CANCER ETIOLOGY, DIAGNOSIS AND TREATMENTS

SIGNALING, GENE REGULATION AND CANCER

CANCER ETIOLOGY, DIAGNOSIS AND TREATMENTS

Additional books in this series can be found on Nova's website under the Series tab.

Additional e-books in this series can be found on Nova's website under the e-book tab.

CANCER ETIOLOGY, DIAGNOSIS AND TREATMENTS

SIGNALING, GENE REGULATION AND CANCER

SHREE RAM SINGH
AND
MANOJ K. MISHRA
EDITORS

New York

For permission to use material from this book please contact us:
Telephone 631-231-7269; Fax 631-231-8175
Web Site: http://www.novapublishers.com

NOTICE TO THE READER

Additional color graphics may be available in the e-book version of this book.

LIBRARY OF CONGRESS CATALOGING-IN-PUBLICATION DATA

Signaling, gene regulation, and cancer / editors, Shree Ram Singh and Manoj K. Mishra.
p. cm.
Includes bibliographical references and index.
ISBN 978-1-61942-088-5 (hardcover)
1. Cancer--Genetic aspects. 2. Cancer--Molecular aspects. 3. Cellular signal transduction. 4. Cancer genes. 5. Genetic regulation. I. Singh, Shree Ram. II. Mishra, Manoj K.
RC268.4.S527 211
616.99'4042--dc23

2011043205

Published by Nova Science Publishers, Inc. † New York

CONTENTS

PREFACE

Although there has been tremendous progress in cancer research and treatment, the mortality caused by this disease is still very high. Cancer is the leading cause of death worldwide and second leading cause of death in the United States of America. The American Cancer Society estimates that approximately 1,596,670 new cancer cases are expected to be diagnosed in 2011 and about 571,950 Americans are expected to die of cancer. They also estimated that by 2030, globally about 21.4 million new cancer cases will arise and 13.2 million cancer deaths are expected to occur. It has been found that uncontrolled growth and migration of abnormal cells forms cancer, which may lead to death if not treated on time. Both external and internal factors are responsible for cancer development and progression. Cancer cells once metastasize, move away from the primary tumor, and colonize in other places in the body. Metastasis is the primary cause of morbidity and mortality in cancer patients.

Metastasis is the result of breach in cell communication (signaling). Cell communication is critical in development, tissue repair, immune responses, and homeostasis. Cell communication between cells and the contact and release of signaling factors are essential to maintain the tissue homeostasis. Any defect in cell communication leads to cancer, diabetes and other developmental abnormalities. There are several signaling pathways known to regulate the physiology and homeostasis of our body. Understanding the role of these signaling molecules will help in the effective treatment of several diseases.

Complex interactions between carcinogens and the host's genome affect the heritability of cancer. Genetic defects in cancer affect two group of genes i.e. oncogenes and tumor suppressor genes. The activation of oncogenes results in defects in programmed cell death, uncontrolled growth and division, and loss of cell polarity. Oncogenes inactivate the tumor suppressor genes in cancerous cells that cause abnormal DNA replication, uncontrolled cell cycle, orientation and adhesion defects within the tissues, and abnormalities in immune system. These changes are caused not only by genetic but also by epigenetic mechanisms.

Cancer is a disease of disordered gene expression. Alteration in gene expression in cancer cells is due to disruption in the machinery required for the production or activity of proteins. In the last few years, tremendous progress has been made in cancer research that dramatically changed our view about the determinants of cancer. In addition, the break-through discoveries in the field of cancer provide clear epigenetic mechanisms such as hypomethylation, hypermethylation, loss of imprinting, and chromatin modification in cancer development and clearance. Thus, an understanding of the mechanism involved in the biology of cancer metastasis will provide novel therapeutic strategies to fight against this devastating disease.

In *Signaling, Gene Regulation and Cancer*, a panel of highly skilled experts and frontrunners in their field describe an in-depth analysis of recent developments in field of cell Signaling, regulation of genes and cancer. The book covers the topic including the role of various signaling pathway in development, regulation of cell fate, tumor angiogenesis, duodenal neoplasias, breast, colorectal and prostate cancer, cancer development and progression, microRNA in cancer and epigenetic regulation of cancer.

We would like to thank Nova Science Publisher and their staff for the invitation and editorial guidance throughout the preparation of this work. We are indebted to the authors who graciously accepted their assignments and shared their profound knowledge of specialized subjects to cell signaling, gene regulation and cancer community. We hope that this book will provide valuable information for cell biologist, geneticists, cancer biologists, clinicians, and students involved in various aspects of cell Signaling, regulation of genes and cancer.

We would like to thank our family members for their continued support.

Shree Ram Singh
Manoj K. Mishra

In: Signaling, Gene Regulation and Cancer
Editors: Shree Ram Singh and Manoj K. Mishra
ISBN: 978-1-61942-088-5

Chapter 1

CELL DEATH AND SURVIVAL SIGNALING IN CANCER THERAPY

Simone Fulda*
Institute for Experimental Cancer Research in Pediatrics,
Goethe-University Frankfurt, Komturstr. 3a, Frankfurt, Germany

ABSTRACT

Most therapies that are currently used for the treatment of human cancer, including chemotherapy, radiotherapy or immunotherapy, primarily exert their anticancer activities by inducing cell death in cancer cells. Therefore, the failure to trigger cell death pathways or the ability of cancer cells to evade the induction of cell death pathways may lead to resistance of cancers to current treatment protocols. In addition to insufficient activation of cell death programs in cancer cells, aberrant activation of cell survival cascades can also lead to treatment resistance. Also, cell death and survival signaling pathways are intricately intertwined. The phosphatidylinositol 3'-kinase (PI3K)/Akt/mammalian target of rapamycin (mTOR) pathway presents one of the most prominent prosurvival pathways that is frequently activated in multiple human cancers. Aberrant signaling via PI3K/Akt/mTOR fosters proliferation and cell growth and simultaneously also interferes with the execution of cell death programs. This calls for a better understanding of the underlying mechanisms that regulate cell death and survival pathways in cancers and in response to cancer therapy in order to develop molecular-targeted therapies that tackle resistance mechanisms. This approach is expected to lead to more effective and individualized treatment strategies that selectively target deregulated signaling pathways in a tumor type- and patient-specific manner.

1. INTRODUCTION

Programmed cell death or apoptosis is an intrinsic death program that is present in all mammalian cells and that exerts an important regulatory function in various physiological and

* Tel.: +49 69 67866557, Fax: +49 69 6786659157, Email: simone.fulda@kgu.de.

pathological conditions [1]. For example, tissue homeostasis is maintained by a delicate balance of both proliferation and cell death [2]. Tipping this balance towards too little or too much apoptosis can contribute to either tumor formation or inappropriate tissue loss [3]. Moreover, defects in apoptosis programs can lead to resistance to chemo-, radio- or immunotherapy, since the efficacy of these treatment approaches largely depends on their ability to engage cell death programs in cancer cells [4-6]. This implies that a better understanding of the molecular components that regulate cell death and survival pathways in human cancers in response to current therapies will likely provide a rational basis for the design of new strategies to selectively target signaling pathways in cancer cells. This approach opens new perspectives to overcome treatment failures in a variety of human cancers.

2. Apoptosis Pathways

There are two principal pathways of apoptosis, i.e. the death receptor (extrinsic) and the mitochondrial (intrinsic) pathway [5]. Activation of either pathway eventually leads to activation of caspases, a family of proteases that act as common death effector molecules in various forms of cell death [7].

Stimulation of death receptors of the tumor necrosis factor (TNF) receptor superfamily, for example CD95 (APO-1/Fas) or TRAIL receptors lead to activation of caspase-8 in a multimeric protein complex called the death-inducing signaling complex (DISC) [8]. Once activated, caspase-8 can transmit the apoptosis signal either by direct cleavaging downstream effector caspases such as caspase-3 or by cleaving Bid [9]. Bid is a proapoptotic, BH3-only domain containing protein of the Bcl-2 family that translocates to mitochondria upon its cleavage where it initiates mitochondrial outer membrane permeabilization, thus linking the death receptor to the mitochondrial pathway [10]. The mitochondrial pathway is initiated by the release of apoptogenic factors such as cytochrome c or second mitochondrial activator of caspases (Smac) from the mitochondrial intermembrane space [10]. Upon its release into the cytosol cytochrome c initiates caspase-3 activation via the cytochrome c/Apaf-1/caspase-9-containing apoptosome complex, while Smac enhances caspase activation by antagonizing Inhibitor of Apoptosis (IAP) proteins [10]. IAPs proteins are a family of endogenous caspase inhibitors that bind to and inhibit caspases such as caspase-3, -7 and -9, thereby blocking apoptosis at a key node [11].

In addition to apoptosis, various non-apoptotic modes of cell death have in recent years been delineated, for example necrosis, mitotic catastrophe and autophagy [12]. Since these alternative modes of cell death depend on different molecular signaling events, they may present an alternative strategy to trigger cell death in apoptosis-resistant types of human cancers [13].

Since the inappropriate initiation of cell death pathways can have detrimental effects on cell survival, cell death programs are tightly controlled by pro- and antiapoptotic regulatory mechanisms. The antiapoptotic blocks that restrain cell death have also been implied in mediating resistance of tumor cells to drug treatment.

3. The PI3K/Akt/mTOR Pathway

The PI3K/Akt/mTOR pathway presents one of central mediators that promote cell survival, proliferation and growth [14, 15]. It integrates growth signals from the extracellular and intracellular space, e.g. in response to growth factors stimulation, extracellular matrix proteins or cell-cell interactions [15, 16]. Upon its activation, PI3K is recruited to the plasma membrane upon activation of transmembrane receptor tyrosine kinases (RTKs), cytokine or antigen receptors and in turn recruits Akt via the production of phospholipids [17]. There are several negative regulatory loops to control PI3K activity, for example the tumor suppressor gene *phosphatase and tensin homologue deleted on chromosome 10 (PTEN)* which terminates PI3K/Akt signaling by dephosphorylating phosphoinositoles [18].

There are three isoforms of the mammalian Akt protein, i.e. Akt1, Akt2, and Akt3 [15, 19]. Akt exerts various functions, including cell proliferation, transcription, cell migration, angiogenesis, glucose metabolism and inhibition of apoptosis [15, 19]. Akt supports the survival of cancer cells by modulating energy metabolism or by activating mTOR signaling [19]. Also, Akt can block apoptosis via direct phosphorylation and inhibition of apoptosis signaling molecules such as X-linked inhibitor of apoptosis (XIAP) [19]. In addition, Akt can indirectly interfere with apoptosis, e.g. by altering the activity of transcription factors such as NF-κB, thereby upregulating expression levels of antiapoptotic proteins [19]. Two mTOR complexes have been identified that involve distinct interaction partners of mTOR, i.e. mTOR complex 1 (mTORC1) mTORC1 that comprises regulatory associated protein of mTOR (Raptor) and mTORC2 that contains rapamycin-insensitive companion of mTOR (Rictor) [15].

4. Aberrant Regulation of Signaling Pathways Contributes to Tumorigenesis, Tumor Progression and Therapy Resistance

Evasion of apoptosis belongs to the hallmarks of human cancers. From a mechanistic point, blockage of cell death programs can result from abnormal activation of antiapoptotic programs, by deletion, mutation or loss of function of proapoptotic molecules or via enhanced activation of survival pathways, for example the PI3K/Akt/mTOR pathway. These alterations can also accumulate and act in concert to promote tumor growth. Depending on the type of lesion and its impact on the signaling network such alterations can differentially affect cancer cell biology. In the following paragraphs, examples of critical alterations in apoptosis signaling pathways are given that frequently occur in human cancers.

5. Apoptosis and Cancer Biology

Typically, human cancers have developed strategies to escape apoptosis in order to progressively grow. Signaling to cell death can in principle be blocked by a decrease or defective function of proapoptotic proteins and/or alternatively by increased levels of anti-apoptotic proteins.

5.1. Alterations in the Death Receptor (Extrinsic) Pathway in Human Cancers

In the death receptor (extrinsic) pathway, surface expression of death receptors may be decreased or may even be completely absent in resistant forms of cancers. For example, downregulation of CD95 expression was reported in leukemia or neuroblastoma cells that were resistant towards a variety of drugs [20, 21]. This suggests that critical levels of CD95 expression may be an important parameter for drug sensitivity. It has also been described in colon carcinoma that defective transport of the apoptosis-inducing TRAIL receptors TRAIL-R1 and -R2 from intracellular stores, e.g. the endoplasmatic reticulum, to the cell surface where they exert their proapoptotic function can lead to resistance towards TRAIL [22]. Mutations of the CD95 gene were detected in several types of tumors, including hematological malignancies as well as in solid cancers [23]. Similar to the CD95 system, also loss of expression of the agonistic TRAIL receptors TRAIL-R1 and -R2 has been described to account for TRAIL resistance. Both TRAIL receptors are located on chromosome 8p, a region which is known to be frequently affected by loss of heterozygosity (LOH) in human cancers [24]. In addition, deletions or mutations leading to loss of both copies of TRAIL-R1 or -R2 occur in a small percentage of different types of cancers, including non-Hodgkin's lymphoma, colorectal, breast, head and neck cancer, osteosarcoma or lung carcinoma [25, 26]. Moreover, death receptor signaling can be blocked by different decoy receptors, e.g. decoy receptor 3 (DcR3) or TRAIL-R3. Genetic amplification or overexpression of DcR3 was described in lung or colon carcinoma and glioblastoma as a mechanism to interfere with CD95-triggered apoptosis via competitive binding to CD95 ligand [27, 28]. Furthermore, TRAIL-R3, a decoy receptor for TRAIL, has been shown to be overexpressed in certain cancers, for example in gastric carcinoma [29].

Besides genetic alterations, epigenetic changes such as CpG-island hypermethylation of gene promoters may contribute to impaired surface expression of death receptors [30, 31]. It is important to note that insufficient expression of CD95 expression, for example by epigenetic changes, may favor immune escape. To this end, treatment with histone deacetylase inhibitors has been reported to restore the expression of CD95 in tumors with epigenetically silenced CD95 and to sensitize for chemotherapy in an NK cell-dependent manner [32].

Death receptor signaling may also be hampered by molecules that associate with their cytoplasmatic domains and interfere with the recruitment of procaspase-8 or -10 to the DISC, for example FLIP or phosphoprotein enriched in diabetes/phosphoprotein enriched in astrocytes-15kDa (PED/PEA-15) [33, 34]. High FLIP expression is a characteristic feature of many tumors and has been correlated with resistance to death receptor- and also to chemotherapy-induced apoptosis [35, 36].

Epigenetic silencing can also lead to inactivation of caspase-8, a critical mediator of death receptor-mediated apoptosis. Interestingly, caspase-8 expression has been reported to be downregulated by hypermethylation of regulatory sequences of the caspase-8 gene in a variety of cancers including neuroblastoma, malignant brain tumors, Ewing tumor, retinoblastoma, rhabdomyosarcoma or small lung cell carcinoma both *in vitro* and also *in vivo* in primary tumor samples [37-41]. Of note, co-methylation of caspase-8 and FLIP, an antiapoptotic protein that blocks caspase-8 signaling, as well as co-methylation of the pairs of agonistic TRAIL receptors TRAIL-R1/-R2 or antagonistic TRAIL receptors TRAIL-R3/ -R4 have been described [42]. This indicates that certain apoptosis regulatory genes are regulated by hypermethylation in a concerted fashion, but not randomly.

5.2. Defects in the Mitochondrial (Intrinsic) Pathway in Human Cancers

Besides changes in the death receptor pathway, alterations in the mitochondrial pathway are frequently found in human cancers and can contribute to treatment resistance. One prominent example is the overexpression of Bcl-2, an antiapoptotic protein of the Bcl-2 family. Bcl-2 is present in approximately 85% of human follicular lymphoma as the result of chromosomal translocation of the Bcl-2 oncogene into the immunglobulin heavy chain gene locus [43]. Overexpression of Bcl-2 in transgenic mice has shown that Bcl-2 promotes the neoplastic transformation of B and T lymphocytes as well as of myeloid cells [44, 45]. In addition to this genetic cause of aberrant Bcl-2 expression, Bcl-2 is frequently upregulated in cancers as the result of oncogenic activation of survival pathways such as PI3K/Akt signaling. Furthermore, somatic mutations that impair the *bax* gene, a proapoptotic protein of the Bcl-2 family, have been detected in solid tumors as well as in leukemias. For example, single nucleotide substitution or frameshift mutations of the *bax* gene have been found in mismatch repair-deficient (MMR) colon cancer and in hematopoietic malignancies [46, 47]. Moreover, BH3-only proteins, a group of protein of the Bcl-2 family that harbor a BH3-only domain, can contribute to the suppression of malignant transformation, since they have been shown to function as *bona fide* tumor suppressors. Loss of a single allele of the BH3-only protein Bim was shown to promote tumor formation in a mouse model of B cell lymphoma that is induced by expression of a *c-myc* oncogene [48]. Homozygous deletions in the chromosomal region of the *bim* gene were also detected in patients with mantle cell lymphoma [49], pointing to the potential clinical relevance of genetic alterations of Bim in certain tumors. Mice lacking the *bid* gene, another BH3-only protein, spontaneously develop a myeloproliferative disorder that can progress to a chronic myelomonocytic form of leukemia [50]. In addition to these alterations in Bcl-2 family proteins, there are alternative mechanisms to impair the mitochondrial pathway of apoptosis in human cancers. At the postmitochondrial level, apoptosis programs can be blocked in melanoma and leukemia by decreased or absent activity of Apaf-1, an important component of the apoptosome complex that drives caspase-3 activation [51, 52].

Moreover, aberrant expression of IAP proteins can contribute to defective signaling via either the death receptor or the mitochondrial pathway in human cancers. This may be caused by increased mRNA or protein expression of IAP proteins or alternatively by loss of endogenous antagonists such as XAF1 [53-55]. XIAP stability has also been described to be subject to regulation, for example in response to phosphorylation by Akt [56]. Overexpression of IAP proteins can further block apoptosis signaling by binding to Smac, thereby freeing XIAP to inhibit effector caspases [57, 58]. Of note, survivin is the fourth most common transcriptome of the human genome in gene profiling studies, whereas survivin was detected at very low levels in normal adult tissues implicating survivin in the malignant phenotype of cancer cells [59]. The *cIAP2* gene can be deregulated by chromosomal translocation, for example by the t(11;18)(q21;q21) translocation that frequently occurs in mucosa-associated lymphoid tissue (MALT) lymphoma [60].

6. Apoptosis and Cancer Therapy

Since the escape of apoptosis presents a characteristic feature of a variety of human cancers that play an important role in promoting tumor formation and progression, there has been much interest to design strategies to target the apoptotic machinery in cancer cells. In principle, cell death pathways can be activated by agents that directly trigger apoptosis pathways. Alternatively, apoptosis-targeted therapies can be used to increase the responsiveness of human cancers towards classical treatment approaches that are currently used in clinical therapies, e.g. chemo-, radio- or immunotherapy, as these therapies primarily exert their anti-tumor activity by inducing apoptosis in cancer cells.

6.1. Strategies to Target the Death Receptor Pathway

The most promising strategy to target death receptors in human cancers is currently the use of agents that activate the proapoptotic TRAIL receptors. Recombinant soluble TRAIL or monoclonocal TRAIL receptor antibodies have been shown in a variety of preclinical studies to predominantly kill cancer cells, while sparing normal cells [61]. For example, no toxicity upon intravenous infusion of even high doses of TRAIL was reported in studies in chimpanzees and cynomolgus monkeys that were used as non-human primates [62]. Similarly, various normal human cells of different lineages including fibroblasts, endothelial cells, smooth muscle cells, epithelial cells or astrocytes exhibited no detectable cytotoxic activity upon exposure to TRAIL [63]. The underlying mechanisms for this differential sensitivity of malignant versus normal cells for TRAIL to undergo apoptosis are not exactly known despite extensive research, but are likely not simply due to the differential expression of agonistic versus antagonistic TRAIL receptors.

It is interesting to note that TRAIL can promote proliferation and survival in certain conditions. For example, it has been reported that TRAIL stimulates proliferation rather than apoptosis in a NF-κB-dependent manner in cancer cells that are resistant towards TRAIL-mediated apoptosis [64]. These findings suggest that the death-inducing ligand TRAIL might paradoxically increase tumor growth in certain tumors, e.g. those that are resistant towards TRAIL.

In preclinical studies using a broad range of cancer cell lines as well as *in vivo* in several human cancer xenograft models recombinant soluble TRAIL was shown to be a potent trigger of apoptosis [61]. Similarly, treatment with monoclonal antibodies that are directed against the proapoptotic TRAIL receptors TRAIL-R1 or -R2 resulted in suppression of tumor growth [65, 66]. In addition to its cytotoxic effect against cancer cells, TRAIL-R2 antibody-based therapy was also shown to be an efficient strategy to trigger tumor-specific T cell memory, which resulted in sustained protection from tumor recurrence [67]. Besides recombinant soluble TRAIL ligand or TRAIL receptor antibodies, there are several gene therapy approaches in order to specifically target tumor cells. For example, an adenoviral vector-based system using the hTERT promoter resulted in high expression levels of TRAIL and tumor-specific induction of apoptosis [68]. By comparison, only little transgene expression and associated toxicity was found in non-malignant human primary mammary epithelial cells

[68]. Furthermore, intralesional injection of adenoviral TRAIL inhibited the growth of human breast cancer xenografts and tumor-free survival of mice [68].

In addition to studies testing TRAIL as a monotherapy, there have been tremendous efforts to develop TRAIL-based combination therapies together with chemo-, radio- or immunotherapy or targeted therapeutics. The rationale for such combination approaches is the fact that many tumors show resistance towards TRAIL-induced apoptosis despite the expression of both agonistic TRAIL receptors. Synergistic interaction between TRAIL receptor agonists and chemo- or radiotherapy has been reported in a large variety of solid cancers and leukemia both *in vitro* and also in mouse models of human cancers [69-76]. This cooperative interaction has been attributed to various molecular mechanisms in a context-dependent manner, e.g. influenced by the cancer, cell type or stimulus. Examples include transcriptional upregulation of the agonistic TRAIL receptors TRAIL-R1 and -R2 in a p53-dependent or alternatively p53-independent manner [77, 78]. In mismatch repair-deficient colorectal cancer cells harboring Bax mutations, p53 turned out to be critical for chemotherapy-mediated sensitization to TRAIL via transcriptional upregulation of TRAIL-R2 [79]. In addition, chemotherapy has been described to promote the assembly of death receptors such as CD95 or TRAIL receptors upon exposure to anticancer drugs, e.g. in colon carcinoma cells [80]. Recombinant TRAIL and fully human monoclonal antibodies for TRAIL-R1 or -R2 are currently under evaluation in early clinical trials either alone or in combination with e.g. anticancer drugs [81-86].

6.2. Cancer Therapeutics Targeting Bcl-2 Family Proteins

Antiapoptotic proteins of the Bcl-2 family are frequently expressed at high levels in human cancers and can confer resistance to current treatment approaches by interfering with the activation of the mitochondrial pathway of apoptosis [3]. Therefore, several strategies have been developed to antagonize the cytoprotective functions of antiapoptotic Bcl-2 proteins.

One of the most prominent examples is the development of small molecule inhibitors, which target the protein-protein interaction site between the antiapoptotic Bcl-2 proteins and the multidomain proteins Bax or Bak[87]. These inhibitors bind to the surface groove of Bcl-2, Bcl-X_L and Bcl-w that otherwise interacts with the BH3 domain of Bax or Bak [87]. ABT-737 was the first generation compound that resulted from this development program [87]. ABT-737 has been reported to directly trigger apoptosis in susceptible cell lines or alternatively increase apoptosis sensitivity of cancer cells in combination protocols [87]. To this end, ABT-737 acted in concert with various classical anticancer drugs to elicit apoptosis [87-89]. One potential drawback of ABT-737 lies in the fact that it does not neutralize Mcl-1, one of the antiapoptotic Bcl-2 proteins that is involved in the regulation of apoptosis and frequently expressed at high levels in cancers. Thus, high expression of Mcl-1 has been linked to resistance to ABT-737 and is discussed as a key determinant of ABT-737 sensitivity [88, 89]. To circumvent this Mcl-1-conferred resistance, both pharmacologic and genetic approaches have been evaluated. These comprise proteasome inhibitors, which cause upregulation of the BH3-only domain protein Noxa that specifically antagonizes Mcl-1, CDK inhibitors (e.g. roscovitine, flavopiridol, seliciclib) or Raf/Mek inhibitors (e.g. sorafenib), which have all been shown to potentiate ABT-737-induced cytotoxicity [88-92]. In addition to cell lines, also primary tumor cells derived from patients, e.g. leukemic blasts from AML patients, proved to respond to treatment with ABT-737 [87-89].

Another example to target the cytoprotective action of Bcl-2 family proteins are Bcl-2 antisense oligonucleotides that downregulate Bcl-2 mRNA expression [93]. Bcl-2 antisense oligonucleotides have been under evaluation in a number of clinical trials as single agent or in combination with chemotherapy for both hematological malignancies as well as solid tumors [93].

In addition to small molecule inhibitors, BH3 peptides, which mimic BH3-only domain proteins in their action to trigger the proapoptotic Bax and Bak proteins, have been developed [94]. Together, this preclinical and first clinical evidence indicates that strategies to target antiapoptotic Bcl-2 family proteins present a promising approach to reactivate the intrinsic apoptosis pathway in cancer cells.

6.3. Cancer Therapeutics Targeting "Inhibitor of Apoptosis" (IAP) Proteins

Since cancer cells often have an intrinsic drive to undergo apoptosis, e.g. because of oncogenic mutations, adaptive alterations have to be in place in order to prevent these intrinsic cell death programs. To this end, high levels of IAP proteins were found to be expressed concomitantly with high basal levels of caspases without any signs of apoptosis in cancer cell lines and tumor samples [95]. This suggests that high levels of IAP proteins may counteract the high basal caspase activity in cancer cells thereby conferring resistance.

Various different approaches have been used to neutralize the abnormally high expression of IAPs in cancers. The binding groove of the BIR3 domain of XIAP has served as a scaffold for the design of small molecule inhibitors that act in a similar fashion as Smac, the native antagonist of IAP proteins [96]. Smac binds to this groove of the BIR3 domain of XIAP after its release from the mitochondria into the cytosol.

Smac peptides harboring the N-terminal part of Smac that is essential for binding of Smac to XIAP were reported to promote caspase activation and to sensitize various tumor cell lines and also primary patients'-derived tumor cells for apoptosis induced by death-receptor ligation or cytotoxic drugs [97]. Importantly, Smac peptides even enhanced the antitumor activity of TRAIL *in vivo* in an intracranial malignant glioma xenograft model [70]. Also, Smac peptides potentiated the antitumor activity of TRAIL and epothilone B derivative BMS 247550 in T cell leukemia cells [98]. To facilitate intracellular delivery Smac peptides were linked to a carrier, for example the protein transduction motif of the HIV Tat protein, the Drosophila antennapaedia penetratin sequence or a polyarginine stretch [97, 99, 100]. Moreover, on the basis of the three-dimensional structure of Smac in complex with XIAP BIR3, Smac peptidomimetics were designed, which bind to one or several of the XIAP-BIR3, cIAP1-BIR3, cIAP2-BIR3 or livin-BIR domains and cooperated with TRAIL, TNFα or anticancer drugs to trigger apoptosis in tumor cells [101-107]. Recently, IAP antagonists were reported to kill cancer cells by inducing autoubiquitination of c-IAPs, NF-κB activation, and TNFα-dependent apoptosis [108-110].

Besides Smac mimetics, virally delivered Smac has been reported to enhance the antitumor activity of TRAIL against hepatocellular cancer both *in vitro* and also in mice [111]. Also, overexpression of Smac sensitized various cancers to TRAIL-induced apoptosis [112, 113]. Furthermore, an ubiquitin-based fusion construct to express mature, biologically active Smac in the cytosol of transfected cells was successful to deliver full-length Smac to tumor cells [114]. Moreover, the natural product embelin, a cell-permeable, small molecule inhibitor of XIAP derived from traditional herbal medicine was reported to trigger cell death in cancer cells [115]. Non-peptidic XIAP antagonists targeting the BIR2 domain of XIAP were identified via screening of a polyphenylurea library, which showed cytotoxicity as single agents and also potentiated chemotherapy- and TRAIL-induced apoptosis *in vitro* and *in vivo* [116-118].

In addition to small molecule inhibitor, antisense oligonucleotides directed against XIAP showed potent antitumor activity in a variety of tumor models in mice [119]. Also, XIAP antisense oligonucleotides together with chemotherapeutic drugs cooperated to suppress tumor growth *in vivo* [120]. This indicates that XIAP antisense oligonucleotides might best be exploited in combination regimens, e.g. together with chemotherapy. XIAP antisense oligonucleotides are evaluated in early clinical trials both as single agent and in combination with chemotherapy (Table 1) [121]. Taken together, strategies to antagonize IAP proteins present promising novel approaches to induce apoptotic cell death in cancer cells or to lower the threshold for the induction of apoptosis.

CONCLUSION

Apoptotic cell death is a critical effector program that mediates the antitumor effects of various anticancer therapies currently used in clinical oncology. Therefore, further insights into the molecular events that regulate apoptosis signaling pathways in response to anticancer drug treatment are expected to have important implications for the development of molecular targeted therapies. A variety of approaches that trigger apoptosis pathways in cancer cells have already entered the clinical stage. These strategies may open new perspectives for more effective cancer treatments.

ACKNOWLEDGMENTS

Work in the author's laboratory is supported by grants from the Deutsche Forschungsgemeinschaft, the Deutsche Krebshilfe, the Bundesministerium für Forschung und Technologie, Wilhelm-Sander-Stiftung, Else-Kröner-Fresenius Stiftung, Novartis Stiftung, für therapeutische Forschung, the European Community (ApopTrain, APO-SYS), and IAP6/18.

REFERENCES

[1] Taylor RC; Cullen SP; Martin SJ. Apoptosis: controlled demolition at the cellular level. *Nat. Rev. Mol. Cell Biol*. [Research Support, Non-U.S. Gov't Review], 2008 Mar, 9, 231-41.

[2] Evan GI; Vousden KH. Proliferation, cell cycle and apoptosis in cancer. *Nature*, 2001 May 17, 411, 342-8.

[3] Fulda S. Tumor resistance to apoptosis. *Int. J. Cancer*, 2009 Feb 1, 124, 511-5.
[4] Johnstone RW; Ruefli AA; Lowe SW. Apoptosis: a link between cancer genetics and chemotherapy. *Cell*, 2002 Jan 25, 108, 153-64.
[5] Fulda S; Debatin KM. Extrinsic versus intrinsic apoptosis pathways in anticancer chemotherapy. *Oncogene*, 2006 Aug 7, 25, 4798-811.
[6] Makin G; Dive C. Apoptosis and cancer chemotherapy. *Trends Cell Biol.*, 2001 Nov, 11, S22-6.
[7] Logue SE; Martin SJ. Caspase activation cascades in apoptosis. *Biochem Soc. Trans.*, 2008 Feb, 36, 1-9.
[8] Ashkenazi A. Targeting the extrinsic apoptosis pathway in cancer. *Cytokine Growth Factor Rev.*, 2008 Jun-Aug, 19, 325-31.
[9] Kroemer G; Galluzzi L; Brenner C. Mitochondrial membrane permeabilization in cell death. *Physiol. Rev.*, 2007 Jan, 87, 99-163.
[10] Fulda S; Galluzzi L; Kroemer G. Targeting mitochondria for cancer therapy. *Nat. Rev. Drug Discov*, 2010 Jun, 9, 447-64.
[11] Gyrd-Hansen M; Meier P. IAPs: from caspase inhibitors to modulators of NF-kappaB, inflammation and cancer. *Nat. Rev. Cancer*, 2010 Aug, 10, 561-74.
[12] Okada H; Mak TW. Pathways of apoptotic and non-apoptotic death in tumour cells. *Nat. Rev. Cancer*, 2004 Aug, 4, 592-603.
[13] Yuan J; Kroemer G. Alternative cell death mechanisms in development and beyond. *Genes Dev.*, 2010 Dec 1, 24, 2592-602.
[14] White E; Karp C; Strohecker AM; Guo Y; Mathew R. Role of autophagy in suppression of inflammation and cancer. *Curr. Opin. Cell Biol.*, 2010 Apr, 22, 212-7.
[15] Shaw RJ; Cantley LC. Ras, PI(3)K and mTOR signalling controls tumour cell growth. *Nature*, 2006 May 25, 441, 424-30.
[16] Liu P; Cheng H; Roberts TM; Zhao JJ. Targeting the phosphoinositide 3-kinase pathway in cancer. *Nat. Rev. Drug Discov.*, 2009 Aug, 8, 627-44.
[17] Blume-Jensen P; Hunter T. Oncogenic kinase signalling. *Nature*, 2001 May 17, 411, 355-65.
[18] Yin Y; Shen WH. PTEN: a new guardian of the genome. *Oncogene*, 2008 Sep 18, 27, 5443-53.
[19] Datta SR; Brunet A; Greenberg ME. Cellular survival: a play in three Akts. *Genes Dev.*, 1999 Nov 15, 13, 2905-27.
[20] Friesen C; Fulda S; Debatin KM. Deficient activation of the CD95 (APO-1/Fas) system in drug-resistant cells. *Leukemia*, 1997 Nov, 11, 1833-41.
[21] Fulda S; Scaffidi C; Susin SA; Krammer PH; Kroemer G; Peter ME, et al. Activation of mitochondria and release of mitochondrial apoptogenic factors by betulinic acid. *J. Biol. Chem.*, 1998 Dec 18, 273, 33942-8.
[22] Jin Z; McDonald ER, 3rd; Dicker DT; El-Deiry WS. Deficient tumor necrosis factor-related apoptosis-inducing ligand (TRAIL) death receptor transport to the cell surface in human colon cancer cells selected for resistance to TRAIL-induced apoptosis. *J. Biol. Chem.*, 2004 Aug 20, 279, 35829-39.
[23] Fulda S. Inhibitor of apoptosis proteins in hematological malignancies. *Leukemia*, 2009 Mar, 23, 467-76.
[24] Ashkenazi A. Directing cancer cells to self-destruct with pro-apoptotic receptor agonists. *Nat. Rev. Drug Discov.*, 2008 Dec, 7, 1001-12.

[25] Dechant MJ; Fellenberg J; Scheuerpflug CG; Ewerbeck V; Debatin KM. Mutation analysis of the apoptotic "death-receptors" and the adaptors TRADD and FADD/MORT-1 in osteosarcoma tumor samples and osteosarcoma cell lines. *Int. J. Cancer*, 2004 May 1, 109, 661-7.

[26] Pai SI; Wu GS; Ozoren N; Wu L; Jen J; Sidransky D, et al. Rare loss-of-function mutation of a death receptor gene in head and neck cancer. *Cancer Res*, 1998 Aug 15, 58, 3513-8.

[27] Pitti RM; Marsters SA; Lawrence DA; Roy M; Kischkel FC; Dowd P, et al. Genomic amplification of a decoy receptor for Fas ligand in lung and colon cancer. *Nature*, 1998 Dec 17, 396, 699-703.

[28] Roth W; Isenmann S; Nakamura M; Platten M; Wick W; Kleihues P, et al. Soluble decoy receptor 3 is expressed by malignant gliomas and suppresses CD95 ligand-induced apoptosis and chemotaxis. *Cancer Res*, 2001 Mar 15, 61, 2759-65.

[29] Sheikh MS; Huang Y; Fernandez-Salas EA; El-Deiry WS; Friess H; Amundson S, et al. The antiapoptotic decoy receptor TRID/TRAIL-R3 is a p53-regulated DNA damage-inducible gene that is overexpressed in primary tumors of the gastrointestinal tract. *Oncogene*, 1999 Jul 15, 18, 4153-9.

[30] Petak I; Danam RP; Tillman DM; Vernes R; Howell SR; Berczi L, et al. Hypermethylation of the gene promoter and enhancer region can regulate Fas expression and sensitivity in colon carcinoma. *Cell Death Differ.*, 2003 Feb, 10, 211-7.

[31] Van Noesel MM; Van Bezouw S; Salomons GS; Voute PA; Pieters R; Baylin SB, et al. Tumor-specific down-regulation of the tumor necrosis factor-related apoptosis-inducing ligand decoy receptors DcR1 and DcR2 is associated with dense promoter hypermethylation. *Cancer Res.*, 2002 Apr 1, 62, 2157-61.

[32] Maecker HL; Yun Z; Maecker HT; Giaccia AJ. Epigenetic changes in tumor Fas levels determine immune escape and response to therapy. *Cancer Cell*, 2002 Aug, 2, 139-48.

[33] Hao C; Beguinot F; Condorelli G; Trencia A; Van Meir EG; Yong VW, et al. Induction and intracellular regulation of tumor necrosis factor-related apoptosis-inducing ligand (TRAIL) mediated apotosis in human malignant glioma cells. *Cancer Res.*, 2001 Feb 1, 61, 1162-70.

[34] Krueger A; Baumann S; Krammer PH; Kirchhoff S. FLICE-inhibitory proteins: regulators of death receptor-mediated apoptosis. *Mol. Cell Biol.*, 2001 Dec, 21, 8247-54.

[35] Fulda S; Meyer E; Debatin KM. Metabolic inhibitors sensitize for CD95 (APO-1/Fas)-induced apoptosis by down-regulating Fas-associated death domain-like interleukin 1-converting enzyme inhibitory protein expression. *Cancer Res.*, 2000 Jul 15, 60, 3947-56.

[36] Longley DB; Wilson TR; McEwan M; Allen WL; McDermott U; Galligan L, et al. c-FLIP inhibits chemotherapy-induced colorectal cancer cell death. *Oncogene*, 2006 Feb 9, 25, 838-48.

[37] Fulda S; Kufer MU; Meyer E; van Valen F; Dockhorn-Dworniczak B; Debatin KM. Sensitization for death receptor- or drug-induced apoptosis by re-expression of caspase-8 through demethylation or gene transfer. *Oncogene*, 2001 Sep 13, 20, 5865-77.

[38] Teitz T; Wei T; Valentine MB; Vanin EF; Grenet J; Valentine VA, et al. Caspase 8 is deleted or silenced preferentially in childhood neuroblastomas with amplification of

MYCN. *Nat. Med.* [Research Support, Non-U.S. Gov't Research Support, U.S. Gov't, P.H.S.], 2000 May, 6, 529-35.

[39] Hopkins-Donaldson S; Ziegler A; Kurtz S; Bigosch C; Kandioler D; Ludwig C, et al. Silencing of death receptor and caspase-8 expression in small cell lung carcinoma cell lines and tumors by DNA methylation. *Cell Death Differ.* [Research Support, Non-U.S. Gov't], 2003 Mar, 10, 356-64.

[40] Pingoud-Meier C; Lang D; Janss AJ; Rorke LB; Phillips PC; Shalaby T, et al. Loss of caspase-8 protein expression correlates with unfavorable survival outcome in childhood medulloblastoma. *Clin. Cancer Res.*, 2003 Dec 15, 9, 6401-9.

[41] Harada K; Toyooka S; Shivapurkar N; Maitra A; Reddy JL; Matta H, et al. Deregulation of caspase 8 and 10 expression in pediatric tumors and cell lines. *Cancer Res.* [Research Support, U.S. Gov't, P.H.S.], 2002 Oct 15, 62, 5897-901.

[42] Van Noesel MM; Van Bezouw S; Voute PA; Herman JG; Pieters R; Versteeg R. Clustering of hypermethylated genes in neuroblastoma. *Genes Chromosomes Cancer*, 2003 Nov, 38, 226-33.

[43] Tsujimoto Y; Finger LR; Yunis J; Nowell PC; Croce CM. Cloning of the chromosome breakpoint of neoplastic B cells with the t(14;18) chromosome translocation. *Science*, 1984 Nov 30, 226, 1097-9.

[44] McDonnell TJ; Korsmeyer SJ. Progression from lymphoid hyperplasia to high-grade malignant lymphoma in mice transgenic for the t(14; 18). *Nature*, 1991 Jan 17, 349, 254-6.

[45] Traver D; Akashi K; Weissman IL; Lagasse E. Mice defective in two apoptosis pathways in the myeloid lineage develop acute myeloblastic leukemia. *Immunity*, 1998 Jul, 9, 47-57.

[46] Kitada S; Pedersen IM; Schimmer AD; Reed JC. Dysregulation of apoptosis genes in hematopoietic malignancies. *Oncogene*. [Research Support, Non-U.S. Gov't Research Support, U.S. Gov't, P.H.S. Review], 2002 May 13, 21, 3459-74.

[47] Rampino N; Yamamoto H; Ionov Y; Li Y; Sawai H; Reed JC, et al. Somatic frameshift mutations in the BAX gene in colon cancers of the microsatellite mutator phenotype. *Science*, 1997 Feb 14, 275, 967-9.

[48] Egle A; Harris AW; Bouillet P; Cory S. Bim is a suppressor of Myc-induced mouse B cell leukemia. *Proc. Natl. Acad. Sci. USA*, 2004 Apr 20, 101, 6164-9.

[49] Tagawa H; Karnan S; Suzuki R; Matsuo K; Zhang X; Ota A, et al. Genome-wide array-based CGH for mantle cell lymphoma: identification of homozygous deletions of the proapoptotic gene BIM. *Oncogene*, 2005 Feb 17, 24, 1348-58.

[50] Zinkel SS; Ong CC; Ferguson DO; Iwasaki H; Akashi K; Bronson RT, et al. Proapoptotic BID is required for myeloid homeostasis and tumor suppression. *Genes Dev.*, 2003 Jan 15, 17, 229-39.

[51] Soengas MS; Capodieci P; Polsky D; Mora J; Esteller M; Opitz-Araya X, et al. Inactivation of the apoptosis effector Apaf-1 in malignant melanoma. *Nature*. [Research Support, Non-U.S. Gov't Research Support, U.S. Gov't, P.H.S.], 2001 Jan 11, 409, 207-11.

[52] Fu W-N; Bertoni F; Kelsey SM; McElwaine SM; Cotter FE; Newland AC, et al. Role of DNA methylation in the suppression of Apaf-1 protein in human leukaemia. *Oncogene*, 2003 Jan 23, 22, 451-5.

[53] Chakravarti A; Noll E; Black PM; Finkelstein DF; Finkelstein DM; Dyson NJ, et al. Quantitatively determined survivin expression levels are of prognostic value in human gliomas. *J. Clin. Oncol.*, 2002 Feb 15, 20, 1063-8.

[54] Tamm I; Kornblau SM; Segall H; Krajewski S; Welsh K; Kitada S, et al. Expression and prognostic significance of IAP-family genes in human cancers and myeloid leukemias. *Clin. Cancer Res.*, 2000 May, 6, 1796-803.

[55] Byun DS; Cho K; Ryu BK; Lee MG; Kang MJ; Kim HR, et al. Hypermethylation of XIAP-associated factor 1, a putative tumor suppressor gene from the 17p13.2 locus, in human gastric adenocarcinomas. *Cancer Res.*, 2003 Nov 1, 63, 7068-75.

[56] Dan HC; Sun M; Kaneko S; Feldman RI; Nicosia SV; Wang HG, et al. Akt phosphorylation and stabilization of X-linked inhibitor of apoptosis protein (XIAP). *J. Biol. Chem.*, 2004 Feb 13, 279, 5405-12.

[57] Dohi T; Okada K; Xia F; Wilford CE; Samuel T; Welsh K, et al. An IAP-IAP complex inhibits apoptosis. *J. Biol. Chem.*, 2004 Aug 13, 279, 34087-90.

[58] Song Z; Yao X; Wu M. Direct interaction between survivin and Smac/DIABLO is essential for the anti-apoptotic activity of survivin during taxol-induced apoptosis. *J. Biol. Chem.*, 2003 Jun 20, 278, 23130-40.

[59] Velculescu VE; Madden SL; Zhang L; Lash AE; Yu J; Rago C, et al. Analysis of human transcriptomes. *Nat. Genet.*, 1999 Dec, 23, 387-8.

[60] Dierlamm J; Baens M; Wlodarska I; Stefanova-Ouzounova M; Hernandez JM; Hossfeld DK, et al. The apoptosis inhibitor gene API2 and a novel 18q gene, MLT, are recurrently rearranged in the t(11;18)(q21;q21) associated with mucosa-associated lymphoid tissue lymphomas. *Blood*, 1999 Jun 1, 93, 3601-9.

[61] LeBlanc HN; Ashkenazi A. Apo2L/TRAIL and its death and decoy receptors. *Cell Death Differ.*, 2003 Jan, 10, 66-75.

[62] Ashkenazi A; Pai RC; Fong S; Leung S; Lawrence DA; Marsters SA, et al. Safety and antitumor activity of recombinant soluble Apo2 ligand. *J. Clin. Invest.*, 1999 Jul, 104, 155-62.

[63] Lawrence D; Shahrokh Z; Marsters S; Achilles K; Shih D; Mounho B, et al. Differential hepatocyte toxicity of recombinant Apo2L/TRAIL versions. *Nat. Med.*, 2001 Apr, 7, 383-5.

[64] Ehrhardt H; Fulda S; Schmid I; Hiscott J; Debatin KM; Jeremias I. TRAIL induced survival and proliferation in cancer cells resistant towards TRAIL-induced apoptosis mediated by NF-kappaB. *Oncogene*, 2003 Jun 19, 22, 3842-52.

[65] Chuntharapai A; Dodge K; Grimmer K; Schroeder K; Marsters SA; Koeppen H, et al. Isotype-dependent inhibition of tumor growth in vivo by monoclonal antibodies to death receptor 4. *J. Immunol.*, 2001 Apr 15, 166, 4891-8.

[66] Ichikawa K; Liu W; Zhao L; Wang Z; Liu D; Ohtsuka T, et al. Tumoricidal activity of a novel anti-human DR5 monoclonal antibody without hepatocyte cytotoxicity. *Nat. Med.*, 2001 Aug, 7, 954-60.

[67] Takeda K; Yamaguchi N; Akiba H; Kojima Y; Hayakawa Y; Tanner JE, et al. Induction of tumor-specific T cell immunity by anti-DR5 antibody therapy. *J. Exp. Med.*, 2004 Feb 16, 199, 437-48.

[68] Lin T; Huang X; Gu J; Zhang L; Roth JA; Xiong M, et al. Long-term tumor-free survival from treatment with the GFP-TRAIL fusion gene expressed from the hTERT promoter in breast cancer cells. *Oncogene*, 2002 Nov 14, 21, 8020-8.

[69] Rohn TA; Wagenknecht B; Roth W; Naumann U; Gulbins E; Krammer PH, et al. CCNU-dependent potentiation of TRAIL/Apo2L-induced apoptosis in human glioma cells is p53-independent but may involve enhanced cytochrome c release. *Oncogene*, 2001 Jul 12, 20, 4128-37.

[70] Nagane M; Pan G; Weddle JJ; Dixit VM; Cavenee WK; Huang HJ. Increased death receptor 5 expression by chemotherapeutic agents in human gliomas causes synergistic cytotoxicity with tumor necrosis factor-related apoptosis-inducing ligand in vitro and in vivo. *Cancer Res.*, 2000 Feb 15, 60, 847-53.

[71] Gliniak B; Le T. Tumor necrosis factor-related apoptosis-inducing ligand's antitumor activity in vivo is enhanced by the chemotherapeutic agent CPT-11. *Cancer Res.*, 1999 Dec 15, 59, 6153-8.

[72] Keane MM; Rubinstein Y; Cuello M; Ettenberg SA; Banerjee P; Nau MM, et al. Inhibition of NF-kappaB activity enhances TRAIL mediated apoptosis in breast cancer cell lines. *Breast Cancer Res. Treat*, 2000 Nov, 64, 211-9.

[73] Belka C; Schmid B; Marini P; Durand E; Rudner J; Faltin H, et al. Sensitization of resistant lymphoma cells to irradiation-induced apoptosis by the death ligand TRAIL. *Oncogene*, 2001 Apr 19, 20, 2190-6.

[74] Chinnaiyan AM; Prasad U; Shankar S; Hamstra DA; Shanaiah M; Chenevert TL, et al. Combined effect of tumor necrosis factor-related apoptosis-inducing ligand and ionizing radiation in breast cancer therapy. *Proc. Natl. Acad. Sci. USA*, 2000 Feb 15, 97, 1754-9.

[75] Singh TR; Shankar S; Chen X; Asim M; Srivastava RK. Synergistic interactions of chemotherapeutic drugs and tumor necrosis factor-related apoptosis-inducing ligand/Apo-2 ligand on apoptosis and on regression of breast carcinoma in vivo. *Cancer Res.*, 2003 Sep 1, 63, 5390-400.

[76] Ray S; Almasan A. Apoptosis induction in prostate cancer cells and xenografts by combined treatment with Apo2 ligand/tumor necrosis factor-related apoptosis-inducing ligand and CPT-11. *Cancer Res.*, 2003 Aug 1, 63, 4713-23.

[77] Takimoto R; El-Deiry WS. Wild-type p53 transactivates the KILLER/DR5 gene through an intronic sequence-specific DNA-binding site. *Oncogene*, 2000 Mar 30, 19, 1735-43.

[78] Meng RD; El-Deiry WS. p53-independent upregulation of KILLER/DR5 TRAIL receptor expression by glucocorticoids and interferon-gamma. *Exp. Cell Res.*, 2001 Jan 15, 262, 154-69.

[79] Wang S; El-Deiry WS. Requirement of p53 targets in chemosensitization of colonic carcinoma to death ligand therapy. *Proc. Natl. Acad. Sci. USA*, 2003 Dec 9, 100, 15095-100.

[80] Lacour S; Micheau O; Hammann A; Drouineaud V; Tschopp J; Solary E, et al. Chemotherapy enhances TNF-related apoptosis-inducing ligand DISC assembly in HT29 human colon cancer cells. *Oncogene*, 2003 Mar 27, 22, 1807-16.

[81] Tolcher AW; Mita M; Meropol NJ; von Mehren M; Patnaik A; Padavic K, et al. Phase I pharmacokinetic and biologic correlative study of mapatumumab, a fully human monoclonal antibody with agonist activity to tumor necrosis factor-related apoptosis-inducing ligand receptor-1. *J. Clin. Oncol.*, 2007 Apr 10, 25, 1390-5.

[82] Patnaik A; Wakelee H; Mita M; Fitzgerald A; Hill M; Fox N, et al. HGS-ETR2--A fully human monoclonal antibody to TRAIL-R2: Results of a phase I trial in patients with

advanced solid tumors. *Proceedings Am. Soc. Clin. Oncol. 42nd Annual Meeting, A*, 2006 3012.

[83] Herbst RS; Mendolson DS; Ebbinghaus S; Gordon MS; O'Dwyer P; Lieberman G, et al. A phase I safety and pharmacokinetic (PK) study of recombinant Apo2L/TRAIL, an apoptosis-inducing protein in patients with advanced cancer. *J. Clin. Oncol.*, 2006 24, 3013.

[84] Younes A; Aggarwall BB. Clinical implications of the tumor necrosis factor family in benign and malignant hematologic disorders. *Cancer*, 2003 Aug 1, 98, 458-67.

[85] Chow LQ; Eckhardt SG; Gustafson DL; O'Bryant C; Hariharan S; Diab S, et al. HGS-ETR1, an antibody targeting TRAIL-R1, in combination with paclitaxel and carboplatin in patients with advanced solid malignancies: Results of a phase 1 and PK study. *J. Clin. Oncol.*, 2006 24, 2515.

[86] Mom CH; Sleijfer S; Gietema JA; Sneller V; Fox NL; Lo L, et al. A Phase 1 Study Of HGS-ETR1, A Fully Human Agonistic Monoclonal Antibody To The TRAIL-R1 In Combination With Gemcitabine And Cisplatin In Subjects With Advanced Solid Malignancies [abstract C74]. *Clin. Cancer Res.*, 2005 11, 9117S.

[87] Oltersdorf T; Elmore SW; Shoemaker AR; Armstrong RC; Augeri DJ; Belli BA, et al. An inhibitor of Bcl-2 family proteins induces regression of solid tumours. *Nature*, 2005 Jun 2, 435, 677-81.

[88] Konopleva M; Contractor R; Tsao T; Samudio I; Ruvolo PP; Kitada S, et al. Mechanisms of apoptosis sensitivity and resistance to the BH3 mimetic ABT-737 in acute myeloid leukemia. *Cancer Cell*, 2006 Nov, 10, 375-88.

[89] Van Delft MF; Wei AH; Mason KD; Vandenberg CJ; Chen L; Czabotar PE, et al. The BH3 mimetic ABT-737 targets selective Bcl-2 proteins and efficiently induces apoptosis via Bak/Bax if Mcl-1 is neutralized. *Cancer Cell*, 2006 Nov, 10, 389-99.

[90] Tahir SK; Yang X; Anderson MG; Morgan-Lappe SE; Sarthy AV; Chen J, et al. Influence of Bcl-2 family members on the cellular response of small-cell lung cancer cell lines to ABT-737. *Cancer Res.*, 2007 Feb 1, 67, 1176-83.

[91] Chen Q; Gong B; Mahmoud-Ahmed AS; Zhou A; Hsi ED; Hussein M, et al. Apo2L/TRAIL and Bcl-2-related proteins regulate type I interferon-induced apoptosis in multiple myeloma. *Blood*, 2001 Oct 1, 98, 2183-92.

[92] Lin X; Morgan-Lappe S; Huang X; Li L; Zakula DM; Vernetti LA, et al. 'Seed' analysis of off-target siRNAs reveals an essential role of Mcl-1 in resistance to the small-molecule Bcl-2/Bcl-XL inhibitor ABT-737. *Oncogene*, 2007 Jun 7, 26, 3972-9.

[93] Tolcher AW; Chi K; Kuhn J; Gleave M; Patnaik A; Takimoto C, et al. A phase II, pharmacokinetic, and biological correlative study of oblimersen sodium and docetaxel in patients with hormone-refractory prostate cancer. *Clin. Cancer Res.*, 2005 May 15, 11, 3854-61.

[94] Letai A; Bassik MC; Walensky LD; Sorcinelli MD; Weiler S; Korsmeyer SJ. Distinct BH3 domains either sensitize or activate mitochondrial apoptosis, serving as prototype cancer therapeutics. *Cancer Cell*, 2002 Sep, 2, 183-92.

[95] Yang L; Cao Z; Yan H; Wood WC. Coexistence of high levels of apoptotic signaling and inhibitor of apoptosis proteins in human tumor cells: implication for cancer specific therapy. *Cancer Res.*, 2003 Oct 15, 63, 6815-24.

[96] Shiozaki EN; Shi Y. Caspases, IAPs and Smac/DIABLO: mechanisms from structural biology. *Trends Biochem. Sci.*, 2004 Sep, 29, 486-94.

[97] Fulda S; Wick W; Weller M; Debatin KM. Smac agonists sensitize for Apo2L/TRAIL- or anticancer drug-induced apoptosis and induce regression of malignant glioma in vivo. *Nat. Med.*, 2002 Aug, 8, 808-15.

[98] Guo F; Nimmanapalli R; Paranawithana S; Wittman S; Griffin D; Bali P, et al. Ectopic overexpression of second mitochondria-derived activator of caspases (Smac/DIABLO) or cotreatment with N-terminus of Smac/DIABLO peptide potentiates epothilone B derivative-(BMS 247550) and Apo-2L/TRAIL-induced apoptosis. *Blood*, 2002 May 1, 99, 3419-26.

[99] Arnt CR; Chiorean MV; Heldebrant MP; Gores GJ; Kaufmann SH. Synthetic Smac/DIABLO peptides enhance the effects of chemotherapeutic agents by binding XIAP and cIAP1 in situ. *J. Biol. Chem.*, 2002 Nov 15, 277, 44236-43.

[100] Yang L; Mashima T; Sato S; Mochizuki M; Sakamoto H; Yamori T, et al. Predominant suppression of apoptosome by inhibitor of apoptosis protein in non-small cell lung cancer H460 cells: therapeutic effect of a novel polyarginine-conjugated Smac peptide. *Cancer Res.*, 2003 Feb 15, 63, 831-7.

[101] Li L; Thomas RM; Suzuki H; De Brabander JK; Wang X; Harran PG. A small molecule Smac mimic potentiates TRAIL- and TNFalpha-mediated cell death. *Science*, 2004 Sep 3, 305, 1471-4.

[102] Bockbrader KM; Tan M; Sun Y. A small molecule Smac-mimic compound induces apoptosis and sensitizes TRAIL- and etoposide-induced apoptosis in breast cancer cells. *Oncogene*, 2005 Nov 10, 24, 7381-8.

[103] Sun H; Nikolovska-Coleska Z; Chen J; Yang CY; Tomita Y; Pan H, et al. Structure-based design, synthesis and biochemical testing of novel and potent Smac peptido-mimetics. *Bioorg Med. Chem. Lett.*, 2005 Feb 1, 15, 793-7.

[104] Sun H; Nikolovska-Coleska Z; Yang CY; Xu L; Liu M; Tomita Y, et al. Structure-based design of potent, conformationally constrained Smac mimetics. *J. Am. Chem. Soc.*, 2004 Dec 29, 126, 16686-7.

[105] Sun H; Nikolovska-Coleska Z; Yang CY; Xu L; Tomita Y; Krajewski K, et al. Structure-based design, synthesis, and evaluation of conformationally constrained mimetics of the second mitochondria-derived activator of caspase that target the X-linked inhibitor of apoptosis protein/caspase-9 interaction site. *J. Med. Chem.*, 2004 Aug 12, 47, 4147-50.

[106] Zobel K; Wang L; Varfolomeev E; Franklin MC; Elliott LO; Wallweber HJ, et al. Design, synthesis, and biological activity of a potent Smac mimetic that sensitizes cancer cells to apoptosis by antagonizing IAPs. *ACS Chem. Biol.*, 2006 Sep 19, 1, 525-33.

[107] Sun H; Nikolovska-Coleska Z; Lu J; Qiu S; Yang CY; Gao W, et al. Design, synthesis, and evaluation of a potent, cell-permeable, conformationally constrained second mitochondria derived activator of caspase (Smac) mimetic. *J. Med. Chem.*, 2006 Dec 28, 49, 7916-20.

[108] Varfolomeev E; Blankenship JW; Wayson SM; Fedorova AV; Kayagaki N; Garg P, et al. IAP antagonists induce autoubiquitination of c-IAPs, NF-kappaB activation, and TNFalpha-dependent apoptosis. *Cell*, 2007 Nov 16, 131, 669-81.

[109] Vince JE; Wong WW; Khan N; Feltham R; Chau D; Ahmed AU, et al. IAP antagonists target cIAP1 to induce TNFalpha-dependent apoptosis. *Cell*, 2007 Nov 16, 131, 682-93.

[110] Petersen SL; Wang L; Yalcin-Chin A; Li L; Peyton M; Minna J, et al. Autocrine TNFalpha signaling renders human cancer cells susceptible to Smac-mimetic-induced apoptosis. *Cancer Cell*, 2007 Nov, 12, 445-56.

[111] Pei Z; Chu L; Zou W; Zhang Z; Qiu S; Qi R, et al. An oncolytic adenoviral vector of Smac increases antitumor activity of TRAIL against HCC in human cells and in mice. *Hepatology*, 2004 May, 39, 1371-81.

[112] Okano H; Shiraki K; Inoue H; Kawakita T; Saitou Y; Enokimura N, et al. Over-expression of Smac promotes TRAIL-induced cell death in human hepatocellular carcinoma. *Int. J. Mol. Med.*, 2003 Jul, 12, 25-8.

[113] Ng CP; Bonavida B. X-linked inhibitor of apoptosis (XIAP) blocks Apo2 ligand/tumor necrosis factor-related apoptosis-inducing ligand-mediated apoptosis of prostate cancer cells in the presence of mitochondrial activation: sensitization by overexpression of second mitochondria-derived activator of caspase/direct IAP-binding protein with low pl (Smac/DIABLO). *Mol. Cancer Ther*, 2002 Oct, 1, 1051-8.

[114] Hunter AM; Kottachchi D; Lewis J; Duckett CS; Korneluk RG; Liston P. A novel ubiquitin fusion system bypasses the mitochondria and generates biologically active Smac/DIABLO. *J. Biol. Chem.*, 2003 Feb 28, 278, 7494-9.

[115] Nikolovska-Coleska Z; Xu L; Hu Z; Tomita Y; Li P; Roller PP, et al. Discovery of embelin as a cell-permeable, small-molecular weight inhibitor of XIAP through structure-based computational screening of a traditional herbal medicine three-dimensional structure database. *J. Med. Chem.*, 2004 May 6, 47, 2430-40.

[116] Schimmer AD; Welsh K; Pinilla C; Wang Z; Krajewska M; Bonneau MJ, et al. Small-molecule antagonists of apoptosis suppressor XIAP exhibit broad antitumor activity. *Cancer Cell*, 2004 Jan, 5, 25-35.

[117] Wang Z; Cuddy M; Samuel T; Welsh K; Schimmer A; Hanaii F, et al. Cellular, biochemical, and genetic analysis of mechanism of small molecule IAP inhibitors. *J. Biol. Chem.*, 2004 Nov 12, 279, 48168-76.

[118] Carter BZ; Gronda M; Wang Z; Welsh K; Pinilla C; Andreeff M, et al. Small-molecule XIAP inhibitors derepress downstream effector caspases and induce apoptosis of acute myeloid leukemia cells. *Blood*, 2005 May 15, 105, 4043-50.

[119] LaCasse EC; Kandimalla ER; Winocour P; Sullivan T; Agrawal S; Gillard JW, et al. Application of XIAP antisense to cancer and other proliferative disorders: development of AEG35156/ GEM640. *Ann. NY Acad. Sci.*, 2005 Nov, 1058, 215-34.

[120] LaCasse EC; Cherton-Horvat GG; Hewitt KE; Jerome LJ; Morris SJ; Kandimalla ER, et al. Preclinical characterization of AEG35156/GEM 640, a second-generation antisense oligonucleotide targeting X-linked inhibitor of apoptosis. *Clin. Cancer Res.*, 2006 Sep 1, 12, 5231-41.

[121] LaCasse E; Morris S; Ward T; Cumings J; Robson L; Di Tullio K, et al. AEG 35156, a XIAP antisense oligonucleotide, suppresses XIAP levels in targeted tissues isolatedd from pre-clinical models and from patients. *AACR*, 2006 66.

In: Signaling, Gene Regulation and Cancer
Editors: Shree Ram Singh and Manoj K. Mishra
ISBN: 978-1-61942-088-5

Chapter 2

C4ST-1 AND CHONDROITIN SULFATE IN GROWTH FACTOR SIGNALING, DEVELOPMENT, AND CANCER

***Catherine M. Willis*[1,3], *Robert D. Prinz*[3] and *Michael Klüppel**[1,2,3]**
[1]Department of Pediatrics and
[2]Robert H. Lurie Comprehensive Cancer Center
Feinberg School of Medicine, Northwestern University, Chicago, IL, US
[3]Human Molecular Genetics Program,
Children's Memorial Research Center, Chicago, IL, US

ABSTRACT

The glycosaminoglycan chondroitin sulfate consists of unbranched chains of repeating disaccharides. CS chains are attached to core proteins to form CS-proteoglycans, located on the cell surface and in the extracellular matrix. The biological roles of specific chondroitin sulfate forms, as well as enzymes involved in their biosynthesis, in development and disease have recently generated great interest. In this chapter we summarize current knowledge that suggests important biological roles for specific chondroitin sulfate forms, the balance of chondroitin sulfation, and chondroitin sulfotransferase enzymes. In particular, we will discuss recent findings regarding the functional roles of chondroitin-4-sulfotransferase-1 (C4ST-1) and its products in mouse and zebrafish development, signaling pathways and cancer. Future research will undoubtedly provide exciting and novel insights into the functional roles of C4ST-1 and specific chondroitin sulfate forms in development processes, cancer progression, and associated cell signaling events.

* Correspondence to: Michael Klüppel, email: m-kluppel@northwestern.edu.

INTRODUCTION

The glycosaminoglycan (GAG) chondroitin sulfate is a long, unbranched sugar polymer made up of the repeating disaccharide units, glucuronic acid (GlcA) and N-acetlygalactosamine (GalNAc) [1-4] [Figure 1]. Other GAGs include heparin, heparan sulfate, dermatan, keratin, and hyluronic acid. GAGs function biologically as proteoglycans, a major substance of the extracellular matrix (ECM). Proteoglycans (PGs) consist of a protein core with covalently attached glycosaminoglycan side-chains [5, 6]. Mature proteoglycans are found on the cell surface or are secreted into the ECM [7] [Figure 2]. In most cases the functional activities of a PG is dependent on the molecular interactions and affinities of its attached GAG side chains [1]. Proteoglycans have important developmental and normal biological functions such as mediating cell migration, proliferation, survival, and modulating growth factor signaling cascades by binding and harboring growth factors, cytokines, morphogens, and extracellular matrix molecules at the cell surface or in the ECM [8, 9].

Chondroitin sulfates are synthesized and modified in the Golgi apparatus. Chondroitin sulfate biosynthesis starts with several distinct enzymes that covalently attach a xylose, two galactose units, and a glucuronic acid to a selected serine residue on the core proteoglycan protein to form the tetrasacchharide linkage region [4, 6]. Next, the elongation of the sugar polymer begins by the addition of a GalNAc followed by GlcA and GalNAc addition in alternating sequence by chondroitin sulfate synthases. Several families of sulfotransferases modify the elongating sugar chains by transferring a sulfate group from a donor, phosphoadenosine phoshosulfate (PAPS), to specific positions on the GlcA and GalNAc residues [6]. Chondroitin sulfotransferase enzymes create four specific forms of chondroitin sulfate based on their sulfation pattern, CS-A, CS-C, CS-D, and CS-E [Figure 1].

Chondroitin-4-*O*-sulfotransferase-1 (C4ST-1 or CHST11) is one of the enzymes responsible for making 4-sulfated chondroitin sulfate forms, including C4S or CS-A, and CS-E.

A second enzyme, Chondroitin 4-sulfate 6-O-sulfotransferase (GalNAc4S-6ST), is responsible for the 6-sulfation of CS-E. Chondroitin-6-sulfotransferases (C6ST) sulfate the C6 of the GalNAc resisdues to make CS-C. Galactosaminyl uronyl 2-O-sulfotransferase (CS/DS2ST) is responsible for sulfating the C2 residue on GalNAc after C6 sulfation by C6ST enzymes to make CS-D. Chondroitin sulfation balance and the specific sulfation modifications to the sugar moieties on proteoglycans mediated by the sulfotransferases is a tightly regulated process throughout development, injury, and disease [10].

Glycosaminoglycan content can differ in quantity and type in tissues during many biological processes, including development and disease [11]. The biological functions of proteoglycans rely heavily on the sulfation balance of the GAG side chains that are carried on the proteoglycan [9, 10, 12-14]. For example, a loss-of-function mutation in the C4ST-1 gene in mice leads to severe embryonic defects and death shortly after birth [15, 16]. In contrast, C6ST-1 knockout mice did not have any severe defects, indicating that sulfation balance is tightly regulated and that synthesis of certain chondroitin sulfate forms are crucial during mammalian development [17]. Additionally, specific chondroitin sulfate forms and chondroitin sulfate proteoglycans (CSPGs) have been shown to influence cancer progression and be involved in tumorigenic signaling pathways, which demonstrates a role for chondroitin sulfates in tumor biology [9, 13, 18-21]. In this chapter we will focus on chondroitin sulfates and C4ST-1 and how they are involved in growth factor signaling, development, and cancer.

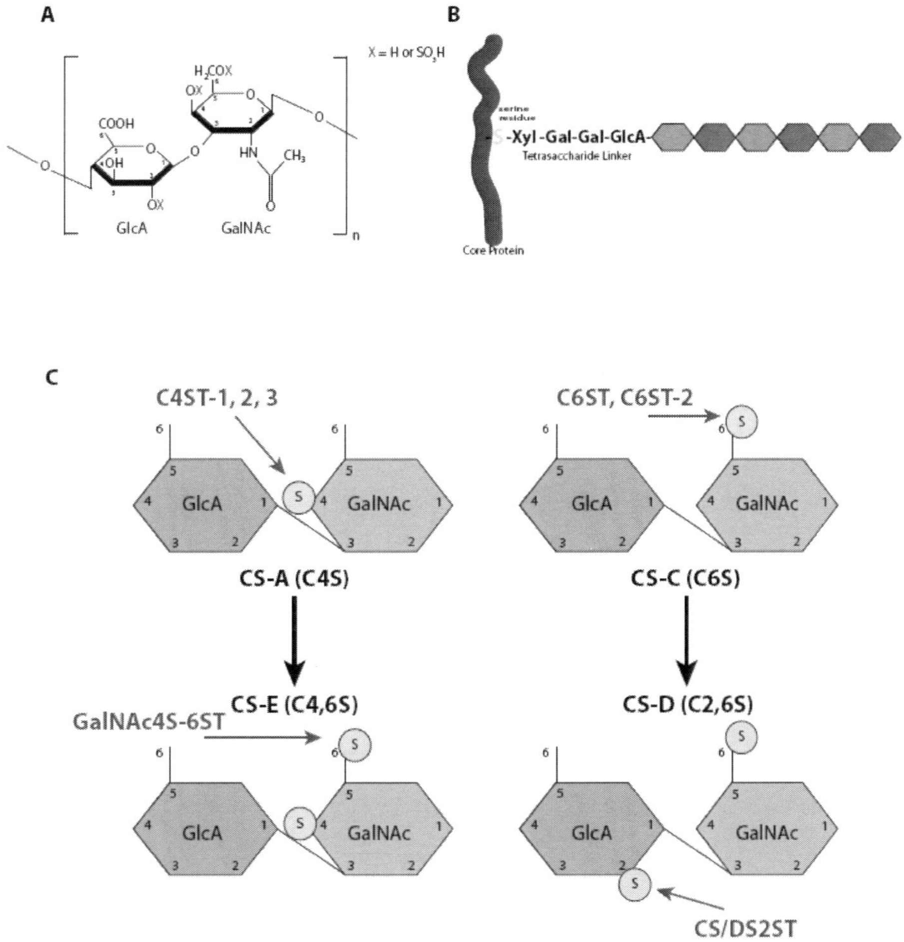

Figure 1. Chondroitin sulfate structure and isoforms. A. Schematic illustration of chondroitin sulfate chemical structure of the disaccharide repeating units [-4GlcAβ1-3GalNAcβ1-]. Chondroitin sulfate consists of repeating glucuronic acid (GlcA) and N-acetylglactosamine (GalNac) sugar residues. The number of individual carbons are indicated. The red 'X' refers to residues that are modified by specific sulfotransferase enzymes. B. Illustration of a chondroitin sulfate proteoglycan (CSPG). The repeating disaccharide units are attached to a selected serine residue on the protein core with a tetrasaccharide linker (Xyl = Xylose; Gal = Galactose; GlcA = Glucuronic acid) followed by the repeating GlcA and GalNac disaccharide units to form the chondroitin sulfate chain on the proteoglycan. C. The four chondroitin sulfate isoforms and the enzymes responsible for their specific sulfation pattern. CS-A (C4S) requires a sulfation on carbon 4 (C4) of GalNAc by the family of chondroitin-4-*O*-sulfotransferases (C4ST-1, 2, 3). CS-E (C4,6S) is double sulfated, and requires sulfation on C4 of GalNAc by C4ST, followed by a subsequent sulfation of C6 of GalNAc by the GalNAc4S-6ST enzyme. CS-C (C6S) is sulfated at the C6 of GalNAc by C6ST enzymes. CS-D (C2,6S) is also double sulfated, and requires C6 sulfationo f GalNAc, followed by C2 sulfation of the GlcA sugar residue by the CS/DS2ST enzyme.

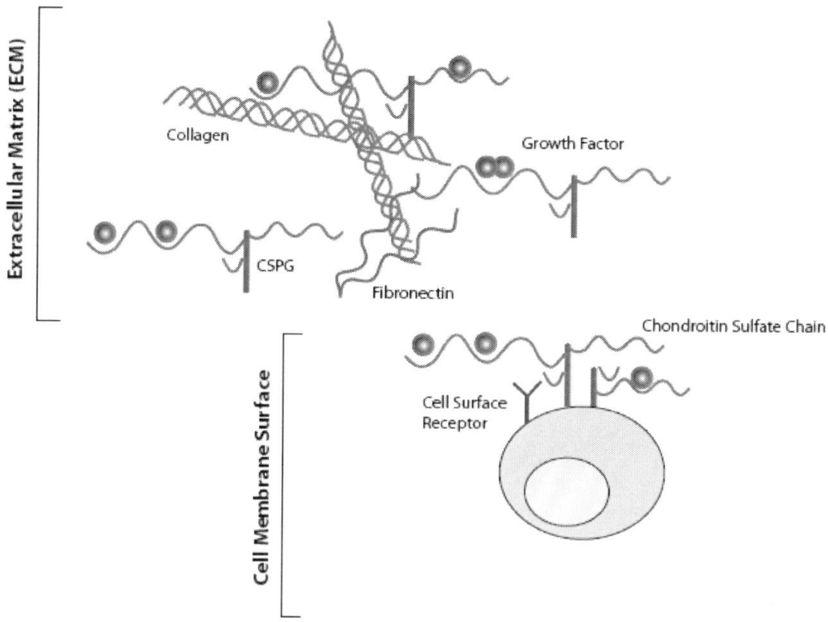

Figure 2. Roles of chondroitin sulfate proteoglycans. Example diagram of the roles of CSPGs. Chondroitin sulfates can be secreted in to the extracellular matrix (ECM) where they can bind ECM molecules like collagen, and harbor growth factors and cytokines and modulate cell signaling. CSPGs are also found on the cell surface and function as cell surface receptors, where they facilitate or inhibit the binding of growth factors/cytokines to the receptor complex.

Chondroitin Sulfates and C4ST-1 in Cellular Signaling Pathways

There are a plethora of studies that describe the role of glycosaminoglycans in cellular signaling pathways. Chondroitin sulfates and CSPGs have been shown to have roles in key signaling pathways that function during several biological processes including, development and tissue morphogenesis, wound repair, and disease through their ability to modulate growth factor activity and morphogenic gradients.

For example, in a study of human mesenchymal stem cells (hMSC) differentiating into an osteoblast lineage, it was found that digestion of cell-surface chondroitin sulfates into disaccharides with the enzyme Chondroitinase ABC (ChABC), leads to an induction of BMP and Wnt signaling in human mesenchymal stem cells, suggesting that CSPGs have a role in modulating these signaling cascades and osteogenic markers [22]. In another study, the effect of chondroitin sulfate in tumor signaling pathways involved in fibrosarcoma progression was investigated. Exogenously added CS-A induced fibrosarcoma cell motility by upregulation of the MAPK pathway (through JNK); increased cell migration, which required tyrosine kinase dependent pathways; and inhibited cell adhesion in a dose-dependent manner [23]. Finally, an example of the involvement of a specific CSPG in cell signaling was shown in a study of melanoma chondroitin sulfate proteoglycan (MCSP) in pathways implicated in melanoma progression. MCSP expression in MCSP-negative melanoma cells induces key pathways important for tumor invasion and growth during melanoma progression [24]. Integrin-mediated cell spreading, focal adhesion kinase phosphorylation, and activation of members of the MAPK pathway – ERK 1/2 were all activated by the expression of MCSP [24].

In contrast, very few studies have investigated the role of C4ST-1 in cellular signaling pathways. Highlighted below are three studies that have investigated a functional role for C4ST-1 in two key signaling pathways, the TGF-β signal transduction pathway and the canonical Wnt/β-catenin pathway.

C4ST-1 as a Target of TGF-β Signaling

One of the first studies to describe C4ST-1's involvement in cellular signaling pathways was a genetic induction gene trap approach, which identified C4ST-1 as a direct target of several members of the TGF-β family of growth factors including TGF-β, bone morphogenetic protein-2 (BMP2), and activin in undifferentiated and differentiated mouse ES cells [15]. Treatment of mouse embryonic explant and whole embryo cultures with BMP2 induced C4ST-1 expression in liver and lung explants, but not kidney explants or whole embryo cultures. This suggests that C4ST-1 gene expression is regulated by TGF-β-like growth factors in a tissue/cell-type specific manner [15].

More recently, a study further investigated the regulation of C4ST-1 expression by the TGF-β signaling cascade [25]. Putative conserved cis-regulatory modules were identified in the mouse and human C4ST-1 gene loci (from -100 to +20kb of the C4ST-1 translational start site). The cis-regulatory modules were tested by luciferase reporter assays to reveal how they are able to control C4ST-1 expression in human embryonic kidney (HEK293T) and mouse normal murine mammary gland (NmuMG) cells. The luciferase data identified a conserved functional promoter just upstream of exon 1, as well as positive regulatory modules at -3.5, +2, and +14 kb, and negative regulatory modules at -51 and +0.1 kb in the C4ST-1 gene locus. Treatment with the TGF-β ligand was able to significantly induce transcriptional activity of the +14kb element in HEK293T cells, and the -51, -3.5, and +2 kb elements in NmuMG cells. These data suggest that activation of the TGF-β signaling cascade regulates C4ST-1 transcription in a cell-type specific manner via multiple long-range cis-regulatory modules [25].

C4ST-1 and CS-E in Wnt/β-Catenin Signaling

Recently, C4ST-1 and its product, CS-E, have been shown to play a role in the Wnt/β-catenin signaling pathway. The Wnt/β-catenin pathway involves the activation, cytoplasmic accumulation and nuclear translocation of intracellular β-catenin where it complexes with transcription factors to stimulate the expression of specific target genes. Activation of the Wnt/β-catenin pathway stimulates biological processes important for embryonic development, tissue homeostasis, and disease [26]. In initial studies, induction of β-catenin accumulation by Wnt3a stimulation was inhibited by a decrease in chondroitin sulfation, or a decrease in chondroitin sulfate levels in articular chondrocytes [27]. Next, studies were performed in mutant sog9 L-cell fibroblasts, which lack the expression of both EXT1 (Extosis-1, required for heparan sulfate biosynthesis) and C4ST-1 genes [28]. These studies showed that the mutant cells had a significant decrease in Wnt3a-stimulated β-catenin accumulation, which can be rescued by stably expressing C4ST-1, but not EXT-1 [28]. In addition, this study showed that the specific chondroitin sulfate form, CS-E, but not the other chondroitin sulfate forms, was able to bind Wnt3a ligand with high affinity. Addition of CS-E to normal L-cells reduced β-catenin levels, much like what was seen in the sog9 mutant L-cells lacking C4ST-1 expression. Together, this data suggests that the C4ST-1, through its ability to make CS-E containing CSPGs, might play a role in the Wnt/β-catenin signaling pathway. The investigators of this study suggest a model in which C4ST-1 expression increases the level of CS-E containing proteoglycans, which can then bind Wnt3a, and facilitate the binding of the ligand to its receptor complex, thus increasing the efficiency of ligand-receptor interactions [28]. In a follow-up study, Nadanaka et al. (2011) show that L-cells stably expressing the Wnt3a ligand had a reduction in C4ST-1 gene expression, and subsequently a change in sulfation balance with a higher concentration of chondroitin sulfate products with low affinity for Wnt3a ligand binding [29]. This allows the Wnt3a ligand to freely diffuse across L-cell fibroblast cultures. Forced expression of C4ST-was suggested to inhibit the diffusion of Wnt3a ligand in L-cell fibroblast cultures, because of the increase in production of CS-E containing proteoglycans [29]. This and their previous study suggest that C4ST-1 expression is able to inhibit Wnt3a diffusion and sustained signaling, but C4ST-1 gene expression is negatively regulated by active Wnt/β-catenin signaling [29].

The Roles of Chondroitin Sulfates and C4ST-1 in Development

Mutations of the core protein, the biosynthetic enzymes or enzymes involved in the GAG modifications, have been used to gather insight into developmental questions of different organisms [16, 30-35]. Studies show that CSPGs are involved in regulating processes in skeletal development in model organisms, and furthermore, show that appropriate chondroitin sulfation balance is crucial for normal development.

Chondroitin Sulfate in Chicken Development

Nanomelia is a lethal autosomal recessive chondrodystrophy of chickens [36]. It results in a skeletal abnormality characterized by shortened and deformed limbs. Nanomelia is caused by a mutation in the chick aggrecan gene, a major CSPG that resides in the ECM of cartilage. The mutation leads to a premature stop codon resulting in a truncated aggrecan core protein precursor that is eventually degraded in the cell. The truncated aggrecan core protein does not go under further maturation processes in the Golgi apparatus, and therefore, it is not secreted into the ECM. This results in an ECM that contains other major components, such as type II collagen, but it lacks aggrecan. Since aggrecan is one of the major ECM components in cartilage, it is no surprise that the nanomelic chicken is phenotypically characterized by severe micromelia and skeletal abnormalities. Specifically, these defects, that progressively exaggerate over time starting on embryonic day 8 (E8), are a reduced trunk and head size, and broad and malformed limbs that are twisted in a cephalad direction away from the body.

Chondroitin Sulfate in Mouse Development

Cartilage Matrix Deficiency (cmd)

Cartilage matrix deficiency (cmd) in mice is caused by an autosomal recessive lethal mutation. Phenotypically, mice homozygous for the cmd-mutation are characterized by an abnormally short trunk, limbs, and snout, an enlarged abdomen, protruding tongue as well as a cleft palate [33]. The cmd/cmd animals die shortly after birth because of respiratory failure [33]. However, heterozygous littermates show a normal phenotype and size. The cmd mutation consists of a 7 base pair deletion in exon 5 of the 18 exons containing murine aggrecan gene. The deletion results in a frame shift leading to premature stop codon within exon 6 resulting in a severely truncated core protein [35]. Again, like the nanomelia in chicks, with the lack of aggrecan in the ECM, these mice have deformed cartilage.

Brachymorphism

Brachymorphism is a nonlethal murine autosomal recessive growth disorder that is characterized by normal CSPG levels, which are significantly undersulfated. This undersulfation is caused by a decreased level of the global sulfate donor PAPS (phosphoadenosine-phoshosulfate) resulting from a missense mutation in PAPS synthetase 2 (PAPSS2) [31]. Biochemical analysis of GAG content in the cartilage of brachymorphic mice showed a preferential undersulfation of chondroitin sulfate chains, but normal sulfation of heparan sulfate chains [37].

While the brachymorphic mice (bm) are phenotypically normal at birth, the differences start to become evident after postnatal day 3 (P3). Over the first 4 weeks of life, skeletal deficits develop. These include a dome-shaped skull, short thick tail, and shortened but not widened limbs, as well as a 50% reduction in limb length and 25% reduction in axial skeletal size compared to their heterozygous and wild type littermates [31].

Detailed analysis of the brachymorphic mouse limbs showed reduced Indian hedgehog (Ihh) signaling in the growth plates as well as irregular Ihh protein distribution in the ECM [37]. The Ihh growth factor is found in prehypertrophic chondrocytes in the growth plate of bones. Ihh is secreted into the ECM and diffuses to its target cells, proliferative chondrocytes within the growth plate, where it stimulates and induces proliferation of these cells. In this study, Cortes et al. show that CSPGs are involved in modulating Ihh signaling. Ihh binds to heparan sulfate and to different forms of chondroitin sulfates (CS-A, CS-C, and unsulfated CS). In particular, Ihh has a high binding affinity to the chondroitin sulfate form, CS-A, and moreover, they show that Ihh binds to the chondroitin sulfate chains of the cartilage CSPG, aggrecan. Cortes et al. propose a model where CSPGs actively modulate the Ihh signaling cascade within the bone growth plate by binding the Ihh growth factor, and facilitating its diffusion across the growth plate to its target cells, where it can bind and stimulate the Ihh receptor, Patched-1. This study suggests that the decreased Ihh concentrations in growth plates seen in brachymorphic mice is due to the severely undersulfated CSPGs and their inability to bind the growth factor and facilitate its diffusion throughout the growth plate, causing a decrease in the number of proliferative chondrocytes. Ultimately, this leads to a reduced growth plate length, which then leads to limb and skeletal defects.

C4ST-1 Gene Trap Mutation (C4ST-1$^{gt/gt}$)

C4ST-1 plays a crucial role in the synthesis of CSPGs, by its involvement in both the sulfation of CS-A and CS-E, and in chondroitin sulfate GAG chain elongation. Therefore, the loss of function mutation of the murine C4ST-1 gene by a gene trap is yet another approach to answer questions concerning the developmental influences of CSPG [16]. This loss of function gene trap mutation shows severe chondrodysplasia. Mice heterozygous for this mutation C4ST-1$^{gt/+}$ are viable and fertile, whereas the homozygous littermates (C4ST-1$^{gt/gt}$) die within 6 hours after birth with severe respiratory distress. Phenotypically, the C4ST-1$^{gt/gt}$ animals show acute dwarfism, cartilage defects, and multiple skeletal abnormalities. Skeletal abnormalities include a smaller rib cage, shorter and thicker iliac bone, femur, tibia, and fibula, as well as misshapen dorsal arches of the vertebrae. Altogether, only bones derived through endochondral ossification are affected by the absence of C4ST-1 and its resulting sulfation imbalance of CSPGs. The growth of bones that develop by intramembraneous ossification are not affected by the loss of C4ST-1 function [16]. The loss of C4ST-1 expression in the C4ST-1$^{gt/gt}$ mice results in a major chondroitin sulfation imbalance, with a switch from the major chondroitin sulfate component, C4S (CS-A) in wild type mice to a predominantly C6S (CS-C) sulfation pattern in the C4ST-1$^{gt/gt}$ cartilage.

Detailed morphological analysis revealed severely shortened, disorganized, and hypocellular growth plates with reductions in size of the columnar and hypertrophic zones. Chondrocyte maturation, negatively regulated by TGFβ signaling and positively regulated by BMP signaling, is affected in C4ST-1$^{gt/gt}$ mice by an increase in TGFβ signaling and a decrease in BMP signaling [16]. In spite of an increase in proliferating cells within the mutant growth plates, an increase in apoptotic cells resulting from an imbalance of apoptotic signals, leads to a net loss of chondrocytes.

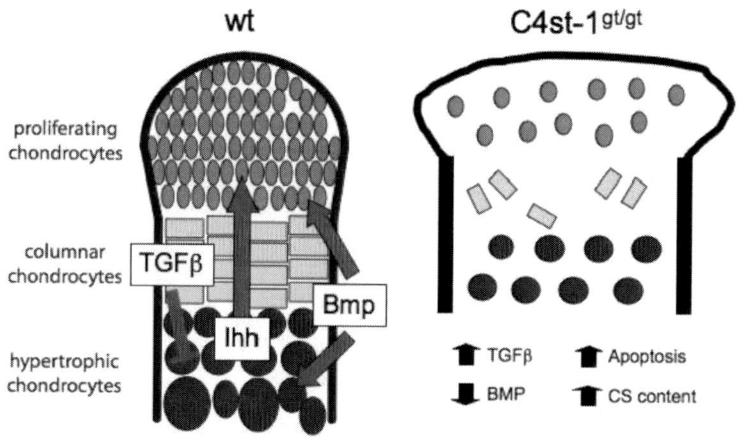

Figure 3. Ihh, TGFβ, and BMP signaling in the developing growth plate. Indian hedgehog (Ihh) is secreted by hypertrophic chondrocytes and acts on the columnar chondrocytes to increase the rate of proliferation. The reduced and abnormal Ihh distribution within the brachymorphic mouse growth plate caused by the undersulfation of CSPGs, results in diminished proliferation and reduced growth plate length. While BMP positively regulates chondrocyte proliferation and hypertrophy, TGFβ functions as a negative regulator. In C4ST-1$^{gt/gt}$ mutants the combination of the down-regulated BMP signaling pathway with the up-regulated TGFβ pathway result in a severe size reduction of the proliferative and hypertrophic zones of the growth plate as well as an increase in apoptosis in the growth plate.

Interestingly, in contrast to the bm mouse [37], no difference in the Ptc-1 expression and Ihh signaling is observed within the columnar and proliferating chondrocytes of C4ST-1$^{gt/gt}$ mice compared to their wild type counterparts. [See Figure 3 for a model of chondroitin sulfates and C4ST-1 and their role in the bone growth plate].

Chondroitin Sulfates in Zebrafish Development

In an in vivo study in zebrafish, antisense morpholino oligonucleotides were injected into the embryos to knockdown C4ST-1 expression [38]. The morphants show morphological abnormalities such as a ventrally bent trunk as well as a curled and/or kinky tail. Both defects are caused by a perturbation of somatic muscle development seen in differential expression patterns of MyoD, a skeletal development marker gene. Additionally, abnormal projections of spinal motor axons (i.e., truncated, misrouted, and side-branched axons) have been observed in these C4ST-1 morphants [38].

Overall, the developmental studies described above show that chondroitin sulfates, CSPGs, and the chondroitin sulfate modifying enzyme, C4ST-1 have an important role in developing organisms. Loss of function and knockdown of C4ST-1, as well as mutations causing severe chondroitin sulfation imbalance, or the lack of aggrecan expression leads to several skeletal abnormalities.

Moreover, some of these mutations are lethal. Understanding the role and function of chondroitin sulfates during development may lead to possible therapeutic actions for cases of developmental skeletal defects and disorders for human patients.

Chondroitin Sulfates and C4ST-1 in Cancer

Expression of chondroitin sulfate and its influence on tumor progression has been the focus of many different cancer studies [9]. It has become increasingly clear that the expression of proteoglycans is altered and typically induced in the tumor stroma during cancer progression, and thus have become new targets for anticancer therapies [39]. Furthermore, it is hypothesized that we can utilize the increased proteoglycan expression in the tumor microenvironment as a potential biological marker for cancer progression and patient survival in certain cancers [39]. The interactions of proteoglycans in the tumor microenvironment, including CSPGs, with growth factors, cytokines, proteinases, and other extracellular molecules secreted by tumor cells can modify the tumor microenvironment, and either facilitate or inhibit cancer cell behavior [21].

CSPGs have been shown to affect such cancer cell behaviors like proliferation, apoptosis, cell signaling, cell adhesion, migration, and angiogenesis. Many studies have shown that there is a differential expression of the specific chondroitin sulfate forms and/or there are changes in sulfation balance in several different cancers compared to the healthy tissue. However, there have been very few studies that describe the role of C4ST-1 in cancer, and therefore the function of chondroitin sulfotransferases in tumorigenesis are not well characterized.

Understanding the mechanisms of chondroitin sulfates and their modifying enzymes in tumor biology can lead to development of beneficial therapies in the treatment of cancer.

Several studies have been summarized below to describe how chondroitin sulfates and C4ST-1 are involved in cancer, and have been categorized into the different types cancer studied. These examples suggest that chondroitin sulfates and the balance of chondroitin sulfation can play important roles in tumor progression.

C4ST-1 Expression in Cancer

There are very few studies describing the functional role of the chondroitin sulfate modifying enzyme, C4ST-1 in cancer. Therefore, it is not well understood how C4ST-1 could influence cancer progression. Here, we will describe the few studies reporting the involvement of C4ST-1 in several cancer types.

One study evaluated the gene expression levels of enzymes involved in chondroitin sulfate and heparan sulfate boiosynthesis and modifications in a comparison of normal bone marrow versus multiple myeloma (MM) cells (malignant plasma cells) [40]. The study showed there was an increase in the expression of C4ST-1, one of nine genes involved in chondroitin sulfate and heparan sulfate biosynthesis that were increased in malignant plasma cells when compared to normal bone marrow [40].

To follow this observation up, the study noted that several known CSPGs were also upregulated in MM compared to normal bone marrow cells. These proteoglycans include, serglycin, testican 2, VCAN, and Col9A2. This suggests that C4ST-1 expression may have an influence on multiple myeloma progression. In another study describing the changes in expression of the genes involved in the glycome of malignant breast tissue, it was shown that C4ST-1 among other sulfotransferases was upregulated in breast cancer compared to normal tissue in a microarray [41]. In contrast, C4ST-1 expression decreased with increasing stage in a study of colon cancer, while C6-sulfation increased [42].

There is one case report that describes a possible role of C4ST-1 in B-cell chronic lymphocytic leukemia (B-CLL) [43]. A patient had a chromosomal translocation of chromosome 14 and chromosome 12 [43]. The breakpoints in chromosome 14 were in the IGH gene locus, and the breakpoints in chromosome 12 were in the C4ST-1 locus.

The breakpoint in chromosome 12 mapped to intron 2 of C4ST-1, and this resulted in the production of two truncated forms of C4ST-1 protein [43]. However, it is unclear if the truncated C4ST-1 expression was involved in the pathogenesis of B-CLL in the patient.

Overall, the few studies listed above suggest that the misregulation of C4ST-1 expression may play role in the progression of several cancer types. However, it remains to be elucidated whether or not C4ST-1 may have a functional role in tumor progression and onset.

Roles of Chondroitin Sulfates and Chondroitin Sulfate Proteoglycans in Several Cancer Studies

Breast Cancer

Chondroitin sulfates have been described as potential regulators of tumor metastasis of multiple cancers, including breast cancer [44]. Some breast cancer cells express cell surface molecules that are recognized by the selectin family of adhesion molecules expressed on the surface of endothelial cells and platelets (E-selectin and P-selectin, respectively). Metastatic breast cancer cells use the interaction with endothelial cells as a mode of transportation through the blood stream and to distant organs. Sialyl Lewis x (sLe^x) and sialyl Lewis a (sLe^a) are the major cell surface ligands of the P- and E-selectin adhesion molecules, and are expressed at high levels on the surface of numerous human cancer cells [45-47]. In a study by Monzavi-Karbassi et al. they found that the 4T1 breast cancer line strongly bind P-selectins on endothelial cells, but this interaction was not dependant on the $sLe^{x/a}$ expression of the breast cancer cell lines [48]. Instead, CSPGs are the major P-selectin ligands expressed on the cell surface of the highly metastatic 4T1 and MDA-MET breast cancer cell lines [44]. Incubating the breast cancer cells with the chondroitin sulfate-digesting enzyme ChABC inhibited P-selectin binding. Furthermore, they show that exogenous treatments of the specific chondroitin sulfate form, CS-E, was able to block the binding of the breast cancer cell lines with P-selectin, all other chondroitin sulfates were unable to do this. This gives evidence that a CS-E containing proteoglycan is the major P-selectin ligand on the metastatic breast cancer cell lines. In this study they show that cell surface CSPGs are important ligands utilized in the metastatic and invasive qualities of tumor cells, and that the chondroitin sulfate form, CS-E can regulate their metastatic potential.

CSPG 4 (CSPG4) is a cell surface proteoglycan that is expressed on the surface of various types of tumors cells, including triple negative breast cancer (TNBC) cells [49]. These breast cancer tumors are considered high risk, and have metastatic potential. Most recently, a study showed that treating TNBC cells with antibodies against CSPG4 inhibited cell growth, migration, and adhesion to ECM molecules *in vitro*. The immunotherapy with the CSPG4 antibody showed the same anti-tumorigenic qualities in orthotopic mouse studies *in vivo* [50]. Moreover, anti-CSPG4 antibody was able to inhibit metastasis to the lungs and prevent re-growth of primary tumors after removal *in vivo*. The study also showed that CSPG4 immunotherapy blocked the activation of several signaling pathways (signaling pathways involving the factors FAK, ERK1/2, PKCα, PI3K/AKT) involved in cell proliferation, migration, and survival in the TNBC tumors (both in vitro and in vivo). This study suggests that CSPG4 may function to positively regulate tumor progression in TNBC cells, the CSPG may be a regulator of several tumorigenic signaling pathways, and anti-CSPG4 immunotherapy could serve as an effect treatment.

The proteoglycan, versican, is highly overexpressed in many solid tumors, including breast cancer [51]. In experiments by Kischel et al. (2010), they set out to determine which of the four isoforms of versican could be targeted for potential anticancer therapies. The four versican isoforms are known as: V0, V1, V2, and V3. In this study, they discovered a new versican isoform that they called, “V4” in human patient breast cancer tissue, expressed specifically in the breast tumor fibroblasts. Kischel et al. discovered that all isoforms (now five isoforms) of versican are highly overexpressed in breast cancer tissue compared to the nearly negligible expression in the adjacent non-malignant mammary tissue. They show that there is not a specific isoform that is expressed more greatly than others in breast cancer. They suggest that the overexpressed versican should be targeted with antibody-based anticancer therapy, with an antibody that targets a domain common to all the versican isoform proteoglycans expressed in breast tumors. They also offer a model to explain the increase in versican production in breast tumor microenvironment. Versican is a known TGFβ responsive gene [52], and their experiments show that recombinant TGFβ and conditioned media from human breast cancer cells can significantly increase the expression of all isoforms of versican in breast tumor fibroblasts. They suggest that TGFβ secretion by the breast cancer cells induces the expression of all five versican isoforms in the tumor stromal cells. This creates an environment suitable for tumor growth.

Colorectal Cancer

In most cancers, CSPG expression levels are altered in the tumor stroma, which can affect the growth and progression of the cancer [53]. In fact, characterization of colon carcinoma is described by an increase in the CSPGs, decorin and versican in the tumor stroma [54]. Furthermore, the GAG side chains of these proteoglycans are altered in colon carcinoma compared to healthy tissue [55]. Even though decorin and versican are present in normal healthy colon tissue, the GAG side chains make-up is generally different, with a predominantly 4-sulfated disaccharide composition in healthy tissue. In the colon carcinoma tumor stroma, the levels of decorin and versican are increased by more than eight-fold compared to healthy tissue, and the sulfation balance is shifted to 6-sulfated and non-sulfated disaccharides [55]. Altered sulfation balance in the colon carcinoma tumor stroma is thought to play a role in inducing changes in cell adhesion, migration, and proliferation of the tumor cells. The change in sulfation balance of proteoglycans (versican and decorin) to an increase in C6-sulfated and non-sulfated chondroitin sulfate chains is not only unique to human colon carcinoma, it is also seen in rectal, pancreatic, liver, and gastric carcinomas [56].

Gastric Carcinoma

In understanding malignant tissue changes versus normal tissue, alterations in the structure and composition of GAGs and proteoglycans have been an area of interest in the tumor biology field. Many studies, including the study by Theocharis et al. (2003), have shown that there are changes in proteoglycan expression and GAG content in cancer. Theocharis et al. showed that the amount of hyluronan, heparan sulfate, chondroitin sulfate, and dermatan sulfate doubled in human gastric carcinoma compared to normal human gastric mucosa [57]. Chondroitin sulfate was the major GAG in human gastric carcinoma, and was expressed 4-fold more when compared to healthy tissue. It was shown that there was a 4-10 fold increase in C6-sulfated and non-sulfated chondroitin sulfate disaccharides in human gastric carcinomas compared to the disaccharide content normal human gastric mucosa. Finally, like many other cancers, the major proteoglycans expressed in human gastric carcinomas were versican and decorin.

Laryngeal Cancer

In a study of laryngeal squamous cell carcinoma (LSCC), increased accumulation of proteoglycans in tumor stroma leads to tumor progression [58]. The CSPGs versican and decorin were shown to have increased accumulation in LSCC tumor stroma in more advanced stages of cancer. Moreover, both proteoglycans had changes in their GAG side chain content compared to normal healthy larynx tissue, and this also was demonstrated in increasing stages of LSCC [58]. In later stages of LSCC, there were more C4-sulfated dermatan structures than compared to healthy tissue.

In contrast to the experiments above, the expression of specific chondroitin sulfates forms were altered in a different way in tissues taken from human laryngeal cancer patients [42, 59-61]. Chondroitin forms CS-C and CS-A have a significantly reduced expression in laryngeal cancer cartilage compared to healthy larynx tissue [62]. Kalathas et al. show in laryngeal cancer tissue there is a misregulation of several chondroitin/dermatan sulfate synthesizing and modifying enzyme expression levels [62]. In healthy tissue, C4ST-1 expression is almost negligible, however, the expression is significantly increased in laryngeal cancer tissue. Moreover, they find the C4ST-1 promoter is hypomethylated in cancer tissue compared to macroscopically normal tissue, which could lead to its increased expression. This in conjunction with differential expression of other modifying enzymes changes the sulfation balance in the cancer tissue to promote tumor progression.

Lung Cancer

The cell surface expression of the highly sulfated chondroitin form, CS-E, has been shown to play a role in the metastatic potential of Lewis lung carcinoma cells lines [19]. Li et al. show that the highly metastatic Lewis lung carcinoma cell line, LM66-H11 expresses a higher amount of CS-E on the cell surface than do P29 cells with low metastatic potential. Removal of chondroitin sulfate chains from the cell surface of LM66-H11 cells with the ChABC enzyme significantly reduced the occurrence of lung metastasis in mice. Exogenous CS-E treatments also had this same effect. This data suggests that endogenous CS-E has a role in positively regulating the metastatic potential of Lewis lung carcinoma cell lines. However, this study did not provide a functional mechanism by which CS-E elicits its effects on lung metastasis.

Melanoma

Several studies have shown that chondroitin sulfates play important roles in the progression of highly metastatic melanomas. The cell surface receptor, melanoma chondroitin sulfate proteoglycan (MCSP) has an important role in tumor adhesion and invasion of melanomas. Iida et al. and others have shown that antibodies against MCSP can inhibit the adhesion of melanoma cells to endothelial cells [59]. Over-expression of MCSP has shown that its involvement in cell adhesion is due to its ability to stimulate α4β1 integrin-mediated cell spreading and focal contact formation in melanoma cells [63]. α4β1 integrin is important for tumor cell adhesion and metastasis because of its ability to bind certain extracellular matrix (ECM) components like fibronectin, and also VCAM-1, an adhesion receptor expressed on the surface of endothelial cells [64]. The α4 component of the α4β1 integrin can specifically bind to chondroitin sulfate chains in MCSP. Elimination of the cell surface chondroitin sulfates with the ChABC enzyme inhibits $\alpha_4\beta_1$ integrin mediated cell adhesion [65].

MCSP was shown to also have a key role in the invasive behavior of metastatic melanomas by forming a complex and regulating the binding and cell surface organization of membrane-type 3-matrix metalloproteinase (MT3-MMP) and pro-MMP-2 [18, 66]. MT3-MMP activates MMP-2 by cleaving its inactive pro-domain. Active MMP-2 degrades ECM molecules like collagen I and IV, and thus is an important regulator in tumor progression by facilitating cellular invasiveness through basement membranes [67]. The MCSP chondroitin chains have a role in the activation of pro-MMP-2 by MT3-MMP. The specific chondroitin form, CS-A, enhances this activation of MMP-2 [18].

Studies have shown that the MCSP expressed in highly metastatic melanomas can be utilized as a tool for treatment mechanisms [68-70]. MCSP antibodies have been used in immunotherapies for melanoma treatment [69]. And more recently, targeting a specific subpopulation of MCSP expressing tumor cells eradicated melanoma lesions and prevented relapse in mice that had subcutaneous transplantation of melanoma cells taken from patient biopsies [68]. These studies show the importance of the expression of specific CSPGs in melanoma tumorigenesis, and possible treatment avenues utilizing the properties of MCSP in cancer.

The CSPG versican, is shown to be upregulated in many cancer types including, melanoma. However, the prognosis for the different cancers overexpressing versican varies based on the roles of versican in the tumor microenvironment. Interestingly, in the case of melanoma, a specific isoform of versican, V3, has been shown to have dual role in tumor growth and metastasis [71]. Overexpression of the V3 versican isoform in human melanoma cells inhibited proliferation in vitro. In vivo, tumors overexpression V3 versican grew much slower than control melanoma tumors. Miquel-Serra et al., found that V3 versican tumors had a low growth rate due to a decrease in proliferation and an increase in apoptosis compared to control melanoma tumors [72]. However, in the same mice overexpressing V3 versican showed an increase in metastasis compared to mice with no overexpression. Miquel-Serra et al. explain this observation as a result of the V3 versican overexpression leads to more hyluronan production by stromal cells, which would induce proliferation of the melanoma cells.

Multiple Myeloma

Serglycin is a unique CSPG, in which it is the only characterized intracellular proteoglycan [39]. It is mainly expressed in hematopoietic and endothelial cells. The core protein contains six GAG attachment sites. Serglycin is packaged into secretory granules or vesicles in hematopoietic cell types (platelets, mast cells, neutrophils, etc.), and is secreted when these cells are activated. It is thought that serglycin uses its GAG chains to serve as a transporter of molecules such as proteases, cytokines, and growth factors during secretion, and the GAG side chains facilitate the binding of the molecules to specific ligands with high affinity. In several hematopoietic tumor cells types it has been found that serglycin is constitutively secreted [73]. In Multiple Myeloma (MM), a B-cell malignancy and essentially a cancer of plasma cells (white blood cells that are responsible for antibody production) in the bone marrow, it was found that there is constitutive secretion of proteoglycans into the ECM. The major PG synthesized and secreted into the ECM by MM cells was serglycin. Serglycin was secreted and associated with the cell surface and also intracellularly in vesicles or granules in MM cells, and the interactions were dependent on its chondroitin sulfate side-chains [73]. The serglycin secreted by MM cells lines had predominantly C4-sulfated chondroitin sulfate side chains, and contained negligible amounts of disulfated, C6-sulfated and non-sulfated chondroitin chains. It is thought that the secretion of serglycin by MM plasma cells accumulates in the bone marrow and inhibits bone mineralization in sites of bone remodeling, which further enhances the MM disease condition in patients [73].

Prostate Cancer

Studies have shown that there is an upregulation of levels of hyluronan and the CSPG versican in the tumor stroma of metastatic prostate cancer. In a more recent study, the functional role of the upregulated levels of hyluronan and versican was analyzed [74]. The study demonstrated a mechanism by which prostate cancer cells can utilize hyluronan and versican secreted by the tumor fibroblasts to create a matrix that facilitates and promotes the prostate cells motility *in vitro* [74]. PC3 and DU145 metastatic prostate cancer cells were able to form a pericellular matrix that incorporated hyluronan and versican. This matrix formation required expression of the hyluronan/CSPG receptor CD-44 and also was dependent on the concentration of versican secreted by the tumor-associated fibroblasts. Together, the results of this study suggest that a complex between CD-44, hyluronan, and versican facilitates the motility of prostate cancer cells.

Osteosarcoma

The murine osteosarcoma cell line, LM8G7, has a high metastatic potential to the liver. In a study to determine the metastatic nature of LM8G7 cells, Basappa et al. (2009), discovered that these cells contained a significantly higher amount of CS-A and CS-E components than do the LM8 murine osteosarcoma cell line, which is preferentially metastatic to the lung [75]. Furthermore, they showed that the increase in CS-E expression in LM8G7 cells has a direct correlation to their metastatic potential. They show that an exogenous pre-injection of CS-E or the GD3G7 antibody against CS-E in the mouse before injection with LM8G7 cells completely inhibited the metastasis to the liver. However, the mice injected with the other chondroitin sulfate forms (CS-A, CS-C, and CS-D) did not have this effect. To further investigate this effect, they showed that exogenous CS-E or the GD3G7 CS-E antibody could inhibit attachment to laminin, invasion, proliferation, and the interaction with VEGF in LM8G7 cells.

This is another demonstration where a specific chondroitin sulfate form is involved in tumor progression and metastasis. However, the specific proteoglycan(s) implicated in tumor progression of osteosarcoma and lung cancer have yet to be determined.

Ovarian Adenocarcinoma

In a study of ovarian adenocarcinomas, it was found that there was high expression of the specific chondroitin form, CS-E, in the extracellular matrix and on the tumor cell surface using the GD3G7 antibody that recognizes CS-E like structures [76]. Normal ovary did not express CS-E like structures at all. The GD3G7 antibody also stained areas of tumor blood vessels and VEGF-rich areas of ovarian adenocarcinomas. In an ELISA assay, it was confirmed that recombinant VEGF protein binds immobilized CS-E with high affinity, whereas CS-A and CS-C did not bind [76].

Additionally, the study showed that different types of ovarian cancers had differential expression of CS-E, which corresponded to the level of gene expression of the chondroitin sulfotransferase partly responsible for CS-E production, GalNAc4S-6ST. This study suggests a specific role for CS-E containing proteoglycans possibly in VEGF biology in ovarian adenocarcinoma. Further exploration is needed to determine the role of CS-E in ovarian tumor biology.

The studies summarized above suggest that CSPGs, chondroitin sulfation balance by differential expression of chondroitin sulfotransferases, and specific chondroitin forms can influence cancer progression in different ways in many types of cancer. Understanding the role of chondroitin sulfates in specific cancers can lead to potential treatment options for patients.

Conclusion

The biological roles of chondroitin sulfates and chondroitin sulfotransferases in development and disease have recently generated great interest. In this chapter we summarized current knowledge that suggest important biological roles for specific chondroitin sulfate forms, the balance of chondroitin sulfation, and chondroitin sulfotransferase enzymes. In particular, C4ST-1 has been shown to be crucial for mouse embryonic skeletal development, and for motor axon guidance in zebrafish. C4ST-1, specific chondroitin sulfate forms, and CSPGs also appear to have important functions in tumor progression. Future research will undoubtedly provide exciting and novel insights into the functional roles of C4ST-1 and chondroitin sulfates in development processes, disease progression, and cell signaling events.

References

[1] Chen RL, Lander AD. Mechanisms underlying preferential assembly of heparan sulfate on glypican-1. *J. Biol. Chem.* 2001; 276: 7507-17.

[2] Habuchi O. [Biological function of chondroitin sulfate and sulfotransferases]. *Seikagaku* 2000; 72: 427-36.

[3] Silbert JE, Sugumaran G. Biosynthesis of chondroitin/dermatan sulfate. *IUBMB Life* 2002; 54: 177-86.

[4] Sugahara K, Kitagawa H. Recent advances in the study of the biosynthesis and functions of sulfated glycosaminoglycans. *Curr. Opin. Struct. Biol.* 2000; 10: 518-27.

[5] Habuchi O. Diversity and functions of glycosaminoglycan sulfotransferases. *Biochim. Biophys. Acta* 2000; 1474: 115-27.

[6] Kusche-Gullberg M, Kjellen L. Sulfotransferases in glycosaminoglycan biosynthesis. *Curr. Opin. Struct. Biol.* 2003; 13: 605-11.

[7] Selleck SB. Proteoglycans and pattern formation: sugar biochemistry meets developmental genetics. *Trends Genet* 2000; 16: 206-12.

[8] Sasisekharan R, Shriver Z, Venkataraman G, Narayanasami U. Roles of heparan-sulphate glycosaminoglycans in cancer. *Nat. Rev. Cancer* 2002; 2: 521-8.

[9] Wegrowski Y, Maquart FX. Chondroitin sulfate proteoglycans in tumor progression. *Adv. Pharmacol.* 2006; 53: 297-321.

[10] Honke K, Taniguchi N. Sulfotransferases and sulfated oligosaccharides. *Med. Res. Rev.* 2002; 22: 637-54.

[11] Dietrich CP. A model for cell-cell recognition and control of cell growth mediated by sulfated glycosaminoglycans. *Braz J. Med. Biol. Res.* 1984; 17: 5-15.

[12] Gama CI, Tully SE, Sotogaku N, et al. Sulfation patterns of glycosaminoglycans encode molecular recognition and activity. *Nat. Chem. Biol.* 2006; 2: 467-73.

[13] Malavaki C, Mizumoto S, Karamanos N, Sugahara K. Recent advances in the structural study of functional chondroitin sulfate and dermatan sulfate in health and disease. *Connect Tissue Res.* 2008; 49: 133-9.

[14] Yamada S, Sugahara K. Potential therapeutic application of chondroitin sulfate/dermatan sulfate. *Curr. Drug Discov. Technol.* 2008; 5: 289-301.

[15] Klüppel M, Vallis KA, Wrana JL. A high-throughput induction gene trap approach defines C4ST as a target of BMP signaling. *Mech. Dev.* 2002; 118: 77-89.

[16] Klüppel M, Wight TN, Chan C, Hinek A, Wrana JL. Maintenance of chondroitin sulfation balance by chondroitin-4-sulfotransferase 1 is required for chondrocyte development and growth factor signaling during cartilage morphogenesis. *Development* 2005; 132: 3989-4003.

[17] Uchimura K, Kadomatsu K, Nishimura H, et al. Functional analysis of the chondroitin 6-sulfotransferase gene in relation to lymphocyte subpopulations, brain development, and oversulfated chondroitin sulfates. *J. Biol. Chem.* 2002; 277: 1443-50.

[18] Iida J, Wilhelmson KL, Ng J, et al. Cell surface chondroitin sulfate glycosaminoglycan in melanoma: role in the activation of pro-MMP-2 (pro-gelatinase A). *Biochem. J.* 2007; 403: 553-63.

[19] Li F, Ten Dam GB, Murugan S, et al. Involvement of highly sulfated chondroitin sulfate in the metastasis of the Lewis lung carcinoma cells. *J. Biol. Chem.* 2008; 283: 34294-304.

[20] Sugahara KN, Hirata T, Tanaka T, et al. Chondroitin sulfate E fragments enhance CD44 cleavage and CD44-dependent motility in tumor cells. *Cancer Res.* 2008; 68: 7191-9.

[21] Wegrowski Y, Maquart FX. Involvement of stromal proteoglycans in tumour progression. *Crit. Rev. Oncol. Hematol.* 2004; 49: 259-68.

[22] Manton KJ, Leong DF, Cool SM, Nurcombe V. Disruption of heparan and chondroitin sulfate signaling enhances mesenchymal stem cell-derived osteogenic differentiation via bone morphogenetic protein signaling pathways. *Stem Cells* 2007; 25: 2845-54.

[23] Fthenou E, Zong F, Zafiropoulos A, Dobra K, Hjerpe A, Tzanakakis GN. Chondroitin sulfate A regulates fibrosarcoma cell adhesion, motility and migration through JNK and tyrosine kinase signaling pathways. *In Vivo* 2009; 23: 69-76.

[24] Yang J, Price MA, Neudauer CL, et al. Melanoma chondroitin sulfate proteoglycan enhances FAK and ERK activation by distinct mechanisms. *J. Cell Biol.* 2004; 165: 881-91.

[25] Willis CM, Wrana JL, Klüppel M. Identification and characterization of TGFbeta-dependent and -independent cis-regulatory modules in the C4ST-1/CHST11 locus. *Genet. Mol. Res.* 2009; 8: 1331-43.

[26] Nusse R. Wnt signaling in disease and in development. *Cell Res.* 2005; 15: 28-32.

[27] Shortkroff S, Yates KE. Alteration of matrix glycosaminoglycans diminishes articular chondrocytes' response to a canonical Wnt signal. *Osteoarthritis Cartilage* 2007; 15: 147-54.

[28] Nadanaka S, Ishida M, Ikegami M, Kitagawa H. Chondroitin 4-O-sulfotransferase-1 modulates Wnt-3a signaling through control of E disaccharide expression of chondroitin sulfate. *J. Biol. Chem.* 2008; 283: 27333-43.

[29] Nadanaka S, Kinouchi H, Taniguchi-Morita K, Tamura J, Kitagawa H. Down-regulation of chondroitin 4-O-sulfotransferase-1 by Wnt signaling triggers diffusion of Wnt-3a. *J. Biol. Chem.*; 286: 4199-208.

[30] Krueger RC, Jr., Kurima K, Schwartz NB. Completion of the mouse aggrecan gene structure and identification of the defect in the cmd-Bc mouse as a near complete deletion of the murine aggrecan gene. *Mamm. Genome* 1999; 10: 1119-25.

[31] Kurima K, Warman ML, Krishnan S, et al. A member of a family of sulfate-activating enzymes causes murine brachymorphism. *Proc. Natl. Acad. Sci. USA* 1998; 95: 8681-5.

[32] Kimata K, Barrach HJ, Brown KS, Pennypacker JP. Absence of proteoglycan core protein in cartilage from the cmd/cmd (cartilage matrix deficiency) mouse. *J. Biol. Chem.* 1981; 256: 6961-8.

[33] Rittenhouse E, Dunn LC, Cookingham J, et al. Cartilage matrix deficiency (cmd): a new autosomal recessive lethal mutation in the mouse. *J. Embryol. Exp. Morphol.* 1978; 43: 71-84.

[34] Li H, Schwartz NB, Vertel BM. cDNA cloning of chick cartilage chondroitin sulfate (aggrecan) core protein and identification of a stop codon in the aggrecan gene associated with the chondrodystrophy, nanomelia. *J. Biol. Chem.* 1993; 268: 23504-11.

[35] Watanabe H, Kimata K, Line S, et al. Mouse cartilage matrix deficiency (cmd) caused by a 7 bp deletion in the aggrecan gene. *Nat. Genet.* 1994; 7: 154-7.

[36] Vertel BM, Walters LM, Grier B, Maine N, Goetinck PF. Nanomelic chondrocytes synthesize, but fail to translocate, a truncated aggrecan precursor. *J. Cell Sci.* 1993; 104 (Pt 3): 939-48.

[37] Cortes M, Baria AT, Schwartz NB. Sulfation of chondroitin sulfate proteoglycans is necessary for proper Indian hedgehog signaling in the developing growth plate. *Development* 2009; 136: 1697-706.

[38] Mizumoto S, Mikami T, Yasunaga D, et al. Chondroitin 4-O-sulfotransferase-1 is required for somitic muscle development and motor axon guidance in zebrafish. *Biochem. J.* 2009; 419: 387-99.

[39] Theocharis AD, Skandalis SS, Tzanakakis GN, Karamanos NK. Proteoglycans in health and disease: novel roles for proteoglycans in malignancy and their pharmacological targeting. *FEBS J.*; 277: 3904-23.

[40] Bret C, Hose D, Reme T, et al. Expression of genes encoding for proteins involved in heparan sulphate and chondroitin sulphate chain synthesis and modification in normal and malignant plasma cells. *Br. J. Haematol.* 2009; 145: 350-68.

[41] Potapenko IO, Haakensen VD, Luders T, et al. Glycan gene expression signatures in normal and malignant breast tissue; possible role in diagnosis and progression. *Mol. Oncol.*; 4: 98-118.

[42] Kalathas D, Theocharis DA, Bounias D, et al. Alterations of glycosaminoglycan disaccharide content and composition in colorectal cancer: structural and expressional studies. *Oncol. Rep.* 2009; 22: 369-75.

[43] Schmidt HH, Dyomin VG, Palanisamy N, et al. Deregulation of the carbohydrate (chondroitin 4) sulfotransferase 11 (CHST11) gene in a B-cell chronic lymphocytic leukemia with a t(12;14)(q23;q32). *Oncogene* 2004; 23: 6991-6.

[44] Monzavi-Karbassi B, Stanley JS, Hennings L, et al. Chondroitin sulfate glycosaminoglycans as major P-selectin ligands on metastatic breast cancer cell lines. *Int. J. Cancer* 2007; 120: 1179-91.

[45] Renkonen R, Mattila P, Majuri ML, et al. In vitro experimental studies of sialyl Lewis x and sialyl Lewis a on endothelial and carcinoma cells: crucial glycans on selectin ligands. *Glycoconj J.* 1997; 14: 593-600.

[46] Renkonen J, Paavonen T, Renkonen R. Endothelial and epithelial expression of sialyl Lewis(x) and sialyl Lewis(a) in lesions of breast carcinoma. *Int. J. Cancer* 1997; 74: 296-300.

[47] Kim YJ, Borsig L, Varki NM, Varki A. P-selectin deficiency attenuates tumor growth and metastasis. *Proc. Natl. Acad. Sci. USA* 1998; 95: 9325-30.

[48] Monzavi-Karbassi B, Whitehead TL, Jousheghany F, et al. Deficiency in surface expression of E-selectin ligand promotes lung colonization in a mouse model of breast cancer. *Int. J. Cancer* 2005; 117: 398-408.

[49] Wang X, Wang Y, Yu L, et al. CSPG4 in cancer: multiple roles. *Curr. Mol. Med.*; 10: 419-29.

[50] Wang X, Osada T, Wang Y, et al. CSPG4 protein as a new target for the antibody-based immunotherapy of triple-negative breast cancer. *J. Natl. Cancer Inst.*; 102: 1496-512.

[51] Kischel P, Waltregny D, Dumont B, et al. Versican overexpression in human breast cancer lesions: known and new isoforms for stromal tumor targeting. *Int. J. Cancer*; 126: 640-50.

[52] Arslan F, Bosserhoff AK, Nickl-Jockschat T, Doerfelt A, Bogdahn U, Hau P. The role of versican isoforms V0/V1 in glioma migration mediated by transforming growth factor-beta2. *Br. J. Cancer* 2007; 96: 1560-8.

[53] Iozzo RV. Matrix proteoglycans: from molecular design to cellular function. *Annu. Rev. Biochem.* 1998; 67: 609-52.

[54] Iozzo RV, Cohen I. Altered proteoglycan gene expression and the tumor stroma. *Experientia* 1993; 49: 447-55.

[55] Theocharis AD. Human colon adenocarcinoma is associated with specific post-translational modifications of versican and decorin. *Biochim. Biophys. Acta* 2002; 1588: 165-72.

[56] Asimakopoulou AP, Theocharis AD, Tzanakakis GN, Karamanos NK. The biological role of chondroitin sulfate in cancer and chondroitin-based anticancer agents. *In Vivo* 2008; 22: 385-9.

[57] Theocharis AD, Vynios DH, Papageorgakopoulou N, Skandalis SS, Theocharis DA. Altered content composition and structure of glycosaminoglycans and proteoglycans in gastric carcinoma. *Int. J. Biochem. Cell Biol.* 2003; 35: 376-90.

[58] Skandalis SS, Theocharis AD, Papageorgakopoulou N, Vynios DH, Theocharis DA. The increased accumulation of structurally modified versican and decorin is related with the progression of laryngeal cancer. *Biochimie* 2006; 88: 1135-43.

[59] Iida J, Meijne AM, Oegema TR, Jr., et al. A role of chondroitin sulfate glycosaminoglycan binding site in alpha4beta1 integrin-mediated melanoma cell adhesion. *J. Biol. Chem.* 1998; 273: 5955-62.

[60] Papadas TA, Stylianou M, Mastronikolis NS, et al. Alterations in the content and composition of glycosaminoglycans in human laryngeal carcinoma. *Acta Otolaryngol.* 2002; 122: 330-7.

[61] Skandalis SS, Stylianou M, Vynios DH, Papageorgakopoulou N, Theocharis DA. The structural and compositional changes of glycosaminoglycans are closely associated with tissue type in human laryngeal cancer. *Biochimie* 2007; 89: 1573-80.

[62] Kalathas D, Triantaphyllidou IE, Mastronikolis NS, et al. The chondroitin/dermatan sulfate synthesizing and modifying enzymes in laryngeal cancer: expressional and epigenetic studies. *Head Neck Oncol.;* 2: 27.

[63] Iida J, Meijne AM, Spiro RC, Roos E, Furcht LT, McCarthy JB. Spreading and focal contact formation of human melanoma cells in response to the stimulation of both melanoma-associated proteoglycan (NG2) and alpha 4 beta 1 integrin. *Cancer Res.* 1995; 55: 2177-85.

[64] Taichman DB, Cybulsky MI, Djaffar I, et al. Tumor cell surface alpha 4 beta 1 integrin mediates adhesion to vascular endothelium: demonstration of an interaction with the N-terminal domains of INCAM-110/VCAM-1. *Cell Regul.* 1991; 2: 347-55.

[65] Iida J, Skubitz AP, Furcht LT, Wayner EA, McCarthy JB. Coordinate role for cell surface chondroitin sulfate proteoglycan and alpha 4 beta 1 integrin in mediating melanoma cell adhesion to fibronectin. *J. Cell Biol.* 1992; 118: 431-44.

[66] Iida J, Pei D, Kang T, et al. Melanoma chondroitin sulfate proteoglycan regulates matrix metalloproteinase-dependent human melanoma invasion into type I collagen. *J. Biol. Chem.* 2001; 276: 18786-94.

[67] Birkedal-Hansen H. Proteolytic remodeling of extracellular matrix. *Curr. Opin. Cell Biol.* 1995; 7: 728-35.

[68] Schmidt P, Kopecky C, Hombach A, Zigrino P, Mauch C, Abken H. Eradication of melanomas by targeted elimination of a minor subset of tumor cells. *Proc. Natl. Acad. Sci. USA*; 108: 2474-9.

[69] Maciag PC, Seavey MM, Pan ZK, Ferrone S, Paterson Y. Cancer immunotherapy targeting the high molecular weight melanoma-associated antigen protein results in a broad antitumor response and reduction of pericytes in the tumor vasculature. *Cancer Res.* 2008; 68: 8066-75.

[70] Burns WR, Zhao Y, Frankel TL, et al. A high molecular weight melanoma-associated antigen-specific chimeric antigen receptor redirects lymphocytes to target human melanomas. *Cancer Res.*; 70: 3027-33.

[71] Miquel-Serra L, Serra M, Hernandez D, et al. V3 versican isoform expression has a dual role in human melanoma tumor growth and metastasis. *Lab Invest.* 2006; 86: 889-901.

[72] Serra M, Miquel L, Domenzain C, et al. V3 versican isoform expression alters the phenotype of melanoma cells and their tumorigenic potential. *Int. J. Cancer* 2005; 114: 879-86.

[73] Theocharis AD, Seidel C, Borset M, et al. Serglycin constitutively secreted by myeloma plasma cells is a potent inhibitor of bone mineralization in vitro. *J. Biol. Chem.* 2006; 281: 35116-28.

[74] Ricciardelli C, Russell DL, Ween MP, et al. Formation of hyaluronan- and versican-rich pericellular matrix by prostate cancer cells promotes cell motility. *J. Biol. Chem.* 2007; 282: 10814-25.

[75] Basappa, Murugan S, Sugahara KN, et al. Involvement of chondroitin sulfate E in the liver tumor focal formation of murine osteosarcoma cells. *Glycobiology* 2009; 19: 735-42.

[76] ten Dam GB, van de Westerlo EM, Purushothaman A, et al. Antibody GD3G7 selected against embryonic glycosaminoglycans defines chondroitin sulfate-E domains highly up-regulated in ovarian cancer and involved in vascular endothelial growth factor binding. *Am. J. Pathol.* 2007; 171: 1324-33.

In: Signaling, Gene Regulation and Cancer
Editors: Shree Ram Singh and Manoj K. Mishra
ISBN: 978-1-61942-088-5

Chapter 3

NFAT SIGNALING AND TRANSCRIPTION IN CANCER DEVELOPMENT AND PROGRESSION

Sandra Baumgart, Elisabeth Glesel, Shiv K. Singh, Garima Singh, Nai-Ming Chen, Kristina Reutlinger and Volker Ellenrieder*
Klinik für Innere Medizin, SP Gastroenterologie
Universitätsklinikum Marburg
Baldingerstrasse, Marburg, Germany

ABSTRACT

The nuclear factor of activated T cells (NFAT) family of transcription factors comprise 4 Ca^{2+}-Calcineurin regulated members which play important roles in the development and function of the immune system. NFAT signaling is induced by Ca^{2+}-Calcineurin mediated dephosphorylation and subsequent nuclear translocation of NFAT proteins which in turn faciliates DNA binding and activation of target genes, notably in cooperation with partner proteins. NFAT activity is tightly regulated by various control mechanisms, such as (re-) phosphorylation by export and maintenance kinases, as well as endogenous inhibitors of calcineurin. Recent studies have discovered that signaling through particularly NFATc1 and NFATc2 is crucially involved in several diseases, including many cancer entities. Basically, various mechanisms of NFAT-driven tumorigenesis have been described which implicate selection of partner proteins or posttranslational modifications, amongst others. Thus, NFAT regulates differentiation, apoptosis, proliferation, migration and angiogenesis to balance normal and uncontroled cell growth dependent on the cellular context. In pancreatic cancer, both NFATc1 and NFATc2 are highly overexpressed and maintain tumor growth via diverse routes: For instance, NFATc2 cooperates with Glycogen synthase kinase-3β (GSK-3β) and transcription factor STAT3 to sustain cell proliferation. Apart from gene activation NFATc2 can act as a repressor, as it silences expression of tumor suppressor gene $p15^{INK4B}$ and consequently enhances tumor cell growth. NFATc1 moreover forms complexes with transcription factor Elk-1 on the DNA to induce local chromatin acetylation and transcription of the oncogene c-myc. Taken together, NFAT signaling has

* Correspondence to: Volker Ellenrieder; Email: ellenrie@med.uni-marburg.de.

emerged as a multifunctional and powerful regulatory principle governing growth, survival and homeostasis of transformed as well as untransformed cells and tissues and this renders them interesting and potentially powerful tools for targeted tumor therapy.

THE NFAT FAMILY OF TRANSCRIPTION FACTORS

Nuclear factor of activated T cells (NFAT) has originally been identified as an inducible nuclear factor which binds to, and upregulates transcription from, the interleukin-2 (IL-2) gene promoter in T cells following stimulation of the T cell antigen receptor [1]. It was soon recognized that the NFAT proteins comprise a family of transcription factors consisting of four closely related proteins, NFATc1 (NFATc, NFAT2), NFATc2 (NFATp, NFAT1), NFATc3 (NFATx, NFAT4) and NFATc4 (NFAT3), as well as a more distant relative, NFAT5 (TonEBP: tonicity element binding protein). Following their initial discovery, a multitude of studies quickly established that NFAT proteins are expressed not only in T cells, but many different cells of the immune system, and that signaling through NFAT proteins is crucial for the development and function of the immune system [for reviews, see 2,3].

The NFAT proteins c1-c4 are at the heart of Ca^{2+} signal transduction in immune cells (Figure 1, left panel). A rise in intracellular calmodulin-(CaM)-bound Ca^{2+} levels activates the heterodimeric phosphatase calcineurin, which dephosphorylates multiple serine residues in the aminoterminal regions of NFAT proteins. Dephosphorylation exposes previously masked nuclear localization sequences (NLS), resulting in the rapid translocation of NFAT proteins from the cytoplasm into the nucleus [4,5]. Since NFAT proteins bind their DNA target sequences only with relatively weak affinity, they regulate many, if not all, of their target gene promoters in cooperation with partnering transcription factors [6-8]. NFAT transcription complexes thus can be seen as integrators of signals from different pathways, where one signal must be Ca^{2+}/calcineurin, while the other signal(s) may be from divers other pathways [9]. This also explains, in part, the broad diversity of cellular responses that Ca^{2+} signaling triggers depending on the cell type or the state of cellular differentiation and/or activation in cells inside or outside of the immune system. Nuclear export, and thus inactivation, of NFAT proteins is induced by re-phosphorylation of the N-terminus of NFAT proteins by specific kinases such as GSK-3 [10]-Phosphorylation-induced conformational changes most likely lead to the exposure of a nuclear export signal (NES), which is responsible for the rapid and Crm-1 mediated translocation to the cytoplasm [11].

The crucial role that NFAT signaling plays in the initiation of a productive immune response is highlighted by the enormous immunosuppressive potential of the calcineurin inhibitors FK506 (tacrolimus) and cyclosporine A (CsA). These microbial agents have revolutionized transplant therapy by their ability to effectively prevent the immune response against allogeneic tissue which normally leads to rejection of the transplant. FK506 and CsA bind to two different intracellular proteins, FKBP and cyclophilin, respectively, and subsequently bind to and inhibit the calcineurin complex [9,12-14]. While calcineurin has been shown to dephosphorylate target proteins other than NFAT proteins, the inhibitory action of CsA and FK506 seem to be surprisingly specific for NFAT-mediated signaling.

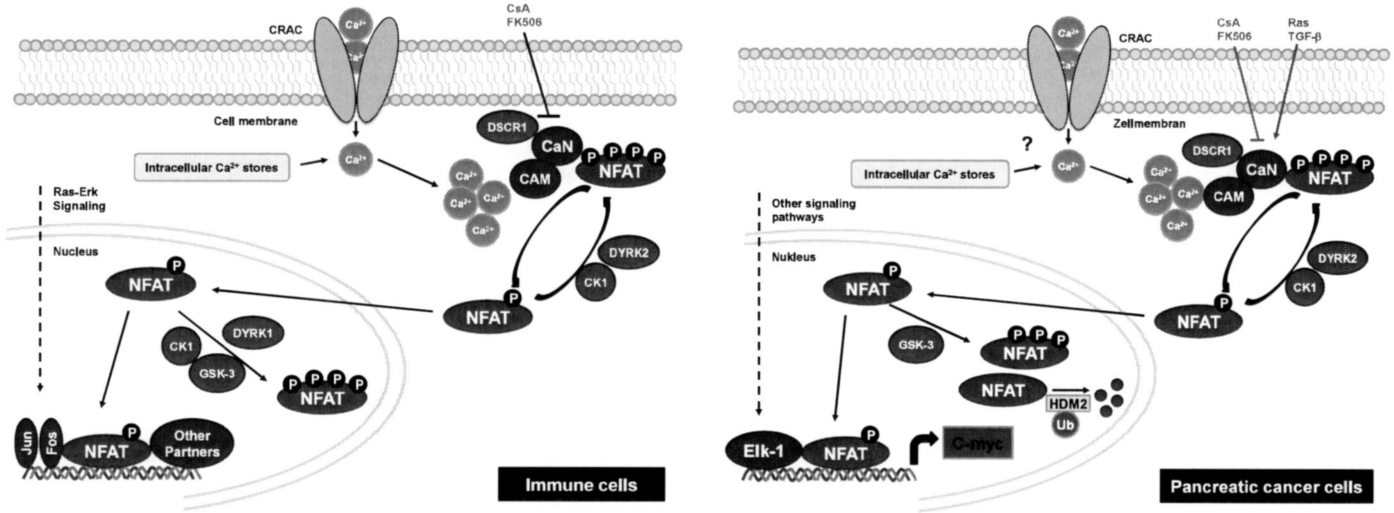

Figure 1. Regulatory circuit of NFAT transcription factors. In Immune cells, activation of intracellular calcium stores opens Ca^{2+} -release activated channels (CRAC) to increase extracellular Ca^{2+} influx. Ca^{2+} binds to calmodulin (CAM) and activates Calcineurin (CaN), which in turn dephosphorylates NFAT proteins. CaN activity is counteracted by endougenous inhibitors such as Down's syndrome candidate region 1 (DSCR1) or maintenance kinases dual-specificity tyrosine-phosphorylation regulated kinase 2 (DYRK2) and Casein Kinase 1 (CK1) which maintain NFAT in a hyperphosphorylated state. Pharmacological inhibitors of CaN are cyclosporin A (CsA) and tacrolimus (FK506). Dephosphorylated NFAT proteins translocate to the nucleus and bind to their GGAAA target sequences on the DNA. Efficient binding requires association with partner proteins, for instance jun-fos complex AP-1 is best described to cooperate with NFAT in immune cells. Termination of DNA binding is achieved by export kinases CK1, DYRK1 and glycogen synthase kinase 3 (GSK-3) which rephosphorylate NFAT in the nucleus and induce re-shuttling to the cytoplasm (left panel). In pancreatic cancer cells (right panel), NFAT activation can be additionally stimulated by Ras and transforming growth factor β (TGF-β) signaling. Upon activation, NFAT associates with partner transcription factor Elk-1 to induce transcription of c-myc and subsequent tumor growth. GSK-3β stabilizes NFAT in the nucleus by phosphorylation of three serines in the SP2 region and consequently prevents a rapid degradation through E3 ligase human double minute 2 (HDM2).

This is for example evidenced by the fact that the phenotype of NFATc3/c4 double knockout mice is nearly identical to that of embryos of mothers treated with CsA during early embryonic development [15]. Both drugs thus represent extremely valuable pharmacological tools for the study of Ca^{2+}/calcineurin/NFAT-mediated signaling events. Recent studies however have demonstrated that the vital role of NFAT signaling is by no means restricted to the immune system. NFAT proteins participate in the regulation of genes influencing the development and differentiation of numerous mammalian cells and tissues. It has been shown, for instance, that NFAT proteins control multiple steps in myogenesis, chondrocyte differentiation and the development of the cardiovascular system [9,16-21]. In addition, recent evidence suggests that NFAT proteins, and in particular NFATc1, regulate important cellular processes such as proliferation and apoptosis in different cell types including epithelial cells, fibroblasts and preadipocytes [22-24]. For instance, growth stimulatory effects of NFATc1 and NFATc2 have been demonstrated in skeletal muscle [25-27] and heart valve development [28,29]. Furthermore, NFAT transcription factors have been shown to control peripheral vascular development during angiogenesis [15,30,31], and to play roles in apoptosis regulation both in immune and non-immune cells [32-35]. Finally, an essential contribution to the regulation of quiescence and self-renewal of stem cells mediated by Ca^{2+}/calcineurin/NFAT signaling has been evidenced [36,37]. One of the reasons why the universal importance of Ca^{2+}/calcineurin/NFAT signaling for the development and homeostasis of many cells and tissues has for a long time gone more or less unnoticed may be the fact that no homologues of the calcium-sensitive NFAT proteins NFATc1-c4 exist in invertebrates [36,38]. The intriguing implication of this is that the evolution of NFATc proteins is tightly linked to the development of fundamental vertebrate-specific features such as a complex nervous system, a vascular system with a complex heart, and a recombinational immune system.

Regulation of NFAT Activity

NFAT – Structural Features

The predicted primary structure of NFAT is schematically illustrated in Figure 2. Two major regions of sequence homology, comprising the Rel-like DNA-binding domain (DBD) and the NFAT homology region (NHR), are represented in both NFATc1 and NFATc2 isoforms. The DNA-binding domain, which is located between amino acid residues ≈400 and ≈700, is highly conserved within the NFAT family and shows moderate sequence similarity to the DNA-binding domains of Rel-family proteins. The ≈300 amino acids spanning NFAT homology region (NHR) is encoded in a single exon in all NFAT proteins and is located N-terminal to the DNA-binding domain [15]. The NHR is heavily phosphorylated in resting cancer cells, with the phosphorylated serine residues distributed among conserved serine-rich sequence motifs termed SRR1-3 and SP1-3 regions [39, 40]. The NHR domain also contains a critical nuclear localization sequence (NLS) and defines two regions responsible for calcineurin binding. The regions outside the DNA-binding and regulatory domains show relatively little sequence conservation among NFAT family members. Noteworthy, the N-terminal ≈100 amino acid region of NFATc2 is not conserved in sequence but works as a transactivation domain (TAD) through its ability to recruit coactivators such as CBP, p300 and JAB1 that enhance NFAT mediated transcription [9].

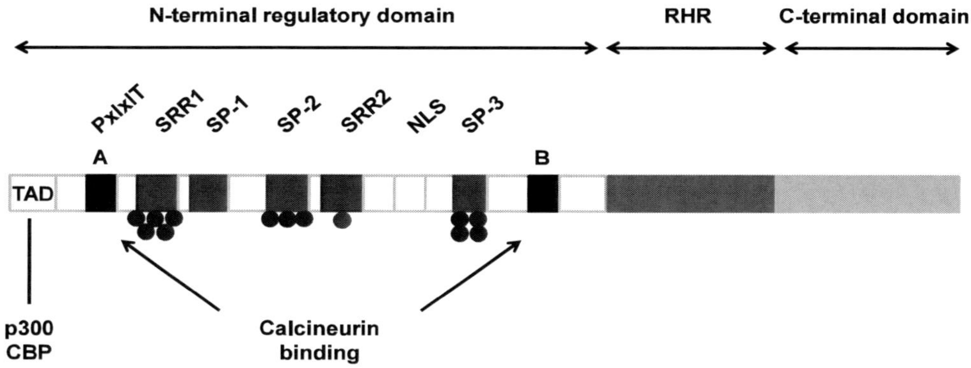

Figure 2. Schematic representation of NFAT primary structure. The domains depicted are retained in NFATc1 and c2 isoforms. The N-terminal regulatory domain comprises the transactivation domain (TAD) which can be acetylated by histone acetyle transferases p300/CBP, calcineurin binding sequences A and B as well as serine-rich regions (SRR) and serine-proline rich motifs (SP). The latter regions contain phosphorylation sites, indicated by blue circles for GSK-3, CK1 and DYRK kinases. The red circle indicates a calcineurin-independent site with unclear function.

Activation of NFAT Signaling

NFAT factors are calcium/calcineurin responsive proteins which are functionally regulated at the level of their subcellular localization (Figure 1). In resting cancer cells, NFAT factors are located in the cytoplasm and in a highly phosphorylated, inactive state. Ligand binding of many receptors results in the activation of phospholipase C (PLC), release of IP_3, and a transient release of Ca^{2+} from intracellular stores through IP_3 receptors. This initial release of Ca^{2+} is not sufficient to activate NFAT target genes in a number of cell types. Rather an influx of Ca^{2+} through specialized Ca^{2+} release activated channels (termed CRAC) is required. CRAC provide the persistent Ca^{2+} signal that is necessary for sufficient calcineurin activation and the subsequent dephosphorylation of cytoplasmic NFAT proteins. Noteworthy, calcineurin activity may also be controlled independently of Ca^{2+}. To date, four cellular inhibitors of calcineurin phosphatase complexes have been identified, all of which are able to block nuclear translocation of NFAT proteins: the calcineurin B homolog, CHP [41], the calcineurin inactivator CABIN [42,43], the AKAP79 scaffold protein [44], and finally, members of the DSCR/MCIP/RCAN family of endogenous calcineurin inhibitors. In case of sustained Ca^{2+} increase and calcineurin activation, the phosphatase targets and dephosphorylates moderately conserved serine rich motifs at the NHR of NFAT to unmask its nuclear localization signals [39,40]. Subsequently, NFAT proteins shuttle into the nucleus where they accumulate on target promoter sites [40, 45-47]. It is not formally known whether dephosphorylation is an ordered process, although mass spectrometry experiments suggest that the SRR1 region, which is immediately adjacent to the major calcineurin docking site (PxIxIT), is preferentially dephosphorylated at low calcineurin activity [40]. Efficient dephosphorylation requires a sucessfull docking interaction between NFAT and calcineurin

[48-50]. The major docking site for calcineurin is located at the N-terminus of the NFAT regulatory domain with the consensus sequence PxIxIT (SPRIEIT in NFATc2, see Figure 2). Current studies suggest that NFAT nuclear localization is accompanied by a relocalization of calcineurin from the cytoplasm into the nucleus [51]. This process has been documented for endogenous calcineurin and NFATc2 in primary keratinocytes [52] and in pancreatic cancer cells as well [53 and Ellenrieder V., unpublished results]. In fact, NFAT and calcineurin display a nuclear co-localisation in tumor cells from early pancreatic cancer stages, and with a higher frequency in tumor cells from advanced stages of the disease [53]. Consistently, NFAT and calcineurin are predominantly expressed in the nuclear fraction of pancreatic cancer cells with sustained Ca^{2+} influx. In the nucleus, calcineurin maintains NFAT proteins in their dephosphorylated form and thus in a transcriptionally active status.

Transcriptional Mechanisms

Upon nuclear translocation, NFAT factors recognize the GGAAA consensus sequence within target promoter elements and bind DNA either as homodimers, heterodimers or through interaction with other transcription factors [54,55]. In fact, NFAT proteins frequently cooperate with other transcription factors to elicit high-affinity binding on common target promoters. The best documented example is the unrelated transcription factor activator protein-1 (AP-1; comprised of Fos–Jun complexes), which forms a quaternary complex with NFAT factors and DNA. AP-1 is the *bonafide* transcriptional partner for NFATs during T-cell activation, but might also be involved in NFAT regulated gene expression in carcinogenesis. NFAT:AP-1 complexes serve as signal integrators for two diametrically different signaling pathways, the Ca^{2+}/calcineurin pathway that activates NFAT and the mitogen activated protein kinase (MAPK) pathway that promotes the synthesis and activation of Fos and Jun family members. In addition to AP-1, NFAT proteins cooperate with numerous other signaling regulated transcription factors with implications in cell activation, differentiation and adaptation such as GATA proteins, FoxP3 and members of the MEF-family. Depending on the partners and cofactors they bind, NFAT factors can either enhance local chromatin acetylation and induce target promoter activation, or interact with histone deacetylases to silence target genes [55-57].

In pancreatic cancer cells, NFATc1 and NFATc2 promote cell cycle progression and growth through complex formation with the ets-like factor Elk-1 (Figure 1, right panel). Sequential chromatin immunoprecipitation analysis revealed that NFAT binding to the serum responsive element stimulates the histone acetyltransferase (HAT) p300 to create a hyperacetylated chromatin state at the proximal c-Myc promoter (Figure 3). This event is required for the consecutive inducible recruitment of Elk-1 and the resulting induction of c-Myc expression following stimulation of the Ca^{2+}/calcineurin pathway [55,57]. Disruption of this transcription complex prevents both c-Myc promoter transactivation and pancreatic cancer growth stimulation *in vitro* and *in vivo*. Recent evidence also indicates that certain NFAT complexes might include transcriptional repressor proteins and members of the HDAC family of histone deactylases that cooperate with the Ca^{2+}/calcineurin pathway to mediate gene silencing in non-transformed cells and in cancer as well. In osteoblasts, for instance,

NFATc1 acts as a transcriptional co-repressor that sustains HDAC3 binding on the proximal region of the osteocalcin promoter in order to silence its activity and transcription [56].

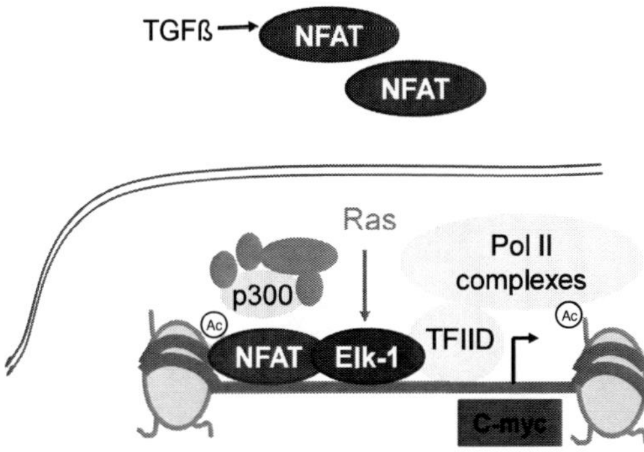

Figure 3. Signal integration by NFAT and partner transcription factors. Activated NFAT factors rapidly bind to and induce transcription from a consensus (GGAAA) promoter sequence localized within the serum responsive element of the c-Myc promoter. NFAT binding to this promoter region enhances the net histone acetylase activity associated with the c-Myc promoter and subsequently allows recruitment of a second Ras-MAPK signaling regulated transcription factor (ELK-1) to recruit Polymerase II complexes that consequently induce c-Myc expression.

Another good example is the NFATc1/C isoform that following sumoylation translocates to promyelocytic leukemia nuclear (PML) bodies, where it interacts with histone deacetylases to induce transcriptionally inactive chromatin at the interleukin-2 promoter [58]. This recent finding by the Serfling's laboratory demonstrated for the first time that the modification by SUMO (small ubiquitin-like modifier) can convert NFATc1/C from an activator to a site-specific transcriptional repressor. In line with these findings, NFATc2-mediated repression of the tumor-suppressor gene p15^{INK4b} could recently been demonstrated [Baumgart S. et al., in press]. Thereby, sequential ChIP analyses have revealed that upon mitogenic stimulation, NFATc2 recruits a complex consisting of histone methyl transferase Suv39H1 and heterochromatin protein 1γ (HP1γ) which in turn bind to and silence the p15^{INK4b} promoter through local Histone 3 Lysine 9 (H3K9) trimethylation (Figure 4). Consequently, facultative heterochromatin formation confers a switch to the repressor function of NFATc2. Thus, NFAT partnering proteins help control the specificity of NFAT promoter binding and the resulting mode of action.

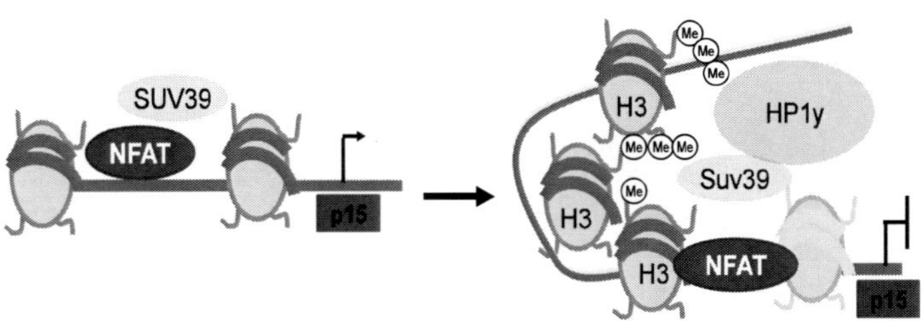

Figure 4. NFAT mediated recruitment of repressor complexes. NFAT factors recruit histone methyl transferase Suv39H1 and subsequently heterochromatin protein 1γ (HP1γ) to reduce trimethylation of histone 3 which induces repression of tumor suppressor p15^{INK4b}.

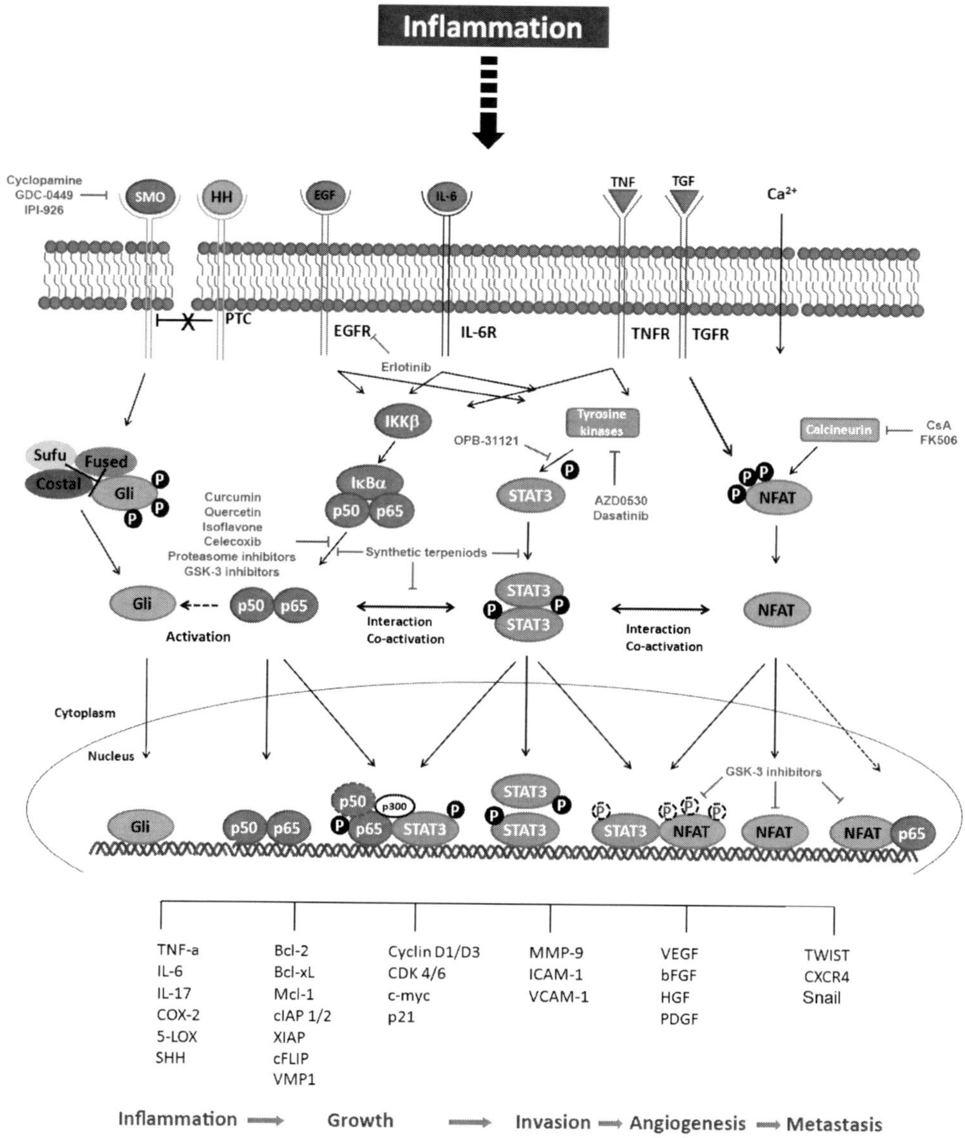

Figure 5. Putative cross-talk between oncogenic transcription factors contributing to pancreatic carcinogenesis. Inflammation induces binding of cytokines and inflammatory ligands to their receptors, which in turn activate the corresponding transcription factor signalling pathways. Binding of HH to the transmembrane protein PTC reverses the inhibitory effect on SMO, which again releases GLI transcription factors from their inhibitory complex and finally initiates nuclear translocation. The activation of the pathway can be blocked by inhibitors of SMO such as cyclopamine and its derivatives currently in clinical studies, GDC-0449 and IPI-926. Interactions with the NFκB signalling pathway are illustrated by the dashed arrowhead, which indicates activation of the HH pathway mainly through the regulation of the expression of its ligand SHH. Binding of EGF or IL-6 activates IKKβ kinase, which stimulates degradation of IκBα. Subsequently, NFκB proteins are released to the nucleus, where they mediate gene transcription alone or in cooperation with Stat proteins. Curcumin, quercetin, isoflavone and proteasome or GSK-3 inhibitors interfere with NFκB signalling mainly through inhibition of NFκB activation. Thereby, p-Stat3 binding to p-p65 induces histone acetyltransferase p300 and retains active

p65 in the nucleus. Synthetic terpenoids can either block isolated NFκB and Stat3 activation or inhibit the interaction of both pathways. A putative binding of p50 to Stat3 is shown by the dashed circle. The Stat3 signalling pathway can also be activated by the aforementioned inflammatory stimuli. Upon binding of the ligands to the transmembrane receptors, (receptor-bound) kinases, such as JAKs, SRC or ABL, are activated and in turn phosphorylate Stat3 proteins. EGFR inhibitor, erlotinib, as well as tyrosine kinase inhibitors, AZD0530 and dasatinib, have been shown to efficiently block Stat3 activation. The subsequent dimerisation leads to nuclear translocation and target gene transcription, partially by interaction with NFAT transcription factors. The interaction on target promoters is not dependent on Stat and NFAT phosphorylation, illustrated by dashed circles. The canonical NFAT signalling pathway is activated by intracellular Ca2+ increases leading to activation of the phosphatase calcineurin and dephosphorylation of NFAT proteins, which shuttle to the nucleus and bind to their target promoters. Calcineurin inhibitors, CsA and FK506, block NFAT dephosphorylation and nuclear translocation. NFAT phosphorylation does not exclude nuclear localisation and binding to partner proteins, shown by dashed white circles. The dashed arrow indicates a possible interaction with p65, which has been demonstrated for other non-pancreatic tissues. GSK-3 inhibitors successfully disrupt NFAT binding to partner proteins such as Stat3 as well as NFAT transcriptional activity in the nucleus. Finally, the network of inflammatory transcription factors regulates numerous target genes mediating inflammation, growth, invasion, angiogenesis and metastasis, thereby contributing to pancreatic carcinogenesis. Bcl-2/xL, B cell lymphoma 2/xL; bFGF, basic fibroblast growth factor; CDK 4/6, cyclin-dependent kinase 4/6; cIAP 1/2, cellular inhibitor of apoptosis; cFLIP, cellular FLICE (FADD-like IL-1b-converting enzyme)-inhibitory protein; COX-2, cyclooxygenase-2; CsA, cyclosporine A; CXCR4, CXC-motif chemokine receptor 4; EGF(R), epidermal growth factor receptor; GSK-3, glycogen synthase kinase- 3; HGF, hepatocyte growth factor; HH, Hedgehog; ICAM-1, intercellular adhesion molecule 1;IKKβ, IκB kinase b; IκBα, inhibitor of NFkB a; IL-6(R), interleukin-6 (receptor); JAK, janus kinase; 5-LOX, 5-lipo-oxygenase; Mcl-1, myeloid cell leukaemia sequence 1; MMP-9, matrix metaloproteinase 9; NFAT, nuclear factor of activated T cells; PDGF, plateletderived growth factor; PTC, Patched; SHH, Sonic HH; SMO, Smoothened; Stat3, signal transducer and activator of transcription 3; Sufu, Suppressor of Fused; TGF(R), transforming growth factor (receptor); TNF(R), tumour necrosis factor (receptor); VCAM-1, vascular cell adhesion molecule 1; VEGF, vascular endothelial growth factor; VMP1, vacuole membrane protein 1; XIAP, X-linked inhibitor of apoptosis protein.

Recently, additional transcription factors have been identified that form stable nuclear complexes with NFATc2 in pancreatic cancer cells [Baumgart S., unpublished results]. This for instance involves the signal transducer and activator of transcription-3 (Stat3) protein, a latent signal-induced transcription factor that has been implicated in numerous biological processes, including cell transformation and tumorigenesis [59-61]. Like NFAT proteins, Stat3 is also primarily regulated at the level of its subcellular localization [59-61]. In normal resting cells, Stat3 resides in a non-phosphorylated version in the cytoplasm. However, following cytokine and growth factor stimulation, Stat3 proteins are inducibly phosphorylated on a critical regulatory tyrosine residue (Y-705) that promotes their homodimerization and subsequent translocation into the nucleus where they control gene transcription. The Clipstone's group has recently shown that Stat3 is activated in pancreatic cancer cells through an NFAT-dependent autocrine factor [62]. Moreover, genetic depletion of Stat3 attenuates the transformation capacity of NFATc1, suggesting a cooperative function of both factors in cancer. In line, we have most recently identified Stat3 as one of a few inflammatory transcription factors that form nuclear complexes with NFATc2. The Stat3-NFAT interaction is tightly regulated on the level of acetylation mediated by GSK-3β and appears to be of outmost importance for NFATc2 dependent cancer progression [63, Baumgart S., unpublished results]. Similar to Stat3, NFAT might in response to inflammatory signals recruit and, most importantly, combine the actions of other signaling regulated transcription factors (e.g. Stat3, NF-kB, and Gli) to integrate pathway-specific signals into Ca^{2+}/calcineurin regulated transcription (Figure 5). Thus, structural and biochemical data indicate that NFAT transcription complexes function as signal integrators and detectors. One signal must be Ca^{2+}/calcineurin [64], while the second can be inflammatory, developmental or oncogenic such as the Ras-MAP kinase pathway [65].

Similar to sumoylation and phosphorylation, additional posttranslational modifications such as ADP-ribosylation can determine the mode and magnitude of NFAT transcription. In an elegant approach Olabisi and coworkers have shown that poly-ADP-ribose polymerase PARP-1 binds to NFAT factors and induces ADP-ribosylation [66]. ADP-ribosylation is a reversible posttranslational modification that transfers ADP-ribose from NAD^+ to Glu, Asp, and/or Arg amino acids of target proteins [66]. ADP-ribosylation modifies target proteins to various masses due to the assorted chain lengths of the ADP-ribose. Poly-ADP-ribose polymerase-1 (PARP-1) is a nuclear enzyme that accounts for the bulk of ADP-ribosylation in vivo. ADP-ribosylation of NFAT provides a molecular switch to positively regulate NFAT-dependent gene transcription through increased DNA binding activity [65]. Similar to phosphorylation, ADP-ribosylation occurs in a reversible fashion, and hydrolysis of ADP-ribose is mediated by PARG. Given that cytoplasmic PARG accounts for most of the cellular PARG activity, it is tempting to speculate that compartmentalization of the molecules involved in NFAT phosphorylation, sumoylation and ADP-ribosylation (PARP-1 and PARG) provides fine-tuning of NFAT-mediated gene transcription.

Thus, NFAT target gene selection and regulation is defined by the nuclear composition and selection of nuclear partner proteins and the presence and activity of NFAT modifying enzymes in the nucleus at a given time.

Termination of Transcription and Nuclear Export

When Ca^{2+} entry is prevented or calcineurin activity is inhibited, nuclear import of the NFATc family is opposed by a highly efficient nuclear export mechanism ($t_{1/2} \approx 15$ min) that requires the sequential actions of a nuclear priming kinase, such as dual-specificity tyrosine kinase 1a (Dyrk1a) [67,68] (Figure 1, left panel). Notably, Dyrk1a phosphorylates nuclear NFAT, priming it for phosphorylation by GSK-3β and subsequent nuclear export. Consequently, NFAT proteins accumulate in the cytosol. Both genetic and biochemical experiments indicate that in addition to Dyrk1a the protein kinase A (PKA) can serve as a priming kinase [10,69]. It is however unclear whether all NFAT proteins are phoshorylated by identical kinases and whether phosphorylation by GSK-3β always results in a negative regulation. It also appears possible that the functional consequence of NFAT phosphorylation depends on the cell-type and the nuclear context at a given time. For instance, others and we have most recently identified a novel and unexpected function of GSK-3β in the regulation of NFATc2 in cancer cells [70,71]. These data implicate a role for GSK-3β-mediated phosphorylation of the serine rich SP2 domain in protein stabilization and life-time regulation of NFATc2. GSK-3β targets nuclear NFATc2 and protects the transcription factor from rapid ubiquitinylation and proteasomal degradation in cancer (Figure 1, right panel). Probably, additional modifications then contribute to successful NFAT export to the cytosol. Once in the cytosol, NFATc retention is controlled through maintenance kinases phosphorylating the unstructured N-terminus [72]. These include c-Jun N-terminal kinase (JNK), extracellular signal-regulated kinase (ERK), p38 mitogen-activated protein kinase (MAPK) and casein kinases [73,74]. The fact that the N-termini are unstructured indicates that many kinases can probably phosphorylate them *in vitro*. To date, there is no gentic evidence that these kinases limit NFAT signaling, and, indeed, activation of MAPKs facilitates NFAT signaling rather than opposing it, as would be expected if they had a role in keeping NFATc in the cytoplasm [75].

NFAT Signaling in Tumor Development and Progression

The broad spectrum of NFAT-mediated processes and the central importance for growth and differentiation of many different cells and tissues strongly implicates a distinct oncogenic potential of NFAT transcription factors. Of special interest in this context is the role of Ca^{2+}/calcineurin/NFAT signaling in cell cycle control, since calcium signaling has been shown to promote cell cycle progression and G1/S phase transition in a variety of normal cells [76-78] as well as transformed cells [79]. Moreover, it has been demonstrated that overexpression of a constitutively active NFATc1 mutant is sufficient to induce a transformed phenotype in pre-adipocyte 3T3-L1 fibroblasts, associated with altered expression of cell-cycle related genes such as cyclin D1, cyclin D2, pRB and, most notably, c-Myc [23] However, direct evidence for prominent roles of NFAT signaling in human cancer is only just beginning to accumulate.

Additional findings for a role of NFAT signaling in the development and the phenotypic behaviour of epithelial tumors come from studies on pancreatic cancer, breast cancer and colon carcinoma, at which the contribution to pancreatic cancer will be discussed below. The Toker laboratory, for instance, had previously discovered that NFATc2 and NFAT5 are downstream mediators of α6β4 integrin signaling and promote cancer invasion [80]. In colon cancer cells, activation of NFATc1 and c2 might contribute to the transcriptional upregulation of the *COX-2* gene, [81] which in turn is a major player in epithelial carcinogenesis. Depending on the tumor type and origin, Cox-2 initiates prostanglandin mediated signaling that finally leads to increased cellular proliferation, promotion of angiogenesis, inhibition of apoptosis and stimulation of invasion/motility. Most recently, the Toker laboratory reported an unexpecting finding in NFAT regulation which identified the Akt kinase as a potential negative regulator of NFATc2 [82]. Akt is a well-known proto-oncogene which has been shown to promote proliferation, growth and survival in a variety of cancer cells. Using constitutively active Akt constructs and Akt-specific siRNA, however, Yoeli-Lerner et al. demonstrated that Akt downregulates the transcriptional activity of NFAT by drastically reducing NFAT protein levels in breast cancer cell lines *via* stimulation of HDM2-mediated NFAT ubiquitination. Akt-induced NFATc2 degradation resulted in significantly impaired migratory and invasive potential of the carcinoma cells, suggesting that while Akt promotes tumor progression through increased cell survival mechanisms, it may at the same time inhibit cell motility and invasion by a mechanism that depends, at least in part, on NFAT proteins [70].

Noteworthy, NFATc1 and NFATc2 might have redundant but also opposing functions in tumorigenesis. Tumor suppressor activities have recently been demonstrated for NFATc2 in fibroblasts, whereas NFATc1 was described in the same study as a strong oncogene able to transform cells [83]. In this recent study by Robbs and colleagues, constitutively active NFATc2 induced cell cycle arrest and inhibited H-Ras induced transformation, whereas introduction of constitutively active NFATc1 showed opposite effects and caused increased proliferation and promotion of cell transformation. In contrast to these findings, a fusion between NFATc2 and ESWR1 genes has been reported to be causal for the development of Ewings sarcoma, an aggressive sarcoma in childhood [84]. These observations underscore the idea that NFATc1 and NFATc2 partially share target specificity but also regulate a non-overlapping subset of transcriptional targets, and therefore have non-redundant functions during cancer development and progression. However, whether or not NFAT proteins exert redundant or opposite effects in cancer cells might also dependent on the origin of the cancer and the nuclear context given in a cell. Accordingly, we and others have reported strong growth promoting functions for both NFAT proteins in pancreatic, colon and breast cancer and in particular in those with a highly invasive and metastatic phenotype [for reviews, see 63,85]. Based on recent key findings, NFAT signaling has thus emerged as a multifunctional and powerful regulatory principle governing growth, survival and homeostasis of transformed as well as untransformed cells and tissues.

NFAT Oncogenic Functions in Pancreatic Cancer

We have recently demonstrated that NFATc1 is ectopically expressed and highly activated in pancreatic cancer cells *in vivo* and *in vitro*. [53] Like NFATc1, the NFATc2 isoform is also frequently overexpressed in solid tumors, including pancreatic adenocarcinoma [Baumgart S. et al., in press], and correlated with increased cell proliferation and invasion. Holzmann and colleagues moreover demonstrated chromosomal amplification of NFATc2 in the majority of pancreatic cancers [86]. In immunohistochemical analyses, 70 % of human pancreatic cancer specimens and all of the seven pancreatic cancer cell lines analyzed displayed strong nuclear staining for NFATc1 and concurrent presence of calcineurin, while staining for both proteins were weak or absent in normal exocrine pancreas or inflammatory control tissues. Proliferation and anchorage-independent growth of cultured pancreatic cancer cells was significantly attenuated by inhibition of Ca^{2+}/calcineurin signaling with CsA or siRNA-mediated knock-down of NFATc1. Intriguingly, however, this was not true for all cell lines under investigation. While ASPC-1 and Panc-1 cells reacted with cell cycle arrest to CsA treatment, no significant effect on proliferation and cell cycle progression was observed for IMIM-PC2 and (murine) TD-2 cells [53]. Incidently, the non-responsive cell lines had previously been shown to harbor genomic amplifications of the c-Myc gene, whereas the responsive Panc-1 and ASPC-1 cell lines lack genomic amplifications of the c-Myc gene, respectively [87,88]. We subsequently demonstrated that in responsive cell lines, activated NFATc1 directly binds to a previously described regulatory element within the proximal c-Myc promoter termed the TGFβ inhibitory element (TIE), which has been shown to harbor binding sites for members of several different families of transcription factors and to play a critical role in the regulation of c-Myc expression during the G1-phase [89,90]. In particular, the TIE is responsible for the downregulation of c-Myc expression in response to activation of the TGFβ signaling cascade, which in turn is a key feature of efficient growth control in normal epithelial cells. Binding of NFATc1 to the TIE strongly stimulated the transcriptional activity of both, the isolated TIE fragment as well as the full length c-Myc promoter. In responsive pancreatic cancer cells, stimulation of c-Myc expression by activated NFATc1 led to accelerated G1/S phase transition as well as significantly increased anchorage-independent growth, which is considered a hallmark of malignant transformation of epithelial cells. Our results thus identify ectopic expression of NFATc1 in conjunction with aberrant activation of the Ca^{2+}/calcineurin signaling cascade as a novel mechanism of trancriptional overexpression of c-Myc in pancreatic cancer [91]. The importance of this novel regulatory mechanism is emphasized by the fact that overexpression of the c-Myc protein is observed in up to 70 % of human pancreatic tumors [91], while genomic amplifications of the *c-Myc* gene locus are found in only ca. 30 % of cases [92]. Our data additionally confirmed that upregulation of c-Myc by NFATc1 and NFATc2 represents a critical step in TGFβ-induced growth stimulation in pancreatic cancer. Mechanistically, TGFβ activates NFAT proteins to displace Smad3 repressor complexes from the TGFβ inhibitory element (TIE element) of the proximal c-Myc promoter and induce transcription of the oncogenic transcription factor [91]. Upon induction, c-Myc feeds back on NFAT to form complexes to induce growth promoting transcriptional programmes in cancer. Upon stimulation, nuclear NFAT factors bind to and induce transcription from key cell cycle regulators such as CDK4, CDK6 and cyclin D resulting in accelerated cell cycle transition [23] on the one hand, and otherwise they bind to tumor suppressor gene $p15^{INK4b}$ promoter to induce silencing thereof [Baumgart S. et al., in press].

As inflammation is one major risk factor for the development of pancreatic cancer, and as NFAT factors mediate inflammatory processes [93], it appears possible that they link inflammation to carcinogenesis in cooperation with other inflammatory transcription factors, such as Stat3 [63]. Our own observations most notably suggest a striking role of NFATc1 in Kras-driven carcinogenesis, accompanied by immune infiltration, using transgenic mouse models of pancreatic cancer. Herein, NFATc1 probably serves as fine-tuning signal integrator mediating differentiation and growth of pancreatic ductal cells and in turn pancreatic cancer development [Ellenrieder V., personal communication]. Together, these recent studies indicate a strong and multifaceted contribution of NFAT transcription factors to the development and maintenance of pancreatic cancer.

Another hallmark of tumorigenesis is the ability of cancer cells to evade cell death. Programmed cell death by apoptosis occurs in virtually all cell types and is precisely regulated at cellular and molecular levels. NFAT proteins have shown to exert anti-apoptotic properties in numerous cancer types [85]. Overexpression of NFATc1 or NFATc2 in adipocytes or lymphocytes protected cells from undergoing apoptosis in response to growth factor withdrawal [23] or following antigen stimulation, respectively [94]. In pancreatic cancer cells, genetic depletion of either factor results in increased sensitivity to gemcitabine mediated cell death [55].

Besides growth stimulation and survival, NFAT proteins play key roles in migration and invasion of breast and pancreatic cancer cells [85]. The mechanisms by which NFAT factors function as pro-invasion transcription factors also lies within the transcriptional programme of genes that are induced in tumor cells. For instance, NFAT proteins induce the transcription of cyclooxygenase 2 (COX2) and of the genes encoding autotaxin, exonucleotide pyrophosphatase and phosphodiesterase 2 (ENPP2). Autotaxin is a secreted protein that converts lysophosphatidylcholine into lysophosphatidic acid (lPA), a potent mitogenic and motogenic factor for cancer cells [85]. COX2 catalyses the synthesis of prostaglandins such as prostaglandin E2 (PGE2) in tumor cells and in endothelial cells as well, leading to increased tumor cell migration and endothelial cell proliferation [95,96]. Moreover, knockdown of COX2 or treatment with COX inhibitors such as non-steroidal anti-inflammatory drugs, rendered pancreatic cancer cells less invasive both *in vitro* and *in vivo* [95,96]. Thus, COX2 has emerged as a key enzyme in tumor cell migration and the metastatic dissemination of most human tumors, including pancreatic cancer cell infiltration to neighboring organ structures. A key step in pancreatic cancer cell migration and invasion is the induction and activation of matrix metalloproteinases such as MMP-2 [97]. In a genetic mouse model using highly metastatic osteosarcoma cells, a role of NFAT proteins in MMP-2 induction has been demonstrated [98]. Increased levels of MMP-2 expression and activity have also been reported in highly invasive and migrating pancreatic cancer cells and in particular in those with increased activation of TGFβ signaling [97]. Since NFATc1 and NFATc2 themselves are TGFβ inducible transcription factors in pancreatic cancer cells, it will be interesting to elucidate their role in TGFβ mediated gene expression during cancer progression. In pancreatic cancer, TGFβ has a dual function, working as a growth inhibitor at early tumor stages and a strong promoter of growth and migration in advanced stages of the disease. TGFβ then stimulates the expression of key cell cycle regulators (e.g. c-Myc and cyclin D) and promotes the induction of an epithelial-to-mesenchymal transdifferentiation, a hallmark of tumor cell migration and invasion [99]. Whether NFAT proteins contribute to TGFβ driven carcinogenesis and whether they work independent or in concert with the Smad signaling molecules to mediate transcription in response to TGFβ is currently under investigation.

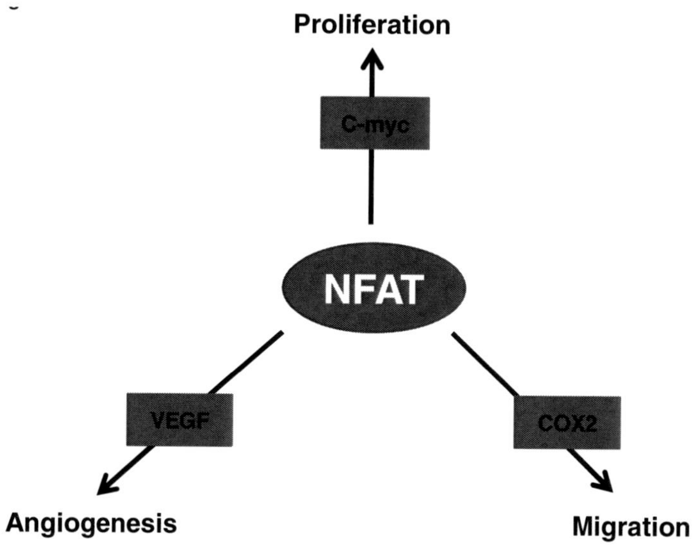

Figure 6. Functional implications of NFAT signaling and transcription during carcinogenesis. NFAT members play major roles in mediation of key oncogenic functions such as proliferation, angiogenesis and tumor cell migration through their abilities to control gene expression. Critical mediators of these functions are vascular endothelial growth factor (VEGF), Cyclooxygenase 2 (COX2) and the transcription factor C-Myc.

Summarily, growing evidence from biochemical, genetic and functional approaches suggest a multifunctional and powerful role for NFAT isoforms in pancreatic carcinogenesis. Therefore, recent research in normal and transformed cell types has uncovered at least three major pathways through which NFAT-mediated Ca^{2+}/calmodulin signaling may influence central aspects of tumor biology: Stimulation of angiogenesis through upregulation of VEGF, stimulation of tumor cell proliferation through upregulation of c-Myc, and stimulation of tumor cell migration through COX2 (Figure 6). Consequently, NFAT proteins have emerged as interesting therapeutic targets in pancreatic cancer therapy.

NFAT Targeting Strategies

Based on their oncogenic potential, one would predict that pharmacological targeting of NFAT factors could be an attractive goal in medical oncology. Two structurally unrelated drugs, CsA and FK506 are potent inhibitors of the calcineurin–NFAT pathway and hence are widely used as immunosuppressive agents in tissue and organ transplant to prevent rejection [100]. Both drugs bind to calcineurin and prevent access of NFAT proteins, thus inhibiting their consecutive dephosphorylation and activation. Inactivation of the calcineurin-NFAT pathway represents a hub of antigen specific T cell activation and differentiation, and therefore inhibition of its activity leads to a complete block of the adaptive immune response, thus preventing transplant rejection. However, the spectrum of calcineurin substrates is not limited to NFAT proteins neither in immune cells nor outside the immune system and most notably, calcineurin can control the activity of other key signaling pathways including the TGFβ-Smad pathway and the mitogenic Ras-MAP kinase cascade [100]. In addition, calcineurin regulates the expression and activation of inflammatory transcription factors such as NF-κB, AP-1, and Elk1 as well as the calcium responsive CREB protein [101,102]. It is therefore not surprising that inhibition of calcineurin through CsA or FK506 does not solely block the expression of NFAT regulated target genes but also targets NFAT independent gene regulation in cancer. This might also explain the toxic side effects that are frequently observed under treatment with CsA or FK506 such as neurotoxicity, nephrotoxicity and the development of high blood pressure [100]. Nonetheless, one would predict that as potent calcineurin–NFAT inhibitors, both CsA and FK506 could work as cancer therapeutics, and in fact pre-clinical studies demonstrate successful tumor growth retardation upon treatment. Somewhat paradoxically, there is actually a significant increase in cancer incidence in patients on long-term immunosuppressive treatments with CsA or FK506, which is most likely due to suppression of local tumor immunosurveillance [85]. Therefore, new treatment strategies that specifically switch off NFAT signaling in the tumor endothelium and the tumor cells themselves, without affecting the local immune response in the tumor environment, are urgently needed. Much effort has been made in the last decade to identify and characterize compounds that work downstream of calcineurin to specifically block NFAT activation and transcription. One such compound is UR-1505, a pentafluoropropoxy derivative of salicylic acid [103]. UR-1505 blocks NFAT-DNA binding and hence antagonizes NFAT mediated transcription in different transformed and non-transformed cell lines. UR-1505 shows no effect on NFAT nuclear-cytoplasmic shuttling and neither interferes with calcineurin activity nor with NFAT independent transcription. Another interesting approach is the use of peptides that specifically interfere with NFAT docking to calcineurin and thus block NFAT dephosphorylation and nuclear translocation. First evidence obtained from cell culture and animal studies suggest antitumor potential of these peptides, called VIVIT, in clinical use. VIVIT proteins have been shown to prolong graft survival in mice and attenuate cancer cell invasion without affecting other known calcineurin downstream substrates [104,105].

Further strategies to block NFAT activity in pancreatic cancer include targeting of endogenous activators of NFAT, such as GSK-3β [63]. Recent studies revealed that the inhibition of GSK-3β kinase activity disrupts Stat3-NFATc2 interaction and consequently reduces NFAT transcriptional activity and cell growth in pancreatic xenograft mouse models [63, Baumgart S., unpublished observations]. Moreover GSK-3β acts to stabilize NFATc2 protein and accordingly stimulates its oncogenic functions [71]. As chronic inflammation represents a risk factor for the development of pancreatic cancer, inflammatory pathways should be considered as critical therapeutic targets. GSK-3β regulates the transcription factors NF-kB, Stat3, NFAT and possibly GLI, in different cells present within the tumor microenvironment [reviewed in 63]. As these transcription factors are also correlated to inflammatory responses, they emerged as potential drug targets in inflammatory diseases. Inflammation is frequently linked to carcinogenesis and as such these proteins primarily represent key modulators of inflammatory processes promoting cancer development. This renders GSK-3β an interesting tool to repress (inflammation-mediated) carcinogenesis.

Finally, some drugs originally introduced for other applications turned out to act as inhibitors of calcineurin/NFAT activation. Among them is the nitrogen containing bisphosphonate zoledronic acid. For instance, we have most recently shown that zoledronic acid blocks cell growth through induction of NFATc2 ubiquitinylation and proteasomal degradation in pancreatic cancer *in vitro* and *in vivo*, through inhibition of GSK-3β kinase activity and induction of HDM2 ligase [71].

REFERENCES

[1] Shaw, JP, Utz, PJ, Durand, DB, Toole, JJ, Emmel, EA, Crabtree, GR. Identification of a putative regulator of early T cell activation genes. *Science* 1988; 241:202-205.

[2] Feske, S, Okamura, H, Hogan, PG, Rao, A. Ca2+/calcineurin signalling in cells of the immune system. *Biochem. Biophys. Res. Commun.* 2003; 311:1117-1132.

[3] Macian, F. NFAT proteins: key regulators of T-cell development and function. *Nat. Rev. Immunol.* 2005; 5:472-484.

[4] Luo, C, Shaw, KT, Raghavan, A, Aramburu, J, Garcia-Cozar, F, Perrino, BA, Hogan, PG, Rao, A. Interaction of calcineurin with a domain of the transcription factor NFAT1 that controls nuclear import. *Proc. Natl. Acad. Sci. USA* 1996; 93:8907-8912.

[5] Shibasaki, F, Price, ER, Milan, D, McKeon, F. Role of kinases and the phosphatase calcineurin in the nuclear shuttling of transcription factor NF-AT4. *Nature* 1996; 382:370-373.

[6] Chen, L, Glover, JN, Hogan, PG, Rao, A, Harrison, SC. Structure of the DNA-binding domains from NFAT, Fos and Jun bound specifically to DNA. *Nature* 1998; 392:42-48.

[7] Garcia-Cozar, FJ, Okamura, H, Aramburu, JF, Shaw, KT, Pelletier, L, Showalter, R, Villafranca, E, Rao, A. Two-site interaction of nuclear factor of activated T cells with activated calcineurin. *J. Biol. Chem.* 1998; 273:23877-23883.

[8] Loh, C, Shaw, KT, Carew, J, Viola, JP, Luo, C, Perrino, BA, Rao, A. Calcineurin binds the transcription factor NFAT1 and reversibly regulates its activity. *J. Biol. Chem.* 1996; 271:10884-10891.

[9] Crabtree, GR, Olson, EN. NFAT signaling: choreographing the social lives of cells. *Cell* 2002; 109 Suppl:S67-79.:S67-S79.

[10] Beals, CR, Sheridan, CM, Turck, CW, Gardner, P, Crabtree, GR. Nuclear export of NF-ATc enhanced by glycogen synthase kinase-3. *Science* 1997; 275:1930-1934.

[11] Neal, JW, Clipstone, NA. Glycogen synthase kinase-3 inhibits the DNA binding activity of NFATc. *J. Biol. Chem.* 2001; 276:3666-3673.

[12] Kissinger, CR, Parge, HE, Knighton, DR, Lewis, CT, Pelletier, LA, Tempczyk, A, Kalish, VJ, Tucker, KD, Showalter, RE, Moomaw, EW. Crystal structures of human calcineurin and the human FKBP12-FK506-calcineurin complex. *Nature* 1995; 378:641-644.

[13] Kawamura, A, Su, MS. Interaction of FKBP12-FK506 with calcineurin A at the B subunit-binding domain. *J. Biol. Chem.* 1995; 270:15463-15466.

[14] Milan, D, Griffith, J, Su, M, Price, ER, McKeon, F. The latch region of calcineurin B is involved in both immunosuppressant-immunophilin complex docking and phosphatase activation. *Cell* 1994; 79:437-447.

[15] Graef, IA, Chen, F, Chen, L, Kuo, A, Crabtree, GR. Signals transduced by Ca(2+)/calcineurin and NFATc3/c4 pattern the developing vasculature. *Cell* 2001; 105:863-875.

[16] Chang, CP, Neilson, JR, Bayle, JH, Gestwicki, JE, Kuo, A, Stankunas, K, Graef, IA, Crabtree, GR. A field of myocardial-endocardial NFAT signaling underlies heart valve morphogenesis. *Cell* 2004; 118:649-663.

[17] Friday, BB, Horsley, V, Pavlath, GK. Calcineurin activity is required for the initiation of skeletal muscle differentiation. *J. Cell Biol.* 2000; 149:657-666.

[18] Graef, IA, Chen, F, Crabtree, GR. NFAT signaling in vertebrate development. *Curr. Opin. Genet. Dev.* 2001; 11:505-512.

[19] Graef, IA, Wang, F, Charron, F, Chen, L, Neilson, J, Tessier-Lavigne, M, Crabtree, GR. Neurotrophins and netrins require calcineurin/NFAT signaling to stimulate outgrowth of embryonic axons. *Cell* 2003; 113:657-670.

[20] Hirotani, H, Tuohy, NA, Woo, JT, Stern, PH, Clipstone, NA. The calcineurin/nuclear factor of activated T cells signaling pathway regulates osteoclastogenesis in RAW264.7 cells. *J. Biol. Chem.* 2004; 279:13984-13992.

[21] Mammucari, C, di Vignano, AT, Sharov, AA, Neilson, J, Havrda, MC, Roop, DR, Botchkarev, VA, Crabtree, GR, Dotto, GP. Integration of Notch 1 and Calcineurin/NFAT signaling pathways in keratinocyte growth and differentiation control. *Dev. Cell* 2005; 8:665-676.

[22] Hogan, PG, Chen, L, Nardone, J, Rao, A. Transcriptional regulation by calcium, calcineurin, and NFAT. *Genes Dev.* 2003; 17:2205-2232.

[23] Neal, JW, Clipstone, NA. A constitutively active NFATc1 mutant induces a transformed phenotype in 3T3-L1 fibroblasts. *J. Biol. Chem.* 2003; 278:17246-17254.

[24] Viola, JP, Carvalho, LD, Fonseca, BP, Teixeira, LK. NFAT transcription factors: from cell cycle to tumor development. *Braz. J. Med. Biol. Res.* 2005; 38:335-344.

[25] Pavlath, GK, Horsley, V. Cell fusion in skeletal muscle--central role of NFATC2 in regulating muscle cell size. *Cell Cycle* 2003; 2:420-423.

[26] Horsley, V, Pavlath, GK. Prostaglandin F2(alpha) stimulates growth of skeletal muscle cells via an NFATC2-dependent pathway. *J. Cell Biol.* 2003; 161:111-118.

[27] Musaro, A, McCullagh, KJ, Naya, FJ, Olson, EN, Rosenthal, N. IGF-1 induces skeletal myocyte hypertrophy through calcineurin in association with GATA-2 and NF-ATc1. *Nature* 1999; 400:581-585.

[28] de la Pompa, JL, Timmerman, LA, Takimoto, H, Yoshida, H, Elia, AJ, Samper, E, Potter, J, Wakeham, A, Marengere, L, Langille, BL, Crabtree, GR, Mak, TW. Role of the NF-ATc transcription factor in morphogenesis of cardiac valves and septum. *Nature* 1998; 392:182-186.

[29] Ranger, AM, Grusby, MJ, Hodge, MR, Gravallese, EM, de la Brousse, FC, Hoey, T, Mickanin, C, Baldwin, HS, Glimcher, LH. The transcription factor NF-ATc is essential for cardiac valve formation. *Nature* 1998; 392:186-190.

[30] Hernandez, GL, Volpert, OV, Iniguez, MA, Lorenzo, E, Martinez-Martinez, S, Grau, R, Fresno, M, Redondo, JM. Selective inhibition of vascular endothelial growth factor-mediated angiogenesis by cyclosporin A: roles of the nuclear factor of activated T cells and cyclooxygenase 2. *J. Exp. Med.* 2001; 193:607-620.

[31] Zaichuk, TA, Shroff, EH, Emmanuel, R, Filleur, S, Nelius, T, Volpert, OV. Nuclear factor of activated T cells balances angiogenesis activation and inhibition. *J. Exp. Med.* 2004; 199:1513-1522.

[32] Benedito, AB, Lehtinen, M, Massol, R, Lopes, UG, Kirchhausen, T, Rao, A, Bonni, A. The transcription factor NFAT3 mediates neuronal survival. *J. Biol. Chem.* 2005; 280:2818-2825.

[33] Chuvpilo, S, Jankevics, E, Tyrsin, D, Akimzhanov, A, Moroz, D, Jha, MK, Schulze-Luehrmann, J, Santner-Nanan, B, Feoktistova, E, Konig, T, Avots, A, Schmitt, E, Berberich-Siebelt, F, Schimpl, A, Serfling, E. Autoregulation of NFATc1/A expression facilitates effector T cells to escape from rapid apoptosis. *Immunity* 2002; 16:881-895.

[34] Iwai-Kanai, E, Hasegawa, K. Intracellular signaling pathways for norepinephrine- and endothelin-1-mediated regulation of myocardial cell apoptosis. *Mol. Cell Biochem.* 2004; 259:163-168.

[35] Kawamura, T, Ono, K, Morimoto, T, Akao, M, Iwai-Kanai, E, Wada, H, Sowa, N, Kita, T, Hasegawa, K. Endothelin-1-dependent nuclear factor of activated T lymphocyte signaling associates with transcriptional coactivator p300 in the activation of the B cell leukemia-2 promoter in cardiac myocytes. *Circ. Res.* 2004; 94:1492-1499.

[36] Oro AE. A new role for an old friend: NFAT and stem cell quiescence. *Cell Stem Cell* 2008; 2: 104-106.

[37] Li X, Zhu L, Yang A, Lin J, Tang F, Jin S, Wei Z, Li J, Jin Y. Calcineurin-NFAT signaling critically regulates early lineage specification in mouse embryonic stem cells and embryos. *Cell Stem Cell* 2011; 8:46-58.

[38] Graef, IA, Gastier, JM, Francke, U, Crabtree, GR. Evolutionary relationships among Rel domains indicate functional diversification by recombination. *Proc. Natl. Acad. Sci. USA* 2001; 98:5740-5745.

[39] Beals CR, Clipstone NA, Ho SN, Crabtree GR: Nuclear localization of NF-ATc by a calcineurin-dependent, cyclosporin-sensitive intramolecular interaction. *Genes Dev.* 1997;11:824-834.

[40] Okamura H, Aramburu J, García-Rodríguez C, Viola JP, Raghavan A, Tahiliani M, Zhang X, Qin J, Hogan PG, Rao A: Concerted dephosphorylation of the transcription factor NFAT1 induces a conformational switch that regulates transcriptional activity. *Mol. Cell* 2000;6:539-550.

[41] Lin X, Sikkink RA, Rusnak F, Barber DL: Inhibition of calcineurin phosphatase activity by a calcineurin B homologous protein. *J. Biol. Chem.* 1999;274:36125-31.

[42] Lai MM, Burnett PE, Wolosker H, Blackshaw S, Snyder SH: Cain, a novel physiologic protein inhibitor of calcineurin. *J. Biol. Chem.* 1998;273:18325-18331.

[43] Sun L, Youn HD, Loh C, Stolow M, He W, Liu JO: Cabin 1, a negative regulator for calcineurin signaling in T lymphocytes. *Immunity* 1998;8:703-711.

[44] Klauck TM, Faux MC, Labudda K, Langeberg LK, Jaken S, Scott JD: Coordination of three signaling enzymes by AKAP79, a mammalian scaffold protein. *Science* 1996;271:1589-1592.

[45] Shaw KT, Ho AM, Raghavan A, Kim J, Jain J, Park J, Sharma S, Rao A, Hogan PG: Immunosuppressive drugs prevent a rapid dephosphorylation of transcription factor NFAT1 in stimulated immune cells. *Proc. Natl. Acad. Sci. U S A* 1995; 92:11205-11209.

[46] Porter CM, Havens MA, Clipstone NA: Identification of amino acid residues and protein kinases involved in the regulation of NFATc subcellular localization. *J. Biol. Chem.* 2000;275: 3543-3551.

[47] Neal JW, Clipstone NA: Glycogen synthase kinase-3 inhibits the DNA binding activity of NFATc. *J. Biol. Chem.* 2001; 276: 3666–3673.

[48] Aramburu J, Garcia-Cózar F, Raghavan A, Okamura H, Rao A, Hogan PG: Selective inhibition of NFAT activation by a peptide spanning the calcineurin targeting site of NFAT. *Mol. Cell.* 1998; 1:627-637.

[49] Aramburu J, Yaffe MB, López-Rodríguez C, Cantley LC, Hogan PG, Rao A: Affinity-driven peptide selection of an NFAT inhibitor more selective than cyclosporin A. *Science* 1999; 285:2129-2133.

[50] Chow CW, Rincón M, Davis RJ: Requirement for transcription factor NFAT in interleukin-2 expression. *Mol. Cell. Biol.* 1999; 19:2300-2307.

[51] Shibasaki F, Price ER, Milan D, McKeon F: Role of kinases and the phosphatase calcineurin in the nuclear shuttling of transcription factor NF-AT4. *Nature* 1996; 382: 370–373.

[52] Al-Daraji WI, Grant KR, Ryan K, Saxton A, Reynolds NJ: Localization of calcineurin/NFAT in human skin and psoriasis and inhibition of calcineurin/NFAT activation in human keratinocytes by cyclosporin A. *J. Invest. Dermatol.* 2002; 118:779-788.

[53] Buchholz M, Schatz A, Wagner M, Michl P, Linhart T, Adler G, Gress TM, Ellenrieder V: Overexpression of c-myc in pancreatic cancer caused by ectopic activation of NFATc1 and the Ca2+/calcineurin signaling pathway. *EMBO J.* 2006; 25:3714-3724.

[54] Buchholz M and Ellenrieder V: An Emerging Role for Ca2+/Calcineurin/NFAT Signaling in Cancerogenesis. *Cell Cycle* 2007; 6:16-19.

[55] Koenig A, Fernandez-Zapico ME, Ellenrieder V. Primers on molecular pathways – the NFAT transcription pathway in pancreatic cancer. *Pancreatology* 2010; 10:416-422

[56] Choo MK, Yeo H, Zayzafoon M: NFATc1 mediates HDAC-dependent transcriptional repression of osteocalcin expression during osteoblast differentiation. *Bone* 2009; 45:579-589.

[57] Koenig A, Linhart T, Schlengemann K, Reutlinger K, Wegele J, Adler G, Singh G, Hofmann L, Kunsch S, Büch T, Schäfer E, Gress TM, Fernandez-Zapico ME, Ellenrieder V: NFAT-induced histone acetylation relay switch promotes c-Myc-dependent growth in pancreatic cancer cells. *Gastroenterology* 2010; 138:1189-1199.

[58] Nayak A, Glöckner-Pagel J, Vaeth M, Schumann JE, Buttmann M, Bopp T, Schmitt E, Serfling E, Berberich-Siebelt F: Sumoylation of the transcription factor NFATc1 leads to its subnuclear relocalization and interleukin-2 repression by histone deacetylase. *J. Biol. Chem.* 2009; 284:10935-10946.

[59] Darnell JE Jr: STATs and gene regulation. *Science* 1997; 277:1630-1635.

[60] Bromberg JF: Activation of STAT proteins and growth control. *Bioessays* 2001; 23:161-169.

[61] Yu H, Jove R: The STATs of cancer--new molecular targets come of age. *Nat. Rev. Cancer* 2004; 4:97-105.

[62] Lagunas L, Clipstone NA: Deregulated NFATc1 activity transforms murine fibroblasts via an autocrine growth factor-mediated Stat3-dependent pathway. *J. Cell. Biochem.* 2009; 108:237-248.

[63] Baumgart S., Ellenrieder V, Fernandez-Zapico ME. Oncogenic transcription factors: cornerstones of inflammation-linked pancreatic carcinogenesis. *Gut* 2011; [Epub ahead of print].

[64] Crabtree GR: Contingent genetic regulatory events in T lymphocyte activation. *Science* 1989; 243:355-361.

[65] Woodrow M, Clipstone NA, Cantrell D: p21ras and calcineurin synergize to regulate the nuclear factor of activated T cells. *J. Exp. Med.* 1993; 178:1517-1522.

[66] Hassa PO, Haenni SS, Elser M, Hottiger MO: Nuclear ADP-ribosylation reactions in mammalian cells: where are we today and where are we going? *Microbiol. Mol. Biol. Rev.* 2006; 70:789-829.

[67] Gwack Y, Sharma S, Nardone J, Tanasa B, Iuga A, Srikanth S, Okamura H, Bolton D, Feske S, Hogan PG, Rao A. A genome-wide Drosophila RNAi screen identifies DYRK-family kinases as regulators of NFAT. *Nature.* 2006; 441:646-650.

[68] Kuhn C, Frank D, Will R, Jaschinski C, Frauen R, Katus HA, Frey N. DYRK1A is a novel negative regulator of cardiomyocyte hypertrophy. *J. Biol. Chem.* 2009; 284:17320-17327.

[69] Arron JR, Winslow MM, Polleri A, Chang CP, Wu H, Gao X, Neilson JR, Chen L, Heit JJ, Kim SK, Yamasaki N, Miyakawa T, Francke U, Graef IA, Crabtree GR: NFAT dysregulation by increased dosage of DSCR1 and DYRK1A on chromosome 21. *Nature* 2006; 441:595-600.

[70] Yoeli-Lerner M, Chin YR, Hansen CK, Toker A. Akt/protein kinase b and glycogen synthase kinase-3beta signaling pathway regulates cell migration through the NFAT1 transcription factor. *Mol. Cancer. Res.* 2009; 7:425-432

[71] Singh SK, Baumgart S, Singh G, König AO, Reutlinger K, Hofbauer LC, Barth P, Gress TM, Lomberk G, Urrutia R, Fernandez-Zapico ME, Ellenrieder V. Disruption of a nuclear NFATc2 protein stabilization loop confers breast and pancreatic cancer growth suppression by zoledronic acid. *J. Biol. Chem.* 2011; 268:28761-29771

[72] Wolfe SA, Zhou P, Dötsch V, Chen L, You A, Ho SN, Crabtree GR, Wagner G, Verdine GL: Unusual Rel-like architecture in the DNA-binding domain of the transcription factor NFATc. *Nature* 1997;385:172-176.

[73] Porter CM, Havens MA, Clipstone NA: Identification of amino acid residues and protein kinases involved in the regulation of NFATc subcellular localization. *J. Biol. Chem.* 2000;275:3543-3551.

[74] Zhu J, Shibasaki F, Price R, Guillemot JC, Yano T, Dötsch V, Wagner G, Ferrara P, McKeon F: Intramolecular masking of nuclear import signal on NF-AT4 by casein kinase I and MEKK1. *Cell* 1998;93:851-861.

[75] Whitehurst CE, Geppert TD: MEK1 and the extracellular signal-regulated kinases are required for the stimulation of IL-2 gene transcription in T cells. *J. Immunol.* 1996;156:1020-1029.

[76] Lipskaia, L, Lompre, AM. Alteration in temporal kinetics of Ca2+ signaling and control of growth and proliferation. *Biol. Cell.* 2004; 96:55-68.

[77] Tomono, M, Toyoshima, K, Ito, M, Amano, H, Kiss, Z. Inhibitors of calcineurin block expression of cyclins A and E induced by fibroblast growth factor in Swiss 3T3 fibroblasts. *Arch. Biochem. Biophys.* 1998; 353:374-378.

[78] Terada, N, Lucas, JJ, Gelfand, EW. Differential regulation of the tumor suppressor molecules, retinoblastoma susceptibility gene product (Rb) and p53, during cell cycle progression of normal human T cells. *J. Immunol.* 1991; 147:698-704.

[79] Mosieniak, G, Pyrzynska, B, Kaminska, B. Nuclear factor of activated T cells (NFAT) as a new component of the signal transduction pathway in glioma cells. *J. Neurochem.* 1998; 71:134-141.

[80] Jauliac, S, Lopez-Rodriguez, C, Shaw, LM, Brown, LF, Rao, A, Toker, A. The role of NFAT transcription factors in integrin-mediated carcinoma invasion. *Nat. Cell. Biol.* 2002; 4:540-544.

[81] Duque, J, Fresno, M, Iniguez, MA. Expression and function of the nuclear factor of activated T cells in colon carcinoma cells: involvement in the regulation of cyclooxygenase-2. *J. Biol. Chem.* 2005; 280:8686-8693.

[82] Yoeli-Lerner, M, Yiu, GK, Rabinovitz, I, Erhardt, P, Jauliac, S, Toker, A. Akt blocks breast cancer cell motility and invasion through the transcription factor NFAT. *Mol. Cell.* 2005; 20:539-550.

[83] Robbs BK, Cruz AL, Werneck MB, Mognol GP, Viola JP: Dual roles for NFAT transcription factor genes as oncogenes and tumor suppressors. *Mol. Cell. Biol.* 2008; 28:7168-7181.

[84] Szuhai K, Ijszenga M, de Jong D, Karseladze A, Tanke HJ, Hogendoorn PC. The NFATc2 gene is involved in a novel cloned translocation in a Ewing sarcoma variant that couples its function in immunology to oncology. *Clin. Cancer. Res.* 2009; 15:2259-2268

[85] Mancini M, Toker A. NFAT proteins: emerging roles in cancer progression. *Nat. Rev. Cancer* 2009; 9:810-820

[86] Holzmann K, Kohlhammer H, Schwaenen C, Wessendorf S, Kestler HA, Schwoerer A, Rau B, Radlwimmer B, Döhner H, Lichter P, Gress T, Bentz M. Genomic DNA-chip hybridization reveals a higher incidence of genomic amplifications in pancreatic cancer than conventional comparative genomic hybridization and leads to the identification of novel candidate genes. *Cancer Res.* 2004; 64:4428-4433

[87] Schreiner, B, Greten, FR, Baur, DM, Fingerle, AA, Zechner, U, Bohm, C, Schmid, M, Hameister, H, Schmid, RM. Murine pancreatic tumor cell line TD2 bears the characteristic pattern of genetic changes with two independently amplified gene loci. *Oncogene* 2003; 22:6802-6809.

[88] Yagi, K, Furuhashi, M, Aoki, H, Goto, D, Kuwano, H, Sugamura, K, Miyazono, K, Kato, M. c-myc is a downstream target of the Smad pathway. *J. Biol. Chem.* 2002; 277:854-861.

[89] Chen, CR, Kang, Y, Massague, J. Defective repression of c-myc in breast cancer cells: A loss at the core of the transforming growth factor beta growth arrest program. *Proc. Natl. Acad. Sci. U S A* 2001; 98:992-999.

[90] Li, YJ, Wei, ZM, Meng, YX, Ji, XR. Beta-catenin up-regulates the expression of cyclinD1, c-myc and MMP-7 in human pancreatic cancer: relationships with carcinogenesis and metastasis. *World J. Gastroenterol.* 2005; 11:2117-2123.

[91] Singh G, Singh SK, König A, Reutlinger K, Nye MD, Adhikary T, Eilers M, Gress TM, Fernandez-Zapico ME, Ellenrieder V. Sequential activation of NFAT and c-Myc transcription factors mediates the TGF-beta switch from a suppressor to a promoter of cancer cell proliferation. *J. Biol. Chem.* 2010; 285:27241-27250

[92] Schleger, C, Verbeke, C, Hildenbrand, R, Zentgraf, H, Bleyl, U. c-MYC activation in primary and metastatic ductal adenocarcinoma of the pancreas: incidence, mechanisms, and clinical significance. *Mod. Pathol.* 2002; 15:462-469.

[93] Müller MR, Rao A. NFAT, immunity and cancer. A transcription factor comes of age. *Nat. Rev. Immunol.* 2010; 10:645-56

[94] Caetano MS, Vieira-de-Abreu A, Teixeira LK, Werneck MB, Barcinski MA, Viola JP: NFATC2 transcription factor regulates cell cycle progression during lymphocyte activation: evidence of its involvement in the control of cyclin gene expression. *FASEB J.* 2002;16:1940-1942.

[95] Juuti A, Louhimo J, Nordling S, Ristimäki A, Haglund C. Cyclooxygenase-2 expression correlates with poor prognosis in pancreatic cancer. *J. Clin. Pathol.* 2006;59:382-386

[96] Eibl G, Takata Y, Boros LG, Liu J, Okada Y, Reber HA, Hines OJ. Growth stimulation of COX-2-negative pancreatic cancer by a selective COX-2 inhibitor. *Cancer Res.* 2005;65:982-990.

[97] Ellenrieder V, Hendler SF, Ruhland C, Boeck W, Adler G, Gress TM: TGF-beta-induced invasiveness of pancreatic cancer cells is mediated by matrix metalloproteinase-2 and the urokinase plasminogen activator system. *Int. J. Cancer* 2001;93:204-211.

[98] Velupillai P, Sung CK, Tian Y, Dahl J, Carroll J, Bronson R, Benjamin T. Polyoma virus-induced osteosarcomas in inbred strains of mice: host determinants of metastasis. *PLoS Pathog.* 2010; 22;:e1000733.

[99] Ellenrieder V, Hendler SF, Boeck W, Seufferlein T, Menke A, Ruhland C, Adler G, Gress TM. Transforming growth factor beta1 treatment leads to an epithelial-mesenchymal transdifferentiation of pancreatic cancer cells requiring extracellular signal-regulated kinase 2 activation. *Cancer Res.* 2001;61:4222-4228.

[100] Sieber M, Baumgrass R. Novel inhibitors of the calcineurin/NFATc hub - alternatives to CsA and FK506? Cell Commun Signal. 2009; 7:1-19.

[101] Frantz B, Nordby EC, Bren G, Steffan N, Paya CV, Kincaid RL, Tocci MJ, O'Keefe SJ, O'Neill EA: Calcineurin acts in synergy with PMA to inactivate I kappa B/MAD3, an inhibitor of NF-kappa B. *Embo J* 1994, 13:861-870.

[102] Sugimoto T, Stewart S, Guan KL: The calcium/calmodulindependent protein phosphatase calcineurin is the major Elk- 1 phosphatase. *J. Biol. Chem.* 1997; 272: 29415-29418.

[103] Roman J, de Arriba AF, Barron S, Michelena P, Giral M, Merlos M, Bailon E, Comalada M, Galvez J, Zarzuelo A, Ramis I: UR-1505 – a new salicylate, blocks T cell activation through nuclear factor of activated T cells. *Mol. Pharmacol.* 2007, 72(2):269-279.

[104] Aramburu J, Garcia-Cozar F, Raghavan A, Okamura H, Rao A, Hogan PG: Selective inhibition of NFAT activation by a peptide spanning the calcineurin targeting site of NFAT. *Mol. Cell* 1998; 1:627-637.

[105] Noguchi, H. *et al.* A new cell-permeable peptide allows successful allogeneic islet transplantation in mice. *Nature Med.* 2004; 10:305–309

In: Signaling, Gene Regulation and Cancer
Editors: Shree Ram Singh and Manoj K. Mishra
ISBN: 978-1-61942-088-5

Chapter 4

JAK/STAT SIGNALING IN CELL FATE AND TUMORIGENESIS

Shree Ram Singh*[*,1] and Manoj K. Mishra[2]
[1]Mouse Cancer Genetics Program, National Cancer Institute,
National Institutes of Health, Frederick, MD, US
[2]Department of Math and Science, Alabama State University,
Montgomery, AL, US

ABSTRACT

The Janus kinase (JAK)/signal transducer and activator of transcription (STAT) pathway plays an important role in a wide variety of biological processes such as cell fate, proliferation, migration, polarity and immune response. JAK/STAT signaling was originally identified in vertebrates and is highly conserved from yeast to human. The misregulation JAK/STAT signaling results in various human diseases, such as immune disorders, metabolic diseases, and tumorigenesis. In this chapter, we presented the role of JAK/STAT signaling on cell fate, proliferation, and tumor formation in *Drosophila*.

Keywords: JAK/STAT; *Drosophila*, stem cell fate, tumor formation

INTRODUCTION

The Janus kinase (JAK)/signal transducer and activator of transcription (STAT) signal transduction pathway was originally identified through studies on the transcriptional activation response to a variety of cytokines and growth factors [1-4]. This pathway is highly conserved from yeast to humans. JAK/STAT pathway plays an essential role in transducing the cytokine signal and regulates various biological processes including fetal development,

* Corresponding author: Shree Ram Singh, Ph. D. Mouse Cancer Genetics Program, National Cancer Institute, National Institutes of Health, Frederick, MD 21702, USA Tel: 301-846-7331 E-mail: singhshr@mail.nih.gov.

cell growth, migration, differentiation, apoptosis, transformation, stem cell fate, inflammation, and immune response.

In the mammalian system, four members of JAK kinase family (JAKs 1–3 and Tyk2) has been identified and are known to associated with cytokine and growth factor receptors. JAK-mediated signaling leads to activation of a signal transduction cascade that involves STAT proteins. The mammalian STAT family of transcription factors contains seven proteins (STATs 1-4, 5A, 5B, and 6), which is tyrosine phosphorylated in response to ligand-induced receptor stimulation. The phosphorylation leads to dimer formation and translocation to the nucleus where STATs bind DNA and positively or negatively regulate transcription [11-13]. In addition to main components, there are several effector proteins have been identified, which are involved in the subset of JAK/STAT signaling events. They are STAMs (signal-transducing adapter molecules), STATIP (STAT-Interacting Protein), and SH2B/Lnk/APS family. In addition to JAK/STAT pathway effectors, there are three major classes of negative regulator: SOCS (suppressors of cytokine signaling), PIAS (protein inhibitors of activated STATs), and PTPs (protein tyrosine phosphatases).

Even though the basic mechanism of JAK/STAT pathway is relatively simple, the biological consequences of this pathway activation are complicated by interactions with other signaling pathways. It is known that JAK/STAT signaling indirectly promotes Ras signaling through the transcriptional activation of SOCS3. In contrast, RTK pathway promotes JAK/STAT signaling by activating EGFR, PDGFR, and MAPK. STATs is also associated with c-Jun, IRF9 (interferon regulatory factor-9), c-Fos, NF-KappaB (nuclear factor-KappaB), SMAD (Sma and MAD (mothers against decapentaplegic), SP1 (transcription factor SP1)) and cofactors like p300, CBP (CREB-binding protein), BRCA1 (breast cancer-1 gene), MCM5, (minichromosome maintenance-5). Hyperactivation of JAK/STAT results in various human diseases such as myocardial hypertrophy leukemia, lymphoma, multiple myeloma, solid neoplasias, and cancer in several adult organs [8-13]. Knockout studies in mice suggest the central role of this signaling in hematopoiesis and regulation of immune functions [14, 15]. Humans with loss-of-function mutations in *Stat1*, *Stat3*, *Tyk2*, or *Jak3* display immunodeficiency and hyper-IgE syndrome because of the requirement of this pathway in the developing blood cells [16,17]. JAK/STAT signaling has been extensively studied in *Drosophila.* In contrast to mammals, only one receptor, Domeless (Dome), one JAK (Hopscotch, Hop), one STAT (Stat92E or Marelle) and three cytokines, Unpaired (Upd), Upd2, and Upd3 have been functionally characterised in *Drosophila* (1, 3,8, 18-25]. There are several negative regulators such as suppressor of cytokine signaling at 36E (Socs36E), ken and barbie (ken), protein inhibitors of activated stats (PIAS), and nucleosome remodeling factor (NURF) also reported (Figure 1).

The JAK/STAT signal transduction pathway regulates various developmental processes in *Drosophila*, including sex determination, eye formation, embryonic segmentation, tracheal formation, male and female germline stem cell proliferation and self-renewal, border cell migration, polar and follicular cell fate determination in the female germline, germ cell proliferation, invasive migration, intestine, kidney, hindgut, stomach and hematopoetic stem cell proliferation and maintenance. In this chapter, we will describe the role of JAK/STAT in stem cell regulation and tumorigenesis in *Drosophila*.

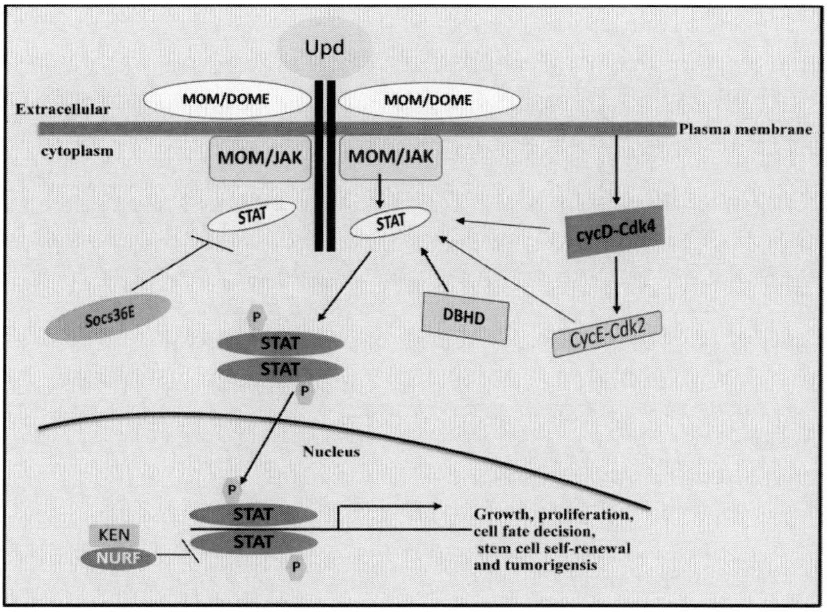

Figure 1. JAK/STAT pathways in *Drosophila.*

JAK/STAT Signaling in Stem Cell Regulation

JAK/STAT pathway is known to regulate stem cell fate in several identified and well-characterized stem cells in *Drosophila* (26-56). In *Drosophila* male germline, JAK/STAT pathway regulates germ line stem cells (GSCs) self-renewal, maintenance, and differentiation [24-26]. The JAK/STAT ligand Upd is expressed in the hub cells and maintains the male GSCs self-renewal by activating the JAK/STAT pathway in GSCs. Upd usually direct the GSCs to undergo self-renewal through the Mom/Hop/Stat92E pathway. Upon phosphorylation Stat92E then enters the nucleus to activate expression of genes that regulate the self-renewal of GSCs and somatic stem cells (SSCs) or cyst progenitor cells (CPCs). Overexpression of the Hop/Stat92E pathway through overexpressing *upd* results in unrestricted stem cell division. However, loss-of-function mutations in the Hop/Stat92E pathway lead to differentiation of GSCs, which suggest that signaling maintains stem cell fate or viability [24-26]. It has been found that early spermatogonia with limited mitotic divisions can repopulate the niche and can be reverted to stem cell identity by conditionally manipulating the Hop/Stat92E signaling [27]. *zfh1*, which was found to be a Stat92E-regulated gene required for SSCs self-renewal [48]. Flaherty et al. [41] found that *Chinmo*, a potential Stat92E downstream effector, expressed in both GSCs and SSCs and required for the maintenance of SSCs. Misexpression of *chinmo* in somatic cells in the testis leads to expansion of both GSCs and SSCs outside of the niche, which is similar to overexpression of JAK/STAT pathway or misexpression of *zfh1*. In addition, they found that *chinmo* does not act through *zfh1* to promote stem cell expansion outside of the niche. Recently, it has been demonstrated that the JAK/STAT signaling also regulates SSCs self-renewal [36,42]. We found that negative regulator, *socs36,* suppress JAK/STAT signaling in somatic cells, and prevent them from out-competing the GSCs. In addition, we demonstrated that the somatic

JAK/STAT signaling is essential for self-renewal and maintenance of both SSCs and GSCs. These results suggest that a single JAK/STAT signal from the niche orchestrate the competitive and dependent co-existence of GSCs and CPCs in the Drosophila testis niche [36].

In the *Drosophila* ovary, JAK/STAT signaling is required for the maintenance and division of GSCs and escort stem cells (ESCs) as well as polar cell specification and follicle formation [40, 45, 52]. Decotto and Spradling [52] found that Upd1 is expressed highly in the terminal filament cells (TFCs), SCCs, and low level in ESCs. Loss of STAT signaling dramatically decreased numbers of GSCs, however, over expression of *upd1* resulted in an increased number of cysts with a small number of ovarioles with tumor phenotype. Further, JAK/STAT signaling in cap cells positively regulates *dpp* expression because in the absence of JAK/STAT signaling, Dpp signaling activity in cap cells is downregulated [50,51].

In addition to the role of JAK/STAT signaling in gemline stem cells regulation, it has been demonstrated that it also regulate stem cells in midgut, malpighian tubules, stomach and hematopoietic stem cells in *Drosophila*. It has been demonstrated that damaged or stressed enterocytes (ECs) produce the cytokines such as Upd, Upd2, and Upd3 that activate the JAK/STAT signaling in *Drosophila* adult intestinal stem cells (ISCs) and enteroblast cells (EBs) by promoting their division and differentiation in normal and regenerating midgut [33-35]. JAK/STAT also controls ISCs proliferation [35] and regulates self-renewal and lineage differentiation of ISCs [31]. Recently, we demonstrated that the JAK/STAT signaling regulates ISCs self-renewal or differentiation [32].

Stem cells are found in *Drosophila* malpighian tubules, a functional homologue of mammalian kidney. We reported that loss of JAK/STAT signaling promotes the differentiation of renal stem cells. In contrast, over-expression of *upd* results in dramatic increase in the number of proliferating cells, mitotically active cells, and renal stem cells. We also found that JAK/STAT ligand Upd and its receptor Dome along with Stat92E protein, are expressed in renal stem cells, which suggests that self-renewal of renal stem cells is controlled by autocrine JAK/STAT signaling and does not require a fixed niche [37].

We recently identified gastric stem cells (GaSCs) in the adult *Drosophila* using the Stat-GFP reporter [38]. We further found that JAK/STAT signaling regulates GaSCs proliferation [38]. JAK/STAT pathway also regulates neuro-epithelial cells in the optic lobe of the brain and hematopoiesis in *Drosophila* [55-58].

JAK/STAT Pathway in Tumorigenesis

JAK/STAT pathway regulates eye development, tumor formation in *Drosophila* and human. It has been found that hyperactivation of mutant of hop^{Tum-} overactivates STAT92E that leads to leukemia-like hematopoietic tumors in *Drosophila* [19,20, 59-61] It has been shown that overexpression of Hop/Stat92E signaling in *Drosophila* induces cell overproliferation and large eye size. It has been demonstrated that Hop/Stat92E signaling can synergize with CycD-Cdk4 and CycE-Cdk2 in melanotic tumors in *Drosophila*, promote the formation of an enlarged eye. These results suggest that the JAK/STAT pathway and Cyclin-Cdk cooperatively regulate tissue outgrowth and tumor formation in *Drosophila*. Further, we found that JAK/STAT signaling interact with *Drosophila* BHD and regulate stem cell and tumorigenesis [62] (Figure 1).

In addition, it has also been demonstrated that overactivation of JAK/STAT leads to deletion of the tumor suppressor genes including PcG and vps25, which results in neoplastic transformation in eye tissue [63-66]. These results suggest that excessive activation of this signaling function as pro-growth, which in turn results in tumor formation in *Drosophila.* Overactivation of JAK/STAT pathway also results in tumor formation in mice and human [67-70]. These studies on hyperactivation of JAK/STAT signaling and tumor formation from *Drosophila*, mice and human suggest that the pro-growth function is conserved.

ACKNOWLEDGMENTS

This research was supported by the Intramural Research Program of the National Institutes of Health, National Cancer Institute.

REFERENCES

[1] Hou XS, Perrimon N. The JAK-STAT pathway in *Drosophila. Trends Genet* 1997; 13:105–10.

[2] Darnell, JE, Jr. STATs and gene regulation. *Science* 1997; 277, 1630–1635.

[3] Hou SX, Zheng Z, Chen X, Perrimon N. The Jak/STAT pathway in model organisms: emerging roles in cell movement. *Dev Cell* 2002; 3,765–78.

[4] O'Shea JJ, Gadina M, Schreiber RD. Cytokine signaling in 2002: new surprises in the JAK/STAT pathway. *Cell* 2002; 109:S121–31.

[5] Schindler, C and Darnell, JE Jr (1995) Transcriptional responses to polypeptide ligands: the JAK–STAT pathway. *Annu. Rev. Biochem.* 64, 621–651

[6] Li, WX (2008) Canonical and non-canonical JAK–STAT signaling. *Trends Cell Biol* 18, 545-551.

[7] Richards, AJ and Stephens JM. Emerging roles of JAK–STAT signaling pathways in adipocytes. *Trends Endocrino Metabol.* 2011; 22, 325-332.

[8] Zeidler MP, Bach EA, Perrimon N. The roles of the *Drosophila* JAK/STAT pathway. *Oncogene* 2000; 19, 2598–606.

[9] Decker T. Introduction: STATs as essential intracellular mediators of cytokine responses. *Cell Mol. Life Sci.* 1999; 55,1505–8.

[10] Luo H, Dearolf CR. The JAK/STAT pathway and *Drosophila* development. *Bioessays.* 2001; 23,1138–47.

[11] Imada K, Leonard WJ. The JAK-STAT pathway. *Mol. Immunol.* 2000; 37:1–11.

[12] Rabinow L. The proliferation of *Drosophila* in cancer research: a system for the functional characterization of tumor suppressors and oncogenes. *Cancer Invest* 2002; 20:531–56.

[13] Bromberg JF, Wrzeszczynska MH, Devgan G, *et al.* STAT3 as an oncogene. *Cell* 1999; 98:295–303.

[14] O'Shea J. J, Murray P. J () Cytokine signaling modules in inflammatory responses. *Immunity* 2008; 28: 477–487.

[15] Dupuis et al., 2003 S. Dupuis, E. Jouanguy, S. Al-Hajjar, C. Fieschi, I.Z. Al-Mohsen, S. Al-Jumaah, K. Yang, A. Chapgier, C. Eidenschenk and P. Eid *et al.*, Impaired response to interferon-alpha/beta and lethal viral disease in human STAT1 deficiency, *Nat. Genet.* 2003; 33, 388–391.

[16] Shuai K, Liu B. Regulation of JAK-STAT signalling in the immune system. Nat Rev Immunol 2003; 3, 900–911.

[17] Minegishi Y., M. Saito, S. Tsuchiya, I. Tsuge, H. Takada, T. Hara, N. Kawamura, T. Ariga, S. Pasic and O. Stojkovic *et al.*, Dominant-negative mutations in the DNA-binding domain of STAT3 cause hyper-IgE syndrome. *Nature* 2007; 448, 1058–1062.

[18] Binari R, Perrimon N. Stripe-specific regulation of pair-rule genes by hopscotch, a putative Jak family tyrosine kinase in *Drosophila*. *Genes Dev.* 1994; 8:300–12.

[19] Hou SX, Melnick MB, Perrimon N. marelle acts downstream of the *Drosophila* HOP/JAK kinase and encodes a protein similar to the mammalian STATs. *Cell* 1996; 84:411–9.

[20] Yan R, Small S, Desplan C, Dearolf CR, Darnell JJ. Identification of a Stat gene that functions in *Drosophila* development. *Cell* 1996; 84, 421–30.

[21] Harrison DA, McCoon PE, Binari R, Gilman M, Perrimon N. *Drosophila* unpaired encodes a secreted protein that activates the JAK signaling pathway. *Genes Dev.* 1998; 12, 3252–63.

[22] Brown S, Hu N, Castelli-Gair Hombría J. Identification of the first invertebrate interleukin JAK/STAT receptor, the *Drosophila* gene domeless. *Curr. Biol.* 2001; 11,1700–5.

[23] Chen HW, Chen X, Oh SW, *et al.* mom identifies a receptor for the *Drosophila* JAK/STAT signal transduction pathway and encodes a protein distantly related to the mammalian cytokine receptor family. *Genes Dev.* 2002; 16, 388–98.

[24] Dearolf CR. JAKs and STATs in invertebrate model organisms. *Cell Mol. Life Sci.* 1999; 55, 1578–84.

[25] Zeidler MP, Perrimon N, Strutt DI. Polarity determination in the *Drosophila* eye: a novel role for unpaired and JAK/STAT signaling. *Genes Dev* 1999; 13, 1342–53.

[26] Kiger AA, Jones DL, Schulz C, Rogers MB, Fuller MT. Stem cell self-renewal specified by JAK-STAT activation in response to a support cell cue. *Science* 2001; 294, 2542–5.

[27] Brawley C, Matunis E. Regeneration of male germline stem cells by spermatogonial dedifferentiation *in vivo*. *Science* 2004; 304, 1331–4.

[28] Fuller MT. Spermatogenesis. In: Bate M, Martinez-Arias A, eds. *The Development of Drosophila melanogaster*, Cold Spring Harbor Laboratory Press, Cold Spring Harbor, NY, 1993; pp 71–147.

[29] Brawley C, Matunis E. Regeneration of male germline stem cells by spermatogonial dedifferentiation *in vivo*. *Science* 2004; 304,1331–4.

[30] Beebe K, Lee WC and Micchelli, CA. JAK/STAT signaling coordinates stem cell proliferation and multilineage differentiation in the *Drosophila* intestinal stem cell lineage. *Dev. Biol.* 2010; 338, 28-37.

[31] Lin G, Xu, N and Xi R. Paracrine unpaired signaling through the JAK/STAT pathway controls self-renewal and lineage differentiation of *Drosophila* intestinal stem cells. *J. Mol. Cell Biol.* 2010; 2, 37-49.

[32] Liu W, Singh SR, and Hou, S X. JAK-STAT is restrained by Notch to control cell proliferation of the *Drosophila* intestinal stem cells. *J. Cell Biochem.* 2010; 109, 992-999.

[33] Buchon, N., Broderick, N. A., Poidevin, M., Pradervand, S. and Lemaitre, B. Drosophila intestinal response to bacterial infection: activation of host defense and stem cell proliferation. *Cell Host Microbe* 2009; 5, 200-211.

[34] Buchon N, Broderick NA, Kuraishi T, and Lemaitre B. *Drosophila* EGFR pathway coordinates stem cell proliferation and gut remodeling following infection. *BMC Biol.* 2010; 8, 152.

[35] Jiang H, Patel PH, Kohlmaier A, Grenley MO, McEwen DG and Edgar BA. Cytokine/Jak/Stat signaling mediates regeneration and homeostasis in the Drosophila midgut. *Cell* 2009; 137, 1343-1355.

[36] Singh SR, Zheng Z, Wang H, Oh, SW, Chen X, and Hou, SX. Competitiveness for the niche and mutual dependence of the germline and somatic stem cells in the *Drosophila* testis are regulated by the JAK/STAT signaling. J. Cell Physiol. 2010; 223, 500-510.

[37] Singh S R, Liu W, and Hou SX. The adult Drosophila malpighian tubules are maintained by multipotent stem cells. Cell Stem Cell, 2007; 1, 191-203.

[38] Singh SR, Zeng, X, Zheng Z, and Hou SX. The adult Drosophila gastric and stomach organs are maintained by a multipotent stem cell pool at the foregut/midgut junction in the cardia (proventriculus) Cell Cycle 2011; 10,1109-1120.

[39] López-Onieva L, Fernández-Miñán A, González-Reyes A. Jak/Stat signalling in niche support cells regulates dpp transcription to control germline stem cell maintenance in the Drosophila ovary. *Development,* 2008; 135, 533-40.

[40] Decotto E. and A.C. Spradling, The *Drosophila* ovarian and testis stem cell niches: similar somatic stem cells and signals, *Dev. Cell* 2005; 9,501–510.

[41] Flaherty MS, Salis P, Evans CJ, Ekas LA, Marouf A, Zavadil J, Banerjee U, Bach EA. chinmo is a functional effector of the JAK/STAT pathway that regulates eye development, tumor formation, and stem cell self-renewal in Drosophila. *Dev. Cell,* 2010; 18, 556-68.

[42] Issigonis MTulina N. de Cuevas M. Brawley C, Sandler L and Matunis E. JAK-STAT signal inhibition regulates competition in the Drosophila testis stem cell niche, *Science* 2009; 326,153–156.

[43] Arbouzova NI and Zeidler MP. JAK/STAT signalling in Drosophila: insights into conserved regulatory and cellular functions, *Development* 2006; 133; 2605–2616.

[44] Decotto and A.C. Spradling, The *Drosophila* ovarian and testis stem cell niches: similar somatic stem cells and signals, *Dev. Cell* 2005, 9 501–510.

[45] L. Gregory, P.J. Came and S. Brown, Stem cell regulation by JAK/STAT signaling in Drosophila, *Semin. Cell Dev. Biol.* 19 (2008), pp. 407–413.

[46] Cronin SJ, *et al.*, Genome-wide RNAi screen identifies genes involved in intestinal pathogenic bacterial infection, *Science* 2009; 325, 340–343.

[47] Jiang H. et al. Cytokine/Jak/Stat signaling mediates regeneration and homeostasis in the *Drosophila* midgut, *Cell* 2009; 137, 1343–1355.

[48] Leatherman JL and Dinardo S. Zfh-1 controls somatic stem cell self-renewal in the *Drosophila* testis and nonautonomously influences germline stem cell self-renewal, *Cell Stem Cell* 2008; 3, 44–54.

[49] Tulina N and Matunis E. Control of stem cell self-renewal in Drosophila spermatogenesis by JAK-STAT signaling, *Science* 2001; 294, 2546–2549.

[50] Lopez-Onieva L, Fernandez-Minan A and Gonzalez-Reyes A. Jak/Stat signalling in niche support cells regulates dpp transcription to control germline stem cell maintenance in the Drosophila ovary, *Development* 2008; 135, 533–540.

[51] Wang L, Li Z and Cai Y, The JAK/STAT pathway positively regulates DPP signaling in the Drosophila germline stem cell niche, *J. Cell Biol.* 2008; 180, 721–728.

[52] Xi R, McGregor JR and Harrison DA. A gradient of JAK pathway activity patterns the anterior–posterior axis of the follicular epithelium. *Dev. Cell* 2003; 4, 167–177.

[53] Baksa K *et al.* The *Drosophila* STAT protein, stat92E, regulates follicle cell differentiation during oogenesis, *Dev. Biol.* 2002; 243,166–175.

[54] McGregor JR, Xi R and Harrison DA. JAK signaling is somatically required for follicle cell differentiation in *Drosophila*, *Development* 2002; 129,705–717.

[55] Gao H, Wu X, and Fossett N. Upregulation of the Drosophila Friend of GATA gene U-shaped by JAK/STAT signaling maintains lymph gland prohemocyte potency. Mol. Cell Biol. 2009; 29,6086-96.

[56] Minakhina S, Tan W, and Steward R. JAK/STAT and the GATA factor Pannier control hemocyte maturation and differentiation in Drosophila. *Dev. Biol.* 2011; 352,308-16.

[57] Ngo KT, Wang J, Junker M, Kriz S, Vo G, Asem B, Olson JM, Banerjee U, and Hartenstein V. Concomitant requirement for Notch and Jak/Stat signaling during neuro-epithelial differentiation in the Drosophila optic lobe. *Dev. Biol.* 2010; 346, 284-95.

[58] Wang W, Li Y, Zhou L, Yue H, and Luo H. Role of JAK/STAT signaling in neuroepithelial stem cell maintenance and proliferation in the Drosophila optic lobe. *Biochem. Biophys. Res. Commun.* 2011; 410, 714-20.

[59] Hanratty WP, and Dearolf CR. The *Drosophila* Tumorous-lethal hematopoietic oncogene is a dominant mutation in the hopscotch locus. *Mol. Gen. Genet.* 1993; 238,33–37.

[60] Harrison DA, Binari R, Nahreini TS, Gilman M, and Perrimon N. Activation of a Drosophila Janus kinase (JAK) causes hematopoietic neoplasia and developmental defects. *EMBO J.* 1995;14, 2857–2865.

[61] Muller P, Kuttenkeuler D, Gesellchen V, Zeidler MP, and Boutros M. Identification of JAK/STAT signalling components by genome-wide RNA interference. *Nature* 2005; 436,871–875.

[62] Singh SR, Zhen W, Zheng Z, Wang H, Oh SW, Liu W, Zbar B, Schmidt LS, and Hou SX. The Drosophila homolog of the human tumor suppressor gene BHD interacts with the JAK-STAT and Dpp signaling pathways in regulating male germline stem cell maintenance. *Oncogene* 2006; 25, 5933-41.

[63] Thompson BJ, Mathieu J, Sung HH, Loeser E, Rorth P, and Cohen SM. Tumor suppressor properties of the ESCRT-II complex component Vps25 in Drosophila. *Dev. Cell* 2005; 9, 711–720.

[64] Sparmann A, van Lohuizen M. Polycomb silencers control cell fate, development and cancer. *Nat. Rev. Cancer* 2006; 6, 846–856.

[65] Martinez AM, Schuettengruber B, Sakr S, Janic A, Gonzalez C, Cavalli G. Polyhomeotic has a tumor suppressor activity mediated by repression of Notch signaling. *Nat. Genet.* 2009; 41, 1076–1082.

[66] Vaccari T and Bilder D. The Drosophila tumor suppressor vps25 prevents nonautonomous overproliferation by regulating notch trafficking. *Dev. Cell* 2005; 9, 687–698.

[67] Calo V, Migliavacca M, Bazan V, Macaluso M, Buscemi M, Gebbia N, and Russo A. STAT proteins: from normal control of cellular events to tumorigenesis. *J. Cell Physiol.* 2003; 197,157–168.

[68] Smirnova OV, Ostroukhova TY, and Bogorad RL. JAK-STAT pathway in carcinogenesis: is it relevant to cholangiocarcinoma progression? *World J. Gastroenterol.* 2007;13,6478–6491.

[69] Constantinescu SN, Girardot M, and Pecquet C. Mining for JAK-STAT mutations in cancer. *Trends Biochem. Sci.* 2008; 33,122–131.

[70] Bromberg JF, Wrzeszczynska MH, Devgan G, Zhao Y, Pestell RG, Albanese C, Darnell JE., Jr Stat3 as an oncogene. *Cell* 1999, 98:295–303.

In: Signaling, Gene Regulation and Cancer
Editors: Shree Ram Singh and Manoj K. Mishra
ISBN: 978-1-61942-088-5

Chapter 5

ROLE OF TGF-β AND REGULATORY T CELLS IN PROSTATE CANCER

Manoj K. Mishra**[*,1], **Erdal Erglou**[2], **Sapna Jain**[2], **Shree R. Singh**[2], **Udai P. Singh**[3], **Renu Bhatt**[4] and **Shree Ram Singh**[5]

[1]Department of Mathematics and Science and
[2]Center for Nanobiotechnology and Life Science Research, Alabama State University, Montgomery, AL, US
[3]Department of Pathology, Microbiology, and Immunology, School of Medicine, University of South Carolina, Columbia, SC, US
[4]Department of Biotechnology, Guru Ghasi Das University, Bilaspur, Chattisgarh, India
[5]Mouse Cancer Genetics Program, National Cancer Institute, Frederick, MD, US

ABSTRACT

Prostate Cancer (PCa) is the most common non-skin malignancy and commonly diagnosed cancer, and second most cancer responsible for death in men in United States. Prior studies in animal models of PCa as well as in human, demonstrated that the prostate produces several growth factors. Addition or blocking of these growth factors can alter the PCa cell proliferation and other important functions. Among these growth factors, transforming growth factor-β1 (TGF-β) plays significant role in tumorigenicity, displays potent immunosuppressive activities, and is required for the conversion of conventional $CD4^+$ T cells to $Foxp3^+$ regulatory T cells (T_R). T_R that expresses the transcription factor Foxp3 is essential for normal immune function. Absence of T_R cells results in multi-organ autoimmunity and death. This chapter reviews the underlying mechanisms involved in the context of TGF-β and/or T_R in prostate cancer development and clearance.

[*] Corresponding Author: Manoj K. Mishra, Department of Mathematics and Science, Alabama State University, Montgomery, AL 36104, Email: mmishra@alasu.edu.

Introduction

Prostate Cancer is the most common non-skin malignancy and most commonly diagnosed cancer in men in United States. This is the second most cancer responsible for death in men. According to an estimate posted on National Cancer Institute's webpage, approximately 240,890 new cases of PCa and 33,720 deaths have been predicted in United States in 2011. According to American Cancer Society, "African-American men have the highest rate of PCa in the world and the lowest rate of survival. Black men in America are 1.5 times more likely to develop PCa and are 2 to 3 times more likely to die of the disease than white men". There are several measures available for early diagnosis and treatment of PCa. Treatment measures include surgery, chemotherapy, radiation therapy, immunotherapy, and vaccine therapy. This chapter summarizes the recent advancement in molecular and cellular mechanisms during prostate tumor development and clearance in the context of TGF-β and T_R [1-3]. This will help to understand the basic components of TGF-β and T_R -dependent control of anti-tumor responses that will eventually assist in design of tumor vaccine and immunotherapy of PCa.

Regulatory T Cells and Cancer

Regulatory T cells (T_R) express forkhead winged helix protein family type transcription factor "Foxp3," which plays significant role in T_R development. Mutation in the Foxp3 gene causes fatal autoimmune disorders called "immune dysregulation, polyendocrinopathy, enteropathy, X-linked (IPEX)" [1, 2, 4]. In this disease, human patients at a very young age show signs of autoimmune disease indicating enlarged lymph nodes, thyroiditis, diabetes mellitus, eczema, food allergies, anemia, as well as several other autoimmune features. T_R cells suppress self or autoreactive cells, thereby protecting the host from autoimmune disease [5, 6]. On the other hand, T_R cells may block anti-tumor immune responses. Particularly in the context of cancer, T_R cell, frequencies, and function are important because increased numbers might favor tumor development or growth and influence the course of the disease. The development of T_R cells during tumor progression has been addressed in a several model systems. For example, T_R development in a fibrosarcoma model in C57BL/6N mice as well as in a colon adenocarcinoma model in BALB/c mice has been shown to have lethal tumor progression [7]. The T_R cells during tumor development have shown to have high suppressor capacity. The suppressive effect of naturally occurring T_R cells against tumor-specific $CD8^+$ T cells was established in a poorly immunogenic B16 melanoma model. T_R cells efficiently suppressed cytotoxic T lymphocyte (CTL)–mediated concomitant immunity against a recalling with the same tumor, demonstrating that precursor T_R cells in naive hosts gave rise to effective suppressors during tumor development. These findings clearly suggest that T_R cells are major regulators of tumor immunity [8]. Further evidence for the interference of T_R cells with $CD8^+$ T cell-mediated anti-tumor immune responses in vivo was established in a transgenic murine colon carcinoma model where T_R cells abrogated $CD8^+$ T cell–mediated tumor rejection by specifically suppressing cytotoxicity of CTLs [9].

Prior studies have revealed that $CD4^+CD25^+$ T_R cells, $CD8^+$ $Foxp3^+$ T_R cells present in prostate tumor–derived tumor infiltrating lymphocytes (TILs) suppress immune responses

[10]. Recent studies indicate that preexisting T_R cells at tumor sites may pose major obstacles to effective cancer immunotherapy, as these cells have a potent ability to suppress host immune responses [11, 12]. Indeed, increased proportions of $CD4^+CD25^+$ T_R cells in the total $CD4^+$ T-cell populations have been documented in patients with different types of cancers, including lung, breast, and ovarian tumors [13, 14]. Our recent findings further show the presence of antigen-specific $CD4^+$ T_R cells at tumor sites, where they induce antigen-specific and local immune tolerance [21-22].

The removal or elimination of T_R cells population with anti-CD25 monoclonal antibody (mAb) treatment results in effective rejection of transplanted tumors in animal models [15] further suggest a functional role for these T_R cells in tumor progression and immunosuppression.

Currently, a number of important questions about the role of TR cells during tumorigenesis are under intense investigation. Prior studies have revealed that increased percentages of $CD4^+CD25^+$ T_R cells in TILs in non–small-cell lung cancer and ovarian cancer. These T_R cells were shown to secrete TGF-β, providing first evidence that T_R cells contribute to immune dysfunction in patients with cancer [13].

In malignant melanoma, an increase of functional $CD4^+CD25^+$ T_R cells was observed [16]. A more recent study demonstrated Foxp3 mRNA expression in the increased subset of $CD4^+$CD25 high T_R cells in patients with melanoma, confirming the previous reports [17]. Studies have demonstrated that $CD4^+CD25^+$ $Foxp3^+$ T_R cells suppress tumor-specific T-cell immunity in ovarian cancer, contribute to tumor growth, and accumulate during progression [14, 18].

TGF-β and Its Role in Prostate Cancer

Uncontrolled proliferation of cells is the essence of cancer. There is a wide range of data available describing how cancer cells escape normal growth pattern and fail to respond normal signals. Cancer up-regulates a family of growth proteins, known as TGF-β, which can inhibit cell growth [19]. Prostate cancer is shown to have enhanced levels of several growth factors that play important roles during tumor progression. The role of these growth factors during cancer has been extensively studied. Among these growth factors, the diverse role of TGF-β has been widely reported [20]. Proper signaling of TGF-β is required for normal functions.

During cancer, some components of TGF-β signaling gets mutated resulting in the cells becoming resistance to the effect of TGF-β and these resistant cells grow without any regulation. Under normal circumstances, TGF-β decreases cell proliferation including cancerous cells. TβR II have an important role in TGF-β production and it has been reported that methylation of TβR II gene in PCa cell lines, may be responsible for mutations that causes cancer. Previous studies have demonstrated higher levels of TGF-β in PCa than normal prostate [21].

TGF-β has several roles such as it enhances tumorigenicity, display several immunosuppressive activities as well as help in conversion of $CD4^+$ T cells to T_R [4]. It is known that T_R is essential for normal immune functions and TGF-β provide help to naïve $CD4^+$ T cells to acquire a T_R phenotype that express Foxp3 [5, 22, 23].

TGF-β and Its Role in Conversion of Naïve CD4$^+$ T Cells to Foxp3$^+$ T_R

Regulatory T cells play unique and important role in maintaining tolerance to self, thereby suppressing the development of autoimmune disease [24]. The best characterized subset of T_R consists of thymically derived CD4$^+$CD25$^+$Foxp3$^+$ T_R that is also referred as naturally arising T_R (nT_R) and T_R that acquire foxp3 phenotype in peripheral organs referred as induced T_R (iT_R) [5, 22]. Foxp3, a transcription factor of forked head winged helix protein family, regulates the development of mature T_R in thymus. Mutation in Foxp3 has been shown to result in massive lymphoproliferation and severe autoimmune disease in human and mouse [25]. After stimulation with TCR and TGF-β, mature CD4$^+$ T cells have been shown to acquire suppressive capabilities and expression of Foxp3. TGF-β induced T_R (iT_R) shows similar phenotype as nT_R [23]. TGF-β has effects on T cell mediated immune suppression and control of autoimmunity. The binding of TGF-β to its receptor complex activates the intracellular kinase domain of TβR II, which leads to phosphorylation and activation of different sets of protein including the Smad family of proteins [26]. TGF-β engages both Smad-dependent and Smad-independent mechanism to exert its action [26], and particularly the Smad3 and Smad2 help in generating T_R cells as well as elevate the expression Foxp3. In conclusion, the role of TGF-β in conversion of naïve CD4$^+$ T cells to Foxp3$^+$ T_R cells could provide an alternative source to generate T_R for cell based immunotherapy. However, a complete understanding of molecular mechanism involved during the conversion of naïve CD4$^+$ T cells to Foxp3 expressing T_R needs more focused investigation.

Significance of TGF-β and T_R in Prostate Cancer Development and Progression

In recent years, studies have established a correlation between TGF-β production and T_R generation during cancer development and clearance [9, 20, 23, 27-29]. T_R is mostly generated in the thymus and has an essential role in maintaining tolerance to self-antigen in peripheral organs [4, 30]. A prime aspect of tumor immunology is immune surveillance in tumor bearing host. Recently it has been shown that tumor cells produces TGF-β that can directly convert naïve CD4$^+$ T cells to T_R suggesting the immune surveillance in tumor bearing host through conversion of T_R cells [31]. Therefore, the role of TGF-β and an understanding between tumor-specific T cells (particularly, naïve CD4$^+$ T cells) and T_R is needed to facilitate the design of novel methodologies to generate a more vigorous anti-tumor response.

Role of TGF-β in Tumorigenesis

Several studies suggest the role of TGF-β as a potent inhibitor of growth there by suppressing proliferation of cells and thus prevent tumor formation in normal signaling condition [19-21, 27, 32]. Cancerous cells become resistant to TGF-β signal and enhance their proliferation, invasion, and metastasis when stimulated by TGF-β. An antagonist, cysteine protease inhibitor cystatin C (CystC), for TGF-β type II receptor has been reported to inhibit TGF-β binding and signaling in normal and cancerous cells [33]. There are several other factors affect the TGF-β signaling. Any diversion from correct signaling of TGF-β enforces cells to become resistant to all endogenous effects of TGF-β. Commonly, a defective ligand binding to cell surface responsible for resistance to activated TGF-β and in some cases lack of binding is associated with the expression of TβR II mRNA (15). Prostate cancer cells produce more TGF-β than normal prostate cells. Cancerous cells, on one hand protect themselves from the effect of TGF-β from being growth inhibited while on the other hand they still maintain some responsiveness to TGF-β that may be linked to their efficacy in tumor growth and metastasis [34].

TGF-β Plays Crucial Role in Conversion of Conventional Naïve $CD4^+$ T Cells to $Foxp3^+$ T_R

In recent years, studies [3, 5, 23, 26, 35] have demonstrated the roles of TGF-β, IL-2, and TCR signaling in the generation of iT_R. To verify, which particular signaling factor has important role in Foxp3 expressing T_R generation, we have shown that TGF-β is a potent inducer of naïve $CD4^+$ T cells to Foxp3 expressing T_R. As we know, TGF-β is a pluripotent cytokine that has vigorous effects on T-cell mediated immune suppression as well as on the control of autoimmunity. This study demonstrated the significant role of TGF-β in constitutive expression of Foxp3 expressing T_R [26].

Converted T_R Suppresses Primary Immune Response to Foreign Antigen

T_R cells constitutively express the IL-2Rα chain (CD25) and a number of other surface markers. However, none of these surface markers is specific for T_R cells [36-38]. This has complicated the *in vivo* identification and analysis of T_R cells. To facilitate the accurate identification and tracking of $Foxp3^+$ T_R cells *in vivo*, mice derived with a bicistronic *Foxp3* locus ($Foxp3^{EGFP}$) that co-expresses the enhanced green fluorescent protein (EGFP) under control of the endogenous *Foxp3* promoter/enhancer elements [4, 30, 39]. EGFP fluorescence accurately and specifically identifies $Foxp3^+$ T_R cells, and the characterization of these mice is consistent with other mice expressing fluorescent proteins under the control of the Foxp3 promoter [40]. When $EGFP^-$ conventional T cells are sorted from $Foxp3^{EGFP}$ mice and activated by TCR cross linking in the presence of TGF-β *in vitro*, 30% of these cells up regulate Foxp3, as demonstrated by both Foxp3 staining and EGFP expression. Adoptive transfer of *in vitro* converted T_R cells suppresses the primary immune response to foreign antigen, further demonstrating the potency of this T_R cell type [5, 22]. Following the similar approach, naïve $CD4^+CD25^-$ T cells can be converted to T_R in the presence of TGF-β and anti CD3/CD28 and be used to identify the role of converted T_R cell population in PCa growth and progression.

ROLE OF TGF-β AND/OR T_R DURING PROSTATE CANCER

Studies using rodent and human prostate tumor cells established that the level of TGF-β is elevated during PCa than the normal prostate. Reduction of tumorigenicity by a murine PCa cell line-(**TR**ansgenic **A**denocarcinoma of the **M**ouse **P**rostate) TRAMP-C3 is also linked to less production of TGF-β (unpublished data). Postulated mechanisms whereby the TRAMP-C3 reduces tumorigenicity include: 1) proper and effective signaling of TGF-β, 2) Interaction with different cellular proteins during TGF-β signaling. However, we propose that the capacity of TRAMP-C3 to elicit a vigorous anti-tumor cellular immune response is an important component of the anti-tumorigenic activity of TRAMP-C3. An understanding of the molecular basis for the immune-mediated anti-tumorigenic activities of TRAMP- C3 will lead to optimize its role in the possible treatment of PCa.

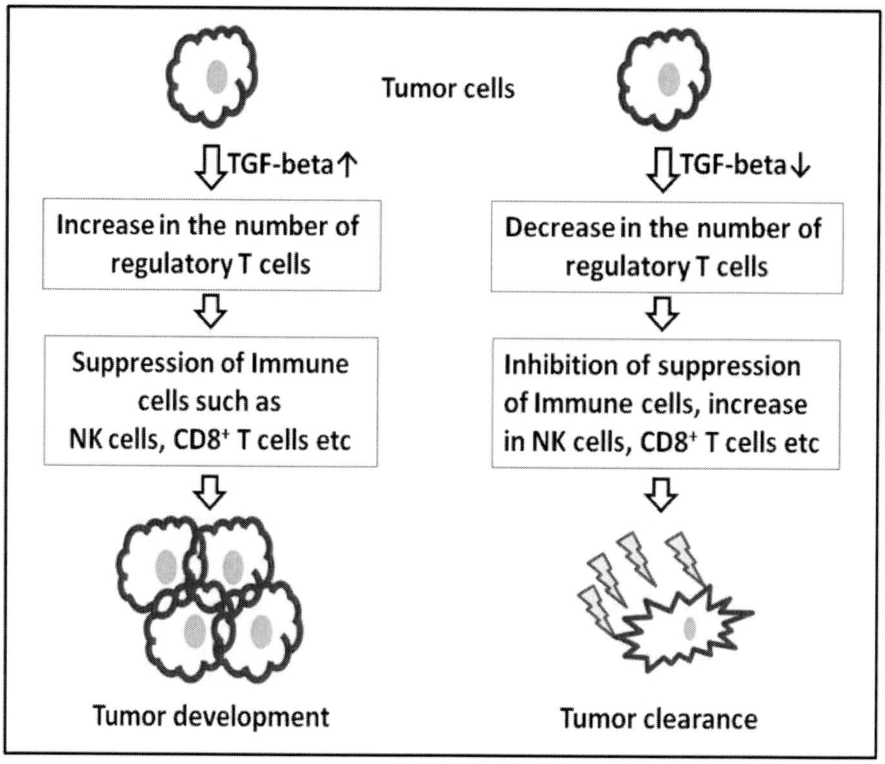

Figure 1. A conceptual diagram demonstrates the role of TGF-β and regulatory T cells during tumor development and clearance.

Experimental evidences demonstrated that tumor-infiltrating immune cells play a major role in fighting cancer and their action links with cancer development and clearance [41, 42]. In most cases, however, tumor-specific T cells ultimately fail to control tumor growth. It is known that conventional $CD4^+$ T cells can be converted to $Foxp3^+CD4^+$ T_R cells if CD3/CD28 ligation occurs in the presence of TGF□-β [6, 23, 24]. Furthermore, adoptively transferring these "converted" $Foxp3^+CD4^+$ T_R cells results in the inhibition of antigen-specific CD4 T cell responses *in vivo* [6, 23, 24]. Based on our preliminary unpublished data and cited literatures, we hypothesizes that: 1) The anti-tumorigenic activity of is mediated, in part, by the inhibition of TGF□-β production by tumor cells; 2) By decreasing the production of TGF□-β, the conversion of conventional $CD4^+$ T cells to $Foxp3^+$ T cells *in vivo* decreases, resulting in an enhanced cell-mediated anti-tumor immune response (Fig. 1). If our hypotheses are correct, studies are needed to exploit this unique activity of tumor cells to enhance antigen-specific, anti-tumor immune responses by inhibiting the generation of "converted" T_R cells *in vivo.*

ACKNOWLEDGMENTS

We acknowledge support from Center for Nanobiotechnology and Life Science Research at Alabama State University. We sincerely apologize to those authors whose works may have been relevant to this review but was not mentioned or cited due to the space limitations.

REFERENCES

[1] Fontenot, J.D., M.A. Gavin, and A.Y. Rudensky, Foxp3 programs the development and function of CD4+CD25+ regulatory T cells. *Nat. Immunol.*, 2003. 4(4): p. 330-6.

[2] Fontenot, J.D., et al., A function for interleukin 2 in Foxp3-expressing regulatory T cells. *Nat. Immunol.*, 2005. 6(11): p. 1142-51.

[3] Marie, J.C., et al., TGF-beta1 maintains suppressor function and Foxp3 expression in CD4+CD25+ regulatory T cells. *J. Exp. Med.*, 2005. 201(7): p. 1061-7.

[4] Rudensky, A.Y., Regulatory T cells and Foxp3. *Immunol. Rev*. 241(1): p. 260-8.

[5] Haribhai, D., et al., A central role for induced regulatory T cells in tolerance induction in experimental colitis. *J. Immunol.*, 2009. 182(6): p. 3461-8.

[6] Haribhai, D., et al., A requisite role for induced regulatory T cells in tolerance based on expanding antigen receptor diversity. *Immunity*. 35(1): p. 109-22.

[7] Peng, L., et al., Tumor-induced L-selectinhigh suppressor T cells mediate potent effector T cell blockade and cause failure of otherwise curative adoptive immunotherapy. *J. Immunol.*, 2002. 169(9): p. 4811-21.

[8] Turk, M.J., et al., Concomitant tumor immunity to a poorly immunogenic melanoma is prevented by regulatory T cells. *J. Exp. Med.,* 2004. 200(6): p. 771-82.

[9] Chen, M.L., et al., Regulatory T cells suppress tumor-specific CD8 T cell cytotoxicity through TGF-beta signals in vivo. *Proc. Natl. Acad. Sci. USA*, 2005. 102(2): p. 419-24.

[10] Kiniwa, Y., et al., CD8+ Foxp3+ regulatory T cells mediate immunosuppression in prostate cancer. *Clin. Cancer Res.*, 2007. 13(23): p. 6947-58.

[11] Wang, H.Y. and R.F. Wang, Antigen-specific CD4+ regulatory T cells in cancer: implications for immunotherapy. *Microbes Infect.*, 2005. 7(7-8): p. 1056-62.

[12] Wang, H.Y., et al., Tumor-specific human CD4+ regulatory T cells and their ligands: implications for immunotherapy. *Immunity*, 2004. 20(1): p. 107-18.

[13] Woo, E.Y., et al., Regulatory CD4(+)CD25(+) T cells in tumors from patients with early-stage non-small cell lung cancer and late-stage ovarian cancer. *Cancer Res.*, 2001. 61(12): p. 4766-72.

[14] Curiel, T.J., et al., Specific recruitment of regulatory T cells in ovarian carcinoma fosters immune privilege and predicts reduced survival. *Nat. Med.*, 2004. 10(9): p. 942-9.
[15] Onizuka, S., et al., Tumor rejection by in vivo administration of anti-CD25 (interleukin-2 receptor alpha) monoclonal antibody. *Cancer Res.*, 1999. 59(13): p. 3128-33.
[16] Javia, L.R. and S.A. Rosenberg, CD4+CD25+ suppressor lymphocytes in the circulation of patients immunized against melanoma antigens. *J. Immunother*, 2003. 26(1): p. 85-93.
[17] Viguier, M., et al., Foxp3 expressing CD4+CD25(high) regulatory T cells are overrepresented in human metastatic melanoma lymph nodes and inhibit the function of infiltrating T cells. *J. Immunol.*, 2004. 173(2): p. 1444-53.
[18] Shevach, E.M., CD4+ CD25+ suppressor T cells: more questions than answers. *Nat. Rev. Immunol.*, 2002. 2(6): p. 389-400.
[19] Massague, J., TGF-beta signal transduction. *Annu. Rev. Biochem.*, 1998. 67: p. 753-91.
[20] Jones, E., H. Pu, and N. Kyprianou, Targeting TGF-beta in prostate cancer: therapeutic possibilities during tumor progression. *Expert Opin. Ther. Targets*, 2009. 13(2): p. 227-34.
[21] Steiner, M.S., et al., Expression of transforming growth factor-beta 1 in prostate cancer. *Endocrinology*, 1994. 135(5): p. 2240-7.
[22] Haribhai, D., et al., Regulatory T cells dynamically control the primary immune response to foreign antigen. *J. Immunol.*, 2007. 178(5): p. 2961-72.
[23] Relland, L.M., et al., Affinity-based selection of regulatory T cells occurs independent of agonist-mediated induction of Foxp3 expression. *J. Immunol.*, 2009. 182(3): p. 1341-50.
[24] Lin, W., et al., Regulatory T cell development in the absence of functional Foxp3. *Nat. Immunol.*, 2007. 8(4): p. 359-68.
[25] Chatila, T.A., Role of regulatory T cells in human diseases. J Allergy Clin Immunol, 2005. 116(5): p. 949-59; quiz 960.
[26] Jana, S., et al., The role of NF-kappaB and Smad3 in TGF-beta-mediated Foxp3 expression. *Eur. J. Immunol.*, 2009. 39(9): p. 2571-83.
[27] Fogli, M., et al., Transforming growth factor beta3 inhibits chronic myelogenous leukemia hematopoiesis by inducing Fas-independent apoptosis. *Exp. Hematol.*, 2000. 28(7): p. 775-83.
[28] Gray, A., et al., Prostate cancer immunotherapy yields superior long-term survival in TRAMP mice when administered at an early stage of carcinogenesis prior to the establishment of tumor-associated immunosuppression at later stages. *Vaccine*, 2009. 27 Suppl 6: p. G52-9.
[29] Taams, L.S. and A.N. Akbar, Peripheral generation and function of CD4+CD25+ regulatory T cells. *Curr. Top Microbiol. Immunol.*, 2005. 293: p. 115-31.
[30] Williams, L.M. and A.Y. Rudensky, Maintenance of the Foxp3-dependent developmental program in mature regulatory T cells requires continued expression of Foxp3. *Nat. Immunol.*, 2007. 8(3): p. 277-84.
[31] Liu, V.C., et al., Tumor evasion of the immune system by converting CD4+CD25- T cells into CD4+CD25+ T regulatory cells: role of tumor-derived TGF-beta. *J. Immunol.*, 2007. 178(5): p. 2883-92.

[32] Freathy, C., et al., Transforming growth factor-beta(1) induces apoptosis in rat FaO hepatoma cells via cytochrome c release and oligomerization of Apaf-1 to form a approximately 700-kd apoptosome caspase-processing complex. *Hepatology*, 2000. 32(4 Pt 1): p. 750-60.

[33] Sokol, J.P., et al., The use of cystatin C to inhibit epithelial-mesenchymal transition and morphological transformation stimulated by transforming growth factor-beta. *Breast Cancer Res.*, 2005. 7(5): p. R844-53.

[34] Blum, R., et al., Molecular signatures of prostate stem cells reveal novel signaling pathways and provide insights into prostate cancer. *PLoS One*, 2009. 4(5): p. e5722.

[35] Haribhai, D., et al., Functional reprogramming of the primary immune response by T cell receptor antagonism. *J. Exp. Med.*, 2004. 200(11): p. 1371-82.

[36] Fehervari, Z. and S. Sakaguchi, Control of Foxp3+ CD25+CD4+ regulatory cell activation and function by dendritic cells. *Int. Immunol.*, 2004. 16(12): p. 1769-80.

[37] Hori, S. and S. Sakaguchi, Foxp3: a critical regulator of the development and function of regulatory T cells. *Microbes Infect.*, 2004. 6(8): p. 745-51.

[38] Sakaguchi, S., Naturally arising CD4+ regulatory t cells for immunologic self-tolerance and negative control of immune responses. *Annu. Rev. Immunol.*, 2004. 22: p. 531-62.

[39] Maynard, C.L., et al., Regulatory T cells expressing interleukin 10 develop from Foxp3+ and Foxp3- precursor cells in the absence of interleukin 10. *Nat. Immunol.*, 2007. 8(9): p. 931-41.

[40] Wan, Y.Y. and R.A. Flavell, Identifying Foxp3-expressing suppressor T cells with a bicistronic reporter. *Proc. Natl. Acad. Sci. USA*, 2005. 102(14): p. 5126-31.

[41] Sato, E., et al., Intraepithelial CD8+ tumor-infiltrating lymphocytes and a high CD8+/regulatory T cell ratio are associated with favorable prognosis in ovarian cancer. *Proc. Natl. Acad. Sci. USA*, 2005. 102(51): p. 18538-43.

[42] Dudley, M.E., et al., Cancer regression and autoimmunity in patients after clonal repopulation with antitumor lymphocytes. *Science*, 2002. 298(5594): p. 850-4.

In: Signaling, Gene Regulation and Cancer
Editors: Shree Ram Singh and Manoj K. Mishra ISBN: 978-1-61942-088-5

Chapter 6

SIGNALING PATHWAYS, GENE REGULATION AND DUODENAL NEOPLASIAS

Carlos A. Rubio*

Gastrointestinal and Liver Pathology Research Laboratory,
Department of Pathology,
Karolinska Institute and University Hospital, Stockholm, Sweden

ABSTRACT

The *duodenum* is located at the crossroads of the three main secretions of the gastrointestinal (GI) tract -gastric, pancreatic and biliary. Despite the detrimental nature of these chemically dissimilar juices upon unprotected cells, the duodenal mucosa remains unharmed, keeping pace with cell renewal. Although anatomists consider the *duodenum* to be part of the small bowel, gastroenterologists regard the *duodenum* - together with the esophagus and the stomach- as part of the upper digestive tract. The latter view may lie on the fact that the *duodenum* has special biological attributes that are at variance with the rest of the small intestine. In this chapter we review signaling pathways affecting the duodenal mucosa, its putative stem cells and their offspring, TAP cells, and their differentiated cells (Paneth cells, goblet cells, enterocytes and neuroendocrine cells) as well as its villous fronds. Classical and more recently discovered stem cell candidate markers are tabulated. In celiac disease intolerance to gluten protein leads to villous atrophy. In contrast to the normal duodenal mucosa that is characterized by short crypts and long villous fronds, villous atrophy is typified by long crypts, high numbers of mitosis and absence of villous formations. We submit that the high numbers of mitosis in enlarged crypts mirrors increased mitotic activity in TAP cells but not in stem cells, as mitoses are not recorded in position +5 cells or in cells at the bottom of the duodenal crypts (the domain of stem cells).

There is an increasing awareness of the supportive role played by Paneth cells in maintaining CBC stem cell homeostasis. Paneth cells contain the potent antimicrobial enzyme lysozyme. Recent developments indicate that in celiac disease many Paneth cells are superseded by lysozyme-producing mucus glands. This metaplastic antimicrobial adaptation might mirror a reprogramming of stem cells, to signals generated by the

* E-mail: Carlos.Rubio@ki.se.

pathogenic duodenal flora recently detected in celiac disease. All nonampullary duodenal adenomas are tubular. We submit that these adenomas are dysplastic aberrations of the crypts. The lack of villous fronds in duodenal adenomas as well as in celiac disease is puzzling. The possible signaling responsible for the abrogation of villous fronds in these two diseases of the duodenal crypts is discussed. Duodenal adenomas often concur with gastric duodenal metaplasia, a non-neoplastic mucosal change triggered by duodenal micro-environmental factors such as high pH and chronic inflammation, and celiac patients develop villous atrophy, but only if gluten protein is present in the duodenal micro-environment. The purported signaling of the duodenal micro-environment to stem cells in villous atrophy and cryptal adenomas, is receiving much attention at this laboratory.

1. INTODUCTION

Herophilos, a Greek physician (300 BC) found that the length of the first portion of the small intestine was equal to 12 finger diameters (approximately 25 cm).

He called this portion δωδεκαδάκτυλον (dodecadaktylon, in Latin, *duodenum*). In spite that anatomists regard the *duodenum* as a portion of the small bowel, gastroenterologists maintain that it is part of the upper digestive tract, together with the esophagus and the stomach. The reason for the latter view may lie on the fact that the *duodenum* has special biological attributes that are at variance with the rest of the small intestine:

i. The duodenum is located at the crossroads of the three main secretions of the gastrointestinal (GI) tract -gastric, pancreatic and biliary. Despite the detrimental nature of these chemically dissimilar juices upon unprotected cells, the duodenal mucosa remains unharmed, keeping pace with cell renewal.
ii. At histology, the duodenal mucosa show particular submucosal glands, the Brunner glands; their secretions buffers the disparate pHs in the duodenal microenvironment.
iii. The bottom of the duodenal crypts (where Paneth cells reside) show the highest expression of Peroxisome proliferator-activated receptor beta or delta (PPAR□β/δ), a nuclear receptor encoded by the *PPARD* gene, than in the rest of the small intestine [1].
iv. The repertoire of Paneth cell antimicrobials differs between duodenum and ileum; cryptdin 4 and cryptdin-related sequences (CRS) 4C peptides are expressed at progressively increasing amounts (10^1- and 10^4-fold, respectively) comparing duodenum and ileum in conventionally reared mice [2].
v. The gene expression of the liver enriched transcription factor HNF6 is more abundant in the *duodenum* than in the jejunum [3].
vi. Lectines cause upregulation of the gene expression of the proinflammatoty cytokines interleukin (IL)-8, tumor necrosis factor in the *duodenum* but not in the ileum in mice [4].
vii. A duodenum-specific enhancer identified in the second intron of the human *ADA* (adenosine deaminase) gene is expressed at high levels only along the villi of the duodenal epithelium. Binding of PDX-1 and GATA-4 is absolutely essential for enhancer function. An additional enhancer binding sites for Cdx factors, for YY1, and for NFI family members has been identified [5].

viii. The *duodenum* is the most common site for extracolonic tumors in patients with familial adenomatous polyposis (FAP) [6].

ix. MUTYH is a mammalian DNA glycosylase that initiates base excision repair by excising adenine opposite 8-oxoguanine and 2-hydroxyadenine opposite guanine, thereby preventing G:C to T:A transversion caused by oxidative stress. In MUTYH-null mice treated with KBrO3, the mean number of small intestinal tumors dramatically increased to 61.88, whereas it was 0.85 in wild-type mice. The tumors developed predominantly in the *duodenum* [7].

In this chapter, pertinent literature regarding putative SCs, TAP cells and villous formations in the duodenal mucosa in health and disease, is presented. Possible differences in signaling and gene regulation between normal donors, patients with villous atrophy in celiac disease (an autoimmune mucosal ailment) and patients with duodenal neoplasias of the crypts, are discussed.

2. Stem Cells: Location in the Crypts and Signaling Pathways

Stem cells are multipotent, self-replicating, with capacity to generate daughter cells committed to terminal differentiation. The estimated number of stem cell population within the crypt, varies from 0.4% to 60% of the crypt [8] with variation from species to species [9, 10].

The position of the stem cells in intestinal crypts has been much debated. For some, stem cells are "placed" at position +4 relative to the crypt bottom [11].

For others, a stem cell zone consists of small, undifferentiated, cycling Crypt Base Columnar (CBC) cells, intercalated with Paneth cells [12]. It has been postulated that CBC cells represent a population of stem cells more ready to respond to stimulating signals generated from adjacent mesenchymal cells such as BMP antagonists like Noggin or Gremlin. There is, however, a mounting awareness that in rapidly renewing adult tissues, the 2 stem cell compartments coexist and work coordinately [13].

Queries and comments

If +4 stem cells and CBC stem cells are two different cell systems, the pertinent question is: How cells, evolving from rapidly dividing CBC reach the TAP cell domain in the crypt? Rationally not by "jumping over" +4 cells, but, by either intercoursing signal information with +4 stem cells, or through by-passing the +4 stem cell niche in the crypt cuff via cellular predetermined corridors, one for CBC cells and one for +4 cells. If the latter is the alternative, then two cellular independent cell columns might exist: one for the off-springs of +4 stem cells and the other for off-springs of CBC stem cells. Other questions: Are TAP cells in these two columns proceeding at the same speed on their route to the villi? If so, do differentiated cells "mingle" in the 2 columns? In this context it should be remembered that six or more independent crypts surround a single villus, resulting in an equal number of parallel columns of epithelial cells running toward the villus tip.

2.1. Signaling Pathways

Several signaling pathways such as Wnt, bone morphogenic protein (BMP), PTEN-controlled PI3K/Akt kinase and Notch are involved in intestinal development and homeostasis, including stem cell regulation, lineage specification and maturation [14].

2.1.1. Wnt Transduction Signaling

In the canonical pathway, the Wnt ligand interacts with its cognate receptor, Frizzled, and a co-receptor from the lipoprotein receptor-related protein 5/6 (LRP 5/6) families [14, 15]. Wnt signalling, which is transduced through β̃catenin/TCF4, maintains the undifferentiated state of intestinal crypt stem/progenitor cells via cell cycle control and inhibition of differentiation, controls migration and localization of epithelial cells along the crypt-villus axis, directs early secretory lineage development and terminal differentiation of Paneth cells [14, 16]. Paradoxically, the terminal differentiation of Paneth cells at the crypt bottom requires Wnt signals. In humans, the canonical Wnt signaling is also involved in the control of the migration/invasion behavior of mesenchymal stem cells (hMSC) [17, 18].

3. Stem Cell Markers

3.1. Detecting Stem Cells in Position+4

Several markers may identify stem cells at +4 position: CD34, cKit, or the Hoechst dye that define Side Population, Musashi-1 [19, 20, 21, 22, 23, 24, 25, 26].

3.2. Detecting CBC Stem Cells

Lgr5/Gpr49 gene, that encodes an orphan G protein-coupled receptor, characterized by a large leucine-rich extracellular domain, labels CBC cells. These cells invariably express the Ki67 cell cycle marker, providing a simple means of their identification amongst noncycling Paneth cells. The average cycling time of CBC cells is about 1 day, ruling out that they are quiescent, while transit-amplifying progenitors (TAP) cells cycle every 12 h [9].

$Lgr5^{hi}$ cells generate all cell types of the small intestine epithelium through life. Using the marker Bmi1 (a gene encoding a component of a Polycomb Repressing Complex 1) long term tracing is observed with kinetics that are surprisingly identical to that of $Lgr5^{hi}$ cells [27]. $Lgr5^{hi}$ and Bmi1 may mark overlapping, if not identical, cell population. Both Bmi-1 and CBC cells produce offspring within days [28].

Recently, PHLDA1 was found expressed in discrete crypt base and some position +4 cells in the human small intestine. Hence, PHLDA1 is a putative epithelial stem cell marker in the human small intestine [29].

Intestinal stem cells receive support from their own specialized progeny Paneth cell markers (lysozyme, defensin A1 and stem-cell markers (Lgr5, Olfm4, Tnfrsf19, Cdca) [30]. Paneth cells serve as multifunctional guardians of stem cells, by providing essential niche signaling support: EGF, Wnt3 and Notch.

Only the direct neighbors of Paneth cells, the Lgr5 stem cells, receive strong Wnt signaling, further increased by R-spondin.

Doublecortin- and Calmodulin Kinase-Like 1 (DCAMKL1), a microtubule-associated protein is expressed in the small intestinal crypt cells [26]. Mushashi-1, a gene that encodes an RNA-binding protein labels both ISCs and progenitor cells [8].

Prominin-1(Prom1)/CD133, a pentaspan trans-membrane glycoprotein identify intestinal CBC cells and TAP cells located above the Paneth cells. Hence, Prom1 marks a much larger stem cell/transit-amplifying progenitor compartment [28, 31]. A summary of the candidate markers of intestinal stem cells is shown in Table 1.

Table 1. Candidate markers of intestinal stem cells

+4 cells marker	CBC cells marker	+4 and CBC cells marker	Crypt Base marker	Crypt base (incl. Paneth cells) marker	Crypt-villous axis marker
Bmi 1 Sox9 mTert PHLDA1	Lgr5 Ascl2 Olfm4	Musashi 1	CD133/ Promin	CD24	Decamkl 1

4. Clonality

Human small bowel crypts are clonal, with all the differentiated lineages originating from a common stem cell, whereas the contribution of several crypts leads to villi being polyclonal. Patches of mutated crypts share the same mtDNA mutation and therefore one parent mutated crypt must have divided by fission, over time, to produce daughter crypts sharing the same mutation [32]. In mice, exposure to a slightly higher dose of radiation (> 1 Gy) yielded 6 additional so-called 'clonogenic' crypt cells with the capacity to regenerate entire lost crypts. On the other hand, as many as 30–40 clonogenic cells are identified per crypt, following higher doses of radiation (8–10 Gy) [33]. Under normal conditions, clonogenic cells are believed to represent early TAP cells. Many more short-lived progenitors rapidly divide before giving rise to a differentiated progeny.

5. Stem Cell Niches

Stem cells are thought to reside in a 'niche' towards the base of the crypt and their activity is regulated by the paracrine secretion of growth factors and cytokines from surrounding mesenchymal cells. [12]. The niche includes all cellular and non-cellular components that interact in order to control the adult stem cell, to physical contact and diffusible factors [34].

Recent studies showed, however, that Lgr5+ cells can repopulate the crypt without a mesechymal niche. Asymmetry of crypt–villus organoids was established by the localized presence of Wnt-producing Paneth cells. [27].

5.1. Integrin Signaling

Integrin-linked kinase (ILK) plays a role in integrin signaling-mediated extracellular matrix (ECM)–cell interactions and also acts as a scaffold protein in functional focal adhesion points. ILK performs crucial roles in the control of human intestinal cell and crypt–villus axis homeostasis— especially with regard to basement membrane fibronectin deposition—as well as cell proliferation, spreading, and migration. The basement membrane directs ECM components, such as type IV collagen, laminins and fibronectin, key players in the establishment and maintenance of tissue morphology [35].

6. Stem Cell Plasticity

ISEMF's are an integral part of the epithelial stem cell niche, and regulate epithelial stem cell function by paracrine secretion. In mice with experimental colitis the number of bone marrow derived ISEMF's may be as high as 70%. Donor-derived cells have been observed forming cellular columns in a crypt to- apical fashion which is morphologically similar to indigenous ISEMF's [35].

7. Progenitor Cells Signaling

3 H-thymidine and BrdU, or antibodies against markers such as Ki67, indicate that the lower two thirds of crypt cells are synthetizing DNA. They are called transit amplifying population (TAP). The asymmetric division of the intestinal stem cells maintains the continuous supply of crypt cells [35, 36]. In mice crypt TA progenitors divide every 12–16 h, generating some 300 cells per crypt every day [38]. Newly formed TAP cells reside within crypts for 48–72 h undergoing up to six rounds of cell division as they migrate apically towards the villous [37]. During this journey up the crypt, TAP cells become committed to specific cell fates, start to differentiate and exit into the villous epithelium as mature absorptive enterocytes, mucus-secreting goblet cells, or hormone-producing enteroendocrine cells [30]. Progenitor cells retain the ability to divide until they terminally differentiate [12]. Short-lived multipotent "Mix" progenitors presumably derive directly from multipotent stem cells and generate different epithelial lineages [39].

7.1. Notch Signaling

Stem cells and TAP cells express Notch receptors (N1 and N2) and Hes1, a Notch target gene implicated in intestinal cell differentiation.

Secretory cells express Notch ligands (Dll1, Dll4, Jag1 and Jag2) [38].

7.2 EphB Signaling

EphB signaling promotes cell-cycle reentry of progenitor cells and accounts for approximately 50% of the mitogenic activity in the adult mouse small intestine. Ephrin ligands and their Eph tyrosine kinase receptors are membrane bound proteins that regulate cell migration [39, 40, 41, 42]. EphB receptors are key coordinators of migration and proliferation in the intestinal stem cell niche [43]. Eph/ephrin signaling regulates cell-to-matrix adhesion EphB2 and EphB3, establishing a boundary between proliferative and differentiated cells [44].

7.3. Cdx2, Math 1(Athol), GATA, ADA Signaling

The transcription factor Cdx2 is related to the differentiation of intestinal epithelial cells. Math1-positive epithelial cells co-expressing Cdx2 are found in normal intestinal mucosa from humans and mice. The basic helix–loop–helix transcription factor Math1, is instrumental in the cell fate decision of enteroendocrine, goblet, and Paneth cells in the intestine [45, 46].

A *duodenum*-specific enhancer is located in intron 2 of the human Adenosine deaminase (ADA) gene. This enhancer has been shown to rely on PDX-1, GATA factors, and Cdx factors for its function. *Duodenal* nuclear extracts contain five distinct DNase I footprints within the enhancer [47, 48, 49].

8. Differentiated Cell Signaling

8.1. Paneth Cells

Paneth cells are terminally differentiated cells characterized by large eosinophilic, refractile granules that contain several anti-microbial compounds, important in immunity and host-defense. These cells are made in pairs and die at the base of the crypts. Paneth cells are the longest-lived differentiated cells within a crypt (in mice estimated to be around 23 days). By extrapolating from the lifespan of a mouse (approximately 2 years) to that of a human (approximately 70 years), the estimated turnover time of human Paneth cells reported is 2.2 years [8].

8.1.1. Wnt Signaling

Paneth cells home towards the source of Wnt signals, that is, the CBC cells domain. Conditional deletion of the Wnt receptor Frizzled-5 abrogates expression of these genes in Paneth cells in the adult intestine. Thus, Wnt signals in the crypt can separately drive a stem-cell/progenitor gene programme and a Paneth-cell maturation programme. In intestinal cancer, both gene programmes are activated simultaneously [50]. Ablation of one of the Wnt receptors, *Fz5*, with a *K19Cre* transgene, caused mislocalization of Paneth cells in the villi [51].

8.1.2. β-Catenin/TCF Signaling

β̃catenin/TCF signalling is essential for maintaining the proliferative/undifferentiated state of intestinal epithelial cells and determining cell positioning along the crypt axis by controlling the expression of EphB and ephrinB genes. β̃catenin and TCF inversely control the expression of the EphB2/EphB3 receptors and their ligand ephrin-B1 along the crypt-villus axis. In EphB2/EphB3 null mice, Paneth cells do not follow their downward migratory path, but scatter along crypt and villus. [52, 53, 54]

8.1.3. Musashi-1 Signaling

Musashi-1 (Msi-1), a RNA-binding protein, known as a putative marker of intestinal stem cells suppresses expression of Paneth cell specific genes, presumably through a pathway independent from Notch or Wnt. Msi-1 is a negative regulator of Paneth cell differentiation, and may contribute to maintain the undifferentiated phenotype of intestinal stem cells [55].

8.1.4. PPARβ/δ Signaling

The peroxisome proliferator–activated receptor (PPAR*β/δ* is a nuclear hormone receptor expressed at the bottom of the crypt of the small intestine. PPARs belong to the superfamily of nuclear receptors and their ligands, most of them fatty acids or derivatives (prostaglandins and leukotrienes). PPAR*β/δ* acts on Paneth cell homeostasis by down-regulating the expression of the *Indian hedgehog (Ihh)* signaling pathway, a signal sent by mature Paneth cells to their precursors, negatively regulating their differentiation. The expression of PPAR*β/δ* is highest in the *duodenal* mucosa [1].

8.1.5. EphB Signaling

Ephrin-B1 and Ephrin-B2 are expressed on differentiated cells along the villus. As cells migrate upwards, along the crypt and farther from the Wnt source, present at the crypt base, EphB expression decreases and Ephrin-B ligand expression increases, thereby preventing downward migration [16]. Paneth cells migrate downward into the crypt base from their site of origin in an EphB3-dependent manner [9]. Because Paneth cells do not express Ephrin-B ligands and only express EphB3, their upward migration is prohibited, and, subsequently, they are forced to remain at the crypt base. Deletion of EphB3 results in aberrantly localized Paneth cells. EphB2, not present on Paneth cells is found to be expressed on CBCs and decreases in expression from just above the Paneth cells toward the crypt top. EphB3 expression is restricted to CBCs and Paneth cells at the crypt base [56].

Queries and Comments

The general consensus is that Paneth cells migrate to the base of the crypts whereas all other differentiated cells migrate upwards towards the tip of the villi. This notion would imply that only +4 cells are able to generate Paneth cells while CBC cells are unable to do so. It appears reasonably to assume, however, that if CBC cells also produce Paneth cells, then at least some Paneth cells would migrate sidewise in the crypts and not downwards.

8.1.6. BMP/Hedgehog Signaling

Members of the BMP and hedgehog (*HH*) families direct the positioning of the crypts in the niche through and interplay between epithelial and subepithelial cells [43, 56, 57, 58, 59, 60, 61].

8.1.7. SOX9 Signaling

SOX9 is a transcription factor found in stem/progenitor cells that maintains the multipotent and proliferative capacity of stem/progenitor populations. Sox-9 expression precedes Fz-5-dependent Paneth cell maturation. Since the *Sox9* conditional knockout mice display a complete absence of Paneth cells, *Sox9* is identified as a crucial Wnt-dependent gene in Paneth cell maturation. Sox9 mutant mice display an increase in crypt cell proliferation as well as an increased number of cells expressing c-MYC and CyclinD1, indicating that Sox9 may play a role in negative feedback of the Wnt pathway. Recently, it was found that CD24 is expressed in multipotent IESCs (Sox9EGFPlo) population and marks IESCs that form organoids in culture. Thus, single IESCs generate crypt-like units in the small intestine without a detectable mesenchymal cell component [62]. Flow cytometry of jejunal epithelial preparations revealed a CD24(+) CD45(-) fraction comprising ~1% of the cells. Flow cytometry with anti-lysozyme antibodies demonstrated that Paneth cells comprise ~30% of the CD24 (lo) subfraction [63].

8.1.8. Other Signaling Pathways

Paneth cells express EGF, TGF-a, Wnt3, the Notch ligand Dll4and CD24+; they are essential signals for stem-cell maintenance in culture. Co-culturing of sorted stem cells with Paneth cells markedly improves organoid formation. Genetic removal of Paneth cells in vivo results in the concomitant loss of Lgr5 stem cells. Lgr5 stem cells compete for essential niche signals provided by a specialized daughter cell, the Paneth cell [64].

8.1.9. Paneth Cells Are Involved in IgA-Mediated Acquired Immunity

It is generally accepted that Paneth cells do not participate in the synthesis of polymeric immunoglobulin receptor (pIgR) or the secretion of immunoglobulin A (IgA) in the small intestine. Polymeric Immunoglobulin Receptor mRNA (pIgR) and IgA are colocalized in the secretory granules of human Paneth cells. These findings suggest that, in addition to their well-recognized role in innate immunity, Paneth cells are involved in IgA-mediated acquired immunity in the intestinal tract [65, 66, 67].

8.2. Goblet Cells

In situ hybridization has been used to study mRNA expression of mucin genes in *duodenum.* The pattern of mucin gene expression in fetal duodenum reiterated the patterns observed during intestinal ontogenesis, with MUC2 and MUC3 expression in the surface epithelium and MUC6 expression associated with the development of Brünner's glands . A regulatory role for mucin gene products in gastroduodenal epithelial cell differentiation was suggested [68]. Atoh1 (Math1 for mouse, Hath1 for human), is a basic loop-helix transcription factor and downstream component of the Notch signaling pathway; it is negatively regulated via Notch signaling and expressed in secretory progenitor and mature secretory cells. Musashi 1 is required for differentiation of the secretory lineage (goblet, enteroendocrine and Paneth cells) [69].

8.3. Enterocytes

8.3.1. Notch Signaling

In concert with the Notch pathway, β-catenin maintains intestinal stem cells and controls their differentiation [70, 71, 72, 73]. Deletion of Math1 causes depletion of the goblet, Paneth, and enteroendocrine cell lineages in the small intestine, suggesting that Math1 is essential for progenitor cell commitment to one of three epithelial adult cell types, and that Math1-negative progenitors become enterocytes [55]. Intestinal stem cells can be distinguished by their enriched expression of the Notch ligand, Delta (Dl), through which ISCs control Notch activity of their differentiating daughters. Notch induces absorptive enterocyte differentiation [15]. High levels of Delta result in enterocytic differentiation. HNF-1α cooperates with Cdx2 and GATA-4 in the modulation of proliferation, acquisition of morphological features, and the initiation of the enterocyte differentiation program [74]. In the *duodenum* GATA factors are necessary for crypt cell proliferation, cell differentiation including absorptive enterocyte gene expression. [75].

8.3.2. Epidermal Growth Factor (EGF) Signaling

Signaling by epidermal growth factor (EGF) is associated with intestinal proliferation. Organoids derived from single stem cells were cultured for more than 8 months. Villin1 mature brush borders and apical alkaline phosphatase demonstrated fully polarized enterocytes [76].

8.3.3. Reelin/ Disabled-1 (Dab1) Signaling

Reelin receptors [apolipoprotein E receptor 2 (ApoER2) and very low density lipoprotein receptor (VldlR)], Disabled-1 (Dab1) protein and integrins alpha3 and beta1 are found in enterocytes. The Disabled-1 and VldlR protein signals are detected in the crypt and villus cells, being particularly abundant in the terminal web domain of the enterocyte. The ApoER2 protein signal is found in the upper half of the villi but not in the crypts [77].

8.4. Enteroendocrine Cells

8.4.1. Neuroendocrine Cell Signaling

Enteroendocrine cell differentiation is controlled positively by the transcription factor Ngn3 [78]. Additional transcription factors such as Pdx1, and Neurod1 are required for terminal differentiation of enteroendocrine cells [79]. Reg IV (RELP), a regenerating protein family member, is constitutively expressed in neuroendocrine cells of the intestinal mucosa. The human helix-loop-helix transcription factor Hath1 regulates the embryonic differentiation of neural and other intestinal secretory lineage cells.

GFI1, a transcriptional repressor that functions downstream of *Math1* to select Paneth/goblet versus entero-endocrine cell fates among intestinal secretory progenitors, is expressed in enteroendocrine cells, but not in goblet or Paneth cells. *Gfi1* is *Gfi1−/−* mice lack Paneth cells, have fewer goblet cells, and supernumerary entero-endocrine cells. Low levels of Delta support the enteroendocrine lineage [78, 79].

9. Patients Lacking Paneth Cells, Goblet Cells and Enteroendocrine Cells

Patients characterized by absence of Paneth, goblet, and enteroendocrine cells in the small bowel have a developmental block in differentiation for secretor cell lineages. [67]. Factors responsible for this process are: anomalies in Wnt/β-catenin signaling; Notch signaling, transcription factors Math1, and Hes1; homeobox transcription factors Cdx1 and Cdx2; Kruppel-like factors KLF4 and KLF5; transcription factor Elf3; platelet-derived growth factor (PDGF) A and its receptor, PDGF-Ra; the winged helix transcription factor Fkh6; the homeodomain transcription factor Nkx2–3; and *Hox* and *ParaHox* cluster genes, *Sonic hedgehog*, and bone morphogenetic proteins. The *duodenum* shows absence of goblet cells in both villi and crypts (confirmed by electron microscopy), with preservation of enterocytes. Mucus cell antigens MUC2, MUC5AC, TFF1, and TFF3; enteroendocrine products CMG, gastrin, and serotonin; or the early Paneth cell marker human defensin 5 (HD5) are all negative [67].

10. Villous Formation Signaling

Villous development depends on the proliferation of mesenchymal cells, which support the villous lining. To transport absorbed nutrients, blood and lymph vessels should be formed within the villi.

Villous development does not rely on the proliferation of intestinal epithelial cells alone and requires a cytokine network including epidermal growth factor (EGF), fibroblast growth factor (FGF), platelet-derived growth factor (PDGF), and bone morphogenetic protein (BMP) 2 and 4 [80].

10.1. Hedgehog Signaling

Sonic hedgehog and *Indian hedgehog* genes encode members of the Hedgehog family of cell signals. Both are expressed in gut endoderm, Because Hedgehog inhibition compromises villus formation, this signal is thought to work as a morphogen within the intestine. [3]. *Sonic hedgehog* mutants display among other somatic anomalies, *duodenal* stenosis. *Indian hedgehog* mutants show reduced epithelial stem cell proliferation and differentiation. Hence, Hedgehog signals are essential for organogenesis in the intestinal tract [59, 80, 81, 82].

11. Villous Atrophy (Celiac Disease)

Celiac disease is a common immune-mediated condition in the proximal small intestine often generated by a permanent intolerance to cereal gluten. In diagnostic biopsies the *duodenal* mucosa may display total villous atrophy (Marsh type IIIC), with enlarged crypts showing increased number of mitosis (TAP cells) and absence of villi. Gluten-free diet restores the normal configuration of the *duodenal* mucosa with short crypts and long villi. In later years much attention has been focused on the abnormal microbiota present in the *duodenum* in patients with celiac disease: *bifido bacterium adolescentis/bifido bacterium animalis lactis, Bacteroides vulgatus, Escherichia coli* and rod-shape bacteria [83, 84, 85, 86].

Comments

Lysozyme is as an innate enzyme with potent antibacterial properties, which is strongly expressed in Paneth cells in normal duodenal crypts. There might exist a critical limit for the number of Paneth cells (intercalated with CBC cells) that can be housed at the base of single crypts. Recently, it was found at the bottom of enlarged crypts in celiac disease that Paneth cells could be absent [87]. Instead, lysozyme-rich mucus glands were found to replace the areas of Paneth cell domain (Figure 1) Apparently, duodenal crypts undergo mucus metaplastic transformation, possibly to increase the demands in antibacterial lysozyme required by the recently detected hostile bacterial flora. The possibility that in celiac disease, stem cells might have been re-programmed, from lysozyme-rich Paneth cells to lysozyme-rich mucus glandular metaplasia, was entertained. The molecular signals behind the abrogation of Paneth cells in duodenal crypts in celiac disease and its substitution by lysozyme-rich mucus metaplastic glands deserves further investigation.

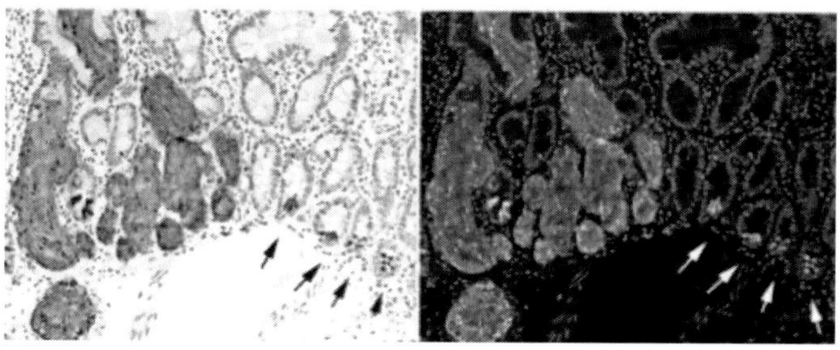

Figure 1. Duodenal mucosa with villous atrophy, antilysozyme immunostain. Left panel: Note normal Paneth cells (at arrows) in the right side of the picture. In the left part, lysozyme-rich mucus metaplasia of the crypts has replaced Paneth cells at the base of the crypts (antilysozyme immunostain x10). Right panel: Same picture using the INVERT function of a Photoshop program (Ps Adobe Photoshop CS3 extended) to highlight the phenomenon.

11.1. Mitosis

Increased number of mitotic figures is a dominant parameter of cell proliferation in patients showing villous atrophy in the duodenal mucosa [91]. The molecular signals that control mitosis in duodenal crypts in villous atrophy remain poorly understood [88].

Comments

The term proliferating cells (from Latin *proles*, offsprings) is here applied for cells in mitosis (M phase) that is, capable to produce daughter cells through cell division and not for cycling cells showing DNA-labeled synthesizing nuclei (S phase). This distinction is crucial, considering that in tumors, a proportion of the DNA synthesizing cells remain arrested in S phase or die, unable to generate daughter tumor cells. In this respect a human nuclear gene (named MTGM) that encodes a novel, small, integral mitochondrial inner-membrane protein is highly expressed in human tumor tissues. Over-expression of MTGM results in inhibition of "cell proliferation", stalling tumor cells in S phase. Hence, a proportion of the DNA synthesizing tumor cells is sterile. Accordingly, these tumor cells will not *proliferate*.

Figures 2, 3 and 4 depict sections from duodenal biopsies in patients with villous atrophy. Sectionion were challenged with the mitosis marker Anti-phospho Histone H3 (Ser 10). Figure 2 reveals enlarged crypts, high numbers of mitotic figures and absence of villous fronds. But the most significant finding is the absence of mitosis in position +5 cells and in cells at the bottom of the crypts (the stem cell domain) (Figures 3 and 4). This observation strongly suggests that proliferation affects exclusively TAP cells and that stem cells are not proliferating at the time of observation. The possibility that stem cells might also proliferate at the time of observation in sections from other patients with villous atrophy cannot be totally rejected. Nevertheless, mitotic cells in position +5 and at the bottom of the crypts were not found in a cohort of 56 biopsy-patients with villous atrophy (Rubio, in preparation).

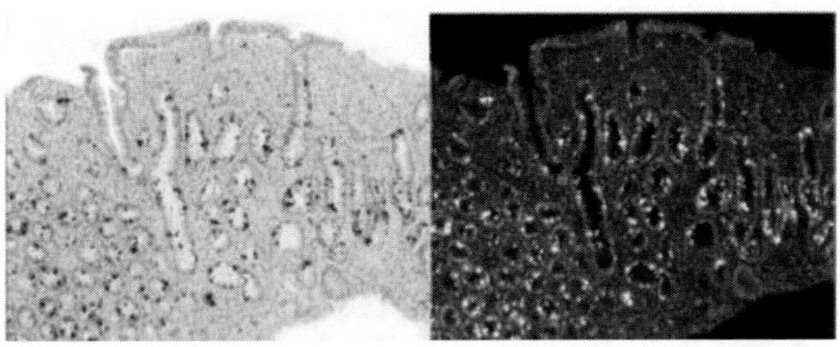

Figure 2. Duodenal mucosa with villous atrophy immunostained with Mitosis marker anti-phospho Histone H3 (Ser 10) (low power view MIT immunostain, x4). Left panel: Note numerous mitotic figures in TAP cells in elongated crypts. Right panel: Same picture using the INVERT function to highlight the phenomenon.

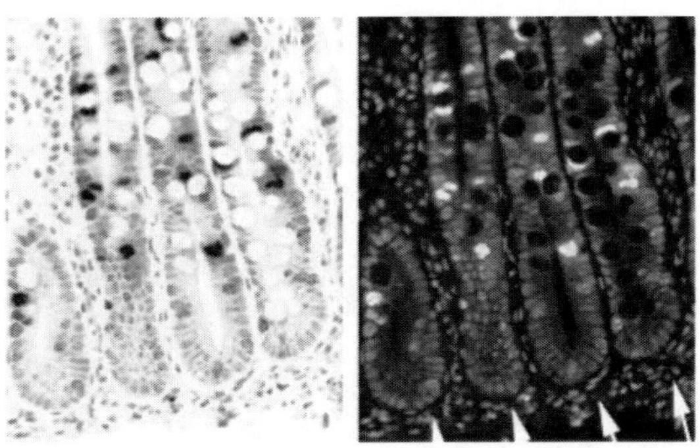

Figure 3. Duodenal mucosa with villous atrophy, immunostained with Mitosis marker anti-phospho Histone H3 (Ser 10) (x20). Left panel: Numerous mitotic figures in TAP cells. Note absence of mitotic figures in the stem cell domain, at basal aspect of the crypts. Right panel: Same picture using the INVERT function to highlight the phenomenon.

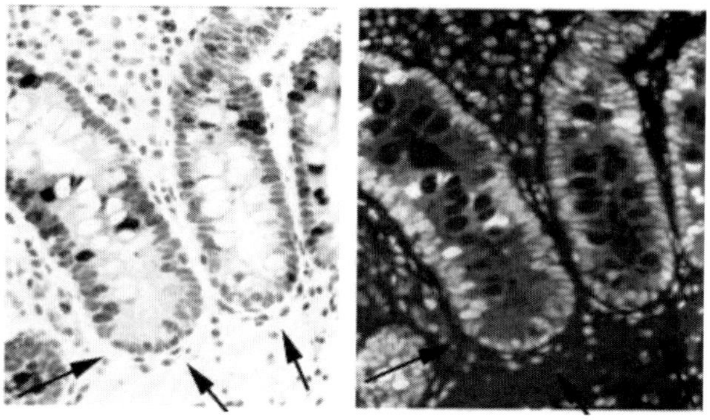

Figure 4. Closer view of duodenal mucosa with villous atrophy, immunostained with Mitosis marker anti-phospho Histone H3 (Ser 10) (x40). Left panel: Mitotic figures are seen in TAP cells. Note absence of mitotic figures in the stem cell domain, at basal aspect of the crypts (arrows). Right panel: Same picture using the INVERT function to highlight the phenomenon.

11.1.2. SUMOylation

One of the key characteristics of stem cells is their capacity to divide for long periods of time in an environment where most of the cells are quiescent. The miRNA pathway might be part of a mechanism that makes stem cells insensitive to environmental signals that normally stop the cell cycle at the G1/S transition [92]. Small ubiquitin-like modifiers (SUMOs) are attached to other proteins to regulate their function (sumoylation). SUMOylation is a downstream target of regulation through Ran, a small GTPase with important functions in both interphase nuclear trafficking and mitotic spindle assembly. Thus, SUMOylation plays a critical role(s) in mitosis [89].

Decreased sumoylation in inducible knockout mice of the E2 enzyme Ubc9 (*Ubc9fl/-/ROSA26-CreERT2* mice) leads to the depletion of the intestinal proliferative compartment and to the rapid disappearance of stem cells. Sumoylation maintains intestinal stem cells and the architecture, mechanical stability. Sox9 in the mutant epithelium at early stage revealed mislocalized Paneth cells throughout the crypt indicating that sumoylation controls allocation of proliferative and differentiated cell populations within the crypt [90]

11.1.3. Feed-back Grow Signaling

The geometric shape of the epithelial tissue in the *duodenum* can be understood as the steady state of a highly dynamic process. Many tissues have an inherent property of growing to a particular size. Self-renewing mucosas seem to ‘‘remember’’ their appropriate sizes, as they accurately regenerate to their original sizes following substantial erosions. Migration of intestinal cells is believed to result from the generation of cells in the crypts that transmits a pressure along the crypt-villus axis. Gaps left by the extruded mature cells at the top of the villi are immediately reoccupied. As a result, epithelial cells attached to each other by cell-to-cell contacts are forced to migrate upward.

The rate of cell renewal might be regulated by a negative feedback control. It has been submitted that size regulation is mediated by secreted factors called *chalones*. *Chalones* are secreted negative feedback factors that control the output of multistage cell lineages. As part of a negative feedback loop, *chalones* repress the proliferation of the cells that secrete the *chalones*, so that when there is a high number or density of the cells, the corresponding high concentration of the *chalone* slows proliferation.

Contrariwise, the exfoliation of *chalone*-rich cells is a signal to stem cells to produce more cells, in order to restore mucosal homeostasis. There is a mounting interest in the mechanisms underlying the execution and regulation of the feedback signaling of intercellular communication. [91, 92, 93]

Comments

In patients with villous atrophy, the duodenal crypts are enlarged (up to 6 times) and the number of mitotic figures, dramatically increased. One conceivable explanation for the latter phenomenon might be that the *chalone*-producing differentiated cells in the villi -that normally refrain stem cells and TAP cells from overproducing mitotic cells- are absent in villous atrophy.

12. Duodenal Adenomas and Carcinomas

12.1. Genetic and Clinical Aspects

Duodenal adenomatosis is found in most patients with familial adenomatous polyposis. Familial adenomatous polyposis (FAP) results from inherited mutation of one APC allele. A 'second hit' mutation of the other allele renders the APC protein inactive and the inability of the destruction complex to degrade cytosolic β-catenin results in the over-transcription of target [12).

About 40% of the duodenal adenomas are sporadic and the remaining 60% from FAP patients [93]. The cumulative incidence of duodenal adenomatosis at age 70 years is 90% [6] and the risk of duodenal cancer in FAP is estimated to be between 100 and 330 times that for the general population, and the lifetime risk of developing duodenal cancer is approximately 4% [94, 95].

Mut Y human homologue-associated polyposis is a recently described colorectal adenomatous polyposis with an autosomal recessive mode of inheritance. Patients affected by Mut Y human homologue-associated polyposis (MAP) have a high prevalence of duodenal polyposis and are at increased risk to develop adenocarcinoma of the duodenum [96].

12.2. Histopathology

Duodenal adenomas in FAP originate in monocryptal adenomas [97]. Early microadenomas show evidence of cellular differentiation; defects in the positioning of Paneth cells suggest disruption of the EphB2:EphB3 receptor system. FAP-associated duodenal epithelium has a significant increase in the number of Paneth cells and endocrine cells per crypt in comparison with controls [97]. Duodenal adenomas spread in a 'bottom-up' fashion with dysplastic cells migrating onto normal villi.

Sporadic nonampullary adenomas exhibit a tubular architecture in all cases [98, 99].

Comments

While reviewing a cohort of 308 duodenal adenomas from patients with familial adenomatous polyposis (FAP), or sporadic, we found in 40.6% (n=125), gastric duodenal metaplasia covering part of the adenomas [100].

Intestinal metaplasia of the esophagus (Barrett´s esophagus), intestinal metaplasia of the stomach, and metaplastic polyposis of the colon are lesions that may antedate neoplastic transformation. It is therefore conceivable that gastric metaplasia in the duodenum might have the same neoplastic proclivity as other metaplastic lesions in the gastrointestinal tract. The known carcinogenic effect of high concentrations of bile acids and pancreatic juices bathing the duodenal mucosa with gastric metaplasia might have set aflame adenomatous neoplastic transformation, either spontaneously (sporadic cases) or reinforcing genetic mutations in patients with FAP. The molecular events telescoping from gastric duodenal metaplasia to dysplastic crypts in duodenal adenomas remains mute. Gastrointestinal pathologists agree that all nonampullar duodenal adenomas are tubular lesions. This is remarkable considering that these tubular neoplasias evolve in a mucosa characterized by villous ornaments. Accordingly, nonampullar duodenal tubular adenomas are dysplastic lesions of the duodenal crypts (Figure 5). The molecular signals responsible for the abrogation of the natural villous formations in tubular adenomas of the *duodenum*, remains elusive.

It should be understood, however, that villous development requires a cytokine network including epidermal growth factor (EGF), fibroblast growth factor (FGF), platelet-derived growth factor (PDGF), and bone morphogenetic protein (BMP) 2 and 4. Whether the lack of villous architecture in tubular adenomas of the *duodenum* is due to a switch-off of EGF, FGF, PDGF and/or BMP signaling during neoplastic transformation, remains to be elucidated.

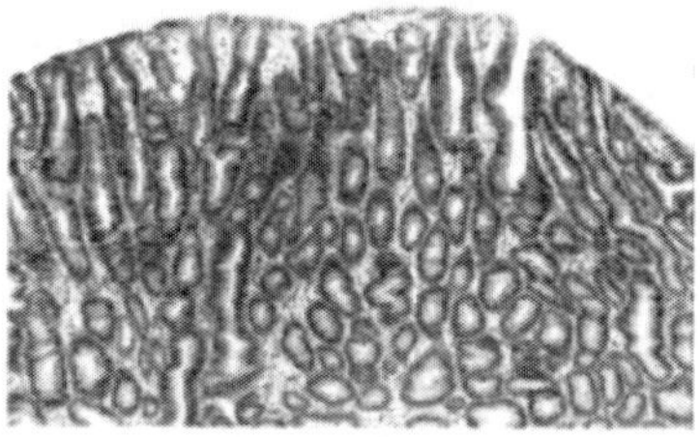

Figure 5. Tubular (cryptal) duodenal adenoma (H and E x 4).

12.3. Duodenal Neoplasia: Mice Models

Duodenal adenomas are also very common in the Min mouse model. Polyps in the small intestine of Min mice develop from cells located around the crypt-villus junction that proliferate inside the mucosa as a disorganized mass that give rise to a tumor. This abnormal migratory behaviour is likely the outcome of the β-catenin/Tcf target gene program autonomously activated in APC mutant cells and results in a further compartmentalization of EphB and ephrin-B expressing cells in the polyp areas. The initial founding polyp cells express high levels of EphB receptors and migrate abnormally inside the villus to avoid the area of high ephrin-B expression at the top positions of the crypts [52]. The morphological similarity of adenomas in the Min mouse and duodenal adenomas in humans suggest the Min mouse is a good model to study carcinogenesis in the proximal small intestine in FAP patients [97].

MUTYH is a mammalian DNA glycosylase that initiates base excision repair by excising adenine opposite 8-oxoguanine and 2-hydroxyadenine opposite guanine, thereby preventing G:C to T:A transversion caused by oxidative stress. MUTYH-null mice are susceptible to spontaneous and oxidative stress–induced intestinal tumorigenesis [7]. In MUTYH-null mice treated with KBrO3, the mean number of small intestinal tumors dramatically increased to 61.88 tumors, (0.85 in wild-type mice). The tumors developed predominantly in the ***duodenum*.** MUTYH suppresses spontaneous tumorigenesis in mammals, thus providing experimental evidence for the association between biallelic germ-line MUTYH mutations and a recessive form of human hereditary colorectal adenoma and carcinoma [7].

12.4. Paneth Cell Translocation in Duodenal Adenomas

Duodenal adenomas from individuals with FAP show Paneth cell differentiation in 92% [30]. In small intestinal adenomas, loss of the wild-type *APC* allele leads to the nuclear localization of *β*-catenin and to the induction of *β*-catenin/tcf-mediated gene transcription. This reduces the expression of Ephrin ligands, leading to Paneth cells losing positional information, being found throughout the adenomatous epithelium. The disruption of Paneth cell positioning tend to be more pronounced in the larger lesions, and this is likely to be associated with the increased disruption of the *β*-catenin–tcf pathways [101].

Queries and Comments

Lysozyme immunostain detects not only differentiated Paneth cells but also Paneth cells precursors. In duodenal adenomas, lysozyme-expressing precursors (undetected in conventionally stained sections (HandE)) are found haphazardly distributed in adenomatous glands, even in the most superficial cell layers [102]. The latter finding suggests that if the innate code that orchestrates the direction of cell migration for Paneth cells is also valid for duodenal adenomas, then the stem cells that normally overly Paneth cells would have been exfoliated into the lumen of the organ [103, 104]. The alternative could be that human duodenal adenomas lack EphB2/EphB3 signaling, as EphB2/EphB3 null mice show misplaced Paneth cells, scattered along crypts and villi. Other explanations could be the ablation of one of the Wnt receptors, *Fz5*, with a *K19Cre* transgene, that causes mislocalization of Paneth cells in the villi or Sox9 mutation that induce mislocalized Paneth cells throughout the crypt.

13. Duodenal Neoplasia Signaling

13.1. Wnt-β-Catenin Signaling

In genetically engineered mice, aberrant Wnt signaling causes tumors predominantly in the small intestine [97, 98]. In adenomas nuclear β-catenin is found in DCAMKL-1-positive cells. Thus, nuclear translocation of β-catenin distinguishes normal and adenoma stem cells within the intestinal tract of mice. Duodenal adenomas are thought to progress to duodenal cancer in a stepwise manner, with the accumulation of genetic mutations (such as *APC*, *k-RAS*, *p53*).

13.2. TGF-β Signaling

Sporadic and FAP-related adenomas show APC and KRAS mutations. BRAF mutations, p53 alterations, and DNA mismatch repair abnormalities are rare. [96, 99]. Regardless of their anatomic location and whether they were sporadic or FAP-related about 75% of the duodenal adenomas show Wnt signaling pathway abnormalities. In fact loss of control of Wnt signaling by the tumour suppressor protein APC is an early event in adenoma formation. Crossing the Lgr5-EGFP mice with APCmin mice allow to examine Lgr5 expression within the premalignant adenomas, which spontaneously arise in their intestine as a result of chronic activation of the Wnt signaling pathway. Lgr5-GFP expression is restricted to a small number of cells within large adenomas. Lgr5 marks the limited population of tumor-initiating/propagating cells thought to exist within cancer, the so-called cancer stem cells [105].

13.3. Eph Signaling

Mice with adenomatous polyposis (APCMin/+) developed invasive adenocarcinoma in the small intestine when EphB signaling is inhibited, thus highlighting the significance of EphB receptors as tumor suppressors. EphB signaling regulates the size of the proliferative compartment without an apparent effect on the number of stem or progenitor cells [43, 106].

13.4. PTEN/PI3K Signaling

The PI3K signaling plays an important role in cell survival, proliferation, growth, and tumorigenesis. Central to this pathway is PI3K, a lipid kinase, composed of both a regulatory subunit, p85, and a catalytic subunit, p110 [107].

13.5. Oncogene Signaling

Disruption of cell polarity is a hallmark of cancer. Basic mechanisms of cell polarity, often targeted by oncogenic signaling pathways, induce deregulation of asymmetric cell divisions of stem or progenitor cells. Cell-polarity mechanisms are important for the diversification of cell shapes and regulation of the asymmetric cell divisions of stem cells. Cell-polarity proteins are known proto-oncogenes or tumor suppressors. [108].

13.6. Mutation Signaling

While mutations in the *p53* and microsatellite instability, occur at approximately similar frequencies in small and large intestinal adenocarcinoma, an inactivation of the adenomatous polyposis coli (*APC*) gene, the gatekeeper mutation in colorectal cancer, is only infrequently seen in small bowel carcinomas [106]. Mutations in *BMPR1*a, *SMAD4* and *PTEN* (phosphatase and tensin homolog -mutated in multiple advanced cancers 1-, give rise to intestinal polyposis syndromes, suggesting a role for BMP and PTEN signaling in the intestinal crypt. BMP signaling is a negative regulator of ISC proliferation [10, 109, 110, 111, 112].

13.7. Methylation

Hypermethylation seems to be a general feature of both FAP-related duodenal carcinomas as well as sporadic duodenal carcinomas with the exception of the PAX6 gene, which is methylated only in FAP-related carcinomas [113].

13.8. Cancer Stem Cells vs. Clonal Evolution

The cancer stem model is highly hierarchical with unique self-renewing cell type at the apex, whereas the clonal evolution model attributes much of the intratumorsal variation to subclonal difference in the mutational profile, and all -except the terminally differentiated cells- may have some self-renewal capacity. Accordingly, in the cancer stem cell model, a phenotypically distinct and generally rare cell type maintains the tumor´s growth, whereas in the clonal evolution model, the dominant subclone(s), sustains it [114].

13.9. Cancer Stem Cell Niche

The cancer stem cell (CSCs) relies on a similar niche, dubbed the "CSC niche," which controls their self-renewal and differentiation [115].

13.10. Cancer in Celiac Disease

Small bowel adenocarcinomas have sporadically been reported in patients with celiac disease [116].

14. NET Tumors

A loss of heterozygosity (LOH) of the MEN1 gene and/or the centromere 11 is demonstrated in approximately 50% of MEN1-associated duodenal NETs. Allelic deletion of the MEN1 gene in microadenomas may reflect a pivotal genetic event in the development of multifocal gastrin and somatostatin cell neoplasms in the duodenum of MEN1 patients [117]. Predominant or exclusive expression of somatostatin was found in 26% of duodenal neuroendocrine tumors (NETs). None of the patients fulfilled the criteria of a somatostatinoma syndrome [118].

15. "Side Population" Cells

The cancer stem cell (CSC) hypothesis suggests that neoplastic clones are maintained exclusively by a small subpopulation of cells that give rise to phenotypically diverse cancer cells [119].

16. Neoplastic Cell Markers

Prominin-1 (CD133). CD133 expression is not restricted to the stem cell zone of the crypt, but expressed all over the intestinal epithelium [28, 31, 103] Direct genetic labeling of adult stem cells in the mouse small intestine, reveal that adenomas arise directly from adult stem cells. Recently, CD 166 was found highly expressed within the endogenous intestinal stem cell niche. CD166-positive cells appear at multiple stages of intestinal carcinoma progression, including benign and metastatic tumors [120]. Prom11/C-L mice containing a Cre-dependent mutant allele of β-catenin show disruption of crypt architecture and a disproportionate expansion of Prom11 cells at the crypt base. The progeny of these cells replace the mucosa of the entire small intestine with neoplastic tissue, characterized by focal high-grade intraepithelial neoplasia and crypt adenoma formation. Although all neoplastic cells arise from Prom11 cells in these mice, only 7% of tumour cells retain Prom1 expression, strongly suggesting that cance, is generated by a rare population of cancer stem cells [121]. *Lgr5*-GFP expression is restricted to a small number of cells within large adenomas, suggesting that *Lgr5* marks not only the normal intestinal stem cells, but also the rare population of tumor-initiating/propagating cell, the so-called cancer stem cells [122].

PHLDA1 [29] is expressed in discrete crypt base and some position +4 cells in the human small (and large) intestine. In small adenomas, PHLDA1 is expressed in a subset of undifferentiated and predominantly Ki-67-negative neoplastic cells, suggesting that a basic hierarchy of differentiation is retained in early tumorigenesis. In large adenomas, carcinomas, and metastases PHLDA1 expression became widespread, with increased expression and nuclear localization at invasive margins. The integrins ITGA2 and ITGA6 were downregulated in response to PHLDA1 suppression, and accordingly cell adhesion to laminin and collagen was significantly reduced. PHLDA1 is a putative epithelial stem cell marker in the human small (and large) intestine [29].

References

[1] Varnat F, Heggeler BB, Grisel P, Boucard N, Corthésy-Theulaz I, Wahli W, Desvergne B. PPARbeta/delta regulates paneth cell differentiation via controlling the hedgehog signaling pathway. *Gastroenterology*. 2006;131:538-53.

[2] Karlsson J, Pütsep K, Chu H, Kays RJ, Bevins CL, Andersson M. Regional variations in Paneth cell antimicrobial peptide expression along the mouse intestinal tract. *BMC Immunol*. 2008; 9: 37-42.

[3] Lehner F, Kulik U, Klempnauer J, Borlak J. Mapping of liver-enriched transcription factors in the human intestine. *World J. Gastroenterol.* 2010;16:3919-27.

[4] Monira P, Koyama Y, Fukutomi R, Yasui K, Isemura M, Yokogoshi H. Effects of Japanese mistletoe lectin on cytokine gene expression in human colonic carcinoma cells and in the mouse intestine. *Biomed. Res.* 2009;30:303-309.

[5] Maier EA, Dusing MR, Wiginton DA Cdx binding determines the timing of enhancer activation in postnatal duodenum. *J. Biol. Chem.* 2005;280:13195-202.

[6] Bülow S, Björk J, Christensen IJ, Fausa O, Järvinen H, Moesgaard F, Vasen HF; DAF Study Group. Duodenal adenomatosis in familial adenomatous polyposis. *Gut.* 2004;53:381-386.

[7] Sakamoto K, Tominaga Y, Yamauchi K, Nakatsu Y, Sakumi K, Yoshiyama K, Egashira A, Kura S, Yao T, Tsuneyoshi M, Maki H, Nakabeppu Y, Tsuzuki T. MUTYH-null mice are susceptible to spontaneous and oxidative stress induced intestinal tumorigenesis. *Cancer Res.* 2007;67:6599-604.

[8] Potten CS, Gandara R, Mahida YR, Loeffler M, Wright NA. The stem cells of small intestinal crypts: where are they? *Cell Prolif.* 2009;42:731-750.

[9] Bjerknes M, Cheng H. Gastrointestinal stem cells. II. Intestinal stem cells. *Am. J. Physiol. Gastrointest Liver Physiol.* 2005;289:G381-7. Review.

[10] Montgomery RK, Breault DT. Small intestinal stem cell markers. *J. Anat.* 2008;213:52-58.

[11] Potten CS, Kovacs L, Hamilton E. Continuous labelling studies on mouse skin and intestine. *Cell Tissue Kinet.* 1974;7:271-283.

[12] Leedham SJ, Brittan M, McDonald SA, Wright NA. Intestinal stem cells. *J. Cell Mol. Med.* 2005;9:11-24. Review.

[13] Scoville DH, Sato T, He XC, Li L. Current view: intestinal stem cells and signaling. *Gastroenterology.* 2008;134:849-864. Review.

[14] Buchert M, Athineos D, Abud HE, Burke ZD, Faux MC, Samuel MS, Jarnicki AG, Winbanks CE, Newton IP, Meniel VS, Suzuki H, Stacker SA, Näthke IS, Tosh D, Huelsken J, Clarke AR, Heath JK, Sansom OJ, Ernst M. Genetic dissection of differential signaling threshold requirements for the Wnt/beta-catenin pathway in vivo. *PLoS Genet.* 2010;6: e1000816.

[15] Nakamura T, Tsuchiya K, Watanabe M. Crosstalk between Wnt and Notch signaling in intestinal epithelial cell fate decision. *J. Gastroenterol.* 2007; 42:705–710. Review.

[16] Huelsken J, Behrens J. The Wnt signalling pathway. *J. Cell Sci.* 2002;115, 3977-3978.

[17] Neth P, Ries C, Karow M, Egea V, Ilmer M, Jochum M. The Wnt signal transduction pathway in stem cells and cancer cells: influence on cellular invasion. Stem Cell Rev. 2007;3:18-29. Review.

[18] Verkaar F, van der Stelt M, Blankesteijn WM, van der Doelen AA, Zaman GJ Discovery of novel small molecule activators of β-catenin signaling.. PLoS One. 2011;6:e 19185.

[19] Kayahara T, Sawada M, Takaishi S, Fukui H, Seno H, Fukuzawa H, Suzuki K, Hiai H, Kageyama R, Okano H, Chiba T. Candidate markers for stem and early progenitor cells, Musashi-1 and Hes1, are expressed in crypt base columnar cells of mouse small intestine. FEBS Lett. 2003;535:131-135.

[20] Potten CS, Booth C, Hargreaves D. The small intestine as a model for evaluating adult tissue stem cell drug targets. Cell Prolif. 2003;36:115-129. Review.

[21] Asai R, Okano H, Yasugi S. Correlation between Musashi-1 and c-hairy-1 expression and cell proliferation activity in the developing intestine and stomach of both chicken and mouse. Dev. Growth Differ. 2005;47:501-510.

[22] Fujimoto K, Beauchamp RD, Whitehead RH. Identification and isolation of candidate human colonic clonogenic cells based on cell surface integrin expression. *Gastroenterology*. 2002;123:1941-1948.

[23] He X et al. BMP signaling inhibits intestinal stem cell self-renewal through suppression of Wnt-b catenin signaling *Nature Genetics* 36: 1117-1122, 2004.

[24] van Es JH, Jay P, Gregorieff A, van Gijn ME, Jonkheer S, Hatzis P, Thiele A, van den Born M, Begthel H, Brabletz T, Taketo MM, Clevers H. Wnt signalling induces maturation of Paneth cells in intestinal crypts. Nat Cell Biol. 2005;7:381-386.

[25] van der Flier LG, Clevers H. Stem cells, self-renewal, and differentiation in the intestinal epithelium. Annu. Rev. Physiol. 2009;71:241-260. Review.

[26] Giannakis M, Stappenbeck TS, Mills JC, Leip DG, Lovett M, Clifton SW, Ippolito JE, Glasscock JI, Arumugam M, Brent MR, Gordon JI. Molecular properties of adult mouse gastric and intestinal epithelial progenitors in their niches. J. Biol. Chem. 2006;281:11292-11300.

[27] Sangiorgi E, Capecchi MR. Bmi1 lineage tracing identifies a self-renewing pancreatic acinar cell subpopulation capable of maintaining pancreatic organ homeostasis. Proc. Natl. Acad. Sci. USA. 2009;106:7101-7106.

[28] Snippert H, van Es J, van der Born M, Begthel H, Stange D, Barker N, Clevers H. Prominin-1/CD133 Marks Stem Cells and Early Progenitors in Mouse Small Intestine. *Gastroenterology*. 2009;36:2187–2194.

[29] Sakthianandeswaren A, Christie M, D'Andreti C, Tsui C, Jorissen RN, Li S, Fleming NI, Gibbs P, Lipton L, Malaterre J, Ramsay RG, Phesse TJ, Ernst M, Jeffery RE, Poulsom R, Leedham SJ, Segditsas S, Tomlinson IP, Bernhard OK, Simpson RJ, Walker F, Faux MC, Church N, Catimel B, Flanagan DJ, Vincan E, Sieber OM. PHLDA1 expression marks the putative epithelial stem cells and contributes to intestinal tumorigenesis. *Cancer Res*. 2011;71:3709-3719.

[30] Lin SA, Barker N 2011 Gastrointestinal stem cells and cancer *Gastroenterology*. 2011;140:1135-1139.

[31] Hou NY, Yang K, Chen T, Chen XZ, Zhang B, Mo XM, Hu JK. CD133+ CD44+ subgroups may be human small intestinal stem cells. Mol. Biol. Rep. 2011;38:997-1004.

[32] Gutierrez-Gonzalez L, Deheragoda M, Elia G, Leedham SJ, Shankar A, Imber C, Jankowski JA, Turnbull DM, Novelli M, Wright NA, McDonald SA. Analysis of the clonal architecture of the human small intestinal epithelium establishes a common stem cell for all lineages and reveals a mechanism for the fixation and spread of mutations. J. Pathol. 2009;217:489-496.

[33] Bach SP, Renehan AG, Potten CS. Stem cells: the intestinal stem cell as a paradigm. Carcinogenesis. 2000;21:469-76. Review.

[34] Walker MR, Patel KK, Stappenbeck TS. The stem cell niche. J. Pathol. 2009;217:169-180. Review.

[35] Benoit YD, Lussier C, Ducharme PA, Sivret S, Schnapp LM, Basora N, Beaulieu JF. Integrin alpha8beta1 regulates adhesion, migration and proliferation of human intestinal

crypt cells via a predominant RhoA/ROCK-dependent mechanism. Biol. Cell. 2009;101:695-708.

[36] Passegué E, Wagers AJ. Regulating quiescence: new insights into hematopoietic stem cell biology. Dev. Cell. 2006;10:415-417.

[37] Potten CS, Gandara R, Mahida YR, Loeffler M, Wright NA. The stem cells of small intestinal crypts: where are they? Cell Prolif. 2009;42:731-750.

[38] Madison BB, Nakagawa H. Delta force in intestinal crypts. Gastroenterology. 2011;140:1135-1139.

[39] Genander M, Halford MM, Xu NJ, Eriksson M, Yu Z, Qiu Z, Martling A, Greicius G, Thakar S, Catchpole T, Chumley MJ, Zdunek S, Wang C, Holm T, Goff SP, Pettersson S, Pestell RG, Henkemeyer M, Frisén J. Dissociation of EphB2 signaling pathways mediating progenitor cell proliferation and tumor suppression. Cell. 2009;139:679-692.

[40] McConnell BB, Kim SS, Yu K, Ghaleb AM, Takeda N, Manabe I, Nusrat A, Nagai R, Yang VW. Krüppel-like Factor 5 is Important for Maintenance of Crypt Architecture and Barrier Function in Mouse Intestine. Gastroenterology. 2011; Jul 13. [Epub ahead of print].

[41] Miao H, Strebhardt K, Pasquale EB, Shen TL, Guan JL, Wang B.Inhibition of integrin-mediated cell adhesion but not directional cell migration requires catalytic activity of EphB3 receptor tyrosine kinase. Role of Rho family small GTPases. *J. Biol. Chem.* 2005;280:923-932.

[42] Furukawa K, Sato T, Katsuno T, Nakagawa T, Noguchi Y, Tokumasa A, Yokote K, Yokosuka O, Saito Y. Smad3 contributes to positioning of proliferating cells in colonic crypts by inducing EphB receptor protein expression. Biochem. Biophys. Res. Commun. 2011;405:521-526.

[43] Holmberg J, Genander M, Halford MM, Annerén C, Sondell M, Chumley MJ, Silvany RE, Henkemeyer M, Frisén J. EphB receptors coordinate migration and proliferation in the intestinal stem cell niche. Cell. 2006;125:1151-1163.

[44] Murai KK, Pasquale EB. Restraining stem cell niche plasticity: a new talent of Eph receptors. Cell Stem Cell. 2010;7:647-648.

[45] Mutoh H, Sakamoto H, Hayakawa H, Arao Y, Satoh K, Nokubi M, Sugano K. The intestine-specific homeobox gene Cdx2 induces expression of the basic helix-loop-helix transcription factor Math1. Differentiation. 2006;74:313-321.

[46] Benoit YD, Paré F, Francoeur C, Jean D, Tremblay E, Boudreau F, Escaffit F, Beaulieu JF. Cooperation between HNF-1alpha, Cdx2, and GATA-4 in initiating an enterocytic differentiation program in a normal human intestinal epithelial progenitor cell line. Am. J. Physiol. Gastrointest Liver Physiol. 2010;298:G504-G517.

[47] Dusing MR, Florence EA, Wiginton DA. High-level activation by a duodenum-specific enhancer requires functional GATA binding sites. Am. J. Physiol. Gastrointest Liver Physiol. 2003;284:G1053-G1065.

[48] Dusing MR, Brickner AG, Lowe SY, Cohen MB, Wiginton DA. A duodenum-specific enhancer regulates expression along three axes in the small intestine. Am. J. Physiol. Gastrointest Liver Physiol. 2000;279:G1080-G1093.

[49] Maier EA, Dusing MR, Wiginton DA. Cdx binding determines the timing of enhancer activation in postnatal duodenum. J. Biol. Chem. 2005;280:13195-13202.

[50] van Es JH, Jay P, Gregorieff A, van Gijn ME, Jonkheer S, Hatzis P, Thiele A, van den Born M, Begthel H, Brabletz T, Taketo MM, Clevers H Wnt signalling induces maturation of Paneth cells in intestinal crypts. *Nat. Cell Biol.* 2005;7:381-386.

[51] Mori-Akiyama Y, van den Born M, van Es JH, Hamilton SR, Adams HP, Zhang J, Clevers H, de Crombrugghe B. SOX9 is required for the differentiation of paneth cells in the intestinal epithelium. Gastroenterology. 2007;133:539-546.

[52] Batlle E, Henderson JT, Beghtel H, van den Born MM, Sancho E, Huls G, Meeldijk J, Robertson J, van de Wetering M, Pawson T, Clevers H. Beta-catenin and TCF mediate cell positioning in the intestinal epithelium by controlling the expression of EphB/ephrinB. *Cell.* 2002;111:251-263.

[53] Wong SY, Chiam KH, Lim CT, Matsudaira P. Computational model of cell positioning: directed and collective migration in the intestinal crypt epithelium. *J. R. Soc. Interface.* 2010;7 Suppl 3:S351-S363.

[54] Camac KS, Thompson FM, Cummins AG. Activation of beta-catenin in the stem cell region of crypts during growth of the small intestine in infant rats. *Dig. Dis. Sci.* 2007;52:1242-1246.

[55] Murayama M, Okamoto R, Tsuchiya K, Akiyama J, Nakamura T, Sakamoto N, Kanai T, Watanabe M. Musashi-1 suppresses expression of Paneth cell-specific genes in human intestinal epithelial cells. *J. Gastroenterol.* 2009;44:173-182.

[56] Cortina C, Palomo-Ponce S, Iglesias M, Fernández-Masip JL, Vivancos A, Whissell G, Humà M, Peiró N, Gallego L, Jonkheer S, Davy A, Lloreta J, Sancho E, Batlle E. EphB-ephrin-B interactions suppress colorectal cancer progression by compartmentalizing tumor cells. Nat. Genet. 2007;39:1376-1383.

[57] Li X, Madison BB, Zacharias W, Kolterud A, States D, Gumucio DL. Deconvoluting the intestine: molecular evidence for a major role of the mesenchyme in the modulation of signaling cross talk. *Physiol. Genomics.* 2007;29:290-301.

[58] Pellegrinet L, Rodilla V, Liu Z, Chen S, Koch U, Espinosa L, Kaestner KH, Kopan R, Lewis J, Radtke F. Dll1- and dll4-mediated notch signaling are required for homeostasis of intestinal stem cells. *Gastroenterology.* 2011;140:1230-1240.

[59] Varum S, Rodrigues AS, Moura MB, Momcilovic O, Easley CA 4th, Ramalho-Santos J, Van Houten B, Schatten G. Energy metabolism in human pluripotent stem cells and their differentiated counterparts. PLoS One. 2011;6(6):e20914.

[60] Nielsen CM, Williams J, van den Brink GR, Lauwers GY, Roberts DJ. Hh pathway expression in human gut tissues and in inflammatory gut diseases. Lab. Invest. 2004;84:1631-1642.

[61] Ishizuya-Oka A, Hasebe T. Sonic hedgehog and bone morphogenetic protein-4 signaling pathway involved in epithelial cell renewal along the radial axis of the intestine. Digestion. 2008;77 Suppl 1:42-47.

[62] Gracz AD, Ramalingam S, Magness ST. Sox9 expression marks a subset of CD24-expressing small intestine epithelial stem cells that form organoids in vitro. *Am. J. Physiol. Gastrointest Liver Physiol.* 2010;298:G590-G600.

[63] von Furstenberg RJ, Gulati AS, Baxi A, Doherty JM, Stappenbeck TS, Gracz AD, Magness ST, Henning SJ Sorting mouse jejunal epithelial cells with CD24 yields a population with characteristics of intestinal stem cells. *Am. J. Physiol. Gastrointest Liver Physio*l. 2011;300:G409-G417.

[64] Sato T, van Es JH, Snippert HJ, Stange DE, Vries RG, van den Born M, Barker N, Shroyer NF, van de Wetering M, Clevers H. Paneth cells constitute the niche for Lgr5 stem cells in intestinal crypts. Nature. 2011;469:415-418.

[65] Santaolalla R, Fukata M, Abreu MT. Innate immunity in the small intestine. Curr Opin Gastroenterol. 2011;27:125-131. Review.

[66] Wershil BK, Furuta GT. Gastrointestinal mucosal immunity. J. Allergy Clin. Immunol. 2008;121(2 Suppl):S380-S383.

[67] Shaoul R, Hong D, Okada Y, Cutz E, Marcon MA. Lineage development in a patient without goblet, paneth, and enteroendocrine cells: a clue for intestinal epithelial differentiation. Pediatr Res. 2005;58:492-498.

[68] Buisine MP, Devisme L, Degand P, Dieu MC, Gosselin B, Copin MC, Aubert JP, Porchet N. Developmental mucin gene expression in the gastroduodenal tract and accessory digestive glands. II. Duodenum and liver, gallbladder, and pancreas. J. Histochem. Cytochem. 2000;48:1667-1676.

[69] Murayama M, Okamoto R, Tsuchiya K, Akiyama J, Nakamura T, Sakamoto N, Kanai T, Watanabe M. Musashi-1 suppresses expression of Paneth cell-specific genes in human intestinal epithelial cells. J. Gastroenterol. 2009;44:173-182.

[70] Fre S, Pallavi SK, Huyghe M, Laé M, Janssen KP, Robine S, Artavanis-Tsakonas S, Louvard D. Notch and Wnt signals cooperatively control cell proliferation and tumorigenesis in the intestine. *Proc. Natl. Acad. Sci. USA*. 2009;106:6309-6314.

[71] Stanger BZ, Datar R, Murtaugh LC, Melton DA. Direct regulation of intestinal fate by Notch. Proc. Natl. Acad. Sci. USA. 2005;102:12443-12448.

[72] Sato T, Vries RG, Snippert HJ, van de Wetering M, Barker N, Stange DE, van Es JH, Abo A, Kujala P, Peters PJ, Clevers H. Single Lgr5 stem cells build crypt-villus structures in vitro without a mesenchymal niche. Nature. 2009;459:262-265.

[73] van Es JH, de Geest N, van de Born M, Clevers H, Hassan BA. Intestinal stem cells lacking the Math1 tumour suppressor are refractory to Notch inhibitors. Nat. Commun. 2010;1:18-23.

[74] Boyle MJ, Seaver EC. Expression of FoxA and GATA transcription factors correlates with regionalized gut development in two lophotrochozoan marine worms: Chaetopterus (Annelida) and Themiste lageniformis (Sipuncula). Evodevo. 2010;1:2-10.

[75] Beuling E, Baffour-Awuah NY, Stapleton KA, Aronson BE, Noah TK, Shroyer NF, Duncan SA, Fleet JC, Krasinski SD.GATA factors regulate proliferation, differentiation, and gene expression in small intestine of mature mice. *Gastroenterology*. 2011;140:1219-1229.

[76] Biteau B, Jasper H. EGF signaling regulates the proliferation of intestinal stem cells in Drosophila. Development. 2011;138:1045-155.

[77] García-Miranda P, Peral MJ, Ilundain AA. Rat small intestine expresses the reelin-Disabled-1 signalling pathway. Exp Physiol. 2010;95:498-507.

[78] Sei Y, Lu X, Liou A, Zhao X, Wank SA. A stem cell marker-expressing subset of enteroendocrine cells resides at the crypt base in the small intestine. Am. J. Physiol. Gastrointest Liver Physiol. 2011;300:G345-G356.

[79] Schonhoff SE, Giel-Moloney M, Leiter AB. Neurogenin 3-expressing progenitor cells in the gastrointestinal tract differentiate into both endocrine and non-endocrine cell types. *Dev. Biol.* 2004;270:443-454.

[80] Ushida K, Yoshida Y, Tsukahara T, Watanabe T, Inoue R. Oral administration of Enterococcus faecalis EC-12 cell preparation improves villous atrophy after weaning through enhancement of growth factor expression in mice. Biomed Res. 2010;31:191-198.

[81] Madison BB, Braunstein K, Kuizon E, Portman K, Qiao XT, Gumucio DL. Epithelial hedgehog signals pattern the intestinal crypt-villus axis. *Development*. 2005;132:279-289.

[82] Zacharias WJ, Madison BB, Kretovich KE, Walton KD, Richards N, Udager AM, Li X, Gumucio DL. Hedgehog signaling controls homeostasis of adult intestinal smooth muscle. *Dev. Biol.* 2011;355:152-162.

[83] Sánchez E, Donat E, Ribes-Koninckx C, Calabuig M, Sanz Y. Intestinal Bacteroides species associated with coeliac disease. J. Clin. Pathol. 2010;63:1105-1111.

[84] Schippa S, Iebba V, Barbato M, Di Nardo G, Totino V, Checchi MP, Longhi C, Maiella G, Cucchiara S, Conte MP. A distinctive 'microbial signature' in celiac pediatric patients. *BMC Microbiol.* 2010;10:175-177.

[85] Forsberg G, Fahlgren A, Hörstedt P, Hammarström S, Hernell O, Hammarström ML. Presence of bacteria and innate immunity of intestinal epithelium in childhood celiac disease. Am. J. Gastroenterol. 2004;99:894-904.

[86] Ou G, Hedberg M, Hörstedt P, Baranov V, Forsberg G, Drobni M, Sandström O, Wai SN, Johansson I, Hammarström ML, Hernell O, Hammarström S. Proximal small intestinal microbiota and identification of rod-shaped bacteria associated with childhood celiac disease. *Am. J. Gastroenterol.* 2009;104:3058-3067.

[87] Rubio CA. Lysozyme-rich mucus metaplasia in duodenal crypts supersedes Paneth cells in celiac disease. *Virchows Arch.* 2011 Jul 18. [Epub ahead of print].

[88] Hatfield SD, Shcherbata HR, Fischer KA, Nakahara K, Carthew RW, Ruohola-Baker H. Stem cell division is regulated by the microRNA pathway. Nature. 2005;435:974-978.

[89] Dasso M. Emerging roles of the SUMO pathway in mitosis. *Cell Div.* 2008;3:5-9.

[90] Demarque MD, Nacerddine K, Neyret-Kahn H, Andrieux A, Danenberg E, Jouvion G, Bomme P, Hamard G, Romagnolo B, Terris B, Cumano A, Barker N, Clevers H, Dejean A Sumoylation by Ubc9 regulates the stem cell compartment and structure and function of the intestinal epithelium in mice. *Gastroenterology.* 2011;140:286-296.

[91] Meinzer HP, Sandblad B. Evidence for cell generation controlled proliferation in the small intestinal crypt. *Cell Tissue Kinet.* 1986;19:581-590.

[92] Bakthavatsalam D, Choe JM, Hanson NE, Gomer RH A Dictyostelium chalone uses G proteins to regulate proliferation *BMC Biol.* 2009;7:44-48.

[93] Lo WC, Chou CS, Gokoffski KK, Wan FY, Lander AD, Calof AL, Nie Q. Feedback regulation in multistage cell lineages. *Math. Biosci. Eng.* 2009;6:59-82.

[94] Johnson MD, Mackey R, Brown N, Church J, Burke C, Walsh RM. Outcome based on management for duodenal adenomas: sporadic versus familial disease. *J. Gastrointest. Surg.* 2010;14:229-235.

[95] Latchford AR, Neale KF, Spigelman AD, Phillips RK, Clark SK. Features of duodenal cancer in patients with familial adenomatous polyposis. *Clin. Gastroenterol. Hepatol.* 2009;7:659-663.

[96] Buecher B, Baert-Desurmont S, Leborgne J, Humeau B, Olschwang S, Frébourg T. Duodenal adenocarcinoma and Mut Y human homologue-associated polyposis. *Eur. J. Gastroenterol. Hepatol.* 2008;20:1024-1027.

[97] Preston SL, Leedham SJ, Oukrif D, Deheregoda M, Goodlad RA, Poulsom R, Alison MR, Wright NA, Novelli M. The development of duodenal microadenomas in FAP patients: the human correlate of the Min mouse. *J. Pathol.* 2008;214:294-301.

[98] Odze RD. Epithelial proliferation and differentiation in flat duodenal mucosa of patients with familial adenomatous polyposis. Mod. Pathol. 1995;8:648-653.

[99] Wagner PL, Immunohistochemical and Molecular Features of Sporadic and FAP-associated Duodenal Adenomas of the Ampullary and Nonampullary Mucosa *Am. J. Surg. Pathol.* 2008;32:1388-1395.

[100] Rubio CA. Gastric duodenal metaplasia in duodenal adenomas. *J. Clin. Pathol.* 2007;60:661-663.

[101] Barker N, van de Wetering M, Clevers H. The intestinal stem cell. *Genes Dev.* 2008;22:1856-1864. Review.

[102] Rubio CA. Stem cells might participate in the cell turnover of duodenal adenomas. *Int. J. Clin. Exp. Pathol.* 2009;2:149-153.

[103] Rubio CA Further studies support the participation of stem cells in the cell turnover of duodenal adenomas. *Anticancer Res.* 2009;29:657-660.

[104] Rubio CA (2011) Putative Stem Cells in Mucosas of the Esophago-Gastrointestinal Tract.Chapter 10. In *Stem Cell, Regenerative Medicine and Cancer* Ed. Shree Ram Singh, Nova Science Publishers, Inc. Haupauge, NY, USA, pp. 281-310.

[105] Zhu L, Gibson P, Currle DS, Tong Y, Richardson RJ, Bayazitov IT, Poppleton H, Zakharenko S, Ellison DW, Gilbertson RJ Prominin 1 marks intestinal stem cells that are susceptible to neoplastic transformation. *Nature.* 2009;457:603-607.

[106] Islam S, Loizides AM, Fialkovich JJ, Grand RJ, Montgomery RK. Developmental expression of Eph and ephrin family genes in mammalian small intestine. *Dig. Dis. Sci.* 2010;55:2478-2488.

[107] Marsh V, Winton DJ, Williams GT, Dubois N, Trumpp A, Sansom OJ, Clarke AR. Epithelial Pten is dispensable for intestinal homeostasis but suppresses adenoma development and progression after Apc mutation. Nat. Genet. 2008;40:1436-444.

[108] Lee M, Vasioukhin V. Cell polarity and cancer--cell and tissue polarity as a non-canonical tumor suppressor. J. Cell Sci. 2008;121:1141-1150.

[109] Breuhahn K, Singh S, Schirmacher P, Bläker H. Large-scale N-terminal deletions but not point mutations stabilize beta-catenin in small bowel carcinomas, suggesting divergent molecular pathways of small and large intestinal carcinogenesis. *J. Pathol.* 2008;215:300-307.

[110] Smalley-Freed WG, Efimov A, Short SP, Jia P, Zhao Z, Washington MK, Robine S, Coffey RJ, Reynolds AB. Adenoma formation following limited ablation of p120-catenin in the mouse intestine. PLoS One. 2011;6:e19880.

[111] Radulescu S, Ridgway RA, Appleton P, Kroboth K, Patel S, Woodgett J, Taylor S, Nathke IS, Sansom OJ. Defining the role of APC in the mitotic spindle checkpoint in vivo: APC-deficient cells are resistant to Taxol. Oncogene. 2010;29:6418-6427.

[112] Haramis AP, Begthel H, van den Born M, van Es J, Jonkheer S, Offerhaus GJ, Clevers H. De novo crypt formation and juvenile polyposis on BMP inhibition in mouse intestine. *Science.* 2004;303:1684-1686.

[113] Berkhout M, Nagtegaal ID, Cornelissen SJ, Dekkers MM, van de Molengraft FJ, Peters WH, Nagengast FM, van Krieken JH, Jeuken JW. Chromosomal and methylation alterations in sporadic and familial adenomatous polyposis-related duodenal carcinomas. *Mod. Pathol.* 2007;20:1253-1262.

[114] Adams JM, Kelly PN, Dakic A, Carotta S, Nutt SL, Strasser A. Role of "cancer stem cells" and cell survival in tumor development and maintenance. Cold Spring Harb. Symp. Quant. Biol. 2008;73:451-459.

[115] Borovski T, De Sousa E Melo F, Vermeulen L, Medema JP. Cancer stem cell niche: the place to be. *Cancer Res.* 2011;71:634-639.

[116] Richir M, Songun I, Wientjes C, Snel P, Dwars B Small Bowel Adenocarcinoma in a Patient with Coeliac Disease: Case Report and Review of the Literature. *Case Rep. Gastroenterol.* 2010;4:416-420.

[117] Anlauf M, Perren A, Klöppel G. Endocrine precursor lesions and microadenomas of the duodenum and pancreas with and without MEN1: criteria, molecular concepts and clinical significance. *Pathobiology.* 2007;74:279-284.

[118] Anlauf M, Perren A, Meyer CL, Schmid S, Saremaslani P, Kruse ML, Weihe E, Komminoth P, Heitz PU, Klöppel G. Precursor lesions in patients with multiple endocrine neoplasia type 1-associated duodenal gastrinomas. Gastroenterology. 2005;128:1187-1198.

[119] Haraguchi N, Utsunomiya T, Inoue H, Tanaka F, Mimori K, Barnard GF, Mori M. Characterization of a side population of cancer cells from human gastrointestinal system. *Stem Cells.* 2006;24:506-513.

[120] Levin TG, Powell AE, Davies PS, Silk AD, Dismuke AD, Anderson EC, Swain JR, Wong MH. Characterization of the intestinal cancer stem cell marker CD166 in the human and mouse gastrointestinal tract. *Gastroenterology.* 2010;139:2072-2082.

[121] Ohashi S, Natsuizaka M, Yashiro-Ohtani Y, Kalman RA, Nakagawa M, Wu L, Klein-Szanto AJ, Herlyn M, Diehl JA, Katz JP, Pear WS, Seykora JT, Nakagawa H. NOTCH1 and NOTCH3 coordinate esophageal squamous differentiation through a CSL-dependent transcriptional network. Gastroenterology. 2010;139:2113-2123.

[122] Sei Y, Lu X, Liou A, Zhao X, Wank SA. A stem cell marker-expressing subset of enteroendocrine cells resides at the crypt base in the small intestine. *Am. J. Physiol. Gastrointest Liver Physiol.* 2011;300:G345-G356.

In: Signaling, Gene Regulation and Cancer
Editors: Shree Ram Singh and Manoj K. Mishra

ISBN: 978-1-61942-088-5

Chapter 7

TARGETING CANCER CELLS IN HEAD AND NECK TUMOURS

C. Bianchini*, A. Ciorba, S. Pelucchi and A.Pastore
ENT Department, University Hospital of Ferrara,
Ferrara, Italy

ABSTRACT

This chapter focuses on the recent advances in understanding the molecular mechanisms at the basis of cancer initiation and progression in Head and Neck, also discussing about the possible development of targeted cellular strategies.

The therapeutic benefits of targeted clinical interventions with increased selectivity and fewer adverse effects hold great promise in the treatment of solid malignancies, both in monotherapy and in combination. Molecular targeted therapies offer increasingly customized solutions based on the targeting of multiple specific pathways essential for cancer development and metastasis, allowing the maintenance of quality of life while efficiently attacking the tumour. Even if in some hematologic malignancies (i.e. non-Hodgkin's lymphomas) antibodies specifically designed to target tumour specific cells have already been introduced, in solid tumours molecular targeted therapy is now entering the clinical practice.

Molecular targeting could achieve specific damage to cancer cells at the same time preserving functionally important tissues: this could offer new prospective in primary and adjuvant treatment also of head and neck tumours.

INTRODUCTION

Head and neck cancer is still the sixth most common cancer type worldwide. Surgery, radiotherapy and chemotherapy, alone or in combination, are considered the main therapeutic approaches. Disappointingly, despite significant advances in head and neck treatments, survival rates and prognosis have only moderately improved through the years [1].

---

* Tel: +39-0532-236383, Fax: +39-0532-237615, Email: chiara.bianchini@unife.it.

Recently studies suggest that several factors can be involved in head and neck cancerogenetic process. The highly tumorigenic "expression" of some cancer types, cancer progression to locally invasive and metastatic states, recurrences are often associated with resistance to treatments: this could be explained with the accumulating genetic and/or epigenic alterations in cancer cells that may contribute to their uncontrolled growth, survival and invasion as well as their intrinsic or acquired resistance to clinical treatments [2]. Particularly it has been proposed the concept of cancer stem cells or cancer-initiating cells for explain cancer initiation and progression to metastatic disease states and resistance to conventional therapies [3, 4, 5].

Still a major problem in the treatment of cancer is the molecular specific targeting to neoplastic cells; ideally, future therapies should act over short distances to minimize damage to healthy cells and target tumoral compartments that have the highest sensitivity.

In this chapter, we discuss the importance of considering these new concepts among the role of cancer progenitor cells and cancer development in order to overcoming resistance to conventional cancer therapies.

Cancer Stem Cells in Head and Neck Tumours

There already are several reports that focus about the possible role of stem cells in cancer occurrence and progression in head and neck tumours [3, 6, 7]. It is likely that carcinogenesis should be considered as a multistep process, in which an accumulation of genetic and epigenetic alterations forms the basis for the development from a normal cell to a cancer one [6]. Particularly, the alteration of two key genes seems to be involved in this process, the p53 and the p16 [8].

Cancer stem cells could represent a minor population of tumour cells that possess the stem cell property of self-renewal; dysregulation of stem cell self-renewal is a likely requirement for the development of cancer [3]. The presence of these cancer progenitors may partially explain the rapid recurrence as well as the high metastatic rate of the most aggressive tumours after current clinical treatments.

Moreover, the reactivation of several developmental signaling cascades [epidermal growth factor (EGF)/EGFR, stem cell factor (SCF)/KIT, platelet-derived growth factor (PDGF)/PDGFR, sonic hedgehog (SHH/PTCH/GLI transcription factor, and/or Wnt/catenin)] combined with the altered DNA repair and mechanisms of multi drug resistance in cancer progenitor cells may take part and may be considered responsible, at least in part, for their resistance to current clinical therapies [2, 9].

Possible Specific Targets in Head and Neck Tumours

Recent advances in basic research, particularly genomics and proteomics, have already improved, but will probably develop more in a near future our understanding of the molecular processes governing head and neck cancer origin and progression.

It has been reported that the simultaneous blockade of several oncogenic cascades activated in cancer progenitor cells during cancer development has been judged essential for improving the current clinical treatments against high-risk metastatic solid tumours, including head and neck cancers [2, 10-18]. Particularly, the molecular targeting of developmental cascades, including hedgehog, Wnt/catenin, Notch, EGFR, PDGFR and KIT pathways and/or oncogenic signalling elements (telomerase, Src, ABL, PI3K/Akt, MYC, NF-KB and survivin), which assume a critical function in regulating the self-renewal, survival and invasion of cancer progenitor cells as well as in drug resistance and disease relapse, has been defined of therapeutic interest [2, 10-19] -see also Table 1-.

So far, the two most promising and advanced strategies are the blockage of growth factor based cellular signalling and the interference with angiogenesis related pathways.

Table 1. Possible sites of intervention for arresting Head and Neck cancer cells progression (47)

Molecular Targets / Possible sites of intervention	Biological Mechanism	References
EGFR in Head and Neck Squamous Cell Carcinoma	Arrest G1 phase cell cycle, stopping cellular proliferation	11,37,38,39
Insuline like growth factor 1 receptor, Telomerase, MYC, BCL2.	Inhibition of intracellular signalling	40,41,42,43
p53	Inducing apoptosis in specific neoplastic cells	44,45
Cox2	Reduction of cellular growth and angiogenesis	46

One of the most comprehensive investigated candidates is the Epidermal Growth Factor Receptor family (EGFR). EGFR is the most prominent candidate for therapeutic targeting due to its more than 90% expression rate in Head and Neck squamous cell carcinoma and its influence on regulation of proliferation, apoptosis, metastasis, angiogenesis and cell differentiation [9]. It has been reported that downstream signalling or inhibition of EGFR may result in the arrest of the G1 phase cell cycle, due to interrupt induced upregulation/accumulation of the cyclin-dependent kinase inhibitors p27kip1 and/or p21cip1/waf1, and this may stop cancer cell proliferation [9]. The extracellular part of the EGFR can be inhibited by monoclonal antibodies, while the intracellular one by more or less specific tyrosine kinase inhibitors (TKIs).

A clinically tested EGFR inhibitor is cetuximab, a monoclonal antibody with a high affinity to EGFR, and there are already several encouraging reports about the use of this drug in treating Head and Neck squamous cell carcinomas [2, 9, 10-19].

Instead, there only are very few data available about the employment of TKIs in head and neck cancer treatment. In particular, it has been shown that TKIs block EGFR downstream signal transduction pathways, inducing an immediate cell cycle arrest, and significantly inhibit angiogenesis by inducing a decrease of angiogenic growth factors [9]. Gefinitib and Erlotinib are the two most advanced drugs available but their use is still experimental [9, 20, 21].

Currently, EGFR inhibitors have been tested as a single agent or in combination with other therapies. As single agent, there are several reports showing that these molecules are moderately active in patients affected by Head and Neck squamous cell carcinoma. In some cases, they have provided disease control with a reasonable therapeutic index in patients who are chemotherapy refractory or are deemed unfit for cytotoxic therapy. The activity of single-agent anti-EGFR therapy, however, in the Head and Neck squamous cell carcinoma setting is largely based on uncontrolled, single-arm phase II trials (See also Table 2].

Table 2. Clinical trials of molecular targeted therapies in recurrent and/or metastatic head and neck squamous cell carcinoma: anti-EGFR molecular-targeted therapies as single agent and in combination (modified from Le Torneau C. and Siu LL., 22, 47)

Agent	Phase	Reference	Treatment	Overall Survival
Single agent				
Cetuximab	II	23	Cetuximab 400 mg / day	6 months
Gefitinib	II	24	Gefitinib 500mg per day	5 months
Gefinitib	III	25	Gefitinib 250mg per day	6 months
Combination				
Platinum + cetuximab	II	26	Platinum + cetuximab	5 months
Platinum + cetuximab	II	27	Platinum + cetuximab	6 months
Platinum + cetuximab	II	28	Platinum + cetuximab	4 months
Cisplatin + erlotinib	II	29	Cisplatin + erlotinib	7 months
Erlotinib and bevacizumab	I, II	30	Erlotinib and bevacizumab	7 months

Few phase II and III trials tested the addition of cetumibax to cisplatin in patients who resulted refractory to platinum based therapy; nonetheless, this combination appears to confer no further benefit over anti-EGFR agents alone, so far (See also Table 2) [22, 23-30].

It is known that angiogenesis plays a key role in the pathogenesis of solid tumours, including Head and Neck. Experimentally it has already been shown that inhibition of vascular endothelial growth factors (VEGF) and its receptors decreases angiogenesis and tumour growth (in mice). Therefore several strategies to target VEGF have been developed and are currently being explored. It is reported that clinical trials with the use of these type of molecules alone or in combination with conventional chemotherapies are already running [17, 18].

The development of new biological agents should focus on inhibitors that are likely to hit multiple targets. The complexity of the aberrant signalling in Head and Neck tumours explains why interfering with only single steps in these pathways have not shown marked clinical responses so far.

Probably, a combination of different agents that target distinct specific pathways is likely to inhibit the escape of tumour cells by alternate mechanisms leading to more effective disease control. Further, different etiological factors and risk habits can result in distinct genetic and epigenetic alterations, which may trigger different signalling pathways, thus impacting differently on the development and progression of Head and Neck tumours. To evaluate the efficacy of these biological agents there is urgent need to identify novel biomarkers that can be used to accurately assess and individualize therapy [31, 32, 33, 34].

FUTURE PERSPECTIVES

Further knowledge, and thus possibly further new therapeutic targets, could arise from the comprehension of head and neck cancer stem cell biology, since recently methods to isolate and characterize the ex vitro and in vivo properties of cancer stem/progenitor cells have been provided and improved. Therefore a re-analysis of the oncogenic gene products specifically activated in cancer progenitor cells will be necessary. Particularly, further investigations by using cancer stem cells isolated from primary cancer patient's malignant tissues at different stages during the cancer progression and metastatic disease should help to identify new biomarkers for the development of more effective diagnostic and prognostic methods and targeted therapies against the aggressive and metastatic cancers. However, the evaluation of the functional and molecular properties of these cells ex vivo must be considered with caution since multiple extrinsic factors may influence the in vitro behaviour of this subset of cells, and more particularly, the experimental culture conditions do not necessarily reflect the local tumoral microenvironment and the intercellular signalling environment. Further investigations of cancer progenitor cells in more vigorous experimental conditions are required, thereby helping in elucidating the molecular mechanisms that provide a critical role for their uncontrolled self-renewal and/or aberrant differentiation capacity. The specific biomarkers and functional properties of cancer progenitor or stem cells clearly should be explored in order to assess select possible targets; this particularly considering that cancer progenitor/stem cells may also exhibit a differential genomic and proteomic pattern, as compared to the bulk mass of further cancer cells [11, 35-47].

In addition, it is also necessary to consider that these molecular markers might allow a very sensitive detection of minimal residual cancer cells or micrometastasis in the various body compartments and/or in lymphonodes, thus facilitating the prognosis definition.

The development of new effective and safe targeted therapies by eradicating the total population of cancer progenitor cells and their further differentiated progenies at the primary and secondary neoplasms should allow us to improve the current cancer treatments, prevent the disease relapse and thereby induce a complete cytogenetic remission and cure of cancer patients in the clinics.

The next years will be exciting also in terms of surgical technique progresses. Newer roadway of nanotechnology and molecular medicine will also lead to technical advancements in areas such as free tissue transfer, robotics, minimally access surgery as well as application of new lasers and optical technologies. Advances in this field could expand our capabilities to repair, restore and preserve organ/tissue function, thus personalizing and tailoring the treatment to the single patient.

Conclusion

A cancer "targeted therapy" actually represents one of the most promising resources in the development of new strategies including in head and neck oncology. A targeted therapy with biological markers in head and neck cancer is a new and fast growing field with interesting aspects concerning decreasing early and late toxicity with better functional outcome and survival.

Since it is likely that carcinogenesis may be considered as a process that arises from the malignant transformation of embryonic or adult stem/progenitor cells into cancer progenitor cells [3, 6, 7], currently, cancer progenitor cells can provide critical functions in cancer initiation and progression into metastatic and recurrent disease states. Based on these observations, the molecular targeting of cancer progenitor cells must be considered for improving the efficacy of the current cancer therapies.

Traditional diagnostic methods such as clinical assessment, histopathological examination and imaging techniques are nowadays considered not sufficient to provide all the information related to prognosis and the treatment choice of head and neck cancer [35,36, 47]. Additional investigations to identify the specific molecular biological characteristics of head and neck tumours should be performed. Only the comprehension of the origin and the molecular behaviour of head and neck cancer will lead us to program and perform a more specific, safe, effective, "personalized", "targeted" therapeutic plan [47].

References

[1] Forastiere A, Koch W, Trotti A, Sidransky D (2001) Medical progress–head and neck cancer. *N. Engl. J. Med.* 345:1890–1900.

[2] Mimeault M., Hauke R., Mehta PP., Batra SK. Recent advances in cancer stem/progenitor cell research: therapeutic implications for overcoming resistance to the most aggressive cancers. *J. Cell. Mol. Med.* 2007: 11, 5, 981-1011.

[3] Bianchini C, Ciorba A, Pelucchi S, Piva R, Pastore A. Head and neck cancer: the possible role of stem cells. *Eur. Arch. Otorhinolaryngol.* 2008 Jan;265(1):17-20.

[4] Graziano A, d'Aquino R, Tirino V, Desiderio V, Rossi A, Pirozzi G. The stem cell hypothesis in head and neck cancer. *J. Cell Biochem.* 2008 Feb 1;103(2):408-12.

[5] Prince ME, Ailles LE. Cancer stem cells in head and neck squamous cell cancer. *J. Clin Oncol.* 2008 Jun 10;26(17):2871-5.

[6] Braakhuis BJ, Leemans CR, BrakenhoV RH (2005) Expanding fields of genetically altered cells in head and neck squamous carcinogenesis. *Semin. Cancer Biol.* 15:113–120.

[7] Hirata Y (1999) Progression of cancer. *Med. Hypotheses* 52(6):51–13.

[8] Gollin SM (2001) Chromosomal alterations in squamous cell carcinomas of the head and neck: window to the biology of disease. *Head Neck* 23:238–253.

[9] Dietz A, Boehm A, Mozet C, Wichmann G, Giannis A. Current aspects of targeted therapy in head and neck tumors. *Eur. Arch. Otorhinolaryngol.* 2008 Jul;265 Suppl 1:S3-12.

[10] Mimeault M, Batra SK. Recent advances on the significance of stem cells in tissue regeneration and cancer therapies. *Stem Cells*. 2006; 24: 2319–45.

[11] Mimeault M, Hauke R, Batra SK. Recent advances on the molecular mechanisms involved in drug resistance of cancer cells and novel targeting therapies. *Clin. Pharmacol. Ther*. 2007; 82: 252–64.

[12] Dean M, Fojo T, Bates S. Tumour stem cells and drug resistance. *Nat. Rev. Cancer*. 2005; 5: 275–84.

[13] Barker N, Clevers H. Mining the Wnt pathway for cancer therapeutics. *Nat. Rev. Drug Discov*. 2006; 5: 997–1014.

[14] Rubin LL, de Sauvage FJ. Targeting the Hedgehog pathway in cancer. *Nat. Rev. Drug Discov*. 2006; 5: 1026–33.

[15] Roberg K, Jonsson AC, Grenman R, Norberg-Spaak L. Radiotherapy response in oral squamous carcinoma cell lines: Evaluation of apoptotic proteins as prognostic factors. *Head Neck*. 2007; 29: 325–34.

[16] Mimeault M, Batra S. Interplay of distinct growth factors during epithelial-mesenchymal transition of cancer progenitor cells and molecular targeting as novel cancer therapies. *Ann. Oncol*. 2007; 18: 1605–19.

[17] Mimeault M, Batra SK. Functions of tumorigenic and migrating cancer progenitor cells in cancer progression and metastasis and their therapeutic implications. *Cancer Metastasis Rev*. 2007; 26: 203–14.

[18] Fodde R, Brabletz T. Wnt/beta-catenin signaling in cancer stemness and malignant behavior. Curr Opin Cell Biol. 2007; 19: 150–8.

[19] Zhang Q, Chen G, Liu X, Qian Q. Monoclonal antibodies as therapeutic agents in oncology and antibody gene therapy. *Cell Res*. 2007; 17: 89–99.

[20] Goerner M, Seiwert TY, Sudhoff H. Molecular targeted therapies in head and neck cancer - An update of recent developements -. *Head Neck Oncol.* 2010 Apr14;2(1):8.

[21] Razak AR, Siu LL, Le Tourneau C. Molecular targeted therapies in all histologies of head and neck cancers: an update. *Curr. Opin. Oncol*. 2010 May;22(3):212-20.

[22] Le Tourneau C., Siu LL. Molecular-targeted therapies in the treatment of squamous cell carcinomas of the head and neck. Current Opinion in *Oncology* 2008, 20:256–263.

[23] Vermorken JB, Trigo J, Hitt R, Koralewski P, Diaz-Rubio E, Rolland F, Knecht R, Amellal N, Schueler A, Baselga J. Open-label, uncontrolled, multicenter phase II study to evaluate the efficacy and toxicity of cetuximab as a single agent in patients with recurrent and/or metastatic squamous cell carcinoma of the head and neck who failed to respond to platinum-based therapy. *J. Clin. Oncol* 2007; 25:2171–2177.

[24] Kirby AM, A'Hern RP, D'Ambrosio C, Tanay M, Syrigos KN, Rogers SJ, Box C, Eccles SA, Nutting CM, Harrington KJ. Gefitinib (ZD1839, Iressa) as palliative treatment in recurrent or metastatic head and neck cancer. *Br. J. Cancer* 2006; 94:631–636.

[25] Stewart JSW, Cohen EEW, Licitra L. A phase III randomized parallel group study of gefitinib (RESSA) versus methotrexate (IMEX) in patients with recurrent squamous cell carcinoma of the head and neck [abstract]. In: American Association for Cancer Research Annual Meeting: Proceedings; 14–18 Apr 2007; Los Angeles. Philadelphia: AACR; 2007. Abstract 3522.

[26] Baselga J, Trigo JM, Bourhis J, Tortochaux J, Cortés-Funes H, Hitt R, Gascón P, Amellal N, Harstrick A, Eckardt A. Phase II multicenter study of the antiepidermal

growth factor receptor monoclonal antibody cetuximab in combination with platinum-based chemotherapy in patients with platinumrefractory metastatic and/or recurrent squamous cell carcinoma of the head and neck. *J. Clin. Oncol* 2005; 23:5568–5577.

[27] Herbst RS, Arquette M, Shin DM, Dicke K, Vokes EE, Azarnia N, Hong WK, Kies MS. Phase II multicenter study of the epidermal growth factor receptor antibody cetuximab and cisplatin for recurrent and refractory squamous cell carcinoma of the head and neck. *J. Clin. Oncol* 2005; 23:5578–5587.

[28] Burtness B, Goldwasser MA, Flood W, Mattar B, Forastiere AA. Phase III randomized trial of cisplatin plus placebo compared with cisplatin plus cetuximab in metastatic / recurrent head and neck cancer: an Eastern Cooperative Oncology Group study. *J. Clin. Oncol* 2005; 23:8646–8654.

[29] Siu LL, Soulieres D, Chen EX, Pond GR, Chin SF, Francis P, Harvey L, Klein M, Zhang W, Dancey J, Eisenhauer EA, Winquist E. Phase I/II trial of erlotinib and cisplatin in patients with recurrent or metastatic squamous cell carcinoma of the head and neck: a Princess Margaret Hospital phase II consortium and National Cancer Institute of Canada Clinical Trials Group Study. *J. Clin. Oncol* 2007; 25:2178–2183.

[30] Cohen EE, Davis DW, Karrison TG, Seiwert TY, Wong SJ, Nattam S, Kozloff MF, Clark JI, Yan DH, Liu W, Pierce C, Dancey JE, Stenson K, Blair E, Dekker A, Vokes EE. Erlotinib and bevacizumab in patients with recurrent or metastatic squamous-cell carcinoma of the head and neck: a phase I/II study. *Lancet Oncol.* 2009 Mar;10(3):247-57. Epub 2009 Feb 7.

[31] Matta A., Ralhan R. Overview of current and future biologically based targeted therapies in head and neck squamous cell carcinoma. *Head and Neck Oncology* 2009, 1:6.

[32] Lamont JP, Kuhn JA, Nemunaitis JJ, McCarty TM. Gene therapy for head and neck cancers. *Oncology* 15(3):303-314, 2001.

[33] Ganly I, Soutar DS, Kaye SB: Current role of gene therapy in head and neck cancer. *Eur. J. Surg. Oncol* 26:338-343, 2000.

[34] Ziober AF, D'Alessandro L, Ziober BL. Is gene expression profiling of head and neck cancers ready for the clinic? *Biomark Med.* 2010 Aug;4(4):571-80.

[35] Mahfouz ME, Rodrigo JP, Takes RP, Elsheikh MN, Rinaldo A, Brakenhoff RH, Ferlito A. Current potential and limitations of molecular diagnostic methods in head and neck cancer. *Eur. Arch. Otorhinolaryngol.* 2009 Dec 27.

[36] Langer CJ. Targeted therapy in head and neck cancer: state of the art 2007 and review of clinical applications. *Cancer.* 2008 Jun 15;112(12):2635-45.

[37] Mimeault M, Batra SK. Recent insights into the molecular mechanisms involved in aging and the malignant transformation of adult stem/progenitor cells and their therapeutic implications. *Ageing Res. Rev.* 2009 Apr;8(2):94-112..

[38] Szakács G, Paterson JK, Ludwig JA, Booth-Genthe C, Gottesman MM. Targeting multidrug resistance in cancer. *Nat. Rev. Drug Discov.* 2006 Mar;5(3):219-34.

[39] Kim S, Grandis JR, Rinaldo A, Takes RP, Ferlito A (2008) Emerging perspectives in epidermal growth factor receptor targeting in head and neck cancer. *Head Neck* 30(5):667–674.

[40] Barnes CJ, Ohshiro K, Rayala SK et al (2007) Insulin-like growth factor receptor as a therapeutic target in head and neck cancer. *Clin. Cancer Res* 13:4291–4299.

[41] Hong DY, Lee BJ, Lee JC, Choi JS, Wang SG, Ro JH. Expression of VEGF, HGF, IL-6, IL-8, MMP-9, Telomerase in Peripheral Blood of Patients with Head and Neck Squamous Cell Carcinoma. *Clin. Exp. Otorhinolaryngol.* 2009 Dec;2(4):186-92.

[42] Wang X, Hao MW, Dong K, Lin F, Ren JH, Zhang HZ. Apoptosis induction effects of EGCG in laryngeal squamous cell carcinoma cells through telomerase repression. *Arch. Pharm. Res.* 2009 Sep;32(9):1263-9.

[43] Kurihara Y, Watanabe Y, Onimatsu H, Kojima T, Shirota T, Hatori M, Liu D, Kyo S, Mizuguchi H, Urata Y, Shintani S, Fujiwara T. Telomerase-specific virotheranostics for human head and neck cancer. *Clin. Cancer Res.* 2009 Apr 1;15(7):2335-43.

[44] Rubin Grandis J, Melhem MF, Gooding WE, Day R, Holst VA, Wagener MM, Drenning SD, Tweardy DJ (1998) Levels of TGF alpha and EGFR protein in head and neck squamous cell carcinoma and patient survival. *J. Natl. Cancer Inst.* 90(11):824–832.

[45] Grandis JR, Chakraborty A, Zeng Q, Melhem MF, Tweardy DJ. Down modulation of TGF-alpha protein expression with antisense oligonucleotides inhibits proliferation of head and neck squamous carcinoma but not normal mucosal epithelial cells. *J. Cell Biochem.* 1998 Apr 1;69(1):55-62.

[46] Lin DT, Subbaramaiah K, Shah JP et al (2002) Cyclooxygenase-2: a novel molecular target for the prevention, treatment of head, neck cancer. *Head Neck* 24(8):792–799.

[47] Bianchini C, Ciorba A, Pelucchi S, Piva R, Pastore A. Targeted therapy in head and neck cancer. Tumori. 2011 Mar-Apr;97(2):137-41.

In: Signaling, Gene Regulation and Cancer
Editors: Shree Ram Singh and Manoj K. Mishra
ISBN: 978-1-61942-088-5

Chapter 8

AKT IN HUMAN COLORECTAL CANCER

Kumar S. Bishnupuri** and Satheesh K. Sainathan
Division of Gastroenterology, Washington University School of Medicine, Saint Louis, Missouri, US

ABSTRACT

Colorectal cancer (CRC) is a commonly diagnosed cancer in both men and women and the leading cause of cancer death in the US and Europe. A better understanding of gene regulation and signal transduction pathways involved in colorectal carcinogenesis is required to substantially improve the treatments of human CRC. Akt/PKB is a serine/threonine protein kinase that plays a key role in mediating signals for cell growth, cell survival (anti-apoptotic), cell-cycle progression, differentiation, transcription, translation, and glucose metabolism. Akt1 also known as Akt is primarily involved in cellular survival and apoptotic pathways. Since it blocks apoptosis, and promotes cell survival, Akt has been implicated as a major factor in many types of cancer. This chapter is focused on defining the regulators of Akt and role of Akt in mediating signaling pathways in human CRC.

Keywords: Carcinoma, EGFR, Insulin, GPCR, GSK-3β, cyclin, cyclin-dependent kinase

INTRODUCTION

Colorectal cancer (CRC) is the third most commonly diagnosed cancer in both men and women and the third leading cause of cancer death in the US [1] and the second in Europe [2, 3]. In spite of substantial progress in the last decade, many new cases of CRC and large number of deaths have been reported in recent years [1]. The American Cancer Society estimates that in 2011 about 141,210 people will be diagnosed with colorectal cancer and about 49,380 people will die of the disease in the US. If diagnosed and treated early human CRC is potentially preventable and curable, however, novel CRC markers for early diagnosis

* Tel: (314) 747-4067, Fax: (314) 362-8959, E-mail: kbishnup@dom.wustl.edu.

and targets for successful treatments are still lacking. A better understanding of gene regulation and signal transduction pathways involved in colorectal carcinogenesis may provide useful tools to establish novel CRC markers and new therapeutic targets to substantially improve the treatments of human CRC. Akt/PKB, a serine/threonine protein kinase is one of the most frequently activated protein kinases in human cancer. Activated Akt plays a key role in mediating signals for resistance to apoptosis, increased cell growth, cell proliferation, and cellular energy metabolism [4]. In humans, Akt family includes three members: Akt1, Akt2, and Akt3. Akt1 also known as Akt, which will be frequently used in this chapter is primarily involved in cellular survival and apoptotic pathways. Akt is also able to induce protein synthesis pathways, and is therefore a key signaling protein in the cellular pathways that lead to skeletal muscle hypertrophy, and general tissue growth. Since it blocks apoptosis, and promotes cell survival, Akt has been implicated as a major factor in many types of cancer. Akt2 is an important signaling molecule in the Insulin signaling pathway and is required to induce glucose transport. The role of Akt3 is less clear, though it appears to be predominantly expressed in brain. It has been reported that mice lacking Akt3 have small brains [5]. This chapter focuses on defining the regulators of Akt and role of Akt in mediating signaling pathways in human CRC.

1. Regulators of Akt

1.1. Growth Factors

A growth factor is a naturally occurring substance regulating a variety of cellular processes including cellular growth, proliferation and differentiation. Some growth factors are important regulators of cell division and proliferation of cancer cells hence influence growth rates of many cancers including human CRC. Growth factors associated with human CRC include Epidermal Growth Factor (EGF), Transforming Growth Factor Beta (TGF-β), and Insulin and Insulin-Like Growth factor (IGF). Akt is a common intermediate kinase mediating signals of these CRC-associated growth factors.

1.1.1. Epidermal Growth Factors

Epidermal growth factors constitutes a family of proteins including epidermal growth factor (EGF), Heparin-binding EGF-like growth factor (HB-EGF), transforming growth factor-α (TGF-α), Amphiregulin (AREG), Epiregulin (EPR), Epigen (EPG), Betacellulin (BTC), neuregulins (1, 2, 3 and 4). Although all EGF famlily members are not associated with cancer growth and development, EGF and TGF-α plays important role inducing signaling pathways involved in the process of carcinogenesis by binding to epidermal growth factor receptor (EGFR). Over the last 20 years, elevated levels of the EGFR and its cognate ligands (which include EGF and TGF-α) have been identified as a common component of numerous cancer types. In many cases aberrant EGFR activation, mediated primarily through changes in gene amplification and autocrine stimulation, appears to be an important factor in tumorigenesis, as well as an essential driving force for the aggressive growth behavior of cancer cells [6]. The epidermal growth factor receptor (EGFR) belongs to the ErbB family of receptor tyrosine kinases (RTK). EGF, TGF-α and EGFR are often overexpressed in many human carcinomas including human CRC. Binding of EGF or TGF-α to EGFR form a complex system in which the interactions occurring between receptors and ligands affect the type and the duration of the intracellular signals that derive from receptor activation. In fact, proteins of the ErbB family form either homo- or hetero-dimers following ligand binding and activates tyrosine kinase activity by phosphorylating tyrosine residues at different position of EGFR (FIGURE 1).

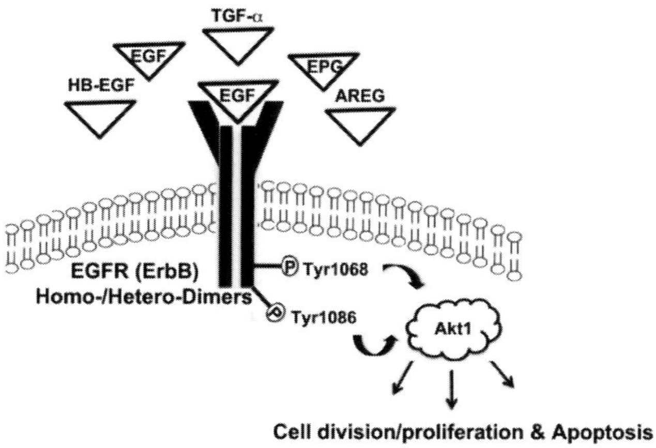

Figure 1. EGFR-Akt Signaling. Binding of EGF family lignads to EGFR proteins of the ErbB family form either homo- or hetero-dimers and activates tyrosine kinase activity by phosphorylating tyrosine residues at different position of EGFR. Phosphorylation of tyrosine residues at positions 1068 and 1086 of EGFR induces Akt kinase activity and is associated with regulation of cell division/proliferation and apoptosis of many human carcinomas including CRC.

Phosphorylation of tyrosine residues at different positions of EGFR induces different protein kinases to mediate their signaling. Phosphorylation of tyrosine residues at 1068 and 1086 positions is associated with induction of Akt kinase activity involved in regulation of cell proliferation and apoptosis of many human carcinomas including CRC. In a recent study we observed Regenerating protein 4 (Reg4)-mediated increase in tyrosine phosphorylation at 1068 position of EGFR and its association with induced Akt activity playing important role in proliferation and apoptosis of human colon adenocarcinoma cells [7].

1.1.2. Transforming Growth Factor Beta (TGF-β)

TGF-βs belong to an important family of cytokines that regulate diverse biologic processes including proliferation, differentiation, apoptosis, and extracellular matrix formation [8, 9]. TGF-β is a potent growth inhibitor or stimulator of apoptosis in many cell types including human CRC cells. The normal colonic epithelium expresses three isoforms of TGF-βs (TGF-βI, II, and III: [10-12]) and TGF-βI serves as a tumor suppressor in the normal colon by inhibiting cell proliferation and inducing apoptosis [13-15]. During late stages of colorectal carcinogenesis, TGF-β serves as tumor promoter [16, 17] and is often overexpressed [18]. A high expression level of TGF-β in the primary tumor is associated with advanced stages [19], tumor recurrence [20], and decreased survival [19]. All three isoforms of TGF-β control cellular processes by signaling through cell surface receptor complexes composed of TGF-β type I and type II receptors [8, 9]. Both receptors are transmembrane serine/threonine kinases, and TGF-β binding activates the type I receptor. When a TGF-ß family ligand binds to a complex of two type I and two type II receptors, the type II receptor phosphorylates the type I receptor through a kinase domain (FIGURE 2).

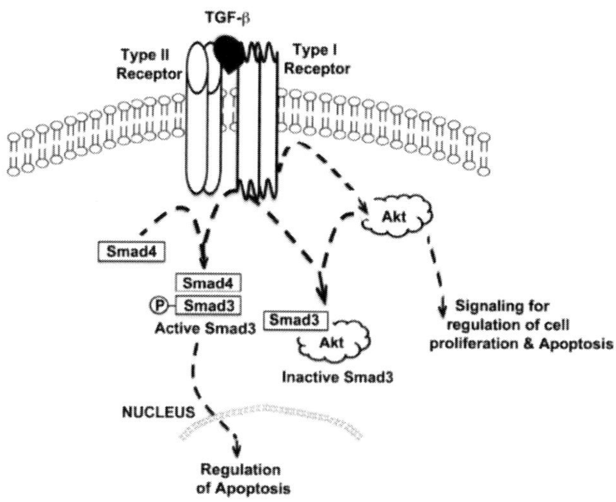

Figure 2. TGF-β-Akt Signaling. Binding of TGF-β ligand to a complex of two type I and two type II receptors phosphorylates the type I receptor through a kinase domain. The phosphorylated type I receptor then phosphorylates the Smad3 and facilitates its binding to Smad4 to form heteromeric complexes. These complexes are translocated into the nucleus to regulate transcription of target genes associated with regulation of proliferation and apoptosis. In addition, Smad3 also physically binds Akt and this interaction underlies Akt inhibitory effects on TGF-β. In certain cases, TGF-β stimulates Akt without involving Smad proteins using multistep PI3 kinase signaling pathway, which is also associated with regulation of proliferation and apoptosis of many cell types including human CRC cells.

The phosphorylated type I receptor then phosphorylates the carboxyl-terminal serine amino acids of two important downstream targets, Smad2 and Smad3, allowing them to bind to Smad4 to form heteromeric complexes [21, 22]. Smad2/3/4 complexes are translocated into the nucleus to regulate transcription of target genes associated with regulation of cell proliferation and apoptosis. In addition, studies have shown that Smad3 physically binds Akt [15, 23]. The interaction between Akt and Smad3 underlie Akt inhibitory effects on TGF-β [23, 24]. In certain cases, TGF-β stimulates a Smad-independent signaling pathway including activation of Akt kinase though multistep induction of PI3 kinase activity and is associated with regulation of proliferation and apoptosis of many cell types including human CRC cells.

1.1.3. Insulin (INS) and Insulin-Like Growth Factor (IGF)

Insulin and the IGF axis each play important and complementary roles in metabolism and growth. Insulin influences metabolism on a short-term basis (e.g., after a meal), whereas the IGF axis exerts a longer-term integrating effect on growth. IGF-1 inhibits apoptosis and is required for cell cycle progression [25]. Both normal and cancer cells of colorectal epithelium express IGF-1 receptors; when activated by IGF-1, the receptor-ligand complex inhibits apoptosis and allows progression through the cell cycle [26-30]. Thus, IGF-1 can influence both premalignant and cancerous stages. Colorectal carcinogenesis results from an accumulation of specific molecular alterations [31], and increased cellular turnover may enhance the rate at which these alterations accumulate. The IGF axis could influence carcinogenesis by being one of the important determinants of cellular turnover. In the colon of acromegalic patients, who are characterized by growth hormone and IGF-1 hypersecretion, epithelial proliferation is increased, the proliferative zone is extended and growth hormone and IGF-1 levels correlate with cell proliferation [32]. Moreover, in colon cancer cell lines, IGF-1 increases production of vascular endothelial growth factor, an angiogenic factor that supports tumor growth [33]. Similar to IGF-1, insulin stimulates growth of normal colonic and carcinoma cells *in vitro* [28-30]. Although colon cancer tissue has both insulin and IGF-1 receptors [27, 34], the mitogenic properties of insulin may be mediated through IGF-1 receptors [35], insulin receptors or a hybrid IGF-1 and insulin receptor [36, 37]. A hypothetical model shown below demonstrates that binding of insulin or IGF-1 to insulin receptor (INSR) or insulin-like growth factor receptor (IGFR) activates Akt kinase activity via PI3K-PIP3-PDK1 pathway and is associated with regulation of cell division/proliferation and apoptosis of carcinomas including human CRC cells (FIGURE 3). In addition, the insulin and IGF-1 signal transduction pathway involved in the regulation of gene expression and mitogenicity is also mediated by activation of the 'Ras' protein in human CRC [38, 39].

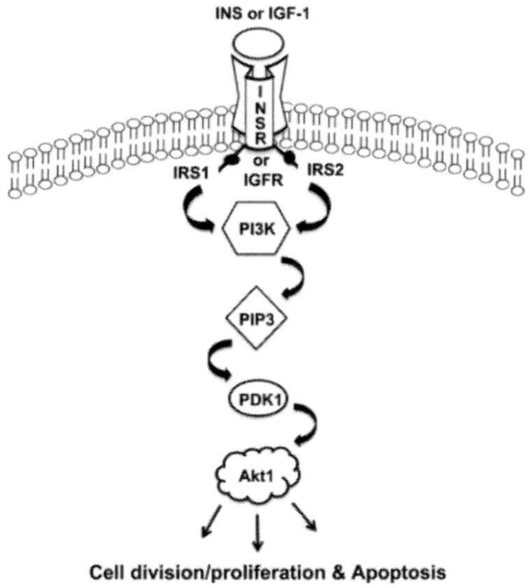

Figure 3. Insulin/IGF-1-Akt Signaling. Binding of INS or IGF-1 to INSR or IGFR activates insulin receptor substrates (IRS1 and 2) - mediated signaling of PI3K-PIP3-PDK1 to induce Akt kinase activity and is associated with regulation of cell division/proliferation and apoptosis of many human carcinomas including CRC.

1.2. G Proteins and G Protein-Coupled Receptors (GPCR)

Guanine nucleotide-binding proteins (G proteins) are a family of proteins involved in transmitting chemical signals outside the cell, and causing changes inside the cell. Receptors coupled to heterotrimeric G proteins are the largest group of integral membrane receptors involved in the transmission of signals from the extracellular environment to the cytoplasm [40].

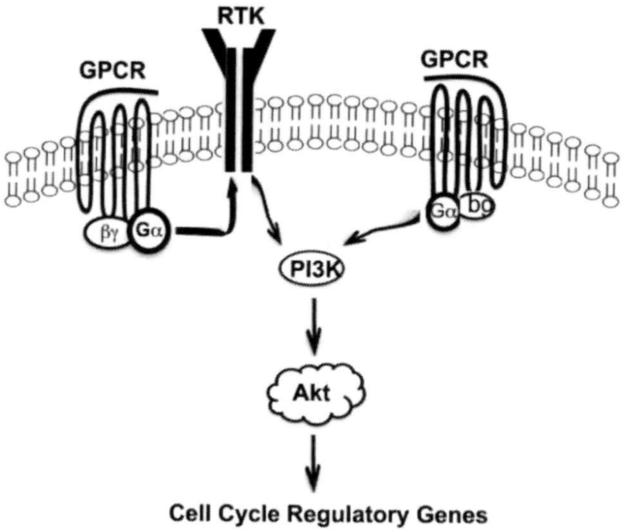

Figure 4. GPCR-Akt signaling. Numerous GPCRs activate PI3K isoforms either through Gaβγ subunits or via RTK transactivation. PI3Ks activate Akt, which is associated with the regulation of cell cycle regulatory gene expression including Cyclin D1.

1.3. PI3 Kinases

Phosphatidylinositol 3-kinases (PI3 kinases or PI3Ks) are a family of enzymes, which play a crucial role in effecting alterations in a broad range of cellular functions in response to extracellular signals.

A key downstream effector of PI3K is the Akt, which in response to PI3K activation, phosphorylates and regulates the activity of a number of targets including kinases, transcription factors and other regulatory molecules. Akt possesses a protein domain known as Pleckstrin Homology Domain (PH domain). In the case of the PH domain of Akt, it binds either PIP_3 (Phosphotidylinositol (3,4,5)-triphosphate, PtdIns(3,4,5)P_3) or PIP_2 (Phosphotidylinositol (3,4,5)-bisphosphate, PtdIns(3,4)P_2) [46]. This is useful for control of cellular signaling because the di-phosphorylated phosphoinositide PIP_2 is only phosphorylated by PI3 kinases, and only upon receipt of chemical messengers, which tell the cell to begin the growth process.

For example, PI3-kinases may be activated by a GPCR or receptor tyrosine kinases (RTKs) such as the EGFR or insulin receptor. Once activated, PI3-kinase phosphorylates PIP_2 to form PIP_3. Once correctly positioned at the membrane via binding of PIP3, Akt can then be phosphorylated by its activating kinases, phosphoinositide dependent kinase 1 (PDK1 at Thr308) and rapamycin complex 2 (mTORC2; at serine 473).

The mTORC2 functionally acts as the PDK2 molecule, although other molecules, including integrin-linked kinase (ILK) and Mitogen-Activated Protein Kinase Activated Protein Kinase-2 (MAPKAPK2) can also serve as PDK2. Phosphorylation by mTORC2 stimulates the subsequent phosphorylation of Akt by PDPK1, which regulates downstream signaling pathways associated with cell survival, proliferation and apoptosis of many human carcinomas including CRC (FIGURE 5).

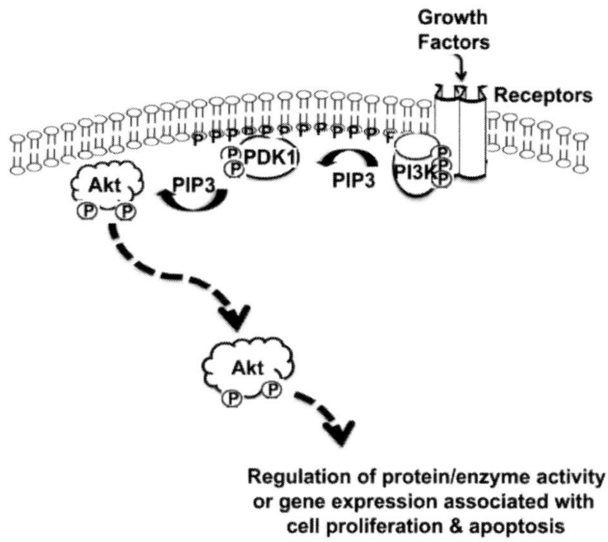

Figure 5. PI3K-Akt signaling. Extracellular signals such as growth factors activate PI3K, which in turn activates the synthesis of PIP3. Akt binds plasma membrane-associated PIP3 via its PH domain, where it undergoes phosphorylation of Thr308 and Ser473 residues. Active Akt then translocates into cytosol or the nucleus, where it interacts with a wide array of downstream effector proteins and regulates protein/enzyme activity or gene expression associated with cell proliferation and apoptosis.

2. Role of Akt in CRC

2.1. Akt in CRC Growth

Akt plays a critical role in the induction and/or progression of a variety of human carcinomas including CRC. Following phosphoryaltion at Thr308 and Ser473, Akt induces multiple signaling pathways associated with cellular survival and apoptosis. A critical downstream effector of Akt, which contributes to tumorigenesis is mTOR. However, the interaction between Akt and mTOR is quite complicated. Nearly a decade after the discovery of mTOR, researchers found that mTOR nucleates a rapamycin- and nutrient-sensitive multiprotein complex (now called mTORC1) [47-49]. Following soon after was the unexpected identification of a second growth-factor-sensitive but nutrient-insensitive mTOR-containing complex called mTORC2 [49-51]. Therefore, the mTOR kinase is considered as the catalytic component of two distinct multiprotein complexes, mTORC1 and mTORC2. A prevailing model suggests that activation of the Akt-mTORC1 pathway plays critical role to drive tumorigenesis (FIGURE 6).

The mTORC1 complex contains mTOR, RAPTOR, mLST8, and PRAS40, whereas the mTORC2 complex contains mTOR and mLST8, but instead of RAPTOR and PRAS40, mTORC2 contains the RICTOR, mSIN1, and PROTOR proteins. Akt-mTORC1 signaling drives cellular growth by controlling numerous processes that regulate protein synthesis and degradation. Diverse positive and negative growth signals influence the activity of mTORC1, many of which converge upon the tuberous sclerosis complex TSC1/2. Akt directly phosphorylates TSC2 to inactivate, which otherwise inhibits mTOR activity [52]. TSC1 and

TSC2 form a heterodimer with GTPase activity that inhibits the activity of Rheb, a small GTPase required for mTOR activation [52]. Currently, the only characterized substrate of mTORC2 is the Akt kinase, which suggests mTORC2 functions downstream in the PI3K pathway to regulate cell growth, proliferation, and survival. In addition, PI3K-Akt signaling directly regulates the activity forkhead box transcription factor (FOX) family. The FOX transcription family includes AFX/FOXO4, FKHR/FOXO1, and FKHR-L1/FOXO3a. Recently, it has been shown that FoxO1 transcription factor, which inhibits cell proliferation, is directly phosphoryalted and inactivated by Akt [53]. As a result of these phosphorylation events, proapoptotic genes induced by FOX family transcription factors are downregulated, which prevents cell death and facilitates cell growth in many human carcinomas.

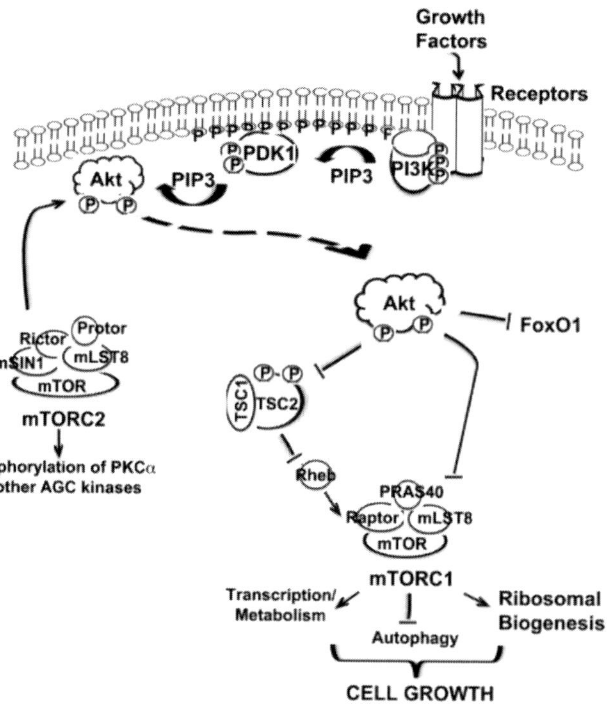

Figure 6. Akt-mTOR signaling in CRC growth. Extracellular signals such as growth factors activate PI3K-PIP3-Akt signaling. Akt binds plasma membrane-associated PIP3, where it undergoes phosphorylation of Thr308 and Ser473 residues. Active Akt then translocates into cytosol, where it interacts with mTOR to regulate CRC cell growth. The mTOR kinase is the catalytic component of two distinct multiprotein complexes, mTORC1 and mTORC2. The mTORC1 complex contains mTOR, RAPTOR, mLST8, and PRAS40. Akt directly phosphorylates tuberous sclerosis complex 2 (TSC2) to inactivate, which otherwise inhibits mTOR activity. Tuberous sclerosis complex 1 (TSC1) and TSC2 form a heterodimer that inhibits the activity of Rheb, a small GTPase required for mTOR activation. The mTORC2 complex contains mTOR and mLST8, RICTOR, mSIN1, and PROTOR proteins. The mTORC2 functions downstream in the PI3K pathway to regulate cell growth, proliferation, and survival. In addition, Akt directly phosphorylate to inactivate FoxO1 transcription factor, which is involved in inhibition of cell proliferation.

Akt has a number of downstream substrates that may contribute to carcinogenesis. Recently, two Akt substrates that are components of the intrinsic cell death machinery have been identified: the Bcl2 family member BAD [54, 55] and the protease Caspase 9 [56]. In both cases, phosphorylation of these proteins by Akt suppresses their proapoptotic function, thereby accounting at least in part for the potent survival effects of Akt.

The phosphorylation of BAD may lead to the prevention of cell death via a mechanism that involves the selective association of the phosphorylated forms of BAD with 14-3-3 protein isoforms.

This BAD/14-3-3 interaction can occur when BAD becomes phosphorylated at Ser-112 and/or Ser-136, and the induced association of BAD with 14-3-3 appears to prevent BAD association with BCL-X_L or BCL-2. It has been postulated that in the absence of BAD phosphorylation, BAD may bind to BCL-X_L or BCL-2 and suppress survival by inducing BAX homodimer formation. The prodeath function of BAD could also be a direct consequence of BAD's heterodimerization with BCL-X_L or BCL-2, and may not involve BAX heterodimerization [57-59].

Phosphorylated BAD may also play an active role in promoting survival. Although there is as yet no direct evidence supporting this hypothesis, an active survival function for BAD is suggested by the finding that the interaction between BAD and 14-3-3 proteins is induced by phosphorylation at BAD Ser-136 (FIGURE 7). Some evidences suggest that overexpression of BAD and Akt can result in the formation of a BAD/Akt complex, and it is possible that BAD brings Akt to the 14-3-3 complex, where Akt may phosphorylate additional signaling molecules to promote survival. Akt is reported to phosphorylate and activate I-kB kinase (IKK) [60, 61], which regulates apoptosis and facilitate cell growth in many human carcinomas.

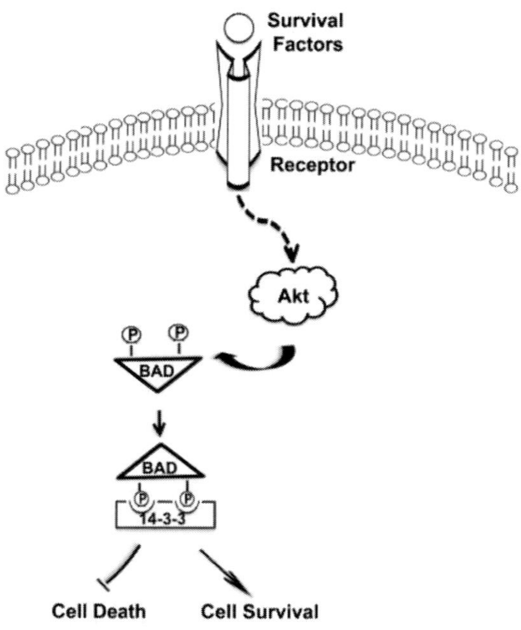

Figure 7. Akt-BAD signaling in CRC cell survival. Akt phosphorylates BAD and facilitates selective association of the phosphorylated forms of BAD with 14-3-3 protein isoforms. This BAD/14-3-3 interaction occurs when BAD becomes phosphorylated at Ser-112 and/or Ser-136. Induced association of BAD with 14-3-3 may prevent association of BAD with BCL-X_L or BCL-2. Phosphorylated BAD may also play an active role in promoting survival.

Activation of IKK causes phosphorylation and degradation of inhibitor of kB (I-kB), which binds to NF-kB transcription factor and inactivates it by localizing NF-kB to the cytosol where it is unable to regulate transcription [62].

IKK-mediated phosphorylation of I-kB targets it for ubiquitination and degradation. In the absence of I-kB, NF-kB's nuclear localization signal is exposed and NF-kB localizes to the nucleus where it is able to induce transcription of antiapoptotic proteins [63-66].

2.2. Akt in CRC Cell Cycle Regulation

Downstream substrate of Akt also includes glycogen synthase kinase-3 beta (GSK-3β) and p21Cip1, which play important role in regulation of cell division cycle of many human carcinomas including human CRC [67]. GSK-3β phosphorylates numerous proteins, such as glycogen synthase, cyclin D, and c-Myc, which maintain these proteins in an inactive state [68]. Activation of PI3K-Akt signaling leads to GSK-3β inactivation and is involved in regulation of expression and activities of cell cycle regulatory genes [69]. Cell division is coordinated by a number of proteins including cyclins and cyclin-dependent kinases (Cdks). Cdks exist as inactive serine/threonine kinase monomers that become activated when bound to specific cyclins. Cyclins are nuclear proteins that are transiently expressed in order to activate its cognate Cdk. These proteins form complexes that facilitate cell cycle progression through various cell cycle stages. The roles that these cell cycle regulatory proteins play in cell cycle progression is modulated by PI3K-Akt signaling. Recently, Akt has been shown to play important role in regulation of the G^2/M transition of the cell cycle [70]. Recent evidences suggest that p21Cip1 (a downstream substrate of Akt) has a complex role in regulating Cdk activity. Under physiological conditions, p21Cip1 can bind to Cdk4/cyclin D and Cdk6/cyclin D complexes, and induce their kinase activities from early G1 until the middle of S phase [71]. In contrast, p21Cip1 can bind to and inhibit kinase activities of Cdk2/cyclin A and Cdk2/cyclin E complexes from the late G1 until the S phase [72, 73]. The function of p21Cip1 during cell cycle progression appears to ensure appropriate Cdk activation. Therefore, p21Cip1 has both positive and negative regulatory effects on cell cycle progression and the PI3K/Akt pathway is required for p21Cip1 induction in human carcinomas. In addition, Akt also inhibits activities of p27 Kip, Myt1 and Wee1, which are associated with cell cycle regulation.

The PI3K/Akt pathway is also a strong activator of cyclin D1, a critical player in cell cycle progression. Akt-mediated-inactivation of GSK-3β plays an important role in regulation of cyclin D1 protein levels. Cyclin D1 degradation occurs mainly through the ubiquitin-dependent proteolysis pathway triggered by phosphorylation [74]. In order to be recognized by ubiquitin-conjugating enzymes for proteolysis, cyclin D1 must be phosphorylated at T286 near its carboxyl terminus [74]. GSK-3β is the primary kinase that phosphorylates cyclin D1 at this residue [74]. Moreover, cyclin D1 phosphorylation at T286 by GSK-3β is also associated with the translocation of cyclin D1 from the nucleus to the cytoplasm. Rapid cyclin D1 degradation induced by GSK-3β is inhibited by activation of PI3K/Akt pathway because Akt directly phosphorylates and inactivates GSK-3β.

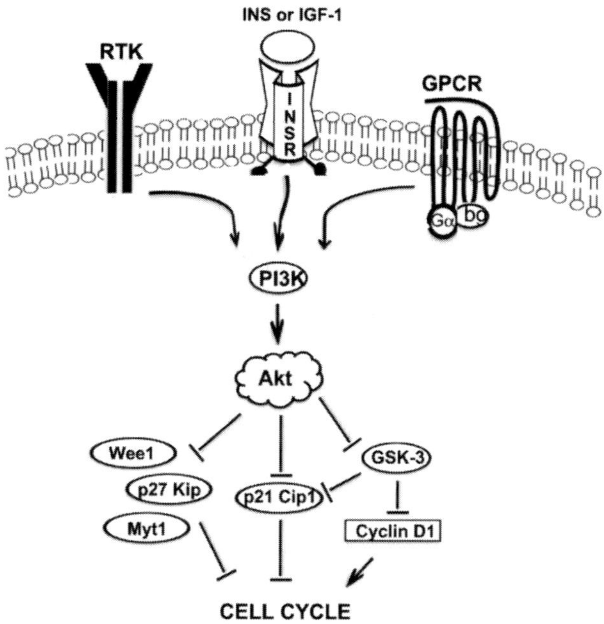

Figure 8. Akt in cell cycle regulation. Downstream substrate of Akt includes glycogen synthase kinase-3 beta (GSK-3β) and p21Cip1. Activated Akt inhibits GSK-3β and and p21Cip1 activity. In addition, Akt also inhibits activities of p27 Kip, Myt1 and Wee1, which are associated with cell cycle regulation. The PI3K/Akt pathway is also a strong regulator of cyclin D1, which is a key cell cycle regulatory gene. Rapid cyclin D1 degradation induced by GSK-3β is inhibited by activation of PI3K/Akt pathway because Akt directly phosphorylates and inactivates GSK-3β.

2.3. Akt in Angiogenesis

Although the contribution of Akt in angiogenesis is still an open question, several lines of evidence suggest a link between Akt and angiogenesis. Thus, Akt has recently been shown to phosphorylate the endothelial nitric oxide synthase (eNOS) leading to a persistent calcium-independent enzyme activation [75, 76]. A large body of literature indicates an essential role of endothelial NO for postnatal neovascularization [77, 78]. Mechanistically, Akt stimulation may enhance endothelial NO synthesis and thereby influence the long-term regulation of vessel growth. The inhibition of the proapoptotic proteins BAD or caspase-9 in addition to the enhanced endothelial NO synthesis, which also inhibits endothelial cell apoptosis, may block apoptosis at several stages in the apoptosis signaling cascade. The question remains what is the importance of the Akt pathway for endothelial cell migration and proliferation, which are prerequisite for angiogenesis. Several *in vitro* studies indicate that endothelial cell migration also depends on NO [79, 80]. Moreover, recent studies suggest that at least the VEGF-induced endothelial cell migration requires the activation of Akt. Therefore, these studies indeed provide evidence that Akt-dependent NO synthesis contributes to endothelial cell migration.

In summary, the activation of Akt is also one of the most frequent alterations observed in human cancer cells. Cancer cells that have constantly active Akt may depend on Akt for survival. Therefore, understanding Akt and associated pathways is important for the creation of better therapies to treat many human cancers including CRC.

References

[1] Jemal, A., et al., Cancer statistics, 2008. CA *Cancer J. Clin.*, 2008. 58(2): p. 71-96.

[2] Bain, N.S., et al., Striking the right balance in colorectal cancer care--a qualitative study of rural and urban patients. *Fam. Pract.*, 2002. 19(4): p. 369-74.

[3] Van Cutsem, E., et al., Adjuvant treatment of colorectal cancer (current expert opinion derived from the Third International Conference: Perspectives in Colorectal Cancer, Dublin, 2001). *Eur. J. Cancer*, 2002. 38(11): p. 1429-36.

[4] Brazil, D.P., Z.Z. Yang, and B.A. Hemmings, Advances in protein kinase B signalling: AKTion on multiple fronts. *Trends Biochem. Sci.,* 2004. 29(5): p. 233-42.

[5] Yang, Z.Z., et al., Physiological functions of protein kinase B/Akt. *Biochem. Soc. Trans*, 2004. 32(Pt 2): p. 350-4.

[6] Salomon, D.S., et al., Epidermal growth factor-related peptides and their receptors in human malignancies. *Crit. Rev. Oncol. Hematol.*, 1995. 19(3): p. 183-232.

[7] Bishnupuri, K.S., et al., Reg IV activates the epidermal growth factor receptor/Akt/AP-1 signaling pathway in colon adenocarcinomas. *Gastroenterology,* 2006. 130(1): p. 137-49.

[8] Derynck, R. and X.H. Feng, TGF-beta receptor signaling. *Biochim. Biophys. Acta*, 1997. 1333(2): p. F105-50.

[9] Massague, J., TGF-beta signal transduction. *Annu. Rev. Biochem.*, 1998. 67: p. 753-91.

[10] Barnard, J.A., G.J. Warwick, and L.I. Gold, Localization of transforming growth factor beta isoforms in the normal murine small intestine and colon. *Gastroenterology*, 1993. 105(1): p. 67-73.

[11] Avery, A., et al., TGF-beta expression in the human colon: differential immunostaining along crypt epithelium. *Br. J. Cancer*, 1993. 68(1): p. 137-9.

[12] Oshima, H., et al., Morphological and molecular processes of polyp formation in Apc(delta716) knockout mice. *Cancer Res*, 1997. 57(9): p. 1644-9.

[13] Ko, T.C., et al., Transforming growth factor-beta inhibits rat intestinal cell growth by regulating cell cycle specific gene expression. *Am. J. Surg.,* 1994. 167(1): p. 14-9; discussion 19-20.

[14] Ko, T.C., et al., Transforming growth factor-beta 1 inhibits cyclin D1 expression in intestinal epithelial cells. *Oncogene*, 1995. 10(1): p. 177-84.

[15] Conery, A.R., et al., Akt interacts directly with Smad3 to regulate the sensitivity to TGF-beta induced apoptosis. *Nat. Cell Biol.*, 2004. 6(4): p. 366-72.

[16] Derynck, R., R.J. Akhurst, and A. Balmain, TGF-beta signaling in tumor suppression and cancer progression. *Nat. Genet.*, 2001. 29(2): p. 117-29.

[17] Akhurst, R.J. and R. Derynck, TGF-beta signaling in cancer--a double-edged sword. *Trends Cell Biol*, 2001. 11(11): p. S44-51.

[18] Derynck, R., et al., Synthesis of messenger RNAs for transforming growth factors alpha and beta and the epidermal growth factor receptor by human tumors. *Cancer Res.*, 1987. 47(3): p. 707-12.

[19] Robson, H., et al., Transforming growth factor beta 1 expression in human colorectal tumours: an independent prognostic marker in a subgroup of poor prognosis patients. *Br. J. Cancer*, 1996. 74(5): p. 753-8.

[20] Friedman, E., et al., High levels of transforming growth factor beta 1 correlate with disease progression in human colon cancer. *Cancer Epidemiol. Biomarkers Prev*, 1995. 4(5): p. 549-54.

[21] Derynck, R., Y. Zhang, and X.H. Feng, Smads: transcriptional activators of TGF-beta responses. *Cell,* 1998. 95(6): p. 737-40.

[22] Miyazono, K., P. ten Dijke, and C.H. Heldin, TGF-beta signaling by Smad proteins. *Adv. Immunol.*, 2000. 75: p. 115-57.

[23] Remy, I., A. Montmarquette, and S.W. Michnick, PKB/Akt modulates TGF-beta signalling through a direct interaction with Smad3. *Nat. Cell Biol.*, 2004. 6(4): p. 358-65.

[24] Song, K., et al., Insulin-like growth factor-I inhibits transcriptional responses of transforming growth factor-beta by phosphatidylinositol 3-kinase/Akt-dependent suppression of the activation of Smad3 but not Smad2. *J. Biol. Chem.*, 2003. 278(40): p. 38342-51.

[25] Aaronson, S.A., Growth factors and cancer. *Science*, 1991. 254(5035): p. 1146-53.

[26] Pollak, M.N., et al., Presence of somatomedin receptors on primary human breast and colon carcinomas. *Cancer Lett.*, 1987. 38(1-2): p. 223-30.

[27] Guo, Y.S., et al., Characterization of insulinlike growth factor I receptors in human colon cancer. *Gastroenterology*, 1992. 102(4 Pt 1): p. 1101-8.

[28] Watkins, L.F., L.R. Lewis, and A.E. Levine, Characterization of the synergistic effect of insulin and transferrin and the regulation of their receptors on a human colon carcinoma cell line. *Int. J. Cancer*, 1990. 45(2): p. 372-5.

[29] Koenuma, M., T. Yamori, and T. Tsuruo, Insulin and insulin-like growth factor 1 stimulate proliferation of metastatic variants of colon carcinoma 26. *Jpn J. Cancer Res*, 1989. 80(1): p. 51-8.

[30] Bjork, J., et al., Growth-regulatory effects of sensory neuropeptides, epidermal growth factor, insulin, and somatostatin on the non-transformed intestinal epithelial cell line IEC-6 and the colon cancer cell line HT 29. *Scand. J. Gastroenterol.*, 1993. 28(10): p. 879-84.

[31] Vogelstein, B., et al., Genetic alterations during colorectal-tumor development. *N. Engl. J. Med.*, 1988. 319(9): p. 525-32.

[32] Cats, A., et al., Increased epithelial cell proliferation in the colon of patients with acromegaly. *Cancer Res*, 1996. 56(3): p. 523-6.

[33] Warren, R.S., et al., Induction of vascular endothelial growth factor by insulin-like growth factor 1 in colorectal carcinoma. *J. Biol. Chem.*, 1996. 271(46): p. 29483-8.

[34] MacDonald, R.S., W.H. Thornton, Jr., and T.L. Bean, Insulin and IGE-1 receptors in a human intestinal adenocarcinoma cell line (CACO-2): regulation of Na+ glucose transport across the brush border. *J. Recept Res.*, 1993. 13(7): p. 1093-113.

[35] Macaulay, V.M., Insulin-like growth factors and cancer. *Br. J. Cancer*, 1992. 65(3): p. 311-20.

[36] Soos, M.A., et al., Receptors for insulin and insulin-like growth factor-I can form hybrid dimers. Characterisation of hybrid receptors in transfected cells. *Biochem. J.*, 1990. 270(2): p. 383-90.

[37] Bach, L.A. and M.M. Rechler, Insulin-like growth factors and diabetes. *Diabetes Metab. Rev.*, 1992. 8(3): p. 229-57.

[38] Burgering, B.M., et al., Insulin stimulation of gene expression mediated by p21ras activation. *EMBO J.*, 1991. 10(5): p. 1103-9.

[39] Jhun, B.H., et al., Insulin and insulin-like growth factor-I signal transduction requires p21ras. *J. Biol. Chem.*, 1994. 269(8): p. 5699-704.

[40] Gudermann, T., T. Schoneberg, and G. Schultz, Functional and structural complexity of signal transduction via G-protein-coupled receptors. *Annu. Rev. Neurosci.*, 1997. 20: p. 399-427.

[41] Neer, E.J., Heterotrimeric G proteins: organizers of transmembrane signals. *Cell,* 1995. 80(2): p. 249-57.

[42] Moule, S.K., et al., Regulation of protein kinase B and glycogen synthase kinase-3 by insulin and beta-adrenergic agonists in rat epididymal fat cells. Activation of protein kinase B by wortmannin-sensitive and -insensitive mechanisms. *J. Biol. Chem.*, 1997. 272(12): p. 7713-9.

[43] Sable, C.L., et al., cAMP stimulates protein kinase B in a Wortmannin-insensitive manner. *FEBS Lett.*, 1997. 409(2): p. 253-7.

[44] Yang, M., et al., G protein-coupled lysophosphatidic acid receptors stimulate proliferation of colon cancer cells through the {beta}-catenin pathway. *Proc. Natl. Acad. Sci. USA*, 2005. 102(17): p. 6027-32.

[45] Liang, J. and J.M. Slingerland, Multiple roles of the PI3K/PKB (Akt) pathway in cell cycle progression. *Cell Cycle*, 2003. 2(4): p. 339-45.

[46] Franke, T.F., et al., Direct regulation of the Akt proto-oncogene product by phosphatidylinositol-3,4-bisphosphate. *Science*, 1997. 275(5300): p. 665-8.

[47] Hara, K., et al., Raptor, a binding partner of target of rapamycin (TOR), mediates TOR action. *Cell*, 2002. 110(2): p. 177-89.

[48] Kim, D.H., et al., mTOR interacts with raptor to form a nutrient-sensitive complex that signals to the cell growth machinery. *Cell,* 2002. 110(2): p. 163-75.

[49] Loewith, R., et al., Two TOR complexes, only one of which is rapamycin sensitive, have distinct roles in cell growth control. *Mol. Cell*, 2002. 10(3): p. 457-68.

[50] Jacinto, E., et al., Mammalian TOR complex 2 controls the actin cytoskeleton and is rapamycin insensitive. *Nat. Cell Biol.*, 2004. 6(11): p. 1122-8.

[51] Sarbassov, D.D., et al., Rictor, a novel binding partner of mTOR, defines a rapamycin-insensitive and raptor-independent pathway that regulates the cytoskeleton. *Curr. Biol.*, 2004. 14(14): p. 1296-302.

[52] Hay, N. and N. Sonenberg, Upstream and downstream of mTOR. *Genes Dev.*, 2004. 18(16): p. 1926-45.

[53] Tran, H., et al., The many forks in FOXO's road. *Sci. STKE*, 2003. 2003(172): p. RE5.

[54] Datta, S.R., et al., Akt phosphorylation of BAD couples survival signals to the cell-intrinsic death machinery. *Cell*, 1997. 91(2): p. 231-41.

[55] del Peso, L., et al., Interleukin-3-induced phosphorylation of BAD through the protein kinase Akt. *Science*, 1997. 278(5338): p. 687-9.

[56] Cardone, M.H., et al., Regulation of cell death protease caspase-9 by phosphorylation. *Science*, 1998. 282(5392): p. 1318-21.

[57] Yang, E., et al., Bad, a heterodimeric partner for Bcl-XL and Bcl-2, displaces Bax and promotes cell death. *Cell*, 1995. 80(2): p. 285-91.

[58] Gajewski, T.F. and C.B. Thompson, Apoptosis meets signal transduction: elimination of a BAD influence. *Cell*, 1996. 87(4): p. 589-92.

[59] Zha, J., et al., Serine phosphorylation of death agonist BAD in response to survival factor results in binding to 14-3-3 not BCL-X(L). *Cell*, 1996. 87(4): p. 619-28.

[60] Romashkova, J.A. and S.S. Makarov, NF-kappaB is a target of AKT in anti-apoptotic PDGF signalling. *Nature*, 1999. 401(6748): p. 86-90.

[61] Madrid, L.V., et al., Akt suppresses apoptosis by stimulating the transactivation potential of the RelA/p65 subunit of NF-kappaB. *Mol. Cell Biol.*, 2000. 20(5): p. 1626-38.

[62] Karin, M. and M. Delhase, The I kappa B kinase (IKK) and NF-kappa B: key elements of proinflammatory signalling. *Semin. Immunol.*, 2000. 12(1): p. 85-98.

[63] Chen, C., L.C. Edelstein, and C. Gelinas, The Rel/NF-kappaB family directly activates expression of the apoptosis inhibitor Bcl-x(L). *Mol. Cell Biol.*, 2000. 20(8): p. 2687-95.

[64] Zong, W.X., et al., The prosurvival Bcl-2 homolog Bfl-1/A1 is a direct transcriptional target of NF-kappaB that blocks TNFalpha-induced apoptosis. *Genes Dev.*, 1999. 13(4): p. 382-7.

[65] Lee, H.H., et al., NF-kappaB-mediated up-regulation of Bcl-x and Bfl-1/A1 is required for CD40 survival signaling in B lymphocytes. *Proc. Natl. Acad. Sci. USA*, 1999. 96(16): p. 9136-41.

[66] Wang, C.Y., et al., NF-kappaB induces expression of the Bcl-2 homologue A1/Bfl-1 to preferentially suppress chemotherapy-induced apoptosis. *Mol. Cell Biol.*, 1999. 19(9): p. 5923-9.

[67] Nicholson, K.M. and N.G. Anderson, The protein kinase B/Akt signalling pathway in human malignancy. *Cell Signal*, 2002. 14(5): p. 381-95.

[68] Cantley, L.C., The phosphoinositide 3-kinase pathway. *Science*, 2002. 296(5573): p. 1655-7.

[69] van Weeren, P.C., et al., Essential role for protein kinase B (PKB) in insulin-induced glycogen synthase kinase 3 inactivation. Characterization of dominant-negative mutant of PKB. *J. Biol. Chem.*, 1998. 273(21): p. 13150-6.

[70] Shtivelman, E., J. Sussman, and D. Stokoe, A role for PI 3-kinase and PKB activity in the G2/M phase of the cell cycle. *Curr. Biol.*, 2002. 12(11): p. 919-24.

[71] Sherr, C.J. and J.M. Roberts, CDK inhibitors: positive and negative regulators of G1-phase progression. *Genes Dev.*, 1999. 13(12): p. 1501-12.

[72] Rank, K.B., D.B. Evans, and S.K. Sharma, The N-terminal domains of cyclin-dependent kinase inhibitory proteins block the phosphorylation of cdk2/Cyclin E by the CDK-activating kinase. *Biochem. Biophys. Res. Commun*, 2000. 271(2): p. 469-73.

[73] Shimizu, T., et al., Complex regulation of CDK2 and G1 arrest during neuronal differentiation of human prostatic cancer TSU-Prl cells by staurosporine. *Anticancer Res.*, 2001. 21(2A): p. 893-8.

[74] Diehl, J.A., et al., Glycogen synthase kinase-3beta regulates cyclin D1 proteolysis and subcellular localization. *Genes Dev.*, 1998. 12(22): p. 3499-511.

[75] Dimmeler, S., et al., Activation of nitric oxide synthase in endothelial cells by Akt-dependent phosphorylation. *Nature*, 1999. 399(6736): p. 601-5.

[76] Fulton, D., et al., Regulation of endothelium-derived nitric oxide production by the protein kinase Akt. *Nature*, 1999. 399(6736): p. 597-601.

[77] Papapetropoulos, A., et al., Nitric oxide production contributes to the angiogenic properties of vascular endothelial growth factor in human endothelial cells. *J. Clin. Invest.*, 1997. 100(12): p. 3131-9.

[78] Murohara, T., et al., Nitric oxide synthase modulates angiogenesis in response to tissue ischemia. *J. Clin. Invest.*, 1998. 101(11): p. 2567-78.

[79] Murohara, T., et al., Role of endothelial nitric oxide synthase in endothelial cell migration. *Arterioscler. Thromb. Vasc. Biol.*, 1999. 19(5): p. 1156-61.

[80] Noiri, E., et al., Podokinesis in endothelial cell migration: role of nitric oxide. *Am. J. Physiol.*, 1998. 274(1 Pt 1): p. C236-44.

In: Signaling, Gene Regulation and Cancer
Editors: Shree Ram Singh and Manoj K. Mishra
ISBN: 978-1-61942-088-5

Chapter 9

DYSREGULATION OF CELLULAR SIGNALING IN COLORECTAL CANCER

William K. K. Wu and Jun Yu*

Institute of Digestive Disease, LKS Institute of Health Sciences and
Department of Medicine and Therapeutics, Faculty of Medicine,
The Chinese University of Hong Kong, Hong Kong

ABSTRACT

Multiple intracellular signaling pathways, such as Wnt/β-catenin signaling, epidermal growth factor receptor/Ras signaling, and p53 signaling are frequently dysregulated in colorectal cancer. Recent evidence also points to the involvement of signaling pathways in the developmental process, including Notch signaling, Hedgehog signaling, and Hippo signaling. Dysregulation of these signaling pathways contribute to the acquisition of malignant phenotypes, including unchecked cell proliferation, evasion of apoptosis, and enhanced invasiveness and metastasis. Understanding their relative importance and cross-talk will provide a rational basis for anticancer drug development.

INTRODUCTION

Colorectal cancer is the third most common cancer and the fourth leading cause of cancer-related death in the world, with an estimated incidence of 1,233,700 new cases and a mortality of 608,700 deaths per year based on the statistics for the year 2008 [1]. While most cases of colorectal cancer arise sporadically, it is estimated that more than one-fifth of colorectal cancer patients may have a familial component. Defined hereditary syndromes such as Lynch syndrome and familial adenomatous polyposis (FAP), however, are only diagnosed in 6% of these patients, indicating the genetic heterogeneity of familial cases. Several intrinsic (*e.g.* age, male gender, family history, and morbidities such as diabetes mellitus,

* Correspondence: Prof. Jun Yu, 7/F, LKS Medical Sciences Building, Prince of Wales Hospital, Shatin, NT, Hong Kong. Email: junyu@cuhk.edu.hk (J. Yu); Tel: +852 3763 6103; Fax: +852 2144 5330.

obesity and inflammatory bowel disease) and extrinsic (*e.g.* sedentary lifestyle, inadequate intake of fiber, high consumption of alcohol and red meat , high-fat diet, and cigarette smoking) factors also heighten the risk for colorectal cancer. Screening and early diagnosis by methods such as fecal occult blood test and colonoscopy remains the desired options to ease the disease burden. Recent advances in molecular diagnostics, such as detections of mutated genes and/or dysregulated microRNAs associated with colorectal cancer, also facilitate the diagnosis and screening of this increasingly prevalent disease. Various treatment approaches have been devised for colorectal cancer, including surgery, chemotherapy and radiation therapy. The advent of targeted cancer therapeutics, such as cetuximab and bevacizumab, also significantly improves the treatment outcomes. Nevertheless, the prognosis of patients with metastatic colorectal cancer remains dismal, highlighting the need to develop novel therapeutics for these patients.

Despite the complex interactions between the host and the environmental factors, the pathogenesis of colorectal cancer converges on the dysregulation of common signaling pathways as consequences of aberrant activation of proto-oncogenes and inactivation of tumor-suppressor genes. These deranged intracellular signals, originated from genetic and epigenetic abnormalities at the molecular level, promote the acquisition of oncogenic properties, including unchecked cell proliferation, evasion of apoptosis, cellular invasiveness and metastasis, angiogenesis, multi-drug resistance, and circumvention of immune surveillance. From a clinical perspective, many components of the dysregulated signaling pathways are amenable to pharmacological inhibition. Further understanding of signaling dysregulation in colorectal cancer will undoubtedly lead to the discovery of novel targets for the development of the next-generation cancer therapeutics. In addition, dysregulation of a particular signaling pathway in cancer tissues may foreshadow a better or worse prognosis or response to treatment, providing clinicians with extra information to make evidence-based judgment. In this chapter, the involvement of several crucial signaling pathways and their biological functions in colorectal carcinogenesis will be discussed.

Cell Cycle Regulators

Progression through the cell cycle is orchestrated by the temporally coordinated expression and interaction among different cyclins, cyclin-dependent kinases (CDKs) and CDK inhibitors. During G_1-S transition, cyclin D and cyclin E, bind to CDK4/6 and CDK2, respectively, to upregulate genes that are required for DNA synthesis. After entering the S phase, the level of cyclin A increases while the expression of cyclin E reduces. At G_2-M transition, a surge of cyclin B/CDK1 activity triggers the mitotic phase. In addition to cyclins, CDK activities are regulated by CDK inhibitors, which can be categorized into p16 family (p15, p16, p18 and p19) and p21 family (p21, p27, p28 and p57). The former binds to and inhibits the activities of CDK4/6, whereas the latter binds to a wide range of cyclin/CDK complexes, resulting in inhibition of all phases of the cell cycle.

Some of the above mentioned cell cycle regulators are dysregulated in colorectal cancer. Aberrant accumulation of cyclin D_1 occurs in about one third of colorectal cancer as an early event during multistage process of colorectal carcinogenesis [2, 3]. Increased expression of cyclin D_1 is also noted in K-Ras-transformed enterocytes [4]. Moreover, cyclin D_1 and CDK4 are upregulated in intestinal adenomas of patients with FAP [5]. Overexpression of cyclin D_1 also predicts disease recurrence in colorectal adenocarcinoma [6]. To this end, antisense to cyclin D_1 impairs the growth and tumorigenicity of human colon cancer cells. These findings suggest that upregulation of cyclin D_1 is crucial to the pathogenesis of colorectal cancer. However, it is noteworthy that the cyclin D_1 gene amplification is rarely detected in colorectal cancer [7]. In this regard, overexpression of cyclin D_1 has been attributed to the aberrant accumulation of β-catenin as a result of APC mutation as well as oncogenic activation of K-Ras, both of which regulate cyclin D_1 expression in a cooperative manner [8]. In contrast with cyclin D_1, the cyclin E gene is amplified in a subset of primary colorectal carcinoma tissues. Interestingly, in more than half of these cyclin E-amplified samples, amplification of CDK2 gene is also noted [9]. Amplification of cyclin E has also been detected in some colorectal cancer cell lines [10]. Moreover, colorectal cancers frequently overexpress cyclin E, whose level is significantly correlated with the proliferative activity and the mutator phenotype of the tumor cells [11-13]. Intriguingly, the expression of cyclin E and CDK2 may be lost when colon cancer cells have metastasized from primary site to the liver [14].

Several members of the p16 and p21 families of CDK inhibitors are downregulated in colorectal cancer tissues. For instance, epigenetic inactivation of p15 and p16 through DNA methylation has been reported in about one-fourth of colorectal cancer cases, in which methylation of p16 gene is associated with K-Ras mutation [15-17]. The expression of p21 is gradually reduced during neoplastic progression from normal colonic mucosa, through adenoma, to adenocarcinoma. The loss of p21 correlates with poor overall and disease-free survival [18, 19]. A strong association between reduced or absent p21 and the development of metastases has also been reported [20]. Inactivation of p21 enhances tumor formation in two different animal models of intestinal tumorigenesis, supporting the tumor suppressing role of p21 [21]. The expression of p27 is also downregulated during colorectal carcinogenesis. The reduction of p27 is mediated through increased proteasome-dependent degradation. Colorectal cancer patients with p27 downregulation exhibit poorer prognosis [22]. An inverse correlation between p27 expression and the degree of tumor aggressiveness has been noted [23, 24]. Overexpression of p27 increases the sensitivity of colon cancer cells to sodium butyrate-induced differentiation [25]. Moreover, targeted inactivation of p27 is sufficient to induce intestinal tumorigenesis in mouse [26], supporting its function as a tumor suppressor. In addition to p21 and p27, the expression of another family member p57 is decreased in colorectal cancer.

Apoptosis

Apoptosis or type I programmed cell death is frequently dysregulated in human cancers. There are two biochemically distinct apoptotic pathways, namely, mitochondria- and death receptor-mediated apoptotic cascades. The former is regulated in part by opposing activities of pro-apoptotic (Bax, Bak, Bik, Bim, Bad, Bid, HRK, NOXA, PUMA, and BNIP3) and anti-apoptotic (Bcl-2, Bcl-xL, Bcl-w, A1, and Mcl-1) members of the Bcl-2 family, whose regulate mitochondrial membrane permeability, cytochrome *c* release, and caspase-9 activation. In contrast, the death receptor-mediated pathway involves ligand-dependent activation of death receptors (Fas, TNF-R1, DR3, TRAIL-R1 (DR4), TRAIL-R2 (DR5),

DR6, p75-NGFR and EDAR) which triggers FADD-dependent activation of caspase-8. Activation of these two pathways leads to activation of a common set of effector caspases, including caspases-3, -6 and -7. The execution of apoptosis can be suppressed by a class of proteins known as inhibitors of apoptosis (IAPs), which include XIAP, survivin, c-IAP1, c-IAP2, NAIP, Bruce, ILP-2, and Livin. These IAPs impair apoptosis through inhibition of the effector caspases. The importance of apoptosis in colorectal carcinogenesis can be highlighted by the finding that reduced rates of apoptosis within the colorectal mucosa are associated with an increased risk for colorectal adenomas [27]. However, neoplastic transformation in early-stage colorectal carcinogenesis is paradoxically associated with increased rates of apoptosis and proliferation. Apoptotic rate decreases afterward during adenoma-to-carcinoma transition [28].

The expression of several Bcl-2 family members is dysregulated in colorectal cancer and may be associated with disease outcomes. In normal and hyperplastic mucosa, the expression of Bcl-2 is restricted to basal epithelial cells. However, Bcl-2 expression extends to the parabasal and superficial regions in dysplastic polyps and carcinomas. Moreover, an inverse correlation is found between Bcl-2 positivity and spontaneous apoptotic rate [29]. Surprisingly, Bcl-2 expression is associated with favorable clinical outcomes, including longer relapse-free survival and smaller tumour size [30, 31]. Moreover, high expression of Bcl-2 is associated with a lower chance of tumour recurrence in colorectal cancer [32]. The expression of Bcl-2 is inversely correlated with p53 expression in adenomas as well as primary and metastatic colorectal carcinomas [33, 34]. Elevated expression of Bcl-xL and Bcl-w together with reduced expression of Bak have been reported in primary colorectal adenocarcinomas [35, 36]. Frameshift mutations in the Bax gene, which leads to absent or very low expression of Bax protein, have been detected in colorectal cancer with the microsatellite mutator phenotype [37, 38]. Bax gene is also frequently mutated in hereditary nonpolyposis colorectal cancers [39]. Moreover, Bax expression decreases significantly from primary tumor to metastasis of colorectal cancer [40]. In this connection, low expression of Bax is correlated with poorer prognosis in patients with hepatic metastases of colorectal cancer [41]. It has been postulated that mutational inactivation of Bax confers colon cancer cells resistance to death receptor-mediated apoptotic signaling and selective advantage during tumor clonal evolution [42, 43]. Elevated expression of pro-apoptotic Bim and Puma proteins in the tumor cytoplasm also predicts better disease-free and overall survival even after adjustment for histologic grade, tumor stage, age, and treatment [44]. The proapoptotic Bad and Bid protein expression is also associated with better survival in stages II and III colon cancers [45].

The death receptor-mediated pathway has been subverted during colorectal carcinogenesis. For instance, the gene encoding DcR3 which acts as a decoy receptor to binds to FasL is amplified in colon tumors [46]. This decoy receptor competes with Fas receptor for Fas ligands to inhibit the apoptotic cascade. Unexpectedly, colon cancer cells express high level of FasL. Upregulation of FasL has been proposed to mediate T cell killing during evasion of immune surveillance and invasion at the margin of colorectal liver metastases [47-49]. FasL upregulation is an early event in colorectal carcinogenesis, being evident at the stage of mild dysplasia, and is not associated with increased apoptosis of tumor cells [50, 51]. Caspase 8 mutations are also detected in a small subgroup of invasive carcinomas but not in adenomas [52].

Besides the Bcl-2 family and molecules involved in the death receptor-mediated pathway, the expression of IAPs are also dysregulated in colorectal cancer. Immunohistochemical staining has revealed that survivin is expressed in about half of cases of colorectal carcinomas while normal colon epithelium cells do not express survivin. Overexpression of survivin is also associated with reduced apoptotic index, which predicts shorter survival rates [53]. In patients with stage II colorectal carcinomas, survivin expression predicts death from recurrent colorectal carcinoma [54]. Survivin expression also correlates with proliferation and angiogenesis during colorectal tumorigenesis [55]. In contrast to survivin, the levels of expression of cIAP1, cIAP2, and NAIP in the cancerous tissues are lower than those found in normal mucosa [56]. Aberrant hypermethylation in the promoter of XAF1, which negatively regulates the function of XIAP, is also associated with the development of colon cancers [57, 58].

Epidermal Growth Factor Receptor (EGFR) and Ras-Raf-MEK-ERK Signaling

Mitogenic signals transduced by receptor tyrosine kinases play a crucial role in oncogenesis. Among all receptor tyrosine kinases, EGFR has perhaps received the greatest attention owing to its frequent upregulation or aberrant activation in solid tumors. EGFR-mediated signaling promotes cell proliferation, confers resistance to apoptosis, induces angiogenesis, and enhances invasion and metastasis. EGFR belongs to the ErbB receptor family which has three additional members, including ErbB2/neu/HER2, ErbB3/HER3 and ErbB4/HER4. These ErbB receptors are activated by their cognate ligands and share an overlapping downstream signaling network. Known ligands for EGFR include EGF, transforming growth factor (TGF)-α, amphiregulin, epiregulin, betacellulin, heparin-binding EGF and epigen. EGFR has also been shown to be transactivated by G-protein-coupled receptor through ectodomain shedding of membrane-bound EGFR ligands. Upon ligand binding, EGFR receptor form homodimers or heterodimers with other ErbB family members, which are subsequently internalized and autophosphorylated on their tyrosine residues. These events trigger diverse downstream signaling pathways including the Ras/mitogen-activated protein kinases (MAPKs), the Janus kinase-signal transducers and activators of transcription (JAK-STAT), the phospholipase Cγ, and the phosphatidylinositol 3-kinase/Akt pathways (Figure 1).

Overexpression or aberrant activation of EGFR and/or its family members have been implicated in the development and progression of many malignancies. Multiple components along the EGFR/erbB signaling pathway, including EGFR, TGFα, heparin-binding EGF, phospho-ErbB2, phospho-extracellular signal-regulated kinase (ERK) and cyclin D_1, are found to be upregulated in aberrant crypt foci, which are putative precursors of colorectal neoplasms [59]. The overexpression of EGFR has also been documented in 30 - 90% of cases of advanced colorectal cancer. Moreover, EGFR mRNA levels in colon cancer cells predict their ability to produce hepatic metastasis while EGFR positivity in colon cancer tissues are associated with worse staging, aneuploidy, local and lymphatic invasion, and poorer survival [60-64]. Overexpression of ErbB2 and ErbB3 has also been noted, in which ErbB3 expression may have correlation with poor survival [65, 66].

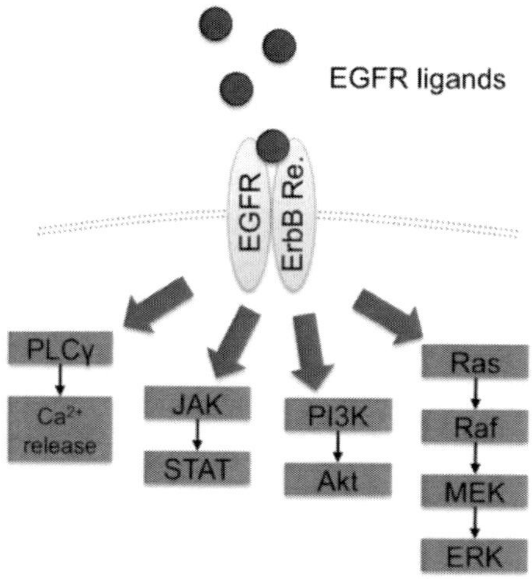

Figure 1. Ligand-dependent activation of EGFR elicits divergent downstream signaling.

Currently, two major approaches, namely, monoclonal antibodies directed toward the extracellular domain of EGFR and small molecule tyrosine kinase inhibitors (TKIs), have been attempted in clinical settings to target EGFR. The former competitively inhibits the binding of ligands and thereby blocking receptor phosphorylation and the downstream signaling while the latter inactivates EGFR intracellular kinase activity. Cetuximab, a FDA-approved EGFR-directed antibody, has shown efficacy as a single agent or in combination with irinotecan in patients with EGFR-expressing tumors. In contrast, EGFR TKIs have demonstrated minimal efficacy in metastatic colorectal cancer [67, 68].

This discrepancy may be explained by the fact that anti-EGFR antibodies but not EGFR TKIs can trigger receptor downregulation on the cell surface through induction of internalization [69, 70]. The specificity of TKI is also in general inferior to that of monoclonal antibody, leading to the occurrence of dose-limiting toxicity in patients treated with TKIs. In this connection, copy number change of EGFR has been shown to predict the sensitivity to cetuximab [71].

Ras, an oncogenic small GTPase, has three isoforms, namely, K-Ras, H-Ras, and N-Ras. Activation and deactivation of Ras is controlled by cycling between the active GTP-bound and inactive GDP-bound forms. Activating mutations of K-Ras, which preferentially occurs in codon 12, occur early in colorectal carcinogenesis and are detected in 27-43% of patients with colorectal cancer [72, 73]. Ras overexpression is also a common event [74]. B-Raf mutations are present in 5-10% of colorectal cancers that do not harbor Ras mutations. Ras mutations are detectable at high frequency in DNA purified from the stool of patients with colorectal tumors, providing a non-invasive method for cancer detection [75]. The oncogenic effect of K-Ras mutation is more prominent in a p53-deficient genetic background [76]. Ras promotes the degradation of p53 through the induction of MDM2, the E3-ubiquitin ligase that targets p53 for destruction [77]. Disruption of activated K-Ras by homologous recombination in colon carcinoma cell lines reduces c-Myc expression and inhibits tumor growth *in vitro* and *in vivo* [78]. Ras also induces cyclin D_1 expression and represses TGF-β signaling [8, 79]. The latter is accomplished through ERK-dependent phosphorylation of Smad2 and Smad3 at specific sites in the region linking the DNA-binding domain and the transcriptional activation domain [79]. Mutation of K-Ras oncogene is also associated with marked up-regulation of vascular endothelial growth factor (VEGF) [80]. In addition, mutant K-Ras or B-Raf upregulates the expression of GLUT-1 to enhance glucose uptake. In this regard, low glucose level in the tumor microenvironment has been proposed to be the driving force for positive selection of K-Ras mutants [81]. The Ras/Raf/MEK/ERK pathway can also be activated by different signaling molecules such as growth factors, cytokines, G-protein-coupled receptors and integrins and impaired by the tumor suppressors PITX1 and Merlin [82, 83]. Since K-Ras is downstream of EGFR, abrogating EGFR signaling by anti-EGFR monoclonal antibodies has limited effect on the downstream oncogenic signaling in K-Ras mutated cells. In this regard, K-Ras mutations are associated with the resistance to EGFR-targeted therapy in colorectal cancer patients [84].

PI3K-AKT-MTOR SIGNALING

The phosphoinositide 3-kinase (PI3K)-Akt-mammalian target of rapamycin (mTOR) cascade is an important signaling pathway involved in the regulation of cell growth, proliferation, differentiation, apoptosis, motility and metabolism. The class I PI3K, whose activity is stimulated by receptor tyrosine kinases or G-protein-coupled receptors and antagonized by the tumor suppressor phosphatase and tensin homolog (PTEN), is an enzyme responsible for the production of phosphoinositides. Upon activation, phosphatidylinositol (3,4,5)-trisphosphate (PIP_3) produced by PI3K leads to the recruitment of Akt, a serine/threonine protein kinase, to the plasma membrane for full activation by phosphoinositide-dependent kinase 1 (PDK1) and PKD2. The phosphorylated Akt in turn activates mTOR complex 1 (mTORC1) through inhibition of TSC1/2, which are negative regulators of mTOR. The mTORC1, which consists of mTOR, regulatory associated protein of mTOR (raptor), mammalian LST8/G-protein β-subunit like protein (mLST8/GβL), and the recently identified partners PRAS40 and DEPTOR, is crucial to the regulation of protein translation and other cellular processes. It acts as a central hub to integrate growth factor and nutrient signals. In growth-permissive conditions, mTORC1 stimulates protein translation through phosphorylation of ribosomal protein S6 kinase 1 (S6K1) at T389 to prime this kinase for full activation by PDK1. Fully activated S6K1 then phosphorylates eIF4B at S422 and promotes its binding to eIF3. On the other hand, mTORC1 inactivates eIF4E-binding protein 1 (4E-BP1), resulting in disinhibition of eIF4E and the formation the eIF4F complex, which is required for the recruitment of the 40S small ribosomal subunit to the mRNA through eIF3. These molecular cascades then enhance translation initiation complex formation and cap-dependent translation.

The PI3K-Akt-mTOR signaling cascade is dysregulated in colorectal cancer. For instances, PIK3CA, the gene encoding the p110α catalytic subunit of PI3K, is somatically mutated in over 25% of colorectal cancers [85, 86]. PIK3CA mutation is also associated with the detection of phosphorylated Akt in colorectal cancer tissues. Surprisingly, phospho-AKT expression in colorectal cancer is associated with low stage and good prognosis [87]. In addition to PIK3CA mutation, germline mutation of PTEN is associated with juvenile polyposis coli, a syndrome characterized by the formation of multiple polyps in the gastrointestinal tract. Biallelic frameshift mutation and loss of heterozygosity of PTEN are also documented in a subset of colorectal cancer [88, 89]. mTOR signaling components, including mTOR, S6K1, and 4EBP1, are also highly activated in glandular elements of colorectal cancer and adenomas with high-grade intraepithelial neoplasia [90].

Activation of the PI3K-Akt-mTOR pathway in cancer cells have multiple biological effects. Induction of K-Ras in intestinal epithelial cells activates the PI3K-Akt-mTOR pathway whose inhibition restricts K-Ras-mediated transformation by inducing G_1 growth arrest [91]. Inhibition of PI3K also significantly enhances enterocyte-like differentiation of cultured colorectal cancer cells [92]. In addition, phosphorylation of Akt is required for suppression of colon cancer cell apoptosis [93]. Activation of the mTOR pathway has been shown to cause translational deregulation and G_1-S acceleration in colon cancer cell lines. These changes are associated with low levels of p27 and activation of cyclin E-Cdk2. Abnormal activation of mTOR signaling is also implicated in chromosomal aberrations in the colonic mucosa of $Apc^{+/\Delta716}$ $Cdx2^{+/-}$ compound mutant mice [94]. Importantly, inhibition of the mTORC1 pathway significantly suppresses intestinal polyp formation and reduces mortality in $Apc^{+/\Delta716}$ mice [95]. Recent findings also indicate that mTOR signaling regulates epithelial-mesenchymal transition, motility, and metastasis of colon cancer cells via RhoA and Rac1 signaling pathways [96]. Moreover, the PI3K-Akt-mTOR pathway modulates drug sensitivity in colon cancer cells. It has been shown that PIK3CA mutations render K-Ras-mutant cells insensitive to the growth inhibitory effect of the MEK inhibitior PD0325901 [97, 98]. eIF4F complex formation also determines the response of colon cancer cells to cetuximab [99].

Wnt Signaling

The canonical Wnt signaling pathway functions by stabilizing β-catenin, whose phosphorylation by glycogen synthase kinase 3β (GSK3β) results in degradation through the ubiquitin-proteasome pathway. The binding of Wnt ligands to a Frizzled receptor represses the function of β-catenin degradation complex (APC, AXIN and GSK3β), allowing the nuclear translocation of stabilized β-catenin followed by transcriptional activation of the Wnt target genes (Figure 2). In normal physiology, Wnt/β-catenin is required for proper cell positioning in the intestinal epithelium by controlling the expression of EphB/ephrinB [100]. Nevertheless, this pathway is aberrantly activated during colorectal carcinogenesis.

Wnt/β-catenin signaling is dysregulated in colorectal cancer through multiple mechanisms. Germline mutations in the human APC gene have been linked to FAP. Inactivating mutations of APC is also reported in up to 70% of sporadic colorectal cancer. The tumor-suppressor activity of mutant APC is compromised due to its inability to be exported from the nucleus to the cytoplasm where it induces β-catenin degradation [101, 102]. Accordingly, axin is abnormally localized in the nucleus in colon cancer cells [103]. Hypermethylation in CpG sites around the CCAAT box of APC promoter have been documented in colorectal cancer cell lines [104]. APC expression may also be deregulated by miR-135a and b in colorectal adenomas and carcinoma [105]. While APC is an important negative regulator of Wnt signaling, a recent study demonstrates that loss of APC by itself may be insufficient for causing β-catenin nuclear localization. To this end, an additional K-

Ras activation is required for nuclear accumulation of β-catenin and proliferation defects [106].

Axin2 and DACT3, both of which are negative regulators of Wnt signaling, are epigenetically silenced in a subset of colorectal carcinoma [107, 108]. The downregulation of Wnt antagonists DKK-1, SOX17 and WIF-1 also contributes to activation of β-catenin in colon cancer cells [109-111]. In FAP patients, β-catenin localizes at the membranes of the cell-to-cell borders in normal epithelial cells but accumulates in the nuclei and cytoplasms in both adenoma and cancer cells, in which the expression level is over three times higher than normal tissues [112]. A subset of colorectal tumors also harbor activating mutations of β-catenin [113-115]. Overexpression of an activated β-catenin causes intestinal dysplasia and adenoma in transgenic mice [116]. In $APC^{-/-}$ colon carcinoma, constitutive transcriptional activation by β-catenin/Tcf-4 complex can be detected [117]. Abnormal activation of Wnt signaling may also be mediated through dysregulation of Wnt ligands. While Wnt-5a is expressed in both nontumorous as well as in colonic tumor tissues, Wnt-2, which expresses at low levels in normal colon, exhibits overexpression in tumor tissues [118]. Moreover, extracellular Wnt antagonists FRP1 and DKK1 reduces β-catenin in colon cancer cell lines, suggesting an autocrine mechanism for constitutive Wnt pathway activation [119]. Furthermore, the transactivation potential of β-catenin/Tcf-4 complex is enhanced by B9L and SHARP, which are aberrantly elevated in colorectal tumors [120, 121]. The colorectal oncoprotein CDK8 also enhances Wnt signaling through repression of E2F1, a potent and specific inhibitor of β-catenin/Tcf-dependent transcription. The negative regulatory effect of E2F1 on β-catenin also explains why colon cancer cells very often retain a functional pRB, an antagonist of E2F1 and a tumor suppressor in other cancer types [122].

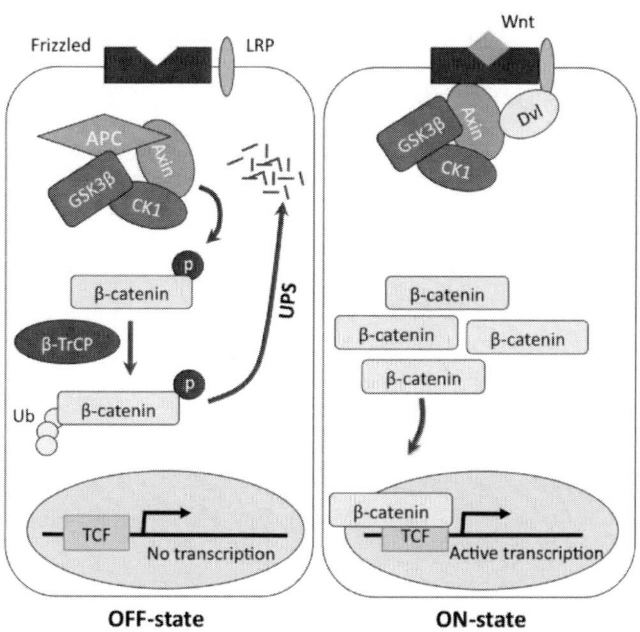

Figure 2. Wnt/β-catenin signaling in mammalian cells. UPS, ubiquitin-proteasome system; CK1, casein kinase I; LRP, Lipoprotein receptor-related proteins; β-TrCP, β-transducing repeat-containing protein.

Activation of Wnt/β-catenin signaling promotes the development of intestinal tumorigenesis through β-catenin-dependent transcription of Wnt target gene. For example, the proto-oncogene c-Myc is activated by β-catenin and repressed by wild-type APC [123]. β-catenin-dependent c-Myc expression imposes a crypt progenitor phenotype on colorectal cancer cells by repressing p21 expression and impairing cellular differentiation [124]. The importance of c-Myc in Wnt/β-catenin signaling is corroborated by the finding that intestine-specific loss of c-Myc rescues the phenotypes of perturbed differentiation, migration, proliferation and apoptosis caused by APC deletion [125]. β-catenin also regulates the expression of cyclin D_1, an important G_1-cyclin, in colon carcinoma cells [8, 126]. Moreover, β-catenin upregulates the expression of c-jun and fra-1, which are important components of the AP-1 transcription complex. β-catenin indirectly increases the expression urokinase-type plasminogen activator receptor (uPAR), whose expression is under the regulation by AP-1. Elevated expression of uPAR is associated with increased tumor invasiveness. The concomitant increase in β-catenin expression and the amount of uPAR have been confirmed in primary colon carcinomas and their liver metastases [127]. Wnt/β-catenin signaling also causes chromosomal instability via up-regulation of conductin [128, 129]. In addition to transcriptional activation, β-catenin mediates transcriptional repression. ZO-1, a gene involved in epithelial polarization, is downregulated by β-catenin [127]. Suppression of β-catenin/Tcf-4-mediated gene transactivation restores epithelial polarization in colorectal cancer cells and promotes colonic epithelial cell differentiation [130, 131]. Loss of APC also induces the expression of HDAC2, a protein responsible for deacetylation of lysine residues on the N-terminal region of the core histones, and thereby promoting transcriptional repression. Inhibition of HDAC2 by valproic acid substantially diminishes adenoma formation in APC^{min} mice [132]. Other colon cancer-related genes that are positively or negatively regulated by β-catenin include matrilysin, WISP-1/2, PPARδ/γ, Drctnnb1a, MDR-1, MCP-3, VEGF, AF17, ENC1, laminin γ2, survivin, axin2, hnkd, HOS, BMP-4, MT1-MMP, HELAD1, c-Met, APCDD1, COX-2, caspases-3/7/9, FGF-18, FGF-BP, EDN1, Tenascin-C, IFITM, ASCL2, CRD-BP, and Fascin [133-161].

Since Wnt/β-catenin signaling is an appealing therapeutic target in colorectal cancer, research effort has been put forth to expedite the discovery of novel compounds that inhibit the Wnt/β-catenin pathway. Lepourcelet *et al.* successfully identified several compounds that disrupt β-catenin/Tcf-4 complexes. These compounds potently antagonize cellular effects of β-catenin-dependent activities, including reporter gene activation, c-Myc or cyclin D_1 expression, and cell proliferation [162]. Emami *et al.* and Huang *et al.* also discovered small molecules that repress β-catenin/Tcf-4 signaling through specific binding to cyclic AMP response element-binding protein and tankyrases, respectively [163, 164]. Tankyrase is an enzyme that interacts with axin and stimulate its degradation through the ubiquitin-proteasome pathway [164]. Besides direct targeting of Wnt signaling, a recent study reported a therapeutic approach that exploits the molecular defect associated with aberrant activation of Wnt signaling. Zhang *et al.* demonstrated that c-FLIP, an endogenous caspase-8 antagonist, is repressed in APC-deficient intestinal cells via activation of c-Myc. c-FLIP repression renders these cells susceptible to death receptor-mediated apoptosis. Co-administration of TRAIL and all-trans-retinyl acetate, an agent that upregulates TRAIL-sensitive death receptors and downregulates decoys receptors, induces apoptosis in APC-deficient premalignant cells without affecting normal cells in intestinal polyps of APC^{Min} mice and human [165]. This approach represents a novel strategy for chemoprevention of colorectal cancer by specific targeting APC-deficient cells for apoptosis. Whether the above findings can be translated into clinical benefits in colorectal cancer patients, however, awaits further clinical evaluation.

p53 Signaling

The TP53 tumor-suppressor gene, the guardian of human genome, is one of the most frequently mutated genes in human cancers, including colorectal cancer. It encodes the 53 kDa phosphoprotein p53 that accumulates in cells in response to DNA damage, oncogene activation and other stresses. Activation of p53 induces DNA repair genes and arrests the cell cycle to allow enough time for fixation of DNA damage. Nevertheless, if DNA damage is beyond repair, p53 induces apoptosis to prevent accumulation of potentially oncogenic mutations. TP53 mutations are found in more than one-third of colorectal tumors and associated with lymphatic invasion and worse prognosis for patients treated with chemotherapy [166]. Higher frequencies of TP53 mutations are observed in distal colon and rectal tumors and lower frequencies in proximal tumors and those with the microsatellite instability [167].

In the early stage of colorectal carcinogenesis, the developing tumor cells experience 'oncogenic stress' caused by aberrant activation of oncogenes that deregulate DNA replication. Such ‘oncogenic stress’ results in stimulation of an ATR/ATM-regulated DNA damage response network in which p53 is a major component. Mutations undermining the function of p53 disrupt this checkpoint and allow cell proliferation, survival, and tumour progression [168]. p53 also inhibits tumorigenesis by inhibiting multipolar mitosis and structural chromosomal instability in cancer cells [169, 170]. Moreover, p53 downregulates VEGF expression and inhibits angiogenesis in human colon cancer [171]. Deficiency of p53 renders colon cancer cells less responsive to antiangiogenic therapy and 5-fluorouracil-based chemotherapy [172]. Activation of p53 induces either cell cycle arrest or apoptosis. The former is mediated by transcriptional activation of p21, whereas the latter is mediated by transcriptional activation of PUMA, PIG3 and TRAIL-sensitive DR5. It has been demonstrated that c-Myc level determines the outcome of p53 response. In this regard, high expression of c-Myc inhibits DNA-bound p53 from activating p21 transcription, shifting p53 response in favor of cell death [173]. p53 has also been shown to counteract the oncogenic Wnt/β-catenin signaling. In transgenic mice, substantial activation of p53 occurs in response to Wnt activation caused by deficiency of casein kinase Iα, a component of the β-catenin-destruction complex. Genetic ablation of casein kinase Iα triggers massive Wnt activation without causing intestinal tumorigenesis in a wild-type p53 genetic background. However, inactivation of both casein kinase Iα and either p53 or its target gene p21 leads to high-grade dysplasia. p21 suppresses the transcription of a set of genes that are associated with cell invasiveness caused by Wnt activation. These findings suggest that p53 inhibits the invasive phenotype caused by aberrant activation of Wnt/β-catenin signaling through p21 [174]. While mutations abrogate the tumor suppressing activity of p53, mutant p53 proteins may acquire new biochemical and biological properties which contribute to tumor malignancy. Such action is mediated through the inhibition of Id2 expression [175]. Depletion of mutant p53 reduces cell proliferation, in vitro and in vivo tumorigenicity, and resistance to anticancer drugs [176].

Restoring p53 activity is an attractive approach for cancer therapy and has led to the identification of a number of molecules, including CP-31398 and Nutlin, both of which impair the ubiquitin-dependent proteasomal degradation of p53. Compounds that activate a p53 response by increasing p73 (a protein related to the p53) expression or inducing a high p53-responsive transcriptional activity in the absence of p53 also exert potent antitumor effects in p53-deficient human colon tumor xenografts [177].

Notch Signaling

Notch signaling has emerged as a crucial signaling pathway in carcinogenesis through its ability to regulate multiple cellular processes, including proliferation, apoptosis, and differentiation. Notch signaling is instigated by the binding of ligands such as Jagged-1 and DLL-4 to the membrane Notch receptors, which are sequentially cleaved by matrix metalloproteinase and γ-secretase upon activation. Cleaved intracellular domains of Notch receptors then translocate to the nucleus and activate the transcription factor Rbpj to modulate gene expression.

Notch signaling is abnormally activated in colorectal cancer. Notch1 activity is increased in colorectal cancers as a result of β-catenin-mediated upregulation of the Notch ligand Jagged-1 [178]. Notch1 is also overexpressed in colon cancer tissues, and the level of Notch1 expression is correlated with the pathologic grade, progression, and metastasis of colon cancers. Notch1 overexpression also promotes tumorsphere formation and inhibits apoptosis in colon cancer cells [179]. Wnt signaling also upregulates the expression of Notch2. In $APC^{Min/+}$ mice, the mRNA levels of the Notch downstream target gene Hes-1 is significantly upregulated in intestinal tumors compared to the normal intestinal mucosa [180]. Notch signaling is rapidly induced and maintained throughout tumorigenesis [181]. Activation of Notch signaling may also be accomplished by the loss of the tumor suppressors FBW7 and Aes. The former mediates the destruction of Notch1 as well as other proto-oncogenes such as c-Myc, cyclin E, and c-Jun while the latter inhibits Notch signaling to repress liver metastasis by converting active Rbpj transcription complexes into repression complexes [182-184]. Tumor invasion and intravasation are also suppressed by Notch signaling inhibition in $Aes^{-/-}$ $APC^{\Delta716}$ double mutant mice [184]. Notch signaling may be activated by the tumor microenvironment. For instance, hypoxia has been shown to activate Notch1 to interfere with colon cancer stem cell differentiation. Notch inhibition by dibenzazepine allows colon cancer stem cell to differentiate into goblet cells, suggesting an important role of Notch signaling in regulating the differentiation of intestinal epithelial cells [185]. However, inactivation of Notch signaling do not always lead to inhibition of intestinal tumorigenesis. It has been demonstrated that genetic inactivation of Notch signaling fails to rescue unrestricted proliferation and impaired differentiation caused by β-catenin hyperactivation. This phenomenon can be explained by the finding that β-catenin can repress the expression of Atoh1, a key secretory determinant factor downstream of Notch [181]. In this respect, Atoh1 is required for the growth and differentiation effects of Notch inhibitors in normal and cancerous intestinal epithelial cells [186].

Notch1 and Notch2 may have different roles in colorectal carcinogenesis. Negative correlation between Notch1 and Notch2 has been observed in colorectal cancer. In addition, elevated levels of Notch1 and Notch2 are associated with opposite clinical outcomes, in which Notch1 positivity and Notch2 negativity predict poor overall survival [187]. Notch3 is upregulated in metastatic colorectal tumors. Altered expression of Notch3 together with Jagged-1 and DLL-4 is associated with aggressive phenotype of colorectal cancer xenograft. Stimulation with DLL-4 or enforced expression of an active form of Notch3 accelerates tumour growth whereas repression of Notch3 levels perturbs cell cycle and substantially reduces cell proliferation, clonogenic capacity, and tumour growth. These findings suggest that Notch3 may modulate the tumourigenic properties of colorectal cancer cells [188]. The Notch ligand DLL-1 may also mediate signaling independent of Notch receptor in colon cancer cells. In this regard, the intracellular domain of DLL-1 can enhance the expression of CTGF gene via binding to Smad2/3 and Tcf-4 to augment TGF-β and Wnt signaling, respectively [189]. Since Notch signaling is involved the colorectal carcinogenesis, blockade of Notch signaling may constitute a novel molecular therapy for cancer. In this regard, the γ-secretase inhibitor DAPT has been shown to inhibit the proliferation and invasion of a goblet cell-like colorectal cancer cell line. Inhibition of γ-secretase is associated with downregulation of Hes1 [190]. Anti-DLL4 antibodies also inhibit the growth of colon tumor xenograft and reduces tumor-initiating cell frequency in colorectal tumor cells irrespective of the mutation status of KRAS [191].

Eicosanoid Signaling

Epidemiological evidence suggests that long-term use of non-steroidal anti-inflammatory drugs (NSAIDs) reduces the risk for colorectal cancer by about 40-50% [192]. The use of NSAIDs also reduces the size and number of intestinal adenomas in patients with FAP. Cyclooxygenase-2 (COX-2), the major target of NSAIDs, is the rate-limiting enzyme for the conversion of arachidonic acid to prostaglandin (PG) H_2. The latter is subsequently converted to other PGs and thromboxane A_2 by specific PG- and thromboxane-synthases. COX-2 is frequently upregulated in gastrointestinal cancers. Both transcriptional and post-transcriptional regulations have been demonstrated to account for COX-2 overexpression in colorectal cancers [193]. At the transcriptional level, transcription factors, such as NF-κB, C/EBP, CREB, NFAT, AP-1, PPAR and HIFα, have been shown to bind to the *cis*-acting regulatory elements of the *COX-2* gene to upregulate COX-2 expression. At the post-transcriptional level, the RNA stabilization factor HuR, which is often overexpressed in colon cancer, binds to the AU-rich element (ARE) of the 3′ untranslated region of COX-2 transcript to prolong its half-life [194] while miR-101, a microRNA downregulated in a subset of colorectal cancer, targets COX-2 mRNA to repress its expression [195]. The upregulation of COX-2 catalytic activity by nitric oxide has also been demonstrated [196]. Among different COX-2-derived products, PGE_2 reaches high level in neoplastic tissues and affects numerous tumorigenic processes, including proliferation, apoptosis, metastasis, and angiogenesis. The biological effects of PGE_2 are mediated by EP receptors, whose genetic ablations have been shown to repress colorectal carcinogenesis in various animal studies [197-199]. The mechanism by which EP receptor activation is linked to oncogenic phenotypes is manifold.

PGE_2 has been shown to transactivate EGFR through induction of amphiregulin or Src-mediated, matrix metalloproteinase-dependent ectodomain shedding of TGFα [200, 201]. In addition, PGE_2 can stimulate β-catenin/TCF-dependent transcription via G-protein activation. To this end, PGE_2 promotes the interaction between the α subunit of G_s protein to RGS domain of axin, which subsequently promotes the release of GSK3β from the β-catenin degradation complex. Meanwhile, free βγ subunits stimulate the PI3K/Akt signaling cascade, resulting in the phosphorylation and inactivation of GSK3β. These molecular events synergize to promote accumulation of β-catenin, causing its nuclear localization and transcription of growth-promoting genes [202]. The growth promoting effect of PGE_2 in colorectal adenoma growth has also been found to be mediated through transactivation of the nuclear peroxisome proliferator-activated receptor delta [203].

Nuclear Factor-κB Signaling

Nuclear factor-κB (NF-κB) is a family of bipartite transcription factors. The most common form in human cells is the p50/RelA. Other configurations, such as p50/p50, p52/p52, p52/RelA, p50/c-Rel, c-Rel/c-Rel, p52/RelB, and p50/RelB, have also been identified. This transcription factor is normally sequestered in the cytoplasm by its inhibitor IκB. Upon activation by stimulus (e.g. UV radiation, inflammatory cytokines), IκB is phosphorylated by the IκB kinase (IKK) complex and subject to degradation through the ubiquitin-proteasome system. The unopposed transcription factor is then released and translocates into the nucleus where it alters expression of oncogenesis-related genes.

Increased expression of RelA is observed in the transition from colorectal adenoma with low-grade dysplasia to adenocarcinoma [204]. Another study also demonstrates that NF-κB binding is increased in 8 of 10 colorectal cancer tumors compared with adjacent normal mucosa [205]. Constitutively active NF-κB is also correlated with angiogenesis and tumor growth in human colorectal cancer [206, 207]. NF-κB immunoreactivity is associated with poorer response to cetuximab-irinotecan therapy and shortened time-to-progression and overall survival in irinotecan-refractory metastatic colorectal cancer [208]. NF-κB signaling is required for the metastasis of colon cancer cells induced by the cell-neural adhesion molecule L1CAM [209]. NF-κB activity is regulated by both oncogenic and tumor suppressing signaling pathways. For instance, the tumor suppressor p53 downregulates NF-κB activity [210, 211]. In this connection, p53-mediated inhibition of NF-κB synergizes with aspirin to induce apoptosis in human colon cancer cells [211]. Reciprocal binding of p50/RelA NF-κB heterodimers to the p53 promoter has also been reported [212, 213]. The oncogenic β-catenin and Ras also inhibit NF-κB activity [214, 215]. In contrast, the anti-apoptotic BAG-1 promotes colorectal tumour cell survival through increased NF-κB activity [216]. Whether the repression of NF-κB activity is required for the tumorigenic actions of Ras and β-catenin, or it represents an autoregulatory mechanism among different oncogenic signaling pathways, requires further elucidation. NF-κB promotes colorectal carcinogenesis through direct transcriptional activation of its oncogenic target genes [217-219]. For example, curcumin, an AP-1 inhibitor, has been shown to inhibit COX-2 expression and colon cancer cell proliferation through repression of NF-κB activity while lipopolysaccharides stimulate colon cancer cell growth via NF-κB-dependent induction of COX-2 [217, 218]. NF-κB also upregulates the expression of FasL, which is implicated in immune escape and metastasis [220]. Inhibition of NF-κB in cancer cells may convert inflammation-induced, tumor necrosis factor (TNF) α-dependent tumor growth to TRAIL-dependent tumor regression [221]. Suppression of NF-κB activity also induces the expression of PTEN, a tumor suppressor that inhibits the oncogenic PI3K activity [222].

The inhibition of NF-κB mediates, at least in part, the anticancer effects of some therapeutic agents. The anti-proliferative action of sulindac, an NSAID, is associated with the inhibition of NF-κB signaling [223]. Butyrate, a short-chain fatty acid, induces apoptosis or differentiation in colon cancer cells, and inhibits NF-κB activation probably through suppression of proteasome-dependent degradation of IκB [224]. Direct inhibition of NF-κB by small molecule inhibitors also induce adhesion-dependent apoptosis of colon cancer cells in an *in vivo* seeding model of metastasis [225]. NF-κB is related to the development of drug resistance. In this regard, NF-κB confers multi-drug resistance by enhancing the expression of MDR1 [226]. Inhibition of NF-κB activation overcomes chemoresistance to CPT-11 and enhances cytotoxicity of 5-fluorouracil in colon cancer cells [227-229]. The pro- or anti-apoptotic properties of NF-κB may be governed by its subcellular localization. Nucleolar sequestration of RelA, a process regulated by COMMD1-dependent ubiquitination, mediates NF-kappaB-driven apoptosis [230, 231]. Sulindac, sulindac sulfone, and indomethacin also induce apoptosis through inducing nucleolar translocation of the RelA [232]. Moreover, nucleolar translocation of RelA mediates the pro-apoptotic effect of p38-dependent inactivation of cyclin D_1/CDK4 in colorectal cancer cells [231]. PUMA, a pro-apoptotic member of the Bcl-2 family, is directly activated by NF-κB and contributes to TNF-α-induced apoptosis [232].

Transforming Growth Factor-β Signaling

Transforming growth factor (TGF)-β and the related bone morphogenetic proteins (BMPs) regulate proliferation and differentiation in a wide spectrum of cell types. TGF-β/BMP signaling is initiated by the binding of TGF-β or BMPs to type II receptor which, upon activation, recruits and phosphorylates the type I receptor. The type I receptor in turn phosphorylates the cytoplasmic signaling Smad proteins (Smad2/3 for TGF-β; Smad1/5/8 for BMP). Phosphorylated Smad2/3 or Smad1/5/8 then heterodimerize with Smad4, translocating into the nucleus to act as a transcription factor. In general, TGF-β/BMP signaling functions as a tumor-suppressing pathway in early stages of cancer by inhibiting cellular proliferation or by promoting differentiation and apoptosis. Nevertheless, in later stages of cancer, TGF-β/BMP signaling promotes invasion and metastasis, increases angiogenesis, and suppresses immune response.

Mutations of components in the TGF-β signaling pathway are associated with colorectal cancer. For example, genetic polymorphism of TGFB1 is associated with genetic susceptibility to colorectal neoplasia [234]. Inactivation of the type II TGF-β receptor has also been described in colon cancer cells with microsatellite instability [235]. Moreover, germline allele-specific expression of type I TGF-β receptor is associated with an increased risk of colorectal cancer [236]. Functional mutations of Smad2 and Smad4 in colon cancer have also been documented [237].

TGF-β signaling is implicated in both oncogenesis and tumor suppression. On one hand, TGF-β signaling represses a PRL-3-mediated cell survival pathway that promotes metastasis and induces apoptosis through repressing the PI3K/Akt/survivin pathway in colon cancer cells [238, 239]. TGF-β also increases the stability of p21 protein and inhibits CDK2 activity [240]. Loss of TGF-β signaling is characterized by malignant phenotypes. Smad3 deficiency promotes tumorigenesis in the distal colon of $Apc^{Min/+}$ mice whereas mutational inactivation of Smad4 contributes to unresponsiveness to the growth-inhibitory effect of TGF-β and enhances migration, aerobic glycolysis, and resistance to 5'-fluoruracil-mediated apoptosis in colon cancer cells [241-243]. On the other hand, TGF-β stimulation in the absence of functional Smad4 leads to increased level of VEGF [241]. TGF-β may also trigger CDK4-dependent phosphorylation of Smad2/3 at both linker and COOH-terminal regions to enhance cell proliferation by upregulating c-Myc. Moreover, JNK-dependent phosphorylation of Smad2 stimulates cell invasion by upregulating matrix metalloproteinase-9. These findings suggest that CDK4 and JNK may turn the TGF-β cascade into a oncogenic signaling pathway in human colorectal cancer [244].

Similar to TGF-β signaling, BMP signaling plays a dual role in the pathogenesis of colorectal cancer. In some contexts, BMP signaling is tumor-suppressive. For instance, germline mutations of BMPRIA have been associated with juvenile polyposis, a syndrome in which patients are at risk of developing colon cancers [245]. BMP2, which is mainly expressed in mature colonocytes, also inhibits normal and cancerous colonic epithelial cell proliferation and promotes apoptosis and differentiation [246]. Moreover, BMP signaling inhibits intestinal stem cell self-renewal through suppression of Wnt-β-catenin signaling [247]. In addition, BMP4 induces differentiation of colorectal cancer stem cells and increases their response to chemotherapy [248]. BMP signaling also contributes to c-Myc downregulation through induction of RUNX3 [249]. However, in other situations, BMP may promote tumorigenesis. Endogenous BMP signaling has been shown to promote tumor growth, cell survival and proliferation in colon cancer xenografts [250]. Overexpression of bone morphogenetic protein 4 also enhances the invasiveness of Smad4-deficient human colorectal cancer cells [251].

Hedgehog Signaling

Hedgehog signaling is implicated in embryonic development, maintenance of stem cell population, and tumorigenesis. In human, activation of Hedgehog signaling is driven by the binding of hedgehog ligands such as Sonic, Indian and Dessert Hedgehogs to the surface receptor Patched, followed by nuclear translocation of Gli. In the intestine, Hedgehog signaling and Wnt signaling are reciprocally regulated. In this regard, Indian Hedgehog antagonizes Wnt signaling to regulate colonic epithelial cell differentiation while inhibition of Wnt/β-catenin signaling increases the expression of Indian Hedgehog, causing a positive feedback loop. Concordantly, the expression of Indian Hedgehog is downregulated in polyps of individuals with FAP [252].

The role of Hedgehog signaling is paradoxical. On one hand, activation of Hedgehog signaling promotes cell survival. A panel of colorectal adenoma and carcinoma cell lines are found to express Sonic and Indian Hedgehogs. Treating these cells with cyclopamine, a known inhibitor of Hedgehog signaling, induces apoptosis. These data suggest that autocrine Hedgehog signaling may be important to the maintenance of cell survival [253]. Expression of Sonic Hedgehog and its receptor Patched is also upregulated in colon hyperplastic polyps, adenomas, and adenocarcinomas, in which Sonic Hedgehog is associated with the proliferation marker Ki-67. Exogenous Sonic Hedgehog also promotes cell proliferation in colonocytes, suggesting the possible involvement of Hedgehog signalling in colon tumour progression [254, 255]. The expression of HIF, an antagonist of Hedgehog signaling, is also downregulated in colorectal cancer tissues [256]. On the other hand, Gli1 has been shown to be downregulated in colorectal cancer. Enforced expression of Gli1 reduces β-catenin nuclear accumulation, represses c-Myc expression and inhibits cell proliferation [257]. Another study also demonstrates that aberrant activation of the Hedgehog signaling pathway is not common in colorectal cancer cell lines [258]. Such dilemma may be reconciled by the finding that upregulation of Smoothened, a negative regulator of Hedgehog signaling, in colorectal cancer may activate Wnt/β-catenin signaling in a Gli1-independent manner [259], posing the possibility that abnormal upregulation of Hedgehog ligands may stimulate colorectal carcinogenesis through promoting Smoothened-dependent activation of Wnt/β-catenin signaling without involvement of Gli.

Hippo Signaling

Hippo signaling is an evolutionarily conserved protein kinase cascade for organ size control. There are evolving evidence suggesting that dysregulation of this signaling pathway is linked to the development of multiple cancer types, including colorectal cancer. In mammals, the core kinase components Mst1/2, WW45, Lats1/2, and Mob1 phosphorylate the downstream transcriptional co-activators YAP and TAZ. 14-3-3 proteins then sequester the phosphorylated YAP and TAZ in the cytoplasm, prohibiting the transactivation of downstream target genes by Hippo signaling. In the colon, YAP1 expression is restricted to the local tissue stem cell compartment while there is strong and diffuse nuclear and cytoplasmic YAP expression in colonic adenocarcinoma [260]. YAP activation is required for the expansion of progenitor cells and the maintenance of their undifferentiated state [261]. In a transgenic mouse model, YAP overexpression causes intestinal dysplasia through activation of Notch signaling, in which the dysplastic phenotype can be blocked by the gamma-secretase inhibitors. Importantly, human colorectal cancer which highly expresses YAP shares molecular features with the YAP-induced dysplastic growth in the mouse [261]. These findings indicate that there is a potential link between Hippo signaling, local tissue stem cells, and colorectal cancer.

Protein Degradation Pathways

The ubiquitin-proteasome system and macroautophagy are two major intracellular pathways for protein degradation. The former degrades short-lived proteins through an intricate set of ubiquitinating enzymes (i.e. E1 ubiquitin-activating enzyme, E2 ubiquitin-conjugating enzyme, and E3 ubiquitin-ligase) and a multimeric proteolytic protein complex known as the proteasome while the latter degrades long-lived proteins and damaged organelles via a lysosome-dependent manner. During autophagy, intracellular materials subject to degradation will be sequestered by the double membrane-bound autophagosome which, in the later stage of autophagy, fuses with the lysosome for digestion by acidic hydrolases. There are accumulating evidence supporting that both protein degradation pathways are dysregulated in colorectal cancer. For instances, the expression of proteasomal subunits PSMD4 and PSMA5 as well as the proteasomal activity as a whole are increased in colorectal cancer [262]. Moreover, the expression of proteasomal subunit PSMA7 is overexpressed and associated with the liver metastasis of colorectal cancer [263]. The upregulation of proteasomal subunits may be partially explained by the finding that colorectal cancer exhibits high nuclear levels of nuclear factor E2-related factor 2, a transcription factor that is involved in the control of proteasomal subunit protein expression [262]. In contrast to the ubiquitin-proteasome system, the role of macroautophagy in colorectal carcinogenesis is more obscure. On one hand, increased expression of beclin-1 and LC3, two biochemical markers of macroautophagy, have been documented in colorectal cancer [264, 265]. The expression of Atg8, a macroautophagy-related gene, is also higher in tumor tissues, in which increased Atg8 expression is correlated with a low grade of differentiation and shortened overall survival [266]. On the other hand, mutations of several genes essential for the formation of autophagosomes, including *UVRAG*, *ATG2B*, *ATG5*, and *ATG9B*, are frequently detected in a subset of patients with high microsatellite instability [267]. These findings indicate that macroautophagy may harbor both oncogenic and tumor-suppressing functions, perhaps in a stage-dependent manner, in the development of colorectal cancer. Therapeutically, inhibitors of the ubiquitin-proteasome system and macroautophagy have both been shown to inhibit proliferation or induce apoptosis in colon cancer cells, in which inhibition of proteasome may upregulate macroautophgy [268, 269].

CONCLUSION

The pathogenesis of colorectal cancer is characterized by dysregulations of multiple signaling pathways. Activation of these signaling cascades leads to acquisition of malignant phenotypes and may predict disease outcomes in colorectal cancer. Nevertheless, their relative importance and temporal relationship in colorectal carcinogenesis are still not fully understood. Although some of these pathways have been successfully targeted by novel molecular therapeutics, their complex interactions may alter the drug responsiveness. Further understanding of signaling dysregulation and genetic testing are required to select the most responsive patients for pathway-directed cancer therapy. Targeting multiple signaling pathways in combination with conventional chemotherapeutics may also enhance the treatment efficacy.

REFERENCES

[1] Jemal A, Bray F, Center MM, Ferlay J, Ward E, Forman D (2011). Global cancer statistics. *CA Cancer J. Clin.* 61: 69-90.

[2] Arber N, Hibshoosh H, Moss SF, Sutter T, Zhang Y, Begg M *et al* (1996a). Increased expression of cyclin D1 is an early event in multistage colorectal carcinogenesis. *Gastroenterology* 110: 669-74.

[3] Bartkova J, Lukas J, Strauss M, Bartek J (1995). Cyclin D1 oncoprotein aberrantly accumulates in malignancies of diverse histogenesis. *Oncogene* 10: 775-8.

[4] Arber N, Sutter T, Miyake M, Kahn SM, Venkatraj VS, Sobrino A *et al* (1996b). Increased expression of cyclin D1 and the Rb tumor suppressor gene in c-K-ras transformed rat enterocytes. *Oncogene* 12: 1903-8.

[5] Zhang T, Nanney LB, Luongo C, Lamps L, Heppner KJ, DuBois RN *et al* (1997). Concurrent overexpression of cyclin D1 and cyclin-dependent kinase 4 (Cdk4) in intestinal adenomas from multiple intestinal neoplasia (Min) mice and human familial adenomatous polyposis patients. *Cancer Res.* 57: 169-75.

[6] Maeda K, Chung YS, Kang SM, Ogawa M, Onoda N, Nakata B *et al* (1997). Overexpression of cyclin D1 and p53 associated with disease recurrence in colorectal adenocarcinoma. *Int. J. Cancer* 74: 310-5.

[7] Toncheva D, Petrova D, Tzenova V, Dimova I, Yankova R, Yordanov V *et al* (2004). Tissue microarray analysis of cyclin D1 gene amplification and gain in colorectal carcinomas. *Tumour Biol.* 25: 157-60.

[8] Tetsu O, McCormick F (1999). Beta-catenin regulates expression of cyclin D1 in colon carcinoma cells. *Nature* 398: 422-6.

[9] Kitahara K, Yasui W, Kuniyasu H, Yokozaki H, Akama Y, Yunotani S *et al* (1995). Concurrent amplification of cyclin E and CDK2 genes in colorectal carcinomas. *Int. J. Cancer* 62: 25-8.

[10] Leach FS, Elledge SJ, Sherr CJ, Willson JK, Markowitz S, Kinzler KW *et al* (1993). Amplification of cyclin genes in colorectal carcinomas. *Cancer Res.* 53: 1986-9.

[11] Sutter T, Dansranjavin T, Lubinski J, Debniak T, Giannakudis J, Hoang-Vu C *et al* (2002). Overexpression of cyclin E protein is closely related to the mutator phenotype of colorectal carcinoma. *Int. J. Colorectal. Dis.* 17: 374-80.

[12] Wang A, Yoshimi N, Suzui M, Yamauchi A, Tarao M, Mori H (1996). Different expression patterns of cyclins A, D1 and E in human colorectal cancer. *J. Cancer Res. Clin. Oncol.* 122: 122-6.

[13] Yasui W, Kuniyasu H, Yokozaki H, Semba S, Shimamoto F, Tahara E (1996). Expression of cyclin E in colorectal adenomas and adenocarcinomas: correlation with expression of Ki-67 antigen and p53 protein. *Virchows Arch.* 429: 13-9.

[14] Li JQ, Miki H, Ohmori M, Wu F, Funamoto Y (2001). Expression of cyclin E and cyclin-dependent kinase 2 correlates with metastasis and prognosis in colorectal carcinoma. *Hum. Pathol.* 32: 945-53.

[15] Guan RJ, Fu Y, Holt PR, Pardee AB (1999). Association of K-ras mutations with p16 methylation in human colon cancer. *Gastroenterology* 116: 1063-71.

[16] Herman JG, Merlo A, Mao L, Lapidus RG, Issa JP, Davidson NE *et al* (1995). Inactivation of the CDKN2/p16/MTS1 gene is frequently associated with aberrant DNA methylation in all common human cancers. *Cancer Res.* 55: 4525-30.

[17] Ishiguro A, Takahata T, Saito M, Yoshiya G, Tamura Y, Sasaki M *et al* (2006). Influence of methylated p15 and p16 genes on clinicopathological features in colorectal cancer. *J. Gastroenterol. Hepatol.* 21: 1334-9.

[18] Sinicrope FA, Roddey G, Lemoine M, Ruan S, Stephens LC, Frazier ML *et al* (1998). Loss of p21WAF1/Cip1 protein expression accompanies progression of sporadic colorectal neoplasms but not hereditary nonpolyposis colorectal cancers. *Clin. Cancer Res.* 4: 1251-61.

[19] Viale G, Pellegrini C, Mazzarol G, Maisonneuve P, Silverman ML, Bosari S (1999). p21WAF1/CIP1 expression in colorectal carcinoma correlates with advanced disease stage and p53 mutations. *J. Pathol.* 187: 302-7.

[20] Bukholm IK, Nesland JM (2000). Protein expression of p53, p21 (WAF1/CIP1), bcl-2, Bax, cyclin D1 and pRb in human colon carcinomas. *Virchows Arch.* 436: 224-8.

[21] Yang W, Velcich A, Lozonschi I, Liang J, Nicholas C, Zhuang M *et al* (2005). Inactivation of p21WAF1/cip1 enhances intestinal tumor formation in Muc2-/- mice. *Am. J. Pathol.* 166: 1239-46.

[22] Loda M, Cukor B, Tam SW, Lavin P, Fiorentino M, Draetta GF *et al* (1997). Increased proteasome-dependent degradation of the cyclin-dependent kinase inhibitor p27 in aggressive colorectal carcinomas. *Nat. Med.* 3: 231-4.

[23] Fredersdorf S, Burns J, Milne AM, Packham G, Fallis L, Gillett CE *et al* (1997). High level expression of p27(kip1) and cyclin D1 in some human breast cancer cells: inverse correlation between the expression of p27(kip1) and degree of malignancy in human breast and colorectal cancers. *Proc. Natl. Acad. Sci. USA* 94: 6380-5.

[24] Thomas GV, Szigeti K, Murphy M, Draetta G, Pagano M, Loda M (1998). Down-regulation of p27 is associated with development of colorectal adenocarcinoma metastases. *Am. J. Pathol.* 153: 681-7.

[25] Yamamoto H, Soh JW, Shirin H, Xing WQ, Lim JT, Yao Y *et al* (1999a). Comparative effects of overexpression of p27Kip1 and p21Cip1/Waf1 on growth and differentiation in human colon carcinoma cells. *Oncogene* 18: 103-15.

[26] Yang W, Bancroft L, Nicholas C, Lozonschi I, Augenlicht LH (2003). Targeted inactivation of p27kip1 is sufficient for large and small intestinal tumorigenesis in the mouse, which can be augmented by a Western-style high-risk diet. *Cancer Res.* 63: 4990-6.

[27] West NJ, Courtney ED, Poullis AP, Leicester RJ (2009). Apoptosis in the colonic crypt, colorectal adenomata, and manipulation by chemoprevention. *Cancer Epidemiol. Biomarkers Prev.* 18: 1680-7.

[28] Sinicrope FA, Roddey G, McDonnell TJ, Shen Y, Cleary KR, Stephens LC (1996). Increased apoptosis accompanies neoplastic development in the human colorectum. *Clin. Cancer Res.* 2: 1999-2006.

[29] Sinicrope FA, Ruan SB, Cleary KR, Stephens LC, Lee JJ, Levin B (1995b). bcl-2 and p53 oncoprotein expression during colorectal tumorigenesis. *Cancer Res* 55: 237-41.

[30] Ofner D, Riehemann K, Maier H, Riedmann B, Nehoda H, Totsch M *et al* (1995). Immunohistochemically detectable bcl-2 expression in colorectal carcinoma: correlation with tumour stage and patient survival. *Br. J. Cancer* 72: 981-5.

[31] Sinicrope FA, Hart J, Michelassi F, Lee JJ (1995a). Prognostic value of bcl-2 oncoprotein expression in stage II colon carcinoma. *Clin. Cancer Res.* 1: 1103-10.

[32] Ilyas M, Hao XP, Wilkinson K, Tomlinson IP, Abbasi AM, Forbes A *et al* (1998). Loss of Bcl-2 expression correlates with tumour recurrence in colorectal cancer. *Gut* 43: 383-7.

[33] Popescu RA, Lohri A, de Kant E, Thiede C, Reuter J, Herrmann R *et al* (1998). bcl-2 expression is reciprocal to p53 and c-myc expression in metastatic human colorectal cancer. *Eur. J. Cancer* 34: 1268-73.

[34] Watson AJ, Merritt AJ, Jones LS, Askew JN, Anderson E, Becciolini A *et al* (1996). Evidence of reciprocity of bcl-2 and p53 expression in human colorectal adenomas and carcinomas. *Br. J. Cancer* 73: 889-95.

[35] Krajewska M, Moss SF, Krajewski S, Song K, Holt PR, Reed JC (1996). Elevated expression of Bcl-X and reduced Bak in primary colorectal adenocarcinomas. *Cancer Res.* 56: 2422-7.

[36] Wilson JW, Nostro MC, Balzi M, Faraoni P, Cianchi F, Becciolini A *et al* (2000). Bcl-w expression in colorectal adenocarcinoma. *Br. J. Cancer* 82: 178-85.

[37] De Angelis PM, Stokke T, Thorstensen L, Lothe RA, Clausen OP (1998). Apoptosis and expression of Bax, Bcl-x, and Bcl-2 apoptotic regulatory proteins in colorectal carcinomas, and association with p53 genotype/phenotype. *Mol. Pathol.* 51: 254-61.

[38] Rampino N, Yamamoto H, Ionov Y, Li Y, Sawai H, Reed JC *et al* (1997). Somatic frameshift mutations in the BAX gene in colon cancers of the microsatellite mutator phenotype. *Science* 275: 967-9.

[39] Yagi OK, Akiyama Y, Nomizu T, Iwama T, Endo M, Yuasa Y (1998). Proapoptotic gene BAX is frequently mutated in hereditary nonpolyposis colorectal cancers but not in adenomas. *Gastroenterology* 114: 268-74.

[40] Jansson A, Sun XF (2002). Bax expression decreases significantly from primary tumor to metastasis in colorectal cancer. *J. Clin. Oncol* 20: 811-6.

[41] Sturm I, Kohne CH, Wolff G, Petrowsky H, Hillebrand T, Hauptmann S *et al* (1999). Analysis of the p53/BAX pathway in colorectal cancer: low BAX is a negative prognostic factor in patients with resected liver metastases. *J. Clin. Oncol.* 17: 1364-74.

[42] Ionov Y, Yamamoto H, Krajewski S, Reed JC, Perucho M (2000). Mutational inactivation of the proapoptotic gene BAX confers selective advantage during tumor clonal evolution. *Proc. Natl. Acad. Sci. USA* 97: 10872-7.

[43] LeBlanc H, Lawrence D, Varfolomeev E, Totpal K, Morlan J, Schow P *et al* (2002). Tumor-cell resistance to death receptor--induced apoptosis through mutational inactivation of the proapoptotic Bcl-2 homolog Bax. *Nat. Med.* 8: 274-81.

[44] Sinicrope FA, Rego RL, Okumura K, Foster NR, O'Connell MJ, Sargent DJ *et al* (2008). Prognostic impact of bim, puma, and noxa expression in human colon carcinomas. *Clin. Cancer Res.* 14: 5810-8.

[45] Sinicrope FA, Rego RL, Foster NR, Thibodeau SN, Alberts SR, Windschitl HE *et al* (2008). Proapoptotic Bad and Bid protein expression predict survival in stages II and III colon cancers. *Clin. Cancer Res.* 14: 4128-33.

[46] Pitti RM, Marsters SA, Lawrence DA, Roy M, Kischkel FC, Dowd P *et al* (1998). Genomic amplification of a decoy receptor for Fas ligand in lung and colon cancer. *Nature* 396: 699-703.

[47] O'Connell J, O'Sullivan GC, Collins JK, Shanahan F (1996). The Fas counterattack: Fas-mediated T cell killing by colon cancer cells expressing Fas ligand. *J. Exp. Med.* 184: 1075-82.

[48] Ryan AE, Shanahan F, O'Connell J, Houston AM (2005). Addressing the "Fas counterattack" controversy: blocking fas ligand expression suppresses tumor immune evasion of colon cancer in vivo. *Cancer Res* 65: 9817-23.

[49] Yoong KF, Afford SC, Randhawa S, Hubscher SG, Adams DH (1999). Fas/Fas ligand interaction in human colorectal hepatic metastases: A mechanism of hepatocyte destruction to facilitate local tumor invasion. *Am. J. Pathol.* 154: 693-703.

[50] Bennett MW, O'Connell J, Houston A, Kelly J, O'Sullivan GC, Collins JK *et al* (2001). Fas ligand upregulation is an early event in colonic carcinogenesis. *J. Clin. Pathol.* 54: 598-604.

[51] Houston A, Waldron-Lynch FD, Bennett MW, Roche D, O'Sullivan GC, Shanahan F *et al* (2003). Fas ligand expressed in colon cancer is not associated with increased apoptosis of tumor cells in vivo. *Int. J. Cancer* 107: 209-14.

[52] Kim HS, Lee JW, Soung YH, Park WS, Kim SY, Lee JH *et al* (2003). Inactivating mutations of caspase-8 gene in colorectal carcinomas. *Gastroenterology* 125: 708-15.

[53] Kawasaki H, Altieri DC, Lu CD, Toyoda M, Tenjo T, Tanigawa N (1998). Inhibition of apoptosis by survivin predicts shorter survival rates in colorectal cancer. *Cancer Res.* 58: 5071-4.

[54] Sarela AI, Macadam RC, Farmery SM, Markham AF, Guillou PJ (2000). Expression of the antiapoptosis gene, survivin, predicts death from recurrent colorectal carcinoma. *Gut* 46: 645-50.

[55] Kawasaki H, Toyoda M, Shinohara H, Okuda J, Watanabe I, Yamamoto T *et al* (2001). Expression of survivin correlates with apoptosis, proliferation, and angiogenesis during human colorectal tumorigenesis. *Cancer* 91: 2026-32.

[56] Endo T, Abe S, Seidlar HB, Nagaoka S, Takemura T, Utsuyama M *et al* (2004). Expression of IAP family proteins in colon cancers from patients with different age groups. *Cancer Immunol. Immunother* 53: 770-6.

[57] Yu LF, Wang J, Zou B, Lin MC, Wu YL, Xia HH *et al* (2007). XAF1 mediates apoptosis through an extracellular signal-regulated kinase pathway in colon cancer. *Cancer* 109: 1996-2003.

[58] Zou B, Chim CS, Zeng H, Leung SY, Yang Y, Tu SP *et al* (2006). Correlation between the single-site CpG methylation and expression silencing of the XAF1 gene in human gastric and colon cancers. *Gastroenterology* 131: 1835-43.

[59] Cohen G, Mustafi R, Chumsangsri A, Little N, Nathanson J, Cerda S *et al* (2006). Epidermal growth factor receptor signaling is up-regulated in human colonic aberrant crypt foci. *Cancer Res.* 66: 5656-64.

[60] Goldstein NS, Armin M (2001). Epidermal growth factor receptor immunohistochemical reactivity in patients with American Joint Committee on Cancer Stage IV colon adenocarcinoma: implications for a standardized scoring system. *Cancer* 92: 1331-46.

[61] Hemming AW, Davis NL, Kluftinger A, Robinson B, Quenville NF, Liseman B *et al* (1992). Prognostic markers of colorectal cancer: an evaluation of DNA content, epidermal growth factor receptor, and Ki-67. *J. Surg. Oncol.* 51: 147-52.

[62] Khorana AA, Ryan CK, Cox C, Eberly S, Sahasrabudhe DM (2003). Vascular endothelial growth factor, CD68, and epidermal growth factor receptor expression and survival in patients with Stage II and Stage III colon carcinoma: a role for the host response in prognosis. *Cancer* 97: 960-8.

[63] Mayer A, Takimoto M, Fritz E, Schellander G, Kofler K, Ludwig H (1993). The prognostic significance of proliferating cell nuclear antigen, epidermal growth factor receptor, and mdr gene expression in colorectal cancer. *Cancer* 71: 2454-60.

[64] Radinsky R, Risin S, Fan D, Dong Z, Bielenberg D, Bucana CD *et al* (1995). Level and function of epidermal growth factor receptor predict the metastatic potential of human colon carcinoma cells. *Clin. Cancer Res.* 1: 19-31.

[65] D'Emilia J, Bulovas K, D'Ercole K, Wolf B, Steele G, Jr., Summerhayes IC (1989). Expression of the c-erbB-2 gene product (p185) at different stages of neoplastic progression in the colon. *Oncogene* 4: 1233-9.

[66] Kapitanovic S, Radosevic S, Slade N, Kapitanovic M, Andelinovic S, Ferencic Z *et al* (2000). Expression of erbB-3 protein in colorectal adenocarcinoma: correlation with poor survival. *J. Cancer Res. Clin. Oncol.* 126: 205-11.

[67] Cho CD, Fisher GA, Halsey J, Sikic BI (2006). Phase I study of gefitinib, oxaliplatin, 5-fluorouracil, and leucovorin (IFOX) in patients with advanced solid malignancies. *Invest New Drugs* 24: 117-23.

[68] Kuo T, Cho CD, Halsey J, Wakelee HA, Advani RH, Ford JM *et al* (2005). Phase II study of gefitinib, fluorouracil, leucovorin, and oxaliplatin therapy in previously treated patients with metastatic colorectal cancer. *J. Clin. Oncol* 23: 5613-9.

[69] Fan Z, Lu Y, Wu X, Mendelsohn J (1994). Antibody-induced epidermal growth factor receptor dimerization mediates inhibition of autocrine proliferation of A431 squamous carcinoma cells. *J. Biol. Chem.* 269: 27595-602.

[70] Prewett M, Rockwell P, Rockwell RF, Giorgio NA, Mendelsohn J, Scher HI *et al* (1996). The biologic effects of C225, a chimeric monoclonal antibody to the EGFR, on human prostate carcinoma. *J. Immunother Emphasis Tumor Immunol.* 19: 419-27.

[71] Moroni M, Veronese S, Benvenuti S, Marrapese G, Sartore-Bianchi A, Di Nicolantonio F *et al* (2005). Gene copy number for epidermal growth factor receptor (EGFR) and clinical response to antiEGFR treatment in colorectal cancer: a cohort study. *Lancet Oncol.* 6: 279-86.

[72] Bos JL, Fearon ER, Hamilton SR, Verlaan-de Vries M, van Boom JH, van der Eb AJ *et al* (1987). Prevalence of ras gene mutations in human colorectal cancers. *Nature* 327: 293-7.

[73] Forrester K, Almoguera C, Han K, Grizzle WE, Perucho M (1987). Detection of high incidence of K-ras oncogenes during human colon tumorigenesis. *Nature* 327: 298-303.

[74] Gallick GE, Kurzrock R, Kloetzer WS, Arlinghaus RB, Gutterman JU (1985). Expression of p21ras in fresh primary and metastatic human colorectal tumors. *Proc. Natl. Acad. Sci. USA* 82: 1795-9.

[75] Sidransky D, Tokino T, Hamilton SR, Kinzler KW, Levin B, Frost P *et al* (1992). Identification of ras oncogene mutations in the stool of patients with curable colorectal tumors. *Science* 256: 102-5.

[76] Sevignani C, Wlodarski P, Kirillova J, Mercer WE, Danielson KG, Iozzo RV *et al* (1998). Tumorigenic conversion of p53-deficient colon epithelial cells by an activated Ki-ras gene. *J. Clin. Invest* 101: 1572-80.

[77] Ries S, Biederer C, Woods D, Shifman O, Shirasawa S, Sasazuki T *et al* (2000). Opposing effects of Ras on p53: transcriptional activation of mdm2 and induction of p19ARF. *Cell* 103: 321-30.

[78] Shirasawa S, Furuse M, Yokoyama N, Sasazuki T (1993). Altered growth of human colon cancer cell lines disrupted at activated Ki-ras. *Science* 260: 85-8.

[79] Kretzschmar M, Doody J, Timokhina I, Massague J (1999). A mechanism of repression of TGFbeta/ Smad signaling by oncogenic Ras. *Genes Dev* 13: 804-16.

[80] Okada F, Rak JW, Croix BS, Lieubeau B, Kaya M, Roncari L *et al* (1998). Impact of oncogenes in tumor angiogenesis: mutant K-ras up-regulation of vascular endothelial growth factor/vascular permeability factor is necessary, but not sufficient for tumorigenicity of human colorectal carcinoma cells. *Proc. Natl. Acad. Sci. USA* 95: 3609-14.

[81] Yun J, Rago C, Cheong I, Pagliarini R, Angenendt P, Rajagopalan H *et al* (2009). Glucose deprivation contributes to the development of KRAS pathway mutations in tumor cells. *Science* 325: 1555-9.

[82] Jin H, Sperka T, Herrlich P, Morrison H (2006). Tumorigenic transformation by CPI-17 through inhibition of a merlin phosphatase. *Nature* 442: 576-9.

[83] Kolfschoten IG, van Leeuwen B, Berns K, Mullenders J, Beijersbergen RL, Bernards R *et al* (2005). A genetic screen identifies PITX1 as a suppressor of RAS activity and tumorigenicity. *Cell* 121: 849-58.

[84] Rizzo S, Bronte G, Fanale D, Corsini L, Silvestris N, Santini D *et al* (2010). Prognostic vs predictive molecular biomarkers in colorectal cancer: is KRAS and BRAF wild type status required for anti-EGFR therapy? *Cancer Treat Rev.* 36 Suppl 3: S56-61.

[85] Abubaker J, Bavi P, Al-Harbi S, Ibrahim M, Siraj AK, Al-Sanea N *et al* (2008). Clinicopathological analysis of colorectal cancers with PIK3CA mutations in Middle Eastern population. *Oncogene* 27: 3539-45.

[86] Samuels Y, Wang Z, Bardelli A, Silliman N, Ptak J, Szabo S *et al* (2004). High frequency of mutations of the PIK3CA gene in human cancers. *Science* 304: 554.

[87] Baba Y, Nosho K, Shima K, Hayashi M, Meyerhardt JA, Chan AT *et al* (2011). Phosphorylated AKT expression is associated with PIK3CA mutation, low stage, and favorable outcome in 717 colorectal cancers. *Cancer* 117: 1399-408.

[88] Guanti G, Resta N, Simone C, Cariola F, Demma I, Fiorente P *et al* (2000). Involvement of PTEN mutations in the genetic pathways of colorectal cancerogenesis. *Hum. Mol. Genet* 9: 283-7.

[89] Lynch ED, Ostermeyer EA, Lee MK, Arena JF, Ji H, Dann J *et al* (1997). Inherited mutations in PTEN that are associated with breast cancer, cowden disease, and juvenile polyposis. *Am. J. Hum. Genet* 61: 1254-60.

[90] Zhang YJ, Dai Q, Sun DF, Xiong H, Tian XQ, Gao FH *et al* (2009). mTOR signaling pathway is a target for the treatment of colorectal cancer. *Ann. Surg. Oncol.* 16: 2617-28.

[91] Shao J, Evers BM, Sheng H (2004). Roles of phosphatidylinositol 3'-kinase and mammalian target of rapamycin/p70 ribosomal protein S6 kinase in K-Ras-mediated transformation of intestinal epithelial cells. *Cancer Res.* 64: 229-35.

[92] Wang Q, Wang X, Hernandez A, Kim S, Evers BM (2001). Inhibition of the phosphatidylinositol 3-kinase pathway contributes to HT29 and Caco-2 intestinal cell differentiation. *Gastroenterology* 120: 1381-92.

[93] Itoh N, Semba S, Ito M, Takeda H, Kawata S, Yamakawa M (2002). Phosphorylation of Akt/PKB is required for suppression of cancer cell apoptosis and tumor progression in human colorectal carcinoma. *Cancer* 94: 3127-34.

[94] Aoki K, Tamai Y, Horiike S, Oshima M, Taketo MM (2003). Colonic polyposis caused by mTOR-mediated chromosomal instability in Apc+/Delta716 Cdx2+/- compound mutant mice. *Nat Genet* 35: 323-30.

[95] Fujishita T, Aoki K, Lane HA, Aoki M, Taketo MM (2008). Inhibition of the mTORC1 pathway suppresses intestinal polyp formation and reduces mortality in ApcDelta716 mice. *Proc. Natl. Acad. Sci. USA* 105: 13544-9.

[96] Gulhati P, Bowen KA, Liu J, Stevens PD, Rychahou PG, Chen M *et al* (2011). mTORC1 and mTORC2 Regulate EMT, Motility, and Metastasis of Colorectal Cancer via RhoA and Rac1 Signaling Pathways. *Cancer Res.* 71: 3246-56.

[97] Halilovic E, She QB, Ye Q, Pagliarini R, Sellers WR, Solit DB *et al* (2010). PIK3CA mutation uncouples tumor growth and cyclin D1 regulation from MEK/ERK and mutant KRAS signaling. *Cancer Res.* 70: 6804-14.

[98] Wee S, Jagani Z, Xiang KX, Loo A, Dorsch M, Yao YM *et al* (2009). PI3K pathway activation mediates resistance to MEK inhibitors in KRAS mutant cancers. *Cancer Res.* 69: 4286-93.

[99] Zindy P, Berge Y, Allal BC, Filleron T, Pierredon S, Cammas A *et al* (2011). Formation of the eIF4F translation initiation complex determines sensitivity to anti-cancer drugs targeting the EGF and HER2 receptors. *Cancer Res.* doi: 10.1158/0008-5472.CAN-11-0420.

[100] Batlle E, Henderson JT, Beghtel H, van den Born MM, Sancho E, Huls G *et al* (2002). Beta-catenin and TCF mediate cell positioning in the intestinal epithelium by controlling the expression of EphB/ephrinB. *Cell* 111: 251-63.

[101] Munemitsu S, Albert I, Souza B, Rubinfeld B, Polakis P (1995). Regulation of intracellular beta-catenin levels by the adenomatous polyposis coli (APC) tumor-suppressor protein. *Proc. Natl. Acad. Sci. USA* 92: 3046-50.

[102] Rosin-Arbesfeld R, Townsley F, Bienz M (2000). The APC tumour suppressor has a nuclear export function. *Nature* 406: 1009-12.

[103] Anderson CB, Neufeld KL, White RL (2002). Subcellular distribution of Wnt pathway proteins in normal and neoplastic colon. *Proc. Natl. Acad. Sci. USA* 99: 8683-8.

[104] Deng G, Song GA, Pong E, Sleisenger M, Kim YS (2004). Promoter methylation inhibits APC gene expression by causing changes in chromatin conformation and interfering with the binding of transcription factor CCAAT-binding factor. *Cancer Res.* 64: 2692-8.

[105] Nagel R, le Sage C, Diosdado B, van der Waal M, Oude Vrielink JA, Bolijn A *et al* (2008). Regulation of the adenomatous polyposis coli gene by the miR-135 family in colorectal cancer. *Cancer Res.* 68: 5795-802.

[106] Phelps RA, Chidester S, Dehghanizadeh S, Phelps J, Sandoval IT, Rai K *et al* (2009). A two-step model for colon adenoma initiation and progression caused by APC loss. *Cell* 137: 623-34.

[107] Jiang X, Tan J, Li J, Kivimae S, Yang X, Zhuang L *et al* (2008). DACT3 is an epigenetic regulator of Wnt/beta-catenin signaling in colorectal cancer and is a therapeutic target of histone modifications. *Cancer Cell* 13: 529-41.

[108] Koinuma K, Yamashita Y, Liu W, Hatanaka H, Kurashina K, Wada T *et al* (2006). Epigenetic silencing of AXIN2 in colorectal carcinoma with microsatellite instability. *Oncogene* 25: 139-46.

[109] Gonzalez-Sancho JM, Aguilera O, Garcia JM, Pendas-Franco N, Pena C, Cal S *et al* (2005). The Wnt antagonist DICKKOPF-1 gene is a downstream target of beta-catenin/TCF and is downregulated in human colon cancer. *Oncogene* 24: 1098-103.

[110] Taniguchi H, Yamamoto H, Hirata T, Miyamoto N, Oki M, Nosho K *et al* (2005). Frequent epigenetic inactivation of Wnt inhibitory factor-1 in human gastrointestinal cancers. *Oncogene* 24: 7946-52.

[111] Zhang W, Glockner SC, Guo M, Machida EO, Wang DH, Easwaran H *et al* (2008). Epigenetic inactivation of the canonical Wnt antagonist SRY-box containing gene 17 in colorectal cancer. *Cancer Res.* 68: 2764-72.

[112] Inomata M, Ochiai A, Akimoto S, Kitano S, Hirohashi S (1996). Alteration of beta-catenin expression in colonic epithelial cells of familial adenomatous polyposis patients. *Cancer Res.* 56: 2213-7.

[113] Ilyas M, Tomlinson IP, Rowan A, Pignatelli M, Bodmer WF (1997). Beta-catenin mutations in cell lines established from human colorectal cancers. *Proc. Natl. Acad. Sci. USA* 94: 10330-4.

[114] Morin PJ, Sparks AB, Korinek V, Barker N, Clevers H, Vogelstein B *et al* (1997). Activation of beta-catenin-Tcf signaling in colon cancer by mutations in beta-catenin or APC. *Science* 275: 1787-90.

[115] Takahashi M, Fukuda K, Sugimura T, Wakabayashi K (1998). Beta-catenin is frequently mutated and demonstrates altered cellular location in azoxymethane-induced rat colon tumors. *Cancer Res.* 58: 42-6.

[116] Romagnolo B, Berrebi D, Saadi-Keddoucci S, Porteu A, Pichard AL, Peuchmaur M *et al* (1999). Intestinal dysplasia and adenoma in transgenic mice after overexpression of an activated beta-catenin. *Cancer Res* 59: 3875-9.

[117] Korinek V, Barker N, Morin PJ, van Wichen D, de Weger R, Kinzler KW *et al* (1997). Constitutive transcriptional activation by a beta-catenin-Tcf complex in APC-/- colon carcinoma. *Science* 275: 1784-7.

[118] Vider BZ, Zimber A, Chastre E, Prevot S, Gespach C, Estlein D *et al* (1996). Evidence for the involvement of the Wnt 2 gene in human colorectal cancer. *Oncogene* 12: 153-8.

[119] Bafico A, Liu G, Goldin L, Harris V, Aaronson SA (2004). An autocrine mechanism for constitutive Wnt pathway activation in human cancer cells. *Cancer Cell* 6: 497-506.

[120] Adachi S, Jigami T, Yasui T, Nakano T, Ohwada S, Omori Y *et al* (2004). Role of a BCL9-related beta-catenin-binding protein, B9L, in tumorigenesis induced by aberrant activation of Wnt signaling. *Cancer Res.* 64: 8496-501.

[121] Feng Y, Bommer GT, Zhai Y, Akyol A, Hinoi T, Winer I *et al* (2007). Drosophila split ends homologue SHARP functions as a positive regulator of Wnt/beta-catenin/T-cell factor signaling in neoplastic transformation. *Cancer Res.* 67: 482-91.

[122] Morris EJ, Ji JY, Yang F, Di Stefano L, Herr A, Moon NS *et al* (2008). E2F1 represses beta-catenin transcription and is antagonized by both pRB and CDK8. *Nature* 455: 552-6.

[123] He TC, Sparks AB, Rago C, Hermeking H, Zawel L, da Costa LT *et al* (1998). Identification of c-MYC as a target of the APC pathway. *Science* 281: 1509-12.

[124] van de Wetering M, Sancho E, Verweij C, de Lau W, Oving I, Hurlstone A *et al* (2002). The beta-catenin/TCF-4 complex imposes a crypt progenitor phenotype on colorectal cancer cells. *Cell* 111: 241-50.

[125] Sansom OJ, Meniel VS, Muncan V, Phesse TJ, Wilkins JA, Reed KR *et al* (2007). Myc deletion rescues Apc deficiency in the small intestine. *Nature* 446: 676-9.

[126] Shtutman M, Zhurinsky J, Simcha I, Albanese C, D'Amico M, Pestell R *et al* (1999). The cyclin D1 gene is a target of the beta-catenin/LEF-1 pathway. *Proc. Natl. Acad. Sci. USA* 96: 5522-7.

[127] Mann B, Gelos M, Siedow A, Hanski ML, Gratchev A, Ilyas M *et al* (1999). Target genes of beta-catenin-T cell-factor/lymphoid-enhancer-factor signaling in human colorectal carcinomas. *Proc. Natl. Acad. Sci. USA* 96: 1603-8.

[128] Aoki K, Aoki M, Sugai M, Harada N, Miyoshi H, Tsukamoto T *et al* (2007). Chromosomal instability by beta-catenin/TCF transcription in APC or beta-catenin mutant cells. *Oncogene* 26: 3511-20.

[129] Hadjihannas MV, Bruckner M, Jerchow B, Birchmeier W, Dietmaier W, Behrens J (2006). Aberrant Wnt/beta-catenin signaling can induce chromosomal instability in colon cancer. *Proc. Natl. Acad. Sci. USA* 103: 10747-52.

[130] Mariadason JM, Bordonaro M, Aslam F, Shi L, Kuraguchi M, Velcich A *et al* (2001). Down-regulation of beta-catenin TCF signaling is linked to colonic epithelial cell differentiation. *Cancer Res.* 61: 3465-71.

[131] Naishiro Y, Yamada T, Takaoka AS, Hayashi R, Hasegawa F, Imai K *et al* (2001). Restoration of epithelial cell polarity in a colorectal cancer cell line by suppression of beta-catenin/T-cell factor 4-mediated gene transactivation. *Cancer Res.* 61: 2751-8.

[132] Zhu P, Martin E, Mengwasser J, Schlag P, Janssen KP, Gottlicher M (2004). Induction of HDAC2 expression upon loss of APC in colorectal tumorigenesis. *Cancer Cell* 5: 455-63.

[133] Andreu P, Colnot S, Godard C, Laurent-Puig P, Lamarque D, Kahn A *et al* (2006). Identification of the IFITM family as a new molecular marker in human colorectal tumors. *Cancer Res.* 66: 1949-55.

[134] Araki Y, Okamura S, Hussain SP, Nagashima M, He P, Shiseki M *et al* (2003). Regulation of cyclooxygenase-2 expression by the Wnt and ras pathways. *Cancer Res.* 63: 728-34.

[135] Beiter K, Hiendlmeyer E, Brabletz T, Hlubek F, Haynl A, Knoll C *et al* (2005). beta-Catenin regulates the expression of tenascin-C in human colorectal tumors. *Oncogene* 24: 8200-4.

[136] Boon EM, van der Neut R, van de Wetering M, Clevers H, Pals ST (2002). Wnt signaling regulates expression of the receptor tyrosine kinase met in colorectal cancer. *Cancer Res.* 62: 5126-8.

[137] Chen T, Yang I, Irby R, Shain KH, Wang HG, Quackenbush J *et al* (2003). Regulation of caspase expression and apoptosis by adenomatous polyposis coli. *Cancer Res.* 63: 4368-74.

[138] Crawford HC, Fingleton BM, Rudolph-Owen LA, Goss KJ, Rubinfeld B, Polakis P *et al* (1999). The metalloproteinase matrilysin is a target of beta-catenin transactivation in intestinal tumors. *Oncogene* 18: 2883-91.

[139] Fujita M, Furukawa Y, Nagasawa Y, Ogawa M, Nakamura Y (2000). Down-regulation of monocyte chemotactic protein-3 by activated beta-catenin. *Cancer Res.* 60: 6683-7.

[140] Fujita M, Furukawa Y, Tsunoda T, Tanaka T, Ogawa M, Nakamura Y (2001). Up-regulation of the ectodermal-neural cortex 1 (ENC1) gene, a downstream target of the beta-catenin/T-cell factor complex, in colorectal carcinomas. *Cancer Res.* 61: 7722-6.

[141] He TC, Chan TA, Vogelstein B, Kinzler KW (1999). PPARdelta is an APC-regulated target of nonsteroidal anti-inflammatory drugs. *Cell* 99: 335-45.

[142] Hlubek F, Jung A, Kotzor N, Kirchner T, Brabletz T (2001). Expression of the invasion factor laminin gamma2 in colorectal carcinomas is regulated by beta-catenin. *Cancer Res.* 61: 8089-93.

[143] Ishiguro H, Shimokawa T, Tsunoda T, Tanaka T, Fujii Y, Nakamura Y *et al* (2002). Isolation of HELAD1, a novel human helicase gene up-regulated in colorectal carcinomas. *Oncogene* 21: 6387-94.

[144] Jansson EA, Are A, Greicius G, Kuo IC, Kelly D, Arulampalam V *et al* (2005). The Wnt/beta-catenin signaling pathway targets PPARgamma activity in colon cancer cells. *Proc. Natl. Acad. Sci. USA* 102: 1460-5.

[145] Jubb AM, Chalasani S, Frantz GD, Smits R, Grabsch HI, Kavi V *et al* (2006). Achaete-scute like 2 (ascl2) is a target of Wnt signalling and is upregulated in intestinal neoplasia. *Oncogene* 25: 3445-57.

[146] Kawasoe T, Furukawa Y, Daigo Y, Nishiwaki T, Ishiguro H, Fujita M *et al* (2000). Isolation and characterization of a novel human gene, DRCTNNB1A, the expression of which is down-regulated by beta-catenin. *Cancer Res.* 60: 3354-8.

[147] Kim JS, Crooks H, Dracheva T, Nishanian TG, Singh B, Jen J *et al* (2002). Oncogenic beta-catenin is required for bone morphogenetic protein 4 expression in human cancer cells. *Cancer Res.* 62: 2744-8.

[148] Kim TH, Xiong H, Zhang Z, Ren B (2005b). beta-Catenin activates the growth factor endothelin-1 in colon cancer cells. *Oncogene* 24: 597-604.

[149] Lin YM, Ono K, Satoh S, Ishiguro H, Fujita M, Miwa N *et al* (2001). Identification of AF17 as a downstream gene of the beta-catenin/T-cell factor pathway and its involvement in colorectal carcinogenesis. *Cancer Res.* 61: 6345-9.

[150] Noubissi FK, Elcheva I, Bhatia N, Shakoori A, Ougolkov A, Liu J *et al* (2006). CRD-BP mediates stabilization of betaTrCP1 and c-myc mRNA in response to beta-catenin signalling. *Nature* 441: 898-901.

[151] Pennica D, Swanson TA, Welsh JW, Roy MA, Lawrence DA, Lee J *et al* (1998). WISP genes are members of the connective tissue growth factor family that are up-regulated in wnt-1-transformed cells and aberrantly expressed in human colon tumors. *Proc. Natl. Acad. Sci. USA* 95: 14717-22.

[152] Ray R, Cabal-Manzano R, Moser AR, Waldman T, Zipper LM, Aigner A *et al* (2003). Up-regulation of fibroblast growth factor-binding protein, by beta-catenin during colon carcinogenesis. *Cancer Res.* 63: 8085-9.

[153] Shimokawa T, Furukawa Y, Sakai M, Li M, Miwa N, Lin YM *et al* (2003). Involvement of the FGF18 gene in colorectal carcinogenesis, as a novel downstream target of the beta-catenin/T-cell factor complex. *Cancer Res.* 63: 6116-20.

[154] Spiegelman VS, Tang W, Katoh M, Slaga TJ, Fuchs SY (2002). Inhibition of HOS expression and activities by Wnt pathway. *Oncogene* 21: 856-60.

[155] Takahashi M, Fujita M, Furukawa Y, Hamamoto R, Shimokawa T, Miwa N *et al* (2002a). Isolation of a novel human gene, APCDD1, as a direct target of the beta-Catenin/T-cell factor 4 complex with probable involvement in colorectal carcinogenesis. *Cancer Res.* 62: 5651-6.

[156] Takahashi M, Tsunoda T, Seiki M, Nakamura Y, Furukawa Y (2002b). Identification of membrane-type matrix metalloproteinase-1 as a target of the beta-catenin/Tcf4 complex in human colorectal cancers. *Oncogene* 21: 5861-7.

[157] Vignjevic D, Schoumacher M, Gavert N, Janssen KP, Jih G, Lae M *et al* (2007). Fascin, a novel target of beta-catenin-TCF signaling, is expressed at the invasive front of human colon cancer. *Cancer Res.* 67: 6844-53.

[158] Yamada T, Takaoka AS, Naishiro Y, Hayashi R, Maruyama K, Maesawa C *et al* (2000). Transactivation of the multidrug resistance 1 gene by T-cell factor 4/beta-catenin complex in early colorectal carcinogenesis. *Cancer Res.* 60: 4761-6.

[159] Yan D, Wiesmann M, Rohan M, Chan V, Jefferson AB, Guo L *et al* (2001). Elevated expression of axin2 and hnkd mRNA provides evidence that Wnt/beta -catenin signaling is activated in human colon tumors. *Proc. Natl. Acad. Sci. USA* 98: 14973-8.

[160] Zhang T, Otevrel T, Gao Z, Ehrlich SM, Fields JZ, Boman BM (2001a). Evidence that APC regulates survivin expression: a possible mechanism contributing to the stem cell origin of colon cancer. *Cancer Res.* 61: 8664-7.

[161] Zhang X, Gaspard JP, Chung DC (2001b). Regulation of vascular endothelial growth factor by the Wnt and K-ras pathways in colonic neoplasia. *Cancer Res.* 61: 6050-4.

[162] Lepourcelet M, Chen YN, France DS, Wang H, Crews P, Petersen F *et al* (2004). Small-molecule antagonists of the oncogenic Tcf/beta-catenin protein complex. *Cancer Cell* 5: 91-102.

[163] Emami KH, Nguyen C, Ma H, Kim DH, Jeong KW, Eguchi M *et al* (2004). A small molecule inhibitor of beta-catenin/CREB-binding protein transcription [corrected]. *Proc. Natl. Acad. Sci. USA* 101: 12682-7.

[164] Huang SM, Mishina YM, Liu S, Cheung A, Stegmeier F, Michaud GA *et al* (2009). Tankyrase inhibition stabilizes axin and antagonizes Wnt signalling. *Nature* 461: 614-20.

[165] Zhang L, Ren X, Alt E, Bai X, Huang S, Xu Z *et al* (2010). Chemoprevention of colorectal cancer by targeting APC-deficient cells for apoptosis. *Nature* 464: 1058-61.

[166] Russo A, Bazan V, Iacopetta B, Kerr D, Soussi T, Gebbia N (2005). The TP53 colorectal cancer international collaborative study on the prognostic and predictive significance of p53 mutation: influence of tumor site, type of mutation, and adjuvant treatment. *J. Clin. Oncol.* 23: 7518-28.

[167] Iacopetta B (2003). TP53 mutation in colorectal cancer. *Hum. Mutat.* 21: 271-6.

[168] Bartkova J, Horejsi Z, Koed K, Kramer A, Tort F, Zieger K *et al* (2005). DNA damage response as a candidate anti-cancer barrier in early human tumorigenesis. *Nature* 434: 864-70.

[169] Dalton WB, Yu B, Yang VW (2010) p53 suppresses structural chromosome instability after mitotic arrest in human cells. *Oncogene* 29: 1929-40.

[170] Vitale I, Senovilla L, Jemaa M, Michaud M, Galluzzi L, Kepp O *et al* (2010). Multipolar mitosis of tetraploid cells: inhibition by p53 and dependency on Mos. *EMBO J.* 29: 1272-84.

[171] Bouvet M, Ellis LM, Nishizaki M, Fujiwara T, Liu W, Bucana CD *et al* (1998). Adenovirus-mediated wild-type p53 gene transfer down-regulates vascular endothelial growth factor expression and inhibits angiogenesis in human colon cancer. *Cancer Res.* 58: 2288-92.

[172] Yu JL, Rak JW, Coomber BL, Hicklin DJ, Kerbel RS (2002). Effect of p53 status on tumor response to antiangiogenic therapy. *Science* 295: 1526-8.

[173] Seoane J, Le HV, Massague J (2002). Myc suppression of the p21(Cip1) Cdk inhibitor influences the outcome of the p53 response to DNA damage. *Nature* 419: 729-34.

[174] Elyada E, Pribluda A, Goldstein RE, Morgenstern Y, Brachya G, Cojocaru G *et al* (2011). CKIalpha ablation highlights a critical role for p53 in invasiveness control. *Nature* 470: 409-13.

[175] Yan W, Liu G, Scoumanne A, Chen X (2008). Suppression of inhibitor of differentiation 2, a target of mutant p53, is required for gain-of-function mutations. *Cancer Res.* 68: 6789-96.

[176] Bossi G, Lapi E, Strano S, Rinaldo C, Blandino G, Sacchi A (2006). Mutant p53 gain of function: reduction of tumor malignancy of human cancer cell lines through abrogation of mutant p53 expression. *Oncogene* 25: 304-9.

[177] Wang W, Kim SH, El-Deiry WS (2006). Small-molecule modulators of p53 family signaling and antitumor effects in p53-deficient human colon tumor xenografts. *Proc. Natl. Acad. Sci. USA* 103: 11003-8.

[178] Rodilla V, Villanueva A, Obrador-Hevia A, Robert-Moreno A, Fernandez-Majada V, Grilli A *et al* (2009). Jagged1 is the pathological link between Wnt and Notch pathways in colorectal cancer. *Proc. Natl. Acad. Sci. USA* 106: 6315-20.

[179] Zhang Y, Li B, Ji ZZ, Zheng PS (2010). Notch1 regulates the growth of human colon cancers. *Cancer* 116: 5207-18.

[180] Ungerback J, Elander N, Grunberg J, Sigvardsson M, Soderkvist P (2011). The Notch-2 gene is regulated by Wnt signaling in cultured colorectal cancer cells. *PLoS One* 6: e17957.

[181] Peignon G, Durand A, Cacheux W, Ayrault O, Terris B, Laurent-Puig P *et al* (2011). Complex interplay between beta-catenin signalling and Notch effectors in intestinal tumorigenesis. *Gut* 60: 166-76.

[182] Babaei-Jadidi R, Li N, Saadeddin A, Spencer-Dene B, Jandke A, Muhammad B *et al* (2011). FBXW7 influences murine intestinal homeostasis and cancer, targeting Notch, Jun, and DEK for degradation. *J. Exp. Med.* 208: 295-312.

[183] Sancho R, Jandke A, Davis H, Diefenbacher ME, Tomlinson I, Behrens A (2010). F-box and WD repeat domain-containing 7 regulates intestinal cell lineage commitment and is a haploinsufficient tumor suppressor. *Gastroenterology* 139: 929-41.

[184] Sonoshita M, Aoki M, Fuwa H, Aoki K, Hosogi H, Sakai Y *et al* (2011). Suppression of colon cancer metastasis by Aes through inhibition of Notch signaling. *Cancer Cell* 19: 125-37.

[185] Yeung TM, Gandhi SC, Bodmer WF (2011). Hypoxia and lineage specification of cell line-derived colorectal cancer stem cells. *Proc. Natl. Acad. Sci. USA* 108: 4382-7.

[186] Kazanjian A, Noah T, Brown D, Burkart J, Shroyer NF (2010). Atonal homolog 1 is required for growth and differentiation effects of notch/gamma-secretase inhibitors on normal and cancerous intestinal epithelial cells. *Gastroenterology* 139: 918-28, 928 e1-6.

[187] Chu D, Zhang Z, Zhou Y, Wang W, Li Y, Zhang H *et al* (2011). Notch1 and Notch2 have opposite prognostic effects on patients with colorectal cancer. *Ann Oncol.* doi:10.1093/annonc/mdq776.

[188] Serafin V, Persano L, Moserle L, Esposito G, Ghisi M, Curtarello M *et al* (2011). Notch3 signalling promotes tumour growth in colorectal cancer. *J. Pathol.*. doi: 10.1002/path.2895.

[189] Bordonaro M, Tewari S, Atamna W, Lazarova DL (2011). The Notch ligand Delta-like 1 integrates inputs from TGFbeta/Activin and Wnt pathways. *Exp. Cell Res.* 317: 1368-81.

[190] Bu X, Li N, Tian X, Huang P (2011). Gamma-secretase inhibitor, a potential target therapy for MUC2-positive colorectal carcinoma. *Neoplasma* 58: 343-7.

[191] Fischer M, Yen WC, Kapoun AM, Wang M, O'Young G, Lewicki J *et al* (2011). Anti-DLL4 inhibits growth and reduces tumor-initiating cell frequency in colorectal tumors with oncogenic KRAS mutations. *Cancer Res* 71: 1520-5.

[192] Peek RM, Jr. (2004). Prevention of colorectal cancer through the use of COX-2 selective inhibitors. *Cancer Chemother Pharmacol* 54 Suppl 1: S50-6.

[193] Wu WK, Sung JJ, Lee CW, Yu J, Cho CH (2010). Cyclooxygenase-2 in tumorigenesis of gastrointestinal cancers: an update on the molecular mechanisms. *Cancer Lett.* 295: 7-16.

[194] Dixon DA, Tolley ND, King PH, Nabors LB, McIntyre TM, Zimmerman GA *et al* (2001). Altered expression of the mRNA stability factor HuR promotes cyclooxygenase-2 expression in colon cancer cells. *J. Clin. Invest.* 108: 1657-65.

[195] Strillacci A, Griffoni C, Sansone P, Paterini P, Piazzi G, Lazzarini G *et al* (2009). MiR-101 downregulation is involved in cyclooxygenase-2 overexpression in human colon cancer cells. *Exp. Cell Res.* 315: 1439-47.

[196] Kim SF, Huri DA, Snyder SH (2005a). Inducible nitric oxide synthase binds, S-nitrosylates, and activates cyclooxygenase-2. *Science* 310: 1966-70.

[197] Mutoh M, Watanabe K, Kitamura T, Shoji Y, Takahashi M, Kawamori T *et al* (2002). Involvement of prostaglandin E receptor subtype EP(4) in colon carcinogenesis. *Cancer Res.* 62: 28-32.

[198] Shoji Y, Takahashi M, Kitamura T, Watanabe K, Kawamori T, Maruyama T *et al* (2004). Downregulation of prostaglandin E receptor subtype EP3 during colon cancer development. *Gut* 53: 1151-8.

[199] Watanabe K, Kawamori T, Nakatsugi S, Ohta T, Ohuchida S, Yamamoto H *et al* (1999). Role of the prostaglandin E receptor subtype EP1 in colon carcinogenesis. *Cancer Res.* 59: 5093-6.

[200] Pai R, Soreghan B, Szabo IL, Pavelka M, Baatar D, Tarnawski AS (2002). Prostaglandin E2 transactivates EGF receptor: a novel mechanism for promoting colon cancer growth and gastrointestinal hypertrophy. *Nat. Med.* 8: 289-93.

[201] Shao J, Lee SB, Guo H, Evers BM, Sheng H (2003). Prostaglandin E2 stimulates the growth of colon cancer cells via induction of amphiregulin. *Cancer Res.* 63: 5218-23.

[202] Castellone MD, Teramoto H, Williams BO, Druey KM, Gutkind JS (2005). Prostaglandin E2 promotes colon cancer cell growth through a Gs-axin-beta-catenin signaling axis. *Science* 310: 1504-10.

[203] Wang D, Wang H, Shi Q, Katkuri S, Walhi W, Desvergne B *et al* (2004). Prostaglandin E(2) promotes colorectal adenoma growth via transactivation of the nuclear peroxisome proliferator-activated receptor delta. *Cancer Cell* 6: 285-95.

[204] Yu HG, Yu LL, Yang Y, Luo HS, Yu JP, Meier JJ *et al* (2003). Increased expression of RelA/nuclear factor-kappa B protein correlates with colorectal tumorigenesis. *Oncology* 65: 37-45.

[205] Lind DS, Hochwald SN, Malaty J, Rekkas S, Hebig P, Mishra G *et al* (2001). Nuclear factor-kappa B is upregulated in colorectal cancer. *Surgery* 130: 363-9.

[206] Sakamoto K, Maeda S, Hikiba Y, Nakagawa H, Hayakawa Y, Shibata W *et al* (2009). Constitutive NF-kappaB activation in colorectal carcinoma plays a key role in angiogenesis, promoting tumor growth. *Clin. Cancer Res.* 15: 2248-58.

[207] Yu HG, Zhong X, Yang YN, Luo HS, Yu JP, Meier JJ *et al* (2004). Increased expression of nuclear factor-kappaB/RelA is correlated with tumor angiogenesis in human colorectal cancer. *Int. J. Colorectal. Dis.* 19: 18-22.

[208] Scartozzi M, Bearzi I, Pierantoni C, Mandolesi A, Loupakis F, Zaniboni A *et al* (2007). Nuclear factor-kB tumor expression predicts response and survival in irinotecan-refractory metastatic colorectal cancer treated with cetuximab-irinotecan therapy. *J. Clin. Oncol* 25: 3930-5.

[209] Gavert N, Ben-Shmuel A, Lemmon V, Brabletz T, Ben-Ze'ev A (2010). Nuclear factor-kappaB signaling and ezrin are essential for L1-mediated metastasis of colon cancer cells. *J. Cell Sci.* 123: 2135-43.

[210] Bartke T, Siegmund D, Peters N, Reichwein M, Henkler F, Scheurich P *et al* (2001). p53 upregulates cFLIP, inhibits transcription of NF-kappaB-regulated genes and induces caspase-8-independent cell death in DLD-1 cells. *Oncogene* 20: 571-80.

[211] Shao J, Fujiwara T, Kadowaki Y, Fukazawa T, Waku T, Itoshima T *et al* (2000). Overexpression of the wild-type p53 gene inhibits NF-kappaB activity and synergizes with aspirin to induce apoptosis in human colon cancer cells. *Oncogene* 19: 726-36.

[212] Benoit V, Hellin AC, Huygen S, Gielen J, Bours V, Merville MP (2000). Additive effect between NF-kappaB subunits and p53 protein for transcriptional activation of human p53 promoter. *Oncogene* 19: 4787-94.

[213] Hellin AC, Calmant P, Gielen J, Bours V, Merville MP (1998). Nuclear factor -kappaB-dependent regulation of p53 gene expression induced by daunomycin genotoxic drug. *Oncogene* 16: 1187-95.

[214] Cadoret A, Bertrand F, Baron-Delage S, Levy P, Courtois G, Gespach C *et al* (1997). Down-regulation of NF-kappaB activity and NF-kappaB p65 subunit expression by ras and polyoma middle T oncogenes in human colonic Caco-2 cells. *Oncogene* 14: 1589-600.

[215] Deng J, Miller SA, Wang HY, Xia W, Wen Y, Zhou BP *et al* (2002). beta-catenin interacts with and inhibits NF-kappa B in human colon and breast cancer. *Cancer Cell* 2: 323-34.

[216] Clemo NK, Collard TJ, Southern SL, Edwards KD, Moorghen M, Packham G *et al* (2008). BAG-1 is up-regulated in colorectal tumour progression and promotes colorectal tumour cell survival through increased NF-kappaB activity. *Carcinogenesis* 29: 849-57.

[217] Kojima M, Morisaki T, Izuhara K, Uchiyama A, Matsunari Y, Katano M *et al* (2000). Lipopolysaccharide increases cyclo-oxygenase-2 expression in a colon carcinoma cell line through nuclear factor-kappa B activation. *Oncogene* 19: 1225-31.

[218] Plummer SM, Holloway KA, Manson MM, Munks RJ, Kaptein A, Farrow S *et al* (1999). Inhibition of cyclo-oxygenase 2 expression in colon cells by the chemopreventive agent curcumin involves inhibition of NF-kappaB activation via the NIK/IKK signalling complex. *Oncogene* 18: 6013-20.

[219] Wang Y, Dang J, Wang H, Allgayer H, Murrell GA, Boyd D (2000). Identification of a novel nuclear factor-kappaB sequence involved in expression of urokinase-type plasminogen activator receptor. *Eur. J. Biochem.* 267: 3248-54.

[220] Harwood FG, Kasibhatla S, Petak I, Vernes R, Green DR, Houghton JA (2000). Regulation of FasL by NF-kappaB and AP-1 in Fas-dependent thymineless death of human colon carcinoma cells. J. Biol. Chem. 275: 10023-9.

[221] Luo JL, Maeda S, Hsu LC, Yagita H, Karin M (2004). Inhibition of NF-kappaB in cancer cells converts inflammation- induced tumor growth mediated by TNFalpha to TRAIL-mediated tumor regression. *Cancer Cell* 6: 297-305.

[222] Wang Q, Zhou Y, Wang X, Chung DH, Evers BM (2007). Regulation of PTEN expression in intestinal epithelial cells by c-Jun NH2-terminal kinase activation and nuclear factor-kappaB inhibition. *Cancer Res.* 67: 7773-81.

[223] Yamamoto Y, Yin MJ, Lin KM, Gaynor RB (1999b). Sulindac inhibits activation of the NF-kappaB pathway. *J. Biol. Chem.* 274: 27307-14.

[224] Yin L, Laevsky G, Giardina C (2001). Butyrate suppression of colonocyte NF-kappa B activation and cellular proteasome activity. *J. Biol. Chem.* 276: 44641-6.

[225] Scaife CL, Kuang J, Wills JC, Trowbridge DB, Gray P, Manning BM *et al* (2002). Nuclear factor kappaB inhibitors induce adhesion-dependent colon cancer apoptosis: implications for metastasis. *Cancer Res.* 62: 6870-8.

[226] Bentires-Alj M, Barbu V, Fillet M, Chariot A, Relic B, Jacobs N *et al* (2003). NF-kappaB transcription factor induces drug resistance through MDR1 expression in cancer cells. *Oncogene* 22: 90-7.

[227] Cusack JC, Jr., Liu R, Baldwin AS, Jr. (2000). Inducible chemoresistance to 7-ethyl-10-[4-(1-piperidino)-1-piperidino]-carbonyloxycamptothe cin (CPT-11) in colorectal cancer cells and a xenograft model is overcome by inhibition of nuclear factor-kappaB activation. *Cancer Res.* 60: 2323-30.

[228] Uetsuka H, Haisa M, Kimura M, Gunduz M, Kaneda Y, Ohkawa T *et al* (2003). Inhibition of inducible NF-kappaB activity reduces chemoresistance to 5-fluorouracil in human stomach cancer cell line. *Exp. Cell Res.* 289: 27-35.

[229] Wang W, McLeod HL, Cassidy J (2003). Disulfiram-mediated inhibition of NF-kappaB activity enhances cytotoxicity of 5-fluorouracil in human colorectal cancer cell lines. *Int. J. Cancer* 104: 504-11.

[230] Stark LA, Dunlop MG (2005). Nucleolar sequestration of RelA (p65) regulates NF-kappaB-driven transcription and apoptosis. *Mol. Cell Biol.* 25: 5985-6004.

[231] Thoms HC, Dunlop MG, Stark LA (2007). p38-mediated inactivation of cyclin D1/cyclin-dependent kinase 4 stimulates nucleolar translocation of RelA and apoptosis in colorectal cancer cells. *Cancer Res* 67: 1660-9.

[232] Loveridge CJ, MacDonald AD, Thoms HC, Dunlop MG, Stark LA (2008). The proapoptotic effects of sulindac, sulindac sulfone and indomethacin are mediated by nucleolar translocation of the RelA(p65) subunit of NF-kappaB. *Oncogene* 27: 2648-55.

[233] Wang P, Qiu W, Dudgeon C, Liu H, Huang C, Zambetti GP *et al* (2009). PUMA is directly activated by NF-kappaB and contributes to TNF-alpha-induced apoptosis. *Cell Death Differ.* 16: 1192-202.

[234] Saltzman BS, Yamamoto JF, Decker R, Yokochi L, Theriault AG, Vogt TM *et al* (2008). Association of genetic variation in the transforming growth factor beta-1 gene with serum levels and risk of colorectal neoplasia. *Cancer Res.* 68: 1236-44.

[235] Markowitz S, Wang J, Myeroff L, Parsons R, Sun L, Lutterbaugh J *et al* (1995). Inactivation of the type II TGF-beta receptor in colon cancer cells with microsatellite instability. *Science* 268: 1336-8.

[236] Valle L, Serena-Acedo T, Liyanarachchi S, Hampel H, Comeras I, Li Z *et al* (2008). Germline allele-specific expression of TGFBR1 confers an increased risk of colorectal cancer. *Science* 321: 1361-5.

[237] Eppert K, Scherer SW, Ozcelik H, Pirone R, Hoodless P, Kim H *et al* (1996). MADR2 maps to 18q21 and encodes a TGFbeta-regulated MAD-related protein that is functionally mutated in colorectal carcinoma. *Cell* 86: 543-52.

[238] Jiang Y, Liu XQ, Rajput A, Geng L, Ongchin M, Zeng Q *et al* (2011). Phosphatase PRL-3 is a direct regulatory target of TGFbeta in colon cancer metastasis. *Cancer Res.* 71: 234-44.

[239] Wang J, Yang L, Yang J, Kuropatwinski K, Wang W, Liu XQ *et al* (2008). Transforming growth factor beta induces apoptosis through repressing the phosphoinositide 3-kinase/AKT/survivin pathway in colon cancer cells. *Cancer Res.* 68: 3152-60.

[240] Gong J, Ammanamanchi S, Ko TC, Brattain MG (2003). Transforming growth factor beta 1 increases the stability of p21/WAF1/CIP1 protein and inhibits CDK2 kinase activity in human colon carcinoma FET cells. *Cancer Res.* 63: 3340-6.

[241] Papageorgis P, Cheng K, Ozturk S, Gong Y, Lambert AW, Abdolmaleky HM *et al* (2011). Smad4 inactivation promotes malignancy and drug resistance of colon cancer. *Cancer Res.* 71: 998-1008.

[242] Sodir NM, Chen X, Park R, Nickel AE, Conti PS, Moats R *et al* (2006). Smad3 deficiency promotes tumorigenesis in the distal colon of ApcMin/+ mice. *Cancer Res.* 66: 8430-8.

[243] Zhou S, Buckhaults P, Zawel L, Bunz F, Riggins G, Dai JL *et al* (1998). Targeted deletion of Smad4 shows it is required for transforming growth factor beta and activin signaling in colorectal cancer cells. *Proc. Natl. Acad. Sci. USA* 95: 2412-6.

[244] Matsuzaki K, Kitano C, Murata M, Sekimoto G, Yoshida K, Uemura Y *et al* (2009). Smad2 and Smad3 phosphorylated at both linker and COOH-terminal regions transmit malignant TGF-beta signal in later stages of human colorectal cancer. *Cancer Res.* 69: 5321-30.

[245] Howe JR, Bair JL, Sayed MG, Anderson ME, Mitros FA, Petersen GM *et al* (2001). Germline mutations of the gene encoding bone morphogenetic protein receptor 1A in juvenile polyposis. *Nat. Genet.* 28: 184-7.

[246] Hardwick JC, Van Den Brink GR, Bleuming SA, Ballester I, Van Den Brande JM, Keller JJ *et al* (2004). Bone morphogenetic protein 2 is expressed by, and acts upon, mature epithelial cells in the colon. *Gastroenterology* 126: 111-21.

[247] He XC, Zhang J, Tong WG, Tawfik O, Ross J, Scoville DH *et al* (2004). BMP signaling inhibits intestinal stem cell self-renewal through suppression of Wnt-beta-catenin signaling. *Nat. Genet.* 36: 1117-21.

[248] Lombardo Y, Scopelliti A, Cammareri P, Todaro M, Iovino F, Ricci-Vitiani L *et al* (2011). Bone morphogenetic protein 4 induces differentiation of colorectal cancer stem cells and increases their response to chemotherapy in mice. *Gastroenterology* 140: 297-309.

[249] Lee CW, Ito K, Ito Y (2010). Role of RUNX3 in bone morphogenetic protein signaling in colorectal cancer. *Cancer Res.* 70: 4243-52.

[250] Lorente-Trigos A, Varnat F, Melotti A, Ruiz i Altaba A (2010). BMP signaling promotes the growth of primary human colon carcinomas in vivo. *J. Mol. Cell Biol.* 2: 318-32.

[251] Deng H, Ravikumar TS, Yang WL (2009). Overexpression of bone morphogenetic protein 4 enhances the invasiveness of Smad4-deficient human colorectal cancer cells. Cancer Lett. 281: 220-31.

[252] van den Brink GR, Bleuming SA, Hardwick JC, Schepman BL, Offerhaus GJ, Keller JJ *et al* (2004). Indian Hedgehog is an antagonist of Wnt signaling in colonic epithelial cell differentiation. *Nat. Genet.* 36: 277-82.

[253] Qualtrough D, Buda A, Gaffield W, Williams AC, Paraskeva C (2004). Hedgehog signalling in colorectal tumour cells: induction of apoptosis with cyclopamine treatment. *Int. J. Cancer* 110: 831-7.

[254] Monzo M, Moreno I, Artells R, Ibeas R, Navarro A, Moreno J *et al* (2006). Sonic hedgehog mRNA expression by real-time quantitative PCR in normal and tumor tissues from colorectal cancer patients. *Cancer Lett* 233: 117-23.

[255] Oniscu A, James RM, Morris RG, Bader S, Malcomson RD, Harrison DJ (2004). Expression of Sonic hedgehog pathway genes is altered in colonic neoplasia. *J. Pathol.* 203: 909-17.

[256] Olsen CL, Hsu PP, Glienke J, Rubanyi GM, Brooks AR (2004). Hedgehog-interacting protein is highly expressed in endothelial cells but down-regulated during angiogenesis and in several human tumors. *BMC Cancer* 4: 43.

[257] Akiyoshi T, Nakamura M, Koga K, Nakashima H, Yao T, Tsuneyoshi M *et al* (2006). Gli1, downregulated in colorectal cancers, inhibits proliferation of colon cancer cells involving Wnt signalling activation. *Gut* 55: 991-9.

[258] Chatel G, Ganeff C, Boussif N, Delacroix L, Briquet A, Nolens G *et al* (2007). Hedgehog signaling pathway is inactive in colorectal cancer cell lines. *Int. J. Cancer* 121: 2622-7.

[259] Arimura S, Matsunaga A, Kitamura T, Aoki K, Aoki M, Taketo MM (2009). Reduced level of smoothened suppresses intestinal tumorigenesis by down-regulation of Wnt signaling. *Gastroenterology* 137: 629-38.

[260] Steinhardt AA, Gayyed MF, Klein AP, Dong J, Maitra A, Pan D *et al* (2008). Expression of Yes-associated protein in common solid tumors. *Hum. Pathol.* 39: 1582-9.

[261] Camargo FD, Gokhale S, Johnnidis JB, Fu D, Bell GW, Jaenisch R *et al* (2007). YAP1 increases organ size and expands undifferentiated progenitor cells. *Curr. Biol.* 17: 2054-60.

[262] Arlt A, Bauer I, Schafmayer C, Tepel J, Muerkoster SS, Brosch M *et al* (2009). Increased proteasome subunit protein expression and proteasome activity in colon cancer relate to an enhanced activation of nuclear factor E2-related factor 2 (Nrf2). *Oncogene* 28: 3983-96.

[263] Hu XT, Chen W, Zhang FB, Shi QL, Hu JB, Geng SM *et al* (2009). Depletion of the proteasome subunit PSMA7 inhibits colorectal cancer cell tumorigenicity and migration. *Oncol. Rep.* 22: 1247-52.

[264] Ahn CH, Jeong EG, Lee JW, Kim MS, Kim SH, Kim SS *et al* (2007). Expression of beclin-1, an autophagy-related protein, in gastric and colorectal cancers. *APMIS* 115: 1344-9.

[265] Yoshioka A, Miyata H, Doki Y, Yamasaki M, Sohma I, Gotoh K *et al* (2008). LC3, an autophagosome marker, is highly expressed in gastrointestinal cancers. *Int. J. Oncol.* 33: 461-8.

[266] Miao Y, Zhang Y, Chen Y, Chen L, Wang F (2010). GABARAP is overexpressed in colorectal carcinoma and correlates with shortened patient survival. *Hepatogastroenterology* 57: 257-61.

[267] Kang MR, Kim MS, Oh JE, Kim YR, Song SY, Kim SS *et al* (2009). Frameshift mutations of autophagy-related genes ATG2B, ATG5, ATG9B and ATG12 in gastric and colorectal cancers with microsatellite instability. *J. Pathol.* 217: 702-6.

[268] Wu WK, Wu YC, Yu L, Li ZJ, Sung JJ, Cho CH (2008). Induction of autophagy by proteasome inhibitor is associated with proliferative arrest in colon cancer cells. *Biochem. Biophys. Res. Commun.* 374: 258-63.

[269] Wu YC, Wu WK, Li Y, Yu L, Li ZJ, Wong CC *et al* (2009). Inhibition of macroautophagy by bafilomycin A1 lowers proliferation and induces apoptosis in colon cancer cells. *Biochem. Biophys. Res. Commun.* 382: 451-6.

In: Signaling, Gene Regulation and Cancer
Editors: Shree Ram Singh and Manoj K. Mishra
ISBN: 978-1-61942-088-5

Chapter 10

SYSTEMS BIOLOGY OF NUCLEAR HORMONE SIGNALING AND GENE REGULATION

Mei Hui Liu and Edwin Cheung*
Genome Institute of Singapore, Genome, Singapore

ABSTRACT

Nuclear hormone receptors (NR) are sequence-specific signal-regulated transcription factors which control defined patterns of gene expression in a wide variety of physiological and disease processes. The availability of next generation sequencing technologies has generated an extensive amount of information on NR-regulated transcription. Combinatorial approaches in analyzing these datasets has further enhanced our understanding of NR biology and provide alternatives to the prevention, diagnosis and treatment of NR driven diseases. This review will focus on the progress of NR studies through the use of current high-throughput platforms in genomics, transcriptomics and proteomics. We will highlight how the integrative approaches of systems biology have contributed to major progress in our understanding of NR biology through examples from two NR, estrogen receptor α (ERα) and androgen receptor (AR). Together, the large-scale integration of datasets provides us with a network-level understanding of diseases such as cancer that will aid us in predicting drug response and provide therapeutic guides at the general and personalized level.

1. INTRODUCTION

1.1. Nuclear Receptors and their Mode of Transcription

Nuclear receptors (NR) are a large superfamily of evolutionarily related DNA-binding transcription factors that have diverse and crucial roles in the regulation of growth, development and homeostasis. Recent advances in the genome-wide analyses of transcription factor binding, histone modification, and gene expression have provided researchers a global

* Genome Institute of Singapore, #02-01 Genome, Singapore 138672. email: cheungcwe@gis.a-star.edu.sg.

view of transcriptional regulation by NR beyond the gene specific level. As genomic and proteomic technologies continue to mature and data generation becomes more routine, computational analysis of data using a systems biology approach will become increasingly important in the study of NR gene regulation. Such combinatorial approaches will further our understanding of NR biology and provide clinicians alternatives to the prevention, diagnosis and treatment of NR driven diseases.

Members of the NR superfamily share a common structural organization that is well-defined and has specific functions (Figure 1A). The most conserved region is the central DNA Binding Domain (DBD). The DBD allows the binding of ligand activated NR on hormone response elements to elicit a biological response. The largest domain is the moderately conserved ligand binding domain (LBD) which contains four structurally distinct but functionally linked surfaces:

i) a dimerization surface, which mediates interaction with partner LBDs,
ii) a ligand binding pocket, which interacts with diverse lipophilic small molecules,
iii) a co-regulator binding surface which binds regulatory protein complexes that modulate positively (by co-activators) or negatively (by co-repressors) transcriptional activity, and
iv) an activation function helix which mediates ligand dependent transactivation.

Collectively, these properties allow transactivation by members of the NR superfamily to occur in 5 general steps upon hormone stimulation (Figure 1B): 1) ligand binding; 2) stable binding of liganded NR to HRE; 3) co-repressors dismissal and co-activators recruitment; 4) activation of transcription; and 5) either shut down or re-initiation of transcription.

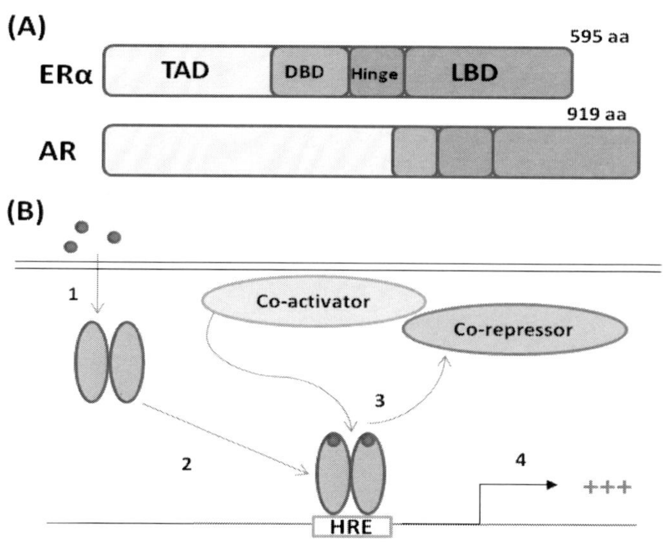

Figure 1. (A) Structural organization of two of the best studied NR, the estrogen receptor and the androgen receptor. Amino acid (aa) length of respective receptor indicated. (B) Schematic representation of transcription activation by NR. Numbers indicate the different steps generally involved, as described in text. TAD, transactivation domain; DBD, DNA binding domain; LBD, ligand binding domain; HRE, hormone response element.

1.2. NR Co-Regulators and Histone Marks

Co-regulators are proteins that repress (co-repressors) or enhance (co-activators) NR transcriptional activity by either bridging the NR to the basic transcription machinery or modifying the structure of chromatin. While chromatin compaction and histone hypoacetylation mainly repress gene activation, remodeling and repositioning of nucleosomes allows binding of transcription factors and formation of the pre-initiation complex. Indeed, regulation of a general transcription unit by NRs requires a vast number of co-regulatory complexes that have various functions and enzymatic activities. Several of these co-regulators can covalently modify histone tails by acetylation, methylation, phosphorylation or ubiquitylation [1]. These modifications can occur in several combinations, often in an inter-dependent manner, and is referred to as the histone code.

Co-activators with histone acetyltransferase (HAT) activity include the homologous CREB binding protein (CBP) and p300 proteins. Transcriptional activation through CBP/p300 recruitment results in increased histone acetylation but also increased contacts with the basal transcription machinery. p300 can also directly acetylate members of the NR [2-3]. The p160 proteins compose a family of three related co-activators (SRC-1, TIF2 and SRC-3). Both SRC-1 and SRC-3 contain an acetyltransferase domain. However, its precise functional targets *in vivo* have not been clarified [4]. Nonetheless, recruitment of p160 proteins by NR bridges the transcriptional complex to the basal transcriptional machinery for gene activation. Other classes of co-activators include the switch/sucrose non-fermenting (SWI/SNF) complexes, which change the location or conformation of nucleosomes and remodel chromatin by using energy from ATP hydrolysis. The mediator (Med1) multi-subunit complex is another class of co-activators that plays an important role in bridging NR to the general transcriptional machinery and in recruiting and activating Pol II. Recent findings from genomic approaches have identified many more groups of co-regulators such as the Forkhead proteins that interact with NR on chromatin. One such example is FOXA1, a co-activator that influences genome-wide chromatin accessibility by its recruitment to enhancers with histone H3 lysine 4 methylation [5].

Histones are characterized by numerous posttranslational modifications that influence gene transcription. The tails of histones can be dynamically modified by acetylation, methylation, phosphorylation, ubiquitylation and sumoylation. These modifications are catalyzed by numerous enzymes in order to activate or repress transcription. For example, CBP/p300 posses histone acetyltransferase activity on the acetylation of H3 and H4 and is correlated with active transcriptional states. Conversely, deacetylation marks transcriptionally repressive states which are catalyzed by histone deacetylase such as HDAC3 through its recruitment by the nuclear co-repressor (NCOR) complex [6]. Histones can also be methylated, for example, on H3 by histone methyltransferases (HMTs). Genomic analysis has shown that an H3 lysine 4 (H3K4) methylation signature correlates with active transcription and accessible chromatin at sites of chromatin remodeling complex occupancy. For example, menin, an integral component of the MLL1/MLL2 histone methyltransferase complexes specific for H3K4 is a transcriptional co-activator of ERα and Vitamin D receptor (VDR) [7]. Conversely, methylation of H3 lysine 27 (H3K27) is known to be associated with inactive genes. Activation of the H3K27 methyltransferase EZH2, a Polycomb group protein, can impose a repressive epigenetic program [8]. In addition, histone lysine demethylases such as LSD1 [9] and JMJD3 [10] work to remove lysine methylation from H3K4 and H3K27, respectively. The contribution of histone modifications to transcription is therefore a complex process involving active and repressive states that are determined by co-regulators with specific enzyme activity and transcription factors, such as NR, which recruits them in a signal responsive manner.

1.3. The Systems Biology Approach

The availability of next generation sequencing technologies has offered new promise for whole genome and transcriptome interrogation. Coupling with proteomics platforms, such as protein microarrays and mass spectrometry platforms, approaches used to decipher regulatory networks in system biology has been constantly improving. Such developments allow for robust and effective system biology approaches to further our understanding in genetic/epigenetic regulatory networks, chromatin states, nuclear architectures, genome variations and the eventual applications to predict drug responses and guide therapeutics and diagnostics.

In this chapter, we will focus on the progress of NR studies through the use of current high-throughput platforms in genomics, transcriptomics and proteomics. We will highlight how the combinatorial approaches of systems biology have contributed to major progress in our understanding of NR biology through mainly examples from two of the best studied NR, ERα and AR. Using ERα, we will highlight how advancements in current technologies have refined our concepts of ERα regulated transcription in the context of three dimensional chromatin architecture. The genome-wide analysis of ERα binding sites has also allowed for the identification of co-operating transcription factors which underlie estrogen signaling, endocrine response and chromatin structure in breast cancer. Further on, we show how using genomics, transcriptomics and proteomics in the study of AR related prostate cancer has allowed for a more comprehensive understanding of its pathogenesis, treatment and diagnostic. Together, the large scale integration of datasets will allow us further understanding of the emergent properties of complex NR related biological systems. Eventually, a network-level understanding of cancer can be used to predict drug responses and guide therapeutics at the general and personalized level [11].

2. Genome-wide NR Binding

ERα and AR are two of the most widely studied members of the NR superfamily. They have important roles in regulating sexual and reproductive development. ERα functions as a transcription factor to regulate cell division and approximately 2/3 of breast cancers are classified as ERα positive and require estrogen for growth and proliferation. One of the main approaches of therapy for ERα-positive breast cancer patients is with anti-estrogens such as tamoxifen. However, many patients will relapse and become resistance to the drug [12]. Therefore, understanding fully how ERα function is essential for developing new strategies for breast cancer treatment.

2.1. Estrogen Receptor Binding

ERα has been the most extensively profiled NR in genomic studies. The first study to enable a partial genome-wide identification of ERα binding was performed with chromatin immunoprecipitation analysis on Affymetrix arrays (ChIP-chip) on arrays covering the entire chromosome 21 and 22 [13]. This resulted in the identification of 57 ERα-binding sites in MCF-7 breast cancer cells. Subsequently, a genome-wide ERα binding map using the same platform identified 3665 ERα binding sites [14]. Using an alternative approach, a ChIP-paired end diTag cloning and sequencing strategy (ChIP-PET) identified 1234 high-confidence ERα-binding sites [15]. More recently using ChIP-seq technology, 10205 ERα-binding sites were identified [16]. Besides estrogen stimulation, the ERα binding profile has also been evaluated in the presence of antagonists, tamoxifen and fulvestrant [16], and growth factors such as EGF [17].

The genome-wide binding of ERα beyond breast cancer cell lines, such as in bone [18] has also examined. Furthermore, post-translational modifications by serine/threonine AKT on genome-wide ERα binding [19] and ERα binding in relation to other NR, such as ERβ, have been investigated [20]. Recently, others have studied the binding of ERα in greater detail by assessing its binding profile in the absence of essential co-regulators (eg. knockdown of a transcription factor, followed by ChIP-seq of ERα) such as FoxA1 [21] and TLE1 [22]. The profiles of ERα binding in the context of subsets of its complexes have also been assessed through sequencing of re-ChIP samples (eg. ChIP of ERα followed by additional ChIP of a separate transcription factor, or vice versa) such as in the RARα complex [23].

In all, genomic patterns of ERα binding have been examined (1) in different species, (2) in various tissues, (3) by multiple signaling pathways, (4) in response to different ligands, (5) with all types of ChIP platforms, (6) in response to the loss or over-expression of its co-regulators and (7) in transcription complexes. Despite the differences, several observations are shared among these studies: (i) ERα binds across the genome, but does so preferentially outside of proximal promoter regions, (ii) ERα binding sites are more significantly associated with estrogen up-regulated, rather than down-regulated genes, and (iii) many regulated genes are associated with multiple ERα binding sites [24].

2.2. Motif Analysis and the Endocrine Response

Besides understanding the global binding characteristics of ERα, many bioinformatics tools have been developed for the identification of transcription factor motifs. These include *de novo* motif discovery tools like MEME [25] or CisFinder [26] and motif scanning programs like CENTDIST [27]. Such approaches fully utilize the data rich information available from the datasets and serve to provide additional clues on the mechanisms of hormone-regulated transcription, such as in relation to the endocrine response.

Using such an approach, motifs for the Forkhead family were shown to be highly enriched at ERα binding sites. The Forkhead transcription factor FOXA1 was subsequently found co-localized at many ERα binding regions across the genome [13, 28-29] and is required for ERα binding events [13, 28], where it functions as a pioneer factor [30-31]. FOXA1 appears to function in a cell-type specific manner through its predominant recruitment at distant enhancers rather than proximal promoters. The demarcations of these lineage specific FOXA1 regulatory regions are determined by the distribution of H3K4 methylation marks. This differential recruitment leads to cell-type specific changes in chromatin structure and is defined by its functional collaboration with lineage-specific transcription factors, such as ERα and AR in breast and prostate cancer cells, respectively [29].

Clinically, FoxA1 expression is a positive prognostic factor among patients with ERα-positive breast tumors and correlates with sensitivity to endocrine therapy [32]. Intriguingly, its effects on the outcome of these patients may be due to FOXA1 influencing the differential interactions between ERα and chromatin. Using cell based assays, it was shown that nearly all ERα chromatin interactions and gene expression changes were FOXA1 dependent [21]. In estrogen-responsive breast cancer cells, tamoxifen bound ERα activity was also FOXA1 dependent. Intriguingly, in tamoxifen-resistant cells, ERα binding was ligand independent but remained FOXA1 dependent. Furthermore, expression of FOXA1 in non-breast cancer cells can also alter ERα binding and function. Thus, these results are in agreement with clinical observations that high FOXA1 level is a positive predictor of response to tamoxifen in ERα-positive patients [32].

Another clinically important co-factor of ERα that may have a role in endocrine response is the paired box 2 gene product (PAX2). PAX2 was shown to be a crucial mediator of ER repression of ERBB2 by the anti-cancer drug tamoxifen [36]. PAX2 and the ER co-activator SRC-3 compete for binding and regulation of ERBB2 transcription and determine tamoxifen response in breast cancer cells. The repression of ERBB2 by ER-PAX2 complex and activation of ERBB2 by ER-SRC3 complex link these two breast cancer subtypes (ERα-positive, ERBB2-negative or positive) and suggests that aggressive ERBB2-positive tumors may originate from ERα-positive tumors that have overcame this repressive mechanism. In addition to FoxA1 and PAX2, genome-wide binding analysis have also helped identify other co-factors of ERα including, AP-1 [14], Oct1 [14], C/EBP [14], GATA3 [33], RUNX1 [34], NKX3-1 [35] and LEF-1 [35].

Through the analysis of genome-wide ERα binding datasets, we have grown to appreciate the role of co-regulators especially in the determination of functional outcomes in ERα binding. Whether it is a pioneering factor or a factor that is recruited through ERα activation, the data suggests that these factors summate to a change at the chromatin level which determines the outcome of transcription. There is also a pressing need to understand NR regulation in the three dimensional chromatin context. Indeed, the global impact of specific factors (e.g. FOXA1) that alter the endocrine response of ERα at the chromatin level may also suggest an influence on the general organization of the chromatin in context.

2.3. Transcription in a Three Dimensional Context

The numerous intergenic NR binding sites mapped from genome-wide ChIP studies is in concordance with the idea that the genome is more than a series of linear sequences. Indeed, the genome exists as elaborate physical structures and their functional properties are strongly determined by their cellular organization [37]. Collectively, findings from genome-wide studies suggest that distal regulation of transcription is a common mechanism used by NR [24]. Despite this recurring observation, the enhancer function of most NR elements is not well established *in vivo*. Consequently, questions about whether and how some of these distal binding sites might communicate with their putative target genes via long distance intra-chromosomal and inter-chromosomal interactions remain intriguing. Importantly, understanding how the genome is organized by NR at the chromatin level may allow us to predict through computational approaches the impact of its related transcription factors in response to drug perturbations.

2.3.1. NR Mediated Chromosomal Interactions

Recently, several groups have demonstrated the role of NR in the regulation of long-range transcription through the use of chromosome conformation capture (3C) assay and its variants such as ChIP-3C. Specifically, ERα can mediate long-range chromatin at several specific gene locus such as the RET [38], PR [39], ERBB2 [36], CTSD [40], CA12 [41], Greb1 [42] and TFF1 [13, 43]. Other NR can also mediate long-range chromatin interactions including GR at the Ciz1-Lcn2 locus [44], LXR at the ABCG1 locus [45], VDR at the p21 locus [46], and AR at the PSA [47], TMPRSS2 [48], UBE2C [49] and FKBP5 [50] loci. Most long-range chromatin interactions reported thus far for NR are intra-chromosomal, however, inter-chromosomal interaction was recently reported for ERα in mediating the interaction between the enhancer of TFF1 on chromosome 21 and the promoter of Greb1 on chromosome 8 [51].

3C based assays have provided numerous options for the detection of both NR intra-chromosomal and inter-chromosomal interactions. However, despite growing information on the role of NR in the mediation of long-range chromatin interaction, progress towards a better understanding on how distal elements interact at the genome-wide level remain restricted by the technical limits of 3C assays. While each of these methods has its strength and weakness relative to others, all 3C assays are constrained by their inability to provide an unbiased, protein specific genome-wide analysis of the chromatin architecture. Although a global view of glucocorticoid receptor (GR) related chromatin dynamics has been carried out using 4C [52], the results does not directly reveal GR-specific role in chromatin architecture after induction. A protein specific genome-wide long-range detection method is therefore required to fully annotate the relationship between individual elements and how they work together to form complex protein mediated chromatin architectures.

2.3.2. The ChIA-PET Method

Chromatin Interaction Analysis by Paired-End Tag Sequencing (ChIA-PET) [53] was a strategy developed to address this challenge. In this method, formaldehyde cross-linked DNA-protein complexes are first sonicated and enriched by ChIP. Through proximal ligation, tethered DNA fragments of each chromatin complexes are joined to each other. A linker sequence in the junction of two DNA fragments allows linker tag ligation products to be extracted and products are analyzed through ultra-high-throughput pair-end sequencing. When mapped onto the reference genome, the sequences are read out to detect the relationship between the two paired DNA sequences.

ChIA-PET is an unbiased, whole-genome, and *de novo* approach for long-range chromatin interaction analysis and is capable of providing two global datasets: the protein factor binding sites (equivalent to ChIP-PET) and the interactions among the binding sites. Furthermore, because of the involvement of ChIP, it adds specificity to chromatin interactions bound by specific factors of interest and reduces the complexity for genome-wide analysis. Due to the relatively harsher conditions of sonication, as compared to restriction enzyme digestion, random chromatin attachment from specific chromatin interaction complexes are reduced, thus decreasing interaction noise. Hence, ChIA-PET is applicable to many different protein factors involved in transcriptional regulation or chromatin structural conformation.

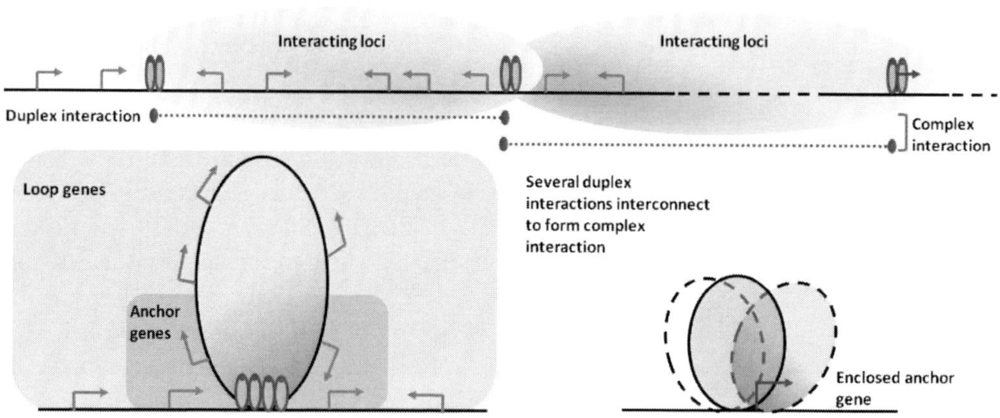

Figure 2. Schematic summary of ER intra-chromosomal interactions identified by ChIA-PET. Upper panel shows an example of ER binding in the linear view. Interacting loci, indicated by ChIA-Pet (pink), formed by two ER binding regions compose a duplex interaction. This duplex interaction may be a subset of a complex interaction (lower right panel). Lower left panel represents a simple duplex interaction loop formed by an ER interacting loci. Genes with their transcription start site within 20kb of anchors are categorized as anchor genes (orange) and the others that fall beyond are classified as loop genes (grey). Anchor genes with both 5' and 3' ends within interaction boundaries are considered enclose anchor genes (red, lower right panel).

Using this method, a total of 1452 ERα intra-chromosomal interactions were identified from the MCF7 genome [53]. Like the direct looping of previously described ERα mediated loops, these 'duplex interactions', on a linear scale, has a basic chromatin structure which features two ERα binding regions (interacting loci) and a loop (the intermediate genomic span between the two binding sites) (Figure 2).

The genomic spans of most duplex interactions are less than 100 kb (86%), about 13% are from 100 kb to 1 Mb, and less than 1% are more than 1 Mb. Structurally, many of these duplex interactions are subsets of bigger chromatin complexes. Many nearby duplex interactions are interconnected, linking three or more anchors into 'daisy-chain' aggregated complex interactions. In all, 689 ERα-bound chromatin interaction regions were identified.

The functionality of these 689 ERα-bound chromatin interactions can be correlated according to the positions and expressions of the genes they regulate. Genes close to any interaction anchors (within 20 kb of the TSS) are considered 'anchor genes', and genes in the interaction loop regions and far away from anchors are classified as 'loop genes'. Altogether, 1,575 'anchor genes' and 3,767 'loop genes' were assigned to interaction regions. Within the anchor gene category, many gene entries have 5' and 3' ends within interaction boundaries. Such entries, are further classified as 'enclosed anchor genes' which frequently occupy the entirety of short interaction loops, engage multiple anchor sites around or within the gene, tend to have intense RNA polymerase II (Pol II) marks covering the entire gene and are preferentially associated with Pol II marks and gene up-regulation as indicated by expression microarrays. These observations suggest that direct ERα binding might be initiated primarily at one or multiple distal sites, which then subsequently recruit other binding sites as anchors to form an interaction complex to ultimately engage the transcriptional machinery at gene promoters. Functionally, most 'anchor genes' are up-regulated (60%), particularly at early time points, in comparison with 'loop genes' (48%), indicating that 'anchor genes' are significantly associated with gene up-regulation.

In summary, the majority of high binding ERα binding sites are involved in ERα-bound chromatin interactions, wherein distal ERα binding sites loop toward gene promoters through connection with proximal ERα binding sites. Genes with promoters near interaction anchor regions (anchor genes) are more likely to be active in transcription and up-regulated by estrogen and ERα binding. Distant genes could be brought together by such interactions for coordinated transcription activation. ERα-bound interactions could have heterogeneous looping structures for different functions. Small gene-centric loops could enhance transcription efficiency, and large multi-genic loops could push partition genes residing in loop regions (loop genes) away from interaction anchor centers in transcriptional hubs. More detailed studies are needed to conclude whether NR can mediate an exclusive looping mechanism or they belong to a subset of a more complex and multi-step mechanism. Regardless, mechanistic insights have began to shed light on how NR and their related co-regulators fit into the bigger picture of long-range chromatin interaction.

2.4. Interrogating the ER Cistrome and Interactome

The regulatory long-range chromatin interactions identified by ERα ChIA-PET data are likely to be mediated by ERα. However, ERα may be a necessary factor but it is unlikely the sole factor mediating the looping interaction. Pol II occupancy at distal ERα binding sites has been suggested to represent the "footprint" of looping interaction with an active promoter [53]. Further investigations on key transcription factors such as Pol II will shed light on the exact roles they play in mediating chromatin loop formation. The role of non-NR in the regulation of long-range chromatin interactions such as the scaffold protein SATB1 [54] and the insulator protein CTCF [55] should also be investigated in the context of NR-mediated long-range interactions.

Data from the ERα interactome have demonstrated the role of a specific protein in the mediation of higher order organization of chromosomal structures. However, the identities of other long-range chromatin mediating factors are only emerging and we are just beginning to define the specific roles they play. One of such factor is FOXA1 [13]. FOXA1 depletion was shown to cause the down regulation of genes mediated by ERα long-range chromatin interaction. Although the direct impact of FOXA1 knockdown on chromatin architecture was not assessed, the data suggest a role for FOXA1 in chromatin looping. However, further investigation is needed to firmly establish its exact role in long-range chromatin interaction besides being differentially recruited at distal enhancers [29] either experimentally through ChIA-PET or through analysis on the accumulating wealth of data on the ERα cistrome and interactome.

Advancement in high-throughput sequencing technologies and the consequential increase in the number of transcription factor binding maps available have allowed us to attribute some of these factors to not just being mere co-activators of ERα. The cistrome and interactome of ERα mapped in breast cancer cells has now revealed the importance of spatial organization in estrogen mediated transcription. Considering the complexity of transcriptional regulation, it is likely there are many more transcription factors that are required for co-ordinating ERα function. In an effort to determine the factors that are involved in regulating ERα binding and long-range chromatin interactions as well as the underlying mechanism(s) mediating these processes, our group used a combination of molecular, genomic, and computational approaches to identify and functionally characterize in detail a novel collaborative factor of ERα. Our results suggest AP-2γ, a transcription factor involved in breast cancer oncogenesis, facilitates ERα binding, long-range chromatin interactions, and transcription by working together with FOXA1 [38]. AP-2γ is pre-bound across the genome before estrogen stimulation, with striking similar binding characteristics as the pioneer factor, FOXA1. Specifically, we demonstrate that most ERα binding sites associated with long-range chromatin interactions are co-localized with AP-2γ and FOXA1, suggesting that collaborative action of these two factors can promote recruitment of ERα which is necessary for the subsequent formation of long-range chromatin interactions.

Long-range transcriptional regulation has been increasingly described in nuclear hormone signaling and the factors that are important in modulating the dynamics are beginning to emerge. One such protein is LSD1. This histone lysine demethylase is required for ligand induced interactive loci to associate with distinct inter-chromatin granules [51]. While a loss of LSD1 did not affect inter-chromosomal interaction of TFF1 and Greb1 gene loci in mammary cells, it was shown that such inter-chromosomal interaction did not associate with transcription factories after LSD1 loss. This work further suggests that the organization of the chromatin structure is a multi-step process which involves specific proteins at different steps and occurs in a rapid manner. This discovery is further accompanied by the implications of motor proteins such as NMI and DLC1 whose canonical role have not been traditionally associated with transcription regulation.

It remains controversial whether the organization of chromatin interaction upon ligand induction is a rapid process as contradicting results have been reported [52-53, 56]. Notably, while ERα mediated inter-chromosomal interaction between two gene loci, Greb1 and TFF1, was demonstrated previously [51], recent work reported no such long-range interaction or chromosome movements [56].

This inter-chromosomal interaction was not detected in our ChIA-PET and fluorescence *in situ* hybridization (FISH) dataset [53]. Furthermore, rapid chromosomal movements upon ligand induction were not observed in response to the ligand activation of another NR, GR, through 4C [52]. The discrepancy in observations for ERα may be primarily due to a difference in the treatment of cells. FISH data suggests that the frequency of the co-localization event in normal and cancer cells were not the same [51]. This implies that the co-localization event may be cell-type dependent and sensitive to the presence of certain regulatory factors such as the protein levels of ER. Different chemical treatments and the duration of these treatments may also contribute to a dissimilar outcome. Nonetheless, the formation of long-range chromatin interactions is likely a multi-step process [51] and there is now an increasing need to clarify the roles of proteins in each step. Emerging roles of other groups of proteins related to the cytoskeleton, DNA repair and cell cycle have open up various potential hypothesis and new questions. Eventually, the aim of investigating transcription factors in long-range chromatin interactions are to differentiate what are the factors that require chromatin looping to occur and what are the factors that mediate the chromatin looping event.

2.5. Integrating the Interactome to Understand Human Diseases

Higher order genome organization is functionally relevant for physiological and pathological processes [57]. Defects in how chromatin is globally organized are emerging as key players in human diseases such as cancer. Datasets such as the ERα interactome will assists to further our understanding on estrogen responsive pathways. Taking a genomics approach, Hsu et al., recently characterized the influence of estrogen signaling on long-range epigenetic silencing (LRES) [58]. LRES is a process where concurrent silencing of multiple loci in large chromosome regions is observed. Through analysis of the transcriptome, methylome, and ERα binding datasets from normal breast epithelia and breast cancer cells, the group uncovered 11 large repressive zones (range, 0.35~5.98 megabases), including a 14-gene cluster located on 16p11.2 which was further characterized by 3C. In normal cells, estrogen signaling induced transient formation of multiple DNA loops in the 16p11.2 region by bringing 14 distant loci to focal ERα-docking sites for coordinate repression, much like what was observed previously [53]. However, the plasticity of this free DNA movement was reduced in breast cancer cells. With these observations, they conclude that an inflexible DNA scaffold may be a novel determinant used by breast cancer cells to reinforce estrogen-mediated repression through the acquisition of DNA methylation and repressive chromatin modifications. Assuming that the ERα status is comparable, it would be interesting to see whether the three dimensional chromatin architecture mediated by ERα differs between normal and cancer cells via ChIA-PET.

Genome-wide association studies have identified numerous gene loci that have been associated with many types of cancer. A better understanding on the mechanisms behind how these loci evolve to affect certain diseases can be explored through the systems biology approach. For example, the 8q24 gene desert contains risk loci for multiple epithelial cancers, including colon, breast, and prostate. It has been shown that each risk locus bears epigenetic marks consistent with enhancer elements and forms a tissue-specific long-range chromatin loop with the MYC proto-oncogene located several hundred kilobases away. Thus, it was proposed that the 8q24 risk loci operate through a common mechanism as tissue-specific enhancers of MYC [59]. It would be intriguing to see whether the integration of genomics, transcriptomics and proteomics data related to ERα in breast cancer and AR in prostate cancer can shed better light on the tissue specific mechanism involved. Indeed, there are numerous NR binding sites within this risk locus and both receptors have been shown to play a role in mediating long-range chromatin interactions [53, 60]. Furthermore, co-factors such as FOXA1 has been shown to control distinct groups of AR and ERα binding sites in a tissue specific manner [21]. The integrative approach may thus provide a framework to understand complex gene switching events during disease progression and development at the mechanistic level.

Besides ChIA-PET, different types of chromatin maps developed on the sequencing platform have allowed us unprecedented access to understand the function of chromatin. Such analysis will include using data on chromatin structure and nucleosomes positioning by MNase-Seq, DNaseI-Seq and FAIRE-Seq, as well as, data on genomic DNA methylation. In addition, genome-wide maps of various transcription factors and histones [61] will contribute to further our understanding. The large amount of data has given us the opportunity for a wide range of hypothesis to be built through systems biology. Although much of the data will likely be cell line specific, it will provide us with the grounds for hypothesis building. The theories generated can be further tested and will go beyond the constraints of individual cell lines. For example, genome-wide analysis has provided us with further mechanistic insights into the molecular basis of ERα transcriptional regulation, as well as, endocrine resistance in breast cancer and will continue to shed light on NR biology as we adapt increasingly to an integrative approach.

3. Global Wide Gene Expression Studies

Different components of the genome can come together to robustly process exogenous and endogenous signals for a coordinated response. Although it is important to establish the characteristics of transcription factor binding, the binding patterns may reflect only the current status of the chromatin and are sometimes limited in providing further information on how such signals translate to function. Hence, pairing transcription factor binding data to global wide gene expression results, such as those obtained through the use of microarray, may complement one another. Such large global wide gene expression datasets have also been useful for the identification, qualification, and application of diagnostic and prognostic biomarkers as it remains one of the final goals of current -omics paradigm [62]. However, the search for biomarkers has not been very fruitful compared to the amount of investment thus far [63]. While numerous reasons are related to this, an integrative approach to analyzing microarray data has been applied successfully for the detection of recurrent fusion genes in AR positive prostate cancer [64].

3.1. A Systems Biology Approach to the Discovery of Fusion Genes in Prostate Cancer

AR is paramount for the maintenance of the differentiated prostate epithelial phenotype [65] and is essential for the initiation and progression of prostate cancer [66]. Androgen-deprivation therapies have been widely used for the treatment of prostate cancer. While such therapies initially lead to disease regression in general, advanced prostate cancer ultimately progresses to an androgen-independent late stage that is refractory to current therapies [67]. Indeed, AR is expressed in the vast majority of both types of prostate cancer and decreasing levels of AR protein expression reduces both types of growth in model systems [66]. It has been shown that AR plays a critical role in both by controlling distinct signaling pathways [49].

Prostate specific antigen (PSA) testing is the current method for the detection of early prostate cancer [68]. However, this method cannot distinguish nonaggressive from aggressive prostate cancers [69]. As a consequence, issues concerning over-diagnosis and treatment arises [70]. Given the limitations of PSA as a screening test, new paradigms for prostate cancer detection are needed. Indeed, classical genetic techniques have identified several candidate genes involved in the pathogenesis of prostate cancer, however, a clear picture of its development has not emerged [68]. To indentify the most promising candidates from the wealth of genomic, transcriptomic and proteomic data being generated, integrative methods must be developed [71].

The development of systems biology has provided flexibility on the types of analysis carried out to allow full utilization of the massive amount of data available publicly. With the correct command lines, the scientist can now pose a question from a different angle and use numerous different approaches to find the answer. Taking advantage of this flexibility, the Chinnaiyan group developed a different strategy in their search for biomarkers in prostate cancer. This method searches for marked over-expression in only a subset of cases and allows for the discovery of factors that are significantly over-represented in only a subset. They reasoned that in the majority of cancer types, heterogeneous patterns of oncogene activation have been observed; thus, traditional analytical methods that search for common activation of genes across a class of cancer samples (e.g. **t** test or signal-to-noise ratio) will serve to allow for the discovery of general biomarkers only. Towards this end, they developed a method termed cancer outlier profile analysis (COPA) to accentuate and identify outlier profiles. They applied COPA to microarray datasets and used a bioinformatics approach to discover candidate oncogenic chromosomal aberrations on the basis of outlier gene expression [64].

From their approach, two ETS transcription factors, ERG and ETV1 [64], were identified as outliers in prostate cancer. Recurrent gene fusions of the 5′ untranslated region of TMPRSS2 to ERG or ETV1 in prostate cancer tissues with outlier expression were identified and validated. Cell line experiments also suggest that the androgen-responsive promoter elements of TMPRSS2 mediate the over-expression of ETS family members in prostate cancer. These results have implications on the molecular diagnosis and treatment of prostate cancer as it was subsequently shown that the pharmacological inhibition of PARP1 selectively inhibits ETS-positive prostate cancer xenograft growth [72]. Hence, the discovery of biomarkers of cancer has paved the way to better target and understand the disease. Recently, using a long-span, paired-end deep sequencing strategy, others have also comprehensively identified fusion transcripts in five primary tumours and three breast cancer cell line [73]. However, due to the lack of a dominant fusion gene in breast cancer, it remains to be seen whether a similar approach can lead to alternative treatments in breast cancer.

Gene fusions are a hallmark of cancer development [74], but the mechanisms underlying their genesis and cell-type specificities are unclear. The systems biology approach in the discovery of fusion proteins has allowed a better understanding of the development of prostate carcinomas and the role that AR plays. Indeed, the Chinnaiyan group subsequently showed that androgen signaling induces intra-chromosomal proximity of the TMPRSS2 and ERG genomic loci. Subsequent exposure of the cells to gamma irradiation, which causes DNA double-strand breaks, facilitates the formation of the TMPRSS2-ERG gene fusion. Such observations may help explain why TMPRSS2-ERG fusions are restricted to the prostate, which is dependent on androgen signaling [60]. These results build on earlier work documenting the role of chromosomal proximity in gene fusions [75] and provide a conceptual framework for the genesis of gene fusions in hormone-regulated epithelial cancers. The impact of hormone-induced proximity may be extended to other NR. Estrogen was shown to induce rapid chromosomal movements that bring together ERα-bound genes on different chromosomes [51] and within chromosomes [53]. Given the broad similarities between estrogen and androgen signaling, it may be reasonable to speculate that other NR might likewise induce chromosomal movements and bring together specific genes, thereby facilitating the formation of gene fusions and other global alterations in chromatin structure.

The integrative approach to the discovery of gene fusions in prostate cancer has encouraged the development of other systems biology approaches to answer specific questions on prostate cancer pathogenesis. To further understand the formation of non-ETS fusion events in prostate cancer, Rubin and colleagues used next-generation transcriptome sequencing (RNA-seq) to explore the whole transcriptome of 25 human prostate cancer samples for the presence of chimeric fusion transcripts. They also developed a novel computational approach Fusion-seq [76] to identify novel gene fusion candidates with high confidence. In total, they discovered and characterized an additional seven new cancer-specific gene fusions, an oncogene and a tumor suppressor gene [77].

Progress in our understanding of prostate cancer is achievable by using an integrative approach. Furthermore, the data indicate that some of the genomic lesions are AR mediated. Future information on the interactome of AR may facilitate the understanding of fusion protein formation.

3.2. Integrating the Genome, the Transcriptome and the Proteome

To begin understanding prostate cancer progression from a systems biology perspective, we need to characterize and integrate the different molecular components involved [78]. A number of groups have used gene expression microarrays to profile prostate cancer tissues at the transcriptome level. Other recent or improved high through-put methods such as RNA-seq [77, 79] and LongSAGE [80], has also contributed to the transcriptome datasets available. In addition, a systematically mapped genomic landscape of AR, ERG, FOXA1, and Pol II, along with eight critical histone marks in multiple prostate cancer cell lines as well as in one prostate tumor specimen was also recently created [8]. In this work, a compendium of 57 genome-wide ChIP-seq experiments and a large set of paired microarray expression profiling data were established [8]. An integrative approach to the analysis of the results shows that TMPRSS2-ERG plays a critical role in cancer progression by disrupting androgen signaling, inhibiting normal prostate differentiation, and turning on EZH2 expression to induce an ES cell-like de-differentiation program. In addition, by inhibiting AR signaling, TMPRSS2-ERG may exert a selective pressure for the development of prostate cancer that is resistant to hormone deprivation therapies. Together, the compendium provided a wealth of data to be analyzed at the genome and transcriptome levels. Further on, advances in high throughput proteomics may provide additional input for analysis.

Proteins represent the functional effectors of cancer progression and can serve as therapeutic targets and markers of disease. Proteomics approaches facilitate the identification of proteins and biochemical pathways involved in tumor development. Multiple technologies have been used to identify or compare proteomic alterations in the serum of prostate cancer patients including protein microarray and SELDI [81-82]. Other groups have used isobaric tags for relative and absolute quantitation (iTRAQ) based quantification to assess global alterations in the proteome using tissues from prostate cancer patients and found 65 proteins to be differentially regulated between paired benign and tumor samples [83]. A high throughput application of an 8-plex iTRAQ approach was also used to profile the proteomes of two distinct panels of isogenic prostate cancer cells lines, with varying growth and metastatic potentials, to quantify global changes associated with the prostate cancer proteome in the stages of progression from organ-confined to metastatic disease [84]. Among the candidates identified, ERp5 showed a significant higher immune-expression in pre-malignant lesions compared with non-malignant epithelium and in high Gleason grade (4-5) versus low grade (2-3) cancers when verified through immunostaining of prostate tissue microarrays.

Numerous proteomic studies using human plasma or serum with a focus on prostate cancer research have been reported to date. Although these efforts led to a list of potential protein biomarkers that have been partly characterized, most remain to be validated with a statistically meaningful study cohort size [85].

More recent developments have employ relatively specific and homogeneous cells obtained by laser capture micro-dissection, seminal fluid and urine for the discovery of prostate-specific secreted proteins [85]. This may allow a better understanding of the molecular changes in the cancer microenvironment more proximal to where cancers arise. Secreted proteins from cell lines have also been targeted for biomarker discovery by using conditioned media from three human prostate cancer cell lines of different origin, PC3 (bone metastasis), LNCaP (lymph node metastasis), and 22Rv1 (localized to prostate) with LC–MS/MS (liquid chromatography tandem mass spectrometry) [86]. Follistatin (FST), chemokine ligand 16 (CXCL16), pentraxin 3 (PTX3), and spondin 2 (SPON2) were suggested as putative biomarkers and subject to further validation using 42 serum samples from patients with or without prostate cancer. Each of these three candidate biomarkers showed a significant difference in up-regulation of patients with prostate cancer.

Due to the non-invasive nature of sample collection and reduced complexity of the proteome over serum, analyzing urine samples has become increasingly popular in recent years [85]. Extensive analysis of urine by proteomics reveal that more than 1500 proteins constitute the normal urine proteome [87]. A large-scale urinary proteomics study was previously carried out to distinguish between different grades of prostate cancer from 407 patient samples by using MALDI-TOF. Using this strategy, uromodulin and semenogelin I (SEMG1) isoform b preproprotein were proposed as two biomarkers that enable distinction between benign and prostate cancer. Unfortunately, further validation indicated that the biomarkers were not specific for prostate cancer [88].

Although gene and protein expression have been extensively profiled in human tumors, little is known about the global metabolomic alterations that characterize cancer progression [89]. Metabolomics is one of the latest biomarker discovery approaches to identify and separate metabolites using technology similar to proteomics [85]. These small molecules are often the final products of biochemical activity in molecular pathways and so may be used within a perturbed biological system to distinguish normal from cancer. Using a combination of high-throughput LC/MS, more than 1126 metabolites were recently identified across prostate cancer related clinical samples of tissue, urine and plasma [89]. Notably, these unbiased metabolomic profiles allowed distinctions to be made between benign prostate, clinically localized prostate cancer and metastatic disease through non-invasive measurements of a distinct metabolite, sarcosine, in the urine. At the molecular level, the master transcriptional regulators of prostate cancer progression, AR and the ETS gene fusions, seem to directly regulate sarcosine by means of transcriptional control of its regulatory enzymes. By profiling the metabolomic alterations of prostate cancer progression, sarcosine was shown to be a potentially important metabolic intermediary of cancer cell invasion and aggressiveness. Components of the sarcosine pathway may have potential as biomarkers of prostate cancer progression and serve as new avenues for therapeutic intervention. Together with accumulating high throughput datasets available, the integrative approach in understanding AR driven prostate cancer seems most poised than other NR mediated diseases to be fully exploited.

3.3. Biomarkers from MicroRNA (miRNA)

Despite these vast amounts of dataset available, the field has only just begun to tap on these resources. Even so, there remain several avenues for exploration. Recently, the proteomic alterations in a cohort of prostate-derived tissues from adjacent benign prostate, clinically localized prostate cancer, and metastatic disease from distant sites were interrogated using iTRAQ and MS [78]. Although over 1000 proteins were found to be differentially regulated, enrichment analysis of the prostate cancer-specific proteomic signature also revealed involvement of miR-128-a/b as a potentially important negative regulator of prostate cancer cell invasion. This finding was further validated in an independent set of 15 clinical specimens. Indeed, the potential role of miRNAs in human cancers has been indicated by several studies which suggest that aberrations in miRNA expression in cancers may be involved in tumor genesis and progression [90]. Abnormal expression of miRNAs has been found in various cancers, and profiling of miRNAs has been shown to be a more accurate method of classifying cancer subtypes than profiling protein-coding genes. In the case of prostate cancer, several miRNA expression profiles have been reported and at least 18 different miRNAs and their targets have been verified biologically in prostate cancer [91].

The global expression patterns of miRNA targets in human prostate cancer using several published microarray datasets were recently explored [92]. Interestingly, in contrast to other mRNAs with altered expression in prostate cancer relative to benign prostate tissue, the transcript levels of miRNA targets were significantly lower. It was also found that the transcript levels of miRNA targets were lower in androgen-independent prostate cancer than in androgen-dependent prostate cancer. Furthermore, when the global analysis was extended to other types of cancers, significant differences in transcript levels between miRNA targets and other genes were found. Importantly, the data suggests that miRNAs have the ability to produce a unique transcriptome signature that may allow differentiation between various class and types of cancer.

The role of miRNA regulation by NR is being established in NR-dependent cancers. NR regulated miRNAs and their specific targets may play crucial roles in establishing phenotypes at different stages of cancer progression. A comprehensive list on the identities of miRNAs that are directly regulated by NR has just begun to emerged through high throughput transcriptome datasets such as those from Global Run-on sequencing (GRO-seq) datasets in breast [93] and prostate [94] cancers. Further work with tissue samples of various cancer stages or matched samples of patient's normal and cancer tissue will clarify whether these hormone sensitive miRNA can serve as prognostic and diagnostic markers for cancer progression beyond the cell line.

Recently, miRNAs profiling from 11 normal breast tissues, 17 non-invasive, 151 invasive breast carcinomas, and 6 cell lines by barcoded Solexa-sequencing were carried out [95]. A systems biology approach allowed for an unbiased clustering and comparative analysis of these miRNA data. Through different miRNA profiles, distinctions between normal breast samples from most non-invasive ductal carcinoma in situ and invasive carcinomas could be made. Furthermore, unique miRNA markers for triple negative breast carcinomas were established. Minimally invasive methods and accurate markers of tumors could improve early diagnosis, reduce treatment costs and the mortality rate of the cancers [91]. Recently, the detection of cell-free nucleic acids (cfNA) in plasma or serum could serve as a 'liquid biopsy' which would be useful for numerous diagnostic applications and would avoid the need for tumour tissue biopsies [96]. Increasing attention has been paid to cfNAs such as DNA, mRNA and miRNA. The miRNAs present in plasma were reported to be stable against varying conditions and endogenous RNases [97]. Thus, understanding such interactions can provide prognostic utility as well as the potential for therapeutic intervention in the future.

4. PERSPECTIVES

4.1. Combination of Genomics and Proteomics

Systems biology has and will continue to allow us to develop a network-level understanding of diseases such as cancer. With advances in sequencing technologies, proliferation of high throughput datasets, and increased capabilities in systems biology, we will begin to move beyond candidate gene identification and progress into the analysis of network signatures and a compendium of molecular concepts [98] to develop a comprehensive understanding of cancer pathogenesis. This will allow us to better predict drug response and guide therapeutics in the future. Examples in the computational strategies used in the study of AR mediated prostate cancer have highlighted the effectiveness and usefulness of this approach (Figure 3).

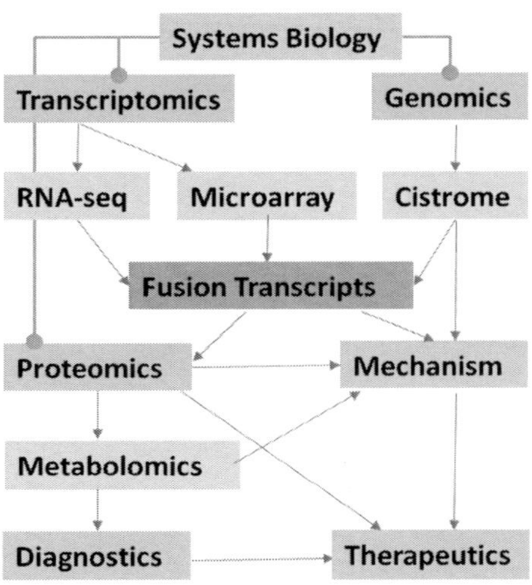

Figure 3. The Systems Biology approach to understand AR driven prostate cancer utilizes different nodes of high throughput data. COPA analysis of microarray data allowed the discovery of fusion transcripts. Other fusion transcripts were identified with RNA-seq and through the development of new computational approaches. Genomics analysis of ChIP-seq data offer insights into transcription factor cistromes and provide hypothesis building of the molecular mechanisms involved. Proteomics allowed the development of diagnostics through metabolomics and also help identify novel interacting proteins of the fusion transcripts. The identification of new interacting partners provided new therapeutic targets for clinical intervention.

Current analysis is mainly driven by the disease it represents. Hence, integrative data analysis is usually disease-centric. However, analyzing the data from a NR based point of view will likely also expand our understanding since NR are drug-able targets which will allow for the manipulation of the genome, transcriptome and proteome through its cognate ligands. In the future, data analysis using high throughput RNAi screening and small molecule perturbations to discover key network components and collaborative factors will yield further results.

4.2. Personalized Medicine and Networks

The constant lowering of cost in sequencing has not only fueled the increasing amount of data available for analysis, it has also allowed the possibility of sequencing an individual's genome to materialize. With the increasing ability to manipulate, query and reconstruct functional molecular networks [11], it may mean that one can predict the outcome of treatments via data input from an individual's genome. Furthermore, systems biology approaches identifying molecular drivers and biomarkers will lead to the implementation of smaller, shorter, cheaper, and individualized clinical trials [99].

Acknowledgments

We would like to thank those scientists in the field whose work, both cited and not cited in this review, have helped to advance our understanding of these fundamentally important biological processes. We also like to thank Kern Rei Chng for reading and providing comments on the chapter. In addition, we would like to acknowledge the generous support of our research from Agency for Science, Technology and Research, Singapore.

References

[1] S.C. Biddie, S. John, and G.L. Hager. Genome-wide mechanisms of nuclear receptor action. *Trends Endocrinol. Metab.*, 21(1):3-9, 2010.

[2] M.Y. Kim, E.M. Woo, Y.T. Chong, D.R. Homenko, and W.L. Kraus. Acetylation of estrogen receptor alpha by p300 at lysines 266 and 268 enhances the deoxyribonucleic acid binding and transactivation activities of the receptor. *Mol. Endocrinol.*, 20(7):1479-1493, 2006.

[3] M. Fu, C. Wang, A.T. Reutens, J. Wang, R.H. Angeletti, L. Siconolfi-Baez, V. Ogryzko, M.L. Avantaggiati, and R.G. Pestell. p300 and p300/cAMP-response element-binding protein-associated factor acetylate the androgen receptor at sites governing hormone-dependent transactivation. *J. Biol. Chem.*, 275(27):20853-20860, 2000.

[4] B. York and B.W. O'Malley. Steroid receptor coactivator (SRC) family: masters of systems biology. *J. Biol. Chem.*, 285(50):38743-38750, 2010.

[5] A.A. Serandour, S. Avner, F. Percevault, F. Demay, M. Bizot, C. Lucchetti-Miganeh, F. Barloy-Hubler, M. Brown, M. Lupien, R. Metivier, G. Salbert, and J. Eeckhoute. Epigenetic switch involved in activation of pioneer factor FOXA1-dependent enhancers. *Genome Res.*, 21(4):555-565, 2011.

[6] X.F. Liu and M.K. Bagchi. Recruitment of distinct chromatin-modifying complexes by tamoxifen-complexed estrogen receptor at natural target gene promoters in vivo. *J. Biol. Chem.*, 279(15):15050-15058, 2004.

[7] K.M. Dreijerink, K.W. Mulder, G.S. Winkler, J.W. Hoppener, C.J. Lips, and H.T. Timmers. Menin links estrogen receptor activation to histone H3K4 trimethylation. *Cancer Res.*, 66(9):4929-4935, 2006.

[8] J. Yu, R.S. Mani, Q. Cao, C.J. Brenner, X. Cao, X. Wang, L. Wu, J. Li, M. Hu, Y. Gong, H. Cheng, B. Laxman, A. Vellaichamy, S. Shankar, Y. Li, S.M. Dhanasekaran, R. Morey, T. Barrette, R.J. Lonigro, S.A. Tomlins, S. Varambally, Z.S. Qin, and A.M. Chinnaiyan. An integrated network of androgen receptor, polycomb, and TMPRSS2-ERG gene fusions in prostate cancer progression. *Cancer Cell*, 17(5):443-454, 2010.

[9] Y. Shi, F. Lan, C. Matson, P. Mulligan, J.R. Whetstine, P.A. Cole, and R.A. Casero. Histone demethylation mediated by the nuclear amine oxidase homolog LSD1. *Cell*, 119(7):941-953, 2004.

[10] F. Lan, P.E. Bayliss, J.L. Rinn, J.R. Whetstine, J.K. Wang, S. Chen, S. Iwase, R. Alpatov, I. Issaeva, E. Canaani, T.M. Roberts, H.Y. Chang, and Y. Shi. A histone H3 lysine 27 demethylase regulates animal posterior development. *Nature*, 449(7163):689-694, 2007.

[11] D. Pe'er and N. Hacohen. Principles and strategies for developing network models in cancer. *Cell*, 144(6):864-873, 2011.

[12] S. Ali, L. Buluwela, and R.C. Coombes. Antiestrogens and their therapeutic applications in breast cancer and other diseases. *Annu. Rev. Med.*, 62:217-232, 2011.

[13] J.S. Carroll, X.S. Liu, A.S. Brodsky, W. Li, C.A. Meyer, A.J. Szary, J. Eeckhoute, W. Shao, E.V. Hestermann, T.R. Geistlinger, E.A. Fox, P.A. Silver, and M. Brown. Chromosome-wide mapping of estrogen receptor binding reveals long-range regulation requiring the forkhead protein FoxA1. *Cell*, 122(1):33-43, 2005.

[14] J.S. Carroll, C.A. Meyer, J. Song, W. Li, T.R. Geistlinger, J. Eeckhoute, A.S. Brodsky, E.K. Keeton, K.C. Fertuck, G.F. Hall, Q. Wang, S. Bekiranov, V. Sementchenko, E.A. Fox, P.A. Silver, T.R. Gingeras, X.S. Liu, and M. Brown. Genome-wide analysis of estrogen receptor binding sites. *Nat. Genet.*, 38(11):1289-1297, 2006.

[15] C.Y. Lin, V.B. Vega, J.S. Thomsen, T. Zhang, S.L. Kong, M. Xie, K.P. Chiu, L. Lipovich, D.H. Barnett, F. Stossi, A. Yeo, J. George, V.A. Kuznetsov, Y.K. Lee, T.H. Charn, N. Palanisamy, L.D. Miller, E. Cheung, B.S. Katzenellenbogen, Y. Ruan, G. Bourque, C.L. Wei, and E.T. Liu. Whole-genome cartography of estrogen receptor alpha binding sites. *PLoS Genet.*, 3(6):e87, 2007.

[16] W.J. Welboren, M.A. van Driel, E.M. Janssen-Megens, S.J. van Heeringen, F.C. Sweep, P.N. Span, and H.G. Stunnenberg. ChIP-Seq of ERalpha and RNA polymerase II defines genes differentially responding to ligands. *EMBO J.*, 28(10):1418-1428, 2009.

[17] M. Lupien, C.A. Meyer, S.T. Bailey, J. Eeckhoute, J. Cook, T. Westerling, X. Zhang, J.S. Carroll, D.R. Rhodes, X.S. Liu, and M. Brown. Growth factor stimulation induces a distinct ER(alpha) cistrome underlying breast cancer endocrine resistance. *Genes Dev.*, 24(19):2219-2227, 2010.

[18] S.A. Krum, G.A. Miranda-Carboni, M. Lupien, J. Eeckhoute, J.S. Carroll, and M. Brown. Unique ERalpha cistromes control cell type-specific gene regulation. *Mol. Endocrinol.*, 22(11):2393-2406, 2008.

[19] P. Bhat-Nakshatri, G. Wang, H. Appaiah, N. Luktuke, J.S. Carroll, T.R. Geistlinger, M. Brown, S. Badve, Y. Liu, and H. Nakshatri. AKT alters genome-wide estrogen receptor alpha binding and impacts estrogen signaling in breast cancer. *Mol. Cell Biol.*, 28(24):7487-7503, 2008.

[20] Y. Liu, H. Gao, T.T. Marstrand, A. Strom, E. Valen, A. Sandelin, J.A. Gustafsson, and K. Dahlman-Wright. The genome landscape of ERalpha- and ERbeta-binding DNA regions. *Proc. Natl. Acad. Sci. USA,* 105(7):2604-2609, 2008.

[21] A. Hurtado, K.A. Holmes, C.S. Ross-Innes, D. Schmidt, and J.S. Carroll. FOXA1 is a key determinant of estrogen receptor function and endocrine response. *Nat. Genet.*, 43(1):27-33, 2011.

[22] K.A. Holmes, A. Hurtado, G.D. Brown, R. Launchbury, C.S. Ross-Innes, J. Hadfield, D.T. Odom, and J.S. Carroll. Breast Cancer Special Feature: Transducin-like enhancer protein 1 mediates estrogen receptor binding and transcriptional activity in breast cancer cells. *Proc. Natl. Acad. Sci. USA,* 2011.

[23] C.S. Ross-Innes, R. Stark, K.A. Holmes, D. Schmidt, C. Spyrou, R. Russell, C.E. Massie, S.L. Vowler, M. Eldridge, and J.S. Carroll. Cooperative interaction between retinoic acid receptor-alpha and estrogen receptor in breast cancer. *Genes Dev.*, 24(2):171-182, 2010.

[24] E. Cheung and W.L. Kraus. Genomic analyses of hormone signaling and gene regulation. *Annu. Rev. Physiol.*, 72:191-218, 2010.

[25] T.L. Bailey, N. Williams, C. Misleh, and W.W. Li. MEME: discovering and analyzing DNA and protein sequence motifs. *Nucleic Acids Res.*, 34(Web Server issue):W369-373, 2006.

[26] A.A. Sharov and M.S. Ko. Exhaustive search for over-represented DNA sequence motifs with CisFinder. *DNA Res.*, 16(5):261-273, 2009.

[27] Z. Zhang, C.W. Chang, W.L. Goh, W.K. Sung, and E. Cheung. CENTDIST: discovery of co-associated factors by motif distribution. *Nucleic Acids Res.*, 2011.

[28] J. Laganiere, G. Deblois, C. Lefebvre, A.R. Bataille, F. Robert, and V. Giguere. Location analysis of estrogen receptor alpha target promoters reveals that FOXA1 defines a domain of the estrogen response. *Proc. Natl. Acad. Sci. USA,* 102(33):11651-11656, 2005.

[29] M. Lupien, J. Eeckhoute, C.A. Meyer, Q. Wang, Y. Zhang, W. Li, J.S. Carroll, X.S. Liu, and M. Brown. FoxA1 translates epigenetic signatures into enhancer-driven lineage-specific transcription. *Cell*, 132(6):958-970, 2008.

[30] L.A. Cirillo, C.E. McPherson, P. Bossard, K. Stevens, S. Cherian, E.Y. Shim, K.L. Clark, S.K. Burley, and K.S. Zaret. Binding of the winged-helix transcription factor HNF3 to a linker histone site on the nucleosome. *EMBO J*, 17(1):244-254, 1998.

[31] L.A. Cirillo, F.R. Lin, I. Cuesta, D. Friedman, M. Jarnik, and K.S. Zaret. Opening of compacted chromatin by early developmental transcription factors HNF3 (FoxA) and GATA-4. *Mol. Cell*, 9(2):279-289, 2002.

[32] S. Badve, D. Turbin, M.A. Thorat, A. Morimiya, T.O. Nielsen, C.M. Perou, S. Dunn, D.G. Huntsman, and H. Nakshatri. FOXA1 expression in breast cancer--correlation with luminal subtype A and survival. *Clin. Cancer Res.*, 13(15 Pt 1):4415-4421, 2007.

[33] J. Eeckhoute, E.K. Keeton, M. Lupien, S.A. Krum, J.S. Carroll, and M. Brown. Positive cross-regulatory loop ties GATA-3 to estrogen receptor alpha expression in breast cancer. *Cancer Res.*, 67(13):6477-6483, 2007.

[34] J.D. Stender, K. Kim, T.H. Charn, B. Komm, K.C. Chang, W.L. Kraus, C. Benner, C.K. Glass, and B.S. Katzenellenbogen. Genome-wide analysis of estrogen receptor alpha DNA binding and tethering mechanisms identifies Runx1 as a novel tethering factor in receptor-mediated transcriptional activation. *Mol. Cell Biol.*, 30(16):3943-3955, 2010.

[35] K.A. Holmes, J.S. Song, X.S. Liu, M. Brown, and J.S. Carroll. Nkx3-1 and LEF-1 function as transcriptional inhibitors of estrogen receptor activity. *Cancer Res.*, 68(18):7380-7385, 2008.

[36] A. Hurtado, K.A. Holmes, T.R. Geistlinger, I.R. Hutcheson, R.I. Nicholson, M. Brown, J. Jiang, W.J. Howat, S. Ali, and J.S. Carroll. Regulation of ERBB2 by oestrogen receptor-PAX2 determines response to tamoxifen. *Nature*, 456(7222):663-666, 2008.

[37] T. Misteli. Beyond the sequence: cellular organization of genome function. *Cell*, 128(4):787-800, 2007.

[38] S.K. Tan, Z.H. Lin, C.W. Chang, V. Varang, K.R. Chng, Y.F. Pan, E.L. Yong, W.K. Sung, and E. Cheung. AP-2gamma regulates oestrogen receptor-mediated long-range chromatin interaction and gene transcription. *EMBO J.*, 2011.

[39] J. Boney-Montoya, Y.S. Ziegler, C.D. Curtis, J.A. Montoya, and A.M. Nardulli. Long-Range Transcriptional Control of Progesterone Receptor Gene Expression. *Mol. Endocrinol.*, 2009.

[40] N. Bretschneider, K. Sara, M. Seifert, G. Reid, F. Gannon, and S. Denger. E2-mediated cathepsin D (CTSD) activation involves looping of distal enhancer elements. *Mol. Oncol.*, 2(2):182-190, 2008.

[41] D.H. Barnett, S. Sheng, T.H. Charn, A. Waheed, W.S. Sly, C.Y. Lin, E.T. Liu, and B.S. Katzenellenbogen. Estrogen receptor regulation of carbonic anhydrase XII through a distal enhancer in breast cancer. *Cancer Res.*, 68(9):3505-3515, 2008.

[42] J. Deschenes, V. Bourdeau, J.H. White, and S. Mader. Regulation of GREB1 transcription by estrogen receptor alpha through a multipartite enhancer spread over 20 kb of upstream flanking sequences. *J. Biol. Chem.*, 282(24):17335-17339, 2007.

[43] Y.F. Pan, K.D. Wansa, M.H. Liu, B. Zhao, S.Z. Hong, P.Y. Tan, K.S. Lim, G. Bourque, E.T. Liu, and E. Cheung. Regulation of estrogen receptor-mediated long range transcription via evolutionarily conserved distal response elements. *J. Biol. Chem.*, 283(47):32977-32988, 2008.

[44] O. Hakim, S. John, J.Q. Ling, S.C. Biddie, A.R. Hoffman, and G.L. Hager. Glucocorticoid receptor activation of the Ciz1-Lcn2 locus by long range interactions. *J. Biol. Chem.*, 284(10):6048-6052, 2009.

[45] T. Jakobsson, N. Venteclef, G. Toresson, A.E. Damdimopoulos, A. Ehrlund, X. Lou, S. Sanyal, K.R. Steffensen, J.A. Gustafsson, and E. Treuter. GPS2 is required for cholesterol efflux by triggering histone demethylation, LXR recruitment, and coregulator assembly at the ABCG1 locus. *Mol. Cell*, 34(4):510-518, 2009.

[46] A. Saramaki, S. Diermeier, R. Kellner, H. Laitinen, S. Vaisanen, and C. Carlberg. Cyclical chromatin looping and transcription factor association on the regulatory regions of the p21 (CDKN1A) gene in response to 1alpha,25-dihydroxyvitamin D3. *J. Biol. Chem.*, 284(12):8073-8082, 2009.

[47] Q. Wang, J.S. Carroll, and M. Brown. Spatial and temporal recruitment of androgen receptor and its coactivators involves chromosomal looping and polymerase tracking. *Mol. Cell*, 19(5):631-642, 2005.

[48] Q. Wang, W. Li, X.S. Liu, J.S. Carroll, O.A. Janne, E.K. Keeton, A.M. Chinnaiyan, K.J. Pienta, and M. Brown. A hierarchical network of transcription factors governs androgen receptor-dependent prostate cancer growth. *Mol. Cell*, 27(3):380-392, 2007.

[49] Q. Wang, W. Li, Y. Zhang, X. Yuan, K. Xu, J. Yu, Z. Chen, R. Beroukhim, H. Wang, M. Lupien, T. Wu, M.M. Regan, C.A. Meyer, J.S. Carroll, A.K. Manrai, O.A. Janne, S.P. Balk, R. Mehra, B. Han, A.M. Chinnaiyan, M.A. Rubin, L. True, M. Fiorentino, C. Fiore, M. Loda, P.W. Kantoff, X.S. Liu, and M. Brown. Androgen receptor regulates a distinct transcription program in androgen-independent prostate cancer. *Cell*, 138(2):245-256, 2009.

[50] H. Makkonen, M. Kauhanen, V. Paakinaho, T. Jaaskelainen, and J.J. Palvimo. Long-range activation of FKBP51 transcription by the androgen receptor via distal intronic enhancers. *Nucleic Acids Res.*, 37(12):4135-4148, 2009.

[51] Q. Hu, Y.S. Kwon, E. Nunez, M.D. Cardamone, K.R. Hutt, K.A. Ohgi, I. Garcia-Bassets, D.W. Rose, C.K. Glass, M.G. Rosenfeld, and X.D. Fu. Enhancing nuclear receptor-induced transcription requires nuclear motor and LSD1-dependent gene networking in interchromatin granules. *Proc. Natl. Acad. Sci. USA,* 105(49):19199-19204, 2008.

[52] O. Hakim, M.H. Sung, T.C. Voss, E. Splinter, S. John, P.J. Sabo, R.E. Thurman, J.A. Stamatoyannopoulos, W. de Laat, and G.L. Hager. Diverse gene reprogramming events occur in the same spatial clusters of distal regulatory elements. *Genome Res..*, 21(5):697-706, 2011.

[53] M.J. Fullwood, M.H. Liu, Y.F. Pan, J. Liu, H. Xu, Y.B. Mohamed, Y.L. Orlov, S. Velkov, A. Ho, P.H. Mei, E.G. Chew, P.Y. Huang, W.J. Welboren, Y. Han, H.S. Ooi, P.N. Ariyaratne, V.B. Vega, Y. Luo, P.Y. Tan, P.Y. Choy, K.D. Wansa, B. Zhao, K.S. Lim, S.C. Leow, J.S. Yow, R. Joseph, H. Li, K.V. Desai, J.S. Thomsen, Y.K. Lee, R.K. Karuturi, T. Herve, G. Bourque, H.G. Stunnenberg, X. Ruan, V. Cacheux-Rataboul, W.K. Sung, E.T. Liu, C.L. Wei, E. Cheung, and Y. Ruan. An oestrogen-receptor-alpha-bound human chromatin interactome. *Nature*, 462(7269):58-64, 2009.

[54] H.J. Han, J. Russo, Y. Kohwi, and T. Kohwi-Shigematsu. SATB1 reprogrammes gene expression to promote breast tumour growth and metastasis. *Nature*, 452(7184):187-193, 2008.

[55] D. Schmidt, P.C. Schwalie, C.S. Ross-Innes, A. Hurtado, G.D. Brown, J.S. Carroll, P. Flicek, and D.T. Odom. A CTCF-independent role for cohesin in tissue-specific transcription. *Genome Res.*, 20(5):578-588, 2010.

[56] S. Kocanova, E.A. Kerr, S. Rafique, S. Boyle, E. Katz, S. Caze-Subra, W.A. Bickmore, and K. Bystricky. Activation of estrogen-responsive genes does not require their nuclear co-localization. *PLoS Genet*, 6(4):e1000922, 2010.

[57] T. Misteli. Higher-order genome organization in human disease. *Cold Spring Harb Perspect Biol.*, 2(8):a000794, 2010.

[58] P.Y. Hsu, H.K. Hsu, G.A. Singer, P.S. Yan, B.A. Rodriguez, J.C. Liu, Y.I. Weng, D.E. Deatherage, Z. Chen, J.S. Pereira, R. Lopez, J. Russo, Q. Wang, C.A. Lamartiniere, K.P. Nephew, and T.H. Huang. Estrogen-mediated epigenetic repression of large chromosomal regions through DNA looping. *Genome Res.*, 20(6):733-744, 2010.

[59] N. Ahmadiyeh, M.M. Pomerantz, C. Grisanzio, P. Herman, L. Jia, V. Almendro, H.H. He, M. Brown, X.S. Liu, M. Davis, J.L. Caswell, C.A. Beckwith, A. Hills, L. Macconaill, G.A. Coetzee, M.M. Regan, and M.L. Freedman. 8q24 prostate, breast, and colon cancer risk loci show tissue-specific long-range interaction with MYC. *Proc. Natl. Acad. Sci. USA,* 107(21):9742-9746, 2010.

[60] R.S. Mani, S.A. Tomlins, K. Callahan, A. Ghosh, M.K. Nyati, S. Varambally, N. Palanisamy, and A.M. Chinnaiyan. Induced chromosomal proximity and gene fusions in prostate cancer. *Science*, 326(5957):1230, 2009.

[61] R. Joseph, Y.L. Orlov, M. Huss, W. Sun, S.L. Kong, L. Ukil, Y.F. Pan, G. Li, M. Lim, J.S. Thomsen, Y. Ruan, N.D. Clarke, S. Prabhakar, E. Cheung, and E.T. Liu. Integrative model of genomic factors for determining binding site selection by estrogen receptor-alpha. *Mol. Syst. Biol.*, 6:456, 2010.

[62] M.S. Abu-Asab, M. Chaouchi, S. Alesci, S. Galli, M. Laassri, A.K. Cheema, F. Atouf, J. VanMeter, and H. Amri. Biomarkers in the age of omics: time for a systems biology approach. *OMICS*, 15(3):105-112, 2011.

[63] Nature. Valid concerns. *Nature*, 463(7280):401-402, 2010.

[64] S.A. Tomlins, D.R. Rhodes, S. Perner, S.M. Dhanasekaran, R. Mehra, X.W. Sun, S. Varambally, X. Cao, J. Tchinda, R. Kuefer, C. Lee, J.E. Montie, R.B. Shah, K.J. Pienta, M.A. Rubin, and A.M. Chinnaiyan. Recurrent fusion of TMPRSS2 and ETS transcription factor genes in prostate cancer. *Science*, 310(5748):644-648, 2005.

[65] M.E. Wright, M.J. Tsai, and R. Aebersold. Androgen receptor represses the neuroendocrine transdifferentiation process in prostate cancer cells. *Mol. Endocrinol.*, 17(9):1726-1737, 2003.

[66] C.A. Heinlein and C. Chang. Androgen receptor in prostate cancer. *Endocr. Rev.*, 25(2):276-308, 2004.

[67] J.D. Debes and D.J. Tindall. Mechanisms of androgen-refractory prostate cancer. *N. Engl. J. Med.*, 351(15):1488-1490, 2004.

[68] S.A. Strope and G.L. Andriole. Prostate cancer screening: current status and future perspectives. *Nat. Rev. Urol.*, 7(9):487-493, 2010.

[69] W.J. Catalona, D.S. Smith, T.L. Ratliff, K.M. Dodds, D.E. Coplen, J.J. Yuan, J.A. Petros, and G.L. Andriole. Measurement of prostate-specific antigen in serum as a screening test for prostate cancer. *N. Engl. J. Med.*, 324(17):1156-1161, 1991.

[70] F.H. Schroder, J. Hugosson, M.J. Roobol, T.L. Tammela, S. Ciatto, V. Nelen, M. Kwiatkowski, M. Lujan, H. Lilja, M. Zappa, L.J. Denis, F. Recker, A. Berenguer, L. Maattanen, C.H. Bangma, G. Aus, A. Villers, X. Rebillard, T. van der Kwast, B.G. Blijenberg, S.M. Moss, H.J. de Koning, and A. Auvinen. Screening and prostate-cancer mortality in a randomized European study. *N. Engl. J. Med.*, 360(13):1320-1328, 2009.

[71] S.A. Tomlins, M.A. Rubin, and A.M. Chinnaiyan. Integrative biology of prostate cancer progression. *Annu. Rev. Pathol.*, 1:243-271, 2006.

[72] J.C. Brenner, B. Ateeq, Y. Li, A.K. Yocum, Q. Cao, I.A. Asangani, S. Patel, X. Wang, H. Liang, J. Yu, N. Palanisamy, J. Siddiqui, W. Yan, X. Cao, R. Mehra, A. Sabolch, V. Basrur, R.J. Lonigro, J. Yang, S.A. Tomlins, C.A. Maher, K.S. Elenitoba-Johnson, M. Hussain, N.M. Navone, K.J. Pienta, S. Varambally, F.Y. Feng, and A.M. Chinnaiyan. Mechanistic Rationale for Inhibition of Poly(ADP-Ribose) Polymerase in ETS Gene Fusion-Positive Prostate Cancer. *Cancer Cell*, 19(5):664-678, 2011.

[73] K. Inaki, A.M. Hillmer, L. Ukil, F. Yao, X.Y. Woo, L.A. Vardy, K.F. Zawack, C.W. Lee, P.N. Ariyaratne, Y.S. Chan, K.V. Desai, J. Bergh, P. Hall, T.C. Putti, W.L. Ong, A. Shahab, V. Cacheux-Rataboul, R.K. Karuturi, W.K. Sung, X. Ruan, G. Bourque, Y. Ruan, and E.T. Liu. Transcriptional consequences of genomic structural aberrations in breast cancer. *Genome Res.*, 21(5):676-687, 2011.

[74] J.D. Rowley. The critical role of chromosome translocations in human leukemias. *Annu. Rev. Genet.*, 32:495-519, 1998.

[75] M.N. Nikiforova, J.R. Stringer, R. Blough, M. Medvedovic, J.A. Fagin, and Y.E. Nikiforov. Proximity of chromosomal loci that participate in radiation-induced rearrangements in human cells. *Science*, 290(5489):138-141, 2000.

[76] A. Sboner, L. Habegger, D. Pflueger, S. Terry, D.Z. Chen, J.S. Rozowsky, A.K. Tewari, N. Kitabayashi, B.J. Moss, M.S. Chee, F. Demichelis, M.A. Rubin, and M.B. Gerstein. FusionSeq: a modular framework for finding gene fusions by analyzing paired-end RNA-sequencing data. *Genome Biol.*, 11(10):R104, 2010.

[77] D. Pflueger, S. Terry, A. Sboner, L. Habegger, R. Esgueva, P.C. Lin, M.A. Svensson, N. Kitabayashi, B.J. Moss, T.Y. MacDonald, X. Cao, T. Barrette, A.K. Tewari, M.S. Chee, A.M. Chinnaiyan, D.S. Rickman, F. Demichelis, M.B. Gerstein, and M.A. Rubin. Discovery of non-ETS gene fusions in human prostate cancer using next-generation RNA sequencing. *Genome Res.*, 21(1):56-67, 2011.

[78] A.P. Khan, L.M. Poisson, V.B. Bhat, D. Fermin, R. Zhao, S. Kalyana-Sundaram, G. Michailidis, A.I. Nesvizhskii, G.S. Omenn, A.M. Chinnaiyan, and A. Sreekumar. Quantitative proteomic profiling of prostate cancer reveals a role for miR-128 in prostate cancer. *Mol. Cell Proteomics*, 9(2):298-312, 2010.

[79] S. Nacu, W. Yuan, Z. Kan, D. Bhatt, C.S. Rivers, J. Stinson, B.A. Peters, Z. Modrusan, K. Jung, S. Seshagiri, and T.D. Wu. Deep RNA sequencing analysis of readthrough gene fusions in human prostate adenocarcinoma and reference samples. *BMC Med. Genomics*, 4:11, 2011.

[80] T.L. Romanuik, G. Wang, O. Morozova, A. Delaney, M.A. Marra, and M.D. Sadar. LNCaP Atlas: gene expression associated with in vivo progression to castration-recurrent prostate cancer. *BMC Med. Genomics*, 3:43, 2010.

[81] E.F. Petricoin, 3rd, D.K. Ornstein, C.P. Paweletz, A. Ardekani, P.S. Hackett, B.A. Hitt, A. Velassco, C. Trucco, L. Wiegand, K. Wood, C.B. Simone, P.J. Levine, W.M. Linehan, M.R. Emmert-Buck, S.M. Steinberg, E.C. Kohn, and L.A. Liotta. Serum proteomic patterns for detection of prostate cancer. *J. Natl. Cancer Inst.*, 94(20):1576-1578, 2002.

[82] A. Sreekumar, B. Laxman, D.R. Rhodes, S. Bhagavathula, J. Harwood, D. Giacherio, D. Ghosh, M.G. Sanda, M.A. Rubin, and A.M. Chinnaiyan. Humoral immune response

to alpha-methylacyl-CoA racemase and prostate cancer. *J. Natl. Cancer Inst.*, 96(11):834-843, 2004.

[83] S.D. Garbis, S.I. Tyritzis, T. Roumeliotis, P. Zerefos, E.G. Giannopoulou, A. Vlahou, S. Kossida, J. Diaz, S. Vourekas, C. Tamvakopoulos, K. Pavlakis, D. Sanoudou, and C.A. Constantinides. Search for potential markers for prostate cancer diagnosis, prognosis and treatment in clinical tissue specimens using amine-specific isobaric tagging (iTRAQ) with two-dimensional liquid chromatography and tandem mass spectrometry. *J. Proteome Res.*, 7(8):3146-3158, 2008.

[84] A. Glen, C.A. Evans, C.S. Gan, S.S. Cross, F.C. Hamdy, J. Gibbins, J. Lippitt, C.L. Eaton, J. Noirel, P.C. Wright, and I. Rehman. Eight-plex iTRAQ analysis of variant metastatic human prostate cancer cells identifies candidate biomarkers of progression: An exploratory study. *Prostate*, 70(12):1313-1332, 2010.

[85] Y.A. Goo and D.R. Goodlett. Advances in proteomic prostate cancer biomarker discovery. *J. Proteomics*, 73(10):1839-1850, 2010.

[86] G. Sardana, K. Jung, C. Stephan, and E.P. Diamandis. Proteomic analysis of conditioned media from the PC3, LNCaP, and 22Rv1 prostate cancer cell lines: discovery and validation of candidate prostate cancer biomarkers. *J. Proteome Res*, 7(8):3329-3338, 2008.

[87] J. Adachi, C. Kumar, Y. Zhang, J.V. Olsen, and M. Mann. The human urinary proteome contains more than 1500 proteins, including a large proportion of membrane proteins. *Genome Biol.*, 7(9):R80, 2006.

[88] A.E. M'Koma, D.L. Blum, J.L. Norris, T. Koyama, D. Billheimer, S. Motley, M. Ghiassi, N. Ferdowsi, I. Bhowmick, S.S. Chang, J.H. Fowke, R.M. Caprioli, and N.A. Bhowmick. Detection of pre-neoplastic and neoplastic prostate disease by MALDI profiling of urine. *Biochem. Biophys. Res. Commun.*, 353(3):829-834, 2007.

[89] A. Sreekumar, L.M. Poisson, T.M. Rajendiran, A.P. Khan, Q. Cao, J. Yu, B. Laxman, R. Mehra, R.J. Lonigro, Y. Li, M.K. Nyati, A. Ahsan, S. Kalyana-Sundaram, B. Han, X. Cao, J. Byun, G.S. Omenn, D. Ghosh, S. Pennathur, D.C. Alexander, A. Berger, J.R. Shuster, J.T. Wei, S. Varambally, C. Beecher, and A.M. Chinnaiyan. Metabolomic profiles delineate potential role for sarcosine in prostate cancer progression. *Nature*, 457(7231):910-914, 2009.

[90] A. Esquela-Kerscher and F.J. Slack. Oncomirs - microRNAs with a role in cancer. *Nat. Rev. Cancer*, 6(4):259-269, 2006.

[91] S. Sevli, A. Uzumcu, M. Solak, M. Ittmann, and M. Ozen. The function of microRNAs, small but potent molecules, in human prostate cancer. *Prostate Cancer Prostatic Dis.*, 13(3):208-217, 2010.

[92] R. Sun, X. Fu, Y. Li, Y. Xie, and Y. Mao. Global gene expression analysis reveals reduced abundance of putative microRNA targets in human prostate tumours. *BMC Genomics*, 10:93, 2009.

[93] N. Hah, C.G. Danko, L. Core, J.J. Waterfall, A. Siepel, J.T. Lis, and W.L. Kraus. A rapid, extensive, and transient transcriptional response to estrogen signaling in breast cancer cells. *Cell*, 145(4):622-634, 2011.

[94] D. Wang, I. Garcia-Bassets, C. Benner, W. Li, X. Su, Y. Zhou, J. Qiu, W. Liu, M.U. Kaikkonen, K.A. Ohgi, C.K. Glass, M.G. Rosenfeld, and X.D. Fu. Reprogramming transcription by distinct classes of enhancers functionally defined by eRNA. *Nature*, 2011.

[95] T.A. Farazi, H.M. Horlings, J. Ten Hoeve, A. Mihailovic, H. Halfwerk, P. Morozov, M. Brown, M. Hafner, F. Reyal, M. van Kouwenhove, B. Kreike, D. Sie, V. Hovestadt, L.F. Wessels, M.J. van de Vijver, and T. Tuschl. MicroRNA sequence and expression analysis in breast tumors by deep sequencing. *Cancer Res*, 2011.

[96] H. Schwarzenbach, D.S. Hoon, and K. Pantel. Cell-free nucleic acids as biomarkers in cancer patients. *Nat. Rev. Cancer*, 2011.

[97] P.S. Mitchell, R.K. Parkin, E.M. Kroh, B.R. Fritz, S.K. Wyman, E.L. Pogosova-Agadjanyan, A. Peterson, J. Noteboom, K.C. O'Briant, A. Allen, D.W. Lin, N. Urban, C.W. Drescher, B.S. Knudsen, D.L. Stirewalt, R. Gentleman, R.L. Vessella, P.S. Nelson, D.B. Martin, and M. Tewari. Circulating microRNAs as stable blood-based markers for cancer detection. *Proc. Natl. Acad. Sci. USA,* 105(30):10513-10518, 2008.

[98] S.A. Tomlins, R. Mehra, D.R. Rhodes, X. Cao, L. Wang, S.M. Dhanasekaran, S. Kalyana-Sundaram, J.T. Wei, M.A. Rubin, K.J. Pienta, R.B. Shah, and A.M. Chinnaiyan. Integrative molecular concept modeling of prostate cancer progression. *Nat. Genet.*, 39(1):41-51, 2007.

[99] A.M. Gonzalez-Angulo, B.T. Hennessy, and G.B. Mills. Future of personalized medicine in oncology: a systems biology approach. *J. Clin. Oncol.*, 28(16):2777-2783, 2010.

In: Signaling, Gene Regulation and Cancer
Editors: Shree Ram Singh and Manoj K. Mishra
ISBN: 978-1-61942-088-5

Chapter 11

TARGETING BREAST CANCER

Ashok Jain*
Department of Natural Sciences, Albany State University,
Albany, GA, US

ABSTRACT

Breast cancer is the most common cancer related death among women worldwide. It is a complex genetic disease which exhibits distinct entities with extremely different biological characteristics and clinical behavior. Routine management based on clinical and pathological procedure has proven insufficient to effective outcome. Recent advances in gene expression profile are helpful to differentiate the various types of breast cancer, which is characterized by multiple molecular markers and expression of hormone receptors. The qualitative and quantitative analysis of variable expression of these multiple markers and receptors among the patients makes this disease highly heterogeneous and complex. Therefore, classification between various sub types of breast cancers is not easy. There are seven biologically and clinically meaningful groups that emerge from the molecular complexity, which are known as Luminal A, Luminal B, Basal like, HER2+ and Normal breast like. These sup-types of breast cancer based on molecular markers aid in predicting prognosis and response to therapy. There is two basic approaches (i) local therapy or (ii) systemic therapy for treating the breast cancer. Different treatment modalities can be adopted for various kinds of breast cancer. Different treatment options for breast cancer include surgery, radiation, hormonal, biological and chemotherapy. This review describes the details of breast cancer sub-type based on gene expression profiling and different approaches for the treatments.

INTRODUCTION

The global incidence for cancer now exceeds 1.1 million cases each year, resulting in more than 400,000 deaths annually [1-4]. In both the U.S. and developing nations, overall breast cancer is the second leading cause of cancer-related deaths among women. The

* Email: ashok.jain@asurams.edu.

statistical analysis reveals that in united states, although fewer African American women develops the breast cancer the mortality associated with this disease in them is twice as compared to the native white women. Since 1989, deaths from breast cancer have decreased by five percent for all women. However, the death rate remains disproportionately high for African-American women. Tumors are found at a later, more advanced stage, so there are fewer treatment options. There are several factors that contribute to breast cancer that include: age, family history (genetics), lack of physical activity, obesity, drinking alcohol, physiological factors (like early age menstrual cycle or late age menopause) and environmental factors [5, 6] includes diet containing food carcinogen (heterocyclic amines) and/or excess saturated fats. Breast cancer can develop for a many number of reasons. The risk factors only serve as a guide. It is estimated that one in every eight American women will develop breast cancer in their lifetime and latest analysis shows that 220,000 American women will develop breast cancer in 2011. Symptoms of breast cancer include: discharge from the nipple (clear or bloody), persistent tenderness of the breast, pain in the nipple, swelling of the breast, swelling or mass in the armpit (lymph node), inverted or scaly nipples, and a lump that can be felt no matter how small it is. Sometimes there are no symptoms of breast cancer when it is in the early stages.

Anatomy of Breast and Classification of Breast Cancer Based on Localization

The female breast is made up of: (i) lobules (milk producing glands), (ii) stroma (fatty tissue and connective tissue that surrounds the lobules and ducts as well as blood and lymphatic vessels), and (iii) ducts (passageways for milk to travel from the lobules to the nipple). Traditionally, breast cancer is being classified based on its location in the tissue. Most cases of breast cancer originate in the cells that line the ducts known as ductal carcinoma which accounts for 85 to 90% of all breast cancer cases. Other times, a type of lobular breast cancer is observed where cancerous cells will develop in the lobules which accounts for 8% of all cases [7]. Cancer can also develop in other tissues of the breast as well.

Ductal Carcinoma (which is referred to Ductal Carcinoma In Situ – DCIS) is a condition in which ductal cancer cells grow only inside the ducts of the breast. It is often referred to as a precancerous condition. In situ is a Latin phrase meaning *in place,* or in this instance, it means the abnormal cells are contained in one place (not spread into other tissue.) Since the cancer cells in DCIS do not usually form lumps, there are often no signs or symptoms associated with this condition. Occasionally DCIS will form a small lump or will cause nipple discharge. DCIS is often associated with deposits of calcium called microcalcifications, which can be detected through X-ray. DCIS is not life threatening as it is not invasive. Common treatment options for DCIS are hormonal therapies, lumpectomy (removal of infected ductal tissue by surgery) and chemotherapy [7].

Lobular Carcinoma (which is referred to Lobular Carcinoma In Situ – LCIS) is associated with the hollow glands or lobules where milk accumulates in the breast; they sometimes fill with abnormal cells. This is a precancerous condition but does not directly lead to breast cancer. However, women with LCIS stand a greater risk of developing breast cancer at some point in their lives.

Unfortunately, there are no signs or symptoms normally associated with LCIS such as lumps or even abnormalities on a mammogram. The only way LCIS can be diagnosed is through biopsy. Common treatment options for LCIS are either a mastectomy or a lumpectomy. Regular mammograms are recommended for early detection of LCIS.

Molecular Classification of Breast Cancer

Breast cancer comprises an extraordinarily diverse group of diseases in terms of presentation, morphology, immunohistochemical features, molecular profile and clinical behavior (response to therapy) [8-12]. For many decades, invasive breast carcinomas were only classified according to histological type, grade and expression of hormone receptors [8-12]. Breast cancer is a prevalent disease and one of the main causes of death from cancer. The main prognostic factor is the lymph node status followed by size of the tumor and other parameters such as hormone-receptor status, Human Epidermal Growth factor-2 (HER-2) expression, grade of differentiation, or patient's age [13-17]. In principle, these factors should guide therapeutic decisions, but in practice they are not powerful enough and offer too little help to identify people who benefit from adjuvant treatments. Therefore, most women with early-stage disease undergo adjuvant chemotherapy, radiotherapy and hormonal therapy. However, only a small percentage will experience the benefit [18-20]. Gene expression analysis has identified molecular classes of breast cancer that are biologically and clinically distinct. Breast cancers are classified by whether or not they express the genes for estrogen receptor (ER), progesterone receptor (PR) or Human epidermal growth factor 2 (Her2/neu). Expressions of these three receptors are known to help in the cancer development. Although morphology is often associated with the pattern of molecular aberrations in breast cancers, it is also clear that the tumors of the same histological nature exhibit remarkably different clinical behavior. Multiple studies have reproducibly identified the intrinsic breast cancer subtypes, which include several luminal subtypes characterized by expression of hormone receptor-related genes and two hormone receptor-negative subtypes - the HER2-positive/ER-negative subtype and the "basal-like" subtype. Contrary to the luminal subtypes, the basal-like subtype is characterized by low expression of ER- and HER2-related genes and clinically is usually, but not always, ER/PR-negative and lack HER2 overexpression, thereby constituting the "triple-negative" phenotype.

Multiple studies have demonstrated that the intrinsic subtypes vary by prognosis, with inferior outcomes illustrated among the two hormone receptor-negative subgroups as compared to the luminal subtypes [21]. They may also differ in other important ways. The gene expression microarray based class discovery studies pioneered by Stanford group have led to the identification of at least five sub-groups of breast cancer [7, 22-28]. These subgroups correspond reasonably well to clinical characterization based on HER2 and ER status as well as proliferation markers or histologic grade. These seven subgroups have been identified as (i) *Luminal-A* cancers, which are mostly HER2 negative, ER-positive and histologically low-grade (nuclear antigen Ki-67), (ii) *Luminal-B* (HER2-) cancers which are mostly ER-positive, HER2 negative but may express low levels of hormone receptors and are often high-grade (Ki-67), (iii) Some *Luminal-B* (HER2+) cancers express ER positive, HER2 is overexpressed in these tumors regardless of Ki-67, the tumor is classified as luminal B

HER2 positive, (iv) *HER2-positive* cancers which show amplification and high expression of the *ERBB2* gene and several other genes of the *ERBB2* amplicon (17q11) including HER2, GRB7, GATA4. These are mostly ER-, however some tumors may be of ER+. ER negative tumors (30 to 45%) are characterized by HR expression and low or absent expression of different luminal markers. The ER negative is further classified in to three different sub types. These share common features like HR and GATA3 negatativity, however, they are different biologically and molecularly, (v) *Normal breast like* cancers characterized mainly by high expression of genes that has characteristics of parenchymal basal epithelial cells and adipose/mesenchymal stromal cells with low expression of luminal epithelial cells. Tumors with normal breast like have prognosis that appears to be better than basal-like tumors [23, 29] and seems do not respond to neoadjuvant chemotherapy as other ER negative type tumors. Up to 10 % of all breast cancer is normal breast like cancers, (vi) *Basal-like* breast cancer (BLBC) differs from normal breast like. These tumors express genes characteristic of basal/myoepithelial cells such as CK5 and CK17, integrin 4, laminin, c-KIT, α6-integrin, metallothiopnein, fatty acid binding protein, P-cadherin, NF–κB and EGFR [7, 22, 23, 27, 28]. It is identified that tumors from patients carrying BRCA1 mutations fall within the basal-like subgroup [24, 30, 31, 32], (vii) *Triple negative* breast cancer (TNBC) is defined as that with negative expression of estrogen and progesterone receptors and HER2. TNBC accounts for 11% of invasive breast cancers associated with poor outcome which lack the benefit of targeted therapy and are frequently resistant to standard hormonal and chemotherapeutic regimens. Most triple negative breast cancers possess a basal phenotype and show varying degrees of basal cytokeratin and myoetithelial marker expression [33]. These tumors tend to express in premenopausal women [34 – 36] and members of specific ethnic groups [36 – 38]. These tumors are of high histological grade and have more aggreesive clinical behavior [35 - 37]. The aggressiveness is best illustrated by the fact that the risk of reoccurrence is between first and third year and most patients die within five years [34, 35].

This taxonomy was devised based on available literature outcome of various clinical trials has proven to be of prognostic significance. It has captured the attention of oncologists, pathologists and scientists as a 'gold standard' for the classification and importance of microarray-based gene expression profiling in breast cancer (Figure 1). Recently, in the last five to six years, several papers have reported triple-negative breast cancer [33]. Triple negative breast cancer (TNBC) and basal-like breast cancer (BLBC) are associated with poor outcome and lack the benefit of targeted systemic therapy. It is widely perceived that TNBC and BLBC tumors are synonymous. However, not all BLBC and TNBC have a poor outcome. The majority of basal-like breast cancers have a triple negative phenotype and the vast majority of triple negative cancers display a basal-like transcriptome, however the two are not synonymous [11, 33, 39]. Therefore, TNBC and BLBC are categorized separately in this chapter.

Microarray studies clearly differentiate between ER positive and ER negative breast cancers originating from distinct cell types and they point to biologic processes that govern metastatic progression. Moreover, such profiling has uncovered molecular signatures that could influence clinical care. Luminal types of tumors express high amounts of luminal cytokeratins, ER and genes associated with an active ER pathway [4, 22, 23, 40, 41]. At least two subgroups of luminal tumors have been identified: luminal A, which is usually of low histological grade, have an excellent prognosis and show high levels of expression of ER-activated genes; and luminal B, which is more often of higher histological grade, have higher proliferation rates and a significantly worse prognosis than luminal A tumors [4, 22, 23, 28, 41].

Normal breast-like cancers are rather poorly characterized; one of the defining features of these tumors is that they consistently cluster together with samples of fibroadenomas and normal breast. The clinical significance of normal breast-like tumors is yet to be determined [4, 22, 23, 28, 41]. Some have suggested that this subgroup may be a mere artifact of expression profiling (i. e. disproportionally high content of stromal cells).

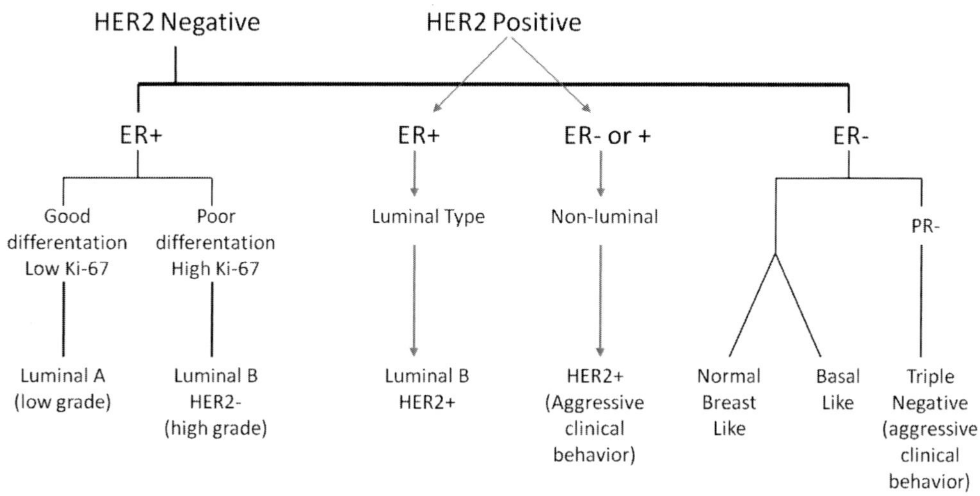

Figure 1. Molecular classification and clinicopathological features of breast cancer. HER2 - Human epidermal growth factor 2: determined by immunohistochemical (IHC) analysis; ER – Estrogen receptor; PR - Progesterone receptor; Ki-67 – Nuclear antigen Ki-67.

HER2 tumors are usually ER-negative and characterized by overexpression of the human epidermal growth factor receptor type 2 (HER2) and genes associated with HER2 pathway and/ or HER2 amplicon on 17q12. HER2 cancers have a very aggressive clinical behavior. However, they are amenable to novel tailored therapies using either monoclonal antibodies against HER2 or HER2 tyrosine kinase inhibitors [4, 22, 23, 28, 41 – 45]. HER2 negative tumors are associated with poor prognosis [22, 25]. Not all tumors that are HER2 amplified fall into the HER2 cluster by expression arrays analysis. There is also evidence to suggest that some HER2 amplified, ER-positive cancers fall within the luminal B subtype rather than the HER2-microarray subtype [15]. Basal-like cancers are another group of ER-negative cancers which are named so because the neoplastic cells of this tumor type consistently express genes usually found in normal basal/myoepithelial cells of the breast, including cytokeratins-5 and cytokeratins-17. It should be emphasized that basal-like breast cancers, unlike 'basal'/ myoepithelial cells of normal breast, almost uniformly express cytokeratins 8 and/ or 18 ER and genes associated with an active ER pathway [4, 22, 23, 41].

The characteristics of basal-like breast cancer have been extensively reviewed [27, 46 – 48]. Some groups have employed microarray-based expression profiling to define basal-like breast cancers, whereas others have used immunohistochemical surrogates. Despite using distinct definitions for basal-like breast cancers, it is documented that basal-like tumors have distinctive clinical presentations [49], histological features [50, 51] response to chemotherapy

[27, 41, 52 – 56] and outcome [22, 27]. In brief, basal-like tumors comprise a heterogeneous group of cancers that account for up to 15% of all breast cancers. These tumors affect younger patients which are more prevalent in African-American women and more often present as interval cancers. Histologically the vast majority of basal-like breast cancers is IDC-NSTs of high histological grade and characterized by high mitotic indices, the presence of central necrotic zones, pushing borders, conspicuous lymphocytic infiltrate and typical/atypical medullary features [30, 50, 51]. At the immunohistochemical level, the majority of basal-like breast cancers lack or display low levels of ER and PR which lack HER2 gene amplification and express genes usually found in 'basal'/ myoepithelial cells of the normal breast, including: high molecular weight cytokeratins (5/6, 14 and 17) [50, 57 – 59], Pcadherin [60], caveolins-1 and caveolins-2 [61, 62], nestin [63], αB crystalline [64, 65], and epidermal growth factor receptor (EGFR) [57]. Alterations of the pRB and p16 G1/S cell cycle checkpoint are remarkably prevalent in these cancers [66, 67] and TP-53 gene mutations are present in up to 85% of cases [27, 33]. In fact, a recent study demonstrated that approximately 30% of basal-like breast cancers concurrently display lack of pRB expression, p16 overexpression and p53 immunoreactivity whereas this profile was rarely seen in tumors of other molecular subgroups [66]. Basal-like display remarkably high proliferation indices as defined by mitotic counting or by MIB1 (Ki67) labeling index [27, 33]. In fact, some studies indicate that expression of basal keratins is a prognostic factor independent of tumor size, grade and lymph node status [68]. However, when compared to either ER-negative nonbasal-like cancers [69] or to grade-matched non-basal-like cancers [67]. The carcinoma with a basal like phenotype does not seem to be associated with a poorer outcome. Basal-like cancers have been shown to have a more aggressive clinical behavior [57, 68, 70].

Triple negative breast cancers (TNBC) are defined as tumors that lack ER, PR and HER2 expression. Depending on the thresholds used to define ER, PR positivity and the methods for HER2 assessment, TNBC account for 10-17% of all breast carcinomas [33 - 38, 54, 56, 71, 72]. It is important to understand that different studies have used diverse methods and thresholds to define lack of expression of these markers. Regardless of the issues, the interest in triple negative cancer persists from the lack of tailored therapies for this group of breast cancer patients and the overlap with the profiles of basal-like cancers. The main characteristics of TNBC that have emerged from the recent studies illustrate the similarities between basal-like and triple negative tumors, including the fact that they more frequently affect younger patients (<fifty years) [35 – 38, 54, 56, 71, 72], are more prevalent in African-American women [36 – 38], often present as interval cancers and are significantly more aggressive than tumors pertaining to other molecular subgroups [35 – 37, 54, 56, 71]. This aggressiveness is best translated by the fact that the peak risk of recurrence is between the first and third year and the majority of deaths occur in the first five years following therapy [35, 71]. Interestingly, patients with triple negative cancers [37, 71] or basal-like cancer show significantly shorter survival rates following the first metastatic event as compared to those with non-triple negative or non-basal-like controls. Pathologically, the majority of TNBC are of high histological grade (grade III) [34, 35, 71] with exceptions indicating 10% with low grade (grade I) in one study [71].

Certain basal-like cancers exhibit a high expression of "basal" cytokeratins such as CK5 and a variety of growth factor receptors including high levels of epidermal growth factor receptor, c-kit (a tyrosine kinase in breast epithelium) and growth factors such as insulin growth factor and hepatocyte growth factor [73, 74]. Immunohistochemical methods for defining basal-like cancers [57, 62] pose difficulties because there is limited feasibility of combining five or more immunohistochemical markers in routine clinical practice which inhibits the ability of these methods to function as perfect molecular makers. Therefore, these methods have not gained wide acceptance in routine tests.

ROLE OF BRCA GENES

Discovery of BRCA genes and identification of their link with breast cancer was an historic milestone. Mutations in BRCA1 and BRCA2 play critical role in differentiating the sporadic tumors and luminal-like tumors. BRCA1 and BRCA2 were the first of many breast cancer genes to be identified and are associated with hereditary breast cancer, as well as ovarian cancer, prostate cancer and colon cancer [75]. BRCA1 is a gene that, when healthy, prevents breast tumors from forming. BRCA1 makes a protein that regulates cell growth and division of the cells that line the milk ducts in breast. If BRCA1 gene is mutated, the protein that it makes is unable to regulate cell division, leaving breast cells susceptible to the growth of cancer. There are more than 600 known mutations of the BRCA1 gene, most of which increase the risk of developing breast cancer [76]. BRCA2 is a gene that, when healthy, prevents breast tumors from forming. BRCA2 makes a protein that is involved in DNA repair and genetic stability. If BRCA2 is mutated, the protein that it makes is unable to repair broken DNA, leaving breast cells susceptible to the growth of cancer [77]. There are more than 450 known mutations of the BRCA2 gene, most of which increase the risk of developing breast cancer. BRCA2 was identified in 1995 and is associated with hereditary breast cancer, as well as ovarian, male breast, pancreatic, prostate, laryngeal, and ocular cancers [78].

There is increasingly more coherent evidence to suggest a link between BRCA 1 pathway and basal-like breast cancers [79, 80]. In fact, the vast majority of tumors arising in BRCA 1 germline mutation carriers, in particular those diagnosed before 50 years of age have morphological features similar to those described in basal-like cancers [81, 82] and display a basal-like phenotype as defined by immunohistochemistry [31, 32] or expression arrays [24].

The immunohistochemical similarities between BRCA 1 tumors and basal-like breast carcinomas are deeper as compared to expression of high molecular weight cytokeratins and other proteins expressed by myoepithelial cells. In fact, both basal-like breast cancers and tumors arising in BRCA 1 germline mutation carriers show a peculiar pattern of cell cycle protein expression [81, 83 – 85]; both rarely harbor CCND1 gene amplification [83, 84]. However they express significantly lower levels of p27 [81, 85] and higher levels of Skp2 [81, 85], cyclin E [81, 85] and caspase-3 [85] when compared to sporadic breast carcinomas and BRCA2 mutation tumors. BRCA 1 protein expression levels have been shown to be significantly lower in tumors of high histological grade, lacking ER and PR expression and of basal-like phenotype [86, 87]. The majority of basal-like breast cancers shows a dysfunctional BRCA 1 pathway [30, 79, 80] and harbor TP53 gene mutations [23, 24, 86, 87]. Interestingly, inactivation of BRCA1 and TP53 in transgenic mouse model has been shown the development of tumors whose morphological and phenotypic characteristics are remarkably similar to basal-like breast cancers [88, 89]. These findings provide circumstantial evidence to suggest that despite the cell of origin, BRCA1 inactivation may lead to basal-like breast cancers. BRCA1 has been shown to play a pivotal role in the regulation of ER expression and that RNA interference-mediated silencing of BRCA 1 in breast cancer cell lines leads to a marked reduction in expression of endogenous ER protein levels [89].

Table 1. Substitute definitions of intrinsic subtypes of breast cancer: current approach uses immunohistochemical definition of estrogen and progesterone receptor, the detection of overexpression and/or amplification of the human epidermal growth factor receptor 2 (HER2) oncogene, and Ki-67 labeling index, a marker of cell proliferation, as the means of identifying tumor subtypes

Intrinsic subtype (Perou et al.1)	Clinicopathological definition	Comments	Treatment options
Luminal A	ER and/or PR positive, HER2 negative, and Ki-67 low (<14%)	Cut point for Ki-67 labelling was based on comparison with PAM50 intrinsic subtyping, Ki-67 staining quality control is important	Endocrine therapy alone, Less responsive to chemotherapy,
Luminal B	(a) Luminal B (HER2 negative): ER and/or PR positive, HER2 negative, and Ki-67 high	Multiple gene array indicating genes with higher proliferation are markers of poor prognosis. In case Ki-67 measurement is not reliable, alternate measurement of tumor proliferation like grade may be used to differentiate 'Luminal A' and 'Luminal B (HER2 negative'	Endocrine ± Cytotoxic therapy. Routine Chemotherapy should include Anthracycline and taxans.
	(b) Luminal B (HER2 positive): ER and/or PR positive, HER2 over-expressed or amplified, and any Ki-67	Endocrine and anti-HER2 therapy may be recommended	Cytotoxics + anti HER2 = endocrine therapy. Routine Chemotherapy should include Anthracycline and taxans.
HER2 positive (non luminal)	Erb-B2 over-expression: HER2 over-expressed or amplified ER and PR absent	Both basal-like and HER2 tumors share certain additional features such as high levels of p53 mutation, aggressive clinical behavior, poor prognosis	Cytotoxics + anti HER2. No definite and routine chemotherapy recommendation; use 'trastuzumab' as standard adjuvant treatment for 1 year only.
Normal Breast Like	ER negative tumors, Characterized by high expression of genes characteristic of parenchymal basal epithelial cells and adipose/mesenchymal stroma cells with low expression of genes characteristic of luminal epithelial cells.	These tumors have a good prognosis that seems to be better than that of basal like cancers.	Does not respond to neoadjuvant chemotherapy

Intrinsic subtype (Perou et al.1)	Clinicopathological definition	Comments	Treatment options
Basal –like	Ductal: ER negative, PR absent, and HER2 negative	Express characteristic of basal/myoepithelialcells such as CK5 and CK17, integrin 4, laminin, c-KIT, α6-integrin, metallothiopnein, fatty acid binding protein, P-cadherin, NF –KB and EGFR	Routine Chemotherapy should include Anthracycline, taxans, and an alkylatting agent (typically cyclophosphamide). Should not use cisplatin or carboplatin routinely; avoid antiangiogenic therapies.
	Triple negative: ER, PR, and HER2 negative	Most triple negative breast cancers possess a basal phenotype and show varying degrees of basal cytokikeratin and myoetithelial marker expression. These tumors are of high histological grade and have more agreesive clinical behavior	Does not respond to hormonal therapy

Dysfunction of the BRCA1 pathway caused by BRCA1 gene promoter methylation or BRCA1 transcriptional inactivation or both is another feature that differentiates sporadic basal-like tumors from luminal-like tumors [79, 90 – 92]. BRCA1 expression is important in DNA repair, activation of cell-cycle checkpoints, maintenance of chromosomal stability and perhaps differentiation of ER-negative stem or progenitor cells into ER-positive luminal cells [93]. These findings are in line with suggestions of a link between the basal-like phenotype and germ-line mutation of BRCA1 [31, 81]. Indeed, almost all breast cancers that are associated with a BRCA1 mutation; whether sporadic or hereditary, have a basal-like triple-negative phenotype [40, 94]. Tumors associated with the BRCA2 mutation have the distribution of phenotypes encountered in the general population.

TREATMENTS OF BREAST CANCER

Meaning of ER, PR or HER2 Positivity or Negativity for Treatment

Estrogen and Progesterone receptor status tests will show whether or not one or both of those hormones fuel the tumor. Cancer that is hormone-sensitive is slightly slower growing and has a better chance of responding to hormone-suppression treatment, than cancer that is hormone receptor negative. Hormone-negative cancer will respond to other kinds of treatment, and hormone-suppression may not be needed.

A score of Estrogen Receptor positive (ER+) means that estrogen is causing the tumor to grow, and that the cancer should respond well to hormone suppression treatments. If the score is Estrogen Receptor negative (ER-) then the tumor is not driven by estrogen. The results need to be evaluated along with other tests such as HER2 status to determine the most effective treatment.

If the score is Progesterone Receptor positive (PgR+ or PR+) the cancer should respond well to hormone suppression treatments. If the score is Progesterone Receptor negative (PgR- or PR-) then the tumor is not driven by progesterone and another test is needed for the HER2 status to determine the most effective treatment. HER2 positive breast cancer responds well to Herceptin treatment. The molecular classification, gene expression signatures based on microarray analysis indicating the gene expression profile of tumor differentiation and cell proliferation (cell cycle gene expression) predicts clinical outcome and prognostic strategies (Figure 2).

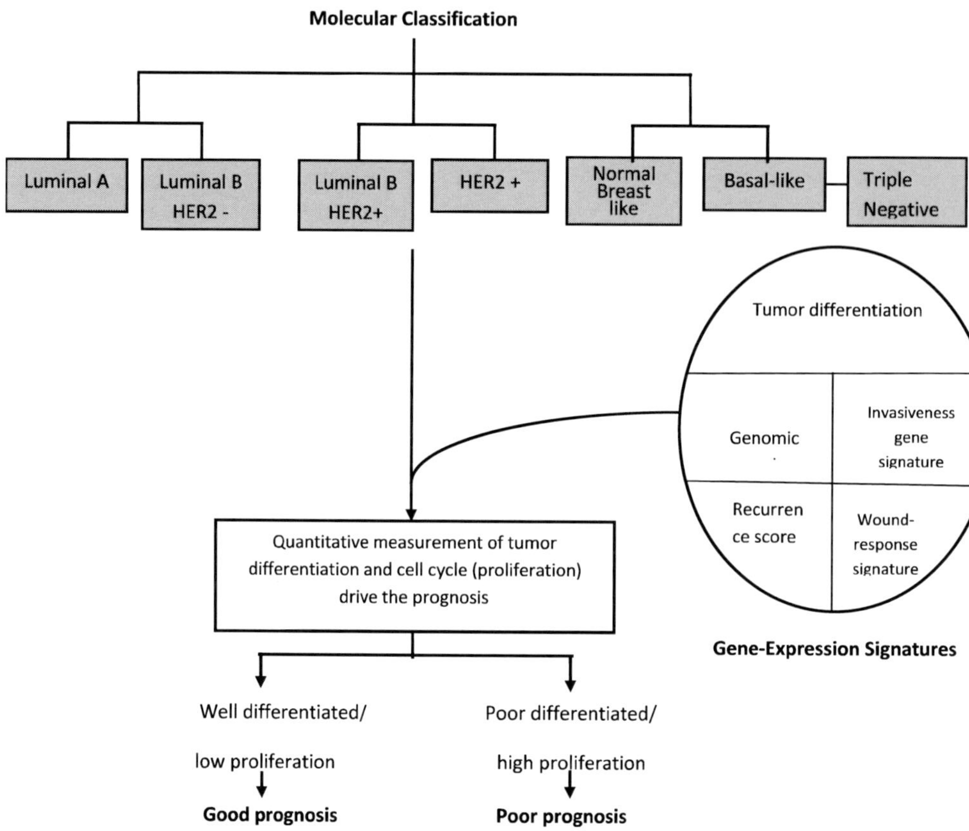

Figure 2. Molecular Classification and Gene-Expression Signatures predicts Clinical Outcome.

As discussed above, breast cancer is expressed as different types depending upon the affected genes. The gene expression profile of a breast cancer patient is important for surgeons and physicians to make correct decisions and appropriate therapy. Despite significant scientific advances in breast cancer managements, most of the world faces resource constraints, which limit our capabilities to early detection, diagnosis, and treatment of breast cancer. Cancer Treatment is either local therapy or systemic therapy. Two recent reviews compiles and illustrate current therapies and targets in treating breast cancer [95] and implications for patient care [12]. The information compiled in this article is primarily based on these review articles. Figure 3 illustrate the current treatments options to target and manage breast cancer.

1. Local Therapy: Surgery and radiation therapy are considered local treatment. They remove or destroy the tumor. If the cancer has spread to other parts of the body, local therapy may be used to control the disease in affected areas specifically.
   a) Surgery: The most common treatment for breast cancer is surgery. There are three common types.
      i. Breast-sparing surgery: An operation to remove the cancer but not the breast is breast-sparing surgery. It is also called breast-conserving surgery, lumpectomy, segmental mastectomy, and partial mastectomy. Sometimes an excisional biopsy serves as a lumpectomy because the surgeon removes the whole lump. The surgeon often removes the underarm lymph nodes as well.
      ii. Axillary surgery: An incision is made in this procedure for axillary lymph node dissection. It shows whether cancer cells have entered the lymphatic system. Most women after breast-sparing surgery receive radiation therapy to the breast. This treatment destroys cancer cells that may remain in the breast [96]. Isolated tumor cells, and even metastases up to 2 mm (micrometastases) in a single sentinel node are not considered to constitute an indication for axillary dissection regardless of the type of breast surgery. It is recommended that in such cases the better options in such cases are lumpectomy and radiation therapy [97].
      iii. Mastectomy: Removal of the breast (or as much of the breast tissue as possible) is a mastectomy. In total mastectomy, the surgeon removes the whole breast and some lymph nodes under the arm may be removed. The modified radical mastectomy involves removal of the whole breast and most or all of the lymph nodes under the arm. Most of the time the lining over the chest muscles and the small chest muscle is removed so it is easier to remove the lymph nodes [98, 99].
   b) Radiation Therapy: X-Ray Radiation therapy (XRT) is used to "sterilize" the remaining breast. XRT destroys cells by fracturing their DNA sequence through free radical creation and release. Normally the whole-breast radiotherapy needs to be avoided (patients with extensive vascular invasions are exceptions) due to the side effects. Partial breast irradiation (PBI) is recommended for patients above the age of seventy. Post-mastectomy radiation therapy is supported for patients younger than forty-five years with one to three positive nodes. Inclusion of anthracyclines and Texans and alkylating agents is recommended. However, routine use of cisplatin or carboplatin is being discouraged. Complications of radiation therapy include fatigue, breast erythema and edema, ipsilateral extremity edema long term (5-17%) and rib fractures [100, 101].
2. Systemic Therapy: Systemic therapy includes treatments that enter through the bloodstream and destroy or control cancer throughout the body. Hormone therapy, biological therapy and chemotherapy are systemic treatments. These therapies can be

initiated either before the local therapy such as surgery as in neoadjuvant therapies or after surgery as in adjuvant therapies.

a) Hormonal Therapy: Estrogen and progesterone are the major growth hormone which interacts through the receptors. In most cases, cancer cells have higher numbers of receptors than the normal cells that allow circulating estrogen or progesterone to attach to the tumor cell and providing food for growth. It is desirable to have the receptor which makes the tumor estrogen receptor (ER)-positive and/or progesterone receptor (PR) positive [102]. This type of tumor cell is potentially responsive to anti-hormonal systemic treatment. If a tumor is ER and/or PR negative then it is unlikely that the anti-hormonal drugs will be used to treat the breast cancer. If a tumor is ER or PR positive, depending on the stage of the cancer, often both anti-hormonal therapy and chemotherapy are used. Sequential endocrine therapy continues as long as the patient remains hormone sensitive. Once disease becomes hormone-resistant, chemotherapy is the alternative. There are five types of anti-hormonal treatments:
   i. *Selective estrogen receptor modulators* (SERMs): Tamoxifen (NovaldexR) is one of the oldest used SERMs. It inhibits the growth of breast tumors by competitive antagonism of estrogen at its receptor site [103, 104]. It also exhibits partial estrogen-agonist effects. Tamoxifen is recommended as adjuvant therapy after surgery for patients with early ER positive and/or PR positive breast cancer. It is also specify as adjuvant therapy in patients with metastatic condition with chemotherapy. Tamoxifen is the only available endocrine option. There are some newer SERMs such as raloxifene (EvistaR), which has lower risk of osteoporosis [105].
   ii. *Progestins:* PR-positive advanced breast tumors can respond to the use of synthetic progesterone-like drugs such as megesterol acetate (MegaceR) which was shown to reduce the frequency of hot flushes in postmenopausal breast cancer patients [106]. Progestins are usually restricted to second or third line therapies following aromatase inhibitors and/or antiestrogens.
   iii. *Lutenising Hormone Releasing Hormone agonists (LHRH):* Goserelin (ZoladexR) and luprolide (LupronR) are a group of drugs that suppress ovarian estrogen production down to postmenopausal levels by inducing a potentially reversible medical ovarian ablation. LHRH analogs are mainly effective in ER-positive early breast cancer premenopausal patients [107].
   iv. *Aromatase Inhibitors (AIs):* Estrogen is synthesized in non-ovarian peripheral tissues in postmenopausal women. Synthesis of estrogen is mainly because of conversion by aromatase of the androgenic substrates androstenedione and testosterone to estrone and estradiol in the adrenal glands and adipose tissue. Aromatase inhibitors are of no value in premenopausal patients where the ovaries are the primary sites of estrogen production [108]. There are two types of aromatase inhibitor which mainly differ structurally: 1) steroidal, substrate analogs such as 4-hydroxyandrostenedione (formestane) and exemestane, and 2) reversible nonsteroidal imidazole-base inhibitors (e.g. anastrozole (ArimidexR) and letrozole (FemaraR) [109]. These are type I and type II aromatase inhibitors

respectively. Use of third-generation AIs inhibits total body aromatization and plasma estradiol levels by more than 95% [110].

v. *Estrogen receptor antagonists* (ERAs): The main purpose of ERAs is to block the active site of estrogen receptors so that the estrogen hormone will not interact with receptor. If the majority of cancer cell's ER is blocked by ERAs then the estrogen will fail to trigger its action. An important addition to the armamentarium of endocrine therapies is the selective estrogen-receptor antagonist fulvestrant (FaslodexR) also termed "an estrogen receptor down-regulator [111]. As a steroidal analog of 17Â-estradiol, fulvestrant has a chemical structure that is similar to that of estradiol but distinct from tamoxifen and other nonsteroidal hormonal agents. Both tamoxifen and fulvestrant competitively inhibit the binding of estradiol to the ER. In contrast to tamoxifen, fulvestrant has no agonist effect and downregulates the expression of the ER. With estradiol as the comparator, fulvestrant's ER binding affinity (0.89) is greater than tamoxifen's (0.025) [112, 113].

b) Biological therapy: Biological therapy is a type of treatment that is designed to boost the patient's immune system and suppressing the side effects of other drugs or treatments such as chemotherapy. Vaccines are the classical example of biological therapy. Normally vaccines are given before the infection such that body produces antibodies and ready to fight Human epidermal growth factor receptor 2 (HER2) gene, also known as HER2/neu is located on chromosome 17q and belongs to the human epithelial receptor (HER) family of genes. It encodes a 185 kDa transmembrane tyrosine kinase growth factor receptor, which mediates signaling for cell proliferation and survival) [114]. HER2 has been found to be an important prognostic and predictive marker of treatment response in women with breast cancer in the adjuvant setting and advanced disease. The HER2 gene is amplified and the HER2 protein is overexpressed in 20% to 25% of breast cancer patients that results poor prognosis and overall shortened survival [115]. HER2 gene amplification and resultant protein overexpression are associated with a more aggressive clinical outcome. Several murine monoclonal antibodies have been identified against the extracellular domain of the HER2 protein. These antibodies are capable to inhibit proliferation of cells overexpressing HER2 [116]. Gentech Inc. (San Francisco, CA) has developed monoclonal antibody to minimize the immunogenicity. The antibody is developed by fusing the antigen-binding region of one of the effective antibodies to the framework region of the human IgG leading to trastuzumab that binds to the HER2. United States Food and Drug Administration (FDA) approved trastuzumab in 1998 for the treatment of HER2-positive metastatic breast cancer (MBC) in combination with paclitaxel or as monotherapy for patients who had received prior chemotherapy for HER2 positive MBC. Trastuzumab is now predominantly used in combination with chemotherapy in the first-line setting of metastatic disease due to its evident benefit in improving clinical outcome [117, 118].

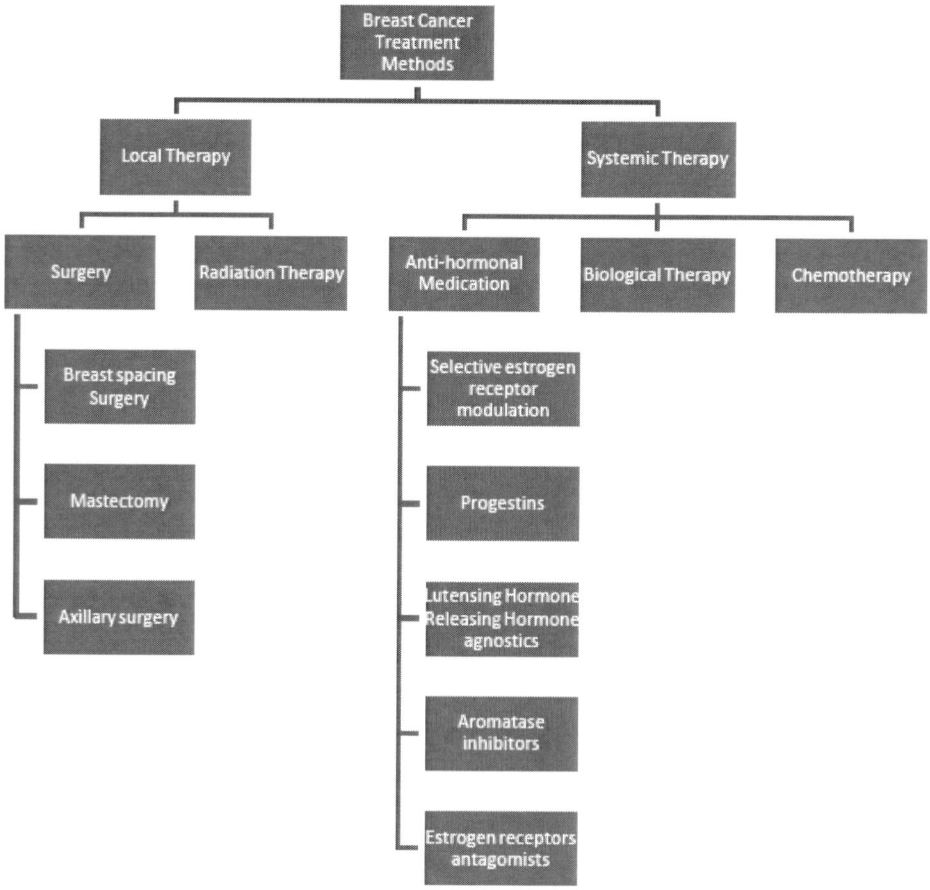

Figure 3. Current Therapies for treating breast cancer.

c) Chemotherapy: Chemotherapy treatment uses medicine to destroy and weaken cancer cells of the tumor or the cancer cells that may have spread in other parts of the body causing metastatic cancer (cancer that has spread beyond the breast). Chemotherapy is used for both before (neoadjuvant) or after the surgery (adjuvant chemotherapy). When treating early-stage invasive breast cancer mostly it is adjuvant to destroy cells that may be left after the surgery. For advanced-stage breast cancer, it is neoadjuvant to destroy cancer cells as much as possible and shrink the tumor so that less tissue has to be removed. Chemotherapy medicines are given in combination with other medicines/treatments and such combination is known as 'chemotherapy regimens'. Chemotherapy treatments are tailored specifically for each person's unique situation like tumor stage, hormone receptor status, HER2 status and menopausal status [119]. Patients with ER+ or PR+ tumors, symptomatic visceral metastasis or hormone refractory disease should receive chemotherapy. Several medicines are used for chemotherapy. The preferred first-line single agents for breast cancer are: Capecitabine, Doxorubicin, Epirubicin, Paclitazel, Pegylated liposomal Daxorubicin and Vinorelbine. Other active agents include cisplatin, carboplatin, paclitaxel protein-bound particles for injectable suspension, etoposide, vinblastine, and fluorouracil by continuous infusion. Adjuvant combination chemotherapy offers higher response rates and longer time to disease progression [120 -122]. Chemotherapy medicines come in many forms and can be given in many ways: intravenously (IV) injection; by mouth (orally); as a pill or capsule; through a port (under this treatment a small disc made of plastic or metal about the size of a quarter is inserted in the patient chest during a short outpatient surgery just under the skin) and through a catheter where a soft thin tube (catheter) is inserted into a large vein during a short outpatient surgery and medicines are injected through the cather.

Conclusion and Future Prospects

Microarray and molecular analysis representing the gene-expression profiling have clearly changed our vision of breast cancer and provided new tools to understand and diagnose the disease. Molecular diagnosis contains information's relevant to classification, prognosis and treatment prediction and potential to be of direct benefit to disease management. Grouping of tumors based on the expression of multiple genes and/or proteins is more powerful than assessment of one or two genes for unraveling the complexity of the tumor tissue. Identification of novel sub-type of breast cancer based on gene profiling studies, such as HER2 positive or negative; ER positive or negative; the normal breast like or basal-like subgroup or triple negative type; luminal A or Luminal B has re-emphasized the clinical outcome. For example ER positive breast cancer have better clinical outcome than the ER negative type of breast cancer.

The rapid advances in the technology available to access tissues have led to significant activities aimed at further refining the sub-classifications. There are still major limitations in assigning consistently a molecular class to new cases of breast cancer. Future work will further determine how many molecular classes of breast cancer are there and how many subtypes of breast tumors can be treated reliably. Currently, several gene signatures whose expression profiles have successfully predicted clinical outcome and survival in patients with breast cancer have been acknowledged. However, it remains to be determined whether these gene signatures will be used individually or in conjunction with other pathological characteristics or there will be a new gene set that will rule them. The new technology assures that it will help the disease management and override better clinical outcome. It is likely that data generated from current and forthcoming clinical trials and applications of genetic techniques in prognosis of disease will have an important impact on the future biomarker and drug discovery process; treating the breast cancer patients and survival of patients.

Acknowledgments

The research is supported by DOD CDMRP Grant No. W81XWH-10-1-1042; and NIH RIMI Grant No. 2P20MD001085-08.

REFERENCES

[1] Global Cancer Facts and Figures *2nd Edition,* American Cancer Society. Cancer facts and figures 2011, No.861811. publication of the American Cancer Society, Atlanta, Georgia., p 1-58. http://www.cancer.org/acs/groups/content/@epidemiology surveilance/ documents/document/acspc-027766.pdf.

[2] J. Ferlay, H.R Shin, F. Bray, D. Forman, C.D. Mathers, D. Parkin. Cancer Incidence and Mortality Worldwide: IARC CancerBase No.10. *International Agency for Research on Cancer.* 2010. http://globocan.iarc.fr.

[3] Salynn Boyles. Global Cancer Deaths to Double by 2030 – Report Predicts Poor Nations Will See Biggest Increase in Cancer; WebMD Health News December 8, 2008, http://www.webmd.com/cancer/news/ 20081209/global-cancer-deaths-double-2030.

[4] N. Howlader, A.M. Noone, M. Krapcho, N. Neyman, R. Aminou, W. Waldron, S.F. Altekruse, C.L. Kosary, J. Ruhl, Z. Tatalovich, H. Cho, A. Mariotto, M.P. Eisner, D.R. Lewis, H.S. Chen, E.J. Feuer, K.A. Cronin, B.K. Edwards (eds). SEER Cancer Statistics Review, 1975-2008, National Cancer Institute. Bethesda, MD, based on November 2010 SEER data submission, posted to the SEER web site, 2011. http://seer.cancer.gov/csr/1975_2008/.

[5] R.T. Chlebowski, Z. Chen, G.L. Anderson, et al. Ethnicity and breast cancer: factors influencing differences in incidence and outcome. *J. Natl. Cancer Inst.* 2005; 97: 439-448.

[6] WCRF and AICR. Food, nutrition and the prevention of cancer: a global perspective.WCRF and American Institute for Cancer Research, Washington DC 1997, 1-670.

[7] G.H. Sakorafas and A.G.H. Tsiotou. Ductal carcinoma in situ (DCIS) of breast evolving perspectives. *Cancer Treatment Review,* 26 (2):103-125; 2000.

[8] J.S. Reis-Filho, P.T. Simpson, T. Gale, S.R. Lakhani. The molecular genetics of breast cancer: the contribution of comparative genomic hybridization. *Pathol. Res. Pract.* 201(11): 713–725, *2005.*

[9] M. Lacroix, R.A. Toillon, G. Leclercq. Stable 'portrait' of breast tumors during progression: data from biology, pathology and genetics. *Endocr. Relat. Cancer,* 11*:* 497–522; 2004.

[10] P.T. Simpson, J.S. Reis-Filho, T. Gale, S.R. Lakhani. Molecular evolution of breast cancer. *J. Pathol.,* 205(2):248–254, 2005.

[11] C.Sotiriou, L.Pusztai, and Lajos Pusztai. Gene expression signatures in breast cancer. *The New England Journal of Medicine*, 360:790-800, 2009.

[12] A. Goldhirsch, W. C. Wood, A.S. Coates, R.D. Gelber, B. Thurlimann, H.-J. Senn, and Panel members. Strategies for subtypes – dealing with the diversity of breast cancer: highlights of the St Gallen International Expert Consensus on the Primary Therapy of Early Breast Cancer 2011. *Annals of Oncology Advance Access*, 2011.

[13] C.L. Carter, C. Allen, and D.E. Henson. Relation of tumor size, lymph node status, and survival in 24,740 breast cancer cases. *Cancer* 63(1):181-187, 1989.

[14] M.H. Galea, R.W. Blamey, C.E. Elston, et al. The Nottingham Prognostic Index in primary breast cancer. *Breast Cancer Res. Treat.*, 22:207-219, 1992.

[15] A.N. Mirza, N.Q. Mirza, G. Vlastos, et al. Prognostic factors in node-negative breast cancer: A review of studies with sample size more than 200 and follow-up more than 5 years. *Ann. Surg.*, 235:10-26, 2002.

[16] M.F. Press, L. Bernstein, P.A. Thomas, et al. HER-2/neu gene amplification characterized by fluorescence in situ hybridization: Poor prognosis in node-negative breast carcinomas. *J. Clin. Oncol.*, 15:2894-2904, 1997.

[17] R.B. Weiss, S.H. Woolf, E. Demakos, et al. Natural history of more than 20 years of node-positive primary breast carcinoma treated with cyclophosphamide, methotrexate, and fluorouracil-based adjuvant chemotherapy: A study by the Cancer and Leukemia Group B. *J. Clin. Oncol.*, 21(9):1825-1835, 2003.

[18] R.E Coleman. Current and future status of adjuvant therapy for breast cancer. *Cancer* 97:880-886, 2003.

[19] B.A. Mincey, F.M. Palmieri, and E.A. Perez. Adjuvant therapy for breast cancer: Recommendations for management based on consensus review and recent clinical trials. *Oncologist* 7:246-250, 2002.

[20] E.P. Winer, C. Hudis, H.J. Burstein, et al. American Society of Clinical Oncology technology assessment working group update: Use of aromatase inhibitors in the adjuvant setting. *J. Clin. Oncol.*, 21:2597-2599, 2003.

[21] B. Weigelt, H.M. Horlings, B. Kreike, M.M. Hayes, M. Hauptmann, L.F. Wessels, O. de Jong, Van de Vijver MJ, Van't Veer LJ, Peterse JL. Refinement of breast cancer classification by molecular characterization of histological special types. *J. Pathol.*, 216(2): 141-50, 2008.

[22] C. M. Perou,T. Sorlie, M. B. Eisen, M. van de Rijn, S. S. Jeffrey, C. A. Rees, J. R. Pollack, D. T. Ross, H. Johnsen, L. A. Akslen, O. Fluge, A. Pergamenschikov, C. Williams, S. X. Zhu, P. E. Lonning, A. L. Borresen-Dale, P.O. Brown, D. Botstein. Molecular portraits of human breast tumours. *Nature*, 406:747-752, 2000.

[23] T. Sorlie, C. M. Perou, R. Tibshirani, T. Aas, S. Geisler, H. Johnsen, T. Hastie, M. B. Eisen, M. van de Rijn, S. S. Jeffrey, T. Thorsen, H. Quist, J. C. Matese, P.O. Brown, D. Botstein, P. Eystein Lonning, A. L. Borresen-Dale. Gene expression patterns of breast carcinomas distinguish tumor subclasses with clinical implications. *Proc. Natl. Acad. Sci. USA*, 98(19):10869-10874, 2001.

[24] T. Sorlie, R. Tibshirani, J. Parker, T. Hastie, J. S. Marron, A. Nobel, S. Deng, H. Johnsen, R. Pesich, S. Geisler, J. Demeter, C. M. Perou, P. E. Lonning, P. O. Brown, A. L. Borresen-Dale, D. Botstein. Repeated observation of breast tumor subtypes in independent gene expression data sets. *Proc. Natl. Acad. Sci. USA*; 100:8418-8423, 2003.

[25] C. Sotiriou, S.Y. Neo, L.M. McShane, et al. Breast cancer classification and prognosis based on gene expression profiles from a population-based study. *Proc. Natl. Acad. Sci. USA*; 100:10393-10398, 2003.

[26] Z. Hu, C. Fan, O. S. Oh, J. S. Marron, X. He, B. F. Qaqish, C. Livasy, L. A. Carey, E. Reynolds, L. Dressler, A. Nobel, J. Parker, M. G. Ewend, L. R.Sawyer, J. Wu, Y. Liu, R. Nanda, M. Tretiakova, A. Ruiz Orrico, O. Dreher0,J. P. Palazzo, L. Perreard, E. Nelson, M. Mone, H. Hansen, M. Mullins, J. F. Quackenbush, M. J. Ellis, O.L. Olopade, P. S. Bernard, C. M. Perou. The molecular portraits of breast tumors are conserved across microarray platforms. *BMC Genomics*; 7:96-96, 2006.

[27] Emad A. Rakha, Jorge S. Reis-Filho, Ian O. Ellis. Basal-like breast cancer: a critical review. *Journal of Clinical Oncology*, 26(15): 2568-81, 2008.

[28] J.S Reis-Filho, C. Westbury, J-Y Pierga. The impact of expression profiling on prognostic and predictive testing in breast cancer. *J. Clin. Patho.*, 59(3): 225-31, 2006.

[29] C. Fan, D.S. Oh, L. Wessels, B. Weigelt, D.S. Nuyten, A.B. Nobel, L.J. van't Veer, and C.M. Perou. Concordance among gene-expression-based predictors for breast cancer. *N. Engl. J. Med.*, 355: 560-9, 2006.

[30] N.C. Turner, J.S. Reis-Filho, A.M. Russell, R.J. Springall, K. Ryder, D. Steele, K. Savage, C.E. Gillett, F.C. Schmitt, A. Ashworth, and A.N. Tutt. BRCA 1 dysfunction in sporadic basallike breast cancer. *Oncogene*, 26: 2126-32, 2007.

[31] W.D. Foulkes, I.M. Stefansson, P.O. Chappuis, L.R. Begin, J.R. Goffin, N. Wong, M. Trudel, and L.A. Akslen. Germline BRCA 1 mutations and a basal epithelial phenotype in breast cancer. *J. Natl. Cancer Inst*, 95(19):1482-5, 2003.

[32] S.R. Lakhani, J.S. Reis-Filho, L. Fulford, F. Penault-Llorca, M. van der Vijver, S. Parry, T. Bishop, J. Benitez, C. Rivas, Y. J. Bignon, J. Chang-Claude, U. Hamann, C. J. Cornelisse, P. Devilee, M. W. Beckmann, C. Nestle-Kramling, P. A. Daly, N. Haites, J. Varley, F. Lalloo, G. Evans, C. Maugard, H. Meijers-Heijboer, J. G. Klijn, E. Olah, B. A. Gusterson, S. Pilotti, P. Radice, S. Scherneck, H. Sobol, J. Jacquemier, T. Wagner, J. Peto, M. R. Stratton, L. McGuffog, D. F. Easton. Prediction of BRCA 1 status in patients with breast cancer using estrogen receptor and basal phenotype. *Clin. Cancer Res.*, 11(14):5175-80, 2005.

[33] J.S. Reis-Filho and A.N.J. Tutt. Triple negative tumors: A critical review. *Histopathology*, 52(1):108-118, 2008.

[34] E.A. Rakha, M.E. El-Sayed, A.R. Green, A.H. Lee, J.F. Robertson, and I.O. Ellis. Prognostic markers in triple-negative breast cancer. *Cancer*, 109(1):25-32, 2007.

[35] M. Tischkowitz, J.S. Brunet, L.R. Begin, D.G. Huntsman, M.C. Cheang, L.A. Akslen, T.O. Nielsen, and W.D. Foulkes. Use of immunohistochemical markers can refine prognosis in triple negative breast cancer. *BMC Cancer*, 7:134, 2007.

[36] K.R. Bauer, M. Brown, R.D. Cress, C.A. Parise, and V. Caggiano. Descriptive analysis of estrogen receptor (ER)-negative, progesterone receptor (PR)-negative, and HER2negative invasive breast cancer, the so-called triple-negative phenotype: a populationbased study from the California cancer Registry. *Cancer*, 109(9):1721-8, 2007.

[37] L.N. Harris, G. Broadwater, N.U. Un, A. Miron, S.J. Schnitt, D. Cowan, J. Lara, I. Bleiweiss, D. Berry, M. Ellis, D.F. Hayes, E.P. Winer, and L. Dressler. Molecular subtypes of breast cancer in relation to paclitaxel response and outcomes in women with metastatic disease: results from CALGB 9342. *Breast Cancer Res.*, 8(6):R66, 2006.

[38] G.J. Morris, S. Naidu, A.K. Topham, F. Guiles, Y. Xu, P. McCue, G.F. Schwartz, P.K. Park, A.L. Rosenberg, K. Brill, and E.P. Mitchell. Differences in breast carcinoma characteristics in newly diagnosed African-American and Caucasian patients: a single-institution compilation compared with the National Cancer Institute's Surveillance, Epidemiology, and end results database. *Cancer*, 110(4):876-84, 2007.

[39] E.A. Rakha, M. E. El-Sayed, J.S. Reis-Filho, and I.O. Ellis. Expression profiling technology: its contribution to our understanding of breast cancer. *Histopathology*, 52(1):67-81, 2008.

[40] Emad A. Rakha, Maysa E. El-Sayed, Andrew R. Green, et al. Biologic and clinical characteristics of breast cancer with single hormone receptor positive phenotype. *J. Clin. Oncol.*, 25(30):4772-4778, 2007.

[41] James D. Brenton, Lisa A. Carey, Ahmed Ashour Ahmed, and Carlos Caldas. Molecular classification and molecular forecasting of breast cancer: ready for clinical application? *J. Clin. Oncol.*, 23(29): 7350-60, 2005.

[42] C.E. Geyer, J. Forster, D. Lindquist, S. Chan, C.G. Romieu, T. Pienkowski, A. Jagiello-Gruszfeld, J. Crown, A. Chan, B. Kaufman, D. Skarlos, M. Campone, N. Davidson, M. Berger, C. Oliva, S.D. Rubin, S. Stein, and D. Cameron. Lapatinib plus capecitabine for HER2-positive advanced breast cancer. *N. Engl. J. Med.*, 355: 2733-43, 2006.

[43] H. Joensuu, P-L. Kellokumpu-Lehtinen, P. Bono, T. Alanko, V. Kataja, R. Asola, T. Utriainen, R. Kokko, A. Hemminki, M. Tarkkanen, T. Turpeenniemi-Hujanen, S. Jyrkkiö, M. Flander, L. Helle, S. Ingalsuo, K. Johansson, A-S. Jääskeläinen, M. Pajunen, M. Rauhala, J. Kaleva-Kerola, T. Salminen, M. Leinonen, I. Elomaa, and J. Isola. Adjuvant docetaxel or vinorelbine with or without trastuzumab for breast cancer. *N. Engl. J. Med.*, 354: 809-20, 2006.

[44] M.J. Piccart-Gebhart, M. Procter, B. Leyland-Jones, A. Goldhirsch, M. Untch, I. Smith, L. Gianni, J. Baselga, R. Bell, C. Jackisch, D. Cameron, M. Dowsett, C.H. Barrios, G. Steger, C-S. Huang, M. Andersson, M. Inbar, M. Lichinitser, I. Láng, U. Nitz, H. Iwata, C. Thomssen, C. Lohrisch, T.M. Suter, J. Rüschoff, T. Sütő, V. Greatorex, C. Ward, C. Straehle, E. McFadden, M.S. Dolci, and R.D. Gelber. Trastuzumab after adjuvant chemotherapy in HER2-positive breast cancer. *N. Engl. J. Med.*, 353: 1659-72, 2005.

[45] D.J. Slamon, B. Leyland-Jones, S. Shak, H.Fuchs, V. Paton, A. Bajamonde, T. Fleming, W. Eiermann, J Wolter, M. Pegram, J. Baselga, and L. Norton. Use of chemotherapy plus a monoclonal antibody against HER2 for metastatic breast cancer that overexpresses HER2. *N. Engl. J. Med.*, 344:783-92, 2001.

[46] E.A. Rakha, M.E. El-Sayed, J. Reis-Filho and I.O. Ellis. Patho-biological aspects of basal-like breast cancer. *Breast Cancer Res. Treat.*, 113(3):411-22, 2008.

[47] E.A. Rakha, J.S. Reis-Filho, I.O. Ellis. Impact of basal-like breast carcinoma determination for a more specific therapy. *Pathobiology*, 75(2):95-103, 2008.

[48] B. Gusterson. Do 'basal-like' breast cancers really exist? *Nat Rev Cancer*, 9:128-134, 2009.

[49] L.A. Carey, C.M. Perou, C.A. Livasy, L.G. Dressler, D. Cowan, K. Conway, G. Karaca, M.A. Troester, C.K. Tse, S. Edmiston, S.L. Deming, J. Geradts, M.C. Cheang, T.O. Nielsen, P.G. Moorman, H.S. Earp, and R.C. Millikan. Race, breast cancer subtypes, and survival in the Carolina Breast Cancer Study. *JAMA* 295(21):2492-502, 2006.

[50] L.G. Fulford, D.F. Easton, J.S. Reis-Filho, A. Sofronis, C.E. Gillett, S.R. Lakhani, and A. Hanby. Specific morphological features predictive for the basal phenotype in grade 3 invasive ductal carcinoma of breast. *Histopathology*, 49(1):22-34, 2006.

[51] C.A. Livasy, G. Karaca, R. Nanda, M.S. Tretiakova, O.I. Olopade, D.T. Moore, C.M. Perou. Phenotypic evaluation of the basal-like subtype of invasive breast carcinoma. *Mod. Pathol.*, 19(2):264-71, 2006.

[52] S. Banerjee, J.S. Reis-Filho, S. Ashley, D. Steele, A. Ashworth, S.R. Lakhani, I.E. Smith. Basal-like breast carcinomas: clinical outcome and response to chemotherapy. *J. Clin. Pathol*, 59(7):729-35, 2006.

[53] K. Altundag, H. Harputluoglu, S. Aksoy, and I.H. Gullu. Potential chemotherapy options in the triple negative subtype of breast cancer. J. Clin. Oncol., 25(10):1294-5, 2007; author reply 1295-6.

[54] B.G. Haffty, Q. Yang, M. Reiss, T. Kearney, S.A. Higgins, J. Weidhaas, L. Harris, W. Hait, and D. Toppmeyer. Locoregional relapse and distant metastasis in conservatively managed triple negative early-stage breast cancer. *J. Clin. Oncol.*, 24(36): 5652-7, 2006.

[55] C. Liedtke, C. Mazouni, K.R. Hess, F. Andre, A. Tordai, J.A. Mejia, W.F. Symmans, A.M. Gonzalez-Angulo, B. Hennessy, M. Green, M. Cristofanilli, G.N. Hortobagyi, and L. Pusztai. Response to neoadjuvant therapy and long-term survival in patients with triple-negative breast cancer. *J. Clin. Oncol.*, 26(8): 1275-81, 2008.

[56] L.A. Carey, E.C. Dees, L. Sawyer, L. Gatti, D.T. Moore, F. Collichio, D.W. Ollila, C.I. Sartor, M.L. Graham, and C.M. Perou. The triple negative paradox: primary tumor chemosensitivity of breast cancer subtypes. *Clin. Cancer Res.*, 13(8): 2329-34, 2007.

[57] T.O. Nielsen, F.D. Hsu, K. Jensen, et al. Immunohistochemical and clinical characterization of the basal-like subtype of invasive breast carcinoma. *Clin. Cancer Res.*, 10: 5367-74, 2004.

[58] M.C.U. Cheang,D. Voduc, C. Bajdik, S. Leung, S. McKinney, S.K. Chia, C.M. Perou, and T.O. Nielsen. Basal-like breast cancer defined by five biomarkers has superior prognostic value than triple-negative phenotype. *Clin. Cancer Res.*, 14(5): 1368-76, 2008.

[59] L.G. Fulford, J.S. Reis-Filho, K. Ryder, C. Jones, C.E. Gillett, A. Hanby, D.Easton, and S.R. Lakhani. Basal-like grade III invasive ductal carcinoma of the breast: patterns of metastasis and long-term survival. *Breast Cancer Res*, 9: R4, 2007.

[60] J.B. Arnes, J.S. Brunet, I. Stefansson, L.R. Begin, N. Wong, P.O. Chappuis, L.A. Akslen, and W.D. Foulkes. Placental cadherin and the basal epithelial phenotype of BRCA 1-related breast cancer. *Clin. Cancer Res.*, 11(11): 4003-11, 2005.

[61] S.M. Pinilla, E. Honrado, D. Hardisson, J. Benitez, and J. Palacios. Caveolin-1 expression is associated with a basal-like phenotype in sporadic and hereditary breast cancer. *Breast Cancer Res. Treat.*, 99(1):85-90, 2006.

[62] K. Savage, S. Leung, S.K. Todd, L.A. Brown, R.L. Jones, D. Robertson, M. James, S. Parry, S.M. Rodrigues Pinilla, D. Huntsman, and J.S. Reis-Filho. Distribution and significance of Caveolin 2 expression in normal breast and invasive breast cancer: an immunofluorescence and immunohistochemical analysis. *Breast Cancer Res. Treat.*, 110(2):245-56, 2007.

[63] S. Parry, K. Savage, C. Marchio, and J.S. Reis-Filho. Nestin is expressed in basal-like and triple negative breast cancers. *J. Clin. Pathol*, 61(9):1045-50, 2008.

[64] J.V. Moyano, J.R. Evans, F. Chen, M. Lu, M.E. Werner, F. Yehiely, L.K. Diaz, D. Turbin, G. Karaca, E. Wiley, T.O. Nielsen, C.M. Perou, and V.L. Cryns. AlphaB-crystallin is a novel oncoprotein that predicts poor clinical outcome in breast cancer. *J. Clin. Invest*, 116(1):261-70, 2006.

[65] S.M. Sitterding, W.R. Wiseman, C.L. Schiller, C. Luan, F. Chen, J.V. Moyano, W.G. Watkin, E.L. Wiley, V.L. Cryns, and L.K. Diaz. AlphaB-crystallin: a novel marker of invasive basal-like and metaplastic breast carcinomas. *Ann. Diagn Pathol.*, 12(1):33-40, 2008.

[66] A.P. Subhawong, T. Subhawong, H. Nassar, N. Kouprina, S. Begum, R. Vang, W.H. Westra, P. Argani. Most Basal-like Breast Carcinomas Demonstrate the Same Rb-/p16+ Immunophenotype as the HPV-related Poorly Differentiated Squamous Cell Carcinomas Which They Resemble Morphologically. *Am. J. Surg. Pathol.*, 33(2):163-75, 2009.

[67] M.L. Gauthier, H.K. Berman, C. Miller, K. Kozakeiwicz, K. Chew, D. Moore, J. Rabban, Y.Y. Chen, K. Kerlikowske, and T.D. Tlsty. Abrogated response to cellular stress identifies DCIS associated with subsequent tumor events and defines basal-like breast tumors. *Cancer Cell*, 12(5): 479-91, 2007.

[68] J.S. Reis-Filho and A.N. Tutt. Triple negative tumours: a critical review. *Histopathology*, 52(1):108-18, 2008.

[69] D.M. Abd El-Rehim, S.E. Pinder, C.E. Paish, J. Bell, R.W. Blamey, J.F. Robertson, R.I. Nicholson, and I.O. Ellis. Expression of luminal and basal cytokeratins in human breast carcinoma. *J. Pathol.*, 2004; 203(2):661-71, 2004.

[70] M. Jumppanen, S. Gruvberger-Saal, P. Kauraniemi, M. Tanner, P.O. Bendahl, M. Lundin, M. Krogh, P. Kataja, A. Borg, M. Ferno, and J. Isola. Basal-like phenotype is not associated with patient survival in estrogen-receptor-negative breast cancers. *Breast Cancer Res.*, 9(1):R16, 2007.

[71] R. Dent, M. Trudeau, K.I. Pritchard, W.M. Hanna, H.K. Kahn, C.A. Sawka, L.A. Uckley, E. Rawlinson, P. Sun, and S.A. Narod. Triple-negative breast cancer: clinical features and patterns of recurrence. *Clin. Cancer Res.*, 13(15 Pt 1):4429-34, 2007.

[72] B.P. Schneider, E.P. Winer, W.D. Foulkes, J. Garber, C.M. Perou, A. Richardson, G.W. Sledge, and L.A. Carey. Triple-negative breast cancer: risk factors to potential targets. *Clin. Cancer Res.*, 14(24): 8010-8, 2008.

[73] M. Van De Rijn, C.M. Perou, R. Tibshirani, et al. Expression of cytokeratins 17 and 5 identifies a group of breast carcinomas with poor clinical outcome. *Am. J. Pathol.*, 161(6): 1991-96, 2002.

[74] I. Matos, R. Dufloth, M. Alvarenga, L.C. Zeferino, and F. Schmitt F. p63, cytokeratin 5, and P-cadherin: three molecular markers to distinguish basal phenotype in breast carcinomas. *Virchows Arch.*, 447(4); 688-694, 2005.

[75] N.C. Turner and J.S. Reis-Filho. Basal-like breast cancer and the BRCA1 phenotype. *Oncogene*, *25*; 5846-53, 2006.

[76] C. Chabalier, C. Lamare, C. Racca, et al. BRCA1 downregulation leads to premature inactivation of spindle checkpoint and confers paclitaxel resistance. *Cell Cycle*, 5(9):1001-7, 2006.

[77] K. Schlacher, N. Christ, N. Siaud, A. Egashira, H. Wu, and M. Jasin. Double-strand break repair-independent role for BRCA2 in blocking stalled replication fork degradation by MRE11. *Cell*, 145 (4):529-542, 2011.

[78] A. Meindl, N. Ditsch, K. Kast, K. Rhie, and R.K. Schmutzler. Hereditary breast and ovarian cancer: new genes, new treatments, new concepts. *Dtsch Arzteb Int.*, 108(19):323-30, 2011.

[79] N. Turner, A. Tutt, and A. Ashworth. Hallmarks of 'BRCAness' in sporadic cancers. *Nat. Rev. Cancer*, 4:814-9, 2004.

[80] N.C. Turner and J.S. Reis-Filho. Basal-like breast cancer and the BRCA 1 phenotype. *Oncogene*, 25: 5846-53, 2006.

[81] W.D. Foulkes, J.S. Brunet, I.M. Stefansson, O. Straume, P.O. Chappuis, L.R. Begin, N. Hamel, J.R. Goffin, N. Wong, M. Trudel, L. Kapusta, P. Porter, and L.A. Akslen. The prognostic implication of the basal-like (cyclin E high/p27 low/p53+/glomeruloid-microvascularproliferation+) phenotype of BRCA1-related breast cancer. *Cancer Res.*, **64**:830-835, 2004.

[82] S.A. Vaziri, L.M. Krumroy, P. Elson, G.T. Budd, G. Darlington, J. Myles, R.R. Tubbs, and G. Casey. Breast tumor immunophenotype of BRCA1-mutation carriers is influenced by age at diagnosis. *Clin. Cancer Res.*, 7(7):1937-45, 2001.

[83] J.S. Reis-Filho, K. Savage, M.B. Lambros, M. James, D. Steele, R.L. Jones, and M. Dowsett. Cyclin 01 protein overexpression and CCND1 amplification in breast carcinomas: an immunohistochemical and chromogenic in situ hybridisation analysis. *Mod. Pathol.*, 19(7):999-1009, 2006.

[84] S.A. Vaziri, R.R. Tubbs, G. Darlington, and G. Casey. Absence of CCND1 gene amplification in breast tumours of BRCA1 mutation carriers. *Mol. Pathol.*, 54(4):259-63, 2001.

[85] J. Palacios, E. Honrado, A. Osorio, A. Cazorla, D. Sarrio, A. Barroso, S. Rodriguez, J.C. Cigudosa, O. Diez, C. Alonso, E. Lerma, J. Dopazo, C. Rivas, J. Benitez. Phenotypic characterization of BRCA 1 and BRCA2 tumors based in a tissue microarray study with 37 immunohistochemical markers. *Breast Cancer Res. Treat.*, 90(1):5-14, 2005.

[86] D.M. Abd El-Rehim, G. Ball, S.E. Pinder, E. Rakha, C. Paish, J.F. Robertson, D. Macmillan, R.W. Blamey, and I.O. Ellis. High-throughput protein expression analysis using tissue microarray technology of a large well-characterised series identifies biologically distinct classes of breast cancer confirming recent cDNA expression analyses. *Int. J. Cancer*, 116(3):340-50, 2005.

[87] S. Calza, P. Hall, G. Auer, J. Bjohle, S. Klaar, U. Kronenwett, E.T. Liu, L. Miller, A. Ploner, J. Smeds, J. Bergh, and Y. Pawitan. Intrinsic molecular signature of breast cancer in a population-based cohort of 412 patients. *Breast Cancer Res.*, 8(4): R34, 2006.

[88] A. McCarthy, K. Savage, A. Gabriel, C. Naceur, J.S. Reis-Filho, and A. Ashworth. A mouse model of basal-like breast carcinoma with metaplastic elements. *J. Pathol.*, 211(4):389-98, 2007.

[89] A.M. Hosey, J.J. Gorski, M.M. Murray, J.E. Quinn, W.Y. Chung, G.E. Stewart, C.R. James, S.M. Farragher, J.M. Mulligan, A.N. Scott, P.A. Dervan, P.G. Johnston, F.J. Couch, P.A. Daly, E. Kay, A. McCann, P.B. Mullan, and D.P. Harkin. Molecular basis for estrogen receptor alpha deficiency in BRCA 1-linked breast cancer. *J. Natl. Cancer Inst*, 99(22):1683-94, 2007..

[90] E. Matros, Z.C. Wang, G. Lodeiro, A. Miron, J.D. Iglehart, and A.L. Richardson. BRCA1 promoter methylation in sporadic breast tumors: relationship to gene expression profiles. *Breast Cancer Res. Treat.*, 91(2):179-86, 2005.

[91] N.C. Turner, J.S. Reis-Filho, A.M. Russell, et al. BRCA1 dysfunction in sporadic basal-like breast cancer. *Oncogene*, 26(14):2126-32, 2007.

[92] I. Hedenfalk, D. Duggan, Y. Chen, et al. Gene-expression profiles in hereditary breast cancer. *N. Engl. J. Med.*, 344:539-548, 2001.

[93] S. Liu, C. Ginestier, E. Charafe-Jauffret, et al. BRCA1 regulates human mammary stem/progenitor cell fate. *Proc. Natl. Acad. Sci. USA*, 105(5):1680-5, 2008.

[94] B. Kreike, M. van Kouwenhove, H. Horlings, et al. Gene expression profiling and histopathological characterization of triple-negative/basal-like breast carcinomas. *Breast Cancer Res.*, 9(5):R65, 2007.

[95] H. Kaur, S. Saini, S. Peer, and J. Singh. Current therapies and novel targets in treatment of breast cancer. *Systematic Reviews in Pharmacy*, 1(1), 40 -49, 2010.

[96] M. Morrow. Techniques in surgery: Lumpectomy for palpable and nonpalpable cancer. In: Harris JR. editors. Diseases of the breast. 3 rd ed. 2004. p. 819-25.

[97] Armando E. Giuliano, Kelly K. Hunt, Karla V. Ballman, Peter D. Beitsch, Pat W. Whitworth, Peter W. Blumencranz, Marilyn Leitch, Sukamal Saha, Linda M. McCall, Monica Morrow. Axillary Dissection vs No Axillary Dissection in Women With Invasive Breast Cancer and Sentinel Node Metastasis. *The Journal of the American Medical Association*, 305(6):569-575, 2011. (doi: 10.1001/jama.2011.90).

[98] U. Veronesi, B. Salvadori, A. Luini, M. Greco, R. Saccozzi, M. del Vecchio, et al. Breast conservation is a safe method in patients with small cancer of the breast. Long-term results of three randomised trials on 1973 patients. *Eur. J. Cancer Am.*, 31A(10):1574-9, 1995.

[99] U. Veronesi, N. Cascinelli, L. Mariani, M. Greco, R. Saccozzi, and A. Luini. Twenty-year-follow-up of ran-domized study comparing breast- conserving surgery with radical mastectomy for early breast cancer. *N. Engl. J. Med.*, 347(16):1227-32, 2002.

[100] Early Breast Cancer Trialists Collaborative Group. Favourable and unfavourable effects on long-term survival of radiotherapy for early breast cancer: An overview of the randomised trials. *Lancet*, 355(9217):1757-70, 2000.

[101] A.D. Hamberger, D.H. Roukos, and T.A Buchholz. Radiation therapy for breast cancer. *N. Engl. J. Med.*, 360:1362-3, 2009.

[102] Y.D. James and N.E. Davidson. Estrogen carcinogenesis in breast cancer. *N. Engl. J. Med.*, 354(3):270-82, 2006.

[103] S. Martino, J. Costantino, M. McNabb, J. Mershon, K. Bryant, T. Powles, et al. The role of selective estrogen receptor modulators in the prevention of breast cancer: Comparison of the clinical trials. *Oncologist*, 9(2):116-25, 2004.

[104] B. Fisher, J.P. Costantino, D.L. Wickerham, R.S. Cecchini, W.M. Cronin, and A. Robidoux. Tamoxifen for the prevention of breast cancer: Current status of the National Surgical Adjuvant Breast and Bowel Project P-1 study. *J. Natl. Cancer Inst.*, 97(22):1652-62, 2005.

[105] E. Barrett-Connor, L. Mosca, P. Collins, M.J. Geiger, D. Grady, and M. Kornitzer. Effects of raloxifene on cardiovascular events and breast cancer in postmenopausal women. *N. Engl. J. Med.*, 355(2):125-37, 2006.

[106] A.N. Wymenga and D.T. Sleijfer. Management of hot flushes in breast cancer patients. *Acta Oncol.*, 41(3):269-75, 2002.

[107] G. von Minckwitz, E. Graf, M. Geberth, W. Eirmann, W. Jonat, B. Conrads, et al. Goserelin versus CMF as adjuvant therapy for node negative, hormone receptor-positive breast cancer in premenopausal patients: A randomised trial (GABG IV-A-93). *Eur. J. Cancer*, 42(12):1780-8, 2006.

[108] H.S. Rugo. The breast cancer continuum in hormone-receptor-positive breast cancer in postmenopausal women: Evolving management options focusing on aromatase inhibitors. *Ann. Oncol.*, 19(1):16-27, 2008.

[109] S.M. Campos. Aromatase inhibitors for breast cancer in postmenopausal women. *Oncologist*, 9(2):126-36, 2004.

[110] J. Geisler and P.E. Lonning. Endocrine effects of aromatase inhibitors and inactivators in vivo: Review of data and method limitations. *J. Steroid Biochem. Mol. Biol.*, 95(1-5):75-81, 2005.

[111] W.J. Gradishar. Clinical value of fulvestrant in breast cancer. *Clin. Breast Cancer*, 6 Suppl 1:S4, 2005.

[112] A. Howell and P. Abram. Clinical development of fulvestrant ('Faslodex'). *Cancer Treat. Rev.*, 31 Suppl 2:S3-9, 2005.

[113] A. Howell. Is fulvestrant ("Faslodex") just another selective estrogen receptor modulator? *Int. J. Gynecol. Cancer*, 16 Suppl 2:521-3, 2006.

[114] A.D. Seidman, M.N. Fornier, and F.J. Esteva. Weekly trastuzumab and paclitaxel therapy for metastatic breast cancer with analysis of efficacy by HER2 immunophenotype and gene amplification. *J. Clin. Oncol.*, 19(10):2587-95, 2001.

[115] C. Tapia, S. Savic, U. Wagner, R. Schönegg, H. Novotny, B. Grilli, et al. HER2 gene status in primary breast cancers and matched distant metastases. *Breast Cancer Res.*, 9(3):R31, 2007.

[116] H.M. Shepard, G.D. Lewis, and J.C. Sarup. Monoclonal antibody therapy of human cancer: Taking the HER2 protooncogene to the clinic. *J. Clin. Immunol*, 11(3):117-27 1991.

[117] S.G. Gilbert. Trastuzumab in breast cancer. *N. Engl. J. Med.*, 354:640-42006.

[118] F. Petrelli, M. Cabiddu, M.E. Cazzaniga, M. Cremonesi, and S. Barni. Targeted therapies for the treatment of breast cancer in the post-trastuzumab era. *Oncologist*, 13(4):373-81, 2008.

[119] M.N. Levine, V.H. Bramwell, K.I. Pritchard, B.D. Norris, L.E. Shepherd, H. Abu-Zahra, et al. Randomized trial of intensive cyclophosphamide, epirubicin, and fluorouracil chemotherapy compared with cyclophosphamide, methotrexate, and fluorouracil in premenopausal women with node-positive breast cancer. *J. Clin. Oncol.*, 16(8):2651-8, 1998.

[120] W.C. Wood, D.R. Budman, A.H. Korzun, M.R. Cooper, J. Younger, R.D. Hart, et al. Dose and dose intensity of adjuvant chemotherapy for stage II node-positive breast cancer. *N. Engl. J. Med.*, 330(18):1253-9, 1994.

[121] G.N. Hortobagyi. Drug therapy: Treatment of breast cancer. *N. Engl. J. Med.*, 339:974-841998.

[122] M.L. Citron, D.A. Berry, C. Cirrincione, C. Hudis, E.P. Winer, W..J Gradishar, et al. Randomized trial of dose-dense vs. conventionally scheduled and sequential vs. concurrent combination chemotherapy as postoperative adjuvant treatment of node-positive primary breast cancer: First report of Intergroup Trial C9741/Cancer and Leukemia Group B Trial 9741. *J. Clin. Oncol.*, 21(8):1431-9, 2003.

In: Signaling, Gene Regulation and Cancer
Editors: Shree Ram Singh and Manoj K. Mishra
ISBN: 978-1-61942-088-5

Chapter 12

NOTCH SIGNALING REGULATES CANCER STEM CELLS AND TUMOR ANGIOGENESIS

Rashmi K. Ambasta[1,3,*], Pravir Kumar[1,2,3], Archita Sharma and S. Esther Priyadharshini[3]

[1]Associate Professors at School of Bio-Sciences and Technology, VIT, Vellore, India
[2]Adjunct Faculty, Department of Neurology, Tufts University School of Medicine, Boston, MA, US
[3]Center for Medical Engineering, VIT University, Vellore, India

"All substances are poisonous, there is none that is not a poison; the right dose differentiates a poison from a remedy" Paracelsus (Auroleus Phillipus Theostratus Bombastus von Hohenheim), Alchemist and Physician, 1538.

ABSTRACT

Notch signaling pathway plays a crucial role in the regulation of cancer stem and endothelial cells proliferation. Upon activation, notch receptor cleaves it intracellular domain, which translocates to the nucleus and trigger the signals for cell proliferation via its transcription factor Hes and Hey. This chapter describes the role of notch pathways in regulation of cancer stem cell and endothelial cell proliferation. Several evidences suggest that the self-renewal capacities of cancer stem cells and tumor angiogenic features are regulated by notch signaling pathway. However, there are various other signaling pathways such as NADPH Oxidase, Wnt, Sonic hedgehog having influence on Notch signaling pathway and explained briefly in this chapter. Recent reports hints towards the controlling mechanism of cancer stem cell and tumor angiogenesis through

[*] Corresponding Author: Address: Rashmi Ambasta, PhD; Cancer biology and functional genomics lab; School of Bio-science and Technology, VIT; Vellore, TN 632014 (INDIA); E-mail: ambasta@vit.ac.in; rashmiambasta@gmail.com; Phone: +91-4162202512 (lab); +91-9003386751; Co-corresponding author: Pravir.Kumar@tufts.edu; Phone: +91-4162202513 (lab); +91-9003386752.

notch signaling molecules. Therefore, notch could be a better target in cancer therapy compared to other signaling pathway. There are several physiological functions are regulated by notch pathway, hence complete inhibition of notch signal is not recommended. Instead of that, in this chapter, we discussed about the specific anti-cancer drug targeting using nanoparticles mediated drug delivery that is very specific to the tumor site and not to the entire body. In this chapter, we also highlighted how a designed twin nanoparticle carries a specific stem cell marker for delivery to the tumor site? The nanoparticles are further concentrated on the tumor site using a magnet as the particles are magnetic in nature. The notch inhibitor drugs are only released in response to hyperthermia generated by magnetic field so that drug is released only on a tumor site and not to other parts of the body where normal functioning of notch can take place. Therefore, this chapter opens new avenues in cancer therapy using a combination of twin nanoparticle-notch inhibitors in addition to cancer stem cell marker.

1. INTRODUCTION

1.1. Notch Signaling

The conserved notch signaling pathway in multicellular organisms plays an important role in cell-cell communication that controls the gene regulatory mechanism, multiple cell differentiation process and most importantly in the determination of cell-fate, i.e. in self-renewal (stem cells) and self-lineage (differentiation) fashion during embryonic and adult phase. The notch signaling molecules triggers the various pathways that are associated with cell fate decisions such as development of neurons and its function, regulation of arterial endothelium and angiogenesis that involves a crucial communication between endocardium and myocardium. However, the abruptly high or low levels of notch signaling are reported to be lethal in embryonic stage.

Several studies shown that aberrant notch signaling pathway has a great impact on human malignancies and plays a significant role in cancer progression. However, genetic and pharmacological manipulation in the notch signaling pathway has a novel therapeutic approach to combat with certain neoplasm and cancer [1]. There is another usage of notch signaling pathway in various diseases such as T-acute lymphoblastic leukemia, cerebral ailments with subcortical infracts leukoencephalopathy and multiple sclerosis. It has been also shown that the notch signaling is responsible for maintaining a homeostasis between cell proliferation and apoptosis [2] hence, inhibition of Notch-1, elicits anti-proliferative and apoptotic properties in pancreatic cancer cells, T-cells acute lymphoblastic leukemia as well as in the animal model [3,4]. Another evidence show that notch signaling induces the cell cycle arrest in small cell lung cancer cells [5]. There are several ligands, which help in the activation of notch signaling that is responsible for the maintenance of stem cells and the progenitor cell.

1.2. Role of Notch Signaling in Cancer Stem Cell and Tumor Angiogenesis

While notch signaling is involved in the cell proliferation and differentiation of cancer stem cells (CSCs), this pathway has a crucial role in tumorigenesis. It has been shown that upon wnt activation certain notch ligands such as Dll-2, Dll-3 and Dll-4 is upregulated that are required for the tumorigenic phenotypes [6]. A new approach regarding the development of solid tumor came from the idea that, like other normal adult tissues, solid tumor also contains stem cells (called cancer stem cells) and is responsible for the growth of solid tumor. There are several genetic mutations that are encoding for proteins are involved critically in the several signaling pathway viz. Notch, Hedgehog, Wnt and NADPH oxidase that allows stem cells to undergo uncontrolled proliferation and form tumor. Considering the above facts, cancer stem cells are the real driving force behind tumor growth where drugs may be resistant to that tumor and hence more effective ant-solid tumor therapies are required to conquer the battle of cancer. In this present chapter, we have highlighted the molecular basis of the Notch signaling pathway in the context with tumor angiogenesis and cancer stem cell proliferation and differentiation.

1.3. Different Type of Notch Ligand and Receptors

Multicellular organism possess a highly conserved notch signaling system with four different type of notch receptors viz. Notch 1, Notch 2, Notch 3 and Notch 4. The discovery of notch has been seen first in the wings of fruit fly in 1914 while allelic genes were identified by T.H. Morgan in 1917. Notch plays a crucial role in cell signaling mechanism by a transmembrane receptor that coordinate notch signaling pathway.

The importance of notch signaling is not limited to the fruitfly wing outgrowth rather also in other developmental processes ranging from worms to human.

The bifunctional properties of Notch are to receive the extracellular cell signals from the cell surface and regulate the gene expression in the nucleus. In this chapter, we will first focus on understanding the notch pathway and then the role of notch pathway in controlling cancer stem cell and tumor angiogenesis. Table 1 lists the different type of ligand, receptor and transcription factor (CBF1; C promoter binding factor RBP: Recombinant signal binding protein RBPjk and RBPL) in Notch signaling pathway.

Table 1. Name of core components of notch signaling in mammals and Drosophila [7]

Core component	Mammals	Drosophila
Ligands	DLL1, DLL2, DLL3, Jag1 and Jag 2	Delta, Serrate
Receptor	Notch 1, 2, 3 and 4	Notch
Transcription factors	CBF1,RBPjk, RBPL	Suppression of Hairless

1.4. Activation of Notch Signaling

Notch receptor comprises a single-pass transmembrane domain where its large extracellular cognate is hetero-oligomer that associates in a calcium dependent manner while a small extracellular portion is a single transmembrane pass and an intracellular region (Figure 1). Activations of notch proteins are ligand based. Notch-1 and Notch-4 in vertebrate are single pass receptors, which are activated by the Delta (or Delta like) and Jagged/Serrate families. Although, they are intact polypeptide entities, during transportation to plasma membrane, they get cleaved.

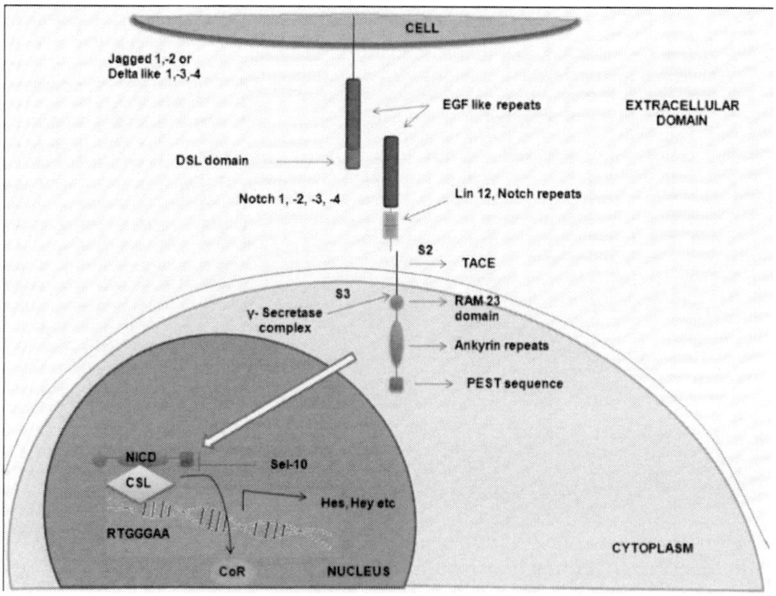

Figure 1. Signaling Pathway of Notch from extracellular domain to nucleus (Adapted and modified from [8].

The other feature of Notch to interact with ligands leads to two additional proteolytic cleavages that liberate the Notch intracellular domain (NICD) from the plasma membrane. The NICD translocates to the nucleus, where it forms a complex with the DNA binding protein CSL, displacing a histone deacetylase (HDAC)-co-repressor (CoR) complex from CSL (CBF1, Suppressor of Hairless and Lag1). Components of an activation complex, such as MAML1 and histone acetyltransferases (HATs), are recruited to the NICD-CSL complex, leading to the transcriptional activation of Notch target genes as shown in Figure 1. The notch target genes are Hes (Hairy Enhancer of Split) and Hey (Hes related with YRPW). The pathway can be regulated by Sel-10 (F-Box protein).

The mechanism of notch signaling is regulated intra- and extracellular. The proteolytic cleavage triggered once the ligand bound to the extracellular domain and then cleaved parts enter the nucleus and modulate the gene expression. Since ligands are also transmembrane proteins hence, receptors are triggered only from direct cell-to-cell contact. This orchestrated process organizes the cells in such a way that, if one cell express certain trait, that cell may switch off the same trait of the neighboring cells by intracellular notch signaling and thus influencing one another to make a large structure. The entire even possess a cascade of events that include, notch, notch ligands as well as intracellular proteins that transmit the notch signal to the cell's nucleus.

In the process of drug designing, the development of new drugs for the treatment of cancer stem cells (CSCs) will also require the same type of understanding between the cellular processes that regulate the cell proliferation. The hallmark feature in hematopoietic stem cells (HSCs) and their transformed counterparts in Leukemia where the origin of CSCs was well understood. It is now becoming increasingly clear that stem cells of many organs share the same cellular pathways as leukemia-derived HSCs. Additionally; a normal stem cell may be transformed into a cancer stem cell through deregulation of the proliferation and differentiation pathways and thus controlling its oncoprotein activities.

2. Cancer Stem Cells (CSCs)

Cancer stem cells (CSCs) possess normal stem cells that have the ability to give rise to all cell types found in a particular cancer sample. The tumorogenic effects of CSCs are well differentiated with non- tumorogenic cancer cells and may generate through the processes of self-renewal and differentiation into multiple cell types. These cells are persisted in tumors as a distinct population and may cause the revival and metastasis by producing new tumor. Therefore, it is necessary to identify a specific therapeutic target at CSCs level for the improvement of survival and quality of life of cancer patients, especially for sufferers of metastatic disease. However, existing cancer therapies are based and derived from the work done on animal models where drug action only helps to reduce in the progression of tumor, but at the same time it is not necessary that the same drug has the equivalent effect on human. Another restriction in the mice model as mice do not exceed two years of life span hence relapse of tumor, and its thorough study is quite difficult to replicate in the human model. Notch signaling has been reported to be playing a crucial role in cancer stem cell [9, 10, 11, 12].

In the initial stage of cancer, the efficacy of cancer treatments is judged by the ablation fraction of tumor mass. Initially, CSCs form a very small proportion of the tumor that may not necessarily select for drugs that act specifically on the stem cells. This theory strengthen the ideas of conventional chemotherapy that indiscriminately kills the differentiated or differentiating cells that is the main constituent of a tumor but unable to generate new cells. This is the reason why a population of CSCs, which gave rise to it, could remain untreated with drug and cause for relapsing of the disease.

Since the beginning of cancer stem cell hypothesis, the cellular heterogeneity of neoplasms was the center of attraction where it originally formulated for hematologic malignancies a precursor of solid tumor. However, the origins of cancer "stem" cells (CSC) or tumor-initiating cells and their method of identification are quite baffling and challenging questions. Nevertheless, the existence of subpopulations of tumor cells with stem-cell like characteristics has own significant therapeutic implications. The stem-cell like phenotype includes self-replication, pluripotency and importantly, resistance to chemotherapeutics. Thus, it is plausible that CSCs, regardless of their origin, may untreatable with standard therapies and cause disease recurrences and/or metastasis after apparently complete remissions. Therefore the idea of selectively targeting of CSCs with novel therapeutics is gaining huge interest. The Notch pathway is one of the most intensively studied putative therapeutic targets in CSC, where several Notch inhibitors have been developed. However, for successful targeting of Notch signaling in CSCs, thorough understandings of Notch regulation with its context-dependent interactions are need to be proven therapeutically along with other signaling molecules. These interactions will enable us to design rational drugs in different combinations that are more likely to safe and effective. Moreover, it also enables us on personalized medicine where an individual can get the best possible treatment of Notch-targeting therapeutics. However, the reliable biomarkers of Notch that measure pathway activities in CSCs from specific tumors will have to be identified and validated. This chapter summarizes the most recent developments in the field of Notch-targeted cancer therapeutics, with emphasis on CSC.

Table 2. Evidences for cancer stem cell promotion by Notch

Notch signaling component	Phenotype or Findings	References
Notch 1	Promotes colony forming capacity and growth in glioma cell. Blockage leads to the reduction in CD133+ cells in glioma. Over expressed in T- cell lymphoblastic leukemia (T-ALL), breast cancer. It induces expression of ABCG2, which is associated with proliferation and drug resistance in cancer stem cells.	[13, 14, 15, 16]
Jagged 1	Increased number of stem cells and self-renewing capacity by recombinant erythropoietin receptors by induction of jagged 1.	[17]
Jagged 2	over expressed in leukemic stem cells	[18]
γ-secretase	Inhibition by DAPT inhibits the growth of leukemic stem cells. Use of GSI has also shown significant anti-proliferative effect on T-ALL cell lines and mouse models either by cell cycle arrest or apoptosis	[18, 19]

In the regenerative medicine, to understand the cell signaling mechanism cellular systems are the best source to understand the etiology of disease that involves signaling mechanism. For example, human pluripotent stem cells (PSCs) derived from a number of different sources, including reprogrammed adult somatic cells that provide the new source which is essential to control the direction and efficiency of human PSCs differentiation.

Although Notch, Wnt, NADPH oxidase and Hedgehog (Hh) signaling pathways have been implicated in the self-renewal/proliferation of hematopoietic stem/progenitor cells, both *in vitro* and *in vivo*, their roles in differentiation processes remain poorly explored. Understanding the individual and collective contributions of Notch, Wnt and Hh signaling for the initial development of hematopoiesis is critical for achieving successful *ex vivo* expansion and differentiation of hematopoietic stem cells (HSCs) from human PSCs that will retain *bona fide* function comparable to adult-derived HSCs.

The baffling issue in nineties was the existence of CSCs in clinical research as it was not successfully established that the difference between normal tissues stem cells and cancer stem cells. However, various theories have been suggested for the genesis of cancer stem cells. These theories tell that may be the stem cells are either mutant in developing stem cells or progenitor cells, adult stem or progenitor cells or mutant cells that acquire stem cell like characteristics.

The propensity of tumor growth from the daughter stem cells upon mutation is very high as mutation is a well-known factor for cancer progression while cancer associated with adult stem cells are those populations where cell turnover is very high. In these tissues, it has long been expected that stem cells are responsible for tumor formation. This is a consequence of the aberrant cell divisions of these stem cells compared to most adult stem cells in conjunction with the extremely long lifespan of adult stem cells.

Combination of these two phenomenon causes an ideal platform for growth and accumulation of cells that ultimately leads to the initiation of cancer. Other possibility to develop cancer is due to the potential de-differentiation of mutated cells such that these cells acquire stem cell like characteristics.

Another related concept is the concept of tumor hierarchy. This concept proposed that a tumor is a heterogeneous population of mutant cells, all of which share some mutations but will vary in specific phenotype. In this model, the tumor is made up of several types of stem cells, one optimal to the specific environment and several fewer successful lines. These secondary lines can become more successful in some environments, allowing the tumor to adapt to its environment, including the methods by which tumors can be treated. If this situation is accurate, it has severe repercussions on the realism of a cancer stem cell specific treatment regime.

2.1. Cancer Stem Cell Marker

However, it is quite difficult to pinpoint the origin of cancer stem cell but recognition of cancer stem cells can be achieved by using specific markers/proteins expressed that are present in the cancer stem cells while absent on normal stem cell and normal tumor cells. Cancer cells must be capable of continuous proliferation and self-renewal to retain the many mutations required for carcinogenesis and for sustainability since differentiated cells cannot divide indefinitely.

However, there are several debates on these cells regarding the population of such cells and how many such cells endowed with stem cell properties. Another confusion regarding the origin of CSCs whether they originate from stem cells that have lost the ability to regulate proliferation, or from more differentiated population of progenitor cells that have acquired abilities to self-renew.

The first conclusive evidence for CSCs was published in 1997 in Nature Medicine where Bonnet and Dick et al isolated a subpopulation of leukemic cells that express a specific surface marker CD34, but lack the CD38 marker. The reports established that the $CD34^+/CD38^-$ subpopulation is capable of initiating tumors in NOD/SCID mouse that is histologically similar to the donor. Table 3 gives the list of cancer stem cell marker until the date.

2.2. Cancer Stem Cell and Tumor Recurrence

Cancer stem cells are that cells population which are responsible for the initiation and recurrence of cancer cells and ultimately leads to the formation of secondary and tertiary tumor. Besides the initiation of the primary tumor, CSCs have also been associated with metastasis formation and cancer relapses.

Table 3. Cancer stem cell markers

Markers	Cancer type	Expression	Role in cancer	References
CD133 (Prominin-1, AC133)	Colorectal, Brain, Lung, Glioblastoma, Prostate, Metastatic pancreatic carcinoma and Medulloblastoma.	Hematopoietic stem cells, endothelial progenitor cells, Glioblastomas, neuronal stem cells and glial stem cells.	It initiates the growth of malignant tumor.	[20, 21, 22]
Lgr5	Colon, Ovarian, Hepatocellular carcinoma and Esophageal Adenocarcinoma	Epithelial cells at the crypt base of the intestine.	Cancer cell proliferation and tumor formation.	[23]
CD44 (H-CAM, HUTCH-1, ECMR III)	Colorectal, Breast, Head and Neck, Pancreatic Prostate,, Ovarian and Gastric cancer.	Lymphocytes and Hematopoietic cells.	It facilitates tumor invasion and metastasis.	[24, 21. 22]
CD166 (ALCAM, Neurolin, SC-1)	Colorectal, Gastric and Pancreatic cancer.	Neurons, activated T-cells and monocytes, epithelium, and fibroblast.	Cancer cell growth, cell survival, motility and invasion.	[24, 21, 25, 26]
ESA (EpCAM, CD326, EGP40, 17-1A antigen, KSA)	Breast, Colorectal, Hepatic and Pancreatic cancer.	Baso- lateral cell surface in the simplest epithelia.	Cancer cell adhesion.	[21, 22]
ALDH 1	Multiple myeloma, Acute myeloid leukemia, Brain and Breast cancer.	Hematopoietic stem cells, neural stem cells and progenitor cells.	Cancer cell survival.	[22]
Integrin α2β	Prostate	Endothelial cells and muscle cells.	Involved cancer cell adhesion, migration and invasion.	[27]
CD24 (BA-1, HAS)	Breast, Brain, Colorectal adenocarcinoma, Prostate and Pancreatic cancer.	B-cell, mature granulocytes and variety of epithelial cell types.	Maintaining cancer cell growth, anchorage-independent proliferation, and survival.	[28, 29]
CD105 (Endoglin)	Renal carcinoma, Colon, Cervical, Laryngeal squamous cell carcinoma, Ovarian and Hepatocellular carcinoma.	Endothelial cells, activated monocytes, tissue macrophages, stromal cells and pre-B cells.	Regulation of tumor angiogenesis.	[30, 31, 32]

Markers	Cancer type	Expression	Role in cancer	References
Nestin	Breast, Glioma, Pancreatic, Prostate and Lung.	Neuroepithelial stem cell, Rostral and Caudal neural tube, radial glial cells, endothelial cells, Myoid cells.	Cancer cell motility and metastasis	[33]
Nanog	Oral squamous cell carcinoma, Cervical and Gastrointestinal cancer.	Neural stem cells and embryonic stem cell.	It promotes proliferation, invasion, and motility of cancer cells	[34, 35]
Musashi	Cervical, Medulloblastoma Glioma, Retinoblastoma, Endometrial carcinoma and Hepatoma.	Small intestinal cells, neural progenitor cells and colon crypt cells.	Increases the cell proliferation.	[35, 36]
4-Oct (POU5F1)	Osteosarcoma, Lung, Prostate and Oral squamous cell carcinoma.	Pluripotent embryonic stem and germ cells, inner cell mass of the blastocyst.	Involved in tumor cell proliferation and migration.	[37, 38, 39, 34]
CD117 (c-Kit, SCFR)	Ovary, Gastrointestinal, Breast, Mammary phyllodes tumor and Thymic carcinoma.	Hematopoietic stem cells, progenitor cells, tissue mast cells and melanocytes. It regulates	Regulates the proliferation of cancer cells and progression of tumor.	[22, 40, 41]
CD34 (gp 105-120)	Acute myeloid leukemia liver and Lung.	Early Hematopoietic stem cells, progenitor cells, small- vessels, endothelial cells and nervous tissue.	Regulates tumor angiogenesis.	[22, 32]

To establish a tumor in animal experimental model researchers sometime induce the animal with tumorous cells followed by the screening and administration of different drugs upon tumor progression. However, efficient tumor formation requires thousands or tens of thousands of cells to be introduced but this condition unable to reflect an exact micro environmental condition as well as the biochemical surroundings inside the cells.

However, several scientists believe that even a small fraction of injected cells of tumor and its CSCs, have the potential to generate a tumor. In other words, heterogeneous population of stem cells possessed multi-differentiate potential, which is a classical hallmark of stem cells.

The existence of leukemic stem cells prompted further research into other types of cancer. CSCs have recently been identified in several solid tumors, including cancers of the brain, breast, colon, ovary, pancreas, prostate, melanoma, etc. Future research in the field of cancer research will throw more light in CSCs and other types of cells found in tumor. Not only, the source of cancer cells necessary for successful treatments, but if current treatments of cancer do not properly destroy enough CSCs, the tumor will reappear.

2.3. Drug Resistance in Cancer Stem Cells

The existence of CSCs has a diversified role in the treatment of cancer that includes identification of disease, selection of drug targets prevention of metastasis and development of new intervention strategies. Normal somatic stem cells are naturally resistant to chemotherapeutic agents- they have various obstacles such as Multi Drug Resistant gene i.e. (MDR) that pump out drugs. CSCs that have mutated from normal stem cells may also express proteins that would increase their resistance towards chemotherapeutic agents. These surviving CSCs then recapitulate the tumor revival. Therefore, it is necessary to identify the CSCs target that enables us to treat a patient, which is suffering from aggressive, non-resectable tumors and thus help us to prevent metastasis. This theory also tells us upon CSC elimination, cancer would regress due to differentiation and/or cell death.

Table 4. Multi drug resistant genes

Gene	Other name	Gene Locus	Size of gene product	Normal expression	Cancer	References
MDR1	ABCB1, P-GP	7q36 [9]	170 KDa	Blood brain barrier, Liver, Kidney, Gut, Testis, and Placenta.	Leukemia, Breast, Colorectal and Lymphoma.	[42, 43, 44]
MRP1	ABCC1	16p13.1 [6]	190 KDa	Brain and many tissues.	Acute myeloid, leukemia and Neuroblastoma.	[45, 46, 47]
MRP2	ABCC2	10q24 [10]	190 KDa	Liver, Kidney, gut and Placenta.	Lung and Renal cell carcinoma.	[48, 49]
MRP3	ABCC3	17q21.3 [11]	190 KDa	Adrenal cortex, Placenta, Gut and Liver.	Lung and Colon cancer.	[49, 50]
MRP4	ABCC4	13q32 [11]	150 KDa	Prostate epithelium, Epidermal and dermal layer of skin.	Leukemia, Ovarian, Breast and Small cell lung cancer.	[51, 52, 53]
MRP5	ABCC5	3q27 [11]	185 KDa	Pancreas and many other tissues.	Pancreatic cancer.	[54, 55]
MRP6	ABCC6	16p13.1 [12]	170 KDa	Liver, Kidney and few other tissues.	Lung cancer	[46, 56]
MRP7	ABCC10	6p12-21 [13]	166 KDa	Wide range of tissues including colon, skin and testis.	Non-small cell lung cancer.	[57, 55]
MRP8	ABCC11	16q 12.1 [14]	150 KDa	Liver, Prostate, Breast and Testes.	Breast cancer.	[58, 59]
BCRP	ABCG2, MXR1, ABCP	4q22 [15]	72.6 Kda	Intestine, Liver, Gut, Breast, endometrium and placenta.	Ovary, Breast, Colon and Gastric Cancer.	[60, 55]

A number of studies have investigated the possibility of identifying specific markers that may distinguish CSCs from the bulk of the tumor (as well as from normal stem cells). However, using the proteomic and genomic approach to identify tumor will greatly enhance our expertise to treat the disease in a better manner.

The rejection of active drugs by any cancerous cells ultimately leads to multiple drug resistance and substantially interferes in chemotherapy. However, recent evidence suggests that the cells which are more rejecting the drug may enriched with stem like cancer cells and that cause recurrence of the cancerous cell's population. Table 4 lists the different type of multi-drug resistant genes identified until the date.

2.4. Multiple Drug Resistance (MDR)

This is a phenomenon where human tumors acquire resistance to one type of therapy and found to be resistant to several other drugs that are often quite different in both structure and mode of action. The better understanding of MDR came with the identification of P-glycoprotein and other related transporters that have been expressed in some cancer cells that recognize and catalyses the efflux of diverse anticancer drugs from cells. Second level of resistance came from the mechanism of programmed cell death or apoptosis, leading to MDR by resistance to anticancer drug-induced apoptosis. The third level of MDR with the finding that the proliferation of human tumors was driven by a small population of self-renewing tumor cells, focusing attention on the MDR properties of these so-called tumor stem cells rather than on the cells that comprised the majority of the tumor population. A fourth advance was the delineation of features of the tumor microenvironment, including immuno-suppression, which essentially provided tumor stem cells with an MDR phenotype. To treat cancer cells over expressing P-glycoprotein (MDR), reports suggest the use of nanoparticles drug to sensitize the tumor stem cells for uptake of drug. There is a great need for therapeutic reagents that are capable of eliminating high drug efflux cancer cells. Several inhibitors of the same cells have been tried over the years, and it is shown that these inhibitors are able to overcome multidrug resistance (MDR) and sensitize high drug efflux cancer cells to chemotherapeutic drugs, or directly reduce the tumorigenicity of lung cancer cells possibly by affecting stem-like cancer cells. These inhibitors may help to improve the efficacy of the chemotherapy to cancer stem cells. Table 5 lists the different type of inhibitors of MDR genes.

Table 5. Chemosensitizers inhibiting MDR genes

MDR Gene	Chemosensitizer	References
Pgp	Alpha-naphthyl isothiocyanates, Agosterol A, tRA 98006, Norverapamil, 5-O-benzoylated taxinine k.	[61, 62, 63, 64, 65]
MRP	Benzyl-, phenethyl- isothiocynates, V-104, Pluronic block copolymer, Quinidine, Chloroquine,	[61, 66, 67, 68].
BCRP	Chrysin, Genistein, Biochanin A, Naligenin, KO143.	[69, 70, 71]

2.5. Cancer Stem Cell and Reactive Oxygen Species

The self-renewal ability of stem cell enables them to maintain their multi-potent differentiation capacity, tissue homeostasis in multicellular organisms. Aerobic metabolism, while enabling efficient energy production, also generates reactive oxygen species (ROS), which damage cellular components. However, the focus in stem cell biology has been on the adverse effects of ROS, particularly the damaging effects of ROS accumulation in aged tissues, development of cancer. Anti-oxidative and anti-stress mechanisms of stem cells have been highlighted. However, it has become increasingly clear that, in some cases, redox conditions play an important role in stem cell maintenance [72], e.g., regulation of the cell-cycle. A redox regulation in various cancer stem cells, the malignant counterparts of normal stem cells are viewed as good targets for cancer therapy. In contrast to cancer cells, where ROS levels are increased, some cancer stem cells maintain low ROS levels, exhibiting redox patterns that are similar to the corresponding normal stem cell. To fully elucidate the mechanisms involved in stem cell maintenance and to effectively target cancer stem cells, it is essential to understand ROS regulatory mechanisms. ROS is known to serve as a second messenger in the intracellular signal transduction pathway for a variety of cellular processes, including inflammation, cell cycle progression, apoptosis, aging and cancer. Recently, ROS has been found to be associated with tumor metastasis involving the processes of tumor cell migration, invasion and angiogenesis. Emerging evidence also suggests that Epithelial-Mesenchymal Transition (EMT), a process that is reminiscent of cancer stem cells, is an important step toward tumor invasion and metastasis, and intimately involved in *de novo* and acquired drug resistance.

In light of recent advances, we are summarizing in this segment of chapter the role of ROS in EMT by cataloging how its deregulation is involved in EMT and tumor aggressiveness. Further attempts have been made to summarize the role of several chemo preventive agents that could be useful for targeted inactivation of ROS, suggesting that many natural agents could be useful for the reversal of EMT, which would become a novel approach for the prevention of tumor progression and/or treatment of human malignancies, especially by killing EMT-type cells that shares similar characteristics with cancer stem cells.

"Everyone Thinks of ROS as Things That Kill Cells, and They Do," Said Dr. Harley Kornblum

Stem cells generally have been thought to maintain low levels of ROS to combat against damage cells. It has been also reported that the neural stem cells were highly responsive to ROS stimulation, which increased their growth and differentiation. Conversely, lower cellular levels of ROS in the neural stem cells interfered with normal cell function in mice, in human and mouse cell lines as well. The body has a system to make ROS when it requires. Some cells, such as immune cells, surround bacteria or viruses and use ROS to kill the invading microbes. Outside influences such as stress and environmental factors such as exposure to radiation can increase ROS levels in cells. Although ROS is produced by all cells in a passive manner as a by-product of normal cell metabolism, some cells also produce ROS in a directed manner using ROS-producing enzymes like NADPH oxidase (NOX).

NOX-generated ROS can act as second messengers in tightly controlled signal transduction pathways for many growth and trophic factors. However, abnormal ROS level damages and ultimately kills cells, so finding the correct balance is vital. Kornblum and his team also found that the ROS-mediated stem cell self-renewal and differentiation of these cells into neurons depended on a cell signaling pathway called PI3K/Akt, which is known to be involved in cellular functions such as cell growth, proliferation, differentiation, motility and survival. NOX-generated ROS affects PI3K/Akt signaling by causing the inactivation of the PTEN protein, an important tumor suppressor and negative regulator of the pathway, by oxidizing a cysteine residue in the protein, which inactivates its function. In brain tumors, if the pathway gets activated in cells already susceptible to becoming cancerous, it may promote the proliferation of brain tumor cells or the propagation of brain tumors. Blocking the pathway may be one way to interrupt the malignant process. Researchers seek to determine whether brain cancer cells use elevated ROS and the PI3K pathway to promote their own growth.

The production of reactive oxygen species (ROS) has traditionally been related to deleterious effects for cells. However, it is now widely accepted that ROS can play an important role in regulating cellular signaling and gene expression. NADPH oxidase mediated ROS production seems to be especially important in this regard [73]. Some lines of evidence suggest that ROS may be important modulators of cell differentiation, including hematopoietic differentiation, in both physiologic and pathologic conditions. In future, the involvement of ROS and NADPH oxidases in cancer, and in particular what is known about the relationship between ROS and haematological malignancies will be interesting to study. Finally, it will be interesting to use ROS generating enzyme like NADPH oxidase as cancer therapeutic targets. However, there are other ROS generating enzymes apart from NADPH Oxidase. NADPH oxidase is a multisubunit enzyme consisting of several subunits, p22phox, Nox isoforms, p47phox, p67phox, rac etc. and assembly of all these subunit leads to activation of the enzyme and generation of ROS. Therefore, targeting NADPH oxidase in cancer stem cell may be one of the attractive ways of cancer therapy. Cancer stem cell is the cause of tumor recurrence, and tumor cannot grow without blood supply. Therefore, in this chapter we focus on understanding the role of notch signaling in controlling both cancer stem cell proliferations as well as tumor angiogenesis. In future, notch mediated therapy can control both tumor relapses caused by cancer stem cell and tumor angiogenesis.

3. Role of Angiogenesis in Tumor Growth

Angiogenesis is the formation of new blood vessels from the pre-existing vasculature. It is an important process as the blood delivers oxygen and nutrients for the survival of cells and their functioning. The link between angiogenesis and cancer formation and metastasis was established by Dr. Judah Folkman in the 1970’s. Inhibition of angiogenesis is thus considered putative target in cancer therapy. He observed that in the absence of blood vessel formation the tumor mass does not grow more than 1-2mm in size and also the cells from the primary tumor cannot escape and metastasize as the newly formed vasculature would not be present for this purpose.

Angiogenesis is controlled by pro-angiogenic and anti-angiogenic factors in the body.

The regulatory factors may be several growth factors like VEGF (Vascular Endothelial Growth Factor), FGF (Fibroblast Growth Factor), PDGF (Platelet Derived Growth Factor),

EGF (Epidermal Growth Factor), placental growth factor, Angiopoietin-1, Angiogenin, Interleukin 8. The natural anti-angiogenic presents in the body are Angostatin, Endostatin, Vasostatin, Prolactin, Angiopoietin-2, Alpha Interferon, Gamma Interferon, Interleukin 12, Fibronectin, Platelet factor-4, etc. a finely tuned equilibrium exists between these pro-angiogenic and anti-angiogenic factors *in vivo* [74].

Angiogenesis is driven due to hypoxic conditions in the tumor, which in turn activates the production of VEGF (Vascular Endothelial Growth Factor) which then binds to the VEGF receptors (VEGFR) which are transmembrane receptor tyrosine kinase on the membrane of endothelial cells and activate them by phosphorylation of tyrosine kinase (TK). The TKs are also present on the receptors of FGF, PDGF, Ang-1, Ang-2, and HGF along with that of VEGF and leads to angiogenesis [74]. In this part of chapter, we will focus on the role of VEGF in tumor progression and targeting it using nanoparticles.

Recently, several groups have identified the importance of Notch signaling in tumor angiogenesis. Notch activity increases, specifically in tumor endothelium and in various tumors types and, in some studies, Notch signaling suppresses angiogenic processes as well as promotes angiogenesis. Because the Notch signaling pathway can mediate communication between various cell types in the tumor microenvironment, interactions between tumor cells and endothelial cells might promote angiogenesis, therefore, targeting the Notch pathway might provide a novel strategy for anti-angiogenic therapies. Notch signaling has also been reported to play a critical role in cancer stem cell proliferation and endothelial cell proliferation. Below we discuss the role of notch in angiogenesis.

Tumor angiogenesis is necessary for tumor growth and progression and, often, solid tumors have an increase in hypoxia, a stimulus for angiogenesis. Low oxygen levels might result from the structural abnormalities in tumor vessels or from the size of the tumor itself, which would lead to inadequate oxygen delivery. Changes in gene expression might help the tumor adapt to its hypoxic environment, and hypoxia might promote angiogenesis by stimulating Notch signaling through HIF1, as well as inducing pro-angiogenic factors.

3.1. Role of Notch Signaling in Angiogenesis

Notch proteins have an important role in the regulation of tumor angiogenesis [75]. Endothelial cells have a low turnover rate and, with a few exceptions, are typically quiescent in the adult. Tumor vasculature, however, is not quiescent; continuously giving rise to new blood vessels to support the growing tumor mass. Tumors secrete soluble pro-angiogenic factors, such as VEGF, to recruit endothelial cells from surrounding host tissues to form their own blood supply. Often, these vessels are abnormally shaped and leaky and can have tumor cells incorporated into their walls. Often, these vessels are abnormally shaped and leaky and can have tumor cells incorporated into their walls. VEGF and bFGF can induce DLL4 expression, both alone and synergistically.

Also VEGF induced the expression of Notch1 and DLL4 through the PI3K–Akt pathway in human arterial endothelial cells. Silencing of DLL4 in endothelial cells led to cell cycle arrest, increased apoptosis, a reduction in tube formation on matrigel and inhibition of VEGF-induced cell migration, all of which have profound effects on the angiogenic process. These studies suggest that pro-angiogenic factors activate Notch signaling to promote angiogenesis.

Table 6. List of reports demonstrating notch promotes angiogenesis

Notch signaling component	Phenotype or Findings	References
DLL4	Upregulated in the vasculature of xenografted breast tumors in mice. Increased in the vasculature of human breast and kidney tumor tissue. Increased under hypoxic conditions and regulated by HIF α, VEGF, bFGF.	[76,77]
Jagged 1	Increased in head and neck squamous cell carcinoma (HNSCC) cells stimulated with growth factors. It induces tube formation of co-cultured endothelial cells (EC) with HNSCC cells.	[78]
Notch 1, 2, 4	Increased activity in Kaposi's sarcoma (KS) cells *in vitro* and *in vivo*. Inhibition of γ-secretase inhibitor reduces KS tumor formation.	[79]
Notch 1	N 1- expressing breast cancer cells induce HIF α transcription. Constitutive activation in EC significantly increases tube formation.	[80]
γ-secretase	Inhibition reduces EC proliferation and sprouting. Inhibition reduces tumor mass and vascularization of xenograft models of lung adenocarcinoma and glioblastoma.	[81]

Anti-angiogenic hypothesis can be a useful tool to inhibit the growth of a tumor by blocking the blood vessels formation and inhibition of several angiogenic markers such as VEGF, Hif1α etc. The well-known drug Avastin, that targets VEGF ligands, has been promising results in colon cancer by attenuation of metastasis. However, anti-VEGF therapies have also been shown to induce tumor drug resistance after treatment, followed by a subsequent activation of other non-VEGF pro-angiogenic markers that promote the blood vessel's formation to support the tumor. Recent studies revealed neovascularization and tumor angiogenesis processes could be induced through the direct interaction of Notch molecules on tumor cells and endothelial cells *in vitro* and *in vivo*. Furthermore, head and neck squamous cell carcinoma (HNSCC) cells that are stimulated with growth factors increase the expression of Jagged-1 in a mitogen-activating protein kinase (MAPK)-dependent manner and hence triggering Notch signaling in co-cultured human dermal microvascular endothelial cells (HDMECs). Therefore, targeting Notch signaling might be beneficial for anti-angiogenic therapies and the crosstalk with VEGF signaling could enhance the efficacy of such an approach, offering a novel strategy in blocking tumor angiogenesis.

Gustafsson *et al.* examined the integration of hypoxic cues with the Notch signaling pathway in neural stem cells, satellite cells and myogenic cells.

3.2. Inhibition of Tumor Angiogenesis by Notch

On contrary, the activation of the Notch signaling pathway also inhibits angiogenesis. In a three-dimensional *in vitro* modeling system, Notch1 signaling suppressed endothelial cell proliferation and inhibited branching at the tip of sprouts in a cell-autonomous manner to regulate the lumen diameter of growing tubules, whereas blocking Notch signaling induced vessel bifurcation and branching. Notch4 inhibited VEGF-induced angiogenesis and endothelial cell sprouting by promoting integrin-mediated collagen adhesion of human endothelial cells.

Table 7. List of reports demonstrating notch inhibits angiogenesis

Notch signaling component	Phenotype or Findings	References
DLL4	Decreases EC proliferation and migration. Reduces VEGFR and neuropllin- 1 expression.	[82]
Jagged	Antisense oligomer increases VEGF induced tube formation and invasion of EC	[83]
Notch 1, 4	Decreases EC VEGFR-2 expression. Decreases VEGF-induced EC proliferation. Over expression induces EC cell cycle arrest through pRb phosphorylation and decreased p21cip1 expression. Blocking with anti-sense oligomers induces EC proliferation and sprouting.	[84, 85, 75]
Notch 4	Inhibits VEGF induced angiogenesis and EC sprouting via integrin mediated matrix adhesion	[86, 87]
HESR 1	Over expression inhibits EC proliferation, migration and tube formation on collagen.	[88]

The overexpression of Notch1 or Notch4 promoted endothelial cell cycle arrest in a mechanism that involved pRb phosphorylation and decreased p21cip1 expression. Activated Notch1 and Notch4 can decrease the expression of VEGFR-2, a target gene of HESR1, resulting in decreased human endothelial cell proliferation in response to VEGF.

HESR1 overexpression in endothelial cells inhibited proliferation, migration and tube formation on collagen. In contrast to studies by Patel *et al.*, Williams *et al.* reported that human endothelial cells transduced with DLL4 had a decrease in VEGF-induced proliferation and migration, as well as a significant reduction in VEGF receptor-2 (VEGFR-2) and its co-receptor neuropilin-1, suggesting that DLL4 is an inhibitor of angiogenesis. Although Notch signaling clearly inhibits sprouting angiogenesis, it is crucial to understand *in vivo* tumor angiogenesis assays to confirm these findings and to determine the physiological relevance in the future.

3.3. Notch Signaling: Pro-Angiogenic or Anti-Angiogenic?

It is still unclear whether activation of Notch signaling can exhibit opposite effects on endothelial cell differentiation and angiogenesis or not. However, one problem could be the overexpression of constitutively activated Notch signaling components. Recently, two studies from the same group demonstrated opposite effects of DLL4 signaling on angiogenesis.

In one study, retroviral transduced overexpression of DLL4 in endothelial cells was used to study the mechanisms of angiogenesis; however, this showed the same phenotype i.e. reduced endothelial cell proliferation, migration and network formation as the knockdown of the endogenous levels by RNAi.

As Patel *et al.* suggest, it is possible that an optimal range of Notch signaling is crucial for endothelial cell differentiation and angiogenesis. This phenomenon has been observed during embryonic development, in which both knockout and overexpression of constitutively active Notch4 led to severe and lethal vascular abnormalities.

Notch function during angiogenesis and tumor genesis enable us to determine if differences in regulation could lead to these seemingly contradictory phenotypes. Furthermore, it is possible that different NICDs have different transcriptional functions and target different genes for expression. Therefore, these activities must be clearly delineated. Similarly, different Notch ligands might transduce different signaling pathways and the identification of which ligands and receptors are expressed on each cell could help determine how all the signals are coordinated to produce different outcomes.

Another reason for the conflicting data seen in these various Notch studies could be due to the cell systems that were used. So far, studies have shown that angiogenesis might be induced or inhibited by Notch signaling (as shown in Figure 2); however, these studies do not address or test the possibility of an increase in vasculogenesis as a mechanism for increased vessel growth in tumors. In early mouse embryonic studies, vasculogenesis was unaffected by the presence or absence of Notch signaling molecules.

However, vasculogenesis in the adult involves bone marrow-derived endothelial precursor cells, and it is unknown if Notch signaling has a role in the recruitment and differentiation of these cells. These progenitor cells have been found in increased levels in the circulation of cancer patients, and it will be interesting to investigate if Notch molecules are involved in the homing of these precursor cells into tumor tissue.

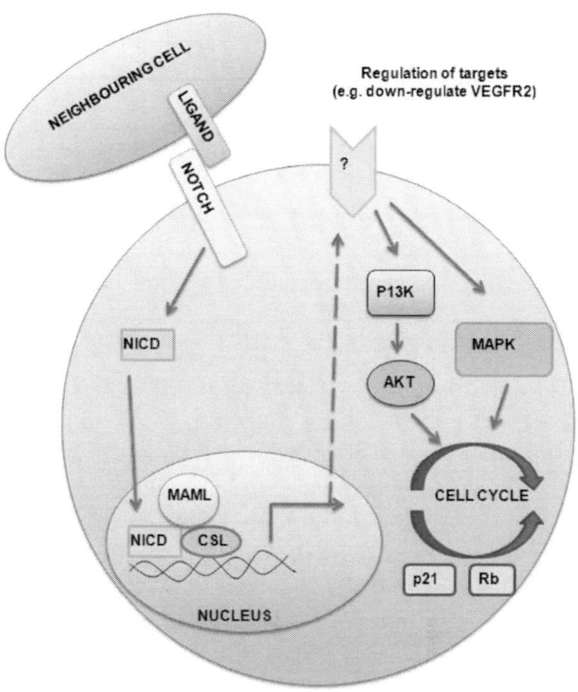

Figure 2. Role of notch signaling pathway in inhibiting VEGF pathway (Redrawn from [89]; Approved by authors).

Because Notch signaling is oncogenic and tumor suppressive in various systems, it is not surprising that Notch can also be pro-angiogenic and anti-angiogenic. In systems in which Notch is oncogenic, it might also have pro-angiogenic roles to support tumor growth. Similarly, in cases in which Notch is a tumor suppressive, perhaps it functions to be anti-angiogenic and contributes to the inhibition of tumor growth. More studies on Notch in tumor angiogenesis are necessary to fully understand its role in tumor growth and progression.

It will be also important to correlate these studies with the oncogenic and tumor suppressor functions of Notch and determine if cell type differences contribute to the dual roles of Notch signaling in tumor angiogenesis. However, with the current understanding about Notch pathway, it is obviously one of the future targets for future cancer therapy.

4. Cancer Stem Cell (CSCS) and Tumor Angiogenesis: Different Pathways Regulating Cell Proliferation

For the drug designing and discovery it is necessary to understand the cellular and molecular phenomenon that regulates the cell proliferation mechanism. The first weel understood mechanism came from Leukemia where haematopoietic stem cells (HSCs) and their transformed counterparts are originated from CSCs.

Also it is now clearer that there are many organs, which share the same cellular pathways as leukemia-derived HSCs. Moreover, the tendency of transformation of cancer stem cells from normal cells via deregulation of the cell cycle that leads to an abnormal proliferation or differentiation pathways that controlling this process or by inducing oncoprotein activities.

There are some important pathways that are CSCs cell proliferation and tumor angiogenesis are mentioned below:

NADPH Oxidase

The NADPH oxidase (nicotinamide adenine dinucleotide phosphate-oxidase) is a membrane-bound enzyme complex. It can be found in the plasma membrane as well as in the membrane of phagosomes. It is made up of six subunits. These subunits are Rac 1/2, gp91 phox, p22phox, p40phox, p47phox, p67phox, etc. as shown in normal circumstances; the complex is latent in neutrophils and is activated to assemble in the membranes during the respiratory burst. It generates superoxide by transferring electrons from NADPH Oxidase inside the cell across the membrane and coupling with two molecular oxygen that produce the superoxide, a reactive free-radical. Superoxide can be produced in phagosomes that contain ingested bacteria and fungi, or it can be produced outside of the cell. In a phagosome, superoxide can spontaneously form hydrogen peroxide that will undergo further reactions to generate reactive oxygen species (ROS). Superoxide is capable of killing bacteria and fungi by mechanisms that are not yet fully understood, but may inactivate critical metabolic enzymes, initiate lipid peroxidation, and liberate redox-active iron, which allows the generation of indiscriminate oxidants such as the hydroxyl radical. It is presumed that superoxide kills bacteria directly, as the virulence of many pathogens is dramatically attenuated when their superoxide dismutase (SOD) genes are deleted.

Major source of ROS in endothelial cells is NADPH oxidase, which consists of different isoforms of Nox (Nox1, Nox2, Nox4, Nox5) p22phox, p47phox and the small G-protein Rac1. NADPH oxidase is activated by various growth factors, including VEGF and angiopoietin-1 as well as hypoxia and ischemia, and ROS derived from this oxidase are involved in VEGFR2. NADPH Oxidase generates ROS and regulates the cell proliferation of both cancer stem cell and endothelial cell.

Sonic Hedgehog

Sonic the hedgehog homolog (SHh) is one of three proteins in the mammalian signaling pathway family called hedgehog, desert hedgehog (DHh) and Indian hedgehog (IHh). SHh is the best studied ligand in the hedgehog signaling pathway.

This plays an important role in the regulation of vertebrate organogenesis, e.g. digit and limb growth while the sonic hedgehog is the best established example of a morphogen as defined by Lewis Wolpert. SHh remains important in the adult where it controls cell division of adult stem cells and has been implicated in development of some cancers. An implication of a Hedgehog (Hh) pathway has widely accepted in a variety of human tumors, and early clinical trials with pathway antagonists have validated Hh signaling as an appropriate anti-cancer target. The SHh ligand binds to the receptor patched, which activates another receptor and downstream pathway via Gli, transcription factor. The Hedgehog-Gli signaling pathway has been reported to play a crucial role in the cancer stem cell proliferation and regulation. Sonic hedgehog has also been reported to induce angiogenesis. Based on above reports, the significance of this signaling pathway is very high for the therapeutic purpose in the treatment of tumor growth and progression.

The Wnt Pathway

The pertinent role of Wnt (which a network of proteins) signaling pathway in embryogenesis and cancer. It consists of a group of secreted lipid-modified signaling proteins of 350-400 amino acids in length. Following the signal sequence, they carry a conserved pattern of 23-24 cysteine residues, on which palmitoylation occurs on a cysteine residue. Further, these proteins activate various pathways in the cell that can be divided into the canonical and non-canonical Wnt pathways. Through these signaling pathways, Wnt proteins play a variety of important roles in embryonic development, cell differentiation and cell polarity generation. This pathway involves a large number of proteins that can regulate the production of Wnt signaling molecules, their interactions with receptors on target cells and the physiological responses of target cells. In this process, cells are exposed to extracellular wnt ligand. Further, wnt binds to the receptor frizzled and activates the cytosolic component which stabilizes beta-catenin and hence enters into the nucleus and activates to wnt target genes. In other circumstances, when wnt is absent, beta-catenin gets degraded by cytosolic complex APC-Axin-GSK-DSh.

The mechanism of action of wnt in such a way that once Wnt receptor complex activated by Wnt binding it inhibits a second complex of proteins that includes axin, GSK-3 and the protein APC. The axin/GSK-3/APC complex normally promotes the proteolytic degradation of the β-catenin intracellular signaling molecule. Furthermore, there are several protein kinases and protein phosphatases have been associated with the ability of the cell-surface Wnt-activation. Here Wnt receptor complex binds axin and disassembles the axin/GSK3 complex. This phosphorylation of the cytoplasmic domain of LRP by CK1 and GSK3 can regulate axin binding on another cellular protein LRP. The protein kinase activity of GSK3 appears to be important for both the formation of the membrane-associated Wnt/FRZ/LRP/DSH/Axin complex and the function of the Axin/APC/GSK3/β-catenin complex. Phosphorylation of β-catenin by GSK3 leads to the destruction of β-catenin. The Wnt Signaling pathway has been reported to be involved in angiogenesis through its molecules and modulators. Even the Cancer stem cell self-renewal capacity has been reported to be driven by Wnt signaling pathway.

Notch

In developmental biology, role of Notch pathway is well known whereas its role in stem cell proliferation has now been demonstrated for several cell types including haematopoietic, neural and mammary stem cells. However, components of the Notch pathway have been proposed to act as oncogenes in mammary and other tumors. This chapter describes in detail the role of notch signaling pathway in regulation of cancer stem cell and endothelial cell proliferation. These developmental pathways are also strongly implicated as stem cell regulators. Both Sonic hedgehog (SHH) and Wnt pathways are commonly hyperactivated and are required to sustain for tumor growth.

Table 8. Notch inhibitor list

Inhibitor	Target	Mode of action	Reference
SEL-10	Notch receptor	Ubiquitin mediated protein degradation	[90]
MK0752, LY450139, R04929097, BMS-708163	Notch ligands, Notch homologs and Other γ-secretase substrates	inhibits notch cleavage by γ-secretase	[91, 92]
MPC-7869	Substrates of γ-secretase	inhibits notch cleavage by γ-secretase	[93]
OMP–21M18, DLL4 antibody	DLL4	Interferes with ligand–receptor interaction	[94, 95, 96]
Negative regulatory region monoclonal antibodies	Notch receptor and ligands	Interferes with ligand-induced Notch subunit separation	[97]
Notch soluble receptor decoys	Relatively specific for Notch homologs.	Interferes with ligand–receptor interaction	[98, 99]
MAML–CSL–Notch, antennapedia/dominant–MAML	Notch homologs and other nuclear transcription factors that target MAML1	Interferes with Notch nuclear co-activator MAML1	[100, 101]
DAPT	γ-secretase	Induces G0-G1 cell cycle arrest and apoptosis	[102]
Dibenzazepine	γ-secretase	inhibits epithelial cell proliferation	[102]

However, the Gli transcription factors that are regulated by SHh take their name from gliomas, where they are commonly expressed at high levels. A degree of crosstalk also exists between these two pathways, and their activation commonly goes hand-in-hand., For instance, in colon cancer hedgehog signaling appears to antagonise Wnt. Sonic hedgehog blockers are available, such as cyclopamine. If it is possible to eliminate the cancer stem cell/angiogenesis from tumor, then a potential cure may be achieved if there are no more CSCs/blood vessels to repopulate a cancer.

5. Cross Talk between Tumor Cells and Endothelial Cell for Tumor Growth Promoted by Angiogenesis Mediated by Notch Pathway

The cross talk between endothelial cell and tumor cell is mediated by Notch signaling pathway (Figure 3). Various growth factors stimulate tumor cell for uncontrolled cell division, and these tumor cells sends a signal to endothelial cell via Notch to activate the notch pathway in endothelial cell, and the transcription factors Hes and Hey activates endothelial cell for the formation of blood vessel network. The increased vascularization of tumor leads to tumor growth and progression. In case of cancer stem cells, it can be speculated that both cancer stem cell and endothelial cell proliferations are controlled and regulated by notch. Therefore, notch is one of the best targets for cancer therapy.

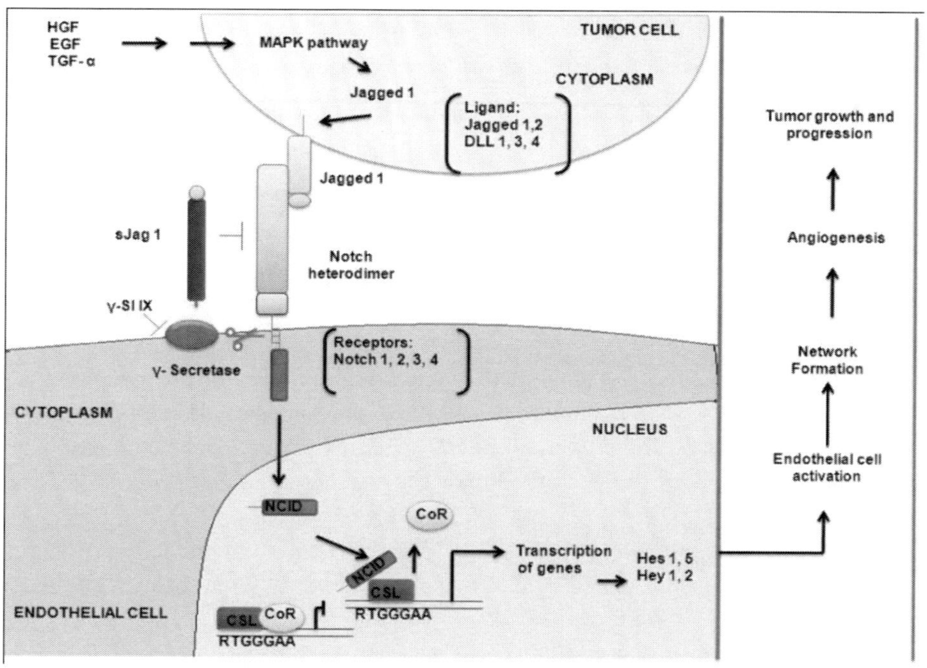

Figure 3. Cross talk between tumor cell and endothelial cell (Adapted from [103] and Trends in Cell Biology; Vol 16 No 6 June 2006 Approved by authors).

6. Cancer Therapy Targeting Notch Signaling Pathway on Cancer Stem Cell and Endothelial Cells of Tumor

Among the variety of cancer-cell subpopulations that make up a tumor, it is thought that only a few cancer stem cells can drive tumor formation. The hierarchy for tumorigenic potential among cancer cells is still not clear. It is also not clear whether any cell within a tumor can become a cancer stem cell or not if so, a cancer stem- cell model can be designed, particularly for cancer treatment: to cure a cancer, essentially all tumor cells, rather than a rare population of potent tumor-initiating cancer stem cells, would have to be killed. In this segment of chapter, we are focusing on cancer therapy targeting a rare population of tumor i.e. cancer stem cell and endothelial cells within the tumor.

6.1. Nanoparticle Mediated Therapy (Using Cancer Stem Cell Marker and Angiogenic Inhibitors)

Nanotechnology is the study of devices that are in the dimensional range of 1-1,000nm. Nanotechnology has been frequently used in oncology and other fatal diseases for the delivery of drugs. The basic principle is that the drug is either attached on or encapsulated within the nanoparticle. These particles may have a polymer or metal coating with magnetic cores. It is possible to attach or immobilize the drug or an antibody on its surface and target it to the desired site. If the nanoparticle is magnetic in nature, then it can be manipulated by applying a magnetic field from an external source. Magnetic drug targeting is an emerging field in the cancer therapy. Gold shell and iron core nanoparticles have been synthesized to carry drugs.

6.2. Targeting of Magnetic Nanoparticle to Tumor Site

Another method of drug delivery is recently developed by the scientists in Brown University, USA is the coupling of magnetic nanoparticles and gold nanoparticles with specific markers and creating a twin nanoparticles. They created twin nanoparticles, which specifically binds to Her-2 receptors and then unloads the drug in the breast cancer cells. This has been successfully demonstrated in cancer cell lines. The nanoparticle can be concentrated in the tumor region by magnetic property, and the specific drug or monoclonal antibody can be targeted and released specifically at the tumor site without affecting any other part of the body.

This shall result in minimum level of side effects and toxicity in the body. Furthermore, the toxic effect of magnetic nanoparticle is countered to a large extent by the gold nanoparticles which are least toxic to the body.

Above figure (Figure 4) demonstrates twin nanoparticles made of iron oxide and gold, which are coupled together. The gold nanoparticles carry an antibody against a specific tumor stem cell marker, and the iron nanoparticle may carry an antibody against Notch monoclonal antibodies or secretase inhibitors absorbed on it. The twin nanoparticle can be mobilized to the tumor region by applying external magnetic field so that it does not affect the whole body.

The heat produced by this force of attraction can mediate the release of Notch monoclonal antibodies or secretase inhibitors in the neighboring endothelial cells of cancer stem cell (after recognition of cancer stem cell marker).

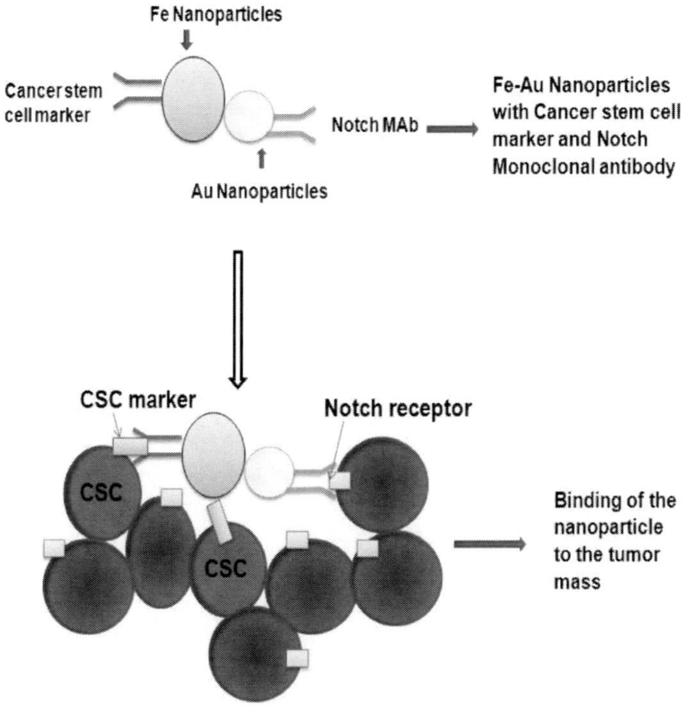

Figure 4. Design of nanoparticle for targeted cancer therapy: Nanoparticle tagged with cancer stem cell marker and notch monoclonal antibody.

CONCLUSION

We conclude that a nanoparticle mediated drug delivery tagging both cancer stem cell marker and Notch inhibitor can be a better model of therapy than a nanoparticle mediated drug delivery, which is tagged with only cancer stem cell marker. The Notch inhibitor used will be more powerful if it is toxic to both endothelial cells as well as cancer stem cell. Alternatively, multidrug resistance gene blocker/sensitizers can also be used. The suggested remote control that releases the magnetic coated nanoparticles is hyperthermia, i.e. elevated temperature as the nanoparticles are magnetic Fe-Au in nature. The idea is to first localize the tumor by specific cancer cell surface marker and then release the Notch monoclonal antibody or inhibitors to inactivate the cell proliferation ability of both cancer stem cell and tumor angiogenesis.

In this way, tumor recurrence problem can also be solved as without cancer stem cell, the tumor mass cannot lead to tumor relapse and even if that is the case the resulting tumor cannot grow in the absence of endothelial cells or non-vascular tumor. Therefore, this chapter on "Notch signaling regulates cancer stem cell and tumor angiogenesis" not only guides the readers to the fundamentals of notch and other signaling pathway controlling cancer stem cell and tumor angiogenesis but also opens new avenues for better targeting of anti-cancer drugs.

REFERENCES

[1] Nickoloff, B J; Osborne, B A and Miele, L. Notch signaling as a therapeutic target in cancer: a new approach to the development of cell fate modifying agents. *Oncogene*, 2003 22, 6598–6608.

[2] Wang, Z; Li, Y; Banerjee, S and Sarkar, F H. Exploitation of the Notch Signaling Pathway as a Novel Target for Cancer Therapy. *Anticancer Research* 2008 28, 3621-3630.

[3] Wang, Z; Zhang, Y; Li, Y; Banerjee, S; Liao, J; Sarkar, FH. Down-regulation of Notch-1 contributes to cell growth inhibition and apoptosis in pancreatic cancer cells. *Mol. Cancer Ther.*, 2006 Mar 5(3), 483-93.

[4] Sharma, VM; Draheim, KM; Kelliher, MA. The Notch1/c-Myc pathway in T cell leukemia. *Cell Cycle*, 2007 Apr 15 6(8), 927-30.

[5] Sriuranpong V, Borges MW, Ravi RK, Arnold DR, Nelkin BD, Baylin SB, Ball DW. Notch signaling induces cell cycle arrest in small cell lung cancer cells. *Cancer Res.,* 2001 Apr 1 61(7), 3200-5.

[6] Ayyanan, A; Civenni, G; Ciarloni, L; Morel, C; Mueller, N; Lefort, K; Mandinova, A; Raffoul, W; Fiche, M; Dotto, GP; Brisken, C. Increased Wnt signaling triggers oncogenic conversion of human breast epithelial cells by a Notch-dependent mechanism. *Proc. Natl. Acad. Sci. USA*, 2006 Mar 7 103(10), 3799-804.

[7] Lai, EC. Notch signaling: control of cell communication and cell fate. *Development,* 2004 Mar 131(5), 965-73.

[8] Rehman, AO and Wang, CY. Notch signaling in the regulation of tumor angiogenesis. *Trends Cell Biol.*, 2006 Jun 16(6), 293-300.

[9] Kim, TH and Shivdasani, RA. Notch signaling in stomach epithelial stem cell homeostasis. *J. Exp. Med.*, 2011 Apr 11 208(4), 677-88.

[10] Hu, YY; Zheng, MH; Cheng, G; Li, L; Liang, L; Gao, F; Wei, YN; Fu, LA; Han, H. Notch signaling contributes to the maintenance of both normal neural stem cells and patient-derived glioma stem cells. *BMC Cancer*, 2011 Feb 22 11, 82.

[11] Pellegrinet, L; Rodilla, V; Liu, Z; Chen, S; Koch, U; Espinosa, L; Kaestner, KH; Kopan, R; Lewis, J; Radtke, F. Dll1- and dll4-mediated notch signaling are required for homeostasis of intestinal stem cells. *Gastroenterology*, 2011 Apr 140(4), 1230-1240.

[12] Pannuti, A; Foreman, K; Rizzo, P; Osipo, C; Golde, T; Osborne, B; Miele, L. Targeting Notch to target cancer stem cells. *Clin. Cancer Res.*, 2010 Jun 15 16(12), 3141-52.

[13] Zhang, XP; Zheng, G; Zou, L; Liu, HL; Hou, LH; Zhou, P; Yin, DD; Zheng, QJ; Liang, L; Zhang, SZ; Feng, L; Yao, LB; Yang, AG; Han, H; Chen, JY. Notch activation promotes cell proliferation and the formation of neural stem cell-like colonies in human glioma cells. Mol. Cell Biochem., 2008 Jan 307(1-2), 101-8.

[14] Fan, X; Matsui, W; Khaki, L; Stearns, D; Chun, J; Li, YM; Eberhart, CG. Notch pathway inhibition depletes stem-like cells and blocks engraftment in embryonal brain tumors. Cancer Res., 2006 Aug 1 66(15), 7445-52.

[15] Diévart, A; Beaulie, N; Jolicoeur, P. Involvement of Notch1 in the development of mouse mammary tumors. *Oncogene,* 1999 Oct 28 18(44), 5973-81.

[16] Patrawala, L; Calhoun,T; Schneider-Broussard, R; Zhou, J; Claypool, K; Tang, DG. Side population is enriched in tumorigenic, stem-like cancer cells, whereas ABCG2+ and ABCG2- cancer cells are similarly tumorigenic. *Cancer Res.,* 2005 Jul 15 65(14), 6207-19.

[17] Phillips, TM; Kim, K; Vlashi, E; McBride, WH; Pajonk, F. Effects of recombinant erythropoietin on breast cancer-initiating cells. *Neoplasia,* 2007 Dec 9(12), 1122-9.

[18] Gal, H; Amariglio, N; Trakhtenbrot, L; Jacob-Hirsh, J; Margalit, O; Avigdor, A; Nagler, A; Tavor, S; Ein-Dor, L; Lapidot, T; Domany, E; Rechavi, G; Givol, D. Gene expression profiles of AML derived stem cells; similarity to hematopoietic stem cells. Leukemia, 2006 Dec 20(12), 2147-54.

[19] Cullion, K; Draheim, KM; Hermance, N; Tammam, J; Sharma, VM; Ware, C; Nikov, G; Krishnamoorthy, V; Majumder, PK; Kelliher, MA. Targeting the Notch1 and mTOR pathways in a mouse T-ALL model. *Blood*, 2009 Jun 11 113(24), 6172-81.

[20] Wang, Q; Chen, Z G; Du, C Z; Wang, H W; Yan, L; Gu, J. Cancer stem cell marker $CD133^+$ tumor cells and clinical outcome in rectal cancer. *Histopathology*, 2009 May 27 55, 284–293.

[21] Camilla, R and Akseli, H. *Virotherapy as an Approach against Cancer Stem Cells Current Gene Therapy*, 2008, Febrauary 13 8.

[22] Takahashi, R; Takeshita,F; Fujiwara, T; Ono, M; Ochiya, T. Cancer Stem Cells in Breast Cancer. *Cancer,* 2011 March 15 3, 1311-1328.

[23] Becker, L; Huang, Q; Mashimo, H. Lgr5, an intestinal stem cell marker, is abnormally expressed in Barrett's esophagus and esophageal adenocarcinoma. *Dis. Esophagus*, 2010 Feb 23(2), 168-74.

[24] David, H; Lydia, K; Jutta, E; Thomas, K; Andreas,Jung. *Prognostic Significance of the Cancer Stem Cell Markers CD133, CD44, and CD166 in Colorectal Cancer Cancer Investigation*, 2009 27,844–850.

[25] Sumiya, I; Shinichi, U; Takaaki, A; Hideo, A; Yasuto, U; Yoshiaki, K; Ken, S; Yuka, N; Hiroshi, K; Masataka, M; Yuko, K; Shoji, N. Clinical implication of cd166 expression in gastric cancer. *Journal of surgical oncology*, 2011 103, 57–61.

[26] Kahlert, C; Weber, H; Mogler, C; Bergmann, F; Schirmacher, P; Kenngott, HG; Matterne, U; Mollberg, N; Rahbari, NN; Hinz, U; Koch, M; Aigner, M; Weitz, J. Increased expression of ALCAM/CD166 in pancreatic cancer is an independent prognostic marker for poor survival and early tumour relapse. *Br. J. Cancer*, 2009 Aug 4 101(3), 457-64.

[27] Chen,W; Cai, SY; Xu, S; Denson, LA; Soroka, CJ; Boyer, JL. Nuclear receptors RXRα: RARα are repressors for human MRP3 expression. Am *J. Physiol. Gastrointest Liver Physiol.*, 2007 292, G1221–G1227.

[28] Christine, F and Charlotte, K. Human breast cancer stem cell markers CD44 and CD24: enriching for cells with functional properties in mice or in man? *Breast Cancer Research*, 2007 May 29 9(303).

[29] Choi, D; Lee, HW; Hur, KY; Kim, JJ; Park, GS; Jang, SH; Song, YS; Jang, KS; Paik, SS. Cancer stem cell markers CD133 and CD24 correlate with invasiveness and differentiation in colorectal adenocarcinoma. *World J. Gastroenterol.*, 2009 May 14 15(18), 2258-2264.

[30] Benedetta,B; Stefania, B; Cristina, G; Ugo F; Giovanni, C; Identification of a tumor-initiating stem cell population in human renal carcinomas. *FASEB J.*, 2008 22, 3696–3705.

[31] Zijlmans, H; Fleuren, GJ; Hazelbag, S; Sier, CF; Dreef, EJ; Kenter, GG; Gorter, A. Expression of endoglin (CD105) in cervical cancer. *Br. J. Cancer,* 2009 May 19 100(10), 1617-26.

[32] Yongzhong, Y; Yiming, P; Jun, C; Xitai, S; Yudong, Q; Yitao, D. Endoglin (CD105) Expression in Angiogenesis of Primary Hepatocellular Carcinomas: Analysis using Tissue Microarrays and Comparisons with CD34 and VEGF. *Annals of Clinical and Laboratory Science*, 2007 37.

[33] Wolfram, K; Steven, B; Matthew, E; Mehsati, H; Ai-Ying, C; Jonathan, I. E; David M; Roles for the Stem Cell–Associated Intermediate Filament Nestin in Prostate Cancer Migration and Metastasis. *Cancer Res.*, 2007 October 1 67(19), 9199–206.

[34] Chiou, SH; Yu, CC; Huang, CY; Lin, SC; Liu, CJ; Tsai, TH; Chou, SH; Chien, CS; Ku, HH; Lo, JF. Positive correlations of Oct-4 and Nanog in oral cancer stem-like cells and high-grade oral squamous cell carcinoma. *Clin. Cancer Res.*, 2008 Jul 1 14(13), 4085-95.

[35] Feng, Y; Caiyun, Z; Qi Cheng; Jiaji, S; Huaizeng, C; Stem-cell-abundant proteins Nanog, Nucleostemin and Musashi1 are highly expressed in malignant cervical epithelial cells. *BMC Cancer*, 2008 April 18 8,108.

[36] Fan, LF; Dong, WG; Jiang, CQ; Xia, D; Liao, F; Yu, QF. Expression of putative stem cell genes Musashi-1 and beta1-integrin in human colorectal adenomas and adenocarcinomas. *Int. J. Colorectal Dis.*, 2010 Jan 25(1), 17-23.

[37] Padraic, P; Sean, V; Thomas, P; David, M; Steven, M; Steven, C; Dennis, A; Parker G. Expression of an Exogenous Human Oct-4 Promoter Identifies Tumor-Initiating Cells in Osteosarcoma *Cancer Res.*, 2009 July 15 69(14), 5648–5655.

[38] Chen, YC; Hsu, HS; Chen, YW; Tsai, TH; How, CK; Wang, CY; Hung, SC; Chang, YL; Tsai, ML; Lee, YY; Ku, HH; Chiou, SH. Oct-4 expression maintained cancer stem-like properties in lung cancer-derived CD133-positive cells. *PLoS One*, 2008 Jul 9 3(7), e2637.

[39] Satyanarayana, R and Lakshmi, N. Oct-4 expression maintained stem cell properties in prostate cancer-derived $CD133^+MDR1^+$ cells. Tropical *Journal of Pharmaceutical Research*, 2009 February 8 (1), 3-9.

[40] Gary, MK; Thomas, C; Philip, CW; Anthony, WI; Richard, A; Bonita, KB; Rooshdiya, K; C Soon Lee. Increased c-kit (CD117) expression in malignant mammary phyllodes tumors. *Modern Pathology*, 2004 March 26 17, 827–831.

[41] Kazuo, N; Yoshihiro, M; Hideo, K; Arafumi, M; Hisao, A; Ryosuke, T. Immunohistochemical KIT (CD117) *Expression in Thymic Epithelial Tumors. Chest*, 2005 128, 140–144.

[42] Baran, Y; Gündüz, U; Ural, A G. Expression of multi drug resistance (MDR1) gene in human promyelocytic leukemia cell line selected with vincristine. *Turk J. Cancer*, 2005 35(2), 88-92.

[43] Yasuhiro, T; Morimasa, W; Kentaro, K; Naoko, K; Taishi, H; Jun, N; Masayuki, N; Seiji, N and Michihiko, K. MDR1 Gene Overexpression and Altered Degree of Methylation at the Promoter Region in Bladder Cancer during Chemotherapeutic Treatment. *Clin. Cancer Res*, 2000 6, 4618-4627.

[44] Vibeke, A; Lene, A; Dorte, J; Mette, O; Mona, S; Julian, H; Elin, K and Ulla, V. The multidrug resistance 1 (MDR1) gene polymorphism G-rs3789243-A is not associated with disease susceptibility in Norwegian patients with colorectal adenoma and colorectal cancer; a case control study. *BMC Med. Gen.*, 2009 10:18.

[45] Zihua, W; Baoshuang, W; Kun, T; Edmund, J.D; Samuel, S.C and Caroline, G.L. A functional polymorphism within the MRP1 gene locus identified through its genomic signature of positive selection. *Hum. Mol. Genet,* 2005 14, 2075–2087.

[46] Dorina, M.V; Edo, V; Anneke, Y; Leonore, N; Hetty, T. B; Gert, J. O; Reinier, A. R; Michael, M; Eva, V.B and Elisabeth, G. E. Deletion of the multidrug resistance protein MRP1 gene in acute myeloid leukemia: the impact on MRP activity. *Blood,* 2000 95, 3514-3519.

[47] Zihua, W; Pui-Hoon, S; Helen, A; Stephen, R; Samuel, S.C; Edmund, JD and Caroline. Nucleotide sequence analyses of the MRP1 gene in four populations suggest negative selection on its coding region. *BMC Genomics*, 2006 7, 111.

[48] Schaub, TP; Kartenbeck, J; König, J; Spring, H; Dörsam, J; Staehler, G; Störkel, S; Thon, WF; Keppler, D. Expression of the MRP2 gene-encoded conjugate export pump in human kidney proximal tubules and in renal cell carcinoma. *J. Am. Soc. Nephrol.*, 1999 Jun 10(6), 1159-69.

[49] Lukasz, P; Katarzyna, K; Marcin, R; Takeshi, U; Michihiko, K; Grzegorz, B. Differential regulation of the human MRP2 and MRP3 gene expression by glucocorticoids. *J. Steroid Biochem. Mol. Biol.*, 2005 96, 229–234.

[50] Chen, W; Wang, G; Liu, Y; Qian, R. Cancer Stem-like Cells in Human Prostate Carcinoma Cells DU145 The Seeds of the Cell Line? *Cancer Biology and Therapy*, 2007 May 6(5).

[51] Rieneke, V; George, L S; Anneke, W. R; Jelle, J. L; Ruud, O; Gerrit, J; Jean-Pierre, G; Joel, N. G; Alexander, P; David, T. C; Rik, J. S and Tanja, D.G: A role for multidrug resistance protein 4 (MRP4; ABCC4) in human dendritic cell migration. *Blood,* 2008 112, 2353-2359.

[52] Murray, D. N; Janice, S; Kara, T; Peter, T; Claudia, F; George, L. S; Peter, W; Susan, L. C; Wendy, B. L; Glenn, M. M; John, D. A and Michelle, H. Expression of multidrug transporter MRP4/ABCC4 is a marker of poor prognosis in neuroblastoma and confers resistance to irinotecan in vitro. *Mol. Cancer Ther*, 2005 4(4), 547–53.

[53] Monika, B; Heiko, T; Dieter, H; Michael, K; Nicola, G and Thorsten, R. Multidrug resistance–associated protein 4 (MRP4) gene polymorphisms and treatment response in adult acute lymphoblastic leukemia. *Blood*, 2009 114, 5400-5401.

[54] Praveen, K. N; Tobias, H; Kathleen, K; Steffen, M; Markus, G; Léa, P; Jérôme, G; Matthias, S; Gabriele, J; Christian, R; Dieter, R; Dariusz, W. K; Heyo, K. K; Frank, U. W; Julia, M; Markus, M. L and Christoph, A. R. Drug efflux transporter MRP5 affects sensitivity of pancreatic cancer cell lines to the nucleoside anticancer drug 5-fluorouracil. *American Society for Pharm and Exp. Ther.*, 2010.

[55] Choi, CH. ABC transporters as multidrug resistance mechanisms and the development of chemosensitizers for their reversal. *Cancer Cell International*, 2005 5:30.

[56] Marcel, K; Marcel, L; Marcel, H; Frank, B and Piet, B. Expression of Human MRP6, a Homologue of the Multidrug Resistance Protein Gene MRP1, in Tissues and Cancer Cells. *Cancer Res.*, 1999 59, 175-182.

[57] Tetsuya, O; Hiroaki, O; Takehiro, U; Yuji, B; Mikinori, M; Ken, M; Hiroyoshi, M; Shigeki, S and Ryuzo, U. MRP7/ABCC10 expression is a predictive biomarker for the resistance to paclitaxel in non-small cell lung cancer. *Mol. Cancer Ther.*, 2008 7, 1150-1155.

[58] Tetsuya, O; Yuji, B; Hiroyuki, A; Hiroaki, O; Ken, M; Hiroyoshi, M; Shigeki, S and Ryuzo, U. MRP8/ABCC11 directly confers resistance to 5-fluorouracil. *Mol. Cancer Ther.*, 2007 6, 122-127.

[59] Yu, T and Toshihisa, I. Pharmacogenomics of Human ABC Transporter ABCC11 (MRP8): Potential Risk of Breast Cancer and Chemotherapy Failure. *Anti-Cancer Agents in Med. Chem.*, 2010 10, 617-624.

[60] Steinbach, D; Sell, W; Voigt, A; Hermann, J; Zintl, F; Sauerbrey, A. BCRP gene expression is associated with a poor response to remission induction therapy in childhood acute myeloid leukemia. *Leukemia*, 2002 16, 1443–1447.

[61] Hu, K; Morris, ME. Effects of benzyl-, phenethyl-, and alphanaphthyl isothiocyanates on P glycoprotein- and MRP1- mediated transport. *J. Pharm. Sci.*, 2004 93, 1901-1911.

[62] Murakami, N; Sugimoto, M; Morita, M; Kobayashi, M. Total synthesis of agosterol A: an MDR-modulator from a marine sponge. *Chemistry,* 2001 7, 2663-2670.

[63] Brooks, T; Minderman, H; O'Loughlin, KL; Pera, P; Ojima, I; Baer, MR; Bernacki, RJ. Taxane-based reversal agents modulate drug resistance mediated by P-glycoprotein, multidrug resistance protein, and breast cancer resistance protein. *Mol. Cancer Ther.*, 2003 2, 1195-1205.

[64] Pauli-Magnus, C; von Richter, O; Burk, O; Ziegler, A; Mettang, T; Eichelbaum, M; Fromm, MF. Characterization of the major metabolites of verapamil as substrates and inhibitors of Pglycoprotein. *J. Pharmacol. Exp. Ther.*, 2000 293, 376-382.

[65] Okumura, H; Chen, ZS; Sakou, M; Sumizawa, T; Furukawa, T; Komatsu, M; Ikeda, R; Suzuki, H; Hirota, K; Aikou, T; Akiyama, SI. Reversal of Pglycoprotein and multidrug-resistance protein-mediated drug resistance in KB cells by 5-O-benzoylated taxinine K. *Mol. Pharmacol.,* 2000 58, 1563-1569.

[66] Evers, R; Kool, M; Smith, AJ; van Deemter, L; de Haas, M. Borst, P. Inhibitory effect of the reversal agents V-104, GF120918 and Pluronic L61 on MDR1 Pgp-, MRP1- and MRP2-mediated transport. *Br. J. Cancer*, 2000 83, 366-374.

[67] Miller, DW; Batrakova, EV; Kabanov, AV. Inhibition of multidrug resistance-associated protein (MRP) functional activity with pluronic block copolymers. *Pharm Res.*, 1999 16, 396-401.

[68] Vezmar, M; Georges, E. Reversal of MRP-mediated doxorubicin resistance with quinoline-based drugs. *Biochem. Pharmacol.*, 2000 59, 1245-1252.

[69] Zhang, S; Yang, X; Morris, ME. Flavonoids are inhibitors of breast cancer resistance protein (ABCG2)-mediated transport. *Mol. Pharmacol.*, 2004 65, 1208-1216.

[70] Imai, Y; Tsukahara, S; Asada, S; Sugimoto, Y. Phytoestrogens/Flavonoids Reverse Breast Cancer Resistance Protein/ABCG2- Mediated Multidrug Resistance. *Cancer Res.,* 2004 64, 4346-4352.

[71] Allen, JD; van Loevezijn, A; Lakhai, JM; van der Valk, M; van Tellingen, O; Reid, G; Schellens, JH; Koomen, GJ; Schinkel, AH. Potent and specific inhibition of the breast cancer resistance protein multidrug transporter in vitro and in mouse intestine by a novel analogue of fumitremorgin C. *Mol. Cancer Ther.* 2002 1, 417-425.

[72] Kobayashi, CI and Suda, T. Regulation of reactive oxygen species in stem cells and cancer stem cells. *J. Cell Physiol.*, 2011 Mar 29.

[73] Ushio-Fukai,M and Alexander, RW. Reactive oxygen species as mediators of angiogenesis signaling: role of NAD (P) H oxidase. *Mol. Cell Biochem.*, 2004 Sep 264(1-2), 85-97.

[74] Fayette, J; Soria, JC; Armand, JP. Use of angiogenesis inhibitors in tumour treatment. *Eur. J. Cancer.* 2005 May 41(8), 1109-16.

[75] Sainson, RC; Aoto, J; Nakatsu, MN; Holderfield, M; Conn, E; Koller, E; Hughes, CC. Cell-autonomous notch signaling regulates endothelial cell branching and proliferation during vascular tubulogenesis. *FASEB J.,* 2005 Jun 19(8), 1027-9.

[76] Mailhos, C; Modlich, U; Lewis, J; Harris, A; Bicknell, R; Ish-Horowicz, D. Delta4, an endothelial specific notch ligand expressed at sites of physiological and tumor angiogenesis. *Differentiation*, 2001 Dec 69(2-3), 135-44.

[77] Patel, NS; Li, JL; Generali, D; Poulsom, R; Cranston, DW; Harris, AL. Up-regulation of delta-like 4 ligand in human tumor vasculature and the role of basal expression in endothelial cell function. *Cancer Res.*, 2005 Oct 1 65(19), 8690-7.

[78] Zeng Q, Li S, Chepeha DB, Giordano TJ, Li J, Zhang H, Polverini PJ, Nor J, Kitajewski J, Wang CY. Crosstalk between tumor and endothelial cells promotes tumor angiogenesis by MAPK activation of Notch signaling. *Cancer Cell.,* 2005 Jul 8(1), 13-23.

[79] Curry, CL; Reed, LL; Golde, TE; Miele, L; Nickoloff, BJ; Foreman, KE. Gamma secretase inhibitor blocks Notch activation and induces apoptosis in Kaposi's sarcoma tumor cells. *Oncogene*, 2005 24, 6333–6344.

[80] Soares, R; Balogh, G; Guo, S; Gärtner, F; Russo, J; Schmitt, F. Evidence for the notch signaling pathway on the role of estrogen in angiogenesis. *Mol. Endocrinol.*, 2004 Sep 18(9), 2333-43.

[81] Paris, D; Quadros, A; Patel, N; Delle Donne, A; Humphrey, J; Mullan, M. Inhibition of angiogenesis and tumor growth by beta and gamma-secretase inhibitors. *Eur. J. Pharmacol.*, 2005 May 2 514(1), 1-15.

[82] Williams, CK; Li, JL; Murga, M; Harris, AL; Tosato, G. Up-regulation of the Notch ligand Delta-like 4 inhibits VEGF-induced endothelial cell function. *Blood*, 2006 Feb 1107(3), 931-9.

[83] Zimrin, AB; Pepper, MS; McMahon, GA; Nguyen, F; Montesano, R; Maciag, T. An antisense oligonucleotide to the notch ligand jagged enhances fibroblast growth factor-induced angiogenesis in vitro. *J. Biol. Chem.*, 1996 Dec 20 271(51), 32499-502.

[84] Taylor, KL; Henderson, AM; Hughes, CC. Notch activation during endothelial cell network formation in vitro targets the basic HLH transcription factor HESR-1 and downregulates VEGFR-2/KDR expression. *Microvasc Res.*, 2002 Nov 64(3), 372-83.

[85] Noseda, M; Chang, L; McLean, G; Grim, JE; Clurman, BE; Smith, LL; Karsan, A. Notch activation induces endothelial cell cycle arrest and participates in contact inhibition: role of p21Cip1 repression. *Mol. Cell Biol.*, 2004 Oct 24(20), 8813-22.

[86] Leong, KG; Hu, X; Li, L; Noseda, M; Larrivée, B; Hull, C; Hood, L; Wong, F; Karsan, A. Activated Notch4 Inhibits Angiogenesis: Role of b1-Integrin Activation. *Mol. Cell. Biol.,* 2002 22, 2830–2841.

[87] MacKenzie, F; Duriez, P; Larrivée, B; Chang, L; Pollet, I; Wong, F; Yip, C; Karsan, A.Notch4-induced inhibition of endothelial sprouting requires the ankyrin repeats and involves signaling through RBP-Jkappa. *Blood*, 2004 Sep 15 104(6), 1760-8.

[88] Henderson, AM; Wang, SJ; Taylor, AC; Aitkenhead, M and Hughes, CC. The basic helix-loop-helix transcription factor HESR1 regulates endothelial cell tube formation. *J. Biol. Chem.*, 2001 276, 6169–6176.

[89] Liu, ZJ; Xiao, M; Balint, K; Soma, A; Pinnix, CC; Capobianco, AJ; Velazquez, OC; Herlyn, M. Inhibition of endothelial cell proliferation by Notch1 signaling is mediated by repressing MAPK and PI3K/Akt pathways and requires MAML1. *FASEB J.,* 2006 May 20(7), 1009-11.

[90] Guangyu, WU; Svetlana, L; Indranil, D; Jinhe, LI; Mark, G; Adele, P; Inca, C; Raymond, J. D and Jan kitajewski. SEL-10 Is an Inhibitor of Notch Signaling That Targets Notch for Ubiquitin-Mediated Protein Degradation. *Mol. and Cell Biol.*, 2001.

[91] Deangelo, D. J; Stone, R M and Silverman, L. A phase I clinical trial of the notch inhibitor MK-0752 in patients with T-cell acute lymphoblastic leukemia/lymphoma (T-ALL) and other leukemias. *J. Clin. Oncol.*, 2006 24.

[92] Peters, J. U; Galley, G; Jacobsen, H; Czech, C; Pierson, P D; Kitasa, E A and Ozmen, L. Novel orally active, dibenzazepinone- based gamma-secretase inhibitors. *Bioorg Med. Chem. Lett.*, 2007 17, 5918–5923.

[93] Kopan, R and Ilagan, M. X. Gamma-secretase: proteasome of the membrane? *Nat. Rev. Mol.Cell Biol.*, 2004 5, 499–504.

[94] Ridgway, J; Zhang, G; Wu, Y; Stawicki, S; Liang, W.C; Chanthery, Y; Kowalski, J; Watts, R J; Callahan, C; Kasman, I; Singh, M; Chien, M; Tan, C; Hongo, J S; Sauvage, F; Plowman, G and Yan, M. Inhibition of Dll4 signaling inhibits tumor growth by deregulating angiogenesis. *Nature,* 2006 444, 1083–1087.

[95] Hoey, T; Yen, W C; Axelrod, F; Basi, J; Donigian, L; Dylla, S; Fitch-Bruhns, M; Lazetic, S; Park, I K; Sato, A; Satyal, S; Wang, X; Clarke, M F; Lewicki, J and Gurney, A. DLL4 blockade inhibits tumor growth and reduces tumor-initiating cell frequency. *Cell Stem. Cell*, 2009 5,168–177.

[96] Noguera-Troise, I; Daly, C; Papadopoulos, N.J; Coetzee, S; Boland, P; Gale, N.W; Lin, H.C; Yancopoulos, G.D and Thurston, G. Blockade of Dll4 inhibits tumor growth by promoting non-productive angiogenesis. *Nature*, 2006 444, 1032–1037.

[97] Li, K; Yucheng, L; Wenjuan, W; Wendy, R.G; David, W.C; Mason, L; Shane, S; Tihui, F; Long, V; Gavin, H; Rachel, M. H; Thomas, D; Sanjaya, S; Stephen, C. B; Zhengbin, Y; Jon, C. A and Bin-Bing, S. Z. Modulation of Notch signaling by antibodies specific for the extracellular negative regulatory region of NOTCH3. *J. Biol. Chem.,* 2006 283, 8046–8054.

[98] Garcés, C; Ruiz-Hidalgo, M.J; Jaime, F; Crystal, P; Lucio, M; Julia, G; Ezio, B; Almudena, P and Jorge, L. Notch-1 controls the expression of fatty acid-activated transcription factors and is required for adipogenesis. *J. Biol. Chem.*, 1997 272, 29729–29734.

[99] Nickoloff, B. J; Qin, JZ and Chaturvedi, V. Jagged-1 mediated activation of notch signaling induces complete maturation of human keratinocytes through NF-kappaB and PPARgamma. *Cell Death Differ.*, 2002 9, 842–855.

[100] Sawyer T. K. AILERON Therapeutics. *Chem. Biol. Drug Des.*, 2009 73, 3–6.

[101] Epenetos, A; Kousparou, C and Stylianou, S. Inhibition of Notch signaling for the treatment of human carcinomas. *AACR Meeting Abstracts* A5502 2009.

[102] Ie-Ming, S and Tian-Li, W. Notch Signaling, γ Secretase Inhibitors, and Cancer Therapy. *Cancer Res.*, 2007 67, 1879-1882.

[103] Li, JL; Harris, AL. Notch signaling from tumor cells: a new mechanism of angiogenesis. *Cancer Cell*, 2005 8(1), 1-3.

In: Signaling, Gene Regulation and Cancer
Editors: Shree Ram Singh and Manoj K. Mishra
ISBN: 978-1-61942-088-5

Chapter 13

CELL CYCLE DEREGULATION IN NEURO-ONCOGENESIS: A FATAL JOURNEY BY CELLS

Pravir Kumar[1,2,3*], Rashmi K. Ambasta[1,3], M. Sonia Angeline Inbakumari[3] and Mandar Bhattacharya[3]

[1]Assistant Director and Associate Professors of Centre for Medical Engineering, VIT, Vellore, India

[2]Adjunct Faculty, Department of Neurology, Tufts University School of Medicine, Boston, MA, India

[3]Functional Genomics and Cancer Biology Laboratory, Centre for Medical Engineering, Vellore Institute of Technology, Vellore, Tamil Nadu, India

ABSTRACT

The hallmark feature in neuro-oncology is an uncontrolled proliferation of the cells leading to the growth of the tumors in the central nervous system and brain. Mutations in the cellular components lead to the cell cycle deregulation. Misregulation of the cell cycle process and its involvement in cancer progression is a proven fact. Several checkpoints, number of molecules, extracellular signals form at the multilevel control system for cell cycle processes. Thereby, it is not surprising that any malfunctioning entity of this network will ultimately lead to a serious disorder, mainly cancer. The cell cycle is controlled by various protein kinases, whose activity is governed by Cyclins. Cyclins combine with cyclin-dependent kinases (CDKs) to enable the cell to pass through each stage of the cell cycle. Important role of cyclin- CDK enzyme is to phosphorylates the retinoblastoma tumor suppressor protein, which inhibits its ability to control the action of transcription factors (E2F family), as these are responsible for inducing the gene expression for proliferation of the cells. Cyclin dependent kinases (CDK) and Cyclins propagates the cell cycle process; on the other hand, several inhibitors are present that diminish the cell cycle rate. Protective proteins like molecular chaperones and degrading

* Corresponding Author Address: Pravir Kumar, PhD, Associate Professor, CBMR-206, Cancer biology and functional genomics lab, School of Bio-science and Technology, VIT Vellore, TN 632014 (INDIA), E-mail: pravirkumar@vit.ac.in; Pravir.Kumar@tufts.edu, Phone: +91-4162202592 (lab); +91-9003386752.

machinery like Ubiquitin-Proteosomal System (UPS) are present in the cells to ensure the physiological homeostasis. Thus overexpression of positive regulators or down regulation, malfunction and mutations of negative regulators are the fundamentals for neoplasm formation. A thorough understanding at the molecular level is needed to design the therapeutic strategy for cancer and to shed some light on the development of cancers of the brain.

1. Introduction

Mounting evidence is suggesting the involvement of post-mitotic cell division in various neurodegenerative disorders and in cancer. In our body, brain is the most important organ that maintains the physiological homeostasis in coordination with other organs. However, the complexity of brain cells and its exact function are quite baffling issues for researchers. One such ailment is a brain tumor and its progression due to the deregulation of cell cycle division. To understand better about neuro-oncogenesis first we should look at the anatomy of brain, which consists of very delicate tissues. Majorities of these tissues are made up of neurons and glial cell (neuroglia). Neurons are structural and functional units of brain while neuroglia is protecting and helping the brain for normal function.

The mitotic cell division is a normal procedure of the body that develops a fully functional organ system. On the other hand, post-mitotic cell division triggers an abnormal increase in the cell number that leads to the formation of a tumor (neoplasm), a mass of cells with an uncontrolled proliferation. Gradually, solid tumors grow bigger and exert a huge amount of pressure on the cells lying below. Due to this, cell experiencing a lack of nutrients and oxygen supply. Initially, these conditions inhibit the cell proliferation but localized cells secrete an extracellular mitogenic and angiogenic factor that compensates and nurtures the deprived tumor cells for further growth and development. Eventually, cells are detached from primary tumor and spread to the adjacent tissues and start moving through the bloodstream or lymphatic system and finally spread in the entire body. This phenomenon is known as metastasis. These metastasized cells produces secondary tumor at a different place from the place of origin which is malignant and lethal in nature.

1.1. Types of Tumors in Neuro-Oncology

Neuro-oncology is basically involved with the various tumors of the central nervous system. The neurons of the brain are not capable of undergoing cell division after their complete growth, unlike the cells in other tissues, which are capable of dividing. The glial cells however, hold back their capacity to proliferate. The human glial tumors are found to show a high potential to proliferate [1]. Human brain tumors have a growth fraction of 20-70%, also considering the size of tumor along with the availability of oxygen and nutrients.

The tumors occurring in the regions of brain and its surrounding structures are found to have been low as well as the high rate of malignancies. Gliomas are the primary tumors of the brain [2]. The most common type of highly malignant tumor among the primary brain tumors is a high- grade glioma (HGG) [3]. Most of the HGG contains a very aggressive subtype of tumors, which are the glioblastomas (GBM). The other kinds of tumors which belong to this group are anaplastic gliomas (AG) [4]. The anaplastic gliomas contain anaplastic astrocytoma (AA), anaplastic oligodendroglioma (AO) and anaplastic oligoastrocytoma (AOA) along with rarer subtypes. These HGGs are cancerous and are not curable hence they are as such related to morbidity and mortality [5].

2. An Overview of Cell Cycle

Cell division is a normal process for the development of an organ. However, once differentiated, cells (e.g. human neuronal cells, human muscle cells) exit the cell cycle and stay at non dividing phase termed as G0 phase. On contrary, terminally differentiated cells are destined to die without dividing again. Cell cycle progression is divided by the different phases. Among those, interphase is a period between divisions of the cytoplasm while G1 phase is responsible for the growth of the newly divided cell. When the growth attains a certain limit, the cell is ready for division. To avoid any aberrant re-entry and abnormality during the cell division, cells have tightly regulated checkpoint's machinery.

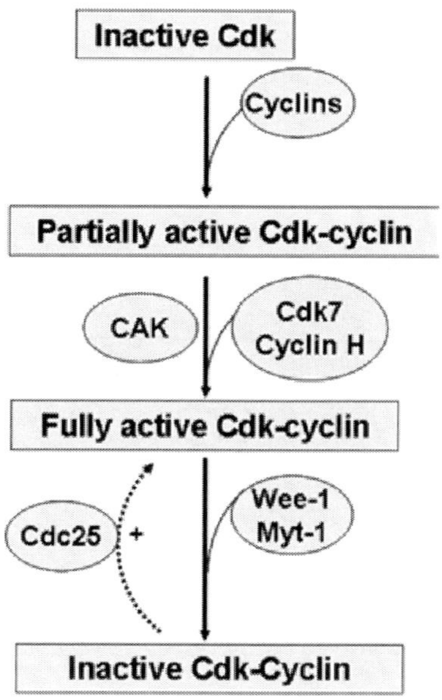

Figure 1. Activation of Cdk-Cyclin complex. CAK: cyclin activating kinase.

There are two main check points that ensure that process is going in a right direction, i.e. G1 checkpoint (restriction point) confirms that environment is favorable and DNA is intact and so that it allows cell to go to the S phase.

A proper stimulation by mitogenic factors can bring a G0 cell to G1 phase or G1 phase cell to proceed to S phase. Lack of mitogenic factors can also make one G1 cell to go to G0 phase.

However, once G1 check point is crossed, the cell must proceed further even if the mitogenic factors are lacking. After transition from G1 to S phase, DNA of the cell (more broadly chromosomes) undergoes replication in G2 phase, which is shorter than the G1 phase,

G2 checkpoint assures that cells do not enter in the mitosis until DNA has been repaired and only after confirmation cell enters into the mitotic phase.

Mitotic phase is divided into four sub phases: prophase, metaphase, anaphase and telophase. During karyokinesis, the chromosomes are shifted towards the polar region. Cytokinesis or division of cytoplasm is the next step where the cell divides at the centre producing two new daughter cells.

2.1. The Cell Cycle Control Mechanism: Cyclin-CDK Interaction

Cyclins are the group of proteins, which govern the activity of protein kinases that enable cell to proceeds in the cell cycle [6]. They are the regulatory subunits that are required for CDK activity. During the cell cycle, cyclins are produced and used up at particular instances, thereby monitoring the kinase activity. In human cells, the loci, that encode for CDKs and cyclins are 13 and 25 respectively [7]. Progress of a cell cycle is controlled by certain complexes of cyclins- CDK. They were found to be the interphase CDKs (CDK2, CDK4 and CDK6) and the mitotic CDK which is CDK1. CDK is capable of interacting with many cyclins and *vice versa*. Chemically, CDKs (Cyclin dependent kinases) are serine threonine proteases (MW 30-40 kD) and present throughout the cell cycle but get activated only after binding with Cyclins.

For instance, G1/S-CDKs activate S-CDKs by destruction of inhibitors and thus inactivate APC. The S-CDKs then phosphorylates the proteins responsible for triggering S phase. As S phase begins G1/S-CDKs promotes self-destruction by degrading G1/S-Cyclins while inactivate M-CDKs are activated by phosphatase and hence progression through G2/M checkpoint. M-CDKs then activate APC.

Two main families of cyclins are the mitotic cyclins and G1 cyclins. Mitotic family cyclins consist of cyclin A and cyclin B, while the G1 cyclins consist of cyclins C, D1-4, and E. Cyclin A is found to show its activity in the S phase and in mitotic phase.

Cyclins are preponderantly located in the nucleus whereas cyclin B is found in the nucleus and cytoplasm as well [8]. Cyclins D and E are needed for progression of G1 phase and to enter S phase [9, 10]. Mutations that are linked to the tumors are mostly caused due to an alteration in the CDK- cyclin complexes, which result in uncontrolled proliferation and cause for aberrant re-entry into the cell cycle [11].

Table 1. Cyclin dependent kinase and their roles in cell cycle progression

Name of CDKs	Role in cell cycle phase	References
Cdk1	M-Phase	[12]
Cdk2	G1/S-phase;S phase;M-phase	[12]
Cdk4	G1 phase	[13]
Cdk6	G1 phase	[13]
Cdk7	Cdk activating kinase	[14]
Cdk7+Cdk8+cdk9	Basal gene transcription by RNA pol II	[15]
Cdk5	Differentiation of Nerve cells	[16]

Table 2. Cdk inhibitors and their role in cell cycle

CDK inhibitors	Role	References
P21cip/kip	Inhibits Cyclin-Cdk2/Cdk4	[22]
P27	Inhibits G1/s and S Cdks	[23]
P57	Inhibits G1/S and S Cdks	[24]
P15 INK4b	Inhibit Cdk4/6	[25]
P16 INK4a	Inhibit Cdk4/6	[25]
P18INK4c	Inhibit Cdk4/6	[26]
P19INK4d	Inhibit Cdk4/6	[26]

Cell cycle defects colligated with tumors due to the modifications in the activity of cyclin-dependent kinase (CDK). Instability occurring at the genomic and chromosomal levels along with an uncontrolled proliferation are mostly caused by the CDKs, which are not regulatable [11, 17]. CDKs are required for impelling every phase of the cell cycle [18] and some specific interphase CDKs are required for proliferation of tumor cells; hence selective inhibitors of CDKs can be an appropriate therapeutic tool to check the tumor progression [19].

Along with other components of cell cycle, Cyclin dependent kinase inhibitors (CDKIs) are also an important group of proteins that regulate the activities of cyclins and CDKs [20, 21].

The CDKIs role is to effectively inhibit the cyclin- CDK complexes and therefore, inhibit the progress of a cell cycle. The CDKIs render the inhibition for cell proliferation. CDKIs are the tumor suppressor genes and divided into two families based on their targets, structure similarity and mechanism adopted to inhibit cell proliferation.

Table 3. Cyclins role in postmitotic condition

Cyclins-Cdks complex	Events at postmitotic condition	References
Cyclin E	Expressed in proliferating neuronal precursor cells and postmitotic neurons; expression is higher in the cytoplasm of adult neurons	[27]
Cyclin D	Loss of import of cyclin D during differentiation in proliferating progenitor cells; inhibition of Cyclin D1-Cdk4 complex formation in the nucleus	[28]
Cdk1	Activates FOXO-1 transcription factor to regulate G2/M transition and stimulate the mitogenic regulator Plk.	[29]
Cyclin 1-Cdk5	MAPK dependent regulation of Bcl-2 and Bcl-XL in mice	[30]
Cyclin C	pRb dependent pathway activates non homologous end joining mechanism for DNA repair	[31]

The CDKIs are INK4 proteins consisting of INK4A, INK4B, INK4C, INK4D, the Cip and Kip family, which contains p21, p27 and p57 [19]. CDK inhibitors (CDKI): Majority of these CDKI inhibits G1 phase CDKs (Table 2).

2.2. Role of Cyclins in Postmitotic Condition

The major events of the postmitotic condition are governed by Cyclins and CDKs, together with different Cyclins D, E, A and C (Table 3). Expression of Cyclin E is found in the proliferating and in differentiating regions. They are expressed in proliferating neuronal precursor cells and postmitotic neurons; while expressions of cyclins are higher in the cytoplasm of adult neurons [27]. In proliferating progenitor cells, loss of Cyclin D function during differentiation is observed accompanied by inhibition of Cyclin D1-Cdk4 complex in the nucleus [28].

2.3. Role of Cyclins in Neurodegeneration

Before explaining the cellular events and aberrant re-entry of cells division under a tumor conditions it is necessary to understand the re-expression of different cyclins (a post mitotic cell division phenomenon) in aged neurons and cause neurodegeneration. In neurodegeneration, Cdk5 undergoes phosphorylation at different serine and threonine residues that leads to accumulation of tau protein, which forms neuro fibrillary tangles (NFT); [32]. Over activation of Cdk5 is monitored by p35 cleavage, which is triggered by Aβ while Cyclin D is degraded by the FBXO31 protein that stops the G1/S regulation. The AD neurons have Cyclin B and Cdc- 2 expressions [33].The roles of different cyclins in neuro-degeneration are depicted in Table 4.

Table 4. Cyclins, CDKs and neurodegeneration

Cyclin/Cdk	Role in neurodegeneration	References
Cdk5	*In vitro* phosphorylation of tau protein in Alzheimer disease	[34]
	Play a role in NFT formation	[32]
	Aβ and PrP trigger Cdk5 over activation, cleavage of p35 plays a major role	[35]
Cyclin D	FBXO31 protein degrades Cyclin D to stop G1/S regulation	[33]
	Up regulation of Cyc D1/Cdk4 in drug induced neuronal cell death in piriform cortex region	[36]
Cyclin E	Nuclear localization of Cyc E has been observed	[37]
Cyclin B	Expression of Cyc B and Cdc-2 in AD neurons	[38]
Cyclin C	RNA level is increased in AD cases, co-localization of Cyc C-Cdk8 in astrocytes	[39]

2.4. Role of Cyclins in Neuro-Oncology

The characteristics features in cancer cells are to overlook the cell cycle check points by re-expression of positive regulators or else loss of negative regulators. Thus aberrant expressions of different cyclins have been detected in various cancers. Cyclin D1 expression in several brain neoplasm is thought to be a main feature in oncogenesis while it expression is significantly correlated with markers like PCNA and MIB-1[40]. Other cyclins such as Cyclin D2hase mixed expression in gliomas and glioblastomas multiformem [41] whereas Cyclin E is associated with advanced stage of the disease and is over expressed as abnormal protein [42,43]. Cyclin A re-expression is correlated with an an anaplastic oligodendroglioma grade [44].

Table 5. Effect of cyclins in cancer

Cyclins	Involvement in Cancer	Reference
Cyclin B1	Topoisomerase I mediated Cyclin B expression in Medulloblastoma	[45]
	CDC2/CycB1 over expression in Glioma cell line(78% CDC expression in grade IV)	[46]
	High expression in the central portion of Glioma	[47]
Cyclin D1	Deregulated Cyclin D in primitive neuroectodermal tumors in 40% cases	[48]
	Overexpression (>50%) and localization in meningioma and glioma	[49]
	Ini1 inactivation mediated CyclinD expression in Rhabdoid tumor	[50]
	Cyclin D1 has an early role in normal GNP development via Shh pathway	[51]
	Boost medulloblastoma formation	[51]
	In grade I astrocytoma and meningioma mitogen induces Cyclin transcription	[52]
	Expression is significantly correlated with markers like PCNA and MIB-1	[40]
	May not be involved in Oligodendroglioma tumorigenesis	[44]
Cyclin D2	Truncated expression in oligodendrogliomas, mixed gliomas and glioblastomas multiforme	[41]
	Grade II astrocytoma shows downregulation of cyclin D2	[52]
Cyclin A	S-phase proliferation marker	[53]
	Potent prognostic factor for low grade astrocytoma	[54]
	Expression is correlated with anaplastic oligodendroglioma grade	[44]
	Sensitive proliferation marker for Meningiomas	[55]
Cyclin E	Correlation with Brg1 and p27 loss of function in pituitary adenoma	[56]
	Cyc E activates Cdc2, compensating the effect of Cdk2 in pituitary tumors	[57]
	Interact with Inhibitor of differentiation 4 (Id4) in glioma cell line	[58]

Over expression of these positive regulators leads to uncontrolled cell proliferation by overlooking the mechanism of normal cell cycle arrest, which is a known characteristic feature of cancer.

3. Family of Cyclin Inhibitors

3.1. Inhibitors of CDK4 (INK) as Negative Growth Regulator

There are around 30% sequences similarities have been found in INK family isoforms (such as p16INK4a, p14ARF, p15INK4b and p18INK4c) that are considered as tumor suppressing proteins. Among these, p16INK4a is thought to be playing the main role in tumor suppression where CDKN2A expresses p16INK4a. These proteins belong to the INK (Inhibitors of CDK4)-family of CKI.

The tumor suppressor gene CDKN2A is located at chromosome 9p21 while genomic studies revealed that there are three exons present at 9p21 that reflects the possibilities of atleast three alternatives spliced variants.

Other inhibitors, p15INK4b and p16INK4a that have four ankyrin repeats whereas p18INK4c and p19INK4d have an additional repeat [26]. These ankyrin repeat to bind to the opposite of Cyclin binding site and cause an allosteric shift hindering cyclin binding and ATP hydrolysis [59]. CDKN2A is induced by vitamin D3 derivatives and by retinoids. P19INK4d knock out cells are sensitive to autophagic cell death [60]. Another Another inhibitor P19INK4d was discovered by Sherr's Laboratory using CDK4 as bait. The main distinctive feature of this protein is its sharp concentration cycling during the cell cycle. Its concentration remains low at G0/G1 onset then reaches at the maximum level at S phase [61]. However, Genetic modification of p19ink4d may not be associated with human neuroglial tumors [62, 63].

CDKN2C gene product was reported through yeast two-hybrid system where human cdk6 used as bait [75]. p18INK4c is needed for normal brain development [76] and has also effect on p18 mouse model. Mice that are lacking p16 as well as p18 are highly susceptible to cancer [77], and it has also been shown that homozygous deletion of p18INK4c in glioblastoma multiforme [66, 67]. One major conclusion was drawn by Solomon et al, that homogeneous deletion of p16INK4a can also trigger the deletion of p18INK18c.

Other proteins that are p15 and p16 interacting with retinoblastoma proteins that severely affected in human pituitary adenoma [64].

There are three major phenomenons in this pathway: (a) mutation in p16, p18 genes, (b) mutation in pRb gene and (c) over expression of CDK4. Correlation between p14, p15 and p16 expression and telomerase activation has also been postulated in meningiomas [65]. In p15INK4b or ARF (mice) double mutant, expression of p16INK4a was observed. Interestingly in p16INK4a/ARF double mutant or single mutant mice, the expression of p15INK4b was found to be increased [78] to compensate the adverse effect. The main cause of p16 gene inactivation is due to the deletion of locus [79] and methylated promoter region of CDKN2A gene. The CDKN2B gene lies near to the region where homozygous deletion of CDKN2A takes place; that may also result in the deletion of CDKN2B.

Table 6. Inhibitors in brain cancer

Brain Cancer	Cdk inhibitors (INK family)	References
Pituitary adenoma	Interaction of p15 and p16 with retinoblastoma, proteins are disrupted	[64]
Meningiomas	Interaction of p14, p15 and p16 with telomerase protein	[65]
Glioblstoma multiforme	Homozygous deletion of p18 gene, alternative splicing mediated p15 in activation (p10 formation)	[65-67]
Astrocytoma	Aberrant hypermethylation p14ARF, p15, pRb and p16; TP53 mutation p16, deletion	[68,69]
Medullo Blastoma	Promoter hypermethylation in p14 p16,	[70]
Oligodendro Glioma	Promoter methylation in p14,p15,p16,p53 and p73	[71]
Chordoma	Homo- or heterozygous loss of CDKN2A loci	[72]
Low grade glioma	Loss of INK4a/ARF associated with BRAF	[73]
Pleomorphic xantho astrocytoma	Homozygous deletion of CDKN2A p14, CDKN2B,	[74]

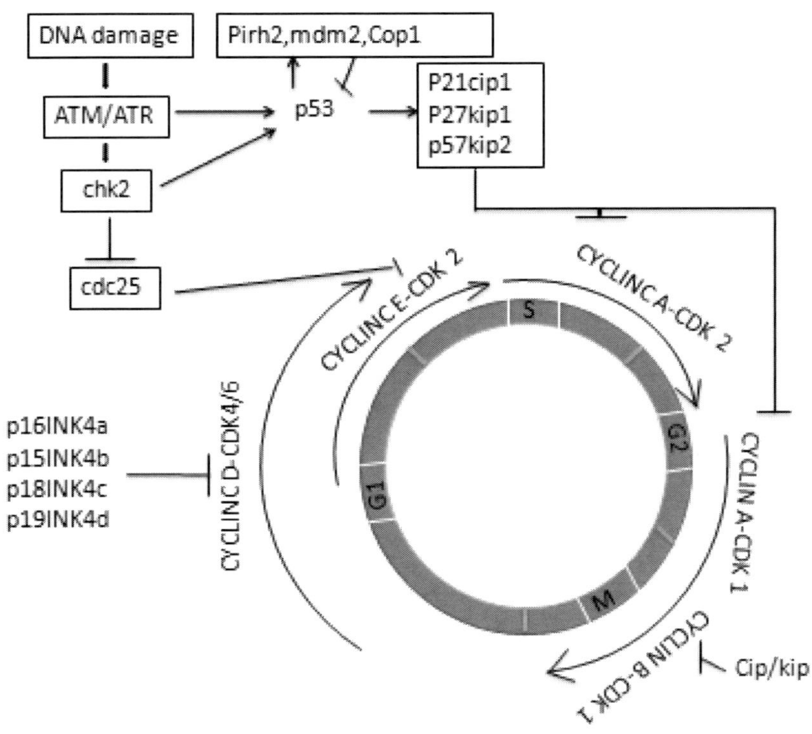

Figure 2. Cell cycle inhibitors and its function in cell cycle progression.

3.2. CIP/KIP FAMILY (p21, p27, p57)

CDKs interacting protein/kinases inhibitor protein (cip/kip) inhibitors are mainly classified of three types-p21cip1, pkip1 and p57kip2. As their name suggests they bind to both CDKs and Cyclin (on contrary, INK interacts only with Cyclin). Unlike INK4 family, cip/kip proteins inhibit wide array of CDKs. All these CKIs share a homology domain which helps them to bind with CDKs. On the other hand, p27 and p57 contains an additional QT domain that helps in recognition of SCF which helps in ubiquitination [80].

CDKN1C (p57kip2) gene resides at the end of chromosome 11p15.5 and its coding sequence consists of four exons gapped by three introns and CpG islands that are present both in upstream and downstream of the transcriptional start site [81]. CDKN1C promoter is a platform for several factor binding sites such as stimulatory proteins, neurofibromin, ΔNp63α, several transcription factors [82-85].

Table 7. Involvement of cip/kif inhibitor family in Brain neoplasms

Brain Cancer	Cdks inhibitor-cip/kip family	References
Neuroblastoma	Presence of inactive native p21 in an probable alternate conformation	[86]
	Inhibition of p21 in deacetylation of histone	[87]
Glioma	Skp2 mediated p27 dowregulation in astrocytic glioma	[88]
Glioblastoma	MicroRNA 221 and 222 mediated p27 inhibition	[89]
	p27-p53 compound mutant mice shows increased glioblastoma latency	[90]
Anaplastic astrocytoma	Methylation of p73/p21 CpG methylation	[91]
	Over expression of p21(waf1/cip1) protein and mRNA in astrocytoma which may be p53-independent	[92]
	p27 confers independent prognostic value; subcellular localization of p27 predicts the prognosis	[93,94]
	Low expression of p27 indicates poor prognosis in high grade astrocytoma	[95]
	p21(waf1/cip1) levels up p27 via IL-4 to arrest astrocytoma growth	[96]
Oligodendrogliomas	Methylation of p73/p21 cpG methylation	[91]
Ependymoma	Methylation of p73/p21 cpG methylation	[91]
Medulloblastoma	Mislocalization of p27 positive role player in cell motility; p27 and patched-1 collaboratively prevent medulloblastoma	[97,98]
	Musashi1(Msi1), the RNA binding protein checks the level of CDKN1A	[99]
Vestibular schwannomas (VS)	Significant p27under expression in aggressive VS than usual VS	[100]
Gemistocytic astrocytomas	p27 immuno-positivity in astrocytes than gametocytes	[101]
Pleomorphic Xantoastrocytoma	Strong P21 immuno-positivity in nucleus may be due to DNA damage	[102]

p21cip1 was identified by Harper et al, 1993 using CDK2 as bait. On the same year El-Diery et al, proposed p21cip1 as a downstream molecule of p53 dependent inactivation while p53 transactivates p21 during DNA damage. Mutation or deletion of p21 gene may not have any effect in human malignancies [103]. In Neuroblastoma NB1643 cell line, native p21 remains inactive and does not bind with CDK2, however recombinant p21 proteins, which are provided from outside, bind to CDKs. However, the endogenous p21 may be linked to a protein or present in an alternative conformation [86]. Treatment with histone deacetylase inhibitors shown that histone acetylation can successfully induce the p21 expression in T98G, A172, U-87MG,U-118 MG, and U-373MG glioma cell lines [87].

The other tumor suppressor gene CDKN1B encodes for a CDK inhibitory protein, p27Kip (27kd). The p27KIP1 is a CDK-binding protein which is activated by TGF-b by contact inhibition or serum starvation [104].

Loss of function of p27kip1 is a quite intriguing topic in cancer research. The level of p27 has been found out to be decreased in almost all brain tumors, including Head and Neck Suamous cell carcinoma [105]. The expression level of p27 is regulated at translation and post translational levels. However, p27kip1 may not be regulated at transcriptional level but inhibition of PI3K pathway is found to mediate the stabilization process [106] Inactivation of p27 is occurred mainly via three processes:

1) Cytoplasmic sequestration of p27kip1 is a main event regarding its loss of function at cancerous condition [107,108,109] where phosphorylation by Src kinase /Akt and not PI3K [106] could be the reason behind this process [110,111].
2) p27kip1 may be ubiquitinated and degraded.
3) Another mechanism of p27kip1 inhibition in glioblastoma cell is the presence of certain micro RNAs. RNAi mediated reduced Dicer can increase p27Kip1 [89].

There is several evidence shown that p27 is related to tumor metastasis and invasion [112,113]. The reason is p27kip1 binds with micro tubules destabilizing protein Stathmin via RhoA-ROCK pathway and inhibit cell motility in astrocytoma, glioblastoma [114-116]. Early evidence suggested that low p27 expression is associated with poor prognosis and increased cell proliferation where heterogeneous expression of p27kip1 could be the reason [117]. p27kip1 is playing an important role in the PTEN-mediated G1 arrest [118].

Another kinase, focal adhesive kinase acts as a catalyst for cell cycle that helps to exit from G1 phase in monolayer cultures of glioblastoma cells. This kinase also enhances the expression of cyclins D1 and E and strives to reduce the expression of p27Kip1 and p21Waf1 [119].

The grade of malignant glioma is associated with p27kip1 expression. The protein p27 is mainly associated with low grade (WHO I-II) glioma than higher one (WHO III-IV) [120].

The role of p57kip2 in a doxycycline regulated system on U373 glioma cells revealed almost 35% decrease in cell mobility suggesting a probable role in metastasis, which is similar to p27 [121].

The mode of action of deleted p57 in cancerous cells is not clear but induction of p57kip2 lowers the level of Retinoblastoma protein family [122]. Like cip/kip family, p57 has also shown to stop the tumor metastasis in *in vitro*. The inhibitory action of p57kip2 as it activates LIM-kinase 1 (but not 2) and deactivates Cofilin, a depolymerization factor of actin filament

[123]. Unlike p27, p57 expressions have shown to be increased with the advancement of the tumor [122].

4. ROLE OF TUMOR SUPPRESSOR PROTEINS

The well-known tumor suppressor gene family, p53 consists of p53, p63 and p73 proteins. Though they are structurally similar but possessing distinct functions [124]. Human p63 has six different isoforms (from 6 mRNAs) and p73 has 28 different isoforms from 35 mRNAs [125] while p53 also have six different isoforms [126].

The function of p53 is to stops G1/S transition if the DNA is damaged. p53 is regulated majorly by E3 ligase MDM2-MDM4 protein [127] while other p53 controlling proteins are pirH2 [128], COP1 [129], MDMX [130] and CHIP [131]. p53 protein interacts with MDM2 via N-terminal During DNA damage the N terminal part of p53 is phosphorylated by ATM/ATR (Ataxia Telangiectasia Mutated, Ataxia Telangiectasia and Rad3 related), CHK1/2 and DNAPK (DNA dependent protein kinase) and p53 translocated to the nucleus acts as a transcription factor.

Several case studies revealed that tumors of the nervous system finally display mutations in the ARF-MDM2-p53 and p16INK4a-CDK4-RB [132] tumor suppressor pathways. Mutation in p53 gene has been observed in astrocytoma, and this event is unlikely in other types of glioma (medulloblastoma, pilocytic grade I astrocytomas, and ependymomas,[133]. In astrocytoma, the p53 mutation has been observed and the mutation is a heterogeneous, even one glioma can carry three mutations.

Another member of p53 protein family, p73 is involved in different types of tumor. Loss of heterozygosity (LOH) is associated with all types of brain tumors [136]. Overe xpression of p73 may not affect p21 or MDM2 expression [137]. Complete CpG island methylation of p73 has been observed in Human Glioma cell line [138].

Table 8. Mutation in cancer

Cancer	Mutation of p53	Grade	References
Anaplastic astrocytoma	67%	WHO Grade III	[134]
Glioblastoma	41%	WHO Grade IV	[134]
Oligodendroglioma	13%, 83%		[135]
Medulloblastoma	11%		[133]
Pilocytic astrocytoma	>5%	WHO Grade I	[133]
Low grade astrocytoma	88%		[135]
Oligoastrocytoma	83%		[135]

4.1. Role of Pocket Proteins

Retinoblastoma protein family consists of three members viz. pRb, p107 and p130. This protein was first reported in the retinal cancer and consists of one pocket domain via which they bind to the transcription factor E2F transcription factors. E2F is a well-known transcription factor that is important for G1/S transition. Each of these transcription factors is associated with a second subunit Dp-1/Dp-2 to form a heterodimeric transcription factor complex. There are eight distinct E2Fs present out of which five are more widely studied- E2F1, E2F2, E2F3, E2F4, and E2F5. pRb interacts with E2F1, E2F2, E2F3, E2F3b and E2F4 while p107 interacts with E2F4 and p130 interacts with E2F4 as well as E2F5. E2F1, E2F2, E2F3 is activators and rest are repressors. E2F4-5 binds to p107/130 and E2F1-3 with pRb. Phosphorylation by CycD-Cdk4 removes the pRb/p107/p130 proteins [139].

E2F1-3 (activator factor) can act freely and transcribe CycA, E gene. p130 located at 16q12.2-13 region which is found to be deleted in several cancers [140]. Mutational gain of chromosomal regions 6p22, 1q31, 2p24-25 and 13q32-q34 or loss of 16q are observed in some cases [141,142]. Deregulation of cyclins (cyclin D/Cdk4 or cyclin E/Cdk2) activity or over-expression of some transcription factors (Myc) inactivates p107 and p130 in human tumors.

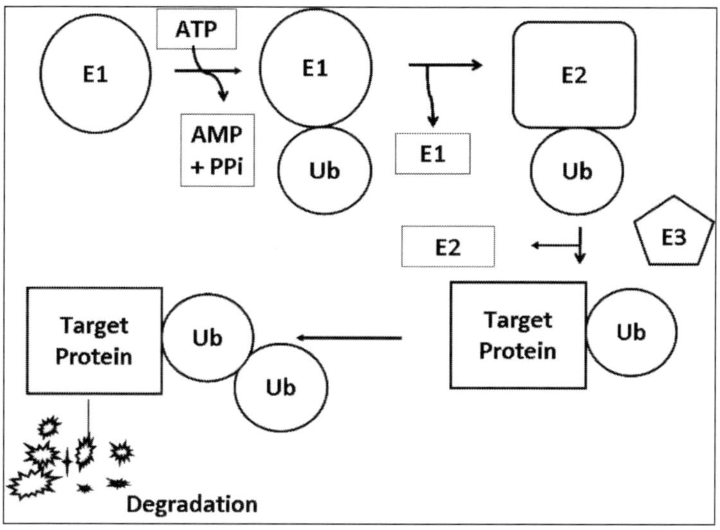

Figure 3. Ubiquitin mediated proteasomal degradation.

5. Ubiquitination, Cell Cycle and Brain Cancer: (UPS Pathway)

Ubiquitination process is regulating the degradation part of cell cycle machinery. Ubiquitin is 8kD in molecular mass that binds to target protein by a reversible mechanism (Figure 3). The binding process needs three distinct enzymes. Ubiquitin gets activated by Ubiquitin activating enzyme (E1) in ATP dependent manner where no transfer of a phosphate group occurs due to the hydrolysis of ATP into AMP and PPi. The E1 enzyme is replaced by

Ubiquitin activating enzyme (E1) in ATP dependent manner where no transfer of a phosphate group occurs due to the hydrolysis of ATP into AMP and PPi. The E1 enzyme is replaced by another enzyme, Ubiquitin conjugating enzyme (E2) and finally Ubiquitin ligase (E3) attaches to the lysine residues of the protein and creating polyubiquitin chain.

There are several other Ubiquitin related proteins that have crucial roles in ubiquitination like small Ubiquitin related modifier (SUMO), NEDD8, ISG15. There are three types of Ubiquitin ligases present 1) N-end rule ubiquitin ligase, 2) HECT type ubiquitin ligase, 3) RING motif containing ubiquitin ligase. Bellow in the table several ubiquitin E3 ligase with their functions have been cited:

Table 9. Cell cycle regulator and ubiquitin E3 ligase

NAME	SUBSTRATE	FUNCTION	BRAIN CANCER	REFERENCES
hBre1 Ubiquitin ligase	Ebp1 tumor suppressor protein (interact with E2F1)	Increase Ebp1 ubiquitation	Primary gliomas	[143]
NARF Ubiquitin ligase	TCF/LEF transcription factors.	Could act as a negative regulator of Wnt/beta-catenin pathways or a positive regulator in colony formation in RING dependent manner.	Glioblastoma multiforme	[144, 145]
E2C/UbcH10 Ubiquitin conjugating enzyme	Cyclin B	Probable proliferation marker. Acts with APC (E3) via N terminal extension. It may induce apoptosis via p53, Bax, G2/M arrest of a cell cycle.	Astrocytoma Glioma	[146-149]
Huwe1 E3 ligase	Histones, Mcl1, p53, c-Myc, cdc6, N-Myc	Suppression of N-DLL3-Myc cascade	Glioblastoma multiforme	[150]
hDM2	p53	Poly ubiquitination of p53		
HAUSP	p53 or Mdm2	stabilizes or destabilize p53	No Report	[151, 152]
SCF-Skp2	p27kip1	Ubiquitination of p27	Glioma	[153]
ISG15 E3 liga		Acts with UBP43 (Usp18) E2 increases in cancer	Glioblastoma	[154]
Nedd4 E3 liagse	ENaC, pTEN, IGF-1R, p63, VEGF-R2, Cbl-b, EPS15, Hrs	Blocking of Akt is extremely studied. FoxM1B mediated regulation.	Astrocytoma	[155]
pVHL E3	HIF-1α	Recognize prolyl hydroxylase mediated proline modification on HIF-1	Hemangioblastoma	[156]

6. Heat Shock Protein and Cell Cycle

The function of molecular chaperones ensures that the nascent polypeptide fold correctly and thcrcby lowering the chance of protein aggregation.

Table 10. HSPs in brain tumor

Cancer type	Role of HSPs	References
Glioblsatoma	HSP90α induces ATP and Ca^{2+} release, helping in cell migration	[158]
Low grade glioma	Dwnregulation of HSP27	[159]
Medulloblastoma	High expression of HSP27,HSP70 and HSP90	[160]

Table 11. Role of HSPs in cell cycle machinery

Cell cycle components	Effect	References
Cdk1	HSP90 helps Cdk-1 in substrate interaction, HSP-Cdk relation is Wee mediated	[164-167]
Cdk2	Client of HSP90 via helix αC	[168]
Cdk4	Cdc37 mediated interaction.	[169]
Cdk6	Client of HSP90	[170]
Myt	DAF21/Hsp90 is needed for proper folding in oogenesis	[171]
Wee	DAF21/Hsp90 is needed for proper Wee-1.3 stability	[171]
Plk-1	Metaphase -anaphase transition regulator	[172]
Aurora B	Client of HSP90 but interaction is not mandatory	[173]
Cdk-1	Cln3-Cdk1 retention at the ER by interaction with the HSP70-related chaperones Ssa1 and Ssa2	[174]
Cdk2	HSP70-2 mediated modulation in mice spermatocytes	[175]
Cdk4	HSP90 client protein (Cdk4) inhibitor induces HSP70 accumulation. Proteosome dependent degradation	[176]
Cdk6	HSP90 client protein (Cdk6) inhibitor induces HSP70 accumulation.	[170]
Survivin	Level decreases in HSP60 knockdown (not cytosolic fraction)	[177]

The main role of chaperone in cell machinery is to give proper folding of nascent protein, which has a conformational alteration upon physiological and pharmacological stress. They were first identified in *Drosophila* in early seventies in response to heat [157] and hence called as Heat Shock Proteins (HSPs).

The nomenclatures of HSPs were given based on their molecular weight. Several publications have highlighted the involvement of HSPs in cell cycle regulation. The levels of chaperones are increased in proliferating cells [161] as well as differentiated cells [162]. A striking similarity is observed between cdk-promoter and Hsp-promoter, which signify the relationship [163].

7. How do HSPs Control the Cell Cycle Machinery?

There are several ways by which HSPs are influencing the cell cycle regulation. These are (a) Upregulation of mitogens, serum, and growth stimulator [178]. (b) Link between DNA synthesis and HSP are well established [179, 180] (c) p53 regulates the transition between G1/S hence regulation of p53 is highly controlled and p53 down regulates several immediate early genes and Hsp70 gene while the oncoproteins are found to stimulate Hsp70 promoters [179] Hsp90, Hsp70 and other co chaperones stabilize and localize p53. Mutated p53 was found to be associated with Hsc70, Hsp90 [181] and Hsp40 [182].

HSF3/c-Myb association is necessary for HSP expression, which is severely disrupted by binding of the p53 to HSF3, resulting in the attenuation of Hsp70 expression. STAT1 interacts with p53 and p73 to induce apoptotic signal [183, 184]. (d) association with CDKs, (e) induction of retinoblastoma proteins, de-phosphorylated form of retinoblastoma protein inhibits cell to progress to S phase, Hsp70 has been found to be directly involved [185] in this phenomenon. (f) interaction with other proteins, p27, as already discussed as a CDK inhibitor where p27 knockout cell culture showed remarkable upregulation of Hsp70 [186]. Interaction of Hsc73 and p27 is also found to be regulated by ATP [187].

As the cancer cells produce some abnormal proteins, they use the chaperone to protect these proteins. Thereby targeting HSPs in anti-cancer therapy is currently being studied.

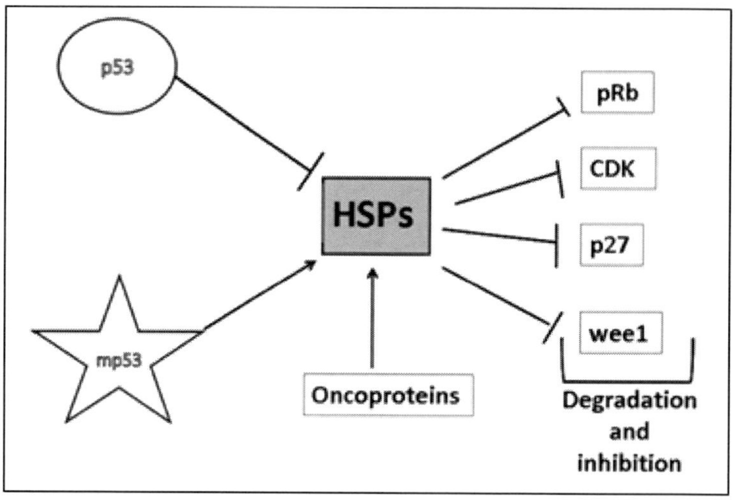

Figure 4. Role of p53 and its mutated form.

8. Therapeutic Applications

Therapeutic approach for the neurooncogeneic conditions has improved to a great extent over the years. Commonly used therapies include surgery, radiation and palliative cytotoxics, which have side- effects along with fewer efficacies. Over the years, the therapeutic applications for treating brain tumors have improved a lot with the effective cure and reduced toxicity.

The glioma patients are subject to multiple treatments immediately after the histological diagnosis. Initial treatment for the tumors is surgery followed by radiation therapy and chemotherapy [188].

8.1. Cell Cycle Therapeutics and Checkpoint Control in Cancer

The cell cycle machinery involves the activity of cyclins, CDKs and CDK inhibitors. CDKs are the most important components which initiate the cell cycle followed by progression and completion, which helps the cells to proliferate.

The malignant cells have to be controlled from proliferating so as to control or prevent the growth of tumor [189]. Specific inhibitors can be developed to inhibit the activity of CDK during the G1 phase. Cdk inhibitors like flavopiridol and UCN-01 [190, 191] are some of the pharmacological inhibitors, being used as therapeutic targets. Replacement of ectopic Cdk inhibitors using adenovirus causes growth arrest [192, 193]. Other treatments also include targeted molecular therapies and anti angiogenic therapies.

8.2. Chaperone and Ubiquitin Proteasome System Mediated Rescue

Chaperone activity is needed in brain tumor like in other cells. Chaperone system is very much needed in the malignant cells as the increased rate of cell division induces the chaperone system to produce the functional proteins inside tumor cells [194]. This leads to upregulation of chaperone protein in tumor cells where they are used as a therapeutic target against malignant tumors, finally leading to death of the tumor [195, 196]. Therapeutic role played by a ubiquitin proteasome system involves the proteasome inhibitors which have anticancer properties [197]. The proven proteasome inhibitor bortezomib (PS-341) has antineoplastic action along with induction of aggresome formation, which leads to apoptosis in multiple myeloma [198]. Some pharmacologic inhibitors like MI-63, an inhibitor of the HDM2 E3 ubiquitin ligase which targets p53, and E3 ligases such as SCF and APC/C has antiproliferative and antineoplastic activity proven in animal models [199]. This anti UPS drugs are effective therapeutic targets against the cancers of brain.

REFERENCES

[1] Hoshino, T; Knebel, KD; Rosenblum, ML; Dougherty, DV; Wilson, CB. Clonogenicity of multiple populations of human glioma cells in vitro sorted by DNA content. *Cancer*, 1982 September 1 50, 997-1002.

[2] Merrell, RT; Eudocia, C Q and Patrick, Y W. Advances in Treatment Options for High-grade Glioma – Current Status and Future Perspectives. European Neurological Review, 2010 5 (1), 49–55.

[3] Behin, A; Hoang-Xuan, K; Carpentier, AF; Delattre, JY. Primary brain tumors in adults. *Lancet,* 2003 361, 323–31.

[4] Louis, DN; Ohgaki, H; Wiestler, OD; Cavenee, WK; Burger, PC; Jouvet, A; Scheithauer, BW; Kleihues, P. The 2007 WHO classification of tumors of the central nervous system. *Acta Neuropathol.*, 2007 Aug 114(2), 97-109.

[5] Wen, PY; Kesari, S. Malignant gliomas in adults. *N. Engl. J. Med.*, 2008 July 21 359 (5), 492–507.

[6] Pines, J. Cyclins and cyclin-dependent kinases: A biochemical view. *Biochem. J.,* 1995 June 15 308 (3), 697-711.

[7] Malumbres, M. and Barbacid, M. Mammalian cyclin-dependent kinases. *Trends Biochem. Sci.*, 2005 30, 630–641.

[8] King, RW; Jackson, PK; Kirschner, MW. Mitosis in transition. *Cell,* 1994 November 18 79(4), 563-571.

[9] Ohtsubo, M; Theodoras, AM; Schumacher, J; Roberts, JM; Pagano, M. WHuman cyclin E, a nuclear protein essential for the G1-to-S phase transition. *Mol. Cell Biol.*, 1995 May 15(5), 2612-24.

[10] Resnitzky, D; Gossen, M; Bujard, H; Reed, SI. Acceleration of the G1/S phase transition by expression of cyclins D1 and E with an inducible system. *Mol. Cell Biol.* 1994 March 14(3), 1669-79.

[11] Malumbres, M; Barbacid, M. To cycle or not to cycle: a critical decision in cancer. Nat. Rev. Cancer. 2001 December 1(3), 222-31.

[12] Morgan, DO. "Cyclin-Dependent Kinase: Engines, Clocks, and Microprocessors." *Annual Review of Cell and Developmental Biology*, 1997 13, 261-291.

[13] Morgan, DO. The Cell Cycle: Principles of Control. *London New Science Press*, 20071st ed.

[14] Nebreda, AR. CDK activation by non-cyclin proteins. *Current Opinion in Cell Biology*, 2006 Apr 18(2), 192-198.

[15] Loyer, P; Trembley, J; Katona, R; Kidd, V; Lahti, J. Role of CDK/cyclin complexes in transcription and RNA splicing. *Cell Signal.*, 2005 Sep 17 (9), 1033–51.

[16] Cicero, S and Herrup, K. Cyclin-Dependent Kinase 5 Is Essential for Neuronal Cell Cycle Arrest and Differentiation. *The Journal of Neuroscience*, 2005 Oct 19 25(42), 9658-9668.

[17] Massague, J. G1 cell-cycle control and cancer. *Nature*, 2004 November 18 432(7015), 298-306.

[18] Kastan, MB; Bartek, J. Cell-cycle checkpoints and cancer. *Nature,* 2004 November 18 432(7015), 316-323.

[19] Malumbres, M; Barbacid, M. Cell cycle, CDKs and cancer: a changing paradigm. *Nat. Rev. Cancer*. 2009 March 9(3), 153-66.

[20] Elledge, SJ; Harper, JW. CDK inhibitors: On the threshold of checkpoints and development. *Curr. Opin. Cell Biol*, 1994 December 6(6), 847-852.

[21] Hunter, T; Pines, J. Cyclins and cancer: II-Cyclin D and CDK inhibitors come of age. *Cell*, 1994 November 18 79(4), 573-582.

[22] Denicourt,, C and Dowdy, SF. Cip/Kip proteins: more than just CDKs inhibitors. *Genes Dev.*, 2004 Apr 15 18(8), 851-5.

[23] Olivier, C. New roles for p21 and p27 cell-cycle inhibitors: a function for each cell compartment? *Trends in Cell Biology*, 2003 Feb 13(2), 65-70.

[24] Matsuoka, S; Edwards, MC; Bai, C; Parker, S; Zhang, P; Baldini, A; Harper, JW; Elledge, SJ. p57KIP2, a structurally distinct member of the p21CIP1 Cdk inhibitor family, is a candidate tumor suppressor gene. *Genes and Development*, 1995 March 19 9(6), 650–62.

[25] Park, KH; Lee, J; Yoo, CG; Kim YW, Han SK, Shim YS, Kim SK, Wang KC, Cho BK, Lee CT. Application of p27 gene therapy for human malignant glioma potentiated by using mutant p27. *J. Neurosurg*., 2004 September 101(3), 505-10.

[26] Brotherton, DH; Dhanaraj, V; Wick, S; Brizuela, L; Domaille, PJ; Volyanik, E; Xu, X; Parisini, E; Smith, BO; Archer, SJ;, Serrano, M; Brenner, SL; Blundell, TL, Laue, ED. Crystal structure of the complex of the cyclin D-dependent kinase Cdk6 bound to the cell-cycle inhibitor p19INK4d. *Nature*, 1998 September 17 395, 244–250.

[27] Ikeda, Y; Matsunaga, Y; Takiguchi, M; Ikeda, MA. Expression of cyclin E in postmitotic neurons during development and in the adult mouse brain. *Gene Expression Patterns*, 2011 Jan-Feb 11(1-2), 64-71.

[28] Sumrejkanchanakij, P; Tamamori-Adachi, M; Matsunaga, Y; Eto, K; Ikeda, MA Role of cyclin D1 cytoplasmic sequestration in the survival of postmitotic neurons. *Oncogene*. 2003 Nov 27, volume 22(54):8723-30.

[29] Yuan, Z; Becker, EB; Merlo, P; Yamada, T; DiBacco, S; Konishi, Y; Schaefer, EM; Bonni, A. Activation of FOXO1 by Cdk1 in Cycling Cells and Postmitotic Neurons. *Science Reports*, 2008 March 21 319(5870), 1665-8.

[30] Brinkkoetter, PT; Olivier, P; Wu, JS; Henderson, S; Krofft, RD; Pippin, JW; Hockenbery, D; Roberts, JM; Shankland, SJ. Cyclin I activates Cdk5 and regulates expression of Bcl-2 and Bcl-XL in postmitotic mouse cells. *J. Clin. Invest*., 2009 October 1 119(10), 3089–3101.

[31] Tomashevski, A; Webster, DR; Grammas, P; Gorospe, M; Kruman, II. Cyclin C-dependent cell cycle entry is required for activation of nonhomologous end joining DNA repair in postmitotic neurons. *Cell Death Differ*., 2010 July 17(7), 1189–1198.

[32] Augustinack, JC; Sanders, JL; Tsai, LH. and Hyman, BT. Colocalization and fluorescence resonance energy transfer between cdk5 and AT8 suggests a close association in pre-neurofibrillary tangles and neurofibrillary tangles. *J. Neuropathol. Exp. Neurol*., 2002 June 61, 557–564.

[33] Santra, MK; Wajapeyee, N; Green, MR. F-box protein FBXO31 mediates cyclin D1 degradation to induce G1 arrest after DNA damage. *Nature,* 2009 Jun 4 459(7247), 722-5.

[34] Hashiguchi, M.; Saito, T.; Hisanaga, S. and Hashiguchi, T. Truncation of CDK5 activator p35 induces intensive phosphorylation of Ser202/Thr205 of human tau. *J. Biol. Chem*., 2002 Nov 277, 44525–44530.

[35] Lopes, JP; Oliveira, CR; Agostinho, P. Role of cyclin-dependent kinase 5 in the neurodegenerative process triggered by amyloid-Beta and prion peptides: implications for Alzheimer's disease and prion-related encephalopathies. *Cell Mol. Neurobiol*., 2007 Nov 27(7), 943-57.

[36] Ino, H; Chiba, T. Cyclin-dependent kinase 4 and cyclin D1 are required for excitotoxin-induced neuronal cell death in vivo. *J Neurosci*., 2001 August 15 21(16), 6086-94.

[37] Hoozemans, J. M. ; Brückner, M. K. ; Rozemuller, A. M.; Veerhuis, R ; Eikelenboom, P ; Arendt, T. Cyclin D1 and Cyclin E Are Co-Localized with Cyclo-Oxygenase 2 (COX-2) in Pyramidal Neurons in Alzheimer Disease Temporal Cortex. *Journal of Neuropathology and Experimental Neurology*, 2002 August 61(8), 678–688.

[38] Nagy, Z; Esiri, M. M; Cato, A.M. and Smith, A. D. Cell cycle markers in the hippocampus in Alzheimer's disease. *Acta Neuropathologica*, 1997 Jul 94(1), 6-15.

[39] Ueberham, U; Hessel, A; Arendt T. Cyclin C expression is involved in the pathogenesis of Alzheimer's disease. *Neurobiol. Aging*, 2003 May-Jun 24(3), 427-35.

[40] Alama, A; Barbieri, F; Spaziante, R; Bruzzo, C; Dadati, P; Dorcaratto, A; Ravetti, JL. Significance of cyclin D1 expression in meningiomas: a preliminary study. *J. Clin. Neurosci.*, 2007 Apr 14(4), 355-8.

[41] Denicourt, C; Legault, P; McNabb, FA; Rassart, E. Human and mouse cyclin D2 splice variants: transforming activity and subcellular localization. *Oncogene.* 2008 Feb 21 27(9), 1253-62.

[42] Keyomarsi, K and Pardee, AB. Redundant cyclin overexpression and gene amplification in breast cancer cells. *Proc. Natl. Acad. Sci. USA,* 1993 90, 1112-1116.

[43] Keyomarsi, K; O'Leary, N; Molnar, G; Lees, E; Fingert, HJ; Pardee, AB. Cyclin E: A potential prognostic marker for breast cancer. *Cancer Res.,* 1994 54,380-385.

[44] Park, SH; Suh, YL. Expression of cyclin A and topoisomerase IIalpha of oligodendrogliomas is correlated with tumor grade, MIB-1 labelling index and survival. *Histopathology,* 2003 Apr 42(4), 395-402.

[45] Janss, AJ; Maity, A; Tang, CB; Muschel, RJ; McKenna, WG; Sutton, L; Phillips, PC. Decreased cyclin B1 expression contributes to G2 delay in human brain tumor cells afte treatment with camptothecin. *Neuro-oncology*, 2001 Jan 3(1), 11-21.

[46] Chen, H; Qiang, H; Jun, D; De-Zhong, Z; Ai-Dong, W and Qing, L. Overexpression of CDC2/CyclinB1 in gliomas, and CDC2 depletion inhibits proliferation of human glioma cells in vitro and in vivo. *BMC Cancer*, 2008 Jan 29 8, 29.

[47] Holtkamp, N; Afanasieva, A; Elstner, A; Landeghem, FK; Könneker, M; Kuhn, SA; Kettenmann, H; Deimling, A. Brain slice invasion model reveals genes differentially regulated in glioma invasion. *Biochem. Biophys. Res. Commun.*, 2005 Nov 4 336(4), 1227-33.

[48] Saab, R; Carlos, RG; Kelly, M; Jerold, ER; Shannon, HB; Joseph, DK; Catherine, B; Geoffrey, N; Kathleen, JH and Stephen, XS. p18Ink4c and p53 Act as Tumor Suppressors in Cyclin D1–Driven Primitive Neuroectodermal Tumor. *Cancer Res.*, 2009 January 15 69, 440.

[49] Ahmad, F; Malin, JA, Jaafar, H. Expressions and localization of cyclin D1 and cyclin-dependent kinase inhibitor p27Kip1 in gliomas and meningiomas. *Biomedical Research* 2009 20 (2), 109-114.

[50] Smith, ME; Cimica, V; Chinni, S; Jana, S; Koba, W; Yang, Z; Fine, E; Zagzag, D; Montagna, C; Kalpana, GV. Therapeutically targeting cyclin D1 in primary tumors arising from loss of Ini1. *Proc. Natl. Acad. Sci. USA*.2011 Jan 4, volume 108(1):319-24.

[51] Pogoriler, J; Millen, K; Utset, M; Du, W. Loss of cyclin D1 impairs cerebellar development and suppresses medulloblastoma formation. *Development.* 2006.133(19):3929-37.

[52] Kheirollahi, M; Mehr-Azin, M; Kamalian, N; Mehdipour, P. Expression of cyclin D2, P53, Rb and ATM cell cycle genes in brain tumors. *Med. Oncol.* 2011 March, volume 28(1):7-14.

[53] Ren, Y; Munyer, P; Kepes, J; Cansler, M; Collins, J; Gayed, B; Tawfik O. Comparison of cyclin A and MIB-1 expression in astrocytic tumors using image-based cell analysis system. *Appl. Immunohistochem Mol. Morphol.*, 2001 Mar 9(1), 14-8.

[54] Hara, A; Saegusa, M; Ichinoe, M; Okayasu, I. Diagnostic and prognostic significance of cyclin A expression in low-grade astrocytomas: comparison with astrogliosis and high-grade tumors. *J. Clin. Pathol.,* 2008 March 61, 287-292.

[55] Korshunov, A; Shishkina, L; Golanov, A. DNA topoisomerase II-alpha and cyclin A immunoexpression in meningiomas and its prognostic significance: an analysis of 263 cases. *Arch. Pathol. Lab. Med.,* 2002 Sep 12 6(9), 1079-86.

[56] Roussel, GA; Bilodeau, S; Vallette, S; Berthelet, F; Lacroix, A; Figarella, BD; Brue, T; Drouin, J. Cooperation between cyclin E and p27 (Kip1) in pituitary tumorigenesis. *Mol. Endocrinol.*, 2010 Sep 24(9), 1835-45.

[57] Aleem, E; Kiyokawa, H; Kaldis, P Cdc2-cyclin E complexes regulate the G1/S phase transition. *Nat. Cell Biol.*, 2005 Aug 7(8), 831-6.

[58] Jeon, HM; Jin, X; Lee, JS; Oh, SY; Sohn, YW; Park, HJ; Joo, KM; Park, WY; Nam, DH; DePinho, RA; Chin, L; Kim, H Inhibitor of differentiation 4 drives brain tumor-initiating cell genesis through cyclin E and notch signaling. *Genes Dev.*, 2008 Aug 1 22 (15), 2028-33.

[59] Russo, AA; Tong, L; Lee, JO; Jeffrey, PD; Pavletich, NP. Structural basis for inhibition of the cyclin-dependent kinase Cdk6 by the tumor suppressor p16INK4a. *Nature,* 1998 September 395, 237–43.

[60] Tavera-Mendoza LE, Wang TT. p19INK4D and cell death, *Cell Cycle.* 2006 March 5(6), 596-8.

[61] Thullberg, M; Bartkova, J; Khan, S; Hansen, K; Ronnstrand, L; Lukas, J; Strauss, M; Bartek, J. Distinct versus redundant properties among members of the INK4 family of cyclin-depen- dent kinase inhibitors. *FEBS Letter*, 2000 March 24 470(2), 161-6.

[62] Zindy, F; van Deursen, J; Grosveld, G; Sherr, CJ; Roussel, MF. INK4d-deficient mice are fertile despite testicular atrophy. *Mol. Cell Biol*, 2000 Jan 20(1), 372-8.

[63] Bouvier-Labit, C; Civatte, M; Bartoli, C; Renaud, W; Pellissier, JF; Figarella-Branger, D. p16INK4a and p19INK4d mRNA expression in neuroglial tumors: Correlation with Ki67 proliferation index. *Neuropathol Appl Neurobiol*, 1999 October 25, 408-416.

[64] Ogino, A; Yoshino, A; Katayama, Y; Watanabe, T; Ota, T; Komine, C; Yokoyama, T; Fukushima, T. The p15 (INK4b)/p16 (INK4a)/RB1 pathway is frequently deregulated in human pituitary adenomas. *J. Neuropathol. Exp. Neuro*. 2005 May 64(5), 398-403.

[65] Simon, M; Park, TW; Köster, G; Mahlberg, R; Hackenbroch, M; Boström, J; Löning, T; Schramm, J. Alterations of INK4a(p16-p14ARF)/INK4b(p15) expression and telomerase activation in meningioma progression. *J. Neurooncol.*, 2001 December 55(3), 149-58.

[66] Solomon, DA; Kim, JS; Jenkins, S; Ressom, H; Huang, M; Coppa, N; Mabanta, L; Bigner, D; Yan, H; Jean, W; Waldman, T. Identification of p18INK4c as a tumor suppressor gene in glioblastoma multiforme. *Cancer Res.*, 2008 April 15 68(8):2564-9.

[67] Wiedemeyer, R; Brennan, C; Heffernan, TP; Xiao, Y; Mahoney, J; Protopopov, A; Zheng, H; Bignell, G; Furnari, F; Cavenee, WK; Hahn, WC; Ichimura, K; Collins, VP; Chu, GC; Stratton, MR; Ligon, KL; Futreal, PA; Chin, L. Feedback circuit among INK4 tumor suppressors constrains human glioblastoma development. *Cancer Cell.*, 2008 Apr 13(4), 355-64.

[68] Watanabe, T; Huang, H; Nakamura, M; Wischhusen, J; Weller, M; Kleihues, P; Ohgaki, H. Methylation of the p73 gene in gliomas. *Acta Neuropathol.*, 2002 October 104(4), 357-62.

[69] Watanabe, T; Katayama, Y; Yoshino, A; Yachi, K; Ohta, T; Ogino, A; Komine, C; Fukushima, T. Aberrant hypermethylation of p14ARF and O6-methylguanine-DNA methyltransferase genes in astrocytoma progression. *Brain Pathol.*, 2007 Jan 17(1), 5-10.

[70] Lindsey, JC; Lusher, ME; Anderton, JA; Bailey, S; Gilbertson, RJ; Pearson, AD; Ellison, DW; Clifford, SC. Identification of tumor-specific epigenetic events in medulloblastoma development by hypermethylation profiling. *Carcinogenesis* 2004 May 25(5), 661-8.

[71] Kuo, LT; Kuo, KT; Lee, MJ; Wei, CC; Scaravilli, F; Tsai, JC; Tseng, HM; Kuo, MF; Tu, YK. Correlation among pathology, genetic and epigenetic profiles, and clinical outcome in oligodendroglial tumors. *Int. J. Cancer.*, 2009 Jun 15 124(12), 2872-9.

[72] Hallor, KH; Staaf, J; Jönsson, G; Heidenblad, M; Vult von Steyern, F; Bauer, HC; Ijszenga, M; Hogendoorn, PC; Mandahl, N; Szuhai, K; Mertens, F. Frequent deletion of the CDKN2A locus in chordoma: analysis of chromosomal imbalances using array comparative genomic hybridisation. *Br. J. Cancer*, 2008 Jan 29 98(2), 434-42.

[73] Robinson, JP; VanBrocklin, MW; Guilbeault, AR; Signorelli, DL; Brandner, S; Holmen, SL. Activated BRAF induces gliomas in mice when combined with Ink4a/Arf loss or Akt activation. *Oncogene*, 2010 Jan 21 29(3), 335-44.

[74] Weber, RG; Hoischen, A; Ehrler, M; Zipper, P; Kaulich, K; Blaschke, B; Becker, AJ; Weber-Mangal, S; Jauch, A; Radlwimmer, B; Schramm, J; Wiestler, OD; Lichter, P; Reifenberger, G. Frequent loss of chromosome 9, homozygous CDKN2A/p14(ARF)/CDKN2B deletion and low TSC1 mRNA expression in pleomorphic xanthoastrocytomas. *Oncogene*, 2007 Feb 15 26(7), 1088-97.

[75] Guan, K; Jenkins, CW; Li, Y; Nichols, MA; Wu, X; O'Keefe, CL; Matera, AG; Xiong, Y. Growth suppression by p18, a p16INK4/MTS1 and p14INK4B/MTS2 related CDK6 inhibitor, correlates with wild-type pRb function. *Genes Dev*, 1994 December 15 8(24), 2939–2952.

[76] Zindy, F; Knoepfler, PS; Xie, S; Sherr, CJ; Eisenman, RN; Roussel, MF. N-Myc and the cyclin-dependent kinase inhibitors p18Ink4c and p27Kip1 coordinately regulate cerebellar development. *Proc. Natl. Acad. Sci. USA*, 2006 Aug 1 103(31), 11579–83.

[77] Ramsey, MR; Krishnamurthy, J; Pei, XH; Torrice, C; Lin, W; Carrasco, DR; Ligon, KL; Xiong, Y; Sharpless, NE. Expression of p16Ink4a compensates for p18Ink4c loss in cyclin-dependent kinase 4/6-dependent tumors and tissues. *Cancer Res.*, 2007 May 15 67(10), 4732-41.

[78] Krimpenfort, P; Ijpenberg, A; Song, JY; Van der Valk, M; Nawijn, M; Zevenhoven, J; Berns, A. p15Ink4b is a critical tumor suppressor in the absence of p16Ink4a. *Nature*, 2007 Aug 23 448(7156), 943-6.

[79] Sabah, M; Cummins, R; Leader, M; Kay, E. Loss of p16INK4A Expression Is Associated With Allelic Imbalance/Loss of heterozygosity of chromosome 9p21 in microdissected malignant peripheral nerve sheath tumors. *Applied Immunohistochemistry and Molecular Morphology*, 2006 March 14(1), 97-102.

[80] Carrano, AC; Eytan, E; Hershko, A; Pagano, M. SKP2 is required for ubiquitin-mediated degradation of the CDK inhibitor p27.*Nat. Cell Biol* 1999 1(4), 193-199.

[81] Kikuchi, T; Toyota, M; Itoh, F; Suzuki, H; Obata, T; Yamamoto, H; Kakiuchi, H; Kusano, M; Issa, JP; Tokino, T; Imai, K. Inactivation of p57KIP2 by regional promoter hypermethylation and histone deacetylation in human tumors. *Oncogene,* 2002 Apr 18 21(17), 2741-2749.

[82] Jia, J; Lin, M; Zhang, L; York, JP; Zhang, P; The Notch signaling pathway controls the size of the ocular lens by directly suppressing p57Kip2 expression. *Mol. Cell Biol.* 2007 October 27(20), 7236-7247.

[83] Alheim, K; Corness, J; Samuelsson, MK; Bladh, LG; Murata, T; Nilsson, T; Okret, S. Identification of a functional glucocorticoid response element in the promoter of the cyclin-dependent kinase inhibitor p57Kip2. *J. Mol. Endocrinol.* 2003 30, 359–68.

[84] Dauphinot, L; De Oliveira, C; Melot, T; Sevenet, N; Thomas, V; Weissman, BE; Delattre O. Analysis of the expression of cell cycle regulators in Ewing cell lines: EWS-FLI-1 modulates p57KIP2and c-Myc expression. *Oncogene* 2001 May 31 20(25), 3258–3265.

[85] Georgia, S; Soliz, R; Li, M; Zhang, P; Bhushan, A. p57 and Hes1 coordinate cell cycle exit with self-renewal of pancreatic progenitors. *Dev Biol*, 2006 October 1 298(1), 22–31.

[86] McKenzie, PP; Danks, MK; Kriwacki, RW; Harris, LC. P21Waf1/Cip1 dysfunction in neuroblastoma: a novel mechanism of attenuating G0-G1 cell cycle arrest. *Cancer Res.* 2003 Jul 1 63(13), 3840-4.

[87] Kamitani, H; Taniura, S; Watanabe, K; Sakamoto, M; Watanabe, T; Eling, T. Histone acetylation may suppress human glioma cell proliferation when p21WAF/Cip1 and gelsolin are induced. Neuro *Oncology*, 2002 April 4(2), 95–101.

[88] Schiffer, D; Cavalla, P; Fiano, V; Ghimenti, C; Piva, R. Inverse relationship between p27/Kip.1 and the F-box protein Skp2 in human astrocytic gliomas by immunohistochemistry and Western blot. *Neurosci Lett.* 2002 August 9, volume 328(2):125-8.

[89] Gillies, JK; Lorimer, IA. Regulation of p27Kip1 by miRNA 221/222 in Glioblastoma. *Cell Cycle*, 2007 August 15 6(16).

[90] Philipp-Staheli, J; Kim, KH; Liggitt, D; Gurley, KE; Longton, G; Kemp, CJ. Distinct roles for p53, p27Kip1, and p21Cip1 during tumor development. *Oncogene.* 2004 January 29, volume 23(4):905-13.

[91] Palani, M; Devan, S; Arunkumar, R; Vanisree, AJ. Frequency variations in the methylated pattern of p73/p21 genes and chromosomal aberrations correlating with different grades of glioma among south Indian population. *Med. Oncol.* 2010 Sep 16.

[92] Xu, QZ; Liu, F; Lu, DH; Yu, SZ; Yang, H. Expression of P21 WAF1/CIP1 in human astrocytomas in correlating with P53, MDM2, and cellular proliferation index. *Zhongguo Yi Xue Ke Xue Yuan Xue Ba.*, 2001 Aug 23(4), 341-5.

[93] Kirla, RM; Haapasalo, HK; Kalimo, H; Salminen, EK. Low expression of p27 indicates a poor prognosis in patients with high-grade astrocytomas. *Cancer*, 2003 February 1 97(3), 644-648.

[94] Hidaka, T; Hama, S; Shrestha, P; Saito, T; Kajiwara, Y; Yamasaki, F; Sugiyama, K; Kurisu, K. The combination of low cytoplasmic and high nuclear expression of p27 predicts a better prognosis in high-grade astrocytoma. *Anticancer Res.*, 2009 Feb 29(2), 597-603.

[95] Nabika, S; Kiya, K; Satoh, H; Mizoue, T; Kondo, H; Katagiri, M; Nishisaka, T; Sugiyama, K; Kurisu, K. Prognostic significance of expression patterns of EGFR family, p21 and p27 in high-grade astrocytoma. *Hiroshima J. Med. Sci.*, 2010 Dec 59(4), 65-70.

[96] Liu, J; Estes, ML; Drazba, JA; Liu, H; Prayson, R; Kondo, S; Jacobs, BS; Barnett, GH; Barna, BP. Anti-sense oligonucleotide of p21(waf1/cip1) prevents interleukin 4-mediated elevation of p27(kip1) in low grade astrocytoma cells. *Oncogene,* 2000 Feb 3 19(5), 661-9.

[97] Bhatia, B; Malik, A; Fernandez-L, A; Kenney, AM p27(Kip1), a double-edged sword in Shh-mediated medulloblastoma: Tumor accelerator and suppressor. *Cell Cycle*, 2010 Nov 9(21), 4307-14.

[98] Ayrault, O; Zindy, F; Rehg, J; Sherr, CJ; Roussel, MF. Two Tumor Suppressors, p27Kip1 and Patched-1, Collaborate to Prevent Medulloblastoma. *Mol. Cancer Res.*, 2009 January 7(1), 33-40.

[99] Sanchez-Diaz, PC; Burton, TL; Burns, SC; Hung, JY; Penalva, LO. Musashi1 modulates cell proliferation genes in the medulloblastoma cell line Daoy. *BMC Cancer*, 2008 Sep 30 8, 280.

[100] Seol, HJ; Jung, HW; Park, SH; Hwang, SK; Kim, DG; Paek, SH; Chung, YS; Sub, LC. Aggressive vestibular schwannomas showing postoperative rapid growth - their association with decreased p27 expression. *J. Neurooncol.*, 2005 Nov 75(2), 203-7.

[101] Avninder, S; Sharma, MC; Deb, P; Mehta, VS; Karak, AK; Mahapatra, AK; Sarkar, C. Gemistocytic astrocytomas: histomorphology, proliferative potential and genetic alterations--a study of 32 cases. *J. Neurooncol.*, 2006 Jun 78(2), 123-7.

[102] Begnami, MD; Rushing, EJ; Evangelista, R; Santi, M; Quezado, M. Evaluation of RB Gene and Cyclin-Dependent Kinase Inhibitors P21 and P27 in Pleomorphic Xantoastrocytoma. *Int. J. Surg. Pathol.*, 2006 Apr 14(2), 113-8.

[103] Shi, Y; Zou, M; Farid, NR; al-Sedairy, ST. Evidence of gene deletion of p21(WAF1/CIP1), a cyclin-dependent protein kinase inhibitor, in thyroid carcinomas. *Br. J. Cancer*. 1996 Nov 74(9), 1336-41.

[104] Toyoshima, H; Hunter, T. p27, a novel inhibitor of G1 cyclin-Cdk protein kinase activity, is related to p21. *Cell.* 1994 July 15, volume 78(1):67-74.

[105] Soo-Yeun Park, Dong Wook Lee1. Effects of p27 Overexpression on Head and Neck Squamous Cell Carcinoma Cell Lines. *The Korean J. Anat.* 2005 38(3), 227-234.

[106] Brandts, CH; Benoit Bilanges, B; Gregor Hare, G; McCormick, F; Stokoe, D. Phosphorylation-independent Stabilization of p27kip1 by the Phosphoinositide 3-Kinase Pathway in Glioblastoma Cells. *J. Biol. Chem.* 2005 Jan 21 280(3), 2012-2019.

[107] Baldassarre, G; Belletti, B; Bruni, P; Boccia, A; Trapasso, F; Pentimalli, F; Barone; MV, Chiappetta, G;, Vento, MT; Spiezia, S; Fusco, A; Viglietto, G. Overexpressed cyclin D3 contributes to retaining the growth inhibitor p27 in the cytoplasm of thyroid tumor cells. *J. Clin. Invest.,* 1999 October, 104:865–874.

[108] Blain, SW; Scher, HI; Cordon-Cardo, C; Koff, A. p27 as a target for cancer therapeutics. *Cancer Cell*, 2003 February 3,111–115.

[109] Rodier, G; Montagnoli, A; Di Marcotullio L; Coulombe, P; Draetta, GF; Pagano, M; Meloche, S. p27 cytoplasmic localization is regulated by phosphorylation on Ser10 and is not a prerequisite for its proteolysis. *EMBO Journal*, 2001 December 3 20(23), 6672-82.

[110] Grimmler, M; Wang, Y; Mund, T; Cilensek, Z; Keidel, EM; Waddell, MB; Jakel, H; Kullmann, M; Kriwacki, RW; Hengst, L. Cdk-inhibitory activity and stability of p27Kip1 are directly regulated by oncogenic tyrosine kinases. *Cell,* 2007 January 26 128(2), 269-280.

[111] Chu, I; Sun, J; Arnaout, A; Kahn, H; Hanna, W; Narod, S; Sun, P; Tan, CK; Hengst, L; Slingerland, J. p27 phosphorylation by Src regulates inhibition of cyclin E-Cdk2. *Cell* 2007 January 26 128(2), 281-294.

[112] Rabbani, F; Koppie, TM; Charytonowicz, E; Drobnjak, M; Bochner, BH; Cordon-Cardo, C. Prognostic significance of p27Kip1 expression in bladder cancer. *BJU Int.*, 2007 August 100(2), 259-63.

[113] Lacoste-Collin, L; Gomez-Brouchet, A; Escourrou, G; Delisle, MB; Levade, T; Uro-Coste E. Expression of p27(Kip1) in bladder cancers: immunohistochemical studyand prognostic value in a series of 95 cases. *Cancer Letter*, 2002 December 1 186(1), 115–20.

[114] Baldassarre, G; Belletti, B; Nicoloso, MS; Schiappacassi, M; Vecchione, A; Spessotto, P; Morrione, A; Canzonieri, V; Colombatti, A. p27 (Kip1)-stathmin interaction influences sarcoma cell migration and invasion. *Cancer Cell*, 2005 January 7, 51–63.

[115] Wu, FY; Wang, SE; Sanders, ME. et al. Reduction of cytosolic p27(Kip1) inhibits cancer cell motility, survival, and tumorigenicity. *Cancer Res.,* 2006 66, 2162–72.

[116] Besson, A; Gurian-West, M; Schmidt, A; Hall, A; Roberts, JM. p27Kip1 modulates cell migration through the regulation of RhoA activation. *Genes Dev.*, 2004 April 15 18, 862–876.

[117] Zagzag, D; Blanco, C; Friedlander, DR; Miller, DC; Newcomb, EW. Expression of p27KIP1 in human gliomas: relationship between tumor grade, proliferation index, and patient survival. *Hum. Pathol.* 2003 Jan 34(1):48-53.

[118] Gottschalk, A. R., Basila, D., Wong, M., Dean, N. M., Brandts, C. H., Stokoe, D., and Haas-Kogan, D. p27Kip1 is required for PTEN-induced G1 growth arrest. *Cancer Research* 2001 March 1 61(5), 2105–2111.

[119] Ding, Q; Grammer, JR; Nelson, MA; Guan, JL; Stewart, JE Jr; Gladson, CL. p27Kip1 and cyclin D1 are necessary for focal adhesion kinase regulation of cell cycle progression in glioblastoma cells propagated in vitro and in vivo in the scid mouse brain. *J. Biol. Chem.* 2005 Feb 25 280(8):6802-15.

[120] Shi Wei, Baibin, Wang Fangru, Liu Chongxiao, and Song Ying P27/Kip1 Expression in Gliomas and Its Clinical Significance. *The Chinese-German Journal of Clinical Oncology Dec.* 2005 4(6), 381–383.

[121] Sakai, K; Peraud, A; Mainprize, T; Nakayama, J; Tsugu, A; Hongo, K; Kobayashi, S; Rutka, JT. Inducible Expression of p57KIP2 Inhibits Glioma Cell Motility and Invasion. *J. Neurooncol.*, 2004 July 68(3), 217-223.

[122] Tsugu, A; Sakai, K; Dirks, PB; Jung, S; Weksberg, R; Fei, YL; Mondal, S; Ivanchuk, S; Ackerley, C; Hamel, PA; Rutka, JT. Expression of p57KIP2 potently blocks the growth of Human Astrocytomas and Induces Cell Senescence. *The American Journal of Pathology*, 2000 September 157(3), 919-932.

[123] Vlachos, P; Joseph, B. The Cdk inhibitor p57Kip2 controls LIM-kinase 1 activity and regulates actin cytoskeleton. *Oncogene*, 26 November 2009 28(47), 4175-4188.

[124] Levrero, M; De Laurenzi, V; Costanzo, A; Gong, J; Wang, JY; Melino, G. The p53/p63/p73 family of transcription factors: overlapping and distinct functions. *J Cell Sci.* 2000 May; volume 113 (Pt 10):1661-70.

[125] Ikawa, S; Nakagawara, A; Ikawa, Y. p53 family genes: structural comparison, expression and mutation. *Cell Death Differ.* 1999 December, volume 6(12):1154-61.

[126] Bourdon, JC; Fernandesm K; Murray-Zmijewskim F; Liu, G; Diot, A; Xirodimas, DP; Saville, MK; Lane, DP. p53 isoforms can regulate p53 transcriptional activity. *Genes Dev.* 2005 Sep 15, volume 19(18):2122-37.

[127] Wade, M; Wang, YV; Wahl, GM. The p53 orchestra: Mdm2 and Mdmx set the tone," *Trends Cell Biol.*, 2010 May 20(5), 299-309.

[128] Leng, RP; Lin, Y; Ma, W; Wu, H; Lemmers, B; Chung, S; Parant, JM; Lozano, G; Hakem, R; Benchimol, S. Pirh2, a p53-induced ubiquitin-protein ligase, promotes p53 degradation. *Cell,* 2003 March 21 112(6), 779-91.

[129] Dornan, D; Wertz, I; Shimizu, H; Arnott, D; Frantz, GD; Dowd, P; O'Rourke, K; Koeppen, H; Dixit, VM. The ubiquitin ligase COP1 is a critical negative regulator of p53. *Nature,* 2004 May 6 429(6987), 86-92.

[130] Finch, RA; Donoviel, DB; Potter, D; Shi, M; Fan, M; Freed, DD; Wang, C; Zambrowicz, BP; Ramirez-Solis, R; Sands, AT; Zhang, N. mdmx Is a negative regulator of p53 Activity in Vivo. *Cancer Research*, 2002 June 1 62, 3221–3225.

[131] Esser, C; Scheffner, M; Höhfeld, J. The Chaperone-associated Ubiquitin Ligase CHIP Is Able to Target p53 for Proteasomal Degradation. *The Journal of Biological Chemistry*, 2005 July 22 280(29), 27443–27448.

[132] Almeida, LO; Custódio, AC; Araújo, JJ; Rey, JA; J.R.W. Almeida, JR; Santos, MJ; Clara, CA; Casartelli, C. Mutational analysis of genes p14ARF, p15INK4b, p16INK4a, and PTEN in human nervous system tumors. *Genetics and Molecular Research.* May 27 2008 7(2), 451-459.

[133] Nozaki, M; Tada, M; Kobayashi, H; Zhang, CL; Sawamura, Y; Abe, H; Ishii, N; Van Meir, EG. Roles of the functional loss of p53 and other genes in astrocytoma tumorigenesis and progression. *Neuro Oncology*, 1999 April 1(2), 124–137.

[134] Tada, M; Iggo, RD; Waridel, F; Nozaki, M; Matsumoto, R; Sawamura, Y; Shinohe, Y; Ikeda, J; Abe, H. Reappraisal of p53 mutations in human malignant astrocytic neoplasms by p53 functional assay: Comparison with conventional structural analyses. *Molecular Carcinogenesis*, 1997 March 18(3), 171–176.

[135] Amatya, VJ; Naumann, U; Weller, M; Ohgaki, HTP53 promoter methylation in human gliomas. *Acta Neuropathol.* 2005 Aug 110 (2), 178-184.

[136] Alonso, ME; Bello, MJ; Gonzalez-Gomez, P; Lomas, J; Arjona, D; de Campos, JM; Kusak, ME; Sarasa, JL; Isla, A; Rey, JA. Mutation analysis of the p73 gene in nonastrocytic brain tumors. *British Journal of Cancer*, 2001July 20 85(2), 204–208.

[137] Kamiya, M; Nakazato, Y. The expression of p73, p21 and MDM2 proteins in gliomas. *J. Neurooncol.*, 2002 September 59(2), 143-149.

[138] Watanabe, T; Huang, H; Nakamura, M; Wischhusen, J; Weller, M; Kleihues, P; Ohgaki, H. Methylation of the p73 gene in gliomas. *Acta Neuropathol,* 2002 October 104(4), 357-62.

[139] Kohn, KW. (1999), Molecular Interaction Map of the Mammalian Cell Cycle Control and DNA Repair Systems. *Mol. Biol Cell.* 1999 Aug 10(8), 2703-34.

[140] Yeung, RS; Bell, DW; Testa, JR; Mayol, X; Baldi, A; Graña, X; Klinga-Levan, K; Knudson, AG; Giordano, A. The retinoblastoma-related gene, RB2, maps to human chromosome 16q12 and rat chromosome 19. *Oncogene.*1993 Dec 8(12), 3465-8.

[141] Potluri, VR; Helson, L; Ellsworth, RM; Reid, T; Gilbert, F. Chromosomal abnormalities in human retinoblastoma. A review. *Cancer.* 1986 August 1, volume 58(3):663-71.

[142] Mairal, A. Pinglier, E; Gilbert, E; Peter, M; Validire, P; Desjardins, L; Doz, F; Aurias, A; Couturier, J. Detection of chromosome imbalances in retinoblastoma by parallel karyotype and CGH analyses. *Genes Chromosomes Cancer.* 2000 Aug;28(4):370-9.

[143] Zhixue Liu, Sang-Muk Oh, Masashi Okada, Xia Liu, Dongmei Cheng, Junmin Peng, Daniel J. Brat, Shi-yong Sun, Wei Zhou, Wei Gu, and Keqiang Ye. *Human BRE1 Is an E3 Ubiquitin Ligase for Ebp1 Tumor Suppressor.* Vol. 20, Issue 3, 757-768, February 1, 2009.

[144] Yamada, M; Ohnishi, J; Ohkawara, B; Iemura, S; Satoh, K; Hyodo-Miura, J; Kawachi, K; Natsume T, Shibuya H. NARF, an Nemo-like Kinase (NLK)-associated Ring Finger Protein Regulates the Ubiquitylation and Degradation of T Cell Factor/Lymphoid Enhancer Factor (TCF/LEF). *J. Biol. Chem.*, 2006 July 28 281(30), 20749-60.

[145] Anderson, TW; Wright, C; Brooks, WS. The E3 Ubiquitin Ligase NARF Promotes Colony Formation in vitro and Exhibits Enhanced Expression Levels in Glioblastoma Multiforme in vivo. *American Journal of Undergraduate Research*, 2010 9, Nos. 2 and 3.

[146] Donato, G; Iofrida, G; Lavano, A; Volpentesta, G; Signorelli, F; Pallante, PL; Berlingieri, MT; Pierantoni, MG; Palmieri, D; Conforti, F; Maltese, L; Tucci, L; Amorosi, A; Fusco, A. Analysis of UbcH10 expression represents a useful tool for the diagnosis and therapy of astrocytic tumors. *Clin. Neuropathol.* 2008 Jul-Aug 27(4), 219-223.

[147] Jiang, L; Huang, CG; Lu, YC; Luo, C; Hu, GH; Liu, HM; Chen, JX; Han, HX. Expression of ubiquitin-conjugating enzyme E2C/UbcH10 in astrocytic tumors. *Brain Res.*, 2008 Mar 27 1201, 161-166.

[148] Summers, MK; Pan, B; Mukhyala, K; Jackson, PK. The Unique N Terminus of the UbcH10 E2 Enzyme Controls the Threshold for APC Activation and Enhances Checkpoint Regulation of the APC. *Mol. Cell.,* 2008 August 22 31(4), 544-56.

[149] Kobirumaki, F; Miyauchi, Y; Fukami, K; Tanaka, H. A Novel UbcH10-Binding Protein Facilitates the Ubiquitinylation of Cyclin B In Vitro. *J. Biochem.*, 2005 February 137 (2), 133-139.

[150] Zhao, XD; Arca, D; Lim, WK; Brahmachary, M; Carro, MS; Ludwig, T; Cardo, CC; Guillemot, F; Aldape, K; Califano, A; Iavarone, A; Lasorella, A. The N-Myc-DLL3 cascade is suppressed by the ubiquitin ligase Huwe1 to inhibit proliferation and promote neurogenesis in the developing brain. *Dev. Cell.*, 2009 Aug 17(2), 210-21.

[151] Masuya, D; Huang, C; Liu, D; Nakashima, ;, Yokomise, H; Ueno, M; Nakashima, N; Sumitomo, S. The HAUSP gene plays an important role in non-small cell lung carcinogenesis through p53-dependent pathways. *J. Pathol.*, 2006 Apr 208(5), 724-32.

[152] Cummins, JM; Vogelstein, B HAUSP is required for p53 destabilization. *Cell Cycle*, 2004 June 3(6), 689-692.

[153] Piva, R; Cancelli, I; Cavalla, P; Bortolotto, S; Dominguez, J; Draetta, GF; Schiffer, D. Proteasome-dependent degradation of p27/kip1 in gliomas. *J. Neuropathol. Exp. Neurol.*, 1999 July 58(7), 691-6.

[154] Desai, SD; Wood, LM; Tsai, YC; Hsieh, TS; Marks, JR; Scott, GL; Giovanella, BC; Liu, LF. ISG15 as a novel tumor biomarker for drug sensitivity. *Mol. Cancer Ther.* 2008 June 7(6), 1430-9.

[155] Dai, B; Pieper, RO; Li, D; Wei, P; Liu, M; Woo, SY; Aldape, KD; Sawaya, R; Xie, K; Huang, S. FoxM1B regulates NEDD4-1 expression, leading to cellular transformation and full malignant phenotype in immortalized human astrocytes. *Cancer Res.* 2010 Apr 1 70(7), 2951-2961.

[156] Butman, JA; Linehan, WM; Lonser, RR. Neurologic manifestations of von Hippel-Lindau disease. *JAMA*.Sep 17 2008 300(11), 1334-42.

[157] Tissières, A; Mitchell, HK; Tracy, UM. Protein synthesis in salivary glands of Drosophila melanogaster: relation to chromosome puffs. *J Mol Biol.* 1974 Apr 15;84(3):389-98.

[158] Thuringer, D; Hammann, A; Benikhlef, N; Fourmaux, E; Bouchot, A; Wettstein, G; Solary, E; Garrido, C. Transactivation of the epidermal growth factor receptor by heat shock protein 90 via Toll-like receptor 4 contributes to the migration of glioblastoma cells. *J. Biol. Chem.*, 2011 Feb 4 286(5), 3418-28.

[159] Shen, G; Liang, S; Xu, Z; Zhou, L; Xiao, S; Xia, X; Li, R; Liao, Y; You, C; Wei, Y. Downregulated expression of HSP27 in human low-grade glioma tissues discovered by a quantitative proteomic analysis. *Proteome Sci.*, 2010 Mar 26 8, 17.

[160] Hauser, P; Hanzély, Z; Jakab, Z; Oláh, L; Szabó, E; Jeney, A; Schuler, D; Fekete, G; Bognár, L; Garami, M. Expression and prognostic examination of heat shock proteins (HSP 27, HSP 70, and HSP 90) in medulloblastoma. *J. Pediatr. Hematol. Oncol.* 2006 Jul 28(7), 461-6.

[161] Helmbrecht, K; Rensing, L. Different constitutive heat shock protein 70expression during proliferation and differentiation of rat C6 glioma cells. *Neurochem. Res.* 1999 Oct 24(10), 1293-9.

[162] García-Bermejo, L; Vilaboa, NE; Pérez, C; de Blas, E; Calle, C; Aller, P. Modulation of HSP70 and HSP27 gene expression by the differentiation inducer sodium butyrate in U-937 human promonocytic leukemia cells. *Leuk Res.* 1995 Oct;19(10):713-8. PubMed PMID: 7500647.

[163] Pechan, PM. Heat shock proteins and cell proliferation. *FEBS Lett.* 1991 Mar 11;280 (1):1-4. Review.

[164] Muñoz, MJ; Jimenez, J. Genetic interactions between Hsp90 and the Cdc2 mitoticb machinery in the fission yeast Schizosaccharomyces pombe. *Mol. Gen. Genet.* 1999 Mar; 261(2):242-50.

[165] Stingl, L; Stühmer, T; Chatterjee, M; Jensen, MR; Flentje, M; Djuzenova, CS. Novel HSP90 inhibitors, NVP-AUY922 and NVP-BEP800, radiosensitise tumor cells through cell-cycle impairment, increased DNA damage and repair protraction. *British Journal of Cancer,* 2010 May 25 102, 1578-1591.

[166] Vassilev, LT; Tovar, C; Chen, S; Knezevic, D; Zhao, X; Sun, H; Heimbrook, DC; Chen, L. Selective small-molecule inhibitor reveals critical mitotic functions of human CDK1. *Proc. Natl. Acad. Sci. US*, 2006 Jul 11 103(28), 10660-5.

[167] Caldas-Lopes, E; Cerchietti, L; Ahn, JH; Clement, CC; Robles, AI; Rodina, A; Moulick, K; Taldone, T; Gozman, A; Guo, Y; Wu, N; de Stanchina, E; White, J; Gross, SS; Ma, Y; Varticovski, L; Melnick, A; Chiosis, G. Hsp90 inhibitor PU-H71, a multimodal inhibitor of malignancy, induces complete responses in triple-negative breast cancer models. *Proc. Natl. Acad. Sci. USA.*2009 May 19 106(20), 8368-73.

[168] Prince, T; Sun, L; Matts, RL. Cdk2: a genuine protein kinase client of Hsp90 and Cdc37. Biochemistry. 2005 Nov 22 44(46), 15287-95.

[169] Vaughan, CK; Gohlke, U; Sobott, F; Good, VM; Ali, MM; Prodromou, C; Robinson, CV; Saibil, HR; Pearl, LH. Structure of an Hsp90-Cdc37-Cdk4 Complex. *Molecular Cell*, 2006. September 1 23(5), 697-707.

[170] Peng, B; Xu, L; Cao, F; Wei, T; Yang, C; Uzan, G; Zhang, D. HSP90 inhibitor, celastrol, arrests human monocytic leukemia cell U937 at G0/G1 in thiol-containing agents reversible way. *Mol. Cancer.* 2010 Apr 16 9:79.

[171] Inoue, T; Hirata, K; Kuwana, Y; Fujita, M; Miwa, J; Roy, R; Yamaguchi, Y. Cell cycle control by daf-21/Hsp90 at the first meiotic prophase/metaphase boundary during oogenesis in Caenorhabditis elegans. *Dev. Growth Differ.* 2006 Jan 48(1), 25-32.

[172] de Cárcer G, do Carmo Avides M, Lallena MJ, Glover DM, González C. Requirement of Hsp90 for centrosomal function reflects its regulation of Polo kinase stability. *EMBO J.* 2001 Jun 1 20(11), 2878-84.

[173] Davies AE, Kaplan KB. Hsp90–Sgt1 and Skp1 target human Mis12 complexes to ensure efficient formation of kinetochore–microtubule binding sites. *Cell Biol.*2010 Apr 19 189(2), 261-74.

[174] Enserink, JM; Kolodner, RD. An overview of Cdk1-controlled targets and processes. *Cell Division*, 2010 May 13 5, 11.

[175] Zhu, D; Dix, DJ; Eddy, EM. HSP70-2 is required for CDC2 kinase activity in meiosis I of mouse spermatocytes. *Development.*, 1997 Aug 124(15), 3007-14.

[176] Zhang, T; Hamza, A; Cao, X; Wang, B; Yu, S; Zhan, CG; Sun, D. A novel Hsp90 inhibitor to disrupt Hsp90/Cdc37 complex against pancreatic cancer cells. *Mol. Cancer Ther.* 2008 Jan; 7(1):162-70.

[177] Ghosh, JC; Dohi, T; Kang, BH; Altieri, DC. Hsp60 Regulation of Tumor Cell Apoptosis. *The Journal of Biological Chemistry*, 2008 February 22 283(8), 5188–5194.

[178] Hansen, LK; Houchins, JP; O'Leary, JJ. Differential regulation of HSC70, HSP70, HSP90 alpha, and HSP90 beta mRNA expression by mitogen activation and heat shock in human lymphocytes. *Exp. Cell Res.* 1991 Feb; 192(2):587-96.

[179] Wu BJ, Morimoto RI. Transcription of the human hsp70 gene is induced by serum stimulation. *Proc. Natl. Acad. Sci. USA.* 1985 Sep 82(18), 6070-4.

[180] Milarski, KL; Morimoto, RI. Expression of human HSP70 during the synthetic phase of the cell cycle. *Proc. Natl. Acad. Sci. USA*. 1986 Dec 83 (24), 9517-21.

[181] Blagosklonny, MV; Toretsky, J; Bohen, S; Neckers, L. Mutant conformation of p53 translated in vitro or in vivo requires functional HSP90. *Proc. Natl. Acad. Sci. USA*. 1996 Aug 6; 93(16):8379-83.

[182] Sugito, K; Yamane, M; Hattori, H; Hayashi, Y; Tohnai, I; Ueda, M; Tsuchida, N; Ohtsuka, K. Interaction between hsp70 and hsp40, eukaryotic homologues of DnaK and DnaJ, in human cells expressing mutant-type p53. *FEBS Lett*. 1995 Jan 23;358(2):161-4.

[183] Townsend, PA; Scarabelli, TM; Davidson, SM; Knight, RA; Latchman, DS; Stephanou, A. STAT-1 interacts with p53 to enhance DNA damage-induced apoptosis. *Journal of Biological Chemistry*, 2004 February 13 279(7), 5811-20.

[184] Soond, SM; Carroll, C; Townsend, PA; Sayan, E; Melino, G; Behrmann, I; Knight, RA; Latchman, DS; Stephanou, A. STAT1 regulates p73-mediated Bax gene expression. *FEBS Lett*., 2007 March 20 581(6), 1217-26.

[185] Hatamoto Keisuke; Torigoe Toshihiko; Mano Yoshinori et al. Disruption of the interaction between retinoblastoma protein and 70 kD heat shock protein leads to growth acceleration of tumor cells. *Tumor Research* 2006, 41, 43-57.

[186] Jinyi Liu, Dongyun Zhang, Xiaoyi Mi, Qing Xia, Yonghui Yu, Zhenghong Zuo, Wei Guo, Xuewei Zhao, Jia Cao, Qing Yang, Angela Zhu, Wancai Yang, Xianglin Shi, Jingxia Li and Chuanshu Huang. p27 Suppresses Arsenite-induced Hsp27/Hsp70 Expression through Inhibiting JNK2/c-Jun- and HSF-1-dependent Pathways. *J. Biol. Chem.* 2010 285: 26058-26065.

[187] Nakamura, S. Tatuno, I; Noguchi, Y; Kitagawa, M; Kohn, LD; Saito, Y; Hirai, A.73-kDa heat shock cognate protein interacts directly with p27Kip1, a cyclin-dependent kinase inhibitor, during G1/S transition. *Biochem. Biophys. Res., Comm.* volume 257, 340, 1999.

[188] Lacroix, M; Abi-Said, D; Fourney, DR; Gokaslan, ZL; Shi, W; DeMonte, F; Lang, FF; McCutcheon, IE; Hassenbusch, SJ; Holland, E; Hess, K; Michael, C; Miller, D; Sawaya, R. A multivariate analysis of 416 patients with glioblastoma multiforme: prognosis, extent of resection, and survival. *J. Neurosurg*, 2001 August 95, 190–198.

[189] Shapiro, GI; Harper, JW. Anticancer drug targets: cell cycle and checkpoint control. *J. Clin. Invest*. 1999 Dec 104(12), 1645-53.

[190] Akiyama, T. et al. G1 phase accumulation induced by UCN-01 is associated with dephosphorylation of Rb and CDK2 proteins as well as induction of Cdk inhibitor p21/Cip1/WAF1/Sdi1 in p53-mutated human epider moid carcinoma A431 cells. *Cancer Res.* 1997 57, 1495–1501.

[191] Sedlacek, H.H., et al. Flavopiridol (L86 8275; NSC 649890), a new kinase inhibitor for tumor therapy. *Int. J. Oncol*. 1996 9, 1143–1168.

[192] Eastham, JA; Hall, SJ; Sehgal, I; Wang, J; Timme, TL; Yang, G; Connell-Crowley, L; Elledge, SJ; Zhang, WW; Harper, JW; et al. In vivo gene therapy with p53 or p21 adenovirus for prostate cancer. *Cancer Res.* 1995 Nov 15; 55(22):5151-5.

[193] Jin, X; Nguyen, D; Zhang, W.W; Kyritsis, A.P and Roth, J.A. Cell cycle arrest and inhibition of tumor cell proliferation the p16INK4 gene mediated by an adenovirus vector. *Cancer Res.* 1995 55, 3250–3253.

[194] Graner, MW and Bigner, DD. Chaperone proteins and brain tumors: Potential targets and possible therapeutics. *Neuro-Oncology* 2005 7, 260–277.

[195] Lee, A.S. The glucose-regulated proteins: Stress induction and clinical applications. *Trends Biochem. Sci.* 2001 26, 504–510.

[196] Mosser, DD; Morimoto, RI. Molecular chaperones and the stress of oncogenesis. *Oncogene*. 2004 Apr 12; 23(16):2907-18.

[197] Lehman, NL. The ubiquitin proteasome system in neuropathology. *Acta Neuropathol.* 2009 Sep; 118(3):329-47.

[198] Richardson, PG; Sonneveld, P; Schuster, M et al. Extended follow-up of a phase 3 trial in relapsed multiple myeloma: final time-to-event results of the APEX trial. *Blood* 2007 110, 3557–3560.

[199] VanderPorten, EC; Taverna, P; Hogan, JN; Ballinger, MD; Flanagan, WM; Fucini, RV. The Aurora kinase inhibitor SNS-314 shows broad therapeutic potential with chemotherapeutics and synergy with microtubule-targeted agents in a colon carcinoma model. *Mol. Cancer Ther.* 2009 Apr 8(4), 930-9.

In: Signaling, Gene Regulation and Cancer
Editors: Shree Ram Singh and Manoj K. Mishra
ISBN: 978-1-61942-088-5

Chapter 14

COMBINATORIAL EFFECT OF HEAT SHOCK PROTEINS AND UBIQUITIN PROTEASOME SYSTEM IN NEURO-ONCOLOGY

Pravir Kumar*[1,2,3,*]*, Rashmi K. Ambasta*[3]*,
***Aditi Sarkar*[3]*, Abhidha Kohli*[3] *and Kushi Anand*[3]**

[1]Centre for Medical Engineering, VIT, Vellore, India
[2]Adjunct Faculty, Department of Neurology, Tufts University School of Medicine, Boston, MA, US
[3]Functional Genomics and Cancer Biology Laboratory, Center for Medical Engineering, Vellore Institute of Technology, Vellore, TamilNadu, India

ABSTRACT

The term *neurooncology* deals with the development of tumor in various parts of the cranium. Heat shock proteins and ubiquitin proteasome systems play a major role in the molecular event occurring in the brain tumors or in any stress situation. Stress response is extremely conserved mechanism in the body that is used for cytoprotection upon physiological and pharmacological insults. Since tumor is a stress condition in a body so it is likely to have an upregulated expression of heat shock proteins to prevent the damage and aggregation of proteins at the protesome level. Little is known about the HSPs cellular dynamics and compartmental localization in brain tumor. Therefore, this disease should intensively be studied and found to have combinatorial action of HSPs and E3 ligases in tumors that are occurring in the central nervous system. Brain's tumors forms 1.4% of all types of cancers revealed by WHO statistical reports, and its mortality rate due to this disease is 2.3% among all cancers. The hallmark features in this disease are deregulation of cell cycle, apoptotic machinery and angiogenesis mechanisms. Tumor of the brain has been found to prevail in varied forms depending upon the part of the brain in which it spurts. Misfolded proteins in tumor tissues are either rectified by

* Corresponding Author Address: Pravir Kumar, PhD. Cancer biology and functional genomics lab, SBST, VIT. E-mail: pravirkumar@vit.ac.in; Pravir.Kumar@tufts.edu. Phone: +91-4162202592 (lab); +91-9003386752.

chaperones activity upto a certain threshold or follow a degradation pathway by proteolytic activity of the ubiquitin-proteasome system. In this chapter, we are highlighting an extensive explorative potential of molecular chaperones in combination with ubiquitin E3 ligase enzymes activities in brain tumors. We have also proposed the therapeutic application of these molecules for clinical amelioration in brain tumor.

1. INTRODUCTION

Brain is the most complex organ and hub of the nervous system that maintains the integrity and physiological homeostasis of our body in highly orchestrated fashion. In last two decades, extensive studies have been carried out by the scientific fraternity to better understand the etiology of brain tumor. However, the molecular mechanism behind tumor formation is well understood [1]. In cancer, the nomenclature has been made based on affected organ; therefore, brain tumor comes under the field of neurooncology. By definition, *"Neurooncology is referred as the branch of medical science dealing with the study of generation and development of tumor in central nervous system (CNS) and spine."*

Brain comprises of well differentiated cells that have been studied for different types of complications related to each part of its structure. Specialized cells of brain mainly known as neurons are the entities which process and transmit the electrical impulses and chemical signals in the entire body. Usually, the nerve cells in adult do not divide, i.e. once they are destroyed, cannot be regenerated. Neurons are the component of the nervous system and perform transmission of signals through synapse formation via the network of nerve cells. There are other cells required to support neurons. The cells of brain have been listed below in Table 1:

Table 1. Various nerve cells and their function

Cells of the brain	Functions	References
Nerve cells	Unit of the brain and it sends a signal from brain to the entire body and *vice versa*.	[2]
Glial cells	Glial cells consist of oligodendrocytes, microglia and astrocytes.	[3]
	Supportive cells to nerve cells.	
Oligodendrocyte	Oligodendrocytes are beads like structure wrapped around every neuron, and they form the myelin sheath.	[4, 5]
	Speed up the signal transduction process.	
	Abnormal functioning of these cells reduces the action potential by 30 times.	
Microglias	Immune cells of brain.	[6, 7]
	Destroys unhealthy or damaged neuron and any foreign particles such as bacteria or virus	
Astrocytes	Tumor prone cells possess star shaped structure.	[8-10]
	Present in between the neurons and act as supportive cells to nerve cells.	
	Each astrocyte interacts with several neurons and thousands of synapses.	
	These cells modify and process the information.	

The word tumor derived from Latin word, which means "*Swelling*" that is the resultant of abnormal cell proliferation and when it formed in any part of the cranium or in the central spinal canal, it is termed as brain tumor. The term neoplasm is used synonymously to the tumor that means abnormal mass of tissue. Tumors can be benign (non-cancerous) or malignant that migrate to other parts of the body due to a process called metastasis. Therefore, based on the origin of neoplasm, brain tumors have been divided into basically two types; primary tumor and secondary tumor. Primary tumors are those which originate in the cells of brain tissue itself. Most commonly they occur in the glial cells and termed as "gliomas." Gliomas can be of various types such as *astrocytoma, brain stem glioma, ependymoma, oligodendroglioma and medulloblastoma*. Other types of primary tumors originating from other than glial cells include *meningioma, schwannoma, craniopharyngioma, germ cell tumor* and *pineal region tumor*. On the other hand, secondary tumors are those which get metastasize from other organs of the body to the brain, e.g. cancerous cells from skin (malignant melanoma), kidney (hypernephroma) and colon (colon carcinoma) migrate to brain via blood flow.

1.1. Classification of (Bailey and Cushing) Brain Tumors

According to world health organization (WHO) tumors of the central nervous system are classified into four different classes viz. I, II, III and IV [11]. This pioneer work of tumor classification was carried out by Bailey and Cushing. The grades reflect the extent of clinical aggressiveness of tumors [12]. Level III and IV are considered as difficult to be cured by treatments like radiation or chemotherapy. This classification categorized tumors based on their resemblance with the cell type present in brain of developing embryo or adult. Based on the above criterion neuropathologists attempted to make a distinction among astrocytoma, oligodendroglioma and neuronal tumors.

Astrocytoma occurs mainly in glail cells. And they have been further divided into four grades of malignancy. *Oligodendroglioma* are the ones consisting of pure oligodendroglial cells or a mixed population of oligodendroglial cells and astrocytic cells. Neuronal tumors are exclusively seen in children, and it comprises *neurocytoma, ganglioglioma* supratentorial primitive neuroectodermal tumors and *medulloblastoma* (MB) [1].

The elements which were considered necessary to make the classification easy and rigid were the anatomical site of tumor, the tissue type, and the degree of malignancy [13]. However, there are certain limitations associated with central nervous system classification given by WHO. The first limitation is ambiguity in histopathological location of a tumor; second the tumors of a central nervous system are of varying morphologies and need to be identified by the particular area representing the characteristic type of tumor as a small area specifying the tumor is difficult to be accurately identified. Third, there are very limited antibodies are available against the peculiar antigens that are characteristics of specific tumor.

1.1.1. Types of Intracranial Neoplasms

The intracranial tumors include all the tumors occurring in brain, spinal cord and region surrounded them. They may be primary brain tumors or secondary brain tumors (Figure 1). There are around 120 entries in the classification of tumors given by WHO. The ones which are most common in children and in adults respectively are *pilocytic astrocytoma, ependymomas* and *medulloblastomas* and diffuse astrocytic tumors, (*astrocytoma*, *anaplastic astrocytoma* and *glioblastomas*) *oligodendrogliomas* and *meningiomas* [14].

Other than these, othertumor types occurring in different parts of the brain include pituitary tumors in pituitary gland, pineal tumors in pineal glands, tumor of nerves, including neuroma, schwannoma and neurofibroma, cystic tumors, tumors in skull base, i.e. chondroma, sarcoma, etc.

Table 2. Characteristic features of Neoplasm

Neoplasms	Characteristics	References
Pilocytic astrocytomas	Occurs in the cerebellum of children.	[14, 15]
	Histopathological observation shows that pilocytic astrocytomas have bipolar cellular areas with Rosenthal fibers and under the microscope, they appear 'hair like' structure.	
	Grow slowly and can be removed surgically.	
Ependymomas	Develops on the surface of ependyma.	[16, 17]
	Occur intracranially in children and in the spinal region in adults, most commonly in the fourth ventricle. They subdivided as-	
	Subependymoma-	
	Subtypes of ependyma that is an intraventricular and malignancy grade, I.	
	Myxopapillary ependymoma-	
	Malignancy grade II	
	Occur in the posterior fossa in the embryonic stages.	
Medulloblastoma (infratentorial PNET)	Malignancy grade IV.	[14, 18-20]
	Originate below the tentorium of the cerebellum.	
	Different types: Astrocytoma, anaplastic astrocytoma and Glioblastoma.	
Astrocyte tumors	Malignancy grade II, III and IV.	[21]
	Arise from the oligodendrocyte cells in the cerebrum.	
Oligo-dendrogliomas	Quite rare and grow slowly.	[22-24]
	Prevalent in middle age adult.	
	Malignancy grade II.	
	It occurs in the meninges.	
Meningiomas	Prevalent in women of age group 30 to 50 years.	[25]
	Benign in nature.	
	It develops from the arachnoidal cells of the meninges.	
Glioma	originates from glial cells	[26]

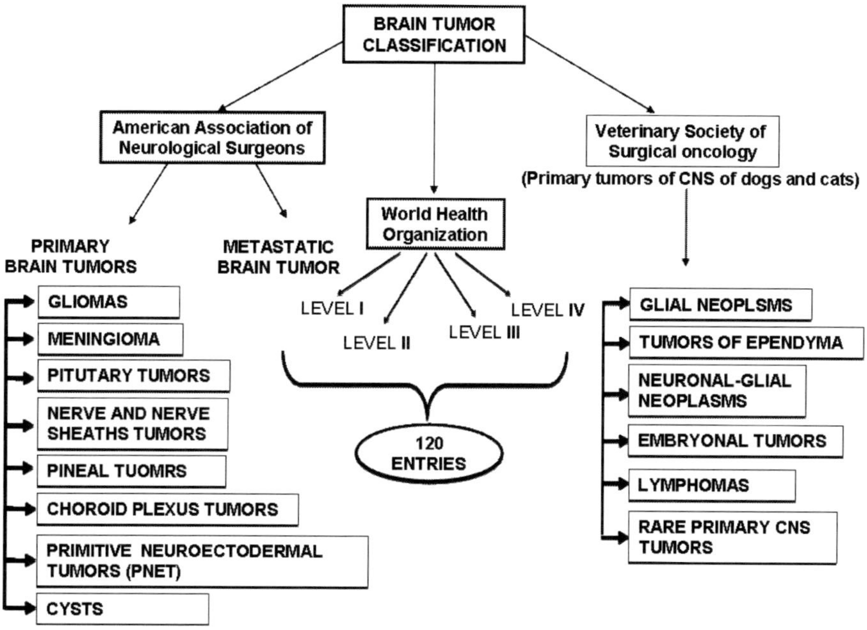

Figure 1. Classification of intracranial tumors.

1.1.2. Epidemiology of Brain Tumors and Possible Treatments

Statistical analysis reveals the prevalence of the fatal disease "brain tumor" in people of different age groups and sexes (Figure2). The cases of brain tumors have increased since past 30 to 50 years and 1.4% of all the cancers are brain cancers today and out of all cancer-related deaths, they form 2.3% [27].

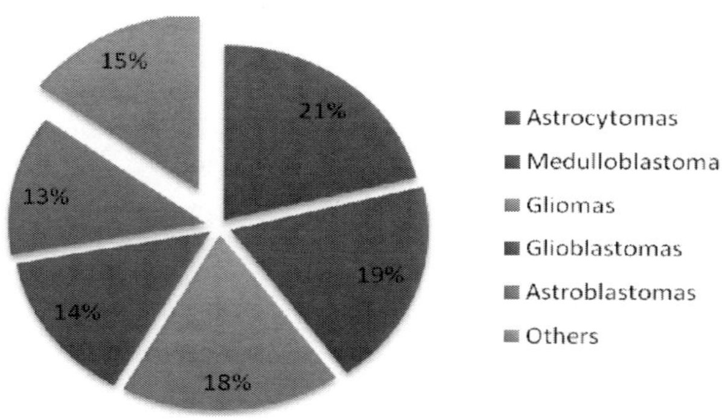

Figure 2. Prevalence of brain tumors.

2. Heat Shock Proteins in Neurooncology

2.1. Introduction to HSPs

Heat shock proteins are most conserved proteins found in the bacteria, yeast, plant and animal cells [28-31]. Heat shock proteins are cytoprotective in nature upon physiological and pharmacological stress situations. These stresses could be ischemia, hyperthermia, hypoxia, tumor, cancer, UV radiation, heavy-metal ions, mental stress, viral infection in the cells, etc. [28]. Cells pre-exposed with these conditions develop tolerance capacity for the above kinds of stresses [32-34] and maintaining the cellular integrity and homeostasis. HSPs are regulated by the activation and trimerization of transcription factor heat shock factors (HSFs). Cells that have undergone these kinds of stress situations and try to combat with an adverse conditions are termed as a stress response. They may be present in the cytoplasm of the cells or some are present in the endoplasmic reticulum and termed as ER stress markers.

2.2. Differential Expression of Heat Shock Proteins in Neurooncology

Heat shock proteins (HSPs) are cytoprotective, antiapoptotic proteins that may predict clinical prognosis in various types of cancer [32]. Here, we describe the role of heat shock proteins in neurooncology mainly overall prognosis for Gliomas, Astrocytomas, Ependymoma, Oligodendroglioma, Medulloblastoma and Meningiomas by differences in HSP expression. The following table explains the concise glimpse of HSPs role in the varied form of intra cranial neoplasm.

Table 3. Preferential expression of heat shock proteins in brain tumor

Brain tumors	Role of HSPs	References
Gliomas	High level of Hsp72 confers drug resistance.	[35, 36]
	Hsp90 helps in the tumor cell survival.	
	High expression of Hsp60 leads to the higher co-expression of proliferating cell nuclear antigen (PCNA).	
Astrocytomas	High level of Hsp27 expression correlates with tumor progression and its grades.	[36, 37]
	High expression of Hsp60 leads the progression of tumor cell survival by increasing cell surviving factors.	
Ependymoma	Significant high expression of Hsp27 and Hsp60 was observed in these kinds of tumors.	[36, 37]
	High level of Hsp27 expression correlates with tumor progression and its grades.	
Oligodendroglioma	Correlation between glial fibrillary acidic protein (GFAP) and Hsp70 have a protective mechanism in the brain tumor.	[38]
Medulloblastoma	Elevated level of Hsp60 leads to the co-expression of proliferating cell nuclear antigen (PCNA)-that is higher in these types of tumors.	[36]
Meningiomas	Hsp27 appears to be constitutively expressed in most meningiomas.	[36, 39]
	Hsp27 seems to participate in the formation of certain inclusion bodies in meningioma.	
	Hsp60 plays a role in tumor progression in indirectly activating cell cycle progressing marker.	

HSPs are known for its multifunctional roles in different signaling molecules that control various important pathways involved in cell cycle regulation [40], cell proliferation, apoptosis

[41, 42] and developmental processes. The superfamily HSPs are classified based on their molecular weight viz. Hsp110, Hsp90, Hsp70, Hsp60, Hsp40, Hsp27 and Hsp10 [43]. Most of the HSPs act as chaperones during the stress conditions to restore normal functions by reversing the misfolding as well as by preventing irreversibly aggregation of proteins [44] and thus balancing the physiological homeostasis. The other important protein with the molecular weight 8 kDa i.e. Ubiquitin, is another vitalfactor, which plays a crucial role in degradation of proteins via the Uiquitin proteasome system (UPS).

2.3. Heat Stress Response (HRS) in Neurooncology

HSPs are integral protein for tumor cell survival and work as molecular chaperones by binding to the unstable or misfolded proteins of tumor cells and prevent their aggregation. It also helps in the destruction of the undesired proteins through the ubiquitination process using ubiquitin proteasome system machinery [45]. The reason for studying HSP expression levels in brain tumor cells is to use them as a potential target for therapeutic application against malignant tumors of the central nervous system. Therefore, it can be understood that if the rescuing chaperones are not available for protein stabilization, the survival of the tumor cells will be obstructed, and thus HSPs can be used as therapeutic agents.

Chaperonic activities are not restricted to the folding or unfolding of target proteins in certain compartment rather it is also responsible for the localization of the denatured or newly synthesized protein to the nucleus as well as in organelle.

There are several transcription factors binding elements in tumor that code for respective proteins and helping in the cell survival pathway. However, slight alteration or deregulation in those binding elements can trigger abrupt cell proliferation and metastasis. For instance, in case of Hif1-alpha translocation, a combinatorial response of Hsp70 and the Hsp60 where Hsp70 helps in translocation of novel protein; foldase Hsp60 performs post-translocation of properly folded proteins [46].

Another major chaperone Hsp90 is considered to be having a different strategy for folding of proteins, which involves the formation of multi-chaperone complex. They basically target the signal transduction proteins and steroid receptors and help in the maintenance of their structures [47]. In tumor cells, inhibition of Hsp90 affects multiple signaling pathways such as angiogenesis, growth signaling, anti-apoptotic pathways, pathways including metastasis and replication pathways [48].

Detailed explanation of each type of heat shock proteins in association with brain tumor and the mechanism of action by these proteins will be discussed in different section of this book chapter (Table 4).

2.4. How Does HSPs Are Rescuing Misfolded Tumor Proteins?

In this section of book chapter, we are highlighting the importance of molecular chaperones that helps in the folding of newly synthesized proteins. There are several pertinent questions like, how these misfolded or damaged proteins are properly folded by heat shock. protein machinery? How these aggregated proteins are sorted out and follow ubiquitin proteasome pathways for destruction?

Table 4. Major role of HSP super family in tumor development

HSP Family	Key function	Function in neuro-oncology	References
Small HSPs	Hsp10 forms mitochondrial chaperonin complex with Hsp60	Hsp27 expression correlates with histological grades of astrocytomas with tumor growth fractions.	[37, 39, 49-51]
	It helps in proper protein folding.	AP-1(activator protein)/Hsp27 co-expression were associated with ascending grade of malignancy.	
	Hsp27 is a folding intermediate.	Hsp27 positive brain tumors impart drug resistant characteristic.	
		Hsp27 participates in the formation of inclusion bodies in meningiomas.	
Hsp 40	Co-chaperone of Hsp70.It helps	Helps in Hsp70 functioning in brain tumors.	[52,Help
	Helps in the renaturation of denatured ptogether with herwith Hsp70 and Hsp104.		
Hsp 60	Largely present in mitochondria and helping in folding of translocated proteins.	High expression of Hsp60 leads to the co-expression of proliferating cell nuclear antigen (PCNA)-higher in glioblastomas, astrocytomas, medulloblastomas.	[36, 54]
Hsp 70	Folding of newly synthesized proteins.	High levels of Hsp72 correlated with drug resistance and high p53 levels in vitro.	[35, 38, 55-57]
	Refolding of misfolded palong with ongwith ATP-controlled interaction in the hydrophobic residues of proteins for prevention of folding during translocation.	Hsp70 is forming a complex with tubulin in brain tumor cells that may stabilize the organization of microtubules and thus protecting the heat-treated cells from Vincristine damage.	
		p53 protein accumulation may be caused as a result of response to stressors, such as Hsp7 3.	
		Glial fibrillary acidic protein (GFAP) and Hsp70 together have a protective mechanism in the bratumor. or	
Hsp 90	Maintenance of signal transduction proteins.	Hsp90 ha vital al role in cell cycle progression and apoptosis in gliomas while th physiological processes isaan absence ce in normal tissue, make it a credible target for cancer therapy.	[47,58-64]
	Hsp90 forming a complex with Hsp70 and chaperonins in cytosol and usa new	Elongation factor-2 (EF-2) imparts survival of human glioblastoma and is chaperoned Hsp90.	
		Increased Hsp90 helps in the cellular survival in brain tumors thus Hsp 90 inhibitors, e.g.	

HSP Family	Key function	Function in neuro-oncology	References
	strategies for folding of proteins.	Geldanamycin, 17-AAG, NVP-AUY922 and NXD30001 are currently havening many beneficial roles in clinical trials.	
		Hsp90 inhibitor 17-AAG is specific for glioblastoma cells and is alsohelpeds in radiosensitization of the tumors	
		Hsp90 is involved in the conformational maturation of a defined group of client proteins, many of which are deregulated in Glioblastoma multiforme	
		Sub-toxic doses of a specific inhibitor of Hsp90, 17-AAG in combination with tumor necrosis factor-related apoptosis-inducing ligand (TRAIL) induces rapid apoptosis in TRAIL-resistant glioma cells reveals the specific target in the brain tumor cells.	
Hsp 110	Imparts heat resistance to the cell.	Hsp105α a novel anti-apoptotic neuroprotective factor in the mammalian brain.	[65, 66]

Small HSPs like alpha-crystallin bind to the misfolded proteins in the abnormal conditions and prevent them from further denaturation or aggregation [67]. It has been also proven that, conformation of HSPs has been changed by the binding of ATP that results in the release of denatured proteins in the cytosol were, for example, Hsp70 helps in renaturation upto a certain threshold [68]. While the processes of proper folding molecular chaperones are forming a oligomeric double ringed structures and allow the misfolded proteins to enter into that and giving proper folding in a sequestered environment. Another example such as refolding of substrates of sHSPs that depends upon the mechanism of disaggregation, which is been mediated by other chaperones [69]. Like the normal tissues, brain tumors also have the extensive misfolded proteins, protein aggregates or partially folded proteins due to physiological stress conditions. It has been observed that in tumor cells, these HSPs work for the survival of the cells and get expressed on the surface of tumor together with co-chaperones [70].

Among the HSP superfamily, Hsp70 proteins have been extensively studied under a stress conditions while its cognates Hsc70 is constitutively expressed in normal cells [71]. Hsp70 family have been found imparting cytoprotective effects by forming a complex with Hsp40 in stressed cells in an ATP-dependent fashion where Hsp70 binds to the hydrophobic segment of the misfolded target molecules [72]. For example, in glioblastoma, synthesis of stress-inducible Hsp70 (Hsp70*i*) enhances the ability of tumor cells to cope with increased concentrations of unfolded or denatured proteins [73]. The Hsp70/Hsc70 complex regulates apoptosis and prevents the formation of aggregated proteins in brain tumors [74]. As mentioned above, Hsp70 family members are found to express on the tumor surfaces that are accompanied with Hsp70-binding protein 1 (HspBP1; [75]). These observations can enable us to use a targeted therapy using antibodies or certain drugs that block the expression of HSPs that have been expressed on the tumor surface [76].

2.5. Angiogenesis, HSPs and Misfolded Proteins in Brain Tumor

The other protective mechanism for tumor regression is an anti-angiogenesis hypothesis. During the development of tumors, various angiogenic markers are upregulated such as VEGF, HIF-1α. Hypoxia inducible factor1 (HIF1) is the master regulator of metabolic adaptation in hypoxic condition [77], which forms the major part in solid tumors. In case of higher-grade gliomas that are distinguished by increased vascular endothelial cell proliferation and peritumoral edema. These are thought to be instigated by vascular endothelial growth factors, which in turn regulated by cellular oxygen tension. Hypoxia inducible factor-1α (HIF-1α) is a main responder of hypoxia and is overexpressed in many human cancers, including glioma, GBM and transitional meningiomas [78] as neovascularization is an important phenomenon in tumor growth and development. Here HSPs and other E3 ligase are working together at the molecular level for resolving aggregated or misfolded proteins for degradation by proteolytic activity of the ubiquitin proteasome system [45].

How HIF1α upregulated and helps in tumor progression? In the normoxia condition, HIF1 prolyl-hydroxylases (PHD; [79]) hydroxylate two proline residues (positioned at pro-402 and pro-564 in human HIF1α) in a reaction that requires molecular oxygen, Fe^{2+} and α-ketoglutarate as co-substrates [80, 81]. Positions of the hydroxylated proline residues are recognized and bound by the Von-Hippel Lindau protein (pVHL) [82-84]. This complex is recognized by components of E3 ubiquitin ligase complex that constitute elongins B and C, cullin2 and Rbx, which conjugates ubiquitin to HIF1α [85-87]. HIF1α regulation is mediated by the ubiquitin E3 ligase via an oxygen-mediated degradation process. After recognition by E3 ubiquitin ligase complex, HIF1α is poly-ubiquitinated and targeted to degradation by 26S proteasomal degradation system, which in turn leads to its lower expression in normoxia cells [88-90]. HIF1 is also regulated by a factor inhibiting hypoxia inducible factor (FIH1) [91] through its oxygen sensing capacity hydroxylating asparagine residue within HIF1α. Another target that modulates HIF expression is Hsp90 that can be used as a target for cancer therapies due to the following reasons [92]:

- Hsp90-mediated regulation of HIF/VEGF signaling leading to the angiogenesis.
- Chaperone-dependent regulation of HIF-independent VEGF-mediated angiogenesis.
- Hsp90 dependent targeting of key pro-angiogenic receptor tyrosine kinases and modulation of drug resistance.
- Consideration of factors in tumor microenvironment that pose several challenges for the clinical efficacy of anti-angiogenic therapy and Hsp90-targeted strategies.

3. Cellular Events in Tumor Progression

The hallmark feature in any cancer is an abnormal and uncontrolled cell proliferation which is controlled by the cell cycle regulating proteins. HSPs play a major role in regulating cell cycle stages, which determine the proliferation rate and fate of the cells. A cell passes. through specific stages which are strictly controlled by different cycle checkpoints to ensure that particular events occur only one time in a cell during cell division and give rise to its daughter cells. Check points are controlled by the catalytic subunit cyclin dependent kinase (cdk) and regulatory subunit, the cyclin-cdk complex.

Formation of this complex is tightly regulated by important mechanisms such as availability of cyclins and phosphorylation [93] of required proteins, inhibitory molecules (INK4, Cip/Kip, p21CIP1, p27KIP1, and p57KIP2) to arrest the cell cycle at the correct point of cell progression [94-101] and the regulatory proteolysis via proteasomal action [102] that in turns inhibiting uncontrolled proliferation. Findings related to tumor and cancer progression have shown deregulation of cyc-cdk complexes because of mutation in any of these proteins which ultimately leads to uncontrolled proliferation and abrogated re-entry cells into the cell cycle [103].

In normal cell division, abnormally dividing cells are arrested at checkpoints by modulating cdk activities that help the cell to recover from deadly alteration. Failure of these cell cycle checkpoints arrest leads to the transmission of altered genetic material to the daughter cells [104], which ultimately triggering a cancerous situation if not undergone for recovery or apoptosis. In case of terminally differentiated neural cells, cell cycle gets arrested in the G0 phase and some key molecules of the cycle are down-regulated. However, one of the most important cyclins in the late G1 phase cyclin D1 is increased in certain models of neural differentiation [105]. Recently, novel functions of p53 in stem cells have been characterized, which includes suppression of pluripotency and inhibition of stem cell's self-renewal [106].

The p53 tumor suppressor gene is frequently mutated or deleted in tumors occurring in humans and is often found to be mutated or lost early in glioma formation [107, 108]. Previous cancer studies have shown that small HSPs such as Hsp27 play a major role in cancer progression. Over expression of Hsp27 is associated with increased anchorage independent growth, increased invasiveness, and resistance to chemotherapeutic drugs and is associated with poor prognosis and reduced disease-free survival [109]. Impact of HSPs (Hsp27, Hsp72, Hsp73, and Hsp90) in glioblastoma showed resistance rather than successful survival even after hyperthermia treatment [35].

Table 5. HSPs and its role in cell cycle check points regulation

HSPs	Cell cycle regulatory molecules	Role in cell cycle	Arrest point	Reference
Hsp90	Cdk1, Cyc-B	Triggering mitosis	G2 check points	[110]
	Plk1	Chromosome functioning	Spindle check points	[111]
	Cyc-D	Rb phosphorylation	G1check points	[112]
	Cyc-E	S phase entry	G1/S check point	[113]
Hsp70	Cdk-4, Wee-1	Nuclear translocation	G1check points	[114]
Hsc 70	p53, pRB	Cell cycle progression, increases nuclear transport capacity	S phase entry, G2/M check points	[115]
Hsp27	P53, p21	suppress senescence/ cell progression	G2/M check points	[116]
Hsp10	Transducin like protein 1	Interact with histone H3	G1/S check point	[117]
	Termination of M phase	Interact with RAS superfamily	G1/S check point	[118, 119]

4. Apoptosis, Heat Shock Proteins and Oncogenesis

Progression of apoptotic signals begins with Fas ligand and tumor necrosis factor (TNF) that leads to the activation of procaspase-8 through molecular interaction between components of the death-inducing signaling complex [120]. Active Caspase-8 cleaves Bid protein and the carboxyl-terminal domain of Bid protein translocates to the mitochondria, and this is responsible for cytochrome C release that in turns triggers the formation of a complex containing Apaf1, a mammalian CED-4 homologue and procaspase-9 that further activates downstream signaling molecules such as procaspase-3.

The processing of these caspases is followed by the cleavage of apoptotic substrates, leading to the disruption of important cellular responses, changes in cellular and nuclear morphology and ultimately cell death [121, 122]. Different stages and signaling molecules that are involved in the apoptosis are tabulated in table 6.

In the case of neuronal apoptosis, cytochrome c, a key factor of apoptotic inducer, is negatively regulated by Hsp27 [123] that has been seen in the case of glioma. Further, Hsp27 activates protein kinase B and Akt, which inhibits cell death by phosphorylating and inactivating procaspase-9 [124]. The expression of Hsp27 in tissue is probably associated with cell differentiation in human gliomas. Rapid up-regulation of Hsp27 against heat shock in in vitro condition may have a temporary potential to increase the survival ability of glioma cells. However, upregulation of Hsp27 has been validated by using SILAC-based quantitative proteomics in cancer situation [125-127].

The correlations of HSPs and apoptotic factors have been satisfactorily proven by the action of Hsp70. The molecular chaperone Hsp70 plays a protective role in the cells and poses interference in the treatment by anticancer drugs, especially in neurooncology [128-130]. Hsp70 blocks several steps of the apoptotic cascade: cytochrome c and apoptosis inducing factor (AIF) releases from mitochondria, nuclear import of AIF, activation of procaspases -9 and caspase -3 and even downstream of active caspase-3. [128, 131-137]. Several cancer studies have shown that heat shock proteins protects cancer cells from TRAIL-induced apoptosis [138]. Apart from this, Hsp70 also actively participates in the regulation of inflammatory and immune responses [139, 140] caused due to apoptosis. Immunological responses due to apoptosis are regulated by Hsp70, and this hypothesis was validated by photodynamic therapy (PDT) [141].

Another key heat shock protein Hsp60 is a well-known stress protein present in mitochondria [142], functions as a chaperone in cell and also plays an important role in the apoptosis mechanism. Cells undergoing apoptosis have Hsp60 in the membrane blebs which then translocate [143] to the surface of the cell [144].

Table 6. Signaling molecules that are involved in apoptosis

Stages	Signaling molecules
Apoptosis inducing signals	TNF-α, Fas (Apo1/CD95), Bid, FADD, JNK/SAPK activation Granzymes, Calpains, Cathepsins, Proteasome.
Programming Phase	Caspase-8, -9, -10, -12, Bax, ROS, MMPT, cytochrome c, apoptosome formation, AIF, endonuclease G, PARP.
Execution phase	Caspase-3, -6, and -7, membrane blebbing, apoptotic body formation, DNA fragmentation

Reduction in Hsp60 expression is inducing the bax localization, release of cytochrome c from the mitochondria, decrease in the protective protein bcl2, activation of caspase 3 by cleavage and fragmentation of DNA [145-147]. This phenomena is sometimes mediated by Toll like receptor 4 (TLR4; [148]). In tumor cells Hsp60 directly associated with cyclophilin D (CypD), a component of the mitochondrial permeability transition pore. This interaction occurs in a multi-chaperonin complex comprising of the Hsp60 and Hsp90, and tumor necrosis factors receptor-associated protein-1 that selectively assembled in tumor but not in normal mitochondria.

In vivo studies on genetic targeting of Hsp60 by siRNA, triggers the CypD-dependent mitochondrial permeability transition, caspase-dependent apoptosis and suppression of intracranial glioblastoma growth. Therefore, Hsp60 is a novel regulator of mitochondrial permeability transition [149], which regulates apoptosis initiation in tumor cells.

5. Ubiquitin Proteasome System and Its Relevance in Neurooncology

The burdens of misfolded proteins in cells are rescued by first through molecular chaperone's machinery that provides proper shape of non functional or misfolded proteins. However, beyond a threshold, molecular chaperones are unable to rescue those defective proteins and hence cells follows the ubiquitin-proteasome mediated destruction pathway for that proteins. In this segment, we would highlight about the ubiquitin proteasome system (UPS) and its malfunctioning that poses a great impact on tumors. Discovery of ubiquitin mediated degradation has been done by various scientists like Drs. Aaron Ciechanover, Avram Hershko, and Irwin A Rose who have been awarded Nobel prize in 2004. Ubiquitination and cell death, differentiation and regulation of cell proteins metabolism are balanced by Ubiquitin Proteasome System (UPS) [150-155].

Ubiquitin is 8 kDa protein, which is present in all parts of the cell. It binds covalently to its target proteins and generates signals, which make them destined for different fates. Gene for ubiquitin is present in multiple copies in the eukaryotic cells. Ubiquitin protein contains seven lysine residues at positions 6, 11, 27, 29, 33, 48 and 63 through which it gets attached to the substrate or to one another.

Ubiquitin is characterized by a β-grasp superfold in which a central α-helix is surrounded by four β-sheets [156]. Degradation of a protein via UPS involves two discrete and successive steps. Proteins that undergo the destruction pathway need a substrate. First, a substrate covalently tagged with multiple ubiquitin molecules to generate a poly ubiquitin chain on the target protein molecule. As destruction pathway proceeds, the first step is mediated by Ubiquitin activating enzyme E1 that activates ubiquitin in an ATP-requiring reaction to generate a high-energy thiol-ester intermediate, E1-Subtrate-ubiquitin. Followed by E1 reaction, there is an addition of ubiquitin-carrier proteins or ubiquitin conjugating enzyme E2, which then transfers the activated ubiquitin from E1 via an additional high-energy thiol-ester intermediate E2-Substrate-ubiquitin, to the substrate that is specifically bound to a member of the ubiquitin-protein ligase family E3. Here Polyubiquitinated substrates are then processed in the 26S proteasome by an energy-consuming manner [157-159]. In UPS system the emergence of another enzyme E4 that contains U-box domain and structurally related to the

E3 RING (Really Interesting New Gene) finger domain. There are two well known E3 ligases CHIP and Parkin that are responsible for the destruction of misfolded proteins in tumorous condition. However, CHIP is known for its dual as a molecular chaperone and E3 ligase, hence it is regarded as connecting link between a molecular chaperone and ubiquitin proteasome systems.

It has been also shown that interaction between CHIP, Hsp90 and Hsp70, CHIP acts as an E3 ligase leading to the ubiquitination and proteasomal degradation of chaperone-bound misfolded proteins [160-162]. The possible molecular mechanism is given in Figure3 where molecular chaperones and Ubiquitin E3 ligase actions are depicted.

Another pertinent functions of ubiquitin E3 ligase in Cyc-cdk complexes, those are the key regulators for the cell cycle progression as described in the earlier sections of this chapter. Transcription of cyclin D is processed by β-catenin/TCF4 where β-catenin is regulated by E3 ligase [164]. Cyclins E1, E2 and A in collaboration with cdk1 and cdk2 helps in transition from S phase to G2 and Cyclin B functions in collaboration with cdk1 in G2 phase and is degraded by the proteasome in late mitosis [165]. Here once again stability of these molecules is maintained by UPS [166].

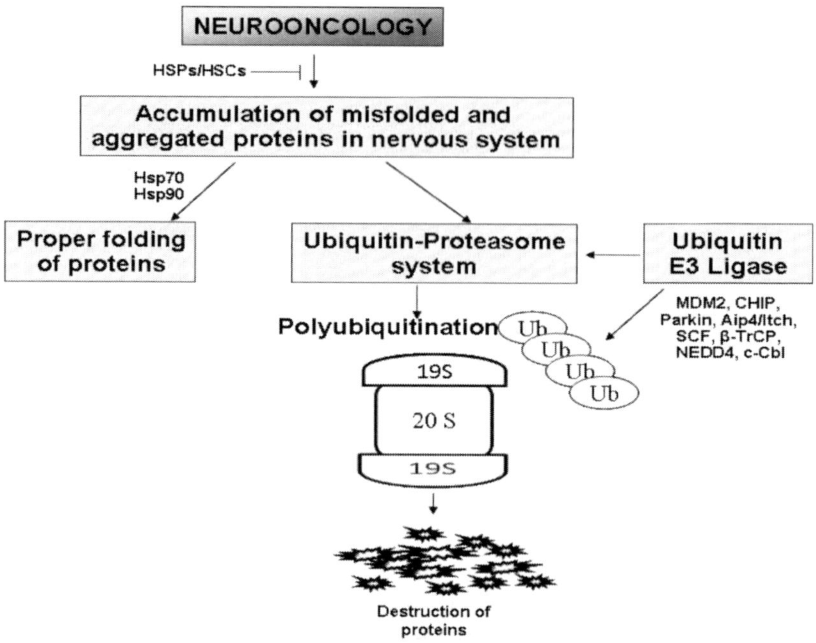

Figure 3. Destruction of misfolded protein via Ubiquitin proteasome pathway: Pictorial representation showing the accumulation of misfolded or aggregated proteins in the cell milieu where molecular chaperones are giving the proper shape but beyond the threshold, the proteins that are trapped in inclusion bodies are intended to proteolysis via a ubiquitin proteasomal pathway [163]. Constitutively expressed HSPs like HSP and HSC70 work in combination with CHIP for Ubiquitination and thereby phosphorylation of tau protein that helps in cell survival [162]. There are several E3 ligases such as. CHIP, Parkin and Atrogin-1 proteins having E3 ligase activity, and they function in co-ordination with heat shock proteins. The protein aggregates can be partially recovered by molecular chaperones depending upon the extent of structural imbalance of the constituent protein and hence avoid the degradation through the ubiquitin -26S proteasome system.

Table 7. Ubiquitin E3 ligases in brain tumor

Ubiquitin E3 ligase	Role of Ubiquitin E3 ligase in intracranial neoplasm	References
MDM2 (Murine double minute 2), HDM2 (Human double minute 2)	Over expressed in gliomas	[171]
	Causes ubiquitination and thereby proteasomal degradation of p53, a tumor suppressor gene in infiltrative astrocytomas.	
CHIP (C-terminus HSC70 interacting protein)	Ubiquitinates and degrades tumor-related proteins via UPS in glioblastoma.	[172]
	Tau proteins responsible for microtubular association in tumor cells are ubiquitinated and degraded by proteasomal action.	
	It regulates levels of Hsp70 in tumor cells.	
Parkin	PARK2 ubiquitinates cyclinE hence causes instability of the cell cycle in glioblastoma.	[173]
Aip4/Itch	Ubiquitinates ErbB4, receptors in EGFR/ErbB family and intends them to get degraded by proteasome in medulloblastomas.	[174, 175]
	Itch belongs to HECT family of E3 ligase, and it ubiquitinates and degrades Gli1.	
hBre	Ebp1, a receptor binding protein, whose repressive effect on E2F1 is suppressed due to polyubiquitination and subsequent proteasomal degradation by hBre-1 in primary gliomas.	[176]
Cullin3-REN complex	Ubiquitinates and subsequently degrades HDAC1, which is involved in deacetylation of Gli1 and Gli2 (glioma associated oncogene) in hedgehog signaling.	[177]
APC/C	Phosphorylation of Human anti apoptotic protein Mcl-1 by CDK1-cyclin B1 and its APC/C (Cdc20)-mediated destruction initiates apoptosis if a cell fails to resolve mitosis, which thereby inhibits glioma formation.	[178, 179]
	It mediates destruction of Cyclin B1 leading to increases mitotic catastrophe, thereby inhibiting glioma development and progression.	
SCF(Skp1-Cullin-F box proteins)	Ubiquitinates and degrades β-catenin, which binds to TCF and LEF transcription factors to regulate neural cell proliferation in medulloblastoma.	[180]
B-TrCP	Ubiquitinates Gli2 and subesquetnly degrades it by proteasomal action.	[175, 181]
NEDD4	NEED4-1 is upregulated by overexpression of FoxM1B transcription factor in Human gliomas.	[182-185]
	Ubiquitinates and degrades phosphatase and tensin homologue (PTEN), a tumor suppressor factor, commonly seen in glioblastoma multiforme (GBM).	
c-Cbl	Enhances invasion of gliomas in extracellular matrix (ECM) through upregulation of Matrix metalloproteinase 2 (MMP2)	[186]

In case of cell cycle, control of p53 mediated pathway is found to be abnormal in all types of brain tumors. Hence, regulation and maintenance of p53 levels in cell are tightly controlled by the constant ubiquitination and proteasomal degradation. Any defect in the E3 ligase activity of mdm2 leads to the accumulation and activation of p53 pathway leading to carcinogenesis [167].

At the same time, DNA damage causes a lethal mutations which is dealt by p53 for determining the cell fate, repairing damage in the cell and continuing cell cycle or cell death [168, 169]. All the above-mentioned steps are highly regulated by E3 ligases (Figure 3).

From previous studies, it has been observed that there is an increased expression of E3 ubiquitin ligase in brain tumor affected patients [170]. Any dysregulation of E3 ubiquitin ligase affects cellular processes and results in carcinogenesis [170]. Therefore, it is necessary to discover a drug that can modulate the efficacy of E3 ligase enzyme in cancer therapeutics. Here is several ubiquitin E3 ligase enzymes that are involved in cancer of the nervous system (Table 7)

6. Future Aspects and Therapeutics

The future of neuro-oncology lies in the progress of improvement in the survival rate of patients suffering from gliomas. There has been much advancement in the treatment of the severe disorder of brain tumour because of novel techniques involved in imaging and localization of the tumors [187]. Few treatments like surgery, radiotherapy or chemotherapy alone are not found to be sufficient enough to deal with complete removal of tumour. Therefore, much research has been done to study the signalling pathways and molecules which play major role in the development of tumour; so that such signalling molecules can be used as potential therapeutic targets. Some of the major inhibitors involved in target therapies for malignant glioma include those molecules which inhibit the regulatory factors, angiogenesis, inhibitors of HSPs, kinase inhibitors, proteasome inhibitors, etc. [188]. Glioblastoma multiform (GBM) arises due to genetic alterations in EGFR (Epidermal Growth Factor Receptor) [189], loss of chromosome 10, amplification or over-expression of MDM2 (Murine Double Minute 2), deletion or mutation in PTEN gene (Phosphatase and tensin) [188]. The inhibitors of these factors and therapies like adjuvant therapy (radiotherapy treatment after the primary surgery) are found to be potentially therapeutic. The therapies like surgery, radiotherapy and stereotactic surgery has been found to be effective in controlling meningioma tumours [190]. Blood-brain barrier resists the delivery of anti-cancerous agents to the central nervous system; therefore, a promising idea of intranasal, noninvasive, rapid delivery to the central nervous system has been enlightened by Hashizume and group [191]. Different treatment therapies have been opted for different types of tumours like surgery and radiotherapy for high-grade glioma, chemotherapy for glioblastoma, anaplastic glioma; besides that targeted drug therapies involve various inhibitors of growth regulators such as erlotinib inhibit EGFR; imatinib, inhibit PDGFR; vandetanib inhibit VEGFR [192]. A novel therapy of using toxins against the surface expressed antigens in tumours has been found as a therapeutic advancement. These toxins are recombinant molecules, which specifically bind to the surface antigens of tumour cells and show the cytotoxic effects. The brain tumour stem cells and the use of nanotechnology assisted therapies have also been referred to be of explorative potential.

Targeting of malignant glioma through a first generations of molecules in proliferative pathways has been disappointing; with relatively rare radiographic responses and no significant prolong "progression-free free survival" (PFS) [193]. An array of ongoing clinical trials is known to combine novel molecular therapies with standard treatments such as radiation and chemotherapy [194]. Among the promising therapeutic targets in early clinical development are fibroblast growth factor receptor (FGFR), heat shock protein 90 (Hsp90), hypoxia-inducible factor 1 alpha (HIF1alpha), cyclin dependent kinases, and many others [194].

REFERENCES

[1] Read, TA; Hegedus, B; Wechsler-Reya, R; Gutmann, DH. The neurobiology of neurooncology. *Annals of Neurology*, 2006 60, 3-11.

[2] Frederico AC. Azevedo, Ludmila RB. Carvalho, Grinberg, LT; Farfel, JM; Renata E.L. Ferretti, Leite, RP; Filho WJ, Lent, R; Herculano-Houzel, S. Equal numbers of neuronal and nonneuronal cells make the human brain an isometrically scaled-up primate brain. *The Journal of Comparative Neurology*, 2009 513(5), 532-541.

[3] Nicola De, AF; Labombarda, F; Deniselle, MC; Gonzalez, SL; Garay, L; Meyer, M; Gargiulo, G; Guennoun, R; Schumacher, M; Progesterone neuroprotection in traumatic CNS injury and motoneuron degeneration. *Front Neuroendocrinol.* 2009 30(2): 173-87.

[4] Vourc'h, P; Dessay, S; Mbarek, O; Marouillat Védrine, S; Müh, JP; Andres, C. The oligodendrocyte-myelin glycoprotein gene is highly expressed during the late stages of myelination in the rat central nervous system. *Brain Res. Dev. Brain Res.* 2003 Sep 10;144(2):159-68.

[5] Simons, M; Trajkovic, K. Neuron-glia communication in the control of oligodendrocyte function and myelin biogenesis. *Journal of cell science,* 2006 119, 4381-4389.

[6] Altman, J. Microglia emerge from the fog. *TINS*, 1994 17, 47-49.

[7] Gehrmann, J; Kreutzberg, GW. Monoclonal antibodies against macrophages/microglia: Immunocytochemical studies of early microglia activation in experimental neuropathology. *Clinical neuropathology,* 1993 12, 301-306.

[8] Goussia, AC; Bruner, JM; Kyritsis, AP; Agnantis, NJ; Fuller, GN; Cytogenetic and molecular genetic abnormalities in primitive neuroectodermal tumors of the central nervous system. *Anticancer Res.* 2000 20(1A) :65-73. Review.

[9] McCall, MA; Gregg, RG, Behringer, RR, Brenner M, Delaney CL, Galbreath EJ, Zhang CL, Pearce RA, Chiu SY, Messing A. Targeted deletion in astrocyte intermediate filament (Gfap) alters neuronal physiology. *Proceedings of the National Academy of Sciences*, 1996 93, 6361-6366.

[10] Giménez Y Ribotta M, Langa F, Menet V, Privat A.Comparative anatomy of the cerebellar cortex in mice lacking vimentin, GFAP, and both vimentin and GFAP. *Glia*, 2000 31(1), 69-83.

[11] Louis, DN ; Ohgaki, H; Wiestler, OD; Cavenee, WK; Burger, PC; Bernd, AJ; Scheithauer, W; Kleihues, P. The 2007 WHO ClassiWcation of Tumors of the Central Nervous System. *Acta Neuropathology*, 2007 114, 97-109.

[12] Huse, JT; Holland, EC. Targeting brain cancer: advances in the molecular pathology of malignant glioma and medulloblastoma. *Nature Reviews Cancer*, 2010 10, 319-331.

[13] Scheithauer, BW.. Development of the WHO classification of tumors of the central nervous system: a historical perspective. *Brain Pathol.* 2009 (4):551-64.

[14] Collins, VP. Brain Tumors: Classification and Genes". *J. Neurol. Neurosurg. Psychiatry,* 2004 75, ii2–ii11.

[15] Jänisch, W. [Classification of gliomas by malignancy grades]. *Zentralbl. Allg. Pathol.* 1985;130(6):447-50.

[16] Chiechi, MV; Smirniotopoulos, JG; Jones, RV. Intracranial subependymomas: CT and MR imaging features in 24 cases. *American Journal of Roentgenology,* 1995 165, 1245-1250.

[17] Wiestler,OD; Schiffer, D. Subependymoma In: World Health Organization Classification of Tumors, *Pathology and Genetics of Tumors of the Nervous System,* 2000, 80-81.

[18] Wells, EM; Khademian, ZP; Walsh, KS; Vezina, G; Sposto, R; Keating, RF; Packer, RJ. Postoperative cerebellar mutism syndrome following treatment of medulloblastoma: neuroradiographic features and origin. *Journal of Neurosurgery: Pediatrics*, 2010 5(4), 329-334.

[19] Höll, T. Kleihues P, Yasargil MG, Wiestler OD. Cerebellar medullomyoblastoma with advanced neuronal differentiation and hamartomatous component. *Acta Neuropathol.* 1991;82(5):408-13.

[20] Schiffer, D; Giordana, MT; Pezzotta, S; Pezzulo, T; Vigliani, MC. Medullomyoblastoma: report of two cases. *Childs Nerv. Syst.* 1992 8(5):268-72.

[21] Tihan, T; Fisher, PG; Kepner, JL; Godfraind, C; McComb, RD; Goldthwaite, PT; Burger, PC. Pediatric astrocytomas with monomorphous pilomyxoid features and a less favorable outcome. *J. Neuropathol. Exp. Neurol.* 1999 58(10):1061-8.

[22] Miller, CR; Dunham, CP; Scheithauer, BW; Perry, A. Significance of necrosis in grading of oligodendroglial neoplasms: a clinicopathologic and genetic study of newly diagnosed high-grade gliomas. *J. Clin. Oncol.* 2006 24(34):5419-26.

[23] van den Bent, MJ; Hegi, ME; Stupp, R. Recent developments in the use of chemotherapy in brain tumours. *Eur. J. Cancer.* 2006 42(5):582-8.

[24] Reifenberger, G; Louis, DN. Oligodendroglioma: toward molecular definitions in diagnostic neuro-oncology. *Experimental Neurology*, 2003 62(2), 111-26.

[25] Yamashima, T. On Arachnoid Villi and Meningiomas: Functional Implication of Ultrastructure, Cell Adhesion Mechanisms, and Extracellular Matrix Composition. *Pathology and oncology research,* 1996 2(3), 144-149.

[26] Perry, A; Aldape, KD; George, DH; Burger, PC. Small cell astrocytoma: an aggressive variant that is clinicopathologically and genetically distinct from anaplastic oligodendroglioma. *Cancer.* 2004 101(10):2318-26.

[27] Fisher, JL; Schwartzbaum, JA; Wrensch, M; Wiemels, JL. Epidemiology of Brain Tumors. *Neurologic Clinics*, 2007 25(4), 867-890.

[28] Lindquist, S. The heat shock response. *Annual Reveiw Biochemistry,* 1986 55, 1151-1191.

[29] Rondinelli, E. Conservation of heat-shock proteins in Trypanosoma cruzi. *Parasitology Today*, 1994 10(5), 172-176.

[30] Rakonczay, Z Jr; Takács, T; Iványi, B; Mándi, Y; Pápai, G; Boros, I; Varga, I; Jost, K; Lonovics, J. Induction of heat shock proteins fails to produce protection against trypsin-induced acute pancreatitis in rats. *Clinical and experimental Medicine*, 2002 2, 89-97.

[31] Katschinski, DM. On heat and cells and proteins. *News Physiol. Sci.* 200419:11-5.

[32] Piacentini, L; Karliner, JS. Altered gene expression during hypoxia and reoxygenation of the heart. *Pharmacology and Therapeutics*, 1999 83, 21-37.

[33] Redaelli, CA; Wagner, M; Kulli, C; Tian, YH; Kubulus, D; Mazzucchelli, L; Wagner, AC; Schilling, MK. Hyperthermia-induced HSP expression correlates with improved rat renal isograft viability and survival in kidneys harvested from non-heart-beating donors. *Transplant international,* 2001 14(6), 351-360.

[34] Stensløkken, KO; Ellefsen, S; Larsen, HK; Vaage, J; Nilsson, GE. Expression of heat shock proteins in anoxic crucian carp (Carassius carassius): support for cold as a preparatory cue for anoxia. *American Journal Pathology - Regular Physiology*, 2010 298(6), R1499-R1508.

[35] Hermisson, M; Strik, H; Rieger, J; Dichgans, J; Meyermann, R; Weller, M. Expression and functional activity of heat shock proteins in human glioblastoma multiforme. *Neurology,* 2000 54(6), 1357-1365.

[36] Kato, S; Kato, M; Hirano, A; Takikawa, M; Ohama, E. The immunohistochemical expression of stress-response protein (srp) 60 in human brain tumor. *Histology and histopathology*, 2001 16, 809-820.

[37] Khalid, H; Tsutsumi, K; Yamashita, H; Kishikawa, M; Yasunaga, A;Shibata, S. Expression of the small heat shock protein (Hsp) 27 in human astrocytomas correlates with histologic grades and tumor growth fractions. *Cellular and Molecular Neurobiology*, 1995 15(2), 257-268.

[38] Lilia, G; Aguilera, R; Cortés-Cuevas, A; Rosario, C; Sumano, H. Circadian serum concentrations of tylosin in broilers after feed or water medication. *Br. Poult. Sci.* 2008 49(5):619-24.

[39] Yokoyama, N; Iwaki, T; Goldman, JE; Tateishi, J; Fukui, M. Small heat-shock protein is expressed in meningiomas and in granulofilamentous inclusion bodies. *Acta neuropathologica*, 1993 85(3), 248-255.

[40] Milarski, KL; Welch, WJ; Morimoto, RI. Cell Cycle-dependent Association of Hsp70 with Specific Cellular Proteins. *The Journal of Cell Biology*, 1989 108, 413-423.

[41] Perotti, C; Liu, R; Parusel, CT; Böcher, N; Schultz, J; Bork, P; Pfitzner, E; Groner, B; Shemanko, CS. Heat shock protein-90-alpha, a prolactin-STAT5 target gene identified in breast cancer cells, is involved in apoptosis regulation. *Breast Cancer Research*, 2008 10:R94.

[42] Gupta, S; Knowlton, AA. Cytosolic Heat Shock Protein 60, Hypoxia, and Apoptosis. *Circulation*, 2002 106, 2727-2733.

[43] Schlesinger, MJ. Heat Shock Proteins. *The Journal of Biological Chemistry* 1990 265(21), 12111-12114.

[44] Benjamin, IJ; McMillan, DR; Stress (Heat Shock) Proteins: Molecular Chaperones in Cardiovascular Biology. *Circulation Research*, 1998 83, 117-132.

[45] Graner, MW; Bigner, DD. Chaperone proteins and brain tumors: Potential targets and possible therapeutics. *Neuro-Oncology,* 2005 7, 260-277.

[46] Hartl, FU. Heat shock proteins in protein folding and membrane translocation. *Seminars in Immunology*, 1991 3(1):5-16.

[47] Young, JC; Moarefi, I, Hartl, FU. Hsp90: a specialized but essential protein-folding tool. *The Journal of cell biology*, 2001 154(2), 267-273.

[48] Neckers, L. Heat shock protein 90: the cancer chaperone. *Journal of Biosciences,* 2007 32, 517-530.

[49] Lin, KM; Lin, B; Lian, IY; Mestril, R; Scheffler, IE; Dillmann, WH. Combined and Individual Mitochondrial Hsp60 and Hsp10 Expression in Cardiac Myocytes Protects Mitochondrial Function and Prevents Apoptotic Cell Deaths Induced by Simulated Ischemia-Reoxygenation. *Circulation*, 2001 103, 1787-1792.

[50] Bausero, MA; Pagea, DT; Osinagab E and Aseaa, A. Surface Expression of Hsp25 and Hsp72 Differentially Regulates Tumor Growth and Metastasis. *Tumor Biology*, 2004 25, 243–251.

[51] Assimakopoulou, M; Varakis, J. AP-1 and heat shock protein 27 expression in human astrocytomas. *Journal of cancer research and clinical oncology,* 2001 127, 727-732.

[52] Sakahira, H; Nagata, S. Co-translational Folding of Caspase-activated DNase with Hsp70, Hsp40, and Inhibitor of Caspase-activated DNase. *The Journal of biological chemistry*, 2002 277(5), 3364-3370.

[53] Glover, JR; Lindquist, S. Hsp104, Hsp70, and Hsp40: A Novel Chaperone System that Rescues Previously Aggregated Proteins. *Cell,* 1998 94, 73-82.

[54] Ghosh, JC; Dohi, T; Kang, BH; Altieri, DC. Hsp60 Regulation of Tumor Cell Apoptosis. *The Journal of biological chemistry*, 2008 283(8), 5188-5194

[55] Mayer, MP; Bukau, B. Hsp70 chaperones: Cellular functions and molecular mechanism. *Cellular and molecular Life Sciences*, 2005 62, 670–684.

[56] Matsumoto, H;Wang, X;Ohnishi, T. Binding between wild-type p53 and hsp72 accumulated after UV and gamma-ray irradiation. *Cancer Letters,* 1995 92, 127-133.

[57] Lee, WC; Lin, KY; Chen, KD; Lai, YK. Induction of Hsp70 is associated with vincristine resistance in heat-shocked 9L rat brain. *British Journal of cancer*, 1992 66, 653-659.

[58] Zhu H, Woolfenden S, Bronson RT, Jaffer ZM, Barluenga S, Winssinger N, Rubenstein AE, Chen R, Charest A. The novel Hsp90 inhibitor NXD30001 induces tumor regression in a genetically engineered mouse model of glioblastoma multiforme. *Mol. Cancer Ther.* 2010 Sep;9(9):2618-26. Epub 2010 Jul 19.

[59] Cruickshanks, N; Shervington, L; Patel, R; Munje, C; Thakkar, D; Shervington, A. Can hsp90alpha-targeted siRNA combined with TMZ be a future therapy for glioma? *Cancer investigation,* 2010 28(6), 608-614.

[60] Shervington, A; Cruickshanks, N; Lea, R; Roberts, G; Dawson, T; Shervington, L. Can the lack of HSP90alpha protein in brain normal tissue and cell lines, rationalise it as a possible therapeutic target for gliomas? *Cancer investigation,* 2008 26(9), 900-904.

[61] Shen, JH; Zhang, Y; Wu, NH; Shen, YF. Resistance to geldanamycin-induced apoptosis in differentiated neuroblastoma SH-SY5Y cells. *Neuroscience letter*, 2007 414(2), 110-114.

[62] Dungey, FA; Caldecott, KW; Chalmers, AJ. Enhanced radiosensitization of human glioma cells by combining inhibition of poly(ADP-ribose) polymerase with inhibition of heat shock protein 90. *Molecular cancer therapy,* 2009 8(8), 2243-2254.

[63] Siegelin, MD; Habel, A; Gaiser, T. 17-AAG sensitized malignant glioma cells to death-receptor mediated apoptosis. *Neurobiology of disorder,* 2009 33(2), 243-249.

[64] Sauvageot, CM; Weatherbee, JL; Kesari, S; Winters, SE; Barnes, J; Dellagatta, J; Ramakrishna, NR; Stiles, CD; Kung, AL; Kieran, MW; Wen, PY Effcacy of the HSP90 inhibitor 17-AAG in human glioma cell lines and tumorigenic glioma stem cells. *Neurooncology*, 2009 11, 109-121.

[65] Oh, HJ; Easton, D; Murawski, M; Kaneko, Y; Subjeck, JR. The Chaperoning Activity of hsp110- identification of functional domains by use of targeted deletions. *The Journal of biological chemistry*, 1999 274(22), 15712-15718.

[66] Yamagishi, N; Saito, Y; Ishihara, K; Hatayama, T. Enhancement of oxidative stress-induced apoptosis by Hsp105alpha in mouse embryonal F9 cells. *European Journal of biochemistry*, 2002 269(16), 4143-4151.

[67] Cashikar, AG; Duennwald, M; Lindquist, SL. A Chaperone Pathway in Protein Disaggregation: Hsp26 alters the nature of protein aggregates to facilitate reactivation by Hsp104s. *Journal of Biological Chemistry,* 2005 280(25), 23869-23875.

[68] Wang, K; Spector, A. ATP causes small heat shock proteins to release denatured protein. *Europian Journal of biochemistry*., 2001 268, 6335-6345.

[69] Mogk, A; Schlieker, C; Friedrich, KL; Scho¨nfeld, HJ; Vierling, E; Bukau, B. (2003), Refolding of Substrates Bound to Small HSPs Relies on a Disaggregation Reaction Mediated Most Efficiently by ClpB/DnaK. *The Journal of biological chemistry*, 2003 278(33), 31033-31042.

[70] Graner, MW; Cumming, RI; Bigner, DD. The Heat Shock Response and Chaperones/Heat Shock Proteins in Brain Tumors: Surface Expression, Release, and Possible Immune Consequences. *The Journal of Neuroscience,* 2007 27(42), 11214-11227.

[71] Boone, AN; Vijayan, MM; Glucocorticoid-mediated attenuation of the hsp70 response in trout hepatocytes involves the proteasome. *Americal Journal Physiological Regular. Integr. Comp. Physiol.* 2002 283(3):R680-7.

[72] Sharma, SK; Christen, P; Goloubinoff, P. Disaggregating Chaperones: An Unfolding Story. *Current Protein and Peptide Science*, 2009 10, 432-446.

[73] Nollen, EA; Brunsting, JF; Roelofsen, H; Weber, LA; Kampinga, HH. In vivo chaperone activity of heat shock protein 70 and thermotolerance. *Molecular and Cellular Biology*, 1999 11, 2069-2079.

[74] Witt, SN. The Fink Blueprint for Hsp70/Hsc70 Molecular Chaperones. *Current Protein and Peptide Science*, 2009 10, 424-431.

[75] Graner, MW; Raynes, DA; Bigner, DD; Guerriero, V. Heat shock protein 70-binding protein 1 is highly expressed in high-grade gliomas, interacts with multiple heat shock protein 70 family members, and specifically binds brain tumor cell surfaces. *Cancer Science*, 2009 100(10):1870-1879.

[76] Radons, J; Multhoff, G. Immunostimulatory functions of membrane-bound and exported heat shock protein 70. *Exercise Immunology Review,* 2005 11, 17-33.

[77] Dikmen, ZG; Gellert, GC; Dogan, P; Yoon, H; Lee, YB; Ahn, CH; Shay, JW. In Vivo and In Vitro Effects of a HIF1α Inhibitor, RX-0047. *Journal of Cellular Biochemistry*, 2008 104:985-994.

[78] Kaynar, MY; Sanus, GZ; Hnimoglu H; Kacira, T; Kemerdere, R; Atukeren P; Gumustas, K; Canbaz, B; Tanriverdi, T. Expression of hypoxia inducible factor-1α in tumors of patients with glioblastoma multiforme and transitional meningioma. *Journal of Clinical Neuroscience,* 2008 15(9), 1036-1042.

[79] Lee, KA; Lynd, JD; O'Reilly, S; Kiupel, M; McCormick, JJ; LaPres, JJ. The biphasic role of the hypoxia-inducible factor prolyl-4-hydroxylase, PHD2, in modulating tumor-forming potential. *Molecular Cancer Research,* 2008 6(5), 829-842.

[80] Jaakkola, P; Mole, DR; Tian, YM; Wilson, MI; Gielbert, J; Gaskell, SJ; Kriegsheim, Av; Hebestreit, HF; Mukherji, M; Schofield, CJ; Maxwell, PH; Pugh, CW; Ratcliffe, PJ. Targeting of HIF-alpha to the von Hippel-Lindau ubiquitylation complex by O2-regulated prolyl hydroxylation. *Science,* 2001 292(5516), 468-472.

[81] Ivan, M; Kondo, K; Yang, H; Kim, W; Valiando, J; Ohh, M; Salic, A; Asara, JM; Lane, WS; Kaelin, WG Jr. HIF-alpha targeted for VHL mediated destruction by prolinehydroxylation: Implications for O2 sensing. *Science,* 2001 292(5516), 464-468.

[82] Maxwell, PH; Wiesener, MS; Chang, GW; Clifford, SC; Vaux, EC; Cockman, ME; Wykoff, CC; Pugh, CW; Maher, ER; Ratcliffe, PJ. The tumor suppressor protein VHL targets hypoxia-inducible facto4rs for oxygen-dependent proteolysis. *Nature*, 1999 399, 271-275.

[83] Hon, WC; Wilson, MI; Harlos, K; Claridge, TD; Schofield, CJ; Pugh, CW; Maxwell, PH; Ratcliffe, PJ; Stuart, DI; Jones, EY. Structural basis for the recognition of hydroxyl proline in HIF-1 alpha by pVHL. *Nature,* 2002 417, 975-978.

[84] Min, JH; Yang, H; Ivan, M; Gertler, F; Kaelin, WG; Pavletich, NP. Structure of an HIF1alpha-pVHL complex: hydroxyproline recognition in signaling. *Science*, 2002 296, 1886-1889.

[85] Cockman, ME; Masson, N; Mole, DR; Jaakkola, P; Chang, GW; Clifford, SC; Maher, ER; Pugh, CW; Ratcliffe, PJ; Maxwell, PH. Hypoxia inducible factor-alpha binding and ubiquitylation by the von Hippel-Lindau tumor suppressor protein. *Journal of Biological Chemistry,* 2000 275, 25733-25741.

[86] Ohh, M; Park, CW; Ivan, M; Hoffman, MA; Kim, TY; Huang, LE; Pavletich, N; Chau, V; Kaelin, WG. Ubiquitination of hypoxia-inducible factor requires direct binding to the beta domain of the von Hippel-Lindau protein. *Nature Cell Biology*, 2000 2, 423-427.

[87] Tanimoto, K; Makino, Y; Pereira, T; Poellinger, L. Mechanism of regulation of the hypoxia-inducible factor-1 alpha by the von Hippel-Lindau tumor suppressor protein. *The EMBO Journal*, 2000 19, 4298-4309.

[88] Salceda, S; Caro, J. Hypoxia-inducible factor 1alpha (HIF1αlpha) protein is rapidly degraded by the ubiquitin-proteasome system under normoxic conditions. *Journal of biological chemistry.*, 1997 272, 22642-22647.

[89] Huang, C; Cheng, H; Hao, S; Zhou, H; Xujia Zhang, X; Gao, J; Sun, Q; Hu, H; Wang, C. Heat Shock Protein 70 Inhibits α-Synuclein Fibril Formation via Interactions with Diverse Intermediates. *Journal of Molecular Biology*, 2006 364,323-336.

[90] Kallio, PJ; Wilson, WJ; O'Brien, S; Makino,Y; Poellinger, L. Regulation of the hypoxia-inducible transcription factor 1alpha by the ubiquitin-proteasome pathway. *The Journal of biological chemistry,* 1999 274, 6519-6525.

[91] Wilkins, SE; Jaana, H; Johana, C; Jeffrey, GJ; Daniel, PJ; Rebecca, B; Peppi, K. Differences in hydroxylation and binding of Notch and HIF1α demonstrate substrate selectivity for factor inhibiting. *Science*, 2009 10, 424-431.

[92] Bohonowych, JE; Gopal, U; Isaacs, JS. Hsp90 as a Gatekeeper of Tumor Angiogenesis: Clinical Promise and Potential Pitfalls". *Journal of Oncology,* 2010 doi:10.1155/2010/412985.

[93] Draetta, GF. Will the real Cdk-activating kinase please stand up. *Current Biology,* 1997 7, R50-R52.

[94] Sherr, CJ; Roberts, JM. CDK inhibitors: positive and negative regulators of G1-phase progression. *Genes and Development*, 1999 13, 1501.

[95] Serrano, M; Hannon, GJ; Beach, D. A new regulatory motif in cell-cycle control causing specific inhibition of cyclin D/CDK4. *Nature*, 1993 366, 704.

[96] Chan, FK; Zhang, J; Cheng, L; Shapiro, DN; Winoto, A. Identification of human and mouse p19, a novel CDK4 and CDK6 inhibitor with homology to p16ink4. *Molecular Cell Biology*, 1995 15, 2682-2688.

[97] Dulic, V; Drullinger, L; Lees, E; Reed, S; Stein, G. Altered regulation of G1 cyclins in senescent human diploid fibroblasts: accumulation of inactive cyclin E-Cdk2 and Cyclin D1-Cdk2 complexes. *Proceedings of the National Academy of Sciences,* 1993 90,11034-11038.

[98] Harper, JW; Adami, GR; Wei, N; Keyomarsi, K; Elledge, SJ. The p21 Cdk-Interacting Protein Cip1 Is a Potent Inhibitor of G1 Cyclin-Dependent Kinases. *Cell,* 1993 75, 805.

[99] Polyak, K; Lee, MH. Erdjument-Bromage H, Koff A, Roberts JM, Tempst P, Massagué J. Cloning of p27Kip1, a cyclin-dependent kinase inhibitor and a potential mediator of extracellular antimitogenic signals. *Cell.* 1994 78(1):59-66.

[100] Toyoshima, H; Hunter, T; p27, a novel inhibitor of G1 cyclin-Cdk protein kinase activity, is related to p21. *Cell.* 1994;78(1):67-74.

[101] Lee, M; Reynisdottir, I; Massague, J. Cloning of p57KIP2, cyclin dependent kinase inhibitor with unique domain structure and tissue distribution. *Genes and Development*, 1995. 9:639-649.

[102] Reed, SI. The ubiquitin proteasome pathway in cell cycle. *Results and Problem in cell differentiation*, 2006 42, 147-181.

[103] Malumbres, M; Barbacid, M. To cycle or not to cycle: a critical decision in cancer. *Nature Rev. Cancer* 2001 1, 222-231.

[104] Bartek, J; Lukas, C; Lukas, J., Checking on DNA damage in S phase. *Nature Reviews Molecular Cell Biology*, 2004 5(10), 792-804.

[105] Van Grunsven, LA; Thomas, A; Urdiales, JL; Machenaud, S; Choler, P; Durand, I; Rudkin, BB. Nerve growth factor-induced accumulation of PC12 cells expressing cyclin D1: evidence for a G1 phase block. *Oncogene*, 1996 12, 855-862.

[106] Zhao, T; Xu, Y. p53 and stem cells: new developments and new concerns. *Trends in cell Biology*, 2010 20(3), 170-175.

[107] Louis, DN. The p53 gene and protein in human brain tumors. *Journal of Neuropathology and Experimental Neurology*, 1994 53(1), 11-21.

[108] Nozaki, M; Tada, M; Kobayashi, H; Zhang, CL; Sawamura, Y; Abe, H; Ishii, N; Van Meir, EG. Roles of the functional loss of p53 and other genes in astrocytoma tumorigenesis and progression. *Neuro-Oncology,* 1999 1(2), 124-137.

[109] Lee, SA; Ndisang, D; Patel, C; Dennis, JH; Faulkes, DJ; D'Arrigo, C; Samady, L; Farooqui-Kabir, S; Heads, RJ; Latchman, DS; Budhram-Mahadeo, VS. Expression of the Brn-3b Transcription Factor Correlates with Expression of Hsp27 in Breast Cancer Biopsies and Is Required for Maximal Activation of the Hsp27 Promoter. *Cancer Research*, 2005 65(8), 3072-3080.

[110] Muñoz, MJ; Jimenez, J. Genetic interactions between Hsp90 and the cdc2 mitotic machinery in the fission yeast Schizo saccharomyces prombe. *Molecular and general genomics*, 1999 261, 241-250.

[111] De Carcer, G. Heat shock protein90 regulate the metaphase-anaphase transition in a polo-like kinase-dependent manner. *Cancer Research,* 2004 64, 5106-5112.

[112] Munster, PN; Srethapakdi, M; Moasser, MM; Rossen, N. Inhibition of heat shock protein 90 Function by ansamycins causes the morphological and functional differentiation of breast cancer cells. *Cancer Research,* 2001 61, 2945-2952.

[113] Bedin, M; Gaben, AM; Saucer, C; Meser, J. Geldenamycin, an inhibitor of the chaperone activity of Hsp90, induces MAPK independent cell cycle arres. *International Journal of cancer*, 2004 109,643-652.

[114] Helmbrecht, K; Zeise, E; Rensing, L. Chaperones in cell cycle regulation and mitogenic signal transduction: a review. *Cell Proliferation* 2000 33(6): 341–365.

[115] Zeise, E; Kühl, N; Kunz, J; Rensing, L. Nuclear translocation of stress protein Hsc70 during S phase in rat C6 glioma cells. *Cell Stress Chaperones*. 1998 Jun; 3(2):94-9.

[116] Venkatakrishnan, CD; Dunsmore, K; Wong, H; Roy, S; Sen, CK; Wani, A; Zweier, JL; Ilangovan, G. HSP27 regulates p53 transcriptional activity in doxorubicin-treated fibroblasts and cardiac H9c2 cells: p21 upregulation and G2/M phase cell cycle arrest. *Am. J. Physiol. Heart Circ. Physiol.* 2008 294(4):H1736-44. 2008.

[117] Uetz, P; Giot, L; Cagney, G; Mansfield, TA; Judson, RS; Knight, J.R; Lockshon, D; arayan, V; Srinivasan, M; Pochart, P;. A comprehensive analysis of proteinprotein interactions in Saccharomyces cerevisiae. *Nature* 2000 403: 623–627.

[118] Levanon D, Eisenstein M, Groner Y. Site-directed mutagenesis supports a three-dimensional model of the runt domain. J Mol Biol. 1998; 277(3):509-12. Erratum in: *J. Mol. Biol.* 1998 282(3):699.

[119] Palaparti, A; Baratz, A; Stifani, S. The Groucho/transducin-like enhancer of split transcriptional repressors interact with the genetically defined amino-terminal silencing domain of histone H3. *J. Biol. Chem.* 1997 272(42):26604-10.

[120] Salvesen, GS; Dixit, VM. Caspases: Intracellular Signaling by Proteolysis. *Cell,* 1997 91, 443-446.

[121] Luo, X; Budihardjo, I; Zou, H; Slaughter, C; and Wang, XD. *Cell* 1998 94, 481–490.

[122] Li, P; Nijhawan, D; Budihardjo, I; Srinivasula, SM; Ahmad, M; Alnemri, ES; Wang, X. Cytochrome c and dATP-dependent formation of Apaf-1/caspase-9 complex initiates an apoptotic protease cascade. *Cell,* 1997 91(4), 479-489.

[123] Bruey, JM; Ducasse, C; Bonniaud, P; Ravagnan, L; Susin, SA; Diaz-Latoud, C; Gurbuxani, S; Arrigo, AP; Kroemer, G; Solary, E; Garrido, C. Hsp27 negatively regulates cell death by interacting with cytochrome c. *Nature Cell Biology*, 2000 2, 645-652.

[124] Mehlen, P; Kretz-Remy, C; Preville, X; Arrigo, AP. Human Hsp27, drosophila Hsp27 and human αB-crystallin expression-mediated increase in glutathione is essential for the protective activity of these proteins against TNF alpha-induced cell death. *The EMBO Journal,* 1996 15, 2695-2706.

[125] Xu, Y; Liang, S; Shen, G; Xu, X; Liu, Q; Xu, Z; Gong, F; Tang, M; Wei, Y. Application of the SILAC (stable isotope labeling with amino acids in cell culture) technique in quantitative comparisons for tissue proteome expression. *Biotechnology and applied biochemistry*, 2009 54, 11-20.

[126] Liang, S; Shen, G; Liu, Q; Xu, Y; Zhou, L; Xiao, S; Xu, Z; Gong, F; Wei, Y. Isoform specific expression and characterization of 14-3-3 proteins in human glioma tissues

discovered by SILAC-based proteomic analysis. *Proteomics Clinical application*, 2009 3, 743-753.

[127] Liang, S; Xu, Y; Shen, G; Liu, Q; Zhao, X; Xu, Z; Xie, X; Gong, F; Li, R; Wei Y. Quantitative protein exprcssion profiling of 14-3-3 isoforms in human renal carcinoma shows 14-3-3 epsilon is involved in limitedly increasing renal cell proliferation. *Electrophoresis,* 2009 30(22), 4152-4162.

[128] Guo, F; Sigua, C; Bali, P; George, P; Fiskus, W; Scuto, A; Annavarapu, S; Mouttaki, A; Sondarva, G; Wei, S; Wu, J; Djeu, J; Bhalla, K. Mechanistic role of heat shock protein 70 in Bcr-Abl-mediated resistance to apoptosis in human acute leukemia cells. *Blood,* 2005 105, 1246-1255.

[129] Kwak, HJ; Jun, CD; Pae, HO; Yoo, JC; Park, YC; Choi, BM; Na, YG; Park, RK; Chung, HT; Chung, HY; Park, WY; Seo, JS. The role of inducible 70-kDa heat shock protein in cell cycle control, differentiation, and apoptotic cell death of the human myeloid leukaemic HL-60 cells. *Cellular Immunology*, 1998 187:1-12.

[130] Li, CY; Lee, JS; Ko, YG; Kim, JI; Seo, JS. Heat Shock Protein 70 Inhibits Apoptosis Downstream of Cytochrome c Release and Upstream of Caspase-3 Activation. *The Journal of Biological Chemistry*, 2000 275(33), 25665-25671.

[131] Daugas, R; Susin, SA; Zamzami, N; Ferri, KF; Irinopoulou, T; Larochette, N; Pre´Vost, MC; Leber, B; Andrews, D; Penninger, J; Kroemer, G. Mitochondrio-nuclear translocation of AIF in apoptosis and necrosis. *The Journal of the Federation of American Societies for Experimental Biology*,. 2000 14, 729-739.

[132] Mosser, DD; Caron, AW; Bourget, L; Denis-Larose, C; Massie, B. Role of the human heat shock protein hsp70 in protection against stress-induced apoptosis. *Molecular and cellular biology*, 1997 17, 5317-5327.

[133] Jaattela, M; Wissing, D; Kokholm, K; Kallunki, T; Egeblad, M. (1998), "Hsp70 exerts its anti-apoptotic function downstream of caspase-3-like proteases. *The EMBO Journal*, 17, 6124-6134.

[134] Beere, HM; Wolf, BB; Cain, K; Mosser, DD; Mahboubi, A; Kuwana, T; Tailor, P; Morimoto, RI; Cohen, GM; Green, DR. Heat-shock protein 70 inhibits apoptosis by preventing recruitment of procaspase-9 to the Apaf-1 apoptosome". *Nature Cell Biology,* 2000 2, 469-475.

[135] Mosser, DD; Caron, AW; Bourget, L; Meriin, AB; Sherman, MY; Morimoto, RI and Massie, B. The chaperone function of hsp70 is required for protection against stress-induced apoptosis. *Molecular Biology Cell,* 2000 20, 7146-7159.

[136] Ravagnan, L; Gurbuxani, S; Susin, SA; Maisse, C; Daugas, E; Zamzami, N; Mak, T; Jaattela, M; Penninger, JM; Garrido, C; Kroemer, G. Heat-shock protein 70 antagonizes apoptosis-inducing factor. *Nature cell biology,* 2001 3, 839-843.

[137] Gurbuxani, S; Schmitt, E; Cande, C; Parcellier, A; Hammann, A; Daugas, E; Kouranti, I; Spahr, C; Pance, A; Kroemer, G; Garrido, C. Heat shock protein 70 binding inhibits the nuclear import of apoptosis-inducing factor. *Oncogene,* 2003 22:6669-6678.

[138] Ozoren, N and El-Deiry, W. Heat shock protects HCT116 and H460 cells from TRAIL-induced apoptosis. *Experimental Cell Research*, 2002 281, 175-181.

[139] Srivastava, P. Interaction of heat shock proteins with peptides and antigen presenting cells: chaperoning of the innate and adaptive immune response. *Annual Review of Immunology*, 2002 20, 395-425.

[140] Pockley, AG. Heat shock proteins as regulators of the immune response. *Lancet,* 2003 362, 469-476.

[141] Zhou, F; Xing, D; Chen, WR. Regulation of Hsp70 on activating macrophages using PDT-induced apoptotic cells. *International Journal of Cancer,* 2009 125, 1380-1389.

[142] Gupta, S; Knowlton, AA. Hsp60, Bax, Apoptosis and the Heart. *Journal of Cellular and Molecular Medicine,* 2005 9(1), 51-58.

[143] Billack, B; Heck, DE; Mariano, TM; Gardner, CR; Sur, R; Laskin, DL; Laskin, JD. (2002), Induction of cyclooxygenase-2 by heat shock protein 60 in macrophages and endothelial cells. *American Journal Physiology Cell Physiology*, 2002 283, C1267-C1277.

[144] Goh, YC; Yap, CT; Huang, BH; Cronshaw, AD; Leung, BP; Lai, PB; Hart, SP; Dransfield, I; Ross, JA. Heat-shock protein 60 translocates to the surface of apoptotic cells and differentiated megakaryocytes and stimulates phagocytosis. *Celluular and Molecular Life Science*, 2010. DOI 10.1007/s00018-010-0534-0

[145] Shan, YX; Liu, TJ; Su, SF; Samsamshariat, A; Mestril, R; Wang, PH. Hsp10 and Hsp60 modulate Bcl-2 family and mitochondria apoptosis signaling induced by doxo. *Journal of mol and Cellular Cardiology,* 2003 35, 1135-1143.

[146] Xanthoudakis, S; Roy, S; Rasper, D; Hennessey, T; Aubin, Y; Cassady, R; Tawa, P; Ruel, R; Rosen, A; Nicholson, DW. Hsp60 accelerates the maturation of pro-caspase-3 by upstream activator proteases during apoptosis. *The EMBO Journal,* 1999 18(8), 2049-2056.

[147] Kirchhoff, SR; Gupta, S; Knowlton, AA. Cytosolic Hsp60, apoptosis, and myocardial injury. *Circulation*, 2002 105, 2899-2904.

[148] Kim, SC; Stice, JP; Chen, L; Jung, JS; Gupta, S; Wang, Y; Baumgarten, G; Trial, J; Knowlton, AA. Extracellular Heat Shock Protein 60, Cardiac Myocytes, and Apoptosis. *Circulatuion Research*, 2009 105:1186-1195.

[149] Ghosh, JC; Siegelin, MD; Dohi, T; Altieri, DC. Heat shock protein 60 regulation of the mitochondrial permeability transition pore in tumor cells. *Cancer Research*, 2010 70(22), 8988-8993.

[150] Hershko, A. Early work on the ubiquitin proteasome system, an interview with Avram Hershko. *Cell Death Differentiation*, 2005 12(9), 1158-1161.

[151] Ciechanover, A. Intracellular protein degradation: from a vague idea thru the lysosome and the ubiquitin–proteasome system and onto human diseases and drug targeting. *Cell Death Differentiation*, 2005 12, 1178-1190.

[152] Melino, G. Discovery of the ubiquitin proteasome system and its involvement in apoptosis. *Cell Death and Differentiation*, 2005 12, 1155-1157.

[153] Bergmann, A. The role of ubiquitylation for the control of cell death in Drosophila. Cell Death and Differentiation, 2010 17, 61-67. Bernstein, M;. An unusual patient request. *Cancer Journal of Neurology Science*. 2009 36(3): 398.

[154] Galba´n, S; Duckett, CS. XIAP as a ubiquitin ligase in cellular signaling. *Cell Death Differentiation*, 2010 17, 54-60.

[155] Mace, PD; Shirley, S; Day, CL. Assembling the building blocks: structure and function of inhibitor of apoptosis proteins. *Cell Death Differentiation*, 2010 17, 46-53.

[156] Welchman, RL, Gordon, C; Mayer, RJ. Ubiquitin and ubiquitin-like proteins as multifunctional signals. *Nature Reviews Molecular Cell Biology*, 2005 6(8), 599-609.

[157] Haas, AL; Warms, JV; Hershko, A; Rose, IA. Ubiquitin-activating enzyme: Mechanism and role in protein-ubiquitin conjugation. *The Journal of biological chemistry,* 1982 257(5), 2543-2548.

[158] Thrower, JS; Hoffman, L; Rechsteiner, M; Pickart, CM. Recognition of the polyubiquitin proteolytic signal. *The EMBO Journal*, 2000 19(1), 94-102.

[159] Risseeuw, EP; Daskalchuk, TE; Banks, TW;Liu, E;Cotelesage, J; Hellmann, H;Estelle, M; Somers, DE; Crosby, WL. Protein interaction analysis of SCF ubiquitin E3 ligase subunits from Arabidopsis. *Plant Journal,* 2003 34(6), 753-767.

[160] Qian, SB; McDonough, H; Boellmann, F; Cyr, DM; Patterson, C. CHIP-mediated stress recovery by sequential ubiquitination of substrates and Hsp70. *Nature.* 2006 440(7083):551-5.

[161] Wyttenbach, A. Role of heat shock proteins during polyglutamine neurodegeneration: Mechanisms and hypothesis. *Jounal of molecular neuroscience*, 2004 23(l-2), 69-96.

[162] Shimura, H; Miura-Shimura, Y; Kosik, K. Binding of tau to heat shock protein 27 leads to decreased concentration of hyperphosphorylated tau and enhanced cell survival. *Journal of biological chemistry*, 2004 279(17), 17957-17962.

[163] Ross, CA; Pickart, CM. The ubiquitin–proteasome pathway in Parkinson's disease and other neurodegenerative diseases. *Trends in cell biology* 2004 14(12), 703-711.

[164] Kitagawa, K; Kotake, Y; Kitagawa, M. Ubiquitin-mediated control of oncogene and tumor suppressor gene products. *Cancer Science*, 2009 100(8), 1374-1381.

[165] Vodermaier, HC. APC/C and SCF: controlling each other and the cell cycle. *Current Biology.,* 2004 14(18), R787-796.

[166] Bornstein, G; Bloom, J; Sitry-Shevah, D; Nakayama, K; Pagano, M; Hershko, A. Role of the SCFSkp2 ubiquitin ligase in the degradation of p21Cip1 in S phase. *Journal of biological chemistry*, 2003 278(28):25752-25757.

[167] Yang, Y; Li, CC; Weissman, AM. Regulating the p53 system through Ubiquitination. *Oncogene,* 2004 23, 2096-2106.

[168] Bader, M; Steller, H. Regulation of cell death by the ubiquitin–proteasome system. *Current Opinion in Cell Biology*, 2009 21, 878-884.

[169] Bassermann, F; Pagano, M. Dissecting the role of ubiquitylation in the DNA damage response checkpoint in G2. *Cell Death Differentiation*, 2010 17, 78-85.

[170] Sun, Y. E3 Ubiquitin Ligases as Cancer Targets and Biomarkers1. *Neoplasia,* 2006 8, 645-654.

[171] Brat, DJ; Mapstone, TB. Malignant Glioma Physiology: Cellular Response to Hypoxia and Its Role in Tumor Progression. *Annals of Internal Medicine.,* 2003 138, 659-668.

[172] Xu, T; Zhou, Q; Zhou, J; Huang, Y; Yan, Y; Li, W; Wang, C; Hu, G; Lu, Y; Chen, J. Carboxyl terminus of Hsp70-interacting protein (CHIP) contributes to human glioma oncogenesis. *Cancer Science*, 2011 doi: 10.1111/j.1349-7006.2011.01888.x

[173] Veeriah, S; Taylor, BS; Meng, S; Fang, F; Yilmaz, E; Vivanco, I; Janakiraman, M; Schultz, N; Hanrahan, AJ; Pao, W; Ladanyi, M; Sander, C; Heguy, A; Holland, EC; Paty, PB; Mischel, PS; Liau, L; Cloughesy, TF; Mellinghoff, IK; Solit, DB; Chan, TA. Somatic mutations of the Parkinson's disease-associated gene PARK2 in glioblastoma and other human malignancies. *Nature Genetic*, 2010 42,77–82.

[174] Omerovic, J; Santangelo, L; Eleonora Maria-Rosaria Puggioni, Marrocco, J; Claudia Dall'Armi, Camilla Palumbo, Belleudi, F; Marcotullio, LD; Frati, L; Torrisi, MR; Cesareni, G; Gulino, A; Alimandi, M. The E3 ligase Aip4/Itch ubiquitinates and targets ErbB-4 for degradation. *The FASEB Journal*, 2007 21, 2849-2862.

[175] Marcotullio, LD; Ferretti, E; Greco, A; Smaele, ED; Screpanti, I; Gulino. A. Multiple Ubiquitin-Dependent Processing Pathways Regulate Hedgehog/Gli Signaling. Implications for Cell Development and Tumorigenesis. *Cell Cycle*, 2007 6(4), 390-393.

[176] Liu, Z; Oh, SM; Okada, M; Liu, X; Cheng, D; Peng, J; Brat, DJ; Shi-yong, Sun; Wei, Zhou, Wei, Gu; Keqiang, Ye. Human BRE1 Is an E3 Ubiquitin Ligase for Ebp1 Tumor Suppressor. *Molecular Biology of the Cell,* 2009 20, 757-768.

[177] Canettieri, G; Marcotullio, LD; Greco, A; Coni, S; Antonucci, L; Infante, P; Pietrosanti, L; Smaele, ED; Ferretti, E; Miele, E; Pelloni, M; Simone, GD; Pedone, EM; Gallinari, P; Giorgi, A; Steinkühler, C; Vitagliano, L; Pedone, C; Schinin, ME; Screpanti, I; Gulino, A. Histone deacetylase and Cullin3–RENKCTD11 Ubiquitin ligase interplay regulates Hedgehog signalling through Gli acetylation. *Nature Cell Biology*, 2010 12, 132-142.

[178] Ji, P; Smith, SM; Wang, Y; Jiang, R; Song, SW; Li, B; Sawaya, R; Bruner, JM; Kuang, J; Yu, H; Fuller, GN; Zhang, W. Inhibition of gliomagenesis and attenuation of mitotic transition by MIIP. *Oncogene,* 2010 29(24), 3501-3508.

[179] Harley, ME; Allan, LA; Sanderson, HS; Clarke, PR. Phosphorylation of Mcl-1 by CDK1-cyclin B1 initiates its Cdc20-dependent destruction during mitotic arrest. *The EMBO Journal*, 2010 29(14), 2407-2420.

[180] Lehman, NL. The ubiquitin proteasome system in neuropathology. *Acta Neuropathology*, 2009 118, 329-347.

[181] Bhatia, N; Thiyagarajan, S; Elcheva, I; Saleem, M; Dlugosz, A; Mukhtar, H; Spiegelman, VS. Gli2 Is Targeted for Ubiquitination and Degradation by β-TrCP Ubiquitin Ligase. *The Journal of biological chemistry,* 2006 281(28), 19320-19326.

[182] Dai, B; Pieper, RO; Li, D; Wei, P; Liu, M; Woo, SY; Aldape, KD; Sawaya, R; Xie, K; Huang, S. FoxM1B regulates NEDD4-1 expression, leading to cellular transformation and full malignant phenotype in immortalized human astrocytes. *Cancer research*, 2010 70(7), 2951-2961.

[183] Wang, X, Trotman, LC, Koppie, T, Alimonti, A, Chen, D,Gao, Z, Wang, J, Erdjument-Bromage, H, Temst, P, Cordon-Cardo, C, Pandolfi, PP, Jiang, X. NEDD$-1 Is a Proto-Oncogenic Ubiquitin ligase for PTEN. *Cell,* 2007 128(1), 129-139.

[184] Kwak, YD; Wang, B; Pan, W; Xu, H; Jiang, X; Liao, FF. Functional interaction of phosphatase and tensin homologue (PTEN) with the E3 ligase NEDD4-1 during neuronal response to zinc. *Journal of biological chemistry*, 2010 285(13), 9847-9857.

[185] Yang, L; Wang, S; Sung, B; Lim, G; Mao, J. Morphine induces ubiquitin-proteasome activity and glutamate transporter degradation. *Journal of biological chemistry*, 2008 283(31), 21703-21713.

[186] Lee, H; Tsygankov, AY. c-Cbl regulates glioma invasion through matrix metalloproteinase 2. *Journal of cellular biochemistry*, 2010 111(5), 1169-1178.

[187] Piepmeier, JM. The future of neuro-oncology. *Acta Neurochirurgica,* 2009 151, 1343-1348.

[188] Reardon, DA; Wen, PY. Therapeutic Advances in the Treatment of Glioblastoma: Rationale and Potential Role of Targeted Agents. *The Oncologist,* 2006 11, 152-164.

[189] Sathornsumetee, S; Rich, JN. New approaches to primary brain tumor treatment. *Anti-Cancer Drugs*, 2006 17, 1003-1016.

[190] Norden, AD; Drappatz, J; Wen, PY. Targeted drug therapy for meningiomas. *Neurosurg. Focus*, 2007 23(4), E12.

[191] Hashizume, R; Ozawa, T; Gryaznov, SM; Bollen, AW; Lamborn, KR; Frey II, WH; Deen, DF. New therapeutic approach for brain tumors: Intranasal delivery of telomerase inhibitor GRN163. *Neuro-Oncology,* 2008 10, 112-120.

[192] Merrell, RT; Quant, EC; Wen, PY. Advances in Treatment Options for High-grade Glioma –Current Status and Future Perspectives. *European Neurological Review*, 2010 5(1), 49-55.

[193] Hwang, Y; Latha, K; Gururaj, A; Rojas, M; Bogler, O. Aberrant EGFR signaling in glioma. *Cancer Drug Discovery and Development*, 2009 2, 441-459.

[194] Erwin, G; Van Meir, CG; Hadjipanayis, AD; Norden, MPH; Hui-Kuo Shu, Patrick, Y; Wen, and Jeffrey, J; Olson.. Exciting New Advances in Neuro-Oncology The Avenue to a Cure for Malignant Glioma. *CA Cancer J. Clin.* 2010 60(3): 166–193.

In: Signaling, Gene Regulation and Cancer
Editors: Shree Ram Singh and Manoj K. Mishra

ISBN: 978-1-61942-088-5

Chapter 15

BIOLOGICAL FUNCTIONS OF IL-4/IL-13, SIGNALING CASCADES AND IMPLICATION IN GROWTH AND DEVELOPMENT OF TUMOR

Praveen Deepak and Arbind Acharya*
Department of Zoology, Faculty of Science,
Banaras Hindu University, Varanasi, India

ABSTRACT

Interleukin-4 and 13 are immunoregulatory cytokines secreted mainly by activated Th2 cells. They regulate multiple biological functions such as proliferation, differentiation and apoptosis in various cell types of hematopoietic and non-hematopoietic origin. Both cytokines are critical to the differentiation of Th0 cells into Th2 cells and simultaneous suppression of Th1 type T cell-mediated immunity thus involved in the amplification of Th2 type of immune responses. Acting on the every immune cell population through their cell surface receptors, they are able to subdue the antitumor function of immunopotent cells and tumor immunosurveillance mechanism(s), and therefore favor unrestricted growth and progression of tumor. Both cytokines have shared biological effects due to common receptor subunits. They have both type I and type II receptor complexes forming at least four receptor complexes viz IL-4Rα-γc, IL-4Rα-IL-13Rα1, IL-4Rα-IL-13Rα2, IL-13Rα2 expressed in many tumor cell types. IL-4 and IL-13 mediate signal through two different pathways IL-4-IRS pathway involving insulin receptor substrate molecules leading to the activation of MAPK and PKC and JAK-STAT pathway resulting in the activation of STAT6. MAPK and PKC activated in response to IL-4 and IL-13 leads to tumor cell growth and survival by activating proliferation and anti-apoptotic mechanism(s), while STAT6 activation leads to the expression of CD23, MHC class II, germ line immunoglobulin ε, CD45, SOCS etc. In addition, signaling through IL-13Rα1 leads to activation of several STAT molecules including STAT1α, 3, 5α, 5β and STAT6, all results in the proliferation of tumor cells and differentiation of Th2 cells. IL-13Rα2, previously known as decoy receptor, also

* Address for correspondence: Prof. Arbind Acharya - Department of Zoology, Faculty of Science, Banaras Hindu University, Varanasi – 221 005, U. P., India. Tel: +91 542 2307140 (Office), 6702518 (Lab), 2368965 (Res). Fax: +91 542 2368174. E-mail: acharya@bhu.ac.in, acharya_arbind@yaoo.in.

transduce signal and leads to the activation of ERK and p38 MAPK which regulate proliferation and metastasis of tumor cells. The signaling cascades induced by IL-4 and IL-13 are modulated by many adaptor and regulatory molecules, which are also known to be associated with many hematological and non-hematological malignancies and therefore they could be an attractive target to develop a therapeutic agent for the treatment of cancer.

INTRODUCTION

Human IL-4 is an 18 to 20 kd glycoprotein composed of 153 amino acids with 50% homology to mouse IL-4 and the human IL-13 is a non-glycosylated protein of 132 amino acids with a molecular weight of 10 kDa containing five cysteine residues ~66 % identical with that of the mouse IL-13. Both are immunoregulatory cytokines expressed and secreted mainly by activated Th2 lymphocytes [1]. Both cytokines are also produced by activated basophils as well as by mast cells derived from nasal or bronchial biopsy specimens from patients with allergic rhinitis and asthma, respectively upon cross-linkage of the high affinity receptor for immunoglobulin E (IgE) [2, 3]. In contrast to IL-4, IL-13 is not produced by eosinophils or neutrophils even after stimulation. Small amount of it is also produced by both T helper (Th1) cells and $CD8^+$ T cells. It has been demonstrated that natural killer T cells are also an important source of IL-13 early in the allergic response [4]. IL-13 transcripts have been found in normal B cells after CD40 cross-linking. IL-13 protein and mRNA has been detected in freshly isolated malignant and Epstein-Barr virus (EBV)–transformed human B lymphocytes, where IL-13 receptors are not expressed.

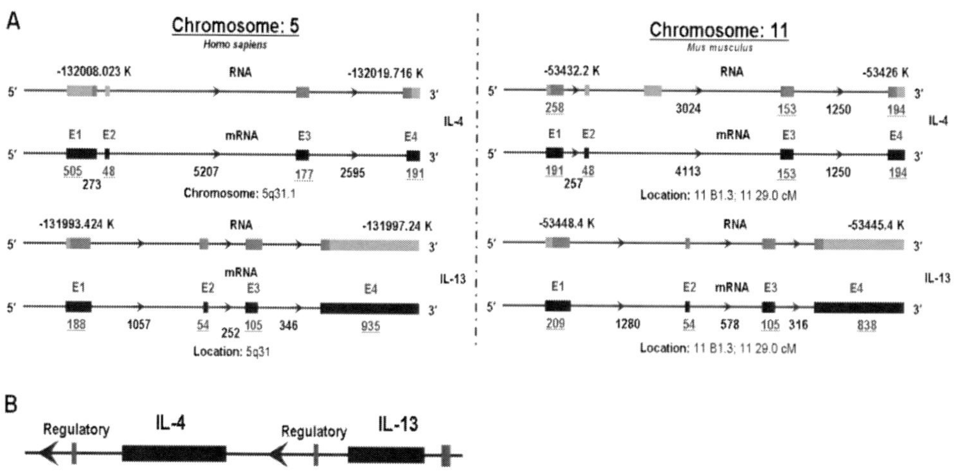

Figure 1. Genomic organization of the **IL-4 and IL-13** gene on chromosome 5 and 11 (not drawn to scale). The genomic organization of the gene was determined from the genomic (Gene ID 16189 for mouse and 3565 for human IL-4, and 16163 for mouse and 3596 for human IL-13). The genes encoding IL-4 and IL-13 are located in tandem within a cluster of cytokine genes on human chromosome 5q and 11b (A). Both genes are located on same locus on chromosome flanked by two regulatory regions (B). IL-4 gene is composed of 8996bp DNA that transcribes into 921bp mRNA in human and 6206bp DNA transcribing for 605bp mRNA in mice. The gene for IL-13 is composed of 2937bp DNA and transcribes into a 1287bp mRNA in human and 3380bp DNA transcribing for 1217bp mRNA in mice. Both genes have four exons as indicated (E1 to E4).

IL-4 was identified in 1982 by Howard et al. as B cell growth factor while much later IL-13 was first discovered in 1989 by Brown et al. by cloning of induction-specific mRNA from a mouse Th2 cell line and in humans by Morgan et al. in 1992 and named as P600 protein (also designated as NC30) [5, 6]. In both human and mouse, the IL-13 gene is closely located to the IL-4 gene being only 25kb upstream to IL-4 gene on the chromosome 5q31 and 11 respectively (Figure 1) [7, 8]. It has been thought that both have originated as duplication event during evolution and therefore they have ~30% homology in the amino acid sequences and shares a conserved hydrophobic structural core [7]. Due to their homology, both of these cytokines show many structural and functional similarities, and are critical in regulating some inflammatory and immune response.

Biological Activities of IL-4 and IL-13

Both have shared biological effects, however at some points both cytokines also exert specific effects, which can be independent to each other. IL-13 elicits many, but not all, of the biologic actions of IL-4 as shown in Figure 2. IL-4 is, however, distinguished from IL-13 by its T cell growth factor activity and its ability to drive differentiation of Th0 precursors toward the Th2 lineage. (Th2 cells secrete IL-4 and IL-5 and lead to a preferential stimulation of humoral immunity. In contrast, Th1 cells which produce IL-2 and interferon [IFN]-γ, lead to a preferential stimulation of cellular immunity.)

Effect of IL-4 And IL-13 on Monocyte/Macrophage System

Both IL-4 and IL-13 have been demonstrated to exhibit pleiotropic effects on human peripheral blood monocytes and mouse granulocyte-macrophage colony stimulating factor (GM-CSF)-derived bone marrow macrophages. They induce morphological and surface antigen changes in human monocytes and mouse GM-CSF-derived bone marrow macrophages [1, 7, 8]. Both induce the formation of long cytoplasmic protrusions that tightly adhere to substratum, giving them a dendritic appearance [8]. Short term treatment of monocytes have been shown to enhance the expression of the surface antigens MHC class I and II, CD11b CD11c, CD18 and CD23, but down regulates the expression of CD16, CD32 and CD64 [9, 10-16]. Both IL-4 and IL-13 inhibits the production of IL-1α and β, IL-6, IL-8, IL-12p35 and p40, GM-CSF, M-CSF, G-CSF, IFN-γ and TNF-α, and enhances the production of IL-10 and IL-1 receptor antagonist (IL-1Ra) synthesis in monocytes and macrophages [9, 13, 16-20]. However, long term treatment of IL-4 or IL-13 for ≥ 20hrs enhances the production of IL-12 and TNF-α in response to LPS or *Staphylococcus aureus* several fold in monocytes and macrophages [21]. It has been demonstrated that these cytokines synergies with IL-10 to inhibit lysosomal enzyme secretion by human monocytes and lamina propria mononuclear cells from patients with inflammatory bowel disease (IBD) [22].

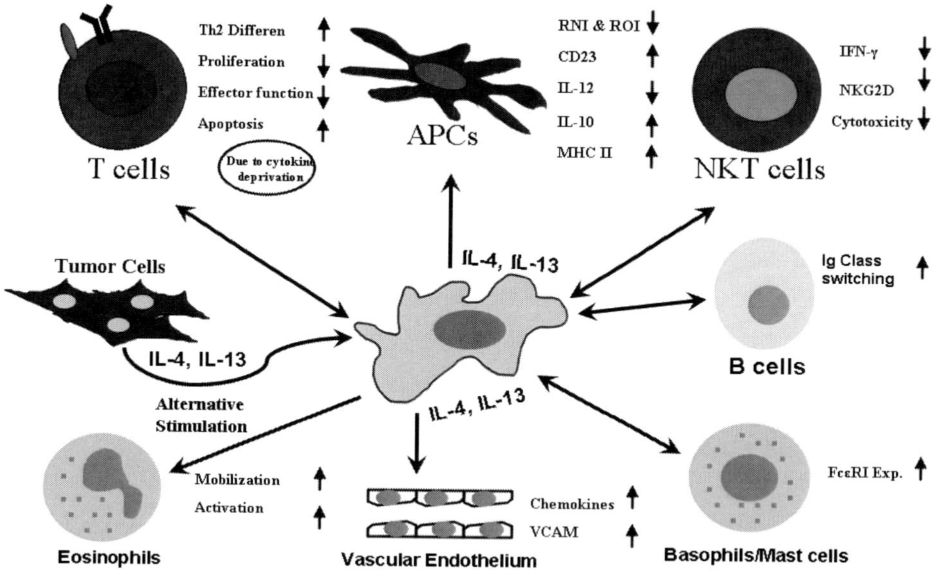

Figure 2. Biological action of IL-4/IL-13 on various hematopoietic/ nonhematopoietic cells. IL-4 and IL-13 are mainly produced by T cells, NKT cells, macrophages and a little bit by B cells in addition to its secretion from tumor cells. In monocytes/macrophages and dendritic cells, they increases CD23 and MHC II expression and inhibit pro-inflammatory cytokine production. They induce the production of Th2 type cytokines in monocytic lineage of cells and in NKT cells and guides the differentiation of NK cells to NK2 type which is immunosuppressive in nature that enhance the differentiation of regulatory T cells (Treg) thereby down regulating tumor-immunosurveillance. They enhance differentiation of TH2 cells and inhibit the differentiation of Th1 cells. Acting on B cells, they induce isotype class switching and enhance the IgE production while acting on mast cells, they enhance the expression of FcεRI and thus increase IgE priming, whereas they enhance the mobilization of eosinophils and therefore increase their recruitment and activation. Amongst nonhematopoietic cells, they exert its effect on fibroblast and enhance collagen synthesis and fibrosis (as in lung fibrosis), where they induce the expression of chemokines in respiratory epithelium and vascular endothelium, and result in mucus hypersecretion and goblet cell hyperplasia in respiratory tract. They enhance VCAM expression in vascular endothelium and in addition, stimulate the proliferation of smooth muscle cells (not shown here).

Though, IL-4 and IL-13 has been shown to up regulate the expression of the mannose-fucose receptor [23], ferritin and transferrin receptor [24] and stimulates chemotaxis in monocytes and macrophages [25]. They modestly decreased the number of Fcγ receptors [26] and potently suppressed the generation of nitric oxide (NO) [23, 24, 26] by down-regulating iNOS expression as observed in resident peritoneal macrophages [27] and in the J774 macrophage cell line [28]. However, no alteration has been found in the iNOS mRNA levels in IL-13 pre-treated mouse bone-marrow derived macrophages and peritoneal exudates macrophages [28, 29]. In contrary to this observation, El-Gayar et al., 2003 did observed the suppression in iNOS protein level after pre-treatment of macrophages with IL-13 for 16hrs [30]. Correspondingly, IL-13 has been shown to inhibit the 12-O-tetradecanoylphorbol-13-acetate (TPA)-triggered reactive oxygen intermediate (ROI) production by stimulating rapid and transient phosphoinositide hydrolysis and intracellular Ca^{2+} mobilization and therefore subsequent accumulation of cAMP [31]. It has been found that IL-13 induces tyrosine phosphorylation of phospholipase Cγ-1 following IRS-2 association in human monocytes, which is resulted into cAMP accumulation and subsequent ROI inhibition [32]. Furthermore, both IL-4 and IL-13 has been shown to inhibit antibody-dependent cellular cytotoxicity (ADCC) activity of macrophages [9].

Both IL-4 and IL-13 have been reported as suppressors of murine osteoclast formation. In contrast to the effect of these molecules on murine osteoclast formation, IL-4 and IL-13 were found to be powerful inducers of osteoclast formation and bone resorption when added to human peripheral blood mononuclear cell (PBMC) cultures containing nonadherent PBMCs for 4 days. In contrast, both molecules significantly suppressed osteoclast formation in lymphocyte-depleted cultures [33]. Therefore, it is clear that both of the cytokines exerts its immunosuppressive effects on monocyte/macrophage system and inhibits vital secretory functions and expression of surface molecules. They act as one of the factor, which is responsible for skewing of M1 phenotype to M2 phenotype of macrophages in tumor-bearing host [34], which play an immunostimulatory role for the growth and progression of tumor rather induction of anti-tumor immunity in the host.

EFFECT OF IL-4 AND IL-13 ON B CELLS AND HUMORAL IMMUNE RESPONSE

IL-4 and IL-13 both have found to play an enhancing effect on the proliferation and differentiation of B lymphocytes. In the presence of costimulatory signals, both cytokines can induce B cell proliferation [7, 35-37] as well as isotype switching from IgM to IgG4 and IgE [35-39]. It enhances proliferation of anti-IgM or anti-CD40-activated human B cells [7, 36, 37] and modulates its surface antigens [1, 7, 35]. However, it has also been reported that **IL-13** enhances the survival not the proliferation of B cells that results in high level of antibody production [40]. A soluble form of IL-13Rα1 has been found to promote IgG2a and IgG2b production by murine germinal center B cells [41]. It has been reported that IL-13 does not stimulate the same protein tyrosine kinase activity in LPS-activated murine vs human B cells [42, 43]. However, one cannot rule out that these differences may be due to the comparison of different B cell populations.

The humoral immune response against an oral immunogen in mice is significantly reduced upon *in vivo* treatment with anti-interleukin-13 antibodies [44]. It has been found that both IL-4 and IL-13 synergizes with the mouse and human CD40L to induce IgM, total IgG, IgG4, and IgE, but not IgA, production by highly purified B cells [45, 46]. Anti-IL-4 antibodies inhibited IL-4 and COS-7/hCD40L induced Ig production by B cells, but had no effect on IL-13 and COS-7/hCD40L induced B cell differentiation, indicating that IL-13 and hCD40L induced immunoglobulin production, including isotype switching to IgE, independently of IL-4. It was also observed that the number of pulmonary metastases decreased more obviously when the mice were intravenously injected with the mixture of wild-type B16 cells and the serum from the mice vaccinated with the IL-4 gene-modified B16 cells [47]. Human B lymphocytes produce significant IL-13, and neutralization of IL-13 impairs IgE synthesis and therefore IL-13 has been suggested as important autocrine growth factor for IgE-producing B lymphocytes [48]. In atopy, persistently stimulated $CD4^{+}45RO^{+}$ memory/effector T cells provide contact-dependent activation signals to B cells, and that these cells may induce IgE switching largely via secretion of IL-13 in atopy [49] and via secretion of both IL-4 and IL-13 in asthma [50]. Both cytokines act at different stages of the

B cell maturation pathway and enhance the expression of CD23/FceRII and class II MHC antigens on resting B cells [37]. It has been reported that both human and mouse B cells express IL-13 receptor subunit IL-13Rα1 that transcend signals to produce antibody in B cells [51-53].

EFFECT OF IL-4 AND IL-13 ON T CELLS AND CELL-MEDIATED IMMUNITY

The role that both cytokines IL-4 and IL-13 play in regulating T cell function is less clear. Both have been reported to exert limited effects on mouse and human T lymphocytes *in vitro* [54]. IL-13 fails to promote the growth of mitogen activated T cells [54, 55]. Alternatively, IL-13 may act directly on T cells under some circumstances as IL-13Rα1 mRNA has been detected in human T cells [56]. Both IL-4 and IL-13 has been reported to induce Stat6 phosphorylation in mitogen activated human T cells [57] and to block the migration of unstimulated human T cells in response to a chemokine gradient [58]. It is possible that T cells express IL-13Rα1 *in vivo* but lose this receptor chain during *in vitro* culture [59] or that increased expression of IL-13Rα2 by T cells inhibits their ability to respond to IL-13 [60].

Though, T cells have been reported to lack IL-4 and IL-13 receptors and therefore do not bind to IL-4 and IL-13, however IL-4 has been reported to bind to murine helper T cell line HT-2 [61], although IL-4 and IL-13 stimulation of anti-CD3-activated splenocytes can modulate the survival, proliferation, and differentiation of T lymphocytes [62, 63] in a STAT6-dependent manner [64]. The explanation of this apparent paradox may be that both of these cytokines act on a cell population in these cultures that promotes Th2 development. Notably, very recent studies suggest that IL-13 regulates the maturation of dendritic cells (DC) [65], which in turn it may modulate T cell function. It has been suggested that IL-13 promoted T cell production of Th2 cytokines and suppression in IFN-γ may indirectly regulate T cells and CTL responses by modulating DC function [66, 67]. It has been found that impaired maturation of dendritic cells leads to low IL-12 level in internal milieu in the neonates. Low IL-12 results in up regulated expression of IL-4Rα, which physically associates with IL-13Rα1 resulting in the formation of a potentially hazardous heteroreceptor on Th1 cells. IL-4 from Th2 cells interacts with these heteroreceptor to signal the death of Th1 cells, leading to the Th2 bias of neonatal immunity [68]. IL-4 has been shown to have a direct inhibitory effect on the development of human Th1 cells *in vitro* [69]. It has been found that IL-13 enhances the production of TGF-β in non-lymphoid cells, which in turn responsible for down regulation of Th1 immunity and CTL response in tumor-bearing host [70]. Furthermore, both IL-4 and IL-13 have been found to play an important role in generating Forkhead box P3-expressing $CD25^+$ regulatory T cells (Tregs) extrathymically in an antigen-dependent manner [71], which is one of the strong factor involved in the tumor-mediated immunosuppression in tumor-bearing host.

Recently, a subpopulation of T cells, Th17 cells have been found to express functional IL-13Rα1 and that IL-13 negatively regulates IL-17A and IL-21 production by decreasing retinoic acid-related-γ T (ReγT) expression and increasing phosphorylation of STAT6 and GATA3 expression. IL-13 blocking attenuates Th17 cell function and secretion of IL-17 [72]. A soluble form of the receptor was immunoprecipitated from the supernatant of activated peripheral T cells, which suggests that T cell IL-13Rα1 might have functions unrelated to the capacity to form a type II IL-4/IL-13R with IL-4Rα [51]. GM-CSF and IL-4 synergistically trigger dendritic cells to acquire retinoic acid-producing capacity. Retinoic acid is responsible for imprinting gut-homing specificity on lymphocytes and further enhances $Foxp3^+$ regulatory T cell differentiation [63].

Role of IL-4 and IL-13 Tumor Immunity and Tumor Immunosurveillance

Tumor cells are known for their genetic instability that results in profound antigenic variation. The genetic instability of tumor cells results in the formation of different tumor variants, which can breach immune response and therefore escape from host's natural immune surveillance mechanism. In escaping tumor from immune surveillance, several immunosuppressive agents and cytokines produced by tumor cells play an indispensable role. Among the cytokines, Th2 cytokines are involved in the negative regulation of immune responses in the tumor-bearing host. As discussed above, IL-4 and IL-13 are also Th2 cytokines and are produced by many tumor types. They impair immune responses generated against the tumor cells.

However, several studies suggest the role of IL-4 in both tumor clearance and tumor tolerance. The presence of IL-4 in the tumor microenvironment enabled the clearance of various tumor types in mice such as melanoma, plasmacytoma, adenocarcinoma, renal carcinoma etc. However, the precise mechanism is yet to know. IL-4 produced by type II polarized $CD4^+$ T cells have been found to confer protection against immunogenic murine melanoma by the recruitment and activity of eosinophils [73]. Adoptive transfer of IL-4 producing tumor-specific $CD8^+$ T cells has been found to restrict growth and metastasis of lung tumor [74]. It has been postulated that inhibitory effect of IL-4 on angiogenesis resulted in the restricted growth of tumor in these studies [75]; therefore it might be assumed that this growth inhibitory effect of IL-4 is restricted to solid tumors. In majority of studies, IL-4 has been shown to impair tumor clearance. Administration of IL-4-expressing mesothelioma cells resulted in accelerated growth of non-IL-4-producing tumor cells [76], which has been thought to occur because of skewing of immune responses to a type 2 polarized T cell responses. Furthermore, mice receiving IL-4 producing tumor cells has been found more prone to develop secondary tumors and also clearance of primary tumor has been found delayed in this study. Altogether, studies suggest that though IL-4 plays a significant protective role in solid tumor, in major cases, it restrict the onset of protective tumor immunity in the host and therefore it has either no role or very least role in immunosurveillance of tumor.

Recently, a role of IL-13 in negative regulation of tumor immunosurveillance through the study in Stat6-deficient knockout mice has been proposed. It has been observed that IL-13 produced by $CD4^+$ CD1d-restricted NKT cells induces TGF-β production by $CD11b^+Gr\text{-}1^+$ myeloid cells, which in turn suppresses the tumor immunosurveillance mechanism by downregulating $CD8^+$ CTL response [67, 77]. Interestingly, in the osteosarcoma model, neither the IL-4R-STAT6 signal pathway, which is induced by IL-13 and/or IL-4, nor TGF-β was necessary for the suppression of tumor immunosurveillance [78]. The blockade of IL-13 has been shown to restore the tumor immunosurveillance in tumor-bearing host also suggesting that IL-13 is involved in negative regulation of tumor immunosurveillance [79, 80].

IL-4/IL-13 RECEPTOR

All these above mentioned biological functions of IL-4 and IL-13 are achieved through receptors present on target cells. Both, IL-4 and IL-13 are type I cytokines and therefore it transmits signal through type I receptors. The type I receptors are characterized by the presence of four conserved cysteine residues, W-S-X-W-S motif, fibronectin type II modules in the extracellular domain and proline rich box regions (Box 1) in the intracellular domain [81]. Proline rich box regions in the intracellular domain which have no intrinsic kinase activaity itself, but are essential for binding of Janus tyrosine kinases (JAK). Proline rich box regions only provides a constitutive association of JAKs [81]. Association of JAK to Box 1 region of intracellular domain leads to the recruitment of downstream signaling molecules and then downward signaling cascade ensues leading to the activation of different transcription regulators necessary for the transcription of IL-4/IL-13 responsive genes. IL-4 and IL-13 have both type I and type II receptor complexes and type I receptor forms heterodimers. There are total of at least four receptors between IL-4 and IL-13 with different receptor subunit combinations and binding (Table 1).

Table 1. IL-4/IL-13 Receptor Complexes

Cytokine	Receptor Complexes	Ligand-Binding Properties	
		IL-4	IL-13
IL-4	IL-4Rα	Binds well	No binding
	IL-4Rα, γc	Binds well with greater affinity	Limited binding; no competition with IL-4
	IL-4Rα, IL-13Rα1	Competes with IL-13	High affinity; competes with IL-4
	IL-4Rα, IL-13Rα2	Limited binding	Strong binding to IL-13Rα2 subunit
IL-13	IL-13Rα1	Low binding	Low binding; Competes with IL-4
	IL-13Rα2	No binding	Strong binding
	IL-4Rα, IL-13Rα1	High affinity; Competes with IL-13	Low affinity; Do not compete with IL-4
	IL-4α, IL-13Rα2	Unpredicted; Limited binding	Unpredicted; High affinity

IL-4 RECEPTOR

IL-4 receptors are composed of two transmembrane protein chains; α chain is a 140kD protein chain having ligand binding domain and gamma chain (γc) shared by other cytokines such as IL-2, IL-7, IL-9, IL-15 and IL-21. The IL-4Rα chain binds IL-4 with high affinity (dissociation constant = 20 to 100 pM) [61] leading to dimerization with another protein to form either a type I or type II receptor. In cells of hematopoeitic lineage, the type I receptor arises by recruitment of a common γc [2, 3, 82]. In non-hemopoietic cells, the type II receptor is formed by interaction of IL-4Rα with IL-13Rα1 instead of γc [3, 82, 83]. The cytoplasmic domain of IL-4Rα chain contains 5 tyrosine (Y) residues at position Y497, Y575, Y603, Y631, and Y713 which are highly conserved among different species and an extracellular W-S-X-W-S motif that is required for ligand binding. The most proximal tyrosine residue (Y497) is critical for generating a proliferating signal, and the second, third and fourth tyrosine residues are within a conserved sequence motif for the activation of STAT6, while the distal tyrosine residue (Y713) is within an ITIM (immunoreceptoe tyrosine-based inhibitory) motif that serves as a docking site for different phosphatases (Figure 3). IL-4Rα receptor is expressed on a variety of hematopoietic cells such as monocytes, macrophages, mast cells, T cells and B cells, and also on non-hematopoietic cells such as fibroblasts, neuroblasts, keratinocytes and hepatocytes. However, it is expressed at relatively low frequencies (100 to 5000 copies per cell) on the non-hematopoietic cells. Alternative splicing of the IL-4R gene can result in the production of a soluble form of IL-4R (sIL-4R) which has no signaling abilities and therefore probably works as a decoy and an antagonist to its ligand [84, 85].

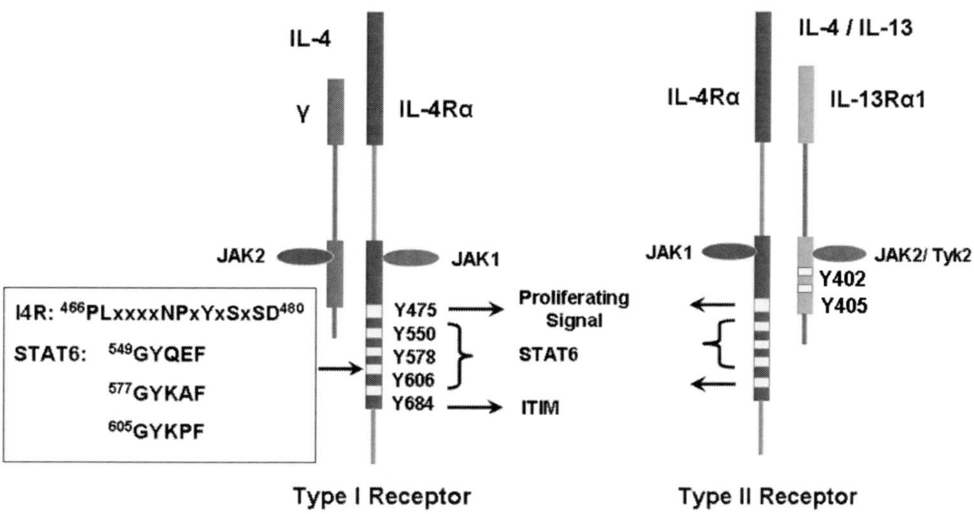

Figure 3. Schematic representation of IL-4 type I and type II receptor. Type I receptor is composed of IL-4Rα and γc subunit and type II receptor is composed of IL-4Rα and IL-13Rα1 subunits. IL-4Rα has five conserved tyrosine residues at the position of 475, 550, 578, 606 and 684, and a praline rich box region (Box-1). JAK1 interacts and binds to Box-1 region of IL-4Rα. IRS1/2 interacts with conserved tyrosine residue at the position of 475 having consensus sequence ^{466}PLxxxxNPxYxSxSD^{480} known as I4R to which PTB domain of IRS molecules associates and transduce mainly IL-4-induced proliferation signal (IL-4/IRS Pathway). Other consecutive three conserved tyrosine residues have STAT6 consensus sequences and therefore binds to STAT6 leading to the activation of STAT6 (JAK-STAT Pathway). While fifth conserved tyrosine residue interacts with ITIM (Immunotyrosine based inhibitory motif)and play a regulatory role in JAK-STAT pathway. In contrast, IL-13Rα1 has one praline rich box region (Box-1) and two conserved tyrosine residues at the position of 402 and 405, which are found to serve the docking sites of STAT3 in vitro, however it might also be possible that these tyrosine residues may serve as the docking sites for additional signaling intermediates.

Studies involving blocking of IL-4Rα with IL-4Rα antibodies or IL-4Rα agonists have shown that both IL-4 and IL-13 responses were inhibited, which suggests that IL-4 uses another receptor subunit that is also utilized by IL-13 for its biological activities [86, 87]. In another studies from γc mutant patients or in the patient where γc is absent have suggested that the biological function of both IL-4 and IL-13 is intact, which is further lead to the speculation that there must be another receptor for IL-4 and IL-13. Later, it has been found that IL-4Rα interacts with IL-13Rα1 to form a heterodimer and drives the IL-4 downstream signaling. IL-4Rα/IL-13Rα1 heterodimer complex is utilized by IL-4 as an alternate receptor, especially in the non-hematopoietic cells, where γc is not expressed.

IL-13 Receptor

IL-13 also binds to two different receptor complexes; IL-13Rα1 and IL-13Rα2. Both IL-13Rα1 and IL-13Rα2 are the members of hematopoietin receptor superfamily and is a type I receptor. Both IL-13 receptors genes mapped to the X chromosome (IL-13Rα1 on Xq24 and IL-13Rα2 on Xq13.1q-28 in human) (Figure 4) and share 37% homology at the amino acids sequences possibly suggesting a role in X-linked immune diseases. The 5′ flanking region of IL-13Rα1 gene (IL-13RA1) has been found to lack both TATA and CCAAT boxes, with a predicted transcription initiation site at -123 base pairs relative to the start codon. Both of these receptors specifically bind to IL-13.

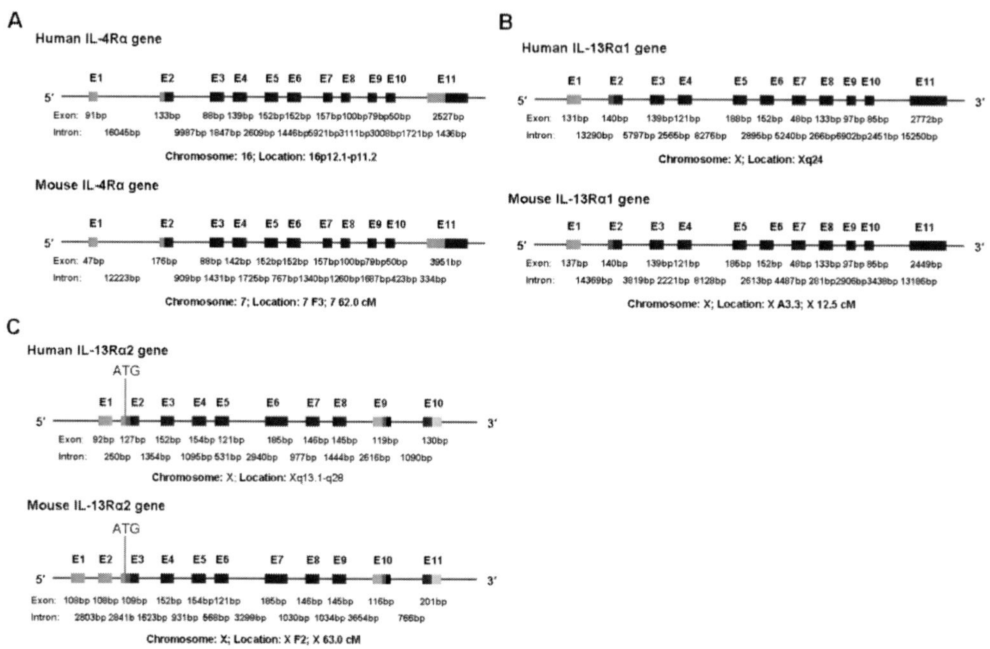

Figure 4. Human and mouse interleukin IL-4Rα, IL-13Rα1 and 13Rα2 gene structures. Gene structures for human and mouse IL-4Rα (A), IL-13Rα1 (B) and 13Rα2 (C) based on mRNA sequences reported in GenBank (NM_000418.2 and NM_001008700.3 for IL-4Rα, NM_001560.2 and NM_133990.4 for IL-13Rα1, NM_000640 and NM_008356 for IL-13Rα2, respectively). The gene structure is well conserved between human and mouse.

IL-13Rα1 (CD213A1), previously known as NR4, is encoded as a protein of 427 amino acids sequences which includes a 21-amino acids signal and a 60 amino acids intra-cellular domain [88]. Extracellular portions of IL-13Rα1 are composed of three fibronectin III domains-D1, D2, and D3-of which the latter two comprise CRH. Amino acids, Leu319 and Tyr321 in the CRH of human IL-13Rα1 are critical residues for binding to IL-13 [89]. The D1 domain is important for binding to IL-13, but not to IL-4. Four positionally conserved cysteine residues in the D2 domain form two disulfide bonds, and a W-S-X-W-S sequence locates in the F'-G' loop in the D3 domain, both of which are critical for the receptors to position correctly and bind to ligands [95]. IL-13Rα1 is a 65-70kD glycosylated protein that binds to IL-13 with low affinity (dissociation constant = 2-10 nM) by itself, but when paired with IL-4Rα, it binds IL-13 with greater affinity (dissociation constant = 400 pM). Pairing of IL-13Rα1 with IL-4Rα forms a functional type II receptor complex for signaling of both IL-4 and IL-13. It has been demonstrated that IL-13 first binds to IL-13Rα1, which then recruits the IL-4Rα forming the type II IL-4R and increases the binding affinity approximately two fold [90]. IL-13 after binding to type II IL-4 receptor, which has tyrosine kinase activity, transmits signal [91]. Because IL-4 and IL-13 both use the type II receptor, the molecular basis of the cytokine' differing functional responses is unclear. However, over expression of γc have been found to decrease the biologic activity of IL-13 by decreasing the expression of IL-13Rα1 and the ration of type I receptor IL-4Rα/γc to the type II IL-4 receptor IL-4Rα/IL-13Rα1 is increased, biologic relevance of which is still unclear [92]. The interaction between IL-4Rα and IL-13Rα1 has been demonstrated to be species specific and therefore human IL-13Rα1 could not interact with mouse IL-4Rα to form a functional receptor, although the open reading frame of human IL-13Rα1 has 81% nucleotides and 76% amino acids similarities with that of murine. [93]. However, the mouse IL-2R γc can interacts and makes a complex with both human IL-4Rα and mouse IL-4, which suggests that the epitope on IL-4Rα is conserved among both the species for associating with IL-2R γc. Soluble IL-13Rα1 has also been detected as a 42kDa protein in activated human T cell supernatant; however biologic significance of this soluble protein has not been investigated yet.

IL-13α2 (CD213A2), on the other hand, is a comparatively short peptide transcribed from *IL-13RA2* gene. The cDNA of human IL-13Rα2, also known as CT19, originally isolated from the Caki-1 renal cell carcinoma cell line encodes a 383 amino acids protein including a 26 amino acid signal sequences, and a 17 amino acids intracellular domain [94]. The extracellular domain of IL-13Rα2 is composed of D1, D2, and D3, of which the latter two comprise CRH. Amino acids, Tyr207, Asp271, Tyr315, and Asp318 in the CRH are critical residues for binding to IL-13 [89]. Tyr315 in IL-13Rα2 is a conserved hydrophobic amino acid residue critical for positioning of the receptor. There exist four positionally conserved cysteine residues in the D2 domain forming two disulfide bonds, and a W-S-X-W-S sequence located in the F'-G' loop in the D3 domain [95, 96]. It has 59% amino acids sequence similarities with that of murine counterpart [97]. N-linked glycosylation in the extra-cellular domain of IL-13Rα2 is essential for binding of IL-13 [98]. Transcripts of IL-13Rα2 have been readily found in the spleen, liver, lung, thymus and brain (Table 2) indicating its wide role in host. Although, it binds to IL-13 with high affinity (dissociation constant 250 pM), but do not seem to transmit signal, which has led to the speculation that IL-13Rα2 is a decoy receptor [99]. IL-13Rα2 is also found free *in vivo* that binds to soluble IL-13 and neutralize them in the internal milieu leading to abrogation in IL-13-mediated downstream signaling

[93]. Presence of very short intracellular domain further supported its non-involvement in signal transduction.

Table 2. Summary of cell types shown to express IL-13 receptor transcripts *

Tissue/Cells	hIL-13Rα1	mIL-13Rα1	hIL-13Rα2	mIL-13Rα2
Brain	+	ND	+	+
Spleen	+	+	+	+
Liver	+	+	+	+
Fetal liver	+	unknown	+	unknown
Thymus	+	+	+	unknown
Heart	+	+	+	unknown
Lung	+	+	+	unknown
Kidney	+	+	unknown	ND
Testis	+	+	+	ND
Stomach	+	+	+	unknown
Skin	+	+	+	unknown
Appendix	+	unknown	+	unknown
PBC	+	unknown	+	unknown
Bone marrow	+	-	+	unknown
Skeletal muscle	+	-	+	unknown
Colon	+	+	unknown	unknown
Small intestine	+	unknown	unknown	unknown
Ovary	+	unknown	unknown	unknown
Prostate	+	unknown	unknown	unknown
Pancreas	+	unknown	+	unknown
B cells	+	unknown	+	unknown
T cells	+	unknown	+	unknown
Endothelial cells	+	unknown	+	unknown

* David J. Matthews and Andrew N.J. McKenzie. ND, Not Detected.

IL-4/IL-13 Receptor Signal Transduction

Consistent with the fact that both cytokines share common receptor subunits, they also share some common signaling pathways and results in the activation of common biological activation programme viz activation and subsequent dimerization of STAT proteins, and activation of STAT targeted gene transcription. It has been further confirmed by a study showing that a mutated IL-4 that blocks the biological activity of IL-4 also antagonizes IL-13 activity [100]. It has been found that in human erythroleukemia cell line TF-1 cells, both cytokines share overlapping binding sites, but the number of sites for IL-13 is half of that for IL-4 [101]. Though, both cytokines have shared receptor complexes, besides the degree of similar signal transduction events, some signaling cascades differ, which results in different biological programme.

The member of hematopoietin receptor family usually lack endogenous kinase activity. Therefore, the IL-4Rα and the associated receptor molecules γc chain and IL-13Rα1, a typical member of the hematopoietin receptor family, also lack endogenous kinase activity. Thus, for the initiation of downstream signaling cascades, intracellular kinases and adaptor proteins have to be recruited. Here, Janus tyrosine kinases (JAKs) are known to play a discrete role in this signaling cascade. The IL-4Rα chain are known to itself recruiting JAK1 and in some cell lines also JAK2 [3]. Binding of JAK1 or JAK2, whichever is applicable, is mediated by a conserved motif of the IL-4Rα chain, located in close proximity to the transmembrane region and containing several acidic amino acids (proline rich box 1 motif) [102, 103]. The γc chain also contains a box1 motif and has been shown to bind JAK3, whereas the IL-13Rα1 chain is associated with JAK2 or the tyrosine kinase 2 (TYK2) [43]. Activation of JAKs or other IL-4R associated kinases results in the phosphorylation of the tyrosine residues of the intracellular domain viz. Y475, Y550, Y578, Y606, and Y684 and thereafter subsequent signaling cascades downstream to receptor subunit is activated. Two major pathways for IL-4R signal transduction have been characterized [3]; one employs insulin receptor substrates (IRS) and transduces mainly IL-4 induced proliferation signals (IL-4/IRS pathway), and other pathway employs signal transducers and activators of transcription (STATs) and mediates gene activation in response to IL-4 (JAK-STAT pathway).

IL-4/IRS Pathway

Signaling via the IRS pathway is initiated by the recruitment of IRS molecules that possess a PTB (phosphotyrosine binding) domain to the phosphorylated consensus motif ^{466}PIxxxxNPxYxSxSD^{480} within the IL-4Rα chain [91] known as insulin-/IL-4R motif (I4R) [104]. Therefore, activation of insulin receptor substrate (IRS) seems to be essentially activated by IL-4Rα subunit of receptor complexes. Four members of IRS family proteins have been identified to date namely IRS 1, 2, 3, and 4 [105]. IRS1, 2 and 3 interact with IL-4Rα. IRS2 is the main family member expressed in hematopoietic cells. IRS1 and IRS2 are cytoplasmic proteins involved in the regulation of cell proliferation and prevention of apoptosis in response to IL-4 and IL-13. The sequence surrounding the proximal tyrosine residues in the cytoplasmic region of the IL-4Rα chain, I4R, is highly homologous to the sequence of the cytoplasmic region of the insulin receptor and insulin-like growth factor (IGF)-1 receptor and shares the same IRS1/2 signaling pathways. IRSs are recruited to the IL-4R complex by the phosphorylation of the tyrosine residues in the IL-4Rα chain I4R domain through an N-terminal PTB domain (phophotyrosine binding domain) and multiply phosphorylated at up to 20 different amino acids by JAKs [106]. After phosphorylation, IRS1/2 acts as adaptor proteins and links other signal transduction molecules to the receptor [107]. JAK1 is critical for IL-4 stimulated induction of IRS1 phosphorylation through the direct action of JAK1 on IRS1 [108]. The phosphorylated IRS proteins provides docking sites for Src homology 2 (SH2) domains-containing signaling proteins such as regulatory subunit p85 of a kinase, phosphoinositide 3-kinases (PI3-K) and a adaptor molecule, growth factor receptor-bound protein 2 (Grb-2) [109]. These interactions lead to the activation of the PI3-K and Ras/Raf signaling pathways respectively [110]. PI3-K consists of 2 subunits, an 85kD regulatory subunit (p85) and an 110kD catalytic subunit (p110). Interaction of the p85 subunit with phosphorylated IRS1/2 molecules leads to conformational changes, which results in activation of the p110 catalytic subunit. Activation of catalytic subunit further activates membrane lipids, as well as serine/threonine residues of proteins. Among the membrane lipids, phosphotidylinositol-(3,4,5)-triphosphate (IP3) and phosphotidylinositol-(3,4)-biphosphate (PIP2) are the most important ones and act as secondary messengers for the activation of downstream kinases, including serine/threonine kinase Akt (protein kinase B) and other molecules involved in cell survival and prevention of apoptosis. In addition, IL-13 has been shown to induce tyrosine phosphorylation of phospholipase Cγ1 following IRS2 association in human monocytes/macrophages leading to the accumulation of cAMP which further enhances cell proliferation [32].

As discussed above, Phosphorylated IRS1/2 interact with adaptor molecule Grb-2 using its SH2 domain. Grb-2 constitutively makes complex with the guanine nucleotide exchange protein, Sos (son of sevenless), which is a putative activator of Ras1, Raf and rolled/MAP kinases. Grb-2 and Sos complex is capable of catalyzing the exchange of GDP for GTP in inactive Ras, producing the active GTP complexed form of Ras. Further, Ras-GTP activates serine/threonine kinase Raf, which then initiate MAPK pathway required for growth and survival, and proliferation of lymphoid cells [3, 111] (Figure 5). Shc through a PTB domain and negative regulation may be mediated by the guanine nucleotide exchange factors (GEFs), p62Dok and FRIPI to the Y1 position in the IL-4Rα [3].

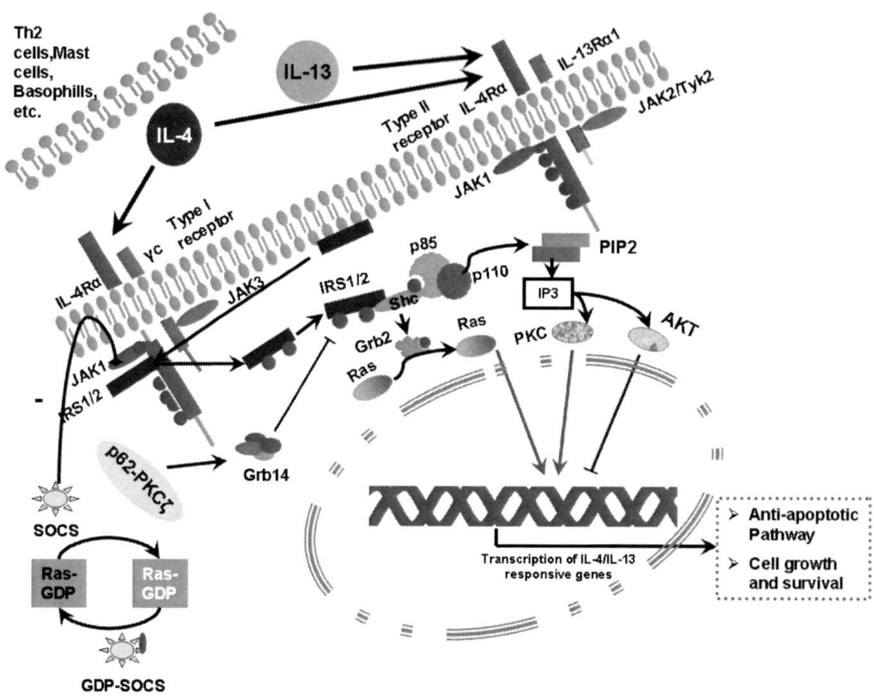

Figure 5. IL-4/IRS pathway in response to IL-4/IL-13 through IL-4Rα subunit. IL-4/IL-13 binding to its receptor complex, IRS1/2 are recruited to the IL-4Rα subunit upon phosphorylation of its first tyrosine residue (Y475). Phosphorylated IRS1/2 then act as a docking site of SH2 domains of signaling proteins such as regulatory subunit of p85 of PI3-K and SHC. P110 then associates with p85, while phosphorylated SHC activates Grb2 by phosphorylation and activates PI3-K and Ras/MAPK signaling pathways respectively. MAPK and PKC are involved in the cell growth and survival whereas activation

of serine/threonine kinase Akt (protein kinase B) prevents apoptosis of cells leading also to increase in the number of cells.

However, recent study on the analysis of cells derived from IRS2-deficient mice leads to a question mark on the importance of the IRS2 to the PI3-K pathway in regulating lymphocyte proliferation and survival [112]. It suggests an additional proliferated signal that might be mediated by signals emanating from the recruitment of the additional adaptor protein. The additional adaptor protein involved in the regulation of IL-4/IRS pathway has been characterized as p62, described under various synonyms. This protein has been identified as a phosphotyrsosine-idependent ligand of the Src homology (Src) 2 domain of $p56^{lck}$ in humans [113], PKCζ interacting protein (ZIP) protein in Ratb [114], and an oxidative stress protein induced in peritoneal macrophages (A170) in mouse [115]. It is also known as signal transduction and adaptor protein (STAP) expressed in murine osteoblast [116]. Amino acid sequences are highly conserved between species, the murine protein shares 97% and 90% homology with rat and human p62, respectively. Proteins share several structural motifs, including an SH2 binding domain, an acidic interaction domain (AID) that binds the atypical PKCζ, a ZZ zink finger, a binding site for TNF receptor associated factor 6 (TRAF6), two PEST sequences that are associated with the rapid degradation of proteins, and a ubiquitin associated (UBA) domain (Figure 7). General mechanism of p62 involves the phosphorylation of Grb14 by PKCζ as a consequence of the interaction of Grb14 and the heterodimeric p62/PKCζ complex. Phosphorylated Grb14 acts as strong inhibitor of insulin signaling and IRS associated IL-4 signaling cascade.

JAK-STAT Pathway

JAK-STAT pathway is regarded as one of the pleiotropic signaling pathway used to transduce signals for multitude of processes from development to homeostasis. Signaling via the JAK-STAT pathway requires the recruitment of cytoplasmic STAT6 transcription factor to the IL-4Rα chain. The binding of IL-4 and IL-13 to its receptor causes conformational changes and heterodimerization of its receptor subunit, which results in the activation of JAKs because two JAKs are brought into close proximity due to receptor heterodimerization. Close interaction of JAKs results in autophosphorylation and transphosphorylation of receptor associated JAKs. Activated JAKs further phosphorylate both the receptor subunits at multiple targets and the major substrates STAT6. JAKs have tandem kinase-homologous domains, one is a non-catalytic regulatory domain and the other is a catalytic domain having tyrosine kinase activity, at the C-terminus, which is required to interact with cytoplasmic domains of receptor subunits. With this respect, JAK tyrosine kinase appears to play a critical role in JAK-STAT signaling. In mammals, the JAK (Just Another Kinases) family comprises four members: JAK1, JAK2, JAK3 and TYK2. All JAKs share common domains, termed as JAK homology (JH) domains 1-7, numbered from the C-terminus to the N-terminus. It is JH7 domain which is associated with the proline rich conserved region of receptor subunit known as Box1. The JH1 domain has catalytic phosphotyrosine kinase activity capable of autophosphorylation [117]. The JH2 domain is pseudokinase domain, which has no catalytic function rather it regulates the basal activity of JAK kinase and receptor induced activation of the catalytic domain. The JH3-JH4 region shares some similarities with SH2 domians but lacks phosphotyrosine binding ability, and the JH4-JH7 region constitutes the FERM domain (Figure 6) (four-point-omne, Ezrin, Radixin, Moesin) also known as 4.1 band [118]. The FERM domain is involved in the interactions between JAKs and other kinases.

As described above, three conserved tyrosine residues Y550 (second), Y578 (third) and Y606 (fourth) of the IL-4Rα chain of receptor complex act as binding sites of the SH2 domain of STAT6. STATs are latent transcription factors that reside in the cytoplasm until they become activated. After binding of IL-4 to its receptor complex, STAT6 is tyrosine phosphorylated by activated JAKs associated with IL-4Rα subunit. Tyrosine residues are conserved in the C-terminus region of the STAT6, phosphorylation of which results in the disengagement of STAT6 from its receptor and homodimerization through interaction with a conserved SH2 domain resided also in the C-terminus region of the STAT6.

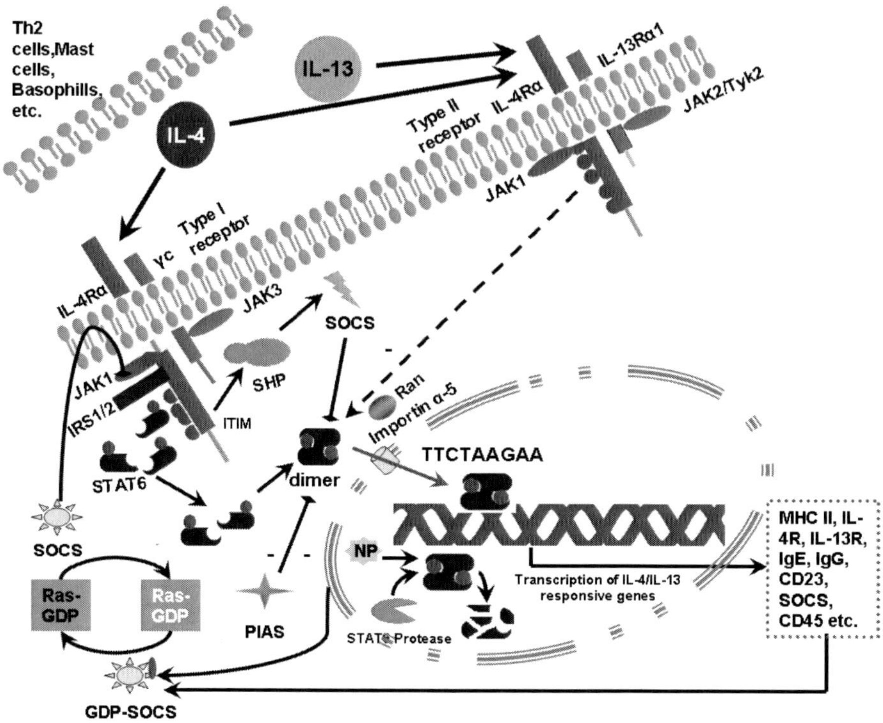

Figure 6. IL-4/IL-13 receptor signaling through JAK-STAT pathway. IL-4/IL-13 binding to its receptor complex, conformational change occur which leads to autophosphorylation of its conserved tyrosine residues in the cytoplasmic region and recruitment of JAK1 to IL-4Rα and JAK3 to γc resulting into the activation of JAK1 and JAK3. STAT6 then binds to phosphorylated tyrosine residues at the position of 550, 578 and 606 through a highly conserved SH2 domain. Binding of STAT6 to its consensus results in its phosphorylation at C-terminal tyrosine residues by activated kinases associated with receptor complex. Upon phosphorylation, STAT6 disengages from the receptor, forms homodimer and translocate into the nucleau through importin-α5 via Ran pathway to bind to specific DNA motifs in the promoter of IL-4 response elements known as GAS-elements of IL-4 inducible genes having a consensus sequence of TTCTAAGAA that include CD23, MHC class II, germline immunoglobulin ε, IL-4Rα chain, SOCS etc. In nucleus, STAT6 is deactivated by nuclear phosphatases (NP) and degraded thereafter by STAT6 proteases. Associated JAK1 activates SOCS protein which results in the formation of Ras-GDP and finally GDP-SOCS. GDP-SOCS plays a regulatory role in the STST6 activation. Another protein, PIAS also involved in the regulation of STAT6. Fifth tyrosine residue of IL-4Rα binds

to ITIM and activates SHP, a tyrosine phosphataes, which activates SOCS proteins and resulted in the inhibition of STAT6.

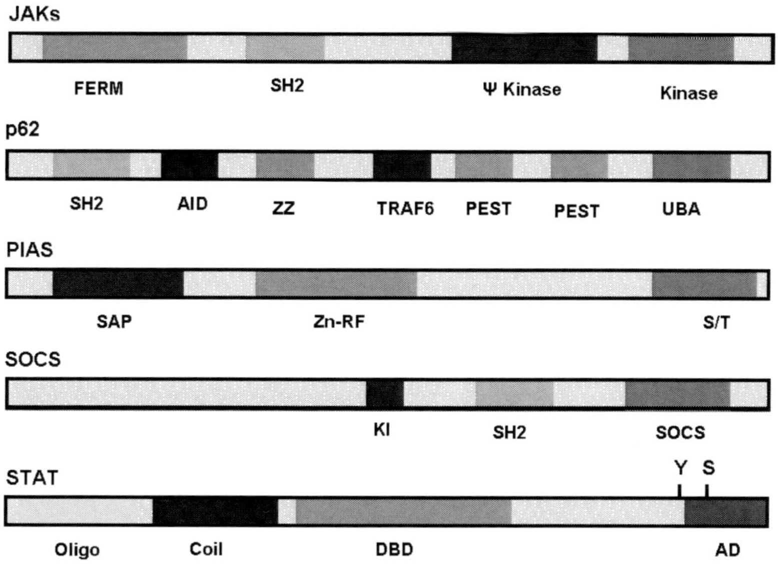

Figure 7. Schematic domain organization of murine JAKs, PIAS, SOCS and STAT proteins. Murine JAKs are composed of four domains; FERM (4.1/ezrin/radixin/moesin domain) which involves in the receptor binding, a SH2 domain that involves phosphotyrosine binding, a ψ kinase which is a non-catalytic kinase like domain and a catalytic kinase domain. P62 is composed of 7 domains including two PEST (a peptide sequence rich in proline (P), glutamic acid (E), serine (S), and threonine (T)) associated with rapid degradation of proteins with a ubiquitin associated domain (UBA), a SH2 binding domain, and acidic interaction domain (AID) that binds to the atypical PKCζ, a ZZ zink finger motif, and a binding sites for TNF receptor associated factor 6 (TRAF6). PIAS protein contains three domains; a SAP (SAF-A (scaffold attachment factor-A)/Acinus/PIAS domain) domain that involves target binding, RING finger domain that involves Sumoylation of target proteins and a serine/threonine rich domain (S/T) also for target binding, a tyrosine phosphatase domain SH2 and a SOCS domain for ubiquitin ligase binding. STAT proteins are composed of four domains; an oligomerization domain (Oligo), a coiled-coil domain for protein interaction, a DNA binding domain (DBD) and a transcriptional activation domain (AD).

Homodimer of phosphorylated STAT6 then translocates into the nucleus by a mechanism requiring importin α-5 (also known as nucleoprotein interactor 1) and the Ran nuclear import pathway to bind to specific DNA motifs in the promoter of IL-4 responsive elements, including CD23, class II MHC or germ line immunoglobulin ε and IL-4Rα chain known as GAS-elements [119, 120]. Activated STAT6 regulates expression of those genes, which are responsible for allergic reaction, isotype class switching and Th2-dominated immune responses [82].

In addition to the principal components of the pathway, at least three adaptor proteins have been identified, which are capable of modulating JAK-STAT pathway by a mechanism(s) yet to fully define. Among them, STAMs (signal-transducing adaptor molecules) is one of the best characterized adaptor proteins, which contain VHS and SH3 domains [121]. Isotype of STAM, STAM1 and STAM2A have been found phosphorylated by JAK1, 2 and 3. Phosphorylation of STAM molecule is dependent on a third domain known as ITAM (inducible tyrosine-based activation motif) that leads to the transcriptional activation target genes including *MYC*. A Second intermediate effector molecule acting as adaptor molecule for JAK-STAT pathway is StIP (STAT-interacting protein). StIP can associate with both JAKs and unphosphorylated STAT6 and serves as a scaffold to facilitate the phosphorylation of STAT6 by JAKs. A third class of adaptor molecule of JAK-STAT signaling is the SH2/Lnk/APS family of protein, which have both pleckstrin homology (PH domain; a protein domain of approximately 120 amino acids) and SH2 domains, and serves as a substrate for JAK phosphorylation. Association of SH2 to JAK results in the facilitation of JAK-STAT signaling, while association of APS results in its inhibition. Therefore, SH2/Lnk/APS family of adaptor molecules acts as effector as well as regulatory molecule of JAK-STAT signaling [122].

Role of IL-13Rα1 in IL-4/IL-13-Mediated Signal Transduction

Although, the role of IL-13Rα1 in the IL-4/IL-13-mediated signal transduction is not clear. It has been described that JAK2/TYK2 binds to the proline rich box region of IL-13Rα1 that might be then involved in recruitment and activation of other signaling intermediates, which are yet to define. IL-13Rα1 has been found to contain 2 tyrosine residues Y402 and Y405 (Figure 3). The tyrosine residue Y402, which is at the +3 position by Gln, has been predicted to be consensus STAT3 recruitment and activation motif and therefore a docking site for STAT3. Recently, it has been demonstrated that the C-terminal tail of IL-13Rα1 is important for activation of JAK1, TYK2, IRS-1 and also STAT6 [123]. However, in a recent stud, other STAT proteins such as STAT1α, STAT5α, and STAT5β have been shown to be activated in response to IL-13 in human monocytes [124] (Figure 8). Activated STAT3 molecules dimerize and translocate to the nucleus and finally serve as transcription factors for various genes such as IRF-1, JunB or glycoprotein 130 [125]. JAK1 associated with IL-4Rα subunit also activates STAT3 and establish a crosstalk between IL-4Rα mediated signal transduction and IL-13Rα1-mediated downstream signal transduction pathway.

Role of IL-13Rα2 in IL-13-Mediated Signal Transduction

Previously, it has been known that in spite of high affinity binding of IL-13Rα2 to its ligand IL-13, it fails to transduce signal, and therefore termed as a decoy receptor for IL-13. This notion is further supported by another observation showing that it lacks Box-1 and Box-2 signaling motifs in its comparatively short intracellular tail region (17 amino acids long in the human), which are necessary components for signal transduction processes. Though, a tyrosine residue has been found in the IL-13Rα2 tail, but the association of this tyrosine residue in the signal transduction in response to IL-13 is yet to clearly define. Recent studies carried out with the help of various metabolic inhibitors suggest that it binds to IL-13 with high affinity, internalizes the receptor-ligand complex dependent on a dileucine motif in the transmembrane domain and transduce signal in cytosolic environment. Recent study reports that short cytoplasmic domain of the IL-13Rα2 binds to the cytoplasmic domain of the IL-4 receptor α chain and forms IL-4Rα-IL-13Rα2 receptor complex thereby involving in the inhibition of IL-4- and IL-13-mediated signal transduction [126].

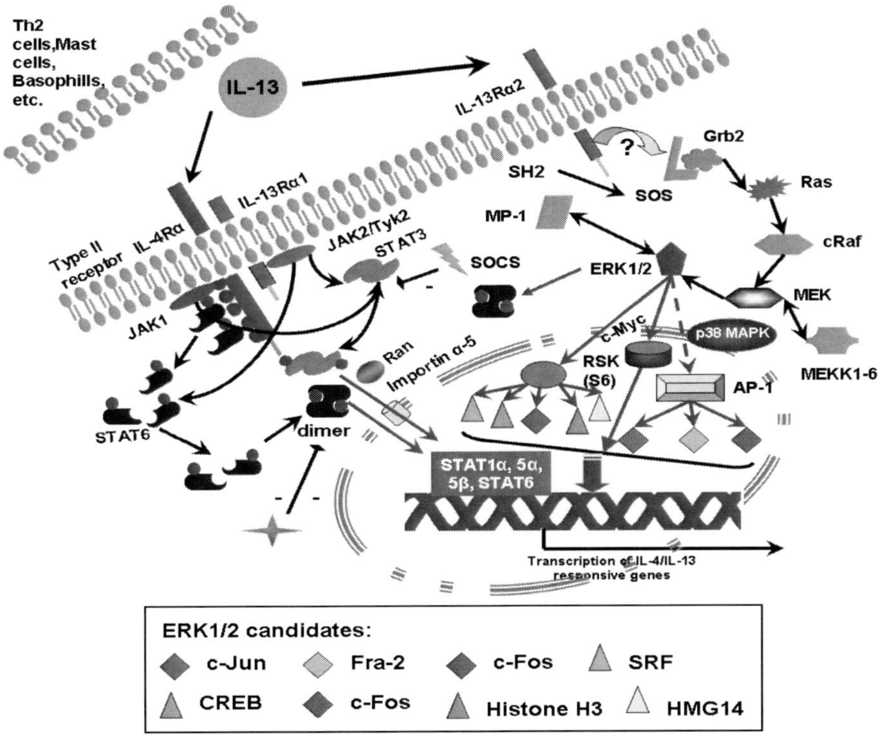

Figure 8. Role of IL-13Rα1 and IL-13Rα2 in IL-4/IL-13 signaling cascade. IL-13 binding to IL-13Rα1 leads to a conformational change and resulting into association of IL-4Rα. Interaction of IL-4Rα and IL-13Rα1 leads to autophosphorylation of their conserved tyrosine residues. Tyrosine phosphorylation of IL-13Rα1 stimulates the recruitment and activation of JAK2/Tyk2, which in one hand, phosphorylates STAT6 resulting in the activation of canonical STAT6 pathway and on the other hand, phosphorylates STAT3, which is dimerized and transported to the nucleus through importin-α5 by Ran pathway in a similar fashion. JAK1 associated with IL-4Rα also phosphorylates STAT3 leading to its activation. IL-13 binding to IL-13Rα2 results in the internalization of receptor-ligand complex and activation of ERK1/2 and AP-1 transcription factor, but how this signal initiated from IL-13Rα2 cytoplasmic tail is still unknown. It has been known that SH2 results in the activation of son of sevenless (SOS), which is phosphorylated and thereafter recruits Grb2. Phosphorylated Grb2 then activates ERK1/2 (extra-cellular regulated kinases) and p38 MAPK (mitogen-activated protein kinase) and results in the gene expression of a number of proteins involved in the regulation of cell survival, growth and proliferation.

However, characterization of the *IL-13R2* promoter results in the identification of several putative transcriptional factor binding sites for nuclear factor of activated T cells-1, AP-1 (c-Jun and c-Fos), AP2, GABP, OCT1, GATA3, PRE, and C-ETS1 [127], which may result in the IL-13-induced biological responses mediated via IL-13Rα2 rather involvement of IL-13Rα2 in the inhibition of the IL-13 signaling cascade. It has been reported that IL-13 mediates its biological response via IL-13Rα2 by activating extracellular signal-related kinases (ERK)1/2 and p38 MAPK in the cells (Figure 8). In macrophages, IL-13Rα2 interaction with IL-13 results in the activation of an AP-1 (activation protein-1) variant containing c-Jun (Jun Oncoprotein) and Fra-2 (Fos-Related Antigen-2) leading to the production of transforming growth factor (TGF)β in a model of murine lung fibrosis [128]. Jun and Fos proteins contain leucine-zipper' motif that allows for dimerization. They dimerize to form AP-1, a transcription factor that binds to the 12-O-tetradecanoylphorbol 13-acetate (TPA) response element (TRE) of several cellular and viral genes including human collagenase, metallothionein IIa, stromelysin, interleukin 2 etc. The transcriptional activity of c-Jun is regulated by phosphorylation at Ser63 and Ser73 through SAPK/JNK, suggesting the involvement of these kinases downstream to IL-13Rα2 in IL-13 signaling [129]. Further, IL-13Rα2 have been shown to up regulate both insulin-like growth factor (IGF-I) and early growth response gene 1 (Egr-1), factors that act as the proximal initiators of the fibrotic program [130]. All together, studies suggest that IL-13 mediates its signal transduction via IL-13Rα2 and performs several biological function specific to IL-13.

Regulation of IL-4/IL-13 Signal Transduction

The detailed mechanism of signal transduction pathway in response to IL-4 and IL-13 in the form of IL-4/IRS pathway and JAK-STAT pathway has been well defined, but the mechanism(s) involved in the regulation of these pathways has been not clearly defined particularly in the case of IL-4/IRS pathway. Recently, the IL-4Rα-associated kinase proto-oncogene tyrosine-protein kinase (PTK) Fes/Fps (feline sarcoma oncogene), a Src-related *fps* PTK member in the regulation of IL-4/IRS pathway whereas a number of regulatory molecules including SH2-containing phosphatases, suppressors of cytokine signaling (SOCS), and protein inhibitors of activated STAT of transcription (PIAS) involved in the regulation of JAK-STAT pathway have been identified.

Regulation of the IL-4/IRS Pathway

IL-4Rα-associated kinase proto-oncogene tyrosine-protein kinase, Fes/Fps has been shown to important regulator of IL-4/IRS signaling induced in response to IL-4 and IL-13 [131]. It is a type of tyrosine kinase expressed mainly on hematopoeitic cells including immature myeloid progenitor cells [132] and lymphocytes [133]. However, it is quite distinct from the cytoplasmic PTK because it lacks a negative regulatory tyrosine phosphorylation domain in the C-terminus region. Fes kinase is supposed to be activated by JAKs especially JAK1, because JAK1-deficient cells has been shown to lack Fes activation. Tyrosine phosphorylation of Fes by activated JAKs in response to IL-4 binding to its receptor components results in the activation of Fes. Activated Fes then blocks the IL-4/IRS pathway. It was further confirmed by a study demonstrating that IL-4–dependent phosphorylation of IRS1/IRS2 is dramatically inhibited in B lymphoid cells expressing a dominant negative kinase-inactive Fes, which significantly correlates with a decrease in IL-4–induced proliferation. In contrast, the IL-4–dependent phosphorylation of JAK1, JAK3, and the STAT6 activation is not affected by expression of this dominant negative Fes construct. It has been found that activated Fes alters the association of JAKs with the p85, which is α subunit of PI3 kinase and also IL-4 inducible activity of $p70^{S6K}$ is reduced [134].

REGULATION OF JAK-STAT PATHWAY

In addition to the inhibitory effect of SH2/Lnk/APS family of adaptor molecules, at least three more molecules have been identified which play a regulatory role in JAK-STAT signaling. The best characterized amongst them is SHP-1 (a class of protein tyrosine phosphatase; product of the mouse *motheaten* gene, previously known as PTP1C), which is characterized by the presence of two SH2 domains N-terminal of the canonical 250 amino acids long tyrosine phosphatase domain. SHP-1 is mainly cytoplasmic and SH2 domain of which binds to either phosphorylated JAKs or phosphorylated receptor subunit to facilitate dephosphorylation leading to the deactivation of receptor subunits and JAKs. CD45 is another tyrosine phosphatase, which has been reported to have involvement in the regulation of JAK-STAT signaling through a receptor subunit. It is expressed in hematopoeitic cells and contains two phosphatase domains in its intracellular region. However, only one phosphatase domain has been demonstrated active, which negatively affect JAK-STAT signaling. Deletion of CD45 gene results in the hyperactivation of JAK1 and JAK3 which is suggestive of its major role in the JAK-STAT regulation. It has been found that both SHP-1 and CD45 do not directly associate with the JAK kinase domain, though there is evidence of dephosphorylation of JAKs, which strongly indicates that some other tyrosine phosphatase might involve in the regulation processes.

Another important regulator of JAK-STAT signaling is SOCS (suppressors of cytokine signaling) protein, which contains an SH2 domain and a SOCS box at the C-terminus [135]. SOCS proteins are a family of at least 8 members, SOCS-1 to SOCS-8, all of which found to be involved in the regulation of JAK-STAT pathway. Activated STAT6 leads to the transcriptional activation of the *SOCS* genes resulting in the expression of SOCS protein, which then binds to phosphorylated JAKs and receptor subunits to turn off the signaling. It also interacts with the elongin BC complex and cullin 2 and promotes the ubiquitination of JAKs due to which stability of JAKs greatly affected. Additionally, SOCS-1 and SOCS-3 regulate JAK-STAT signaling through an inhibitory region located on its N-terminal to the SH2 domain (ITIM). The third class of JAK-STAT regulator is PIAS (protein inhibitors of activated STATs) proteins; PIAS1, PIAS3, PIASx and PIASy. PIAS proteins contain zinc-binding RING-finger domain in the central portion, a well conserved SAP (SAF-A/Acinus/PIAS) domain at the N-terminus, and a less well conserved carboxyl domain [136] (Figure 7). The PIAS proteins bind to activated STAT dimmers through carboxyl domain and prevent its binding to DNA. Furthermore, inhibitory motif (ITIM) of IL-4Rα interacts with the SH2 domains of the tyrosine phosphatase Shp-1, which inhibits the phosphorylation of STAT6 resulting in abrogation of JAK-STAT signaling cascades [137].

Corresponding to the regulation of STAT6 activation, activation of effector substrates and translocation in the case of STAT3 are also negatively regulated by phosphatases (SHPs, SHIP), which generally regulate growth and functional responses of hematopoietic cells through tyrosine phosphorylation of proteins [3]. Regulation is further accomplished by specific inhibitors such as CISI-1/SOCS-1 or PIAS3 [138].

IL-4/IL-13 SIGNALING IN TUMOR DEVELOPMENT, GROWTH AND PROGRESSION

IL-4 and IL-13 have prominent role in tumor growth and progression by directly acting on the tumor cells or indirectly by down regulating anti-tumor immune responses. It has been shown that many tumor types express type II IL-4 receptor including lymphoma [139], renal cell carcinoma, glioblastoma [140], head and neck cancer [140, 141], hepatoma [142], pancreatic cancer, breast cancer [143], prostate cancer [144], and colon cancer [145]. The role of IL-4R receptor is best defined in the Hodgkin's lymphoma/Reed-Sternberg cells (H/RS), where it has been demonstrated to play as an autocrine growth factor *in vitro*. In this study, blocking of IL-13 by IL-13Rα2 induces apoptosis of tumor cells [146]. In a colon cancer cell line, both IL-4 and IL-13 has been found to down regulate the expression of E-cadherin, which is an adhesion molecule responsible for cell-to-cell contact, and carcinoembryonic antigen leading to tumor metastasis [145]. In glioma cell lines, which express IL-4Rα, IL-13Rα1, and IL-13Rα2, IL-13 up regulates VCAM-1 in a dose- and time-dependent manner that leads to metastasis [140]. IL-13 also directly acts on B chronic lymphocytic leukemia (B-CLL) and inhibits IL-2-induced proliferation and spontaneous apoptosis.

As discussed in previous section, both the cytokines are capable of down regulating the phenotypes and functions of various immunopotent hematopoietic cells. Studies indicate that optimal removal of tumor cells needs the activation of type I T cell-mediated host defenses *in vivo*. Therefore, the deviation toward a more prominent Th2 type response has been postulated as a mechanism that might block the rejection of tumor cells and/or promote tumor progression [147]. IL-4 and IL-13 secreted either by tumor cells or by immune cells under the influence of immunosuppressive microenvironment generated by tumor cells, suppress the type I T cell-mediated arm of immune system and substantially promotes Th2 type of immune responses. The expression of FcεR1 by basophils and mast cells due to IL-4/IL-13 exposure, is further enhanced by resulting IgE cross-linking, which provides an additional but powerful mechanism of amplification of Th2 type immune responses by producing large amounts of IL-4 both in mice [148] and in humans [149]. Further, IL-4, through activating STAT6, up regulates the expression of GATA-binding protein 3 (GATA3) that has known to be a master regulator of Th2 cell differentiation [150, 151], which are also involved in the suppression of Th1 and Th17 cell responses, partly through the up regulation of growth factor independent 1 (GFI-1), a transcriptional repressor of IFN-γ and IL-17 production [152]. In turn, loss of Th2 cytokine gene expression has been attributed to be at least a part a consequence of downregulation of GATA-3 expression [150]. IL-4 knockout (KO) mice have been observed not resistant to tumor regrowth. In contrast, IL-4Rα- and STAT6- mice have been found highly resistant to tumor regrwoth [77, 153]. Moreover, STAT5 activation in response to IL-4 or IL-13 through IL-13Rα1 also found to be critical for cytokine-mediated T cell proliferation, survival and Th2 differentiation [154]. It has been already discussed in previous section that both cytokines IL-4 and IL-13 are capable of down regulating the functions of almost every immune cells either directly or through the involvement of intermediate immune cell population such as monocytes/macrophages and promotes the induction of Th2 type immune responses and subsequent down regulation of Th1 immune responses (Figure 2). Recent reports advocate the direct role of IL-13Rα2 in tumor growth, invasion and metastasis. IL-13Rα2 is earlier known as a non-signaling decoy receptor of IL-13, but recently found to transduce signals after binding to IL-13. Expression of IL-13Rα2 has been found to shortens survival time of mice with human pancreatic cancer [155] indicating its role in tumor in tumor growth and invasion.

IL-13 signal transduction through IL-13Rα2 induces many kinases including mitogen activated protein kinases (MAPK) and extracellular-signal regulated kinases (ERKs) (Figure 8). ERKs (ERK1/2) are known to be a main component of the MAPK pathway transducing signals which are mainly responsible for proliferative effects when activated. Abnormal activation of the ERKs cascade found to be associated with metastasis in many human cancers [156-158]. Furthermore, activation of ERK/AP-1 is associated with increased expression of matrix metalloproteinases (MMPs), which are known to be key mediators of cancer invasion and metastasis. It has been demonstrated that IL-13Rα2 positive tumor cells when treated with IL-13 expressed different types of matrix metalloproteinases (MMP 9, MMP 12, MMP 14), which mediates cancer metastasis by degrading extracellular matrices via the AP-1 transcription factor pathway [159, 160]. Knocking down of IL-13Rα2 by RNAi in tumor cells completely abolishes AP-1 signaling cascade and subsequent expression of MMPs. Correspondingly, treatment of tumor cells with Batimastat, a broad spectrum MMP inhibitor, also inhibited the AP-1 transcription pathway and increased invasion mediated by IL-13 confirming the role of IL-13 in tumor cell invasion and metastasis (Unpublished observation). Contrasting results have been found when pancreatic and ovarian cancer cells are activated by IL-13 through IL-13Rα2, increased tissue invasion and metastasis has been found. Therefore, IL-13 modulates tumor cell growth, progression and metastasis by two distinct mechanisms; one mechanism involves ERK and MAPK activation which activates proliferation of tumor cells and suppression of apoptotic pathways while other mechanism involves enhanced expression of MMPs that actively disintegrate tissue matrices.

Table 3. Association of different signaling intermediates and modulators abnormalities in in cancer

Molecular abnormalities	Association in cancer	Reference
PTPN11 mutations	Juvenile myelomonocytic leukaemia (JMML), Myeloplastic syndrome (MDS), Childhood acute myeloid leukaemia (AML)	218
Loss of JAK1 (*v-abl* mutation)	Sarcoma	161
JAK2 (V617F) mutation	Myeloproliferative disorders (MPD), polycythemia vera, essential thrombocythaemia and myelofibrosis	162
Loss of PIAS3 expression	Anaplstic lymphoma kinase-positive T/null-cell lymphoma (ALK^+ TCL)	219
Loss of PIASy expression	Myelodisplastic syndrome (MDS)	220
Mutations of *SOCS-1*	Primary mediastinal B-cell lymphoma	221
Loss of SOCS-1 expression	Primary mediastinal B-cell lymphoma	222
Aberrant SOCS-3 expression	Cutaneous T cell lymphoma (CTCL)	206
c-kit activating mutation	Acute myelogenous leukemia (AML), Gastrointestinal stromal cell tumors (GIST)	178
FLT3 mutations	Acute myelogenous leukemia (AML)	188

Apart from that a number of molecules, which are intermediate to IL-4/IL-13 signaling cascades, have been found associated with tumor growth, invasion and metastasis. Aberrant expression or mutated form of these signaling intermediates are directly linked to specific tumor types as described in Table 3.

ROLE OF JANUS KINASES IN TUMOR GROWTH AND INVASION

The Janus kinase JAK1 has been recently implicated in tumor formation by the *Abelson* onbcogene [161]. It has been found that Jak1 is involved in the transformation of cells by the *v-abl* (viral *Abelson* gene) as well as by the *v-src* (viral *sarcoma* gene) oncogenes. Initial studies showed that STAT5 is constitutively activated in *v-abl*-transformed pre-B cells and transformation by *v-abl* is dependent on its ability to bind and activate JAK1 *in vitro*. Constitutive activation of STAT5 leads to proliferation of *v-abl* transformed cells. Although, it has found that the *v-abl*-infected JAK-1-deficient B cell precursors induce tumor formation in severe combined immunodeficient (SCID) mice at a higher rate and with an earlier onset that JAK-1 wild type cells possibly due to the loss of interferon-γ (IFN-γ) sensitivity associated with the JAK-1 deficiency. This study demonstrates that IFN-γ is key determinant in the progression of tumorigenicity of *v-abl*-transformed B cells. JAKs can also be aberrantly activated by mutations such as JAK2 (V617F) mutation, which has been found in Philadelphia chromosome-negative myeloproliferative disorders (MPD), polycythaemia vera, essential thrombocythaemia and myelofibrosis. In these disorders, JAK2 (V617F) translocates to the nucleus, phosphorylates histone H3 at tyrosine 41 (H3Y41), and disrupts the binding of transcriptional repressor HP1-α (heterochromatin protein 1-alpha). Disruption of HP1-α binding leads to the activation of pro-hematopoietic and pro-lymphopoietic genes such as lmo2 and also causes genomic instability increasing oncogenesis [162]. Moreover, it has been shown that constitutive phosphorylation of JAKs (JAK2 in HTLV tax transformed fibroblasts and JAK-3 in HTLV-1 transformed human T cells) may facilitate tumor growth in both transgenic mouse model and HTLV-1 transformed transformed human T cells [163]. In multiple myeloma (MM), the activation of JAKs and subsequent activation of STAT signaling pathways promotes tumor growth, survival and drug resistance. Due to the involvement of JAKs in the pathogenesis of tumor, possibility of novel approach for the treatment of cancer where JAKs are involvement, has been postulated, that dealt in later.

ROLE OF IRS-1/2 IN TUMOR DEVELOPMENT AND PROGRESSION

The IRS proteins are cytoplasmic docking proteins that function as signaling intermediates in IL-4/IL-13 signaling cascade downstream of activated cell surface receptors and function as adaptor protein [164]. However, recent evidence suggests that it is considerably involved in pathogenesis of breast cancer [165]. Both IRS-1 and IRS-2 have been found to be expressed in breast carcinoma cells [166]. IRS-1 is predominant family member of IRS protein, which is regulated by estrogen and activated by IGF-1 in differentiated estrogen receptor positive (ER+) human breast carcinoma cell lines and not only promotes growth and survival of breast tumor cell lines but also its metastasis [167]. The expression of IRS-2 protein in tumor cells results in highly aggressive nature of tumor. Interestingly, the incidence of lung metastasis have been found to decrease in *Irs-2*-deficient tumors and increase in *IRS-1*-deficient tumors compared with wild type Irs genotype [167] suggesting differential effect of IRS-1 and IRS-2 in tumor progression and invasion. Though, its direct relationship to tumor growth has not been well worked out, it has been assumed that IRS proteins are involved in the growth and progression of mammary carcinoma due to its role in somatic cell growth and proliferation as discussed in signaling section. Apart from its role in breast cancer invasion, its involvement in several other tumor types has been demonstrated. In neuroblastoma and mesothelioma cells, IRS-2 has been found to promote cell motility and invasion [168, 169]. Correspondingly, the elevated expression of IRS-1 in pancreatic and hepatocellular carcinoma indicates its direct or indirect role in the pathogenesis of prostate tumor [170-171].

IL-4/IL-13 Induced STATs and Tumor Development

As described IL-4/IL-13 signaling leads to the activation of many STATs including STAT1α, STAT3, STAT5α, STAT5β and STAT6. Apart from their role in maintaining normal physiology and cellular homeostasis, the STAT proteins have been strongly implicated in cellular transformation and oncogenesis. Amongst them, the involvement of STAT1, 3, 5 and 6 in tumor formation and progression have been well defined. STAT proteins contribute tumor formation through their intimate connection to growth factor, reducing apoptosis and thus increasing tumor cell survival and angiogenesis.

STAT1, in addition to its role in promoting innate immune responses, known also to serve as potent inhibitors of growth and apoptosis leading to increased tumor cell survival. Though, STAT1-deficient mice have not been found to develop spontaneous tumors, they have been found highly susceptible to chemical carcinogenesis [172]. It has been found that STAT1 activation can lead to the up regulated expression of different caspases and cdk (cyclin-dependent kinase) inhibitor p21 in some cell types, which might be implicated in enhanced apoptosis and growth restriction [173]. Interestingly, p21 up regulation by STAT1 in mammary cells appears to involve BRCA1, which is often lost in familial and other forms of breast cancer [174]. It has been found that STAT1 is mutated in some of the human tumors leading to its constitutive activation and therefore unregulated cell transformation. Additionally, it can be assumed that constitutive activation of STAT1 may result in inflammatory state of the host's body that can lead to the tumor development and progression.

Another STAT protein, STAT3 reported to have a role in growth and maintenance of transformed cells. In a study using antisense or dominant negative constructs directed at STAT3, it has been demonstrated that cell proliferation is highly dependent on constitutively activated STAT3 in many tumor-derived cell lines [175]. A mutant form of STAT3, STAT3c is found which does not require tyrosine phosphorylation for the activation, and found to have a positive role in transforming fibroblast cells [176]. In thyroid cancer where Ret receptor tyrosine kinase has been found aberrantly regulated, STAT3 phosphorylation and subsequent activation is required for the transformation of cell mediated by Ret receptor tyrosine kinase

[177]. Correspondingly, an association of c-kit activating mutations and expression of a dominant negative STAT3 in the acute myelogenous leukemia (AML) and gastrointestinal stromal cell tumors (GIST) has been found [178], which strongly suggest its involvement in the tumor development. STAT3 has been shown to be persistently activated in Hodgkin's lymphoma [179], primary prostate cancer and prostate cancer derived cell lines [180], which might play a significant role in the growth and progression of these tumors. Furthermore, persistently activated STAT3 and JAK2 are found in the rare malignancy large granular lymphocyte leukemias [181, 182] suggesting that STAT3 is involved in the development and progression of a large number of tumor types but mostly in hematological malignancies.

STAT5α and β are important transcriptional regulators, also known to frequently activated in a persistent manner in human leukemia/lymphoma cells due to constitutive tyrosine phosphorylation and DNA binding activity. Though, persistent STAT5 activation has been found weakly oncogenic and therefore it leads to late emergence of clonal B cell leukemia/lymphoma, it strongly supports its involvement in tumor development particularly in hematological malignancies. Constitutive STAT5 activation has been frequently observed in different myeloid and lymphoid malignancies [183], and in carcinoma [184]. Its activationshas been directly linked to the activity of oncogenic protein tyrosine kinases that includes BCR-ABL in chronic myeloid leukemia (CML), TEL-JAK2 in acute lymphoid leukemia (ALL) and CML, and mutated FLT3 in AML (in about half of the cases) [185-188]. Activating mutation in JAK2, V617F mutation leads to its constitutive activation as mentioned above, which results in the constitutive activation of STAT5 in Philadelphia chromosome negative myeloproliferative disorders, polycythema vera, essential thrombocythemia and idiopathic myelofibrosis [189, 190]. Deletion of STAT5 prevents HER2/Neu/ErbB2-induced mammary tumor development and its inhibition suppresses tumor cell invasion in human colorectal cancer by inducing G1 cell cycle arrest [191]. These observations are suggestive of profound implication in tumor development and invasion in some hematological tumor cell types; however detailed account of molecular mechanism needs to be studied. In colorectal cancer (CRC), STAT5 is involved in cell growth, cell cycle progression, invasion and migration by regulating the expression of Bcl-2, p16^{ink4A}, p21$^{waf1/cip1}$, p27^{kip1}, E-cadherin, the focal adhesion kinase (FAK), vascular endothelial growth factor (VEGF) and matrix metalloproteinases (MMPs) genes. FAK has been also linked to the activation of STAT3 transcription factor. The presence of STAT5 complex with p44/42 MAPK and SAPK/JNK in colorectal cancer provide an insight into ongoing cross-talk between STAT5 and MAPK pathway which may induce proliferation of CRC cells. Likewise the STAT5-MAPK cross-talk provides aggressiveness to the renal cell carcinoma (RCC) cell lines [192].

Corresponding to these STAT proteins, STAT6 has also been found to aggregate tumor development and invasion.. Mice lacking STAT6 shows enhanced tumor immunity to both primary and metastatic mammary carcinoma [153, 193], and induced spontaneous rejection of implanted tumors [194]. STAT6 is found constitutively activated in several tumor types including prostate cancer [195], cutaneous T cell lymphoma [196], Hodgkin's lymphoma [146] and primary mediastinal large B cell lymphoma [197], where it enhances tumor development, growth and metastasis by inducing resistance to apoptosis and enabling the host to escape immune surveillance. It has been observed that the tumor cells carrying defective STAT6^{null} phenotype show enhanced cell death and simultaneously show decreased invasiveness and metastasis which is correlated with differential expression of surviving,

MDM2, p53 and CAV1 genes [198]. Further, STAT6 has been found to interact with AnnexinA2 which is a multifunctional Ca^{2+}-dependent phospholipids binding protein involved in malignant transformation of several cancers. STAT6-AnxA2 interaction results in STAT6 stabilization in cytosol and its enhanced nuclear entry and subsequent DNA binding in prostate cancer leading to increased metastatic processes [199]. Moreover, $STAT6^{-/-}$ mice have been found to generate M1 macrophages due to blockade of end product activation in response to IL-13 that reflects inoperative IL-4/STAT6 pathway [200] and involvement in tumor progression.

ROLE OF SOCS IN TUMOR DEVELOPMENT AND PROGRESSION

SOCS is a family of proteins involved in the regulation of cytokine signaling cascade including IL-4/IL-13-mediated signal transduction. Though its role in tumorigenesis is not strongly supported in existing literature, SOCS proteins over expression has been found in many tumor types including ductul carcinoma, a relative to mammary carcinoma [201] 9% of breast cancer patients and a significant population of patient with pituitary adenoma showed hypermethylation and silencing of SOCS1 but not SOCS2 and SOCS3 [202]. Recently a correlation between loss of SOCS2 and expression of genes such as p27, Ki-67, cyclin-A, pRb and EGFR has been demonstrated, which actively regulate cell proliferation and migration [203]. SOCS1 downregulation causes a potent growth stimulation of prostate cancer cell lines by up regulating the expression of cyclins D1 and E as well as cyclin-dependent kinases 2 and 4, in turn, its over-expression exerts growth inhibition by down regulating the expression of these cyclins and cdks [204]. In contrast, silencing of SOCS1 inhibits tumor cell migration and invasion. Interestingly, tumor cell proliferation seems to be dependent on the expression of SOCS1 in B16F10 murine melanoma cell line [205]. Since, SOCS proteins are important for regulation of cytokine signaling, its aberrant expression leads to uncontrolled cytokine signaling through ERK-MAPK pathway leading to neoplastic growth and progression. In cutaneous T cell lymphoma (CTCL), aberrantly expressed SOCS3 mediated by constitutive activation of STAT3 appears to play a significant role in CTCL pathogenesis [206]. SOCS-3 also inhibits cell growth by interfering with heterodimer formation between DP-1 playing an important role in controlling JAK-STAT signaling, and E2F [207].

ROLE OF PIAS IN TUMOR DEVELOPMENT AND PROGRESSION

PIAS proteins act as transcriptional coregulators by colocalizing transcriptional regulators at different subnuclear compartments or by modifying the function of transcriptional regulators. It also functions as adaptor proteins through its association with SUMO-modified substrate. With this respect, it seems to function as a regulator of IL-4/IL-13-mediated signal transduction and other signaling cascades. Few reports revealed its association with tumor development and progression. Human prostate cancer cells have been shown to express PIAS1, PIAS3 and PIASy, which found to display distinct effects on androgen receptor-mediated gene activation in prostate cancer cells. In these tumor cells,

PIAS1 and PIAS3 enhance the transcriptional activity of androgen receptor (AR), while PIASy inhibits the transcriptional activity of AR [208].

Therapeutic Approach against Malignancies

Both IL-4 and IL-13 are immunoregulatory cytokines that regulate a variety of functions in hematopoietic and non-hematopoietic cells. They are known to be involved in various immunopathogenesis of diseases including malignancies. Both IL-4 and IL-13 induce Th2 type of immunity with simultaneous suppression of Th1 type T cell-mediated immunity believed to be essential for tumor cell removal and repression. They act either directly or indirectly on every immune cell population and lead to Th2 polarized state resulting in uninterrupted tumor growth and progression as described above. Moreover, signaling intermediates of IL-4/IL-13 signaling pathways have been found associated with many tumor types. Thus, IL-4/IL-13, its receptors and signaling intermediates could be an attractive target for cancer immunotherapy or to develop a therapeutic agent against cancer. Currently, studies are underway to evaluate the clinical potential of IL-4/IL-13 antibody, several IL-4/IL-13 antagonists, soluble IL-13Rα2 and inhibitors of different signaling intermediates and transcriptional regulators.

Targeting IL-4/IL-13

To target, IL-4/IL-13 in those immunopathologic conditions where both cytokines have a major role in disease progression, soluble compounds such as extracellular portion of IL-4Rα and IL-4/IL-13 monoclonal antibody (mAb) have been used in the study involving clinical trials. Extracellular portion of IL-4Rα binds IL-4 and inhibits its function [209]. Initial studies with soluble IL-4R were performed in subjects with mild-to-moderate persistent asthma who were withdrawn from their ICS therapy and then randomly assigned to receive placebo or single dose nebulized IL-4R [209, 210]. Evidence from animal models supported the use of a number of pharmaceutical IL-13 neutralizing antibodies like anrukinzumab, IMA-026, TNX650, QAX576, and MILR1444, GSK-679586 and CAT-354 in phase I /II clinical trials of IL-13-mediated immunopathogenesis. Results suggest that anti-IL-13 monoclonal antibody have efficacious effect in IL-13-induced immunopathologies. However, IL-13 mAb anrukinzumab failed the initial study target, but other IL-13 mAb such as QAX-576 [211] and CAT-354 [212] showed very promising results. Though, most of the studies comprised asthma and other immunopathologies, it provides a way for the designing and development of a therapeutic approach against the malignancies too.

Targeting IL-4/IL-13 Receptor

IL-4/IL-13 receptors are targeted to block the progression of IL-4/IL-13-mediated immunopathologies. Puri et al. developed a chimeric human IL-13 protein (cintredekin besudox), which comprises human IL-13 and a mutated from of the *Pseudomonas* exotoxin

[143], evaluated against pulmonary fibrosis and asthma. Study showed its potential as a therapeutic agent and now National Institute of Health (NIH), the US Food and Drug Administration (FAD) and the University of Michigan licensed the NeoPharm Company to use it as a therapeutic agent for the treatment of idiopathic pulmonary fibrosis and asthma. More studies involving chimeric human IL-13Rα2 and *Pseudomonas* exotoxin have been performed to show its anti-cancer therapeutic potential against many cancer types in the lab of Puri [159, 160]. Recently, a recombinant human IL-4 variant, pitrakinra (AER 001, Aerovant^{TM}), which can inhibit both IL-4R and IL-13R, has entered phase II clinical trial for the treatment of asthma. Two candidates for soluble IL-13R currently exist: IL-13Rα1 and IL-13Rα2. Though, IL-13 has >10 times higher affinity for IL-13Rα2 than that for IL-13Rα1 and therefore IL-13Rα2 is potentially a more robust blocking agent [90]. AMG-317 is fully human monoclonal antibody currently under investigation for its ability to bind IL-4Rα and blocks the function of IL-4 and IL-13 [213]. An antisense drug, AIR645, has been developed, which target mRNA encoding the IL-4Rα subunit and inhibits the appearance of IL-4Rα protein in the cells [214]. Furthermore, Bioengineered version of IL-4, mutein has been developed that acts as IL-4/IL-13 antagonist by binding to the shared IL-4Rα subunit and prevents oligomerization of the subunit [215, 216]. Mutated IL-13 compounds chelated with a cellular toxin such as GB-13 and TRIGLIO evaluated and found very effective in limiting growth and invasion of high-grade astrocytomas including glioblastoma multiforme. GB-13 is a Targepeutics lead 2nd generation compound and TRIGLIO is a Targepeutics lead 3rd generation anti-cancer IL-13-based cytotoxin, yielding both higher specificity and potency and therefore makes them ideal candidates for drugs for the management of brain, pancreas, head and neck, and skin cancers.

TARGETING STAT PROTEINS

Several approaches directed at inhibiting STAT proteins are under investigation, including small-molecule inhibitors, antisense therapy and small interfering RNA (siRNA) and dominant negative peptides. Small molecule inhibitors such as YM-341619 hydrochloride has been developed, which can suppress IL-4-induced STAT-6-dependent reporter gene expression that subsequently leads to the inhibition of IL-4/IL-13 production and T cell differentiation into Th2 cells [217] and therefore resulting in subsequent activation of Th1 type T cell-mediated immunity. Depletion of STAT5 with a small interfering RNA (siRNA) results in the inhibition of expression of Bcl-2, p16^{ink4a}, p21$^{waf1/cip1}$, p27^{kip1}, E-cadherin, FAK, VEGF and MMPs genes responsible for cell growth, cell cycle progression, tumor invasion and metastasis suggesting it as a potential approach for the treatment of cancer [191].

In addition to the above approaches, several studies have been undertaken to target different signaling intermediates of IL-4/IL-13-mediated signal transduction pathways including IRS proteins, JAKs and SOCS proteins. Several JAK inhibitors have been developed for the treatment of malignancies where malignant condition is associated with constitutive activation of JAKs. One of such compound Ruxolitinib have cleared the phase III clinical trial under the study called COMFORT-1 (Controlled MyeloFibrosis Study with Oral JAK Inhibitor Therapy) and found very effective in reducing spleen volume in patients with myelofibrosis (MF). Studies are underway to evaluate the clinical efficacy of several other compounds such as Lestaurtinib, TG101348, Tasocitinib, XL109 (under Phase II clinical trial), AZD1480, CYT387, GLPG0634, INCB028050, ONX0803 (under Phase III clinical trial) and INCB16562 under preclinical studies. Recently, INCB16562, a selective inhibitor of JAK1/2, has been found very efficacious in multiple myeloma (MM) patients [158]. Though, research for the immunotherapeutic or therapeutic approaches against malignancies specifically by targeting IL-4/IL-13, its receptors, IL-4/IL-13 signaling cascades and signaling intermediates are in its infancy, initial studies provides very enthusiastic results that may guide future research in this direction.

Conclusion

IL-4 and IL-13 both are immunoregulatory cytokines that regulate many of the important biological functions such as proliferation, differentiation and immune effector functions. On one hand, they suppress the function of every immune cells through the receptor expressed on their cell surface and on the other hand they enhance Th2 polarization by activating the differentiation of Th0 cells to Th2 cells. Thus, they are involved in negative regulation of anti-tumor immunity and tumor immunosurveillance. Both cytokines activates different STAT proteins and many of the kinases viz RTK, ERK, p38 MAPK and PKC and leads to the amplification of Th2 microenvironment, transmission of proliferating signals and suppression of apoptotic processes in tumor cells that results in unrestricted tumor growth and progression. The involvement of IL-4/IL-13, its receptor complexes and signaling intermediates in the negative regulation of anti-tumor immune response leads to the speculation that they could be targeted for cancer immunotherapy or for designing and developing a therapeutic agent that may lead to effective rejection and repression of tumor cells. Initial studies involving several pharmacological inhibitors, antisense therapy, small interfering RNA and dominant negative peptides against IL-4/IL-13, IL-4/IL-13 receptor subunits, signaling intermediates provide very enthusiastic results that may allow future clinical studies to design and develop novel therapeutic approaches. Interestingly, many of such targeted molecules are under phase II and phase III clinical trial studies and may be adopted in regular clinical practices in future.

References

[1] Minty, A; Chalon, P; Derocq, JM; Dumont, X; Guillemot, J-C; Kaghad, M; Labit, C; Leplatois, P; Liauzun, P; Miloux, B; Minty, C; Casellas, P; Loison, G; Lupker, J; Shire, D; Ferrara, P; Caput, D. Interleukin-13 is a new human lymphokine regulating inflammatory and immune responses. *Nature*, 1993, 362, 248–250.

[2] Chomarat, P; Banchereau, J. Interleukin-4 and interleukin-13: their similarities and discrepancies. *Int. Rev. Immunol*. 1998, 17, 1–52.

[3] Nelms, K; Keegan, AD; Zamorano, J; Ryan, JJ; Paul, WE. The IL-4 receptor: signaling mechanism and biological function. *Annu. Rev. Immunol.,* 1999, 17: 701–738.

[4] Akbari, O; Stock, P; Meyer, E; Kronenberg, M; Sidobre, S; Nakayama, T; Taniguchi, M; Grusby, MJ; DeKruyff, RH; Umetsu, DT. Essential role of NKT cells producing IL-4 and IL-13 in the development of allergen-induced airway hyperreactivity. *Nat. Med.*, 2003, 9, 582–588.

[5] Howard, M; Farrar, J; Hilfiker, M; Johnson, B; Takatsu, K; Hamaoka, T; Paul, WE. Identification of a T cell-derived B cell growth factor distinct from interleukin 2. *J. Exp. Med.,* 1982, 155, 914–923.

[6] Brown, KD; Zurawski, SM; Mosmann, TR; Zurawski, G. A family of small inducible proteins recreated by leucocytes is member of a new super family that includes leukocytes and fibroblast derived inflammatory agents, growth factors, and indicators of various activation processes. *J. Immunol.*, 1989, 142, 679–687.

[7] McKenzie, AN; Li, X; Largaespada, DA; Sato, A; Kaneda, A; Zurawski, SM; Doyle, EL; Milatovich, A; Francke, U; Copeland, NG. Structural comparison and chromosomal localization of the human and mouse IL-13 genes. *J. Immunol.*, 1993, 150, 5436–5444.

[8] Smirnov, DV; Smirnova, MG; Korobko, VG; Frolova, EI. Tandem arrangement of human genes for interlukin-4 and interleukin-13: resemblance in their organization. *Gene,* 1995, 155, 277–281.

[9] de Waal Malefyt, R; Figdor, CG; Huijbens, R; Mohan-Peterson, S; Bennett, B; Culpepper, J; Dang, W; Zurawski, G; de Vries, JE. Effects of IL-13 on phenotype, cytokine production and cytotoxic function of human monocytes: comparison with IL-4 and modulation by IFN-γ or IL-10. *J. Immunol.*, 1993, 151, 6370–6381.

[10] VerceUi, D; Jabara, HH; Lee, BW; Woodland, N; Geha, RS; Leung, DY. Human recombinant interleukin 4 induces FcεR2/CD23 on normal human monocytes. *J. Exp. Med.*, 1988, 167, 1406–1416.

[11] Te Velde, AA; Klomp, JPG; Yard, BA; de Vries, JE; Figdor, CG. Modulation of phenotypic and functional properties of human peripheral blood monocytes by IL-4. *J. Immunol.,* 1988, 140, 1548–1554.

[12] Littman, BH; Dastvan, FF; Carlson, PL; Sanders, KM. Regulation of monocyte/macrophage C2 production and HLA-DR expression by IL-4 (BSF-1) and IFN-γ. *J. Immunol.*, 1989, 142, 520–525.

[13] Te Velde, AA; Rousset, E; Peronne, C; de Vries, JE; Figdor, CG. IFN-γ and IFN-γ have different regulatory effects on IL-4 induced membrane expression of FcεRIIb and release of soluble FcεRIIb by human monocytes. *J. Immunol.*, 1990, 144, 3052–3059.

[14] Becker, S; Daniel, EG. Antagonistic and additive effects of IL-4 and IFN-γ on human monocytes and macrophages: effects on Fc receptors, HLA-D antigens, and superoxide production. *Cell. Immunol.*, 1990, 129, 351–362.

[15] Gerrard, TL; Dyer, DR; Mostowski, HS. IL-4 and granulocyte-macrophage colony stimulating factor selectively increase HLA-DR and HLA-DP antigens, but not HLADQ antigens on human monocytes. *J. Immunol.,* 1990, 144, 4670–4674.

[16] Bogdan, C; Nathan, C. Modulation of macrophage functions by transforming growth factor β, interleukin-4, and interleukin-10. *Ann. NY Acad. Sci.,* 1993, 685, 713–739.

[17] Hart, PH; Vitti, GF; Burgess, DR; Whitty, GA; Piccoli, DS; Hamilton, JA. Potential anti-inflammatory effects of interleukin 4: suppression of human monocyte tumor necrosis factor α, interleukin 1 and prostaglandin E2. *Proc. Natl. Acad. Sci. USA*, 1989, 86, 3803–3807.

[18] Essner, R; Rhoades, K; McBride, WH; Morton, DL; Economou, JS. IL-4 down-regulates IL-1 and TNF gene expression in human monocytes. *J. Immunol.*, 1989, 142, 3857–3861.

[19] Donnelly, RP; Fenton, MJ; Kaufman, JD; Gerrard, TL. IL-1 expression in human monocytes is transcriptionally and posttranscriptionally regulated by IL-4. *J. Immunol.*, 1991, 146, 3431–3436.

[20] Hamilton, JA; Whitty, GA; Royston, AKM; Cebon, J; Layton, JE. Interleukin 4 suppresses granulocyte colony-stimulating factor and granulocyte-macrophage colony stimulating factor levels in stimulated human monocytes. *Immunology*, 1992, 76, 566–571.

[21] D'Andrea, A; Ma, X; Aste-Amezaga, M; Paganin, C; Trinchieri, G. Stimulatory and inhibitory effects of interleukin (IL)-4 and IL-13 on the production of cytokines by human peripheral blood mononuclear cells: priming for IL-12 and tumor necrosis factor α production. *J. Exp. Med.*, 1995, 181, 537–546.

[22] Lugering, N; Kucharzik, T; Stein, H; Winde, G; Lugering, A; Hasilik, A; Domschke, W; Stoll, R. IL-10 synergizes with IL-4 and IL-13 in inhibiting lysosomal enzyme secretion by human monocytes and lamina propria mononuclear cells from patients with inflammatory bowel disease. *Digestive Dis. Sci.*, 1998, 43, 706–714.

[23] Doyle, AG; Herbein, G; Montaner, LJ; Minty, AJ; Caput, D; Ferrara, P; Gordon, S. Interleukin-13 alters the activation state of murine macrophages in vitro: comparison with interleukin-4 and interferon-γ. *Eur. J. Immunol.*, 1994, 24, 1441–1445.

[24] Weiss, G; Bogdan, C; Hentze, MW. Pathways for the regulation of macrophage iron metabolism by the anti-inflammatory cytokines IL-4 and IL-13. *J. Immunol.*, 1997, 158, 420–425.

[25] Lebel-Binay, S; Laguerre, B; Quintin-Colonna, F; Conjeaud, H; Magazin, M; Miloux, B; Pecceu, F; Caput, D; Ferrara, P; Fradelizi, D. Experimental gene therapy of cancer using tumor cells engineered to secrete interleukin-13. *Eur. J. Immunol.*, 1995, 25, 2340–2348.

[26] Doherty, TM; Kastelein, R; Menon, S; Andrade, S; Coffman, RL. Modulation of murine macrophage function by IL-13. *J. Immunol.*, 1993, 151, 7151–7160.

[27] Takeda, K; Kamanaka, M; Tanaka, T; Akira, S. Impaired IL-13-mediated function of macrophages in STAT6-deficient mice. *J. Immunol.*, 1996, 157, 3220–3222.

[28] Bogdan, C; Thüring, H; Dlaska, M; Röllinghoff, M; Weiss, G. Mechanism of suppression of macrophage nitric oxide release by IL-13. *J. Immunol.*, 1997, 159, 4506–4513.

[29] Rutschman, R; Lang, R; Hesse, M; Ihle, JN; Wynn, TA; Murray, PJ. STAT6-dependent substrate depletion regulates nitric oxide production. *J. Immunol.*, 2001, 166, 2173–2177.

[30] El-Gayar, S; Thüring-Nahler, H; Pfeilschifter, J; Röllinghoff, M; Bogdan, C. Translational control of inducible nitric oxide synthase by IL-13 and arginine availability in inflammatory macrophages. *J. Immunol.*, 2003, 171, 4561–4568.

[31] Sozzani, P; Cambon, C; Vita, N; Séguélas, M-H ; Caput, D; Ferrara, P; Pipy, B. Interleukin-13 inhibits protein kinase C-triggered respiratory burst in human monocytes. Role of calcium and cyclic AMP. *J. Biol. Chem.*, 1994, 270, 5084–5088.

[32] Sozzani, P; Hasan, L; Séguélas, M-H; Caput, D; Ferrara, P; Pipy, B; Cambon, C. IL-13 induces tyrosine phosphorylation of phospholipase Cγ-1 following IRS-2 association in

human monocytes: relationship with the inhibitory effect of IL-13 on ROI production. *Biochem. Biophys. Res. Commun.*, 1998, 244, 665–670.

[33] Scopes, J; Massev, HM; Ebrahim, H; Horton, MA; Flanagan, AM. Interleukin-4 and interleukin-13: bidirectional effects on human osteoclast formation. *Bone,* 2001, 29, 203–208.

[34] Sica, A; Schioppa, T; Mantovani, A; Allavena, P. Tumour-associated macrophages are a distinct M2 polarised population promoting tumour progression: Potential targets of anti-cancer therapy. *Eur. J. Cancer,* 2006, 42, 717–727.

[35] Punnonen, J; Aversa, G; Cocks, BG; McKenzie, ANJ; Menon, S; Zurawski, G; de Waal Malefyt, R; de Vries, JE. Interleukin 13 induces interleukin 4-independent IgG4 and IgE synthesis and CD23 expression by human B cells. *Proc. Natl. Acad. Sci. USA*, 1993, 90, 3730–3734.

[36] Cocks, BG; de Waal Malefyt, R; Galizzi, J-P; de Vries, JE; Aversa, G. IL-13 induces proliferation and differentiation of human B cells activated by the CD40 ligand. *Int. Immunol.*, 1993, 5, 657–663.

[37] Defrance, T; Carayon, P; Billian, G; Guillemot, JC; Minty, A; Caput, D; Ferrara, P. Interleukin 13 is a B cell stimulating factor. *J. Exp. Med.,* 1994, 179, 135–143.

[38] Punnonen, J; de Vries, JE. IL-13 induces proliferation, Ig isotype switching, and Ig synthesis by immature human fetal B cells. *J. Immunol.,* 1994, 152, 1094–1102.

[39] Tangye, SG; Ferguson, A; Avery, DT; Ma, CS; Hodgkin, PD. Isotype switching by human B cells is division-associated and regulated by cytokines. *J. Immunol.*, 2002, 169, 4298–4306.

[40] Lai, YH; Mosmann, TR. Mouse IL-13 enhances antibody production in vivo and acts directly on B cells in vitro to increase survival and hence antibody production. *J. Immunol.*, 1999, 162, 78–87.

[41] Poudrier, J; Graber, P; Herren, S; Gretener, D; Elson, G; Bemey, C; Gauchat, J-F; Kosco-Vilbois, MH. A soluble form of IL-13 receptor α1 promotes IgG2a and IgG2b production by murine germinal center B cells. *J. Immunol.*, 1999, 163, 1153–1161.

[42] Welham, MJ; Learmonth, L; Bone, H; Schrader, JW. Interleukin-13 signal transduction in lymphohemopoietic cells: similarities and differences in signal transduction with interleukin-4 and insulin. *J. Biol. Chem.,* 1995, 270, 12286–12296.

[43] Orchansky, P; Ayres, SD; Hilton, DJ; Schrader, JW. An interleukin (IL)-13 receptor lacking the cytoplasmic domain fails to transduce IL-13-induced signals and inhibits responses to IL-4. *J. Biol. Chem.*, 1997, 272, 22940–22497.

[44] Bost, KL; Holton, RH; Cain, TK; Clements, JD. In vivo treatment with anti-interleukin-13 antibodies significantly reduces the humoral immune response against an oral immunogen in mice. *Immunology,* 1996, 87, 633–641.

[45] Ford, D; Sheehan, C; Girasole, C; Priester, R; Kouttab, N; Tigges, J; King, TC; Luciani, A; Morgan, JW; Maizel, AL. The human B cell response to IL-13 is dependent on cellular phenotype as well as mode of activation. *J. Immunol.*, 1999, 163, 3185–3193.

[46] Kuroki, A; Iyoda, M; Shibata, T; Sugisaki, T. Th2 cytokines increase and stimulate B cells to produce IgG4 in idiopathic membranous nephropathy. *Kidney Int.*, 2005, 68, 302–310.

[47] Yu, Y; Cao, X; Zhang, M; Lei, H; Tao, Q. The humoral antitumor responses induced by IL-4 gene-modified tumor vaccine. *Chinese J. Cancer Res.,* 1997, 9, 250–253.

[48] Hajoui, O; Janani, R; Tulic, M; Joubert, P; Ronis, T; Hamid, Q; Zheng, H; Mazer, BD. Synthesis of IL-13 by human B lymphocytes: Regulation and role in IgE production. *J. Allergy Clin. Immunol.*, 2004, 114, 657–663.

[49] Levy, F; Kristofic, C; Heusser, C; Brinkmann, V. Role of IL-13 in CD4 T cell-dependent IgE production in Atopy. *Int. Arch. Allergy Immunol.*, 1997, 112, 49–58.

[50] Ryzhov, S; Goldstein, AE; Matafonov, A; Zeng, D; Biaggioni, I; Feoktistov, I. Adenosine-activated mast cells induce IgE synthesis by B lymphocytes: an A2B-mediated process involving Th2 cytokines IL-4 and IL-13 with implications for asthma. *J. Immunol.*, 2004, 172, 7726–7733.

[51] Graber, P; Gretener, D; Herren, S; Aubry, J-P; Elson, G; Poudrier, J; Lecoanet-Henchoz, S; Alouani, S; Losberger, C; Bonnefoy, J-Y; Kosco-Vilbois, M-H; Gauchat, J-F. The distribution of IL-13 receptor α1 expression on B cells, T cells and monocytes and its regulation by IL-13 and IL-4. *Eur. J. Immunol.*, 1998, 28, 4286–4298.

[52] Ogata, H; Ford, D; Kouttab, N; King, TC; Vita, N; Minty, A; Stoeckler, J; Morgan, D; Girasole, C; Morgan, JW; Maizel, AL. Regulation of interleukin-13 receptor constituents on mature human B lymphocytes. *J. Biol. Chem.*, 1998, 273, 9864–9871.

[53] Hajoui, O; Zheng, H; Guay, J; Letuve, S; Fawaz, LM; Mazer, BD. Regulation of IL-13 receptor α1 expression and signaling on human tonsillar B-lymphocyte subsets. *J. Allergy Clin. Immunol.*, 2007, 120, 1425–1432.

[54] Zurawski, G; de Vries, JE. Interleukin 13, an interleukin-4-like cytokine that acts on monocytes and B cells, but not on T cells. *Immunol. Today,* 1994, 15, 19–26.

[55] de Vries, JE. The role of IL-I 3 and its receptor in allergy and inflammatory responses. *J. Allergy Clin. Immunol.*, 1998, 102, 165–169.

[56] Gauchat, J-F; Schlagenhauf, E; Feng, N-P; Moser, R; Yamage, M; Jeannin, P ; Alouani, S; Elson, G; Notarangelo, ND; Wells, T; Eugster, H-P; Bonnefoy, J-Y. A novel 4-kd interleukin-13 receptor alpha mRNA expressed in human B, T, and endothelial cells encoding an alternate type-II interleukin-4/interleukin-13 receptor. *Eur. J. Immunol.*, 1997, 27, 971–978.

[57] Curiel, RE; Lahesmaa, R; Subleski, J; Cippitelli, M; Kirkeli, RA; Young, HA; Ghosh, P. Identification of a Stat-6-responsive element in the promoter of the human interleukin-4 gene. *Eur. J. lmmunol.*, 1997, 27, 1982–1987.

[58] Tan, J; Deleuran, B; Gesser, H; Maare, H; Deleuran, M; Larsen, CG; Thestrup-Pederson, K. Regulation of human T lymphocyte chemotaxis in vitro by T cell-derived cytokines IL-2, IFN-y, IL-4, IL-10 and IL-13. *J. Immunol.*, 1995, 154, 3742–4752.

[59] McKenzie, GJ; Emson, CL; Bell, SE; Anderson, S; Fallon, P; Zurawski, G; Murray, R; Grencis, R; McKenzie, AN. Impaired development of Th2 cells in IL-I 3-deficient mice. *Immunity,* 1998, 9, 423–432.

[60] Finkelman, FD; Wynn, TA; Donaldson, DD; Urban, JF. The role of IL-13 in helminth-induced inflammation and protective immunity against nematode infections. *Curr. Opin. Immunol.*, 1999, 11, 420–426.

[61] Lowenthal, JW; Castle, BE; Christiansen, J; Schreurs, J; Rennick, D; Arai, N; Ho, YP; Takebe, Y; Howard, M. Expression of high affinity receptors for murine interleukin 4 (BSF-1) on hemopoietic and nonhemopoietic cells. *J. Immunol.*, 1988, 140, 456–464.

[62] Paul, WE; Ohara, J. B-cell stimulatory factor-1/interleukin 4. *Annu. Rev. Immunol.,* 1987, 5, 429–459.

[63] Yokota, T; Arai, N; deVries, J; Spits, H; Banchereau, J; Zlotnik, A; Rennick, D; Howard, M; Takebe, Y; Miyatake, S; Lee, F; Arai, K. Molecular biology of interleukin-4 and interleukin 5 genes and biology of their products that stimulate B cells, T cells and hemopoietic cells. *Immunol. Rev.,* 1988, 102, 137–187.

[64] Kaplan, MH; Schindler, U; Smiley, ST; Grusby, MJ. Stat6 is required for mediating responses to IL-4 and for development of Th2 cells. *Immunity,* 1996, 4, 313–409.

[65] Padilla, J; Daley, E; Chow, A; Robinson, K; Parthasarathi, K; McKenzie, ANJ; Tschernig, T; Kurup, VP; Donaldson, DD; Grunig, G. IL-13 regulates the immune response to inhaled antigens. *J. Immunol.,* 2005, 174, 8097–8105.

[66] Bellinghausen, I; Brand, P; Böttcher, I; Klostermann, B; Knop, J; Sloga, J. Production of interleukin-13 by human dendritic cells after stimulation with protein allergens is a key factor for induction of T helper 2 cytokines and is associated with activation of signal transducer and activator of transcription-6. *Immunology,* 2003, 108, 167–176.

[67] Terabe, M; Park, JM; Berzofsky, JA. Role of IL-13 in regulation of antitumor immunity and tumor growth. *Cancer Immunol. Immunother.*, 2004, 53, 79–85.

[68] Lee, HH; Hoeman, CM; Hardaway, JC; Guloglu, FB; Ellis, JS; Jain, R; Divekar, R; Tartar, DM; Haymaker, CL; Zaghouani, H. Delayed maturation of an IL-12-producing dendritic cells subset explains the early Th2 bias in neonatal immunity. *J. Exp. Med.*, 2008, 205, 2269–2280.

[69] Maggi, E; Parronchi, P; Manetti, R; Simonelli, C; Piccinni, MP; Rugiu, FS; De Carli, M; Ricci, M; Romagnani, S. Reciprocal regulatory effects of IFN-γ and IL-4 on the in vitro development of human Th1 and Th2 clones. *J. Immunol.*, 1992, 148, 2142–2147.

[70] Terabe, M; Matsui, S; Park, J-M; Mamura, M; Noben-Trauth, N; Donaldson, DD; Chen, W; Wahl, SM; Ledbetter, S; Pratt, B; Letterio, JJ; Paul, WE; Berzofsky, JA. Transforming growth factor-β production and myeloid cells are an effector mechanism through which CD1d-restricted T cells block cytotoxic T lymphocyte-mediated tumor immunosurveillance: abrogation prevents tumor recurrence. *J. Exp. Med.,* 2003, 198, 1741–1752.

[71] Skapenko, A; Kalden, JR; Lipsky, PE; Schulze-Koops, H. The IL-4 Receptor α-Chain-Binding Cytokines, IL-4 and IL-13, Induce Forkhead Box P3-Expressing CD25+CD4+ Regulatory T Cells from CD25–CD4+ Precursors. *J. Immunol.*, 2005, 175, 6107–6116.

[72] Newcomb, DC; Zhou, W; Moore, ML; Goleniewska, K; Hershey, GKK; Kolls, JK; Peebles, RS Jr. A functional IL-13 receptor is expressed on polarized murine CD4+ Th17 cells and IL-13 signaling attenuates Th17 cytokine production. *J. Immunol.*, 2009, 182, 5317–5321.

[73] Mattes, J; Hulett, M; Xie, W; Hogan, S; Rothenberg, ME; Foster, P; Parish, C. Immunotherapy of cytotoxic T cell-resistant tumors by T helper 2 cells: an eotaxin and STAT6-dependent process. *J. Exp. Med.,* 2003, 197, 387–393.

[74] Rodolfo, M; Zilocchi, C; Accornero, P; Cappetti, B; Arioli, I; Colombo, MP. IL-4-transduced tumor cell vaccine induces immunoregulatory type 2 CD8 T lymphocytes that cure lung metastases upon adoptive transfer. *J. Immunol.,* 1999, 163, 1923–1928.

[75] Volpert, OV; Fong, T; Koch, AE; Peterson, JD; Waltenbaugh, C; Tepper, RI; Bouck, NP. Inhibition of angiogenesis by interleukin-4. *J. Exp. Med.*, 1998, 188, 1039–1046.

[76] Mukherjee, S; Nelson, D; Loh, S; Van Bruggen, I; Palmer, LJ; Leong, C; Garlepp, MJ; Robinson, BW. The immune anti-tumor effects of GM-CSF and B7-1 gene transfection are enhanced by surgical debulking of tumor. *Cancer Gene Ther.*, 2001, 8, 580–588.

[77] Terabe, M; Matsui, S; Noben-Trauth, N; Chen, H; Watson, C; Donaldson, DD; Carbone, DP; Paul, WE; Berzofsky, JA. NKT cell-mediated repression of tumor immunosurveillance by IL-13 and the IL-4R-STAT6 pathway. *Nat. Immunol.,* 2000, 1, 515–520.

[78] Terabe, M; Khanna, C; Bose, S; Melchionda, F; Mendoza, A; Mackall, CL; Helman, LJ; Berzofsky, JA. CD1d-restricted natural killer T cells can down-regulate tumor immunosurveillance independent of interleukin-4 receptor-signal transducer and activator of transcription 6 or transforming growth factor-β. *Cancer Res.*, 2006, 66, 3869–3875.

[79] Park, JM; Terabe, M; van den Broeke, LT; Donaldson, DD; Berzofsky, JA. Unmasking immunosurveillance against a syngeneic colon cancer by elimination of CD4+ NKT regulatory cells and IL-13. *Int. J. Cancer,* 2004, 114, 80–87.

[80] Fichtner-Feigl, S; Terabe, M; Kitani, A; Young, CA; Fuss, I; Geissler, EK; Schlitt, H-J; Berzofsky, JA; Strober, W. Restoration of tumor immunosurveillance via targeting of interleukin-13 receptor-α2. *Cancer Res.*, 2008, 68, 3467–3475.

[81] Leonard, WJ; Lin, JX. Cytokine receptor signaling pathways. *J. Allergy Clin. Immunol.*, 2000, 105, 877–888.

[82] Pernis, AB; Rothman, PB. JAK-STAT signaling in asthma. *J. Clin. Invest.,* 2002, 109, 1279–1283.

[83] Wynn, TA. IL-13 effector functions. *Annu. Rev. Immunol.*, 2003, 21, 425–456.

[84] Fanslow, WC; Clifford, KN; Park, LS; Rubin, AS; Voice, RF; Beckmann, MP; Widmer, MB. Regulation of alloreactivity in vivo by IL-4 and the soluble IL-4 receptor. *J. Immunol.*, 1991, 147, 535–540.

[85] Kruse, S; Forster, J; Kuehr, J; Deichmann, KA. Characterization of the membrane bound and a soluble form of human IL-4 receptor alpha produced by alternative splicing. *Int. Immunol.,* 1999, 11, 1965–1970.

[86] Lefort, S; Vita, N; Reeb, R; Caput, D; Ferrara, P. IL-13 and IL-4 share signal transduction elements as well as receptor components in TF-1 cells. *FEBS Lett.,* 1995, 366, 122–126.

[87] de Waal Malefyt, R; Abrams, JS; Zurawski, SM; Lecron, JC; Mohan-Peterson, S; Sanjanwala, B; Bennett, B; Silver, J; de Vries, JE; Yssel, H. Differential regulation of IL-13 and IL-4 production by human CD8+ and CD4+ TH0, Th1 and Th2 T cell clones and EBV-transformed B cells. *Int. Immunol.*, 1995, 7, 1405–1416.

[88] Aman, MJ; Tayebi, N; Obiri, NI; Puri, RK; Modi, WS; Leonard, WJ. cDNA cloning and characterization of the human interleukin-13 receptor alpha chain. *J. Biol. Chem.,* 1996, 27, 29265–29270.

[89] Arima, K; Sato, K; Tanaka, G; Kanaji, S; Terada, T; Honjo, E; Kuroki, R; Matsuo; Y; Izuhara, K. Characterization of the interaction between interleukin-13 and interleukin-13 receptors. *J. Biol. Chem.*, 2005, 280, 24915–24922.

[90] Miloux, B; Laurent, P; Bonnin, O; Lupker, J; Caput, D; Vita, N; Ferrara, P. Cloning of the human IL-13 R alpha 1 chain and reconstitution with the IL-4R alpha of a functional IL-4/IL-13 receptor complex. *FEBS Lett.,* 1997, 401, 163–166.

[91] Keegan, AD; Nelms, K; White, M; Wang, L-M; Pierce, JH; Paul, WE. An IL-4 receptor region containing an insulin receptor motif is important for IL-4 mediated IRS-1 phosphorylation and cell growth. *Cell,* 1994, 76, 811–820.

[92] Obiri, NI; Murata, T; Debinski, W; Puri, RK. Modulation of interleukin (IL)-13 binding and signaling by the gamma chain of the IL-2 receptor. *J. Biol. Chem.*, 1997, 272, 20251–20258.

[93] Andrews, R; Rosa, L; Daines, M; Khurana Harshey, G. Reconstitution of a functional human type II IL-4/IL-13 receptor in mouse B cells: demonstration of species specificity. *J. Immunol.*, 2001, 166, 1716–1722.

[94] Caput, D; Laurent, P; Kaghad, M; Lelias, JM; Lefort, S; Vita, N; Ferrara, P. Cloning and characterization of a specific interleukin (IL)-13 binding protein structurally related to the IL-5 receptor α chain. *J. Biol. Chem.,* 1996, 271, 16921–16926.

[95] Bazan, JF. Structural design and molecular evolution of a cytokine receptor superfamily. *Proc. Natl. Acad. Sci. USA*, 1990, 87, 6934–6938.

[96] Guo, J; Apiou, F; Mellerin, MP; Lebeau, B; Jacques, Y; Minvielle, S. Chromosome mapping and expression of the human interleukin-13 receptor. *Genomics*, 1997, 42, 141–145.

[97] Donaldson, DD; Whitters, MJ; Fitz, LJ; Neben, TY; Finnerty, H; Henderson, SL; O'Hara, RM Jr; Beier, DR; Turner, KJ; Wood, CR; Collins, M. The murine IL-13 receptor alpha 2: molecular cloning, characterization, and comparison with murine IL-13 receptor alpha 1. *J. Immunol.,* 1998, 161, 2317–2324.

[98] Kioi, M; Seetharam, S; Puri, RK. N-linked glycosylation of IL-13R alpha2 is essential for optimal IL-13 inhibitory activity. *FASEB J.,* 2006, 20, 2378–2380.

[99] Wood, N; Whitters, MJ; Jacobson, BA; Witek, J; Sypek, JP; Kasaian, M; Eppihimer, MJ; Unger, M; Tanaka, T; Goldman, SJ; Collins, M; Donaldson, DD; Grusby, MJ. Enhanced interleukin (IL)-13 responses in mice lacking IL-13 receptor alpha 2. *J. Exp. Med.*, 2003, 197, 703–709.

[100] Aversa, G; Punnonen, J; Cocks, BG; de Waal Malefyt, R; Vega, E Jr; Zurawski, SM; Zurawski, G; de Vries, JE. An interleukin 4 (IL-4) mutant protein inhibits both IL-4 or IL-13-induced human immunoglobulin G4 (IgG4) and IgE synthesis and B cell proliferation: support for a common component shared by IL-4 and IL-13 receptors. *J. Exp. Med.,* 1993, 178, 2213–2218.

[101] Lefort, S; Vita, N; Reeb, R; Caput, D; Ferrara, P. IL-13 and IL-4 share signal transduction elements as well as receptor components in TF-1 cells. *FEBS Lett.*, 1995, 366, 122–126.

[102] Deutsch, H; Koettnitz, K; Chung, J; Kalthoff, FS. Distinct sequence motifs within the cytoplasmic domain of the human IL-4 receptor differentially regulate apoptosis inhibition and cell growth. *J. Immunol.,* 1995, 154, 3696–3703.

[103] Miyamasu, M; Misaki, Y; Yamaguchi, M; Yamamoto, K; Morita, Y; Matsushima, K; Nakajima, T; Hirai, K. Regulation of human eotaxin generation by Th1-/Th2-derived cytokines. *Int. Arch. Allergy Immunol.,* 2000, 122, 54–58.

[104] Wang, LM; Keegan, AD; Li, W; Lienhard, GE; Pacini, S; Gutkind, JS; Myers, MG; Sun, XJ; White, MF; Aaronson, SA.. Common elements in interleukin 4 and insulin signaling pathways in factor-dependent hematopoietic cells. *Proc. Natl. Acad. Sci. USA*, 1993, 90, 4032–4036.

[105] Lavan, BE; Lane, WS; Lienhard, GE. The 60 kDa phosphotyrosine protein in insulin treated adipocytes is a new member of the insulin receptor substrate family. *J. Biol. Chem.,* 1997, 272, 11439–11443.

[106] Sun, XJ; Rothenberg, P; Kahn, CR; Backer, JM; Araki, E; Wilden, PA; Cahill, DA; Goldstein, BJ; White, MF. Structure of the insulin receptor substrate IRS-1 defines a unique signal transduction protein. *Nature,* 1991, 352, 73–77.

[107] Sun, XJ; Wang, LM; Zhang, Y; Yenush, L; Myers, MG; Glasheen, E; Lane, WS; Pierce, JH; White, MF. Role of IRS-2 in insulin and cytokine signalling. *Nature,* 1995, 377, 173–177.

[108] Wang, HY; Zamorano, J; Yoerkie, JL; Paul, WE; Keegan, AD. The IL-4 induced tyrosine phosphorylation of the insulin receptor substrate is dependent on JAK1 expression in human fibrosarcoma cells. *J. Immunol.,* 1997, 158, 1037–1040.

[109] Pesu, M; Takaluma, K; Aittomaki, S; Lagerstedt, A; Saksela, K; Kovanen, PE; Silvennoinen, O. Interleukin-4-induced transcriptional activation by stat6 involves multiple serine/threonine kinase pathways and serine phosphorylation of stat6. *Blood,* 2000, 95, 494–502.

[110] Denhardt, DT. Signal-transducing protein phosphorylation cascades mediated by Ras/Rho proteins in the mammalian cell: the potential for multiplex signaling. *Biochem. J.,* 1996, 318, 729–747.

[111] Fantin, VR; Keller, SR; Liehard, GE; Wang, LM. Insulin receptor substrate 4 supports insulin- and interleukin 4-stimulated proliferation of hematopoietic cells. *Biochem. Biophys. Res. Commun*., 1999, 260, 718–723.

[112] Wurster, AL; Rodgers, VL; White, MF; Rothstein, TL; Grusby, MJ. Interleukin-4-mediated Protection of Primary B Cells from Apoptosis through Stat6-dependent Up-regulation of Bcl-xL. *J. Biol. Chem*., 2002, 277, 27169–27175.

[113] Joung, I; Strominger, JL; Shin, J. Molecular cloning of a phosphotyrosineindependent ligand of the p56lck SH2 domain. *Proc. Natl. Acad. Sci. USA,* 1996, 93, 5991–5995.

[114] Puls, A; Schmidt, S; Grawe, F; Stabel, S. Interaction of protein kinase C ζ with ZIP, a novel protein kinase C-binding protein. *Proc. Natl. Acad. Sci. USA*, 1997, 94, 6191–6196.

[115] Ishii, T; Yanagawa, T; Kawane, T; Yuki, K; Seita, J; Yoshida, H; Bannai, S. Murine peritoneal macrophages induce a novel 60-kDa protein with structural similarity to a tyrosine kinase p56lck-associated protein in response to oxidative stress. *Biochem. Biophys. Res. Comm*., 1996, 226, 456–460.

[116] Okazaki, M; Ito, S; Kawakita, K; Takeshita, S; Kawai, S; Makishima, F; Oda, H; Kakinuma, A. Cloning, expression profile, and genomic organization of the mouse STAP/A170 gene. *Genomics,* 1999, 60, 87–95.

[117] Leonard, WJ; O'Shea, JJ. JAKs and STATs: biological implications. *Annu. Rev. Immunol*., 1998, 16, 293–322.

[118] Girault, JA; Labesse, G; Mornon, JP; Callebaut, I. Janus kinases and focal adhesion kinases play in the 4.1 band: a superfamily of band 4.1 domains important for cell structure and signal transduction. *Mol. Med.,* 1998, 4, 751–769.

[119] Hou, J; Schindler, U; Henzel, WJ; Ho, TC; Brasseur, M; McKnight, SL. An interleukin-4 induced transcription factor IL-4 *Stat. Science*, 1994, 265, 1701–1706.

[120] Ihle, JN. STATs: signal transducers and activators of transcription. *Science*, 1996, 84, 331–334.

[121] Lohi, O; Lehto, V-P. STAM/EAST/Hbp adaptor proteins-integrators of signaling pathways. *FEBS Lett*., 2001, 508, 287–290.

[122] Greenhalgh, CJ; Hilton, DJ. Negative regulation of cytokine signaling. *J. Leukoc. Biol.*, 2001, 70, 348–356.

[123] Umeshita-Suyama, R; Sugimoto, R; Akaiwa, M; Arima, K; Yu, B; Wada, M; Kuwano, M; Nakajima, K; Hamasaki, N; Izuhara, K. Characterization of IL-4 and IL-13 signals dependent on the human IL-13 receptor alpha chain 1: redundancy of requirement of tyrosine residue for STAT3 activation. *Int. Immunol.*, 2000, 12, 1499–509.

[124] Roy, B; Bhattacharjee, A; Xu, B; Ford, D; Maizel, AL; Cathcart, MK. IL-13 signal transduction in human monocytes: phosphorylation of receptor components, association with Jaks, and phosphorylation/activation of Stats. *J. Leukoc. Biol.*, 2002, 72, 580–589.

[125] O'Brien, CA; Manolagas, SC. Isolation and characterization of the human gp130 promoter. Regulation by STATs. *J. Biol. Chem.,* 1997, 272, 15003–15010.

[126] Rahaman, SO; Sharma, P; Harbor, PC; Aman, MJ; Vogelbaum, MA; Haque, SJ. IL-13Rα2, a decoy receptor for IL-13 acts as an inhibitor of IL-4-dependent signal transduction in glioblastoma cells. *Cancer Res.,* 2002, 62, 1103–1109.

[127] Wu, AH; Low, WC. Molecular cloning and identification of the human interleukin 13 alpha 2 receptor (IL-13Ra2) promoter. *Neuro-Oncol.*, 2003, 5, 179–187.

[128] Fichtner-Feigl, S; Strober, W; Kawakami, K; Puri, RK; Kitani, A. IL-13 signaling through the IL-13alpha(2) receptor is involved in induction of TGF-beta(1) production and fibrosis. *Nature Med.,* 2006, 12, 99–106.

[129] Davis, RJ. Signal transduction by the JNK group of MAP kinases. *Cell,* 2000, 103, 239–252.

[130] IGFapoptosis76,

[131] Jiang, H; Foltenyi, K; Kashiwada, M; Donahue, L; Vuong, B; Hehn, B; Rothman, P. Fes mediates the IL-4 activation of insulin receptor substrate-2 and cellular proliferation. *J. Immunol.,* 2001, 166, 2627–2634.

[132] Greer, P; Maltby, V; Rossant, J; Bernstein, A; Pawson, T. Myeloid expression of the human c-fps/fes proto-oncogene in transgenic mice. *Mol. Cell. Biol.,* 1990, 10, 2521–2527.

[133] MacDonald, I; Levy, J; Pawson, T. Expression of the mammalian c-fes protein in hematopoietic cells and identification of a distinct fes-related protein. *Mol. Cell. Biol.*, 1985, 5, 2543–2551.

[134] Jiang, H; Harris, MB; Rothman, P. IL-4/IL-13 signaling beyond JAK/STAT. *J. Allergy Clin. Immunol.*, 2000, 105, 1063–1070.

[135] Alexander, WS. Suppressors of cytokine signalling (SOCS) in the immune system. *Nat. Rev. Immunol.*, 2002, 2, 410–416.

[136] Rawling, JS; Rosler, KM; Harrison, DA. The JAK/STAT signaling pathway. *J. Cell Sci.,* 2004, 117, 1281–1283.

[137] Hanson, EM; Dickensheets, H; Qu, CK; Donnelly, RP; Keegan, AD. Regulation of the Dephosphorylation of Stat6: Participation of Tyr-713 in the interleukin-4 receptor α, the tyrosine phosphatase SHP-1 and the proteasome. *J. Biol. Chem.*, 2003, 278, 3903–3911.

[138] Chung, CD; Liao, J; Liu, B; Rao, X; Jay, P; Berta, P; Shuai, K. Specific inhibition of Stat3 signal transduction by PIAS3. *Science,* 1997, 278, 1803–1805.

[139] Chaouchi, N; Wallon, C; Goujard, C; Tertian, G; Rudent, A; Caput, D; Ferrera, P; Minty, A; Vazquez, A; Delfraissy, JF. Interleukin-13 inhibits interleukin-2-induced proliferation and protects chronic lymphocytic leukemia B cells from in vitro apoptosis. *Blood*, 1996, 87, 1022–1029.

[140] Kawakami, K; Joshi, BH; Puri, RK. Sensitization of cancer cells to interleukin 13-pseudomonas exotoxin-induced cell death by gene transfer of interleukin 13 receptor alpha chain. *Hum. Gene. Ther.*, 2000, 11, 1829–1835.

[141] Joshi, BH; Kawakami, K; Leland, P; Puri, RK. Heterogeneity in interleukin-13 receptor expression and subunit structure in squamous cell carcinoma of head and neck: differential sensitivity to chimeric fusion proteins comprised of interleukin-13 and a mutated form of Pseudomonas exotoxin. *Clin. Cancer Res.*, 2002, 8, 1948–1956.

[142] Gabay, C; Porter, B; Guenette, D; Billir, B; Arend, WP. Interleukin-4 (IL-4) and IL-13 enhance the effect of IL-1beta on production of IL-1 receptor antagonist by human primary hepatocytes and hepatoma HepG2 cells: differential effect on C-reactive protein production. *Blood*, 1999, 93, 1299–1307.

[143] Kawakami, K; Kawakami, M; Snoy, PJ; Husain, SR; Puri, RK. In vivo overexpression of IL-13 receptor alpha2 chain inhibits tumorigenicity of human breast and pancreatic tumors in immunodeficient mice. *J. Exp. Med.,* 2001, 194, 1743–1754.

[144] Kawakami, K; Husain, SR; Bright, RK; Puri, RK. Gene transfer of interleukin 13 receptor alpha2 chain dramatically enhances the antitumor effect of IL-13 receptor-targeted cytotoxin in human prostate cancer xenografts. *Cancer Gene Ther.*, 2001, 8, 861–868.

[145] Kanai, T; Watanabe, M; Hayashi, A; Nakazawa, A; Yajima, T; Okazawa, A; Yamazaki, M; Ishii, H; Hibi, T. Regulatory effect of interleukin-4 and interleukin-13 on colon cancer cell adhesion. *Br. J. Cancer,* 2000, 82, 1717–1723.

[146] Skinnider, BF; Elia, AJ; Gascoyne, RD; Patterson, B; Trumper, L; Kapp, U; Mak, TW. Signal transducer and activator of transcription 6 is frequently activated in Hodgkin and Reed-Sternberg cells of Hodgkin lymphoma. *Blood*, 2002, 99, 618–626.

[147] Kobayashi, M; Kobayashi, H; Pollard, RB; Suzuki, F. A pathogenic of Th2 cells and their cytokine products on the pulmonary metstasis of murine B16 melanoma. *J. Immunol.,* 1998, 160, 5859–5863.

[148] Min, B; Prout, M; Hu-Li, J; Zhu, J; Jankovic, D; Morgan, ES; Urban, JF Jr; Dvorak, AM; Finkelman, FD; LeGros, G; Paul, WE. Basophils produce IL-4 and accumulate in tissue after infection with a Th2-inducing parasite. *J. Exp. Med.,* 2004, 200, 507–517.

[149] MacGlashan, D Jr; White, JM; Huang, SK; Ono, SJ; Schroeder, JT; Lichtenstein, LM. Secretion of IL-4 from human basophils. The relationship between IL-4 mRNA and protein in resting and stimulated basophils. *J. Immunol.*, 1994, 152, 3006–3016.

[150] Zheng, W-p; Flavell, RA. The transcription factor GATA-3 is necessary and sufficient for Th2 cytokine gene expression in CD4 T cells. *Cell,* 1997, 89, 587–596.

[151] Zhang, DH; Cohn, L; Ray, P; Bottomly, K; Ray, A. Transcription factor GATA-3 is differentially expressed in murine Th21 and Th2 cells and controls Th2-specific expression of the interleukin-5 gene. *J. Biol. Chem.,* 1997, 272, 21597–21603.

[152] Zhu, J; Davidson, TS; Wei, G; Jankovic, D; Cui, K; Schones, DE; Guo, L; Zhao, K; Shevach, EM; Paul, WE. Downregulation of Gfi-1 expression by TGF-β is important for differentiation of Th17 and CD103+ inducible regulatory cells. *J. Exp. Med.*, 2009, 206, 329–341.

[153] Ostrand-Rosenberg, S; Grusby, MJ; Clements, VK. STAT6-deficient mice have enhanced tumor immunity to primary and metastatic mammary carcinoma. *J. Immunol.*, 2000, 165, 6015–6019.

[154] Zhu, J; Cote-Sierra, J; Guo, L; Paul, WE. Stat5 activation plays a critical role in Th2 differentiation. *Immunity,* 2003, 19, 739–748.

[155] Fujisawa, T; Joshi, B; Nakajima, A; Puri, RK. A novel role of interleukin-13 receptor alpha 2 in pancreatic cancer invasion and metastasis. *Cancer Res.,* 2009, 69, 8678–8685.

[156] Ming, J; Liu, N; Gu, Y; Qiu, X; Wang, EH. PRL-3 facilitates angiogenesis and metastasis by increasing ERK phosphorylation and upregulating the elevels and activities of Rho-A/C in lung cancer. *Pathology,* 2009, 41, 118–126.

[157] Fu, H; Hu, Z; Wen, J; Wang, K; Liu, Y. TGF-beta promotes invasion and metastasis of gastric cancer cells by increasing fascin 1 expression via ERK and JNKI signal pathways. *Acta Biochim. Biophys. Sin.* (Shanghai), 2009, 41, 648–656.

[158] Li, LH; Zheng, MH; Luo, Q; Ye, Q; Feng, B; Lu, AG; Wang, ML; Chen, XH; Su, LP; Liu, BY. P21-activated protein kinase 1 induces colorectal cancer metastasis involving ERK activation and phosphorylation by FAK at Ser-910. *Int. J. Oncol.,* 2010, 37, 951–962.

[159] Benbow, U; Brinckerhoff, CE. The AP-1 site and MMP gene regulation: what is all the fuss about? *Matrix Biol.*, 1997, 15, 519–526.

[160] Westermarck, J; Kahari, VM. Regulation of matrix metalloproteinase expression in tumor invasion. *FASEB J.,* 1999, 13, 781–792.

[161] Sexl, V; Kovacic, B; Piekorz, R; Moriggl, R; Stoiber, D; Hoffmeyer, A; Liebminger, R; Kudlacek, O; Weisz, E; Rothammer, K; Ihle, JN. Jak1 deficiency leads to enhanced Abelson-induced B-cell tumor formation. *Blood*, 2003, 101, 4937–4943.

[162] Quintás-Cardama, A; Kantarjian, H; Cortes, J; Verstovsek, S. Janus kinase inhibitors for the treatment of myeloproliferative neoplasias and beyond *Nature Reviews Drug Discovery*, 2011, 10, 127–140.

[163] Xu, X; Kang, S-H; Heidenreich, O; Okerholm, M; O'Shoea, JJ; Nerenberg, MI. Constitutive activation of different JAK tyrosine kinases in human T cell leukemia virus type 1 (HTLV-1) tax protein or virus transformed cells. *J. Clin. Invest.,* 1995, 96, 1548–1555.

[164] White, MF. Regulating insulin signaling and beta-cell function through IRS proteins. *Can. J. Physiol. Pharmacol.*, 2006, 84,725–737.

[165] Mardilovich, K; Pankratz, SL; Shaw, LM. Expression and function of the insulin receptor substrate proteins in cancer. *Cell Commun. Signaling*, 2009, 7, 14–28.

[166] Nagle, JA; Ma, Z; Byrne, MA; White, MF; Shaw, LM. Involvement of insulin receptor substrate 2 in mammary tumor metastasis. *Mol. Cell Biol.,* 2004, 24, 9726–9735.

[167] Nolan, MK; Jankowska, L; Prisco, M; Xu, S; Guvakova, MA; Surmacz, E. Differential roles of IRS-1 and SHC signaling pathways in breast cancer cells. *Int. J. Cancer*, 1997, 72, 828–834.

[168] Kim, B; van Golen, CM; Feldman, EL. Insulin-like growth factor-I signaling in human neuroblastoma cells. *Oncogene*, 2004, 23, 130–141.

[169] Hoang, CD; Zhang, X; Scott, PD; Guillaume, TJ; Maddaus, MA; Yee, D; Kratzke, RA. Selective activation of insulin receptor substrate-1 and -2 in pleural mesothelioma cells: Association with distinct malignant phenotypes. *Cancer Res.,* 2004, 64, 7479–7485.

[170] Kornmann, M; Maruyama, H; Bergmann, U; Tangvoranuntakul, P; Beger, HG; White, MF; Korc, M. Enhanced expression of the insulin receptor substrate-2 docking protein in human pancreatic cancer. *Cancer Res.*,1998, 58, 4250–4254.

[171] Boissan, M; Beurel, E; Wendum, D; Rey, C; Lecluse, Y; Housset, C; Lacombe, ML; Desbois-Mouthon, C. Overexpression of insulin receptor substrate-2 in human and murine hepatocellular carcinoma. *Am. J. Pathol.,* 2005, 167, 869–877.

[172] Kaplan, DH; Shankaran, V; Dighe, AS; Stockert, E; Aguet, M; Old, LJ; Schreiber, RD. Demonstration of an interferon gamma-dependent tumor surveillance system in immunocompetent mice. *Proc. Natl. Acad. Sci. USA,* 1998, 95, 7556–7561.

[173] Chin, YE; Kitagawa, M; Su, W-CS; You, Z-H; Iwamoto, Y; Fu, X-Y. Cell growth arrest and induction of cyclin-dependent kinase inhbitor p21 WAF1/CIP1 mediated by STAT1. *Science,* 1996, 272, 719–722.

[174] Ouchi, T; Lee, SW;; Ouchi, M; Aaronson, SA; Horvath, CM. Collaboration of signal transducer and activator of transcription 1 (STAT1) and BRCA in differential regulation of IFN-gamma target genes. *Proc. Natl. Acad. Sci. USA,* 2000, 97, 5208–5213.

[175] Akira, S. Roles of STAT3 defined by tissue-specific gene targeting. *Oncogene,* 2000, 19, 2607–2611.

[176] Bromberg, J; Wrzeszczynska, MH; Devgan, G; Zhao, Y; Pestell, RG; Albanese, C; Darnell, JE Jr. Stat3 as an oncogene. *Cell,* 1999, 98, 295–303.

[177] Schuringa, JJ; Wojtachnio, Hagens, W; Vellenga, E; Buijs, C; Hofstra, R; Kruijer, W. MEN2A-RET-induced cellular transformation by activation of STAT3. *Oncogene,* 2001, 20, 5350–5358.

[178] Ning, ZQ; Li, J; McGuinness, M; Arceci, RJ. STAT3 activation is required for Asp(816) mutant c-Kit induced tumorigenicity. *Oncogene,* 2001, 20, 4528–4536.

[179] Kube, D; Holtick, U; Vockerodt, M; Ahmadi, T; Haier, B; Behrmann, I; Heinrich, PC; Diehl, V; Tesch, H. STAT3 is constitutively activated in Hodgkin cell lines. *Blood,* 2001, 98, 762–770.

[180] Gao, B; Shen, X; Kunos, G; Meng, Q; Goldberg, ID; Rosen, EM; Fan, S. Constitutive activation of JAK-STAT3 signaling by BRCA1 in human prostate cancer cells. *FEBS Lett.,* 2001, 488, 179–184.

[181] Epling-Burnette, PK; Liu, JH; Catlett-Falcone, R; Turkson, J; Oshiro, M; Kothapalli, R; Li, Y; Wang, JM; Yang-Yen, HF; Karras, J; Jove, R; Loughran, TP Jr. Inhibition of STAT3 signaling leads to apoptosis of leukemic large granular lymphocytes and decreased Mcl-1 expression. *J. Clin. Invest.*, 2001, 107, 351–362.

[182] Epling-Burnette, PK; Zhong, B; Bai, F; Jiang, K; Bailey, RD; Garcia, R; Jove, R; Djeu, JY; Loughran, TP Jr, Wei, S. Cooperative regulation of Mcl-1 by Janus kinase/stat and phosphatidylinositol 3-kinase contribute to granulocytemacrophage colony-stimulating factor-delayed apoptosis in human neutrophils. *J. Immunol.*, 2001, 166, 7486–7495.

[183] Benekli, M; Baer, MR; Baumann, H; Wetzler, M. Signal transducer and activator of transcription proteins in leukemias. *Blood,* 2003, 101, 2940–2954.

[184] Bromberg, J. Stat proteins and oncogenesis *J. Clin. Invest.*, 2002, 109, 1139–1142.

[185] Carlesso, N; Frank, DA; Griffin, JD. Tyrosyl phosphorylation and DNA binding activity of signal transducers and activators of transcription (STAT) proteins in hematopoietic cell lines transformed by Bcr/Abl. *J. Exp. Med.,* 1996, 183, 811–820.

[186] Lacronique, V; Boureux, A; Valle, VD; Poirel, H; Quang, CT; Mauchauffe, M; Berthou, C; Lessard, M; Berger, R; Ghysdael, J; Bernard, OA. A TEL-JAK2 fusion protein with constitutive kinase activity in human leukemia. *Science*, 1997, 278, 1309–1312.

[187] Carron, C; Cormier, F; Janin, A; Lacronique, V; Giovannini, M; Daniel, M-T; Bernard, O. TEL-JAK2 transgenic mice develop T-cell leukemia. *Blood*, 2000, 95, 3891–3899.

[188] Mizuki, M; Fenski, R; Halfter, H; Matsumura, I; Schmidt, R; Muller, C; Grüning, W; Kratz-Albers, K; Serve, S; Steur, C; Büchner, T; Kienast, J; Kanakura, Y; Berdel, WE; Serve, H. Flt3 mutations from patients with acute myeloid leukemia induce transformation of 32D cells mediated by the Ras and STAT5 pathways. *Blood*, 2000, 96, 3907–3914.

[189] James, C; Ugo, V; Le Couedic, JP; Staerk, J; Delhommeau, F; Lacout, C; Garçon, L; Raslova, H; Berger, R; Bennaceur-Griscelli, A; Villeval, JL; Constantinescu, SN; Casadevall, N; Vainchenker, W. A unique clonal JAK2 mutation leading to constitutive signalling causes polycythaemia vera. *Nature,* 2005, 434, 1144–1148.

[190] Levine, RL; Wedleigh, M; Cools, J; Ebert, BL; Wernig, G; Huntly, BJ; Boggon, TJ; Wlodarska, I; Clark, JJ; Moore, S; Adelsperger, J; Koo, S; Lee, JC; Gabriel, S; Mercher, T; D'Andrea, A; Fröhling, S; Döhner, K; Marynen, P; Vandenberghe, P; Mesa, RA; Tefferi, A; Griffin, JD; Eck, MJ; Sellers, WR; Meyerson, M; Golub, TR; Lee, SJ; Gilliland, DG. Activating mutation in the tyrosine kinase JAK2 in polycythemia vera, essential thrombocythemia, and myeloid metaplasia with myelofibrosis *Cancer Cell,* 2005, 7, 387–397.

[191] Xiong, H; Su, W-S; Liang, Q-C; Zhang, Z-G; Chen, H-M; Du, W; Chen, Y-X; Fang, J-Y. Inhbition of STAT5 induces G1 cell cycle arrest and reduces tumor cell invasion in human colorectal cancer cells. *Lab. Invest.,* 2009, 89, 717–725.

[192] Ferbeyre, G; Moriggle, R. The role of STAT5 transcription factor as tumor suppressorss or oncogenes. *Biochimica et Biophysica Acta*, 2011, 1815, 104–114.

[193] Ostrand-Rosenberg, S; Sinha, P; Clements, V; Dissanayake, SI; Miller, S; Davis, C; Danna, E. Signal transducer and activator of transcription 6 (STAT6) and CD1: inhibitors of immunosurveillance against primary tumors and metastatic disease. *Cancer Immunol. Immunother.*, 2004, 53, 86–91.

[194] Kacha, AK; Fallarino, F; Markiewicz, MA; Gejenski, TF. Spontaneous rejection of poorly immunogenic P1 HTR tumors by STAT6-deficient mice. *J. Immunol.,* 2000, 165, 6024–6028.

[195] Ni, Z; Lou, W; Lee, SO; Dhir, R; DeMiguel, F; Grandis, JR; Gao, AC. Selective activation of members of the signal transducers and activators of transcription family in prostate carcinoma. *J. Urol.,* 2002, 167, 1859–1862.

[196] Qin, JZ; Kamarashev, J; Zhang, CL; Dummer, R; Burg, G; Döbbeling, U. Constitutive and interleukin-7 and interleukin-15-stimulated DNA binding of STAT and novel factors in cutaneous T cell lymphoma cells. *J. Invest. Dermatol.*, 2001, 117, 583–589.

[197] Guiter, C; Dusanter-Fourt, I; Copie-Bergman, C; Boulland, ML; Le Gouvello, S; Gaulard, P; Leroy, K; Castellano, F. Constitutive STAT6 activation in primary mediastinal large B-cell lymphoma. *Blood,* 2004, 104, 543–549.

[198] Li, BH; Yang, XZ; Li, PD; Yuan, Q; Liu, XH; Yuan, J; Zhang, WJ. IL-4/Stat6 activities correlate with apoptosis and metastasis in colon cancer cells. *Biochem. Biophys. Res. Commun.*, 2008, 369, 554–560.

[199] Das, S; Shetty, P; Valapala, M; Dasgupta, S; Gryczynski, Z; Vishwanatha, JK. Signal transducer and activator of transcription 6 (STAT6) is a novel interactor of annexin A2 in prostate cancer cells. *Biochemistry*, 2010, 49, 2216–2226.

[200] Sinha, P; Clements, VK; Ostrand-Rosenberg, S. Tumor immunity: a balancing act between T cell activation, macrophage activation and tumor-induced immune suppression. *Cancer Immunol. Immunother.*, 2005, 54, 1137–1142.

[201] Raccurt, M; Tam, SP; Lau, P; Mertani, HC; Lambert, A; Garcia-Caballero, T; Li, H; Brown, RJ; McGuckin, MA; Morel, G; Waters, MJ. Suppressor of cytokine signalling gene expression is elevated in breast carcinoma. *Br. J. Cancer*, 2003, 89, 524–532.

[202] Sutherland, KD; Lindeman, GJ; Choong, DY; Wittlin, S; Brentzell, L; Phillips, W; Campbell, IG; Visvader, JE. Differential hypermethylation of SOCS genes in ovarian and breast carcinomas. *Oncogene*, 2004, 23, 7726–7733.

[203] Farabegoli, F; Ceccarelli, C; Santini, D; Taffurelli, M. Suppressor of cytokine signalling 2 (SOCS-2) expression in breast carcinoma. *J. Clin. Pathol.,* 2005, 58, 1046–1050.

[204] Neuwirt, H; Puhr, M; Santer, FR; Susani, M; Doppler, W; Marcias, G; Rauch, V; Brugger, M; Hobisch, A; Kenner, L; Culig, Z. Suppressor of cytokine signaling (SOCS)-1 is expressed in human prostate cancer and exerts growth-inhibitory function through down-regulation of cyclins and cyclin-dependent kinases. *Am. J. Pathol.*, 2009, 174, 1921–1930.

[205] Scutti, JAB; Matsuo, AL; Pereira, FV; Massaoka, MH; Figueiredo, CR; Moreira, DF; Belizário, JE; Travassos, LR. Role of SOCS-1 gene on melanoma cell growth and tumor development. *Translational. Oncol.*, 2011, 4, 101–109.

[206] Brender, C; Nielsen, M; Kaltoft, K; Mikkelsen, G; Zhang, Q; Wasik, M; Billestrup, N; Ødum, N. STAT3-mediated constitutive expression of SOCS-3 in cutaneous T-cell lymphoma. *Blood,* 2001, 97, 1056–1062.

[207] Masuhiro, Y; Kayama, K; Fukushima, A; Baba, K; Soutsu, M; Kamiya, Y; Gotoh, M; Yamaguchi, N; Hanazawa, S. SOCS-3 inhibits E2F/DP-1 transcriptional activity and cell cycle progression via interaction with DP-1. *J. Biol. Chem.*, 2008, 283, 31575–31583.

[208] Gross, M; Liu, B; Tan, J; French, FS; Carey, M; Shuai, K. Distinct effects of PIAS proteins on androgen-mediated gene activation in prostate cancer cells. *Oncogene,* 2001, 20, 3880–3887.

[209] Borish, LC; Nelson, HS; Lanz, MJ; Claussen, L; Whitmore, JB; Agosti, JM; Garrison, L. Interleukin-4 receptor in moderate atopic asthma. A phase I/II randomized, placebocontrolled trial. *Am. J. Respir. Crit. Care Med.,* 1999, 160, 1816–1823.

[210] Borish, LC; Nelson, HS; Corren, J; Bensch, G; Busse, WW; Whitmore, JB; Agosti, JM. Efficacy of soluble IL-4 receptor for the treatment of adults with asthma. *J. Allergy Clin. Immunol.*, 2001, 107, 963–970.

[211] Kariyawasam, HH; Nicholson, GC; Tan, AJ; Syngal, N; Quinn, D; Boulton, C; Walker, C; Rodman, D; Westwick, J; Kon, OM; Barnes, PJ; Hohlfeld, JM; Badorrek, P; Krug, N; Hansel, TT. Effects of anti-IL-13 (Novartis QAX576) on inflammatory responses following nasal allergen challenge (NAC). *Am. J. Respir. Crit. Care Med.*, 2009, 179, A3642.

[212] Blanchard, C; Mishra, A; Saito-Akei, H; Monk, P; Anderson, I; Rothenberg, ME. Inhibition of human interleukin-13-induced respiratory and oesophageal inflammation by anti-human-interleukin-13 antibody (CAT-354). *Clin. Exp. Allergy*, 2005, 35, 1096–1103.

[213] Vincent, M; Banfield, C; Kakkar, T; Shakib, S; Schicchitano, R; Cheah, TC; Gerson, J. Single-dose, first-in-human study of AMG 317: safety and pharmacokinetics in healthy and asthmatic adults. *Eur. Respir. J.*, 2008, 32, 516s.

[214] Hodges, MR; Castelloe, E; Chen, A; Geary, RS; Karras, JG; Shapiro, D; Yeung, B; Yu, R; Gregory, SA. Randomized, double-blind, placebo controlled first in human study of inhaled AIR645, an IL-4Ra oligonucleotide, in healthy volunteers. *Am. J. Respir. Crit. Care Med.*, 2009, 179, A3640.

[215] Tony, HP; Shen, BJ; Reusch, P; Sebald, W. Design of human interleukin-4 antagonists inhibiting interleukin-4-dependent and interleukin-13-dependent responses in T-cells and B-cells with high efficiency. *Eur. J. Biochem.*, 1994, 225, 659–665.

[216] Fitch, N; Morton, M; Bowden, A; Shanafelt, M-C; Shanafelt, A; Shanafelt, G; Wetzel, G; Moy, J. Preclinical evaluation of BAY 16-9996 a dual Il-4/Il-13 receptor antagonist. *J. Allergy Clin. Immunol.*, 2001, 107, S331–S339.

[217] Ohga, K; Kuromitsu, S; Takezawa, R; Numazaki, M; Ishikawa, J; Nagashima, S; Shimizu, Y. YM-341619 suppresses the differentiation of spleen T cells into Th2 cells in vitro, eosinophilia, and airway hyperresponsiveness in rat allergic models. *Eur. J. Pharmacol.*, 2008, 590, 409–416.

[218] Tartaglia, M; Niemeyer, CM; Fragale, A; Song, X; Buechner, J; Jung, A; Hählen, K; Hasle, H; Licht, JD; Gelb, BD. Somatic mutations in PTPN11 in juvenile myelomonocytic leukemia, myelodysplastic syndromes and acute myeloid leukemia. *Nat. Genet.*, 2003, 34, 148–150.

[219] Zhang, Q; Raghunath, PN; Xue, L; Majewski, M; Carpentieri, DF; Odum, N; Morris, S; Skorski, T; Wasik, MA. Multilevel dysregulation of STAT3 activation in anaplastic lymphoma kinase-positive T/null cell lymphoma. *J. Immunol.*, 2002, 168, 466–474.

[220] Ueda, M; Ota, J; Yamashita, Y; Choi, YL; Ohki, R; Wada, T; Koinuma, K; Kano, Y; Ozawa, K; Mano, H. DNA microarray analysis of stage progression mechanism in myelodysplastic syndrome. *Br. J. Haematol.*, 2003, 123, 288–296.

[221] Melzner, I; Bucur, AJ; Bruderlein, S; Dorsch, K; Hasel, C; Barth, TF; Leithäuser, F; Möller, P. Biallelic mutation of SOCS-1 impairs JAK2 degradation and sustains phospho-JAK2 action in the MedB-1 mediastinal lymphoma line. *Blood,* 2005, 105, 2535–2542.

[222] Melzner, I; Weniger, MA; Bucur, AJ; Bruderlein, S; Dorsch, K; Hasel, C; Leithäuser, F; Ritz, O; Dyer, MJ; Barth, TF; Möller, P. Biallelic deletion within 16p13.13 including SOCS-1 in Karpas1106P mediastinal B-cell lymphoma line is associated with delayed degradation of JAK2 protein. *Int. J. Cancer*, 2006, 118, 1941–1944.

In: Signaling, Gene Regulation and Cancer
Editors: Shree Ram Singh and Manoj K. Mishra
ISBN: 978-1-61942-088-5

Chapter 16

POTENTIAL MOLECULAR TARGETS IN OVARIAN CANCER

Indrajit Chowdhury*, Roland Matthews and Winston E. Thompson

Department of Obstetrics and Gynecology, Reproductive Science Research Program, Morehouse School of Medicine, Atlanta, GA, US

ABSTRACT

Ovarian cancer is the most lethal gynecological cancer and ranked as the fifth leading cause of cancer deaths in the United States than any other type of female reproductive tract cancer. More than 80% of patients have symptoms, even when the disease is still limited to the ovaries. The peak age of occurrence of ovarian cancers is after menopause, however, up to 17% of all ovarian cancers occur in women <40 years of age. Over the past two decades, the prognosis of ovarian cancer has been impacted (only 20-30% of women with overall 5-year survival rates) mainly by improvement in surgical approaches, the introduction of multi-disciplinary care, and the refinement of first line cytotoxic combinations, namely with platinum-based chemotherapy and taxanes. The main purpose of this chapter is to provide an update the current understanding of ovarian cancer signaling and prominent target molecules in treatment of ovarian cancer.

Keyword: Ovary, cancer, signaling, inhibitor

OVARIAN CANCER

Ovarian cancer is the most lethal gynecological cancer and ranked as the fifth leading cause of cancer deaths in the United States than any other type of female reproductive tract cancer (Jemal et al., 2005, 2008, 2009). The annual worldwide incidence and mortality has

* Correspondence: Indrajit Chowdhury, Ph.D., Department Of Obstetrics and Gynecology, Reproductive Science Research Program, Morehouse School of Medicine, 720 Westview Drive Southwest, Atlanta, GA 30310. Tel.: 404-752-1849, Fax: 404-616-4131, e-mail:ichowdhury@msm.edu, indrajitfbs@gmail.com.

been estimated to be 190,000 and 114,000 respectively (Stewart and Kleihues, 2003). Ovarian cancer has a distinctive biology and behavior at the cellular and molecular levels (Bast Jr et al., 2009), although, clinically represented as a complex cystic mass in the pelvis, and termed as the 'silent killer'. More than 80% of patients have symptoms, even when the disease is still limited to the ovaries (Goff et al., 2000). The peak age of occurrence of ovarian cancers is after menopause, however, up to 17% of all ovarian cancers occur in women <40 years of age (Duska et al., 1999). Unlike other cancers at many other sites, ovarian cancer has no anatomical barrier exists to widespread metastasis throughout the peritoneal cavity (Bast Jr et al., 2009). Most frequently, small clusters of cancer cells are shed by the ovary and implant on the peritoneal surface, forming numerous nodules. Thus, metastases can occur through lymphatics to nodes at the renal hilus or through blood vessels to the parenchyma of the liver or lung. Approximately 70% of ovarian cancers are diagnosed at an advanced stage. Survival rates can be predicted by the stage and histological grade of ovarian cancer at the time of diagnosis. In general, women with ovarian cancer confined to the ovary (Stage I, limited to one or both ovaries) have >90% five-year survival rate, whereas those with Stage III (microscopic peritoneal implants outside of the pelvis; or limited to the pelvis with extension to the small bowel or omentum; abdomen)-IV (distant metastases to the liver or outside the peritoneal cavity) tumors have <40% five-year survival rate (SEERS; Gonzalez-Lira et al., 1997; Soslow, 2008) (Table 1). In Stage II ovarian cancer, pelvic extension (pelvic organs) or implants take place.

Table 1. Ovarian cancer classification and current treatment guideline

Stage	Characteristics	Current treatment guideline
1	Tumor limited to ovary or ovaries a. one ovary involve b. both ovaries involve c. tumor limited to ovary or ovaries with ascites, positive peritoneal washings, surface involvement, or rupture present	Cytoreductive surgery Taxane/carboplatin for 3-6 cycles
2	Ovarian tumor with pelvic extension	Cytoreductive surgery Taxane/carboplatin for 3-6 cycles
3	Tumor involving the upper abdomen or lymph nodes	a. Consider neoadjuvent chemotherapy and interval cytoreduction with bulky stage III disease, not appropriate for surgery b. Taxane/carboplatin for 6 cycles with interval debulking as indicated by tumor response and potential respectability. Intraperitoneal chaemotherapy may also be considered, particularly in low volume optimally debulked stage III patients
4	Distant organ involvement includes a malignant pleural effusion and parenchymal involvement of liver or spleen	a. Cytoreductive surgery b. Taxane/carboplatin for 6 cycles with interval debulking as indicated by tumor response and potential respectability. Intraperitoneal chaemotherapy may also be considered, particularly in low volume optimally debulk stage III patients

Over the past two decades (1990-2010), the prognosis of ovarian cancer has been impacted (only 20-30% of women with overall 5-year survival rates) mainly by improvement in surgical approaches, the introduction of multi-disciplinary care, and the refinement of first line cytotoxic combinations, namely with platinum-based chemotherapy in the 1980's and the addition of taxanes in the 1990s (Dinh et al., 2008; Jemal et al., 2008, 2009; Park et al., 2009). The main purpose of this chapter is to provide an update of the current understanding of ovarian cancer signaling and prominent target molecules in treatment of ovarian cancer. Thus, a better understanding of the ovarian signaling pathways would provide targets for the development of therapeutic interventions of ovarian cancer. This chapter focuses on two major areas (a) the molecular pathology in ovarian cancer and (b) the molecular oncogenic signaling molecules and their inhibitors.

MOLECULAR PATHOLOGY IN OVARIAN CANCER

The normal ovary is a complex tissue with several distinct components. Thus, at the cellular and molecular levels, ovarian cancers are remarkably heterogeneous (Table 2). Ovarian cancers can develop from germ cells or granulosa–theca cells. Based on ovarian cancer histology, it is classified into three major categories: epithelial ovarian tumors (surface epithelial stromal tumor), germ cell tumors, and sex cord stromal tumors (Feeley and Wells, 2001; Willner et al., 2007).

Table 2. The two pathway concepts of origins and molecular pathology of ovarian cancer

Type	Histology	Precursor	Molecular features
I (~25% frequency)	Low grade serous carcinoma	Cystadenoma-borderline tumor-carcinoma sequence	Mutation in K-RAS and/or B-RAF (>60%)
	Mucinous carcinoma	Cystadenoma-borderline tumor-carcinoma sequence, metastases from bowl	Mutation in K-RAS, TP53 mutation associated with transition from borderline tumor to carcinoma
	Low grade endometrioid carcinoma	Endometriosis and endometrial-like hyperplacia	Mutation in CTNNB1 (B-catenin gene) and PIK3CA, and PTEN with microsatellite instability (MSI)
	Clear cell carcinoma	Endometriosis	PTEN mutation or loss of heterozygosity, PIK3CA mutation
II (~75% frequency)	High grade serous carcinoma	'de novo' in epithelial inclusion cysts, fallopian tube	TP53 mutation (80%) and PIK3CA mutation (25-40%), BRCA1/2 dysfunction (usually promoter methylation), chromatin instability (CI)
	High grade endometrioid carcinoma	Epithelial inclusion gland or cysts	TP53 and PIK3CA mutation, BRCA1/2 dysfunction (usually promoter methylation), chromatin instability (CI)

However, the exact cause of ovarian cancer or tumourigenesis is unknown. The risk of developing ovarian cancer appears to be affected by several factors including hormonal imbalance, multiple genetic and epigenetic abnormalities [including activation of genes occurs through amplification, point mutation and hypomethylation, whereas genetic inactivation results from deletion or translocation of large chromosomal regions, more limited loss of heterozygosity (LOH) at particular loci and promoter methylation] or life style that lead to the disruption or enhancement of intracellular signalling pathways which control cell proliferation, apoptosis or metastasis (Landen Jr et al., 2008). Over the past several year studies using digital karyotyping counts sequence tags from specific loci distributed throughout the human genome and array based technologies contributed to our understanding of the biology of the ovarian surface epithelium (OSE) and ovarian carcinogenesis in a great detail (Kurman and Shih, 2008; Kurman et al., 2008; Willner et al., 2007).

Many well established and putative tumour suppressor genes have been implicated in ovarian oncogenesis including tumor protein 53 (TP53), also known as protein 53 (p53) (Bast J. et al., 2009; Jcobs et al., 1992). In humans, p53 is encoded by the *TP53* gene located on the short arm of chromosome (17p13.1). p53 is important in regulation of the cell cycle as a tumor suppressor or as a "the guardian of the genome", referring to its role in conserving stability by preventing genome mutation. Mutation and loss of *TP53* function is one of the most frequent genetic abnormalities in ovarian cancer and is observed in 60–80% of both sporadic and familial cases.

Based on the basis of genetic changes in ovarian cancer, it is divided into two groups, a low-grade tumours with mutations in *KRAS* (15%), *BRAF* (12%) and *PIK3CA* (8%), LOH on chromosome Xq, microsatellite instability and expression of amphiregulin; and high-grade tumours with aberrations in *TP53* and potential aberrations in breast cancer 1 (*BRCA1*) and *BRCA2*, as well as LOH on chromosomes 7q and 9p (Bast J. et al., 2009; Jazaeri et al., 2002). In humans, the *KRAS* gene encodes GTPase KRas protein also known as V-Ki-ras2 Kirsten rat sarcoma viral oncogene homolog (KRAS). Like other members of the Ras family, the KRAS protein is a GTPase and is an early player in many signal transduction pathways. The *BRAF* gene encodes serine/threonine-protein kinase b-Raf (b-Raf), also known as proto-oncogene b-Raf or v-Raf murine sarcoma viral oncogene homolog B1, and is involved in sending signals for cell growth. The gene *PIK3CA* encodes phosphatidylinositol 3-kinase, a 85 kDa regulatory subunit and a 110 kDa catalytic subunit that involve in phosphorylation of Phosphatidylinositol (PtdIns), PtdIns4P and PtdIns(4,5)P2 using ATP.

Several growth inhibitory genes are silenced by methylation or imprinting in ovarian cancer including Ras association domain-containing protein 1 (*RASSF1*), deleted in lung and esophageal cancer 1 (*DLEC1*) and opioid binding protein/cell adhesion molecule like (*OPCML*) (Bast J. et al., 2009). The function of imprinted tumour suppressor genes can also be lost in a single step. Approximately 70 human genes are known to be imprinted: only one allele is expressed in each cell of the embryo and adult and silencing of the maternal or paternal allele is inherited despite a normal nucleotide sequence. Of the 16 candidate tumour suppressor genes identified to date in ovarian cancer, 3 are known to be imprinted: *ARHI* (also known as Ras-related GTP-binding protein, member I, *DIRAS3/* Rho-related GTP-binding protein RhoI), pleiomorphic adenoma gene like 1 (*PLAGL1*, also known as *LOT1*) and paternally expressed 3 (*PEG3*). All three genes are down regulated in a substantial fraction of ovarian cancers (Bast J. et al., 2009).

Inheritance of DNA repair defects contributes to as many as 10–15% of ovarian cancers. The lifetime risk of developing ovarian cancer in mutation carriers varies with the genetic defect (for breast cancer 1 *BRCA1* 30–60%, for *BRCA2* 15–30% and for hereditary non-polyposis colon cancer 7%) (Bast J. et al., 2009; Carreira et al., 2009; Turner et al., 2007). At least 15 oncogenes have been implicated in ovarian cancers and 11 of these oncogenes show genomic amplification. DNA copy number abnormalities have also been found in 37% of 283 loci that are known to contain non-coding micro RNAs. In contrast to copy number anomalies, activating mutations in oncogenes are uncommon in the most frequently occurring histotypes (Baldwin et al., 2000; Bast J. et al., 2009).

Changes in cell adhesion and motility also contribute to disease development and metastasis. Adhesion of ovarian cancer cells to the mesothelial cells and to the underlying stroma is mediated by CD44, CA125 and b1 intergrin on the surface of ovarian cancer cells that bind to mesothelin and hyaluronic acid on mesothelial cells, or to fibronectin, laminin and type IV collagen in the underlying matrix (Bast J. et al., 2009).

Molecular Oncogenic Signalling Pathways and Their Inhibitors In Ovarian Cancer

Oncogenesis/Tumorigenesis is a multi-step process that reflects many possible abnormal changes in cellular physiology (Figure 1). These processes include a gain in self-sufficient growth signals, an insensitivity to growth-inhibitory (antigrowth) signals, an ability to evade apoptosis, a potential for unlimited replication, a capacity for sustained angiogenesis, and an ability for tissue invasion and metastases (Chen et al., 2006). At least seven signalling pathways are activated in >50% of ovarian cancers (Bast J. et al., 2009; Ricciardelli and Oehler, 2009). Many cascades, pathways, and molecules are involved and thus, they all represent potential therapeutic targets. Given the complexity of crosstalk between protein signaling pathways, predicting the impact and efficacy of any one signaling inhibitor is difficult. Inhibition of multiple pathways will almost certainly be required to substantially affect ovarian cancer growth. A crucial goal is to identify patients who would benefit from particular targeted therapies. In brief, we are discussing few important signaling molecules and their inhibitors.

1. Epidermal Growth Factor Receptor (EGFR/ ErbB)

The ErbB family comprises four structurally related transmembrane receptor tyrosine kinases (RTK) includes ErbB1 (epidermal growth factor, EGFR/Her1), ErbB2 (Her2/neu), ErbB3 (Her3), and ErbB4 (Her4). All members have an extracellular ligand-binding region, a single membrane-spanning region and a cytoplasmic tyrosine kinase-containing domain, the last being absent in Her-3. In normal cells, the EGFR is activated by a ligand, such as epidermal growth factor (EGF), heparin binding EGF (Hb-EGF), transforming growth factor-alpha (TGFa), amphiregulin and crypto, binding to the extracellular region results in homodimerization or heterodimerization and autophosphorylation to activation of the cytoplasmic kinase domain and phosphorylation of a specific tyrosine. This leads to the activation of intracellular signalling pathways, such as the mitogen-activated protein kinase (MAPK) and the phosphatidylinositol 3-kinase (P13K)-AKT pathways involve in development, cellular proliferation, differentiation, metabolism, motility and survival (Ciardiello, 2005; Salomon et al., 1995).

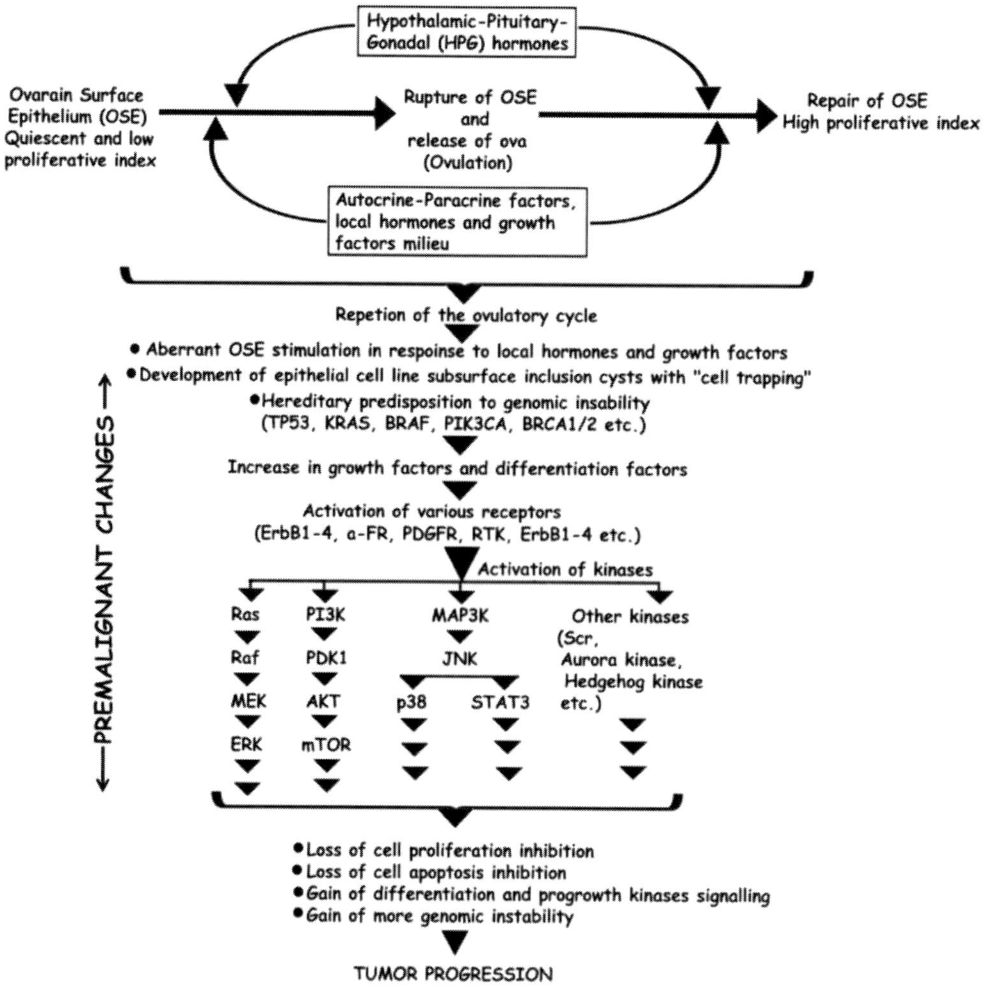

Figure 1. Schematic representation of the mechanism of ovarian carcinogenesis. During the menstrual cycle, ovarian surface epithelium (OSE) cell rupture and repair occurs with each ovulation, which is regulated by hypothalamic-pituitary-gonadal (HPG) hormones, autocrine-paracrine factors, local hormones and growth factor milieu. Prolonged and aberrant exposure of OSE cells to local growth factors lead to a risk of genomic changes, that may lead by 'cell trapping' within epithelial inclusion cysts. Moreover, hereditary predisposition to genomic damage with increase growth and differentiation factors by activation of kinase signaling pathways through autocrine-paracrine fashion causes the increase risk of early-onset (premalignant changes) of ovarian cancer. Ultimately, exposure of various factors may control the exact pattern of homeobox gene expression as in embryonic development and enhance proliferation, differentiation and survival of ovarian cancer cells.

In cancer cells, the EGFR pathway is frequently disrupted resulting in growth, stromal proliferation and angiogenesis. EGFR family overexpression is associated with poor prognosis and nonresponse to chemotherapy in many cancers including ovarian cancer. In ovarian cancer EGFR family is overexpressed in 35–70% of cases (Lafky et al., 2008).

A large number of small-molecule inhibitors including tyrosine kinases inhibitors [gefitinib (ZD1839, Iressa, Astra-Zeneca), erlotinib (OSI-774, Tarceva, Roche), lapatinib (GW572016, Tykerb/Tyverb, GlaxoSmithKline), and CI-1033] (Posadas et al., 2007; Gordon et al., 2005; Campos et al., 2005) and monoclonal antibodies (mAbs) [cetuximab (IMC-C225, Erbitux, Bristol-Myers Squibb), Omnitarg (pertuzumab rhuMab 2C4, Roche/Genetech), trastuzumab, and matuzumab (EMD72000)] (Seiden et al., 2007; Gordon et al., 2005) have been designed for use against members of the EGFR family or their ligands. Some of these inhibitors are now licensed for use in common malignancies (Blagden and Gabra, 2009; Campo and Ghosh, 2010; Dinh et al., 2008). However, in ovarian cancer, monoclonal antibodies (mAbs) or small molecules designed to target the EGFR pathway have only minimal activity, either as single agents or in combination with standard chemotherapy. Moreover, there are inadequate patient selections with inappropriate use of the agents in studies, as it appears that EGFR mutation status does not predict response to treatment as in other tumors (Blagden and Gabra, 2009).

2. Vascular Endothelial Growth Factor Receptor (VEGFR) Family

Angiogenesis is one of the cardinal processes leading to invasion and metastasis of solid tumor. Thus, angiogenesis help in the formation of new vessels in order to supply a growing tumour with adequate oxygen and nutrients. This is dependent on the activation of an 'angiogenic switch' whereby the tumour upregulates expression of proangiogenic growth factors such as vascular endothelial growth factor (VEGF) and fibroblast growth factor, and their receptors, with a corresponding down regulation of antiangiogenic factors such as angiostatin and thrombospondin-1 (Blagden and Gabra, 2009; Oliver and Srinivasan, 2008). The VEGF family of growth factors and receptors are well characterized and significant in their control of angiogenesis. VEGF receptors (VEGFR) are members of a family of receptor of tyrosine kinase (RTK) and are activated by VEGF, an angiogenic factor. Seven VEGF ligands have been identified: VEGF A–E and placental growth factor (PlGF) 1 and 2. They signal through three VEGF receptors, VEGFR 1–3, of which VEGFR 1(FLT-1) and VEGFR 2(KDR/Flk-1) are selectively located on endothelial cells and promote angiogenesis (Ellis and Hicklin, 2008). In ovarian cancer, VEGFR 3 (along with VEGFR 2) has been shown to be required for formation of new but nonfunctional lymphatic vessels (lymphangiogenesis) and is overexpressed in proliferating lymphatic endothelial cells. This contributes to ascites formation, particularly chylous ascites, and is probably regulated by VEGF-A, C and D, and other proinflammatory cytokines [such as interleukin (IL)-6 and tumour necrosis factor alpha (TNF)-a] (Jeon et al., 2008; Hong et al., 2004). Over expression of VEGF ligand correlates with increase microvessel density (Nishida et al., 2004; Raspollini et al., 2004).

A number of agents have been developed against VEGF or VEGFR including mAbs for VEGF [bevacizumab (Avastin, Genentech/Roche), VEGF Trap (AVE005, Aflibercept, Sanofi-Aventis)] and VEGFR [sunitinib (SU11248, Sutent, Pfizer), sorafenib (Nexavar, BAY 43-0006, Onyx), cediranib (AZD2171, Recentin, AstraZeneca), and pazopanib (GW786034,

GlaxoSmithKline) (Blagden and Gabra, 2009; Dinh et al., 2008; Campos and Ghosh, 2010). Inhibition of the VEGF pathway has proven an effective therapeutic strategy in other cancers in which VEGF inhibitors are now routinely used alongside cytotoxic agents (Ellis and Hicklin, 2008). However, in ovarian cancer VEGF and VEGFR are in clinical studies and few studies have shown that these inhibitors can be better antiangiogenic therapies in ovarian cancer.

3. Platelet-Derived Growth Factor Receptor (PDGFR)

Platelet-derived growth factor receptors (PDGF-R) are cell surface tyrosine (RTK) for members of the platelet-derived growth factor (PDGF) family that induces cell growth and survival, transformation, migration, vascular permeability, and wound healing process (Andrae et al., 2008; Schmitt and Matei, 2008). PDGFR activation in cancer occurs as a consequence of gene amplification, chromosomal rearrangements, or activating mutations (Mao and Hamoudi, 2000; Heinrich et al., 2003). PDGFR activation is critical to tumor initiation in addition to functioning as a mediator of connective tissue stroma (Apte et al., 2004). Increase expression of PDGFR has been shown in 50–80% of ovarian tumors (Apte et al., 2004). Several small molecule inhibitors for PDGFR [imatinib mesylate (STI571; Gleevec), sorafenib (Nexavar, BAY 43-0006, Onyx), sunitinib (SU11248, Sutent, Pfizer), dasatinib, 3G3, and CDP 860] are under study or in the clinical trial (Campos and Ghosh, 2010; Schmitt and Matei, 2008).

4. Phosphoinositide 3 Kinase (PI3K)/Protein Kinase B (PKB or AKT)/Mammalian Target of Rapamycin (mTOR)

AKT represents a subfamily of intracellular serine/threonine kinases, upon activation it has multiple roles in the cancer cell such as antiapoptotic, proangiogenic, tumour invasion and metastasis, and cell cycle regulation. In addition, AKT regulates mRNA translation by phosphorylating mTOR (Testa and bellacosa, 2001). Three members of the AKT family have been identified (AKT1, AKT2 and AKT3) so far. AKT1 and AKT2 have been most commonly associated with ovarian cancer (Altomare et al., 2004; Carpten et al., 2007). Phospho-activated AKT2 is upregulated in 40% of primary ovarian cancer cases (Yuan et al., 2000), whereas co-amplification of AKT2 and PIK3CA, the gene encoding the p110a catalytic subunit of the upstream kinase PI3K, is associated with high grade ovarian tumours (Nakayama et al., 2006). The PI3K pathway is activated in approximately 70% of ovarian cancers, and activation of this pathway is associated with resistance to cytotoxic chemotherapy.

The PI3K/AKT/mTOR pathway mediates chemotherapy resistance through its interaction with the apoptosis inhibitor survivin (Fraser et al., 2008; Yuan et al., 2000) and by blocking p53 response to proapoptotic stimuli (Xing et al., 2008). Moreover, activation of the PI3K pathway can be driven by gene amplification (*PIK3CA* and *AKT2*), activating mutations (*PIK3CA*) or inactivating mutations (Phosphatase and tensin homolog, *PTEN*), but in many cancers the pathway is activated by autocrine or paracrine signaling through receptor tyrosine kinase (RTK) (Bast Jr et al., 2009). Therefore, inhibition of the PI3K/AKT/mTOR pathway is not only a potential first-line therapeutic strategy in ovarian cancer but also could be used in recurrent disease as a means of reversing acquired chemotherapy resistance. A number of small molecule inhibitors targeting this pathway are currently in development and early clinical trials including the inhibitor LY29002 (selectively inhibits the catalytic activity of PI3K) (Biomol International) and Perifosine; and potential target for the mTOR pathway include mTOR inhibitors Temsirolimus (CCI-779), everolimus (RAD001), and deforolimus (AP23573) (Campos and Ghosh, 2010; Yap et al., 2008).

5. Rat Sarcoma (Ras)/Raf/Mitogen Activated Protein Kinase (MAPK or MEK)/Extracellular Signal-Regulated Kinases (ERK)

Ras/Raf/MEK/ERK mitogen-activated protein kinase (MAPK) cascade signaling pathway is essential for cellular responses to growth signals and is frequently dysregulated in cancer. This pathway is activated by RTKs, cytokine receptors and integrins, that affect multiple cellular functions by stimulating tumor growth, by producing matrix degrading metalloproteinases (MMPs) and facilitating drug resistance (Weinstein-Oppenheimer et al., 2000). Most commonly, b-Raf (Serine/threonine-protein kinase B-Raf, also known as proto-oncogene B-Raf or v-Raf murine sarcoma viral oncogene homolog B1) and c-Raf (v-Raf murine sarcoma viral oncogene homolog B1) are highly expressed in ovarian cancers, and high b-raf expression appears to be associated with improved survival, while high c-Raf expression is associated with poor prognosis (McPhillips et al., 2001, 2006). Activating mutations have also been documented in Ras in ovarain tumor (Chien and Chow, 1993). In ovarian cancer, b-Raf, K-Ras (GTPase Kras, also known as V-Ki-ras2 Kirsten rat sarcoma viral oncogene homolog) or both are mutated in ~20-32% of tumours, and are predominantly clustered in nonserous and low-grade serous tumours. Moreover, novel and unique MEK1 mutation in ovarian cancer indicates that this pathway may be particularly important for the survival of the malignant phenotype (Estep et al., 2007), and more sensitive to the MEK inhibitor (CI-1040, Pfizer) compared with wild-type sequences (LoRusso et al., 2002; Nakayama et al., 2008). Unfortunately, CI-1040 lacked efficacy in clinical trials. However, recent preclinical evaluation of other MEK inhibitor (PD98059) combined with the PI3K inhibitor (LY294002) are demonstrated synergistic activity and enhanced paclitaxel sensitivity in ovarian cell lines and mouse xenografts (Kigawa et al., 2007). Thus AKT pathway inhibition, alone or in combination with MEK inhibitors, seems to be a promising strategy in order to overcome taxane and platinum resistance in patients with ovarian cancer.

6. Sarcoma (Src) Family

Src is a family of nonreceptor cytoplasmic tyrosine kinases that mediates the signal between growth factor receptors (such as EGFR, c-Met), insulin-like growth factor 1 receptor and downstream signalling cascades (such as focal-adhesion kinase, MAPK and PI3K/AKT/mTOR). Src is activated by a number of RTKs, G-protein coupled receptors and integrins. Activation of Src family involve in many processes of tumorigenesis including proliferation, differentiation, mitosis, adhesion, invasion, motility and angiogenesis (Chen et al., 2006), and more specifically, in the promotion of survival and drug resistance of ovarian cancer cell lines (Pengetnze et al., 2003). Src is overexpressed in late-stage cancers (>50%)(Wiener et al., 2003). In human, over-expression of Src protein facilitates activation of other signaling molecules rather than being oncogenic on its own (Biscardi et al., 1999; LoRusso et al., 2002). Studies in nude mouse models with ovarian cancer have shown that inhibition of Src reduces VEGF expression and retard tumor growth (Wiener et al., 1999). Furthermore, *in vitro* Src inhibition enhances the cytotoxic effects of platinum and paclitaxel treatment and restores chemosensitivity to a chemotherapy-resistant ovarian cancer cell line (Pengetnze et al., 2003). Moreover, orthotopic mouse model of ovarian cancer, the Src-selective inhibitor AP23846 have an antiangiogenic effect by reducing VEGF and IL-8. The effect of these inhibitors are enhanced when the inhibitor are given in combination with docetaxel (Han et al., 2006). Currently several Src inhibitors [Dasatinib (BMS-354825, Sprycel, Bristol-Myers Squibb), bosutinib (SKI-606, Wyeth) andXL-999 (Excelixis)] are now entering the clinical arena (Ma and Adjei, 2009).

7. Aurora Kinase

Aurora kinases are serine-threonine protein kinases that are central mitotic regulators as chromosome segregation notably centromere separation, as well as mitotic spindle formation and cytokinesis processes during mitosis (Harrington et al., 2004; Keen and taylor, 2004). Three aurora kinases (A, B, C) are known. The activity of aurora kinase is cell cycle dependent and active during the gap-2-mitosis (G2M) phase of the cell cycle. Aurora kinases are involved in the progression of a variety of solid tumors (Fu et al., 2006; Harrington et al., 2004; Matthews et al., 2006; Zhou et val., 1998). Aurora A is overexpressed in 50-83% of human epithelial ovarian carcinomas (Gritsko et al., 2003; Landen et al., 2007). Frequently in ovarian cancer, amplification of human chromosome 20q13.2 with Aurora-A take place (Tanner et al., 2000), and phosphorylates the tumor suppressor protein p53, resulting cell cycle progression (Katayama et al., 2004). Thus, Aurora kinase A has been significantly associated with tumor grade and stage FIGO (the International Federation of Obstetricians and Gynecologists), and survival (Landen Jr et al., 2007; Kulkarni et al., 2007).

Several small molecule inhibitors of aurora kinases are in early stage of clinical trials including MK-0457 (a small molecule pan-aurora kinase inhibitor) (Lin et al., 2008), MLN8054 (an orally active small molecule inhibitor of aurora kinase) (Manfredi et al., 2007). Furthermore, MLN8054 is currently being explored in the management of patients with platinum-refractory or resistant epithelial, fallopian, or primary peritoneal carcinoma. Studies from chemoresistant human ovarian cell lines showed that aurora kinase inhibition alone significantly reduced tumor burden. Moreover, combination treatment with docetaxel resulted in significantly improved reduction in tumor growth beyond that afforded by docetaxel alone. The aurora kinase inhibitors synergize with paclitaxel to induce apoptosis in ovarian cancer cells (Scharer et al., 2008).

8. Hedgehog Pathway

Hedgehog signaling plays a role in many processes such as cell differentiation, growth, and proliferation. This pathway is active during embryonic development and remains active in the adult in the maintenance of stem cell populations. The Hedgehog family has several proteins include Sonic hedgehog (Shh), Indian hedgehog (Ihh), and Desert hedgehog (Dhh), which function as signaling molecules (Bhattacharya et al., 2008). There are two Hedgehog receptors PATCHED1 and SMOOTHENED. In the absence of a ligand, PATCHED1 inhibits SMOOTHENED, a transmembrane G-coupled protein. However, when the ligand binds PATCHED1, SMOOTHENED suppression is relieved resulting in transcription of the Hedgehog genes. PATCHED1 or SMOOTHENED receptor mutations or overexpression of the Hedgehog ligand leads to uncontrolled cell proliferation. In ovarian cancer, hedgehog pathway blockers inhibit proliferation of ovarian tumors through down regulation of PATCHED1 (Bhattacharya et al., 2008). Moreover, Hedgehog pathway inhibitor cyclopamine, inhibits the hedgehog molecules (Shh, Dhh, Ptch, Smo, and Gli 1 proteins) in ovarian carcinoma cell lines and inhibits cell proliferation and ovarian cancer (Chen et al., 2007). At present several novel hedgehog signaling pathway blockers are in clinical trial which include IPI-926 (Infinity Pharmaceuticals, Inc., Cambridge, Mass) and GDC-0449 (Genentech, Inc).

9. Alpha Folate Receptor (α-FR)

Folic acid is an essential vitamin and of importance for a single-carbon transfer processes medicated by enzyme systems involved in metabolism for nucleotide synthesis and methylation reactions (Markert et al., 2008). Folic acid is act through α-Folate receptor that is encoded by the *FOLR1 gene* in human, a member of the folate receptor (FOLR) family. α-FR has high affinity for folic acid and for several reduced folic acid derivatives, and mediate delivery of 5-methyltetrahydrofolate to the interior of cells. The expression of α-FR has been increased in various tumor tissues, including ovarian, endometrial, and breast cancer (Elnakat and Ratnam, 2006; Kalli et al., 2008). As a tumor-associated antigen α-FR has been detected in 70% of patients with breast and ovarian cancer (Knutson et al., 2006). Over 90% of nonmucinous ovarian cancers overexpress α-FR (Kalli et al., 2008). Overexpression of α-FR support tumor growth by increasing folate availability to cancer cells (Elnakat and Ratnam, 2006).

Recent studies have demonstrated that α-FR may have a potential as a target for immunotherapeutic approaches in ovarian cancer. Some of these include the use of anti-α-FR antibodies or folic acid conjugates. Further studies have shown that when vaccination regimen created with autologous dendritic cells engineered with mRNA encoded α-FR regress recurrent epithelial ovarian cancer more than 50% lymph node metastasis and a dramatic decrease in CA125 concentrations within 4 weeks after the first vaccination (Hernando et al., 2007). At present several other α-FR inhibitors including Farletuzumab (MORAb-003, monoclonal antibody-dependent cell-mediated cytotoxicity and complement-mediated toxicity) (Ebel et al., 2007) and EC145 (a drug that is specifically designed to enter cancer cells via the folate vitamin receptor, FR) are designed to target ovarian cancer and in clinical trial.

10. Poly-ADP-Ribose Polymerase (PARP)

Poly (ADP-ribose) polymerase (PARP) is a key nuclear enzyme involve in the repair of DNA single-strand breaks (SSBs) using the base excision repair pathway (Tutt et al., 2005). In normal cell, PARP-1 and PARP-2 are the only members of the PARP family known to be activated by DNA damage. PARP inhibition results in the accumulation of DNA SSBs, which may lead to double strand DNA breaks (DSBs). One of the essential components in the repair of DSBs and maintenance of genomic stability is BRCA1 and BRCA2 (Lord and Ashworth, 2008). BRCA1 participates in cell cycle checkpoint control (Mullan et al., 2006), regulation of steroid hormone function (Katiyar et al., 2006) and apoptosis (Gilmore et al., 2003). Most significant role of BRCA1 is within the DNA damage repair pathway, in which BRCA1 exists within the BRCA1-associated genome surveillance complex (BASC) and involve in recognition and repair of DSBs by homologous recombination (Wang et al., 2000). In addition, BRCA1 has been implicated in DNA replication, through repair of 'daughter strand gaps', DNA breaks that occur following stalling of replication forks (Nagaraju and Scully, 2007). Like BRCA1, BRCA2 is also required for DNA replication as well as homologous recombination, via regulation of the double-strand break repair protein, Rad51 (Carreria et al., 2009). The lifetime risk of ovarian cancer for BRCA1 and BRCA2 mutation carriers is estimated at 40–50% and 10–20%, respectively (Sowter et al., 2005). Between 5-10% of all ovarian cancer cases are associated with inheriting a mutation in the BRCA1 or BRCA2 gene (Ford et al., 1994). Thus, PARP inhibitors can be use as a therapeutic tool in BRCA mutation carrier patients. There are several small molecular inhibitor for PARP are in different phases of clinical trial including [Olaparib (AZD2281, KU-0059436; AstraZeneca), AG0146999 (Pfizer), ABT888 (Abbott), BSI-201 (Bipar), INO-1001 (Inotek/Genentech), and MK4827 (Merck)] (Chalmers, 2009).

11. Interleukin-6 (IL-6)/JAK/STAT Pathway

Interleukin-6 (IL-6) is secreted by T cells and macrophages and acts as both a pro-inflammatory and anti-inflammatory cytokine to stimulate immune response to trauma, especially burns or other tissue damage leading to inflammation (Nash et al., 1999). Il6 is overexpressed in most ovarian cancers, providing autocrine stimulation of the Il6 receptor, which activates janus kinase 2 (JAK2), facilitating phosphorylation and nuclear translocation of the signal transducer and activator of transcription 3 (STAT3). Up regulation of STAT3 genes stimulate proliferation, inhibit apoptosis and induce angiogenesis. Nuclear localization of activated phosphorylated STAT3 occurs in more than 70% of ovarian cancers (Rosen et al., 2006). Activated STAT3 also translocates to focal adhesion complexes and stimulates motility along with Src. Currently several small molecular inhibitor and mAbs for Il6, JAK2 and STAT3 are in development or in clinical trial (Burke et al., 2001; Duan et al., 2007; Klampfer et al., 2006; McMurray, 2006).

12. Nuclear Factor Kappa-Light-Chain-Enhancer of Activated B Cells (NF-κB)

NF-κB plays a key role in regulating the immune response to infection. NF-κB is a protein complex that controls the *transcription* of *DNA* in response to various stimuli such as stress, cytokines, free radicals, ultraviolet irradiation, oxidized low-density lipoprotein *(LDL)*, and bacterial or viral antigens. The kappa light are critical components of immunoglobulins. Incorrect regulation of NF-κB has been linked to cancer, inflammatory and autoimmune disease, septic shock, viral infection, and improper immune development. The NFκB transcription factor is constitutively activated in more than half of the ovarian cancers (Lin et al., 2007; Samanta et al., 2004) through signalling initiated by several cytokines (**Interleukin-1,** Il1 and tumor necrosis factor-alpha, TNFα) and growth factors (EGF). Inactive NFκB forms complex with inhibitor of NFκB (IκB) and is activated by the IκB kinase (IKK) complex, which contains two kinase subunits (IKKα and IKKβ) and a regulatory subunit (IKKγ) (Hacker and Karin, 2006). The IKK complex is activated by several kinases including MEKK3 (Yang et al., 2001; Karin, 2006). MEKK3 is overexpressed in >50% of ovarian cancers. The resultant activation of NFκB upregulates antiapoptotic genes including *CFLAR* (encode CASP8 and FADD-like apoptosis regulator protein), antioxidant proteins (superoxide dismutase or ferritin heavy chain), growth regulatory cytokines [(Il6 or growth regulated (GrO1)] and angiogenic factors (Il8) (Karin, 2006). Selective inhibition of NFκB has been difficult to achieve, but gene therapy with adenoviral E1A reduces NFκB signalling and increases sensitivity to paclitaxel in xenograft models, which has prompted a clinical trial of liposomal E1A in combination with paclitaxel in patients with intra-abdominal recurrence of ovarian cancer.

13. Histone Deacetylase (HDAC)

Histone deacetylases (HDAC) (EC number 3.5.1), also known as lysine deacetylases (KDAC) are a class of *enzymes* that remove *acetyl groups* from an ε-N-acetyl lysine amino on a histone or from other numerous non-histone proteins (de Ruijter et al., 2003). Aberrant histone modifications such as hypoacetylation have been associated with malignancy through the transcriptional silencing of tumor suppressor genes in various cancers including ovarian cancer. Currently few histone deacetylase inhibitors are in clinical trial including Belinostat along with combination with carboplatin in patients with recurrent persistent or resistant cancer (Takai and Narahara, 2010).

Conclusion

Ovarian cancers are complex malignancies with significant chromosomal instability which partly occurs as a result of aberrant autocrine/paracrine stimulation and activation of multiple signaling pathways. However, the etiology of ovarian cancer remains poorly understood. A comprehensive compilation of aberrant events and processes in ovarian cancer with understanding of the relationship between histology-specific genetic aberrations and the regulation of tissue histogenesis will increase our understanding of pathophysiology and provide novel markers for early diagnosis and prediction of prognosis. Moreover, a better understanding of the redundant pathways that exist in cell signaling, creative targeting of horizontal and vertical signaling pathway, identification of other predictive markers to better identify a targeted subpopulation of patients that will respond, and an underlying of the mechanisms of resistance will enable the development of more rational, individualized therapies. Furthermore, clinical trials are increasingly including quality of life components in trial designs in efforts to both prolong and improve the quality of life among ovarian cancer patients. In the long term, earlier detection of ovarian cancer is likely to have an important effect on the management of this disease.

Acknowledgment

We are thankful to our library facility for literature survey at Morehouse School of Medicine, Atlanta.

References

Altomare DA, Wang HQ, Skele KL, De Rienzo A, Klein-Szanto AJ, Godwin AK, Testa JR (2004). AKT and mTOR phosphorylation is frequently detected in ovarian cancer and can be targeted to disrupt ovarian tumor cell growth. *Oncogene* 23:5853–5857.

Andrae J, Gallini R, Betsholtz C (2008) Role of platelet-derived growth factors in physiology and medicine. *Genes and Development* 22(10):1276–1312.

Apte SM, Bucana CD, Killion JJ, Gershenson DM, Fidler IJ (2004) Expression of platelet-derived growth factor and activated receptor in clinical specimens of epithelial ovarian cancer and ovarian carcinoma cell lines. *Gynecologic Oncology* 93(1):78–86.

Baldwin RL, Nemeth E, Tran H, Shvartsman H, Cass I, Narod S, Karlan BY (2000). BRCA1 promoter region hypermethylation in ovarian carcinoma: a population-based study. *Cancer Res.* 60:5329–5333.

Bast RC Jr, Hennessy B, Mills GB (2009). The biology of ovarian cancer: new opportunities for translation. *Nat. Rev. Cancer.* 9(6):415-28.

Bhattacharya R, Kwon J, Ali B, Wang E, Patra S, Shridhar V, Mukherjee P (2008). Role of hedgehog signaling in ovarian cancer. *Clinical Cancer Research.*14(23):7659–7666.

Biscardi JS, Tice DA, Parsons SJ (1999). C-Src receptor tyrosine kinases, and human cancer, *Adv. Cancer Res.* 76: 61–119.

Blagden S, Gabra H (2009). Promising molecular targets in ovarian cancer. *Curr. Opin. Oncol.* 21(5):412-9.

Burke WM, Jin X, Lin HJ, Huang M, Liu R, Reynolds RK, Lin J. (2001) Inhibition of constitutively active Stat3 suppresses growth of human ovarian and breast cancer cells. *Oncogene* 20:7925–7934.

Campos S, Hamid O, Seiden MV, Oza A, Plante M, Potkul RK, Lenehan PF, Kaldjian EP, Varterasian ML, Jordan C, Charbonneau C, Hirte H (2005). Multicenter, randomized phase II trial of oral CI-1033 for previously treated advanced ovarian cancer. *J. Clinical Oncol.* 23(24):5597–5604.

Campos SM, Ghosh S (2010). A current review of targeted therapeutics for ovarian cancer. *J. Oncol.* 149362-73.

Carpten JD, Faber AL, Horn C, Donoho GP, Briggs SL, Robbins CM, Hostetter G, Boguslawski S, Moses TY, Savage S, Uhlik M, Lin A, Du J, Qian YW, Zeckner DJ, Tucker-Kellogg G, Touchman J, Patel K, Mousses S, Bittner M, Schevitz R, Lai MH, Blanchard KL, Thomas JE (2007). A transforming mutation in the pleckstrin homology domain of AKT1 in cancer. *Nature* 448:439–444.

Carreira A, Hilario J, Amitani I, Baskin RJ, Shivji MK, Venkitaraman AR, Kowalczykowski SC (2009). The BRC repeats of BRCA2 modulate the DNA-binding selectivity of RAD51. *Cell* 136:1032–1043.

Chalmers AJ (2009). The potential role and application of PARP inhibitors in cancer treatment. *Br. Med. Bull.* 89:23–40.

Chen T, George JA, Taylor CC (2006). Src tyrosine kinase as a chemotherapeutic target: is there a clinical case? *Anti-cancer Drugs* 17 (2):123–131.

Chen X, Horiuchi A, Kikuchi N, Osada R, Yoshida J, Shiozawa T, Konishi I (2007). Hedgehog signal pathway is activated in ovarian carcinomas, correlating with cell proliferation: it's inhibition leads to growth suppression and apoptosis. *Cancer Sci.* 98(1):68–76.

Chien CH, Chow SN (1993). Point mutation of the Ras oncogene in human ovarian cancer. *DNA Cell Biol.* 12 (7): 623–627.

Ciardiello F (2005). Epidermal growth factor receptor inhibitors in cancer treatment, *Future Oncol.* 1 (2):221–234.

de Ruijter AJ, van Gennip AH, Caron HN, Kemp S, van Kuilenburg AB (2003). "Histone deacetylases (HDACs): characterization of the classical HDAC family". *Biochem. J.* 370 (Pt 3): 737–49.

Dinh P, Harnett P, Piccart-Gebhart MJ, Awada A (2008). New therapies for ovarian cancer: cytotoxics and molecularly targeted agents. *Crit. Rev. Oncol. Hematol.* 67(2):103-12.

Duan Z, Bradner J, Greenberg E, Mazitschek R, Foster R, Mahoney J, Seiden MV (2007). 8-benzyl-4-oxo-8-azabicyclo[3.2.1]oct-2-ene-6, 7-dicarboxylic acid (SD-1008), a novel janus kinase 2 inhibitor, increases chemotherapy sensitivity in human ovarian cancer cells. *Mol. Pharmacol.* 72:1137–1145.

Duska LR, Chang YC, Flynn CE, Chen AH, Goodman A, Fuller AF, Nikrui N (1999). Epithelial ovarian carcinoma in the reproductive age group. *Cancer* 85(12):2623-9.

Ebel W, Routhier EL, Foley B, Jacob S, McDonough JM, Patel RK, Turchin HA, Chao Q, Kline JB, Old LJ, Phillips MD, Nicolaides NC, Sass PM, Grasso L (2007). Preclinical evaluation of MORAb-003, a humanized monoclonal antibody antagonizing folate receptor-alpha. *Cancer Immunity* 7:6–13.

Ellis LM, Hicklin DJ (2008). VEGF-targeted therapy: mechanisms of antitumour activity. *Nat. Rev. Cancer* 8:579–591.

Elnakat H, Ratnam M (2006). Role of folate receptor genes in reproduction and related cancers. *Front Biosci.* 11:506–519.

Estep AL, Palmer C, McCormick F, Rauen KA (2007). Mutation analysis of BRAF, MEK1 and MEK2 in 15 ovarian cancer cell lines: implications for therapy. *PLoS ONE* 2:e1279.

Feeley KM, Wells M (2001) Precursor lesions of ovarian epithelial malignancy. *Histopathology* 38:87.

Ford D, Easton DF, Bishop DT, Narod SA, Goldgar DE (1994). Risks of cancer in BRCA1-mutation carriers. *The Lancet* 343(8899):692–695.

Fraser M, Bai T, Tsang BK (2008). Akt promotes cisplatin resistance in human ovarian cancer cells through inhibition of p53 phosphorylation and nuclear function. *Int. J. Cancer* 122:534–546.

Fu S, Hu W, Kavanagh JJ, Bast RC Jr (2006). Targeting aurora kinases in ovarian cancer. *Expert Opin. Ther. Targets* 10 (1): 77–85.

Gilmore PM, Quinn JE, Mullan PB, Andrews HN, McCabe N, Carty M, Kennedy RD, Harkin DP (2003). Role played by BRCA1 in regulating the cellular response to stress. *Biochem. Soc. Trans.* 31(Pt 1):257–262.

Godwin AK, Testa JR, Handel LM, Liu Z, Vanderveer LA, Tracey PA, Hamilton TC (1992). Spontaneous transformation of rat ovarian surface epithelial cells: association with cytogenetic changes and implications of repeated ovulation in the etiology of ovarian cancer. *J. Natl. Cancer Inst.* 84:592–601.

Goff BA, Mandel L, Muntz HG, Melancon CH (2000). Ovarian carcinoma diagnosis. *Cancer.* 89:2068-75.

Gonzalez-Lira G, Escudero-De Los Rios P, Salazar-Martinez E, Lazcano-Ponce EC (1997). Conservative surgery for ovarian cancer and effect on fertility. *Int. J. Gynaecol. Obstet.* 56(2): 155-62.

Gordon AN, Finkler N, Edwards RP, Garcia AA, Crozier M, Irwin DH, Barrett E (2005). Efficacy and safety of erlotinib HCI, an epidermal growth factor receptor (HER1/EGFR) tyrosine kinase inhibitor, in patients with advanced ovarian carcinoma: results from a phase II multicenter study. *Int. J. Gynecol. Cancer* 15:785–792.

Gritsko TM, Coppola D, Paciga JE, Yang L, Sun M, Shelley SA, Fiorica JV, Nicosia SV, Cheng JQ (2003). Activation and overexpression of centosome kinase BTAK/Aurora-A in human ovarian cancer. *Clin. Cancer Res.* 9 (4): 1420–1426.

Häcker H, Karin M (2006). Regulation and function of IKK and IKK-related kinases. *Science STKE.* 357:re 13.

Han LY, Landen CN, Trevino JG, Halder J, Lin YG, Kamat AA, Kim TJ, Merritt WM, Coleman RL, Gershenson DM, Shakespeare WC, Wang Y, Sundaramoorth R, Metcalf CA 3rd, Dalgarno DC, Sawyer TK, Gallick GE, Sood AK (2006). Antiangiogenic and antitumor effects of SRC inhibition in ovarian carcinoma. *Cancer Res.* 66:8633–8639.

Harrington EA, Bebbington D, Moore J, Rasmussen RK, Ajose-Adeogun AO, Nakayama T, Graham JA, Demur C, Hercend T, Diu-Hercend A, Su M, Golec JM, Miller KM (2004). VX-680, a potent and selective small-molecule inhibitor of the Aurora kinases, suppresses tumor growth in vivo. *Nature Medicine* 10(3):262–267.

Heinrich MC, Corless CL, Duensing A, McGreevey L, Chen CJ, Joseph N, Singer S, Griffith DJ, Haley A, Town A, Demetri GD, Fletcher CD, Fletcher JA (2003). PDGFRA activating mutations in gastrointestinal stromal tumors. *Science.* 299(5607):708–710.

Hernando JJ, Park T-W, Fischer H-P, Zivanovic O, Braun M, Pölcher M, Grünn U, Leutner C, Pötzsch B, Kuhn W. (2007). Vaccination with dendritic cells transfected with mRNA-encoded folate-receptor-α for relapsed metastatic ovarian cancer. *The Lancet Oncology* 8(5):451–454.

Hong YK, Shin JW, Detmar M (2004). Development of the lymphatic vascular system: a mystery unravels. *Dev. Dyn.* 231:462–473.

Jacobs IJ, Kohler MF, Wiseman RW, Marks JR, Whitaker R, Kerns BA, Humphrey P, Berchuck A, Ponder BA, Bast RC Jr. (1992). Clonal origin of epithelial ovarian cancer: analysis by loss of heterozygosity, p53 mutation and X chromosome inactivation. *J. Natl Cancer Inst.* 84:1793–1798.

Jazaeri AA, Yee CJ, Sotiriou C, Brantley KR, Boyd J, Liu ET (2002). Gene expression profiles of BRCA1-linked, BRCA2-linked, and sporadic ovarian cancers. *J. Natl. Cancer Inst.* 94:990–1000.

Jemal A, Siegel R, Ward E, Hao Y, Xu J, Thun MJ (2009). Cancer statistics 2009. *CA Cancer J. Clin.* 59(4): 225-49.

Jemal A., Murray T, Ward E, Samuels A, Tiwari RC, Ghafoor A, Feuer EJ, Thun MJ (2005). Cancer statistics, 2005, *CA Cancer J. Cl*in. 55:10–30.

Jemal,A., Siegel R, Ward E, Hao Y, Xu J, Murray T, Thun MJ (2008). Cancer statistics, *CA Cancer J. Clin.* 58:71–96.

Jeon BH, Jang C, Han J, Kataru RP, Piao L, Jung K, Cha HJ, Schwendener RA, Jang KY, Kim KS, Alitalo K, Koh GY. (2008). Profound but dysfunctional lymphangiogenesis via vascular endothelial growth factor ligands from CD11b+ macrophages in advanced ovarian cancer. *Cancer Res.* 68:1100–1109.

Kalli KR, Oberg AL, Keeney GL, Christianson TJ, Low PS, Knutson KL, Hartmann LC (2008). Folate receptor alpha as a tumor target in epithelial ovarian cancer. *Gynecologic Oncology.* 108:619–626.

Karin M (2006). Nuclear factor κB in cancer development and progression. *Nature* 441:431–436.

Katayama H, Sasai K, Kawai H, Yuan ZM, Bondaruk J, Suzuki F, Fujii S, Arlinghaus RB, Czerniak BA, Sen S (2004). Phosphorylation by aurora kinase A induces Mdm2-mediated destabilization and inhibition of p53. *Nature Genetics* 36(1):55–62.

Katiyar P, Ma Y, Fan S, Pestell RG, Furth PA, Rosen EM (2006). Regulation of progesterone receptor signaling by BRCA1 in mammary cancer. *Nucl. Recept. Signal* 4:e006.

Keen N, Taylor S (2004). Aurora-kinase inhibitors as anticancer agents. *Nature Rev. Cancer* 4(12):927–936.

Kigawa J, Kawaguchi W, Itamochi H, Kanamori Y, Oishi T, Shimada M, Sato S, Sato S, Terakawa N (2007). Effect of simultaneous inhibition of MEK and PI3K/Akt pathways on paclitaxel sensitivity in ovarian cancer [abstract #16046]. *J. Clin. Oncol.* 25(18S).

Klampfer L (2006). Signal transducers and activators of transcription (STATs): novel targets for chemopreventative and chemotherapeutic drugs, *Curr. Cancer Drug Targets* 6 (2):107–121.

Knutson KL, Krco CJ, Erskine CL, Goodman K, Kelemen LE, Wettstein PJ, Low PS, Hartmann LC, Kalli KR (2006). T-cell immunity to the folate receptor alpha is prevalent in women with breast or ovarian cancer. *J. Clini. Oncol.* 24(26):4254–4261.

Kulkarni AA, Loddo M, Leo E, Rashid M, Eward KL, Fanshawe TR, Butcher J, Frost A, Ledermann JA, Williams GH, Stoeber K (2007). DNA replication licensing factors and aurora kinases are linked to aneuploidy and clinical outcome in epithelial ovarian carcinoma. *Clinical Cancer Res.* 13(20):6153–6161.

Kurman RJ, Shih Ie M (2008). Pathogenesis of ovarian cancer: lessons from morphology and molecular biology and their clinical implications. *Int. J. Gynecol. Pathol.* 27:151–160.

Kurman, RJ, Visvanathan K, Roden R, Wu TC, Shih IeM. (2008). Early detection and treatment of ovarian cancer: shifting from early stage to minimal volume of disease based on a new model of carcinogenesis, *Am. J. Obstet. Gynecol.* 198:351–356.

Lafky JM, Wilken JA, Baron AT, Maihle NJ (2008). Clinical implications of the ErbB/epidermal growth factor (EGF) receptor family and its ligands in ovarian cancer. *Biochim. Biophys. Acta* 1785:232–265.

Landen CN Jr, Lin YG, Immaneni A, Deavers MT, Merritt WM, Spannuth WA, Bodurka DC, Gershenson DM, Brinkley WR, Sood AK (2007). Overexpression of the centrosomal protein Aurora-A kinase is associated with poor prognosis in epithelial ovarian cancer patients. *Clinical Cancer Res.* 13(14):4098–4104.

Landen Jr. CN, Birrer MJ, Sood AK (2008). Early events in the pathogenesis of epithelial ovarian cancer, *J. Clin. Oncol.* 26:995–1005.

Lin YG, Immaneni A, Merritt WM, Mangala LS, Kim SW, Shahzad MM, Tsang YT, Armaiz-Pena GN, Lu C, Kamat AA, Han LY, Spannuth WA, Nick AM, Landen CN Jr, Wong KK, Gray MJ, Coleman RL, Bodurka DC, Brinkley WR, Sood AK (2008). Targeting Aurora kinase with MK-0457 inhibits ovarian cancer growth. *Clinical Cancer Res.* 14(17):5437–5446.

Lin YG, Kunnumakkara AB, Nair A, Merritt WM, Han LY, Armaiz-Pena GN, Kamat AA, Spannuth WA, Gershenson DM, Lutgendorf SK, Aggarwal BB, Sood AK (2007). Curcumin inhibits tumor growth and angiogenesis in ovarian carcinoma by targeting the nuclear factor-κB pathway. *Clin. Cancer Res.* 13:3423–3430.

Lord CJ, Ashworth A (2008). Targeted therapy for cancer using PARP inhibitors. *Current Opinion Pharm.* 8(4):363–369.

LoRusso, PM, Adjei AA, Meyer MB, Wozniak A, Gadgeel SM, Hanson LJ, Reid JM, Mitchell DY, Bruzek LM, Leopold JS, Herrera R, Van Becelaere K, Carlson T, Packard C, Gulyas SW, Erlichman C (2002). A phase I clinical and pharmacokinetic evaluation of oral CI-1040 administered for 21 consecutive days, repeated every 4 weeks in patients with advanced cancer. *Proc. Am. Soc. Clin. Oncol.* 21 (81a):321a.

Ma WW, Adjei AA (2009). Novel agents on the horizon for cancer therapy. *CA Cancer J. Clin.* 59:111–137.

Manfredi MG, Ecsedy JA, Meetze KA, Balani SK, Burenkova O, Chen W, Galvin KM, Hoar KM, Huck JJ, LeRoy PJ, Ray ET, Sells TB, Stringer B, Stroud SG, Vos TJ, Weatherhead GS, Wysong DR, Zhang M, Bolen JB, Claiborne CF (2007). Antitumor activity of MLN8054, an orally active small-molecule inhibitor of Aurora a kinase. *Proc. National Academy Sci.,* USA 104(10):4106–4111.

Mao X, Hamoudi RA (2000). Molecular and cytogenetic analysis of glioblastoma multiforme. *Cancer Genet. Cytogene* 122(2):87–92.

Markert S, Lassmann S, Gabriel B, Klar M, Werner M, Gitsch G, Kratz F, Hasenburg A (2008). Alpha-folate receptor expression in epithelial ovarian carcinoma and non-neoplastic ovarian tissue. *Anticancer Res.* 28(6):3567–3572.

Matthews N, Visintin C, Hartzoulakis B, Jarvis A, Selwood DL (2006). Aurora A and B kinases as targets for cancer: will they be selective for tumors? *Expert Rev. Anticancer Therapy* 6(1):109–120.

McMurray JS. (2006). A new small molecule Stat 3 inhibitor. *Chem. Biol.* 13:1123–1124.

McPhillips F, Mullen P, MacLeod KG, Sewell JM, Monia BP, Cameron DA, Smyth JF, Langdon SP (2006). Raf-1 is the predominant Raf isoform that mediates growth factor-stimulated growth in ovarian cancer cells, *Carcinogenesis* 27 (4):729–739.

McPhillips F, Mullen P, Monia BP, Ritchic AA, Dorr FA, Smyth JF, Langdon SP (2001). Association of c-Raf expression with survival and its targeting with antisense oligonucleotides in ovarian cancer. *Br. J. Cancer* 85 (11):1753–1758.

Mok SC, Kwong J, Welch WR, Samimi G, Ozbun L, Bonome T, Birrer MJ, Berkowitz RS, Wong KK (2007). Etiology and pathogenesis of epithelial ovarian cancer. *Dis. Markers* 23:367–376.

Mullan PB, Quinn JE, Harkin DP (2006). The role of BRCA1 in transcriptional regulation and cell cycle control. *Oncogene* 25:5854–5863.

Nagaraju G, Scully R (2007). Minding the gap: the underground functions of BRCA1 and BRCA2 at stalled replication forks. *DNA Repair* (Amst) 6:1018–1031.

Nakayama K, Nakayama N, Kurman RJ, Cope L, Pohl G, Samuels Y, Velculescu VE, Wang TL, Shih IeM (2006). Sequence mutations and amplification of PIK3CA and AKT2 genes in purified ovarian serous neoplasms. *Cancer Biol. Ther*. 5:779–785.

Nakayama N, Nakayama K, Yeasmin S, Ishibashi M, Katagiri A, Iida K, Fukumoto M, Miyazaki K. (2008). KRAS or BRAF mutation status is a useful predictor of sensitivity to MEK inhibition in ovarian cancer. *Br. J. Cancer* 99:2020–2028.

Nash MA, Ferrandina G, Gordinier M, Loercher A, Freedman RS (1999). The role of cytokines in both the normal and malignant ovary. *Endocr. Relat. Cancer* 6:93–107.

Nishida N, Yano H, Komai K, Nishida T, Kamura T, Kojiro M (2004). Vascular endothelial growth factor C and vascular endothelial growth factor receptor 2 are related closely to the prognosis of patients with ovarian cancer. *Cancer* 101:1364–1374.

Oliver G, Srinivasan RS (2008). Lymphatic vasculature development: current concepts. *Ann. N Y Acad. Sci*. 1131:75–81.

Park JY, Kim DY, Kim JH, Kim YM, Kim YT, Nam JH (2009). Surgical management of borderline ovarian tumors: The role of fertilitysparing surgery. *Gynecol. Oncol*. 113(1): 75-82.

Pengetnze Y, Steed M, Roby KF, Terranova PF, Taylor CC (2003). Src tyrosine kinase promotes survival and resistance to chemotherapeutics in a mouse ovarian cancer cell line, *Biochem. Biophys. Res. Commun*. 309 (2):377–383.

Posadas EM, Liel MS, Kwitkowski V, Minasian L, Godwin AK, Hussain MM, Espina V, Wood BJ, Steinberg SM, Kohn EC. (2007). A phase II and pharmacodynamic study of gefitinib in patients with refractory or recurrent epithelial ovarian cancer. *Cancer* 109(7):1323–1330.

Raspollini MR, Amunni G, Villanucci A, Baroni G, Boddi V, Taddei GL (2004). Prognostic significance of microvessel density and vascular endothelial growth factor expression in advanced ovarian serous carcinoma. *Int. J. Gynecol. Cancer* 14:815–823.

Ricciardelli C, Oehler MK (2009). Diverse molecular pathways in ovarian cancer and their clinical significance. *Maturitas* 62(3):270-5.

Rosen DG, Mercado-Uribe I, Yang G, Bast RC Jr, Amin HM, Lai R, Liu J (2006). The role of constitutively active signal transducer and activator of transcription 3 in ovarian tumorigenesis and prognosis. *Cancer* 107:2730–2740.

Salomon DS, Brandt R, Ciardiello F, Normanno N (1995). Epidermal growth factor-related peptides and their receptors in human malignancies. *Crit. Rev. Oncol./Hematol.* 19:183–232.

Samanta AK, Huang HJ, Bast RC Jr, Liao W (2004). Overexpression of MEKK3 confers resistance to apoptosis through activation of NF κB. *J. Biol. Chem.* 279:7576–7583.

Scharer CD, Laycock N, Osunkoya AO, Logani S, McDonald JF, Benigno BB, Moreno CS (2008). Aurora kinase inhibitors synergize with paclitaxel to induce apoptosis in ovarian cancer cells. *J. Translational Medic.* 6, article 79:1-13.

Schmitt J, Matei D (2008). Platelet-derived growth factor pathway inhibitors in ovarian cancer. *Clinical Ovarian. Can.* 1(2):120–126.

SEERS. National Cancer Institute. *Cancer Surveillance Epidemiology and End Results* (SEER) Available at: http://seer.cancer. gov/statfacts/html/ corp.html.

Seiden MV, Burris HA, Matulonis U, Hall JB, Armstrong DK, Speyer J, Weber JD, Muggia F (2007). A phase II trial of EMD72000 (matuzumab), a humanized anti-EGFR monoclonal antibody, in patients with platinum-resistant ovarian and primary peritoneal malignancies. *Gyneco. Oncol.* 104(3):727–731.

Soslow RA (2008). Histologic subtypes of ovarian carcinoma: an overview, *Int. J. Gynecol. Pathol.* 27:161–174.

Sowter HM, Ashworth A (2005). BRCA1 and BRCA2 as ovarian cancer susceptibility genes. *Carcinogenesis* 26:1651–1656.

Stewart BW, Kleihues P (2003). Editors, *WHO World Cancer Report*, Lyon, France: IARC Press.

Takai N, Narahara H (2010). Histone deacetylase inhibitor *J. Oncol.* 2010:458431.

Tan DS, Rothermundt C, Thomas K, Bancroft E, Eeles R, Shanley S, Ardern-Jones A, Norman A, Kaye SB, Gore ME. (2008). 'BRCAness' syndrome in ovarian cancer: a case-control study describing the clinical features and outcome of patients with epithelial ovarian cancer associated with BRCA1 and BRCA2 mutations. *J. Clin. Oncol.* 26:5530–5536.

Tanner MM, Grenman S, Koul A, Johannsson O, Meltzer P, Pejovic T, Borg A, Isola JJ (2000). Frequent amplification of chromosomal region 20q12-q13 in ovarian cancer. *Clinical Cancer Research.* 6(5):1833–1839.

Testa JR, Bellacosa A (2001). AKT plays a central role in tumorigenesis. *Proc. Natl. Acad. Sci. U S A* 98:10983–10985.

Turner N, Tutt A, Ashworth A. (2004). Hallmarks of 'BRCAness' in sporadic cancers. *Nat. Rev. Cancer* 4:814–819.

Tutt ANJ, Lord CJ, McCabe N, Farmer H, Turner N, Martin NM, Jackson SP, Smith GC, Ashworth A. (2005). Exploiting the DNA repair defect in BRCA mutant cells in the design of new therapeutic strategies for cancer. *Cold Spring Harbor Symposia on Quantitative Biology.* 70:139–148.

Wang Y, Cortez D, Yazdi P, Neff N, Elledge SJ, Qin J. (2000). BASC, a super complex of BRCA1-associated proteins involved in the recognition and repair of aberrant DNA structures. *Genes Dev.* 14:927–939.

Weinstein-Oppenheimer C.R., Blalock WL, Steelman LS, Chang F, McCubrey JA (2000). The Raf signal transduction cascade as a target for chemotherapeutic intervention in growth factor-responsive tumors, *Pharmacol. Ther.* 88:229–279.

Wiener JR, Windham TC, Estrella VC, Parikh NU, Thall PF, Deavers MT, Bast RC, Mills GB, Gallick GE (2003). Activated SRC protein tyrosine kinase is overexpressed in late-stage human ovarian cancers. *Gynecol. Oncol.* 88:73–79.

Willner, J, Wurz K, Allison KH, Galic V, Garcia RL, Goff BA, Swisher EM (2007). Alternate molecular genetic pathways in ovarian carcinomas of common histological types. *Hum. Pathol.* 38:607–613.

Xing H, Weng D, Chen G, Tao W, Zhu T, Yang X, Meng L, Wang S, Lu Y, Ma D (2008). Activation of fibronectin/PI-3K/Akt2 leads to chemoresistance to docetaxel by regulating survivin protein expression in ovarian and breast cancer cells. *Cancer Lett.* 261:108–119.

Yang J, Lin Y, Guo Z, Cheng J, Huang J, Deng L, Liao W, Chen Z, Liu Z, Su B (2001). The essential role of MEKK3 in TNF-induced NF-κB activation. *Nature Immunol.* 2:620–624.

Yap TA, Garrett MD, Walton MI, Raynaud F, de Bono JS, Workman P (2008). Targeting the PI3K-AKT-mTOR pathway: progress, pitfalls, and promises. *Curr. Opin. Pharmacol.* 8:393–412.

Yuan ZQ, Sun M, Feldman RI, Wang G, Ma X, Jiang C, Coppola D, Nicosia SV, Cheng JQ (2000). Frequent activation of AKT2 and induction of apoptosis by inhibition of phosphoinositide-3-OH kinase/Akt pathway in human ovarian cancer. *Oncogene* 19:2324–2330.

Zhou H, Kuang J, Zhong L, Kuo WL, Gray JW, Sahin A, Brinkley BR, Sen S (1998). Tumour amplified kinase STK15/BTAK induces centrosome amplification, aneuploidy and transformation. *Nature Genet.* 20(2):189–193.

In: Signaling, Gene Regulation and Cancer
Editors: Shree Ram Singh and Manoj K. Mishra
ISBN: 978-1-61942-088-5

Chapter 17

MicroRNAs and Their Involvement in Cancer

Kartz E. Bibb
Department of Mathematics and Science,
Alabama State University, Montgomery, AL, US

Abstract

Our cells' DNA contains many genes, approximately 30,000, but not all are active at once. Genes determine cell characteristics and functions. Genes are actively expressed as mRNA which is produced in the nucleus, transported to the cytoplasm and translated into protein. Only a small percentage of the genome is made up by genes and most of the genome do not code for proteins [1]. Scientist for many years, referred to regions outside of the genes as "junk DNA," but it has been realized that RNA from intergenic regions, code for genes that do not code for protein and are now called micro RNAs (miRNA).

Introduction

The size of most mRNAs is between 2000 and 5000 nucleotides, however the size of miRNAs is approximately 17-21 nts in length [1]. About 3% of human genes encode for miRNAs, and up to 30% of human protein coding genes may be regulated by miRNAs [4].The first miRNA, lin-4 was expressed temporally through a mutant *Caenorhabditis elegans* (roundworm) [1, 12]. Another miRNA let-7 was later found also in *C. elegans*, which also had homologous sequences in several other species including human suggesting that miRNAs are not specific to worms [1]. To date, there has been more than 500 human miRNAs experimentally identified, and up to 30% of the human protein coding genes may be regulated by miRNAs, which makes miRNAs one of the most abundant classes of regulatory genes in humans [18]. miRNAs are experimentally discovered by cloning small RNAs from a particular tissue type, preferentially during a developmental stage and then sequencing them to identify which subgroup of miRNAs they might belong to [20, 33]. The number of annotated miRNAs which increased from less than 300 to more than 4,000 over the last 4 years [4].

BIOGENESIS OF MIRNAS

microRNAs genes are transcribed by RNA polymerase II. The primary transcript (pri-miRNA) is capped at the 5' end and polyadenylated at the 3' end. Pri-miRNAs are cleaved in the nucleus by an Rnase III type enzyme Drosha/DGCR8 and the product of this cleavage is one or more precursor miRNA [1, 14]. Two ribonucleic enzymes, Drosha and Dicer, subsequently process the primary transcripts (or pri-miRNA) to generate mature miRNAs, which contain one or more stem-loop structures of about 70 bases. [14-16, 32]. The nuclear splicing machinery provides pre-miRNA from introns, miRNAs produced by this pathway are appropriately called mitrons [9]. Alternatively, pre-miRNAs are derived directly from size-matched introns. These so-called "mitrons" skip the Drosha-DGCR8 processing step and are spiced out of their host genes [9, 15]. These lariats are de-branched, refolded into the stem-loop structure of typical pre-miRNAs [9]

The ribonuclease Drosha excises the stem-loop structure to form the precursor miRNA. After export into the cytoplasm, the pre-miRNA is cleaved, by the ribonuclease Dicer to generate a short RNA duplex [4, 16-17]. After untwisting, one RNA strand become the mature single-stranded miRNA, while the complementary strand, termed miRNA* is usually rapidly degraded [4, 15].

Dicer cleaves pr-miRNAs into 21-25 nts long miRNA/miRNA* duplexes, each strand of which bears 5'monophosphate, 3' hydroxyl group and a 3' 2 nt overhang [11]. Of a miRNA/miRNA* duplex, only one strand, designated the miRNA strand is selected as the guide of mature RISC, whereas the other strand, the miRNA* strand, is discarded during RISC assembly [11, 16]. Dicer was also responsible for the production of lin-4 and let-7 in *C. elegans* and in fact let was also produced in a Dicer dependant manner in human *Hela* cells [13]. RAN-GTP and Exprotin 5 transport this pre-miRNA into the cytoplasm. A helicase then unwinds the cleaved double stranded RNA in strand specific direction [3, 19].

One of the strands of this duplex is incorporated into a protein complex called RNA induced silencing complex (RISC) [14-16]. The strand incorporated into RISC is th e mature miRNA while the other strand (called miRNA* strand) gets degraded. The mature miRNA guides RISC to mRNAs containing a partially complementary sequence to the miRNA. If the miRNA can anneal to the target site, RISC cleaves the mRNA, degrades the mRNA through de-capping and de-adenylation or suppresses the translation without affecting the mRNA level [17]. Short RNA molecules encoded by the genome that regulate the expression of protein coding genes ensuring normal development and functioning of the body [1, 12, 14].

RISC Assemby

RISC assembly follows a multi-step pathway. During RISC loading, miRNA/miRNA* duplexes are incorporated into Ago proteins in an ATP-dependent active process [11, 16]. It has been reported that Hsc70/Hsp90 chaperone machinery mediates such conformational opening [11, 68, 70]. After RISC loading, the duplex is unwound and the miRNA* strand is discarded form Ago protein [64, 67, 70]. Passenger strand cleavage occurs only if the duplex is extensively base paired especially around position 10-11 of the guide [33]. Most miRNA/miRNA* duplexes have central mismatches and arc unwound by a slicer-independent mechanism [11].

RISC Functions

A target mRNA may bear multiple biding sites of the same miRNA. Not all nucleotides of a miRNA contribute equally to RISC target recognition [16]. This target recognition is largely determined by base-pairing of nucleotides in the seed region and is enhanced by additional base-pairing in the middle of the 3' region [11]. Ago2 is capable of RNA cleavage, but this reaction requires extensive base-pairing bewee the miRNA strand and mRNA target [70]. This base pairing subsequently causes inhibition of protein translation and/or degredation of the messenger RNA [15-17]. In humans, miRNAs mainly inhibit protein translation of their target genes and only infrequently cause degradation cleavage of the messenger RNA [4, 69]. Some miRNAs silence target mRNAs through cleavage. In contrast, the complementarity between the miRNA stand and target mRNA is typically limited, which renders RISC incapable of target cleavage [67, 70]. Ago protein may provide a platform to recruit factors, required for translational repression and mRNA de-adenylation/degradation of target mRNAs [70]. Such slicer-independent silencing can be reconstituted by artificial tethering of an Ago protein or a mRNA [11, 70].

At least six models of translational repression have been proposed; 1. RISC induces de-adenylation causing a decrease of translational efficiency by blocking target mRNA; 2. RISC blocks cap function by interacting with either the cap or eIF4E; 3. RISC blocks a late step in initiation of translation such as recruitment of 60S ribosomal subunit; 4. RISC blocks a post-initiation steps; 5. RISC induces proteolysis of nascent peptides during translation; or 6. RISC recruits target mRNAs to processing bodies, where mRNA is degraded and or stored in a translationally inactive site [11, 67].

Point mutations within miRNAs or their flanking regions can lead to altered transcription levels. When point mutations occur in miRNA binding sites of target mRNAs, these can compromise the ability of small SNAs to silence target mRNA transcripts [11].

Protein components of the miRNA pathway can be compromised by the derailed genome. In human cell, master watchman p53 interacts with p68 and p72 RNA helicases [31, 73]. p53 is a prominent tumor suppressor; its function as a transcription factor for apoptosis or cell-cycle arrest is so central that, in order for most cancers to proliferate, the p53 gene needs to be disabled tor the p53 pathway needs to be disrupted [11, 31]. p68 and p72 RNA helicases are required for efficient maturation of a subse t of miRNas [73]. it has been revealed that their association with the Drosha/FGCR8 complex facilitates the conversion of pri- to pre-mRNAs [11, 15]. p53 mutants affect the interaction between p68/72 and Drosha [11, 31, 73].

Enhancing pre-miRNA conversion could be a mechanism by which TGF-Beta superfamily signaling contributes to oncogenesis [11]. Dicer and TRBP process pre-miRNAs into miRNA/miRNA* duplexes. A recent study found that TRBP is a phophoprotein [11]. Phosphorylation of TRBP is dependent on serine/threonin-selective protein kinases belonging to the MAPK/Erk pathway, a mitosis-inducing signal transduction pathway that relays exracellular-activated signals to the nucleus and promotes cell division [75]. MAPK-mediated activation of TRBP enhances pr-miRNA processing and alters mature miRNA composition [11, 75].

Although the available expression profiling data has documented nearly ubiquitous dysregulation of miRNA expression in human, miRNA expression patterns are highly specific for cell-type and cellular differentiation status [6, 11]. It is likely that much of the aberrant miRNA expression observed in tumors is a secondary consequence of the loss of normal cellular identity that accompanies malignant transformation [48]. The case that a miRNA functions as a tumor suppressor or oncogene is strengthened by at least four types of evidence: (1) widespread dysregulation in diverse cancers, (2) gain or loss of miRNA function in tumors due to deletion, amplification or mutation, (3) direct tumor-suppressing or tumor-promoting activity using animal models and (4) the identification and verification of cancer-relevant targets that illuminate mechanisms through which the miRNA participates in oncogenesis [6, 47, 48].

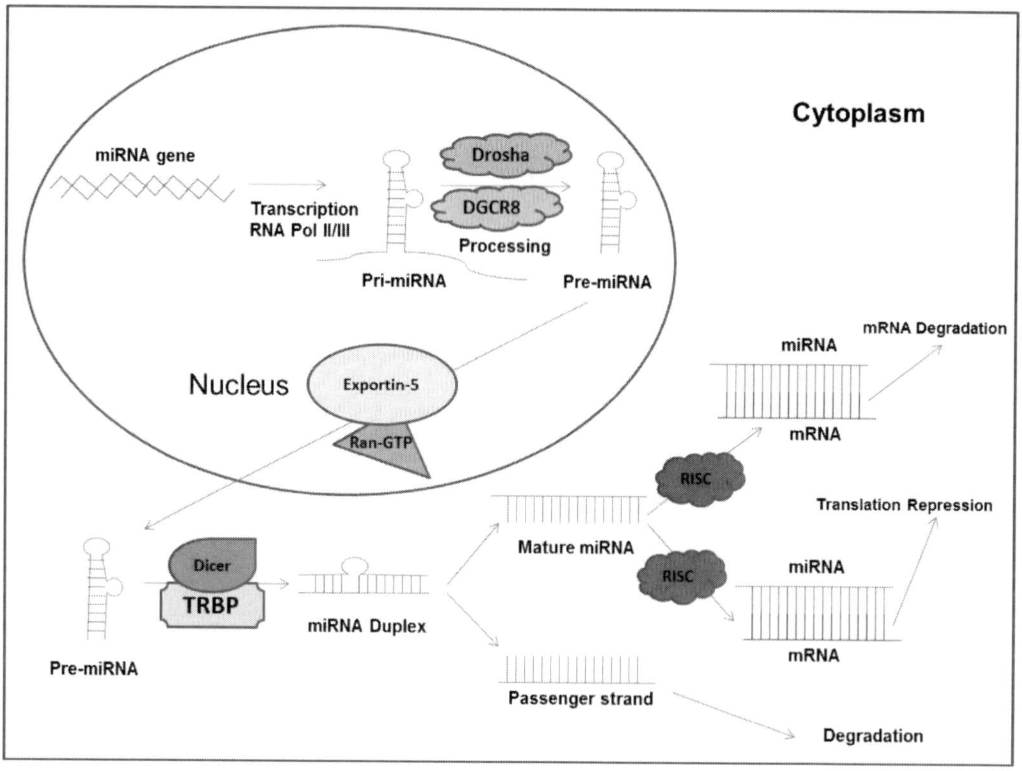

Figure 1 Proposed Model for miRNA Biogenesis: Biogenesis of microRNAs(miRNAs): miRNA genes are transcribed by RNA polymerase II. The pri-miRNA is poly-adenylated. Pri-miRNA molecules contain stem loop structures that are recognized by Drosha/DGCR8 complex. The hairpin structures are trimmed, forming pre-miRNA which is transported to the cytoplasm by Exportin-5/RAN-GTP. Dicer processes the pre-miRNA and one duplex is released from each. The two strands separated by Dicer/TRBP complex and one is incorporated into RISC. The mature miRNA will target mRNAs while the other strand will be degraded.

MicroRNAs as Oncogenes and Tumor Suppressors

A global decrease in miRNA ;levels has been observed in human cancer, indicating that small RNAs may have an intrinsic function in tumor suppression. Four Cancers have shown significant reduction in miRNA expression when a total of 217 human and mouse miRNAs across 334 human cancers, cancer cell lines, and normal tissues [4, 17]. Poorly differentiated tumors had lower miRNA levels compared with more-differentiated tumors [4, 31, 34]. miRNAs can functionto drive terminal differentiation and prevent cell division.

Because the level of miRNAs and their targets are well balanced in healthy cells, altering the level of miRNAs in gene knock-out or over-exprssion may result in disease. Specific miRNAs can either act to bind to and suppress oncogenes, leading to a decreased probability of turmorgenesis, or bind to tumor suppressors, leading to an increased probability of developing cancer [7]. Oncogenes promote cell proliferation and turmorigenesis while turmor suppressor genes express cell division and tumor formation. Different oncogenes ans tumorsuppressor genes are expressed in different tissues [1]. Cancer has five major stages (initiation, promotion, malignant conversion, progression and metastasis) and different genes are important in each stage [19]. As deleterious mutations are accumulated in tumor suppressors, the balance tips towards oncogene activity and cancer development [36].

Aberrant miRNA expression may lead to tumorigenesis in two different manners [7]. First, miRNAs that normally inhibit oncogenes, may be located in a fragile chromosomal region that is deleted within a cell. This deletion would result in a significant decrease in transcription of these miRNAs, leading to a decreased ability to repress their oncogene targets [50]. With a lower abundance of repressive miRNAs, the oncogene transcripts would undergo an increase in translation and protein expression, potentially leading to increased cell proliferation, invasiveness, angiogenesis or decreased levels of apoptosis, ultimately leading to tumor formation [7, 43].

The second way the irregular expression of miRNAs could lead to tumorigenesis, would be by the amplification of a region of a chromosome that encompasses miRNAs involved in negatively regulating the transcript of a known tumor suppressor [44]. Therefore a magnification of this region of the chromosome would result in an increased expression of this particular miRNA or miRNA cluster, leading to a greater translational repression of its tumor suppressor transcript target [7, 50]

Reduced expression of let-7 showed good correlation with shortened postoperative survivals. In let-7 mutants, stem cells can fail to exit the cell cycle and terminally differentiate at the correct time, so that they continue to divide which is an indication of cancer [3]. Increased expression of let-7 in lung adenocarcinoma cell lines suppressed cell proliferation. Tumor suppressor activity of let-7 was elucidated by validating two predicted target genes [1]. miRNA miR-17-92 cluster demonstrates that they can be oncogenes [12]. This cluster contains 14 similar miRNAs and present in three copies in the human genome. The expression of this cluster is strongly up-regulated in lung cancer that is the most aggressive activity of these miRNAs their predicted target are PTEN and RB2, two known tumor suppressor genes [25]. Their over-expression in lung cancer was shown to be c-myc-dependent [72]. The pathology of tumors indicated lower rates of apoptosis as compared to tumors with Myc overexpression alone. The miR-17-92 cluster which is induced by c-Myc, in contrast, inhibit E2F1 expression [32]. This suggests that c-Myc fine-tunes gene expression by activating the transcription of target genes and by simultaneously inducing inhibitory miRNAs that reduced their translation [4, 32].

Mutations in the RAS oncogene are present in approximately 15-30% of all human cancer, and over-expression of the RAS oncogne is common in lung cancer [4]. It has been confirmed that let-7 can inhibit RAS expression in human cancer cell lines [4, 25]. Loss or reduction of let-7 in lung cancer leads to RAS over-expression, thus, promoting cellular growth and contributing to tumorgenesis, thereby acting as a tumor suppressor [28, 34].

Recent studies have identified a miRNA, miR-34 to be regulated by the p53 transcription factor [31]. The p53 protein, regulates the cellular response to stress and cancer-initiating events such as DNA damage. Expression of miR-34 induces cell cycle arrest and thereby acts together with other effectors of the p53 tumor suppressor network to inhibit inappropriate cell proliferation [41, 59]. Another group independently demonstrated that miR-34 is up-regulated by p53 upon DNA damage and promotes apoptosis [4]. Over-expressionof miR-34a induces apoptosis and alters the expression of several genes related to cell cycle progression, apoptosis, DNA repair, and angiogenesis [10]. miR-34 activation recapitulates tumor suppressor events associated with p53 activity and the down regulation of miR-148a, miR-9-2 and miR-34 is part of a unique 20 miRNA metastasis signature in liver tumors [8].

It has been found while studying a region at chromosome 13q14, that is frequently deleted in chronic lymphocytic leukemia (CLL), the region contains two miRNA genes miR-15a and miR 16-1. These two miRNAs derive from the same primary transcript and down regulation of this transcript is associate d with the 13q14 deletion [43]. The proposed tumor suppressor role of miR-15a and miR-16-1 was supported by the discovery of germ-line mutation in two CLL patients that decreased the accumulation othe the two miRNAs [44]. Both miR-15 and miR-16-2 recognize target sites on the 3'UTR of BCL2 an ant-apoptotic oncogene [45]. These miRNAs control the level of BCL2 in normal cells but the expression of the oncogene is increased when the miRNAs are deleted leading to CLL [1, 15, 16, 50]. A germ-line mutation in the miR-16-1 primary transcript seven nucleotides away from the base of the miRNA hairpin was found in two out of 75 CLL patients [43, 44]. It is not clear how a mutation at this position would impact miRNA processing, however the presence of this variant correlated with diminished expression of this miRNA in patient-derived CLL cells and prevented normal miR-16-1 maturation after heterologous expression in cell lines [6, 48].

Given that loss-of-function of miR-15a and miR-16-1 is so common in CLL, and loss of 13Q14 occurs frequently in other malignancies including mantle cell lymphoma, multiple myeloma and prostate cancer, these miRNAs are likely to be tumor suppressor genes [6]. The tumor suppressing properties of these miRNAs appears to be mediated at least in part through their ability to down regulate the anti-apoptotic protein Bcl2 [6, 48]

Data suggest that aberrant expression of miR-21 may promote tumorigenesis by inhibiting apoptosis [6]. Over-expression of this miRNA in diverse tumors including those derived from breast, colon, lung, pancreas, stomach and prostat [6, 30]. miR-31 is another example of an miRNA with oncogenic properties (termed an oncomir) in lung cancer [9, 37]. Data showed that miR-31 functions as an oncomir by directly repressing the tumor/suppressors LATS2 and PPP2R2A [9, 59]

One study used quantitative reverse transcription PCR (qRT-PCR) to determine the levels of a restricted range of 156 miRNAs in adult gonadal germ cell tumors (GCT), compared with three cases of normal testis [38, 39]. This work suggested that the miR-371-373 gene cluster is highly expressed in adult malignant GCTs [5]. Data indicates that the mir-371-372 and miR-302 clusters, of which six members (miR-372-373 and miR-302a-302d) share the identical 2-7 nucleotide seed region AAGUGC, have a fundamental role in the pathogenesis of malignant GCTs by down-regulating functionally significant target genes [5, 40].

Oncogenic activation of Ras in primary cells induces an irreversible growth arrest termed cellular senescence [6]. Loss of p53, allows cancer cells to overcome this response and continue proliferating. To identify miRNAs that when over-expressed could substitute for p53 loss and allow continued proliferation in the context of Ras activation [28]. In addition to blocking Ras-induced cellular senescence, these miRNAs were shown to potentiate Ras-mediated cellular transformation [28]. These miRNAs do not function by simply down-regulating p53 as miR-372 and miR-373 –expressing cells show normal induction of p53 and its downstream target p21 in response to Ras activation [23]. Thus, these miRNAs render cells insensitive to the effects of a normally functioning p53 allele [31]. Based on these observations, the authors hypothesized that tumors which are associated with normal p53 function might show pathologic elevation of miR-372 and miR-373 [40]. miR-372 and miR-373 are two additional examples of oncogenic miRNAs that are shown to promote cell proliferation and tumor development by neutralizing p53-mediated CDK inhibition, possibly through direct targeting of the tumor suppressor gene LATS2 [10, 54]. Testicular germ cell tumors, which rarely exhibit p53 loss-of-function, frequently exhibit miR-372/373 over-expression [6].

miRNA profiling showed that three miRNAs (miR-221, miR-222 and miR-146) accumulate at a much higher level in thyroid tumors than in matching healthy tissues [1]. miR-143 and miR-145 identified in colon tissue, showed reduced expression in different stages of colorectal neoplasia [1]. Accumulation of the mature miRNAs suggested, that the altered level of mature miRNAs is due to differential processing in normal and cancer tissues [1]. Gliobastoma multiforme (GBM) is the most common form of brain cancer in human but its development [62]. GBM samples show higher accumulation of miR221 and lower accumulation of miR-181 than in normal tissues [62]. The Apo21/tumor necrosis factor (TNF)-related apoptosis-inducing ligand (TRAIL) is a member of the TNF family known to induce apoptosis in a variety of cancers. Garofalo et al showed that miR-221 and miR-222 contribute to lung cancer resistance to TRAIL therapy by down-regulating PTEN and TIMP3 tumor suppressors [9, 58, 60].

The BIC gene has been linked to several types of lymphomas. It does not encode for a protein and it was considered as a non-coding RNA with oncogene activity [1, 14]. A study of miR-155 accumulation in B-cell derived lymphomas, which found a consistently high level of miR-155 expression [1]. miRNA profile of clinical samples (76 neoplastic and 10 normal breast tissues) was established using microarrays containing probes against all known miRNAs [46]. Several showed lower accumulation in cancer samples miR-125b, miR-145 miR-21 and miR-155 [1, 53, 55]. Decreased expression of miR-143and miR-145 has been demonstrated to be a frequent feature of colorectal tumors [6]. Down regulation of miR-145 is also very common in breast carcinomas and breast cancer cell lines [6, 26].

It is now known that BIC is a pre-miRNA and miR-155 is located in the final exon of this transcript [6]. Therefore, BIC/miR-155 was the first miRNA transcript definitely shown to possess tumor-promoting activity [45]. Over-expression of BIC/miR-155 has consistently been observed in diverse tumor types including B-cell lymphoma, and breast, lung, colon and thyroid cancers [46]. Available evidence strongly implicates BIC/miR-155 as an oncogene.

The mechanisms through with BIC/miR-155 promotes tumorigenesis remain unknown but it has been demonstrated that the angiotensin II type I receptor is a miR-155 target [6, 48].

It is indicated that DNA methylation-associated silencing of tumor suppressor miRNAs contributes to the development of human cancer metastasis [8]. Metastasis is responsible for 90% of death in patients with solid tumors. The metastatic process consists of sequential, interrelated steps by which primary tumor cells acquire the capacity to invade the adjacent tissue, enter the systemic circulation, translocate through the vasculature, arrest in distant capillaries, move into the surrounding tissue parenchyma, and finally proliferate from micro-metastasis into macroscopic secondary tumors [54]. Three metastasis-promoting miRNAs have been described; miR-10b promotes cell migration and invasion through homeobox D10 inhibition, whereas miR-373 and miR520c [8]. MiRNAs like miR-126 and miR-335 can act as metastasis suppressor, the latter by targeting oncogene SOX4 and tenacin C [8].

The transcription of a miRNA gene, mir-124a, was shown to be inactivated by hypermethylation of its promoter in various human tumors [32]. The miRNA gene mir-127 is usually expressed in normal cells but not in cancer cells [63]. It was highly induced in cultured human cancer cells after treatment with demethylating drugs, suggesting that it is subject to epigenetic silencing through promoter hypermethlylation [4]. There is evidence that miRNAs are regulated indirectly through control of their processing enzymes. Down regulation of miRNAs in human cancer was not associated with reduced levels of the primary miRNA transcripts [4, 74].

miR-148, miR-34b/c and miR-9 were found to undergo specific hypermethylation-associated silencing in cancer cells compared with normal tissues [8, 57, 60]. The reintroduction of miR-148 and miR-34 b/c in cancer cells with epigenetic inactivation inhibited their motility, reduced tumor growth, and inhibited metastasis formation in xenograft models, with an associated down regulation of the miRNA oncogenic target genes, such as C-MYC, E2F3, CDK6 and TGIF2 [8, 55].Several recent studies have uncovered a relationship between the EGFR signaling pathway and miRNAs. miR-128b is a direct regulator of EGFR in which loss-of-heterozygosity is frequently found [61]. MiR-7, an miRNA frequently downregulated in lung cancer, has been shown to suppress EGFR and Raft mRNA expression [9].

Signalling

MicroRNAs are critical post-transcriptional regulators that may collectively control a majority of animal genes [2]. The major signal transduction cascades that control animal development are highly dose-sensitive and frequently altered in human disorders [22]. There is evidence that miRNA targeting of the major signaling pathways and implications for haw aberrant miRNA activity might underlie human disease and cancer [2]. The miRNA-mediated regulatory network has fueled great speculation about the potential relevance of miRNAs to human disease [1].

The précise level of transcriptional output of cell signaling pathways such as Notch (N), TGFBeta, Receptor tyrosine kinase (RTK), nuclear receptor, jak/STAT, and Hippo, is essential for normal cell specification and tissue growth (21). The fundamental control of cell division, growth and apoptosis by core cell signaling pathways makes many of their components potent oncogenes and tumor suppressors [2, 24]. miRNA targeting of transduction-limiting repressors could increase pathway activity or they might repress signal-regulated events by targeting positively-acting pathway components [2].

In worms, miR-61 is a direct transcriptional target of lin-12/Notch, and proposed to function as a Notch effector that inhibits EGFR/MAPK signaling via the vav-1 target gene [2, 21]. It was recently shown tht eendogenous miR-14 regulates nuclear receptor signaling by the steroid hormone Ecdysone, where it limits the expression o f the Ecdysone Receptor [2]. miR-14 was isolated from Drosophila as a putative cell death-related gene [2]. Two additional Drosophila miRNAs, bantam and miR-278, have both anti-apoptotic and growth promoting activity, dual functions that suggest oncogene-like function [25]. miR-278 similarly inhibits apoptisis and promotes growth, although in this case its loss-of function phenotype is manifest in energy homeostasis [1, 2, 25]. The graded expression of these miRNAs in Xenopus due to TGFbeta signaling may be a consequence of repression by dorsal Wnt signaling [2]. It was proposed that miR-15/16 functions at the crossroads for these signaling pathways [2].

miRNAs are integrated into vast regulatory networks that affect biological events. miRNAs represent a layer above transcriptional control for both fine-tuning and dramatically altering the activity and output of cell signaling [22]. miRNAs may also serve as points of crosstalk between signaling pathways by integrating transcriptional inputs or by their functional regulatory output on different pathways [2, 75].

MICRORNAS AND CANCER

A cancerous cell is one in which the normal machinery malfunctions, potentially resulting in an increase in proliferation, evasion of cell death, and insensitivity to anti-growth signals, among other possible consequences [7]. Molecular signature of miRNAs in different malignancies suggests that these are not only actively involved in the pathogenesis of human cancer(3). Oncogenes promote cell proliferation and turmorigenesis while turmor suppressor genes express cell division and tumor formation [12]. Different oncogenes and tumorsuppressor genes are expressed in different tissues [15]. Cancer has five major stages (initiation, promotion, malignant conversion, progression and metastasis) and different genes are important in each stage [1]. As deleterious mutations are accumulated in tumor suppressors either by aging, because of carcinogens, from the envioronment or through genetics, the balance tips towards oncogene activity and cancer development [1, 17]. If the miRNAs are associated with the same cancer type with similar dysregulation patterns, then their target genes may share common cellular and functional characteristics and the miRNAs may have common target onco/tumor suppressor genes [3]. About 50% of miRNA genes are localized in cancer associated genomic regions (CAGR) or fragile sites. MiRNAs located near fragile sties could be possible targets of genomic alterations [3]. In various cancer malignancies approximately 200 miRNAs have been identified which are dysregulated significantly [3].

Malignant tumourigenesos causes cell death and enhances cell division such as the ability to spread. Changes in the interaction between tumor and surrounding tissues that allow sustained angiogensis and metastasis allows for this phenomenon [52]. The differential expression of seversl miRNAs between health and cancer samples was in line with the predicted function of these miRNAs [1, 51].

Germ cell tumors (GCT) are clinically and pathologically complex neoplasms that occur from the neonatal period up to lat adulthood [35]. The germ cell theory of tumorigenesis states that all GCTs arise from totipotent primodial germ cells [42]. Mi RNAs regulate gene expression via translational repression and mRNA destabilization [5].

In recent years a number of miRNAs are reported to be involved in various type of leukemia [3, 37]. The five most highly expressed miRNAs in ALL reported in are miR-128b, miR-204, miR-218, miR-331, and miR-181b-1 [3, 42]. In CLL, the five most highly expressed miRNAs are miR-331, miR-29a, miR-195, miR-34a, and miR-29c [3, 49]. It has been shown that miR-181a family miRNAs are expressed at relatively low levels in acute myeloid leukemia(AML) [3, 29, 46]. In another experiment, it was revealed that miR-23b is repressed and miR-221, miR-222 and miR-34a are significantly up-regulate in AML specimens compared to normal bone marrow and CD34+ haematopoietic progenitor cell [3, 48].

Prostate cancer is the most common form of cancer among men [3]. Current diagnostic test for prostate cancer is the prostate specific antigen (PSA) test, which suffers from high false positive rate. Recently researchers have begun analyzing the miRNA expression profiles of normal and tumor samples for the diagnosis of prostate cancer [45]. MiR-21 is over-expressed in various types of cancer indicating that miR-21 is involved in tumor growth, invasion and metastasis [62]. MiR-21 could promote apoptosis resistance, motility and invasion in prostate cancer cells and these events of miR-21 may be partly due to its regulation of target genes [3].

The expression of miR-15 and miR-16 was analyzed in blood samples from patients with CLL. Both miRNAs were absent or downregulated in the majority (68%) of cases when compared to normal tissue or lymphocytes [4]. Specific miRNAs may have a role beyond the tumor initiating event and directly participate in tumor progression and metastasis [4].

In order to block miRNA functions, the use of synthetic mRNA containing multiple pairing sites for endogenous miRNA may be employed [10]. Such synthetic mRNAs, referred as miRNA sponges, function by sucking up the target miRNAs and thus blocking their regulatory function in the cell]66[. Recent studies have also employed an approach of using target protectors (miR-masks). miR-masks are like AMOs, however, they bind to the pairing site of miRNAs in the 3' UTR of the target mRNA by fully complementary mechanism [71]. These studies suggest that silencing or restoration of miRNA function can serve as a promising therapeutic strategy for cancer treatment [10].

MicrRNAs as Diagnostics and Therapeutics

Cancer cells have different miRNA profiles compared with normal cells, which strongly suggest that miRNA profiles could be used for diagnosis. It has been established that the miRNA profiles of 17 poorly differentiated tumors and 12 samples were correctly diagnosed based on miRNA signatures [1]. miRNAs may act as potential diagnostic marker of specific cancer type. In some cases miR-221 is only over-expressed in papillary thyroid tumor tissue and is over-expressed in normal thyroid tissue which is adjacent to papillary tyroid carcinoma, suggesting that this might be an important tyroid cancer marker [3].

It has been hypothesized that miRNAs have a correlation with patient's survival [19]. The expression level of Dicer is reduced in a fraction of lung cancers and have a prognostic impact of the survival of the surgically treated patients [3]. Eight miRNAs (miR-155, miR-17-3p, miR-106a, miR-93, let—7a-2, miR-145, let-7b and miR-21) were related to the adenocarcinoma patients survival [27]. Patients with high expression of either miR-155, miR-17-3p, miR-106a, miR-93 or miR-21 and low expression of either let-7a-2, let-7bor miR-145 were found to have a significantly worse prognosis [3, 28].

The most intuitive choice of molecules to correct altered miRNA-messenger RNA interactions are RNA olighnucleotides. Modified antisense oligonucleotides are already being developed to utilize the intrinsic RNAi pathway for delivery of gene therapy. If the delivery problem can be overcome, then miRNA therapies may also be possible [4].

Anti-miRNA oligonucleotides (AMOs) can be used to suppress miRNA activity if the AMOs can bind strongly to the miRNA and are stable enough in physiological conditions [64]. 2'-Omethylation of RNA nucleotides and locked nucleic acid (LNA) DNA nucleotides are modification to miRNAs to attain strong binding [65]. These modifications are also used in the opposite approach were modified oligonuceotides are delivered into cells that under-express miRNAs with tumour suppressor activity [69]. miRNAs are now known to regulate pathways controlled by classic tumor suppressors and oncogenes including p53, MYC and RAS [6]. These findings emphasize the need to integrate the study of miRNA expression and function into ongoing molecular analyses of cancer pathogenesis in order to achieve a complete understanding of the disorders [48, 56]. Technologies that allow the systemic delivery of small RNA mimics or inhibitors to whole animals, hold great promise for the development of therapeutic strategies based on miRNA overexpression or inhibition [6, 44].

Synthetic miRA mimics have also been used to restore miRNA function within the tumor cell [10]. For oncogenic miRNAs, use of anti-miRA oligonuleotides (AMOs), small molecule inhibitors, miRNA sponges and miRNA masking, have been utilized or to achieve their downregulation [10]. AMOs can block the interactionsbetween miRNA andits targe mRNAs through competitive inhibion of base pairing [10].

Some recent studies have also employed an approach of using target protectors (miR-masks) [10]. miR-masks are like AMOs, however, they do not directly interact with its target miRNA but instead bind to the pairing site of miRNAs in the 3' UTR of the target mRNA by fully complementary mechanism [71]. These studies suggest that silencing or restoration of miRNA function can serve as a promising therapeutic strategy for cancer treatment. [10].

CONCLUSION

It is evident that microRNAs play a significant role in regulation of human diseases such as cancer. Experimental and computational data has indicated that within the cell, microRNAs may function normally as tumor suppressors and/or abnormally as oncogenes. They are able to carry out these functions by binding to mRNAs to affect their post-transcriptional or pre-translational activities. Finally miRNAs may be targeted for means of cancer prevention or therapeutic intervention, possibly by using antagomirs/miRNA mimics to manipulate the levels of particular miRNAs [7]. Synthetic miRA mimics have also been used to restore miRNA function within the tumor cell [10]. For oncogenic miRNAs, use of anti-miRA oligonuleotides (AMOs), small molecule inhibitors, miRNA sponges and miRNA masking, have been utilized or to achieve their downregulation [64]. AMOs can block the interactions between miRNA andits targe mRNAs through competitive inhibion of base pairing [10].

ACKNOWLEDGMENTS

Thanks to Dr. Manoj Mishra for an invitation to contribute this chapter on gene regulation in cancer. Thanks to Mr. Jerald Palmer for assisting with the design of the proposed model of MicroRNA Biogenesis.

REFERENCES

[1] Dalmay, T. MicroRNAs and cancer. *Journal of Internal Me*dicine 2008; 263: 366-375.

[2] Hagan, J.W. and Lai, E. C. microRNA control of cell-cell signaling during development and disease. *Cell Cycle,* 2008 August; 7(15): 2327-2332.

[3] Bandyopadhyay, S., Mitra, R. Maulik, U. and Zhang, M. Development of the human cancer microRNA network. Silence: *A Journal of RNA Regulation,* 2010, 1:6.

[4] Sassen, S., Miska, E., and Caldas, C. MicroRNA-implications for cancer. *Virchows Arch*. 2008;452: 1-10.

[5] Palmer, R.D., Murray, M. J., Saini, H.K., et al.Malignant Germ Cell Tumors Display Common MicroRNA Profiles Resulting in Global Changes in Expression of Messenger RNA Targets. *Cancer Research,* 2010;70:2911-2923.

[6] Kent, O.A. and Mendell, J.T. A small piece in the cancer puzzle: microRNAs as tumor suppressors and oncogenes. *Oncogene* (2006) 25, 6188-6196.

[7] Armstrong, L. *Micro RNA involvement in tumorigenesis*. AbCam 1998-2011. http://www.abcam.com/index.html?pageconfig=resourceandrid=1 2971andpid=10628.

[8] Lujambio, A., Calin, G., Villanueva, A., Ropero, S., Sanchez-Cespedes, M.., Blanco, D., Montuenga, L., Rossi, S., Nicoloso, M., Faller, W., Gallagher, W., Eccles, S., Croce, C., and Esteller, M. A microRNA DNA methylation signature for human cancer metastasis. *PNAS*, 2008 September;105:36:13556-13561.

[9] Lin, P.Y. and Yang, P.C. Minireview:MicroRNA in lung cancer. *British Journal of Cancer*, 2010;103,1144-1148.

[10] Bhardwaj, A., Singh, S., and Singh, A. MicroRNA-based Cancer Therapeutics: Big Hope from Small RNAs. *Molecular Cell Pharmacology*, 2010;2(5):213-219.

[11] Kwak, P., Shintaro, I., and Tomari, Y. The microRNA pathway and cancer. *Cancer Science*, 2010 November;101:11:2309-2315.

[12] Lee RC, Feinbaum RL, Ambros V. the C. elegans heterochronic gene lin-4 encodes small RNAs with antisense complementarity to lin-4. *Cell* 1993;75: 843-54.

[13] Fire A, Xu s, Montgomery MK, Kostas SA, Driver, Mel CC. Potent and specific genetic interference by double-stranded RNA in C. elegans. *Nature* 1998; 391:806-11.

[14] Lee Y, Kim M, Han JJ et al. MicroRNA genes are transcribed by RNA polymerase II *EMBO J.* 2004; 23: 4051-60.

[15] Lee Y. Jeon K, Lee JT, Kim S, Kim VN. MicroRNA maturation: stepwise Processing and subcellular localization. *EMBO J.* 2002; 21: 4663-70.

[16] Lee Y, Ahn C, Han J et al. The nuclear RNase III Drosha initiates microRNA processing. *Nature* 2003; 425: 415-9.

[17] Lim LP, La NC, Garrett-Egele P et al. Microarray analysis shows that some microRNAs down regulate large numbers of target mRNAs. *Nature* 2005; 433: 769-73.

[18] Lewis BP, Burge CB, Bartel DP (2005) Conserved seed pairing, often flanked by adenosines, indicates that thousands of human genes are microRNA, targets. *Cell* 120:15-20.

[19] Volina S, Calin GA, Liu C, Ambs S, Climmino A, Pruett R, Yanaihara N, Lanza G, Scarpa A, Vecchione A et al. A microRNA expression signature of human solid tumors defines cancer gene targets. *Proc. Natl. Acad. Sci. USA* 2006, 103:2257-2261.

[20] Griffiths-Jones S, Grocock RJ, van Dongen S., Bateman A, Enright AJ (2006) miRBase: microRNA sequences, targets and gene nomenclature. *Nucleisc. Acids Res.* 34:DD140-D144.

[21] Lai EC, Tam B, Rubin GM. Pervasive regulation of Drosophila Notch target genes by GY=box-. Brd-box and K-box-class microRNAs. *Genes Dev.* 2005;19:1067-80.

[22] Chen K, Rajewsky N. he evolution of gene regulation by transcription factors and microRNAs. *Nat. Rev. Genet.* 2007;8:93-103.

[23] Voorhoeve PM, le Sage C, Schrier M, Gillis AJ, Stoop H, Nagel R, Liu YP, van Duijse J, et al. A genetic screen implicates miRNA-372 and miRNA-373 as oncogenes in testicular germ cell tumors. *Cell* 2006;124169-81.

[24] Flynt AS, Li, Thatcher EJ, Solnica-Krezel L, Patton JG. Zebrafish miR-214 modulates Hedgehog signaling to specify muscle cell fate. *Nat. Genet.* 2007;39:259-6.

[25] Nolo R, Morrison CM, Tao C, Zhang X, halder G. The bantam microRNA is a target of the hippo tumor-suppressor pathway. *Curr. Biol.* 2006;16:1895-904.

[26] Lorio MV Fercn M, Liu C Broese A, Spizo R, Sabbioni S, et al. micro RNA gene expression deregulation in human breast cancer. *Cancer Research* 2005,65:7065-7070.

[27] Yanahara N, Caplen N, Bowman E, Seike m, Kumamoto K, Yi M, Stephens RM, et al. Unique microRNA molecular profiles in lung cancer diagnosis and prognosis. *Cancer Cell* 2006, 9:189-198.

[28] Johnson SM, Grosshans H, Shingara J, Byrom M, Jarvis R, Cheng A, Labourier E, Reinert KL, Brown D, and Slack FJ. RAS is regulated by the let-7 microRNA family. *Cell* 2005¸120:635-647.

[29] Marcucci G, Radmacher M, maharry K, Mrzek K, Ruppert A, et al. MicroRNA expression in cytogenetically normal acute myeloid leukemia. *The New England Journal of Medicine* 2008, 358:1919-1928.

[30] Li T, Li D, Sha J, Sun P, Huang Y. MicroRNA-21 directly targets MARCKS and promotes apoptosis resistance and invasion in prostate cancer cells. *Biochem. Biophys. Res. Commun.* 2009: 1171-1182.

[31] Chang TC, Wentzel EA, Kent OA, Ramachandran K, Mullendore M, et al. (2007) Transactivation of miR-34a by p53 broadly influences gene expression and promotes apoptosis. *Mol. Cell* 26:745-752.

[32] Brennecke J, Stark A, Russell RB, Cohen SM (2005) Principles of microRNA-target recognition. *PLoS Biol.* 3:e85.

[33] Lagos-Quintana M, Rauhut R, lendeckel W, Tuschl T (2001) Identification of novel genes coding fro small expressed RNAs. *Science* 294:853-858.

[34] Lu J, Getz G, Miska EA, Alvarex-Saavedra E, Lamb J, Peck D, et al. (2005) MicroRNA expression profiles classify human cancers. *Nature* 435:834-838.

[35] Palmer RD, Nicholson JC, Hale JP. Mangement of germ cell tumors in childhood. *Curr. Paediatr.* 2003; 13:213-20.

[36] Zhang B, Pan X, Cobb GP, Anderson TA. microRNAs as oncogenes and tumor suppressors. *Dev. Biol.* 2007;302:1-12.

[37] Esquela-Kersher A, Slack FJ. Oncomirs-microRNAs with a role in cancer. *Nat. Rev. Cancer* 2006;6:259-69.

[38] Kent, OA, Mendell JT. A small piece in the cancer puzzle: microRNAs as tumor suppressors and oncogenes. *Oncogene* 2008;25:6188-96.

[39] Gillis AJ, Stoop HJ, Hersmus R, et al. High-through-put microRNAome analysis in human germ cell line tumors. *J. Pathol.* 2007;213:319-28.

[40] Voorhoweve PM, le Sage C, Shrier M, et al A genetic screen implicates miRNA-372 and microRNA 373 as oncogenes in testicular germ cell tumors. *Cell* 2006;124:169-81.

[41] Wang Y, Baskerville S, Shenoy A, Babiarz JE, Baehner L, Blelloch R. Embryonic stem cell-specific microRNAs regulate the G1-S transition and promote rapid proliferation. *Nat. Genet.* 2008;40:1478-83.

[42] Wang Y, Blelloch R. Cell cycle regulation by microRNAs in embryonic stem cells. *Cancer Res.* 2009;69:4093-6.

[43] Calin GA, Dumitru CD, Shimizu M, Bichi R, Zupo S, Noch E et al. Frequent deletions and down-regulation of microRNA genes miR-15 and miR-16 at 13q14 in chronic lymphocytic leukemia. *Proc. Natl. Acad. Sci. USA* 2002;99:15524-15529.

[44] Calin GA, Liu G, Sevignani C, Ferracin M, Felli N, Dumitru CD et al. MicroRNA profiling reveals distinct signatures in B cell chronic lymophocytic leukemias. *Proc. Natl. Acad. Sci. USA* 2004;101:11755-11760.

[45] Dong JT, Boyd JC, Fierson JR HF. Loss of heterozygosity at 13q14 in high grade, *high stage prostate cancer.* 2001;49:166-171.

[46] Eis PS, Tam W un L, Chadburn A, Li Z, Gomez MF et al. Accumulation of miR-155 and BIC RNA in human B cell lymphomas. *Proc. Nat. Acad. Sci. USA* 2005;102:3627-3632.

[47] Krutzfeldt J, Rajewsky N, Braich R, Rajeev KG, Tuschl T, Manoharan M et al. Silencing of microRNAs in vivo with 'antagomirs' 2005; *Nature* 438:685-689.

[48] Volinia S, Calin GA, Liu CG, Ambs S, Cimmino A, Petrocca F et al. A microRNA expression signature of human solid tumors defines cancer gene targets. *Proc. Natl. Acad. Sci. USA* 2006;103:2257-2261.

[49] Calin GA, Ferracin M, CinoA, I Leva G, Shimiu M, Wojcik SE, Iorio MV, Visone R, et al. A microRNA signature associated with prognosis and progression in chronic lymphocytic leukemia. *New England Journal of Medicine* 2005;353(17):1793-801.

[50] Rassenti L, Alder H, Bolinia S, Liu CG, Kipps TJ, Negrini M, Croce CM. miR-15 and miR-16 induce apoptosis by targeting BCL2. *Proc. Natl. Acad. Sci. USA* 2005;102(39):1394-409.

[51] Lu J, Getz G, Miska EA, Alvarez-Saavedra E, Lamb J, Peck D, Sweet-Cordero A, et al. MicroRNA expression profiles classify human cancers. *Nature* 2005;435(7043):834-8.

[52] Ma L, Teruya-Feldstein J, Weinberg RA. Tumor invasion and metastasis initiated bymicroRNA-10 in breast cancer. *Nature* 2007;49:682-688.

[53] Scott GK, Goga A, Bhaumik D, Berger CE, Sullivan CS, Benz CC. Coordinate suppression of ERBB2 and ERBB3 by enforced expression of microRNA miR-125a or miR-125b *J. Biol. Chem.* 2007;282:1479-1486.

[54] Huang Q et al. The microRNAs miR-373 and miR-502c promote tumor invasion and metastasis. *Nat. Cell Biol.* 2008;10:202-210.

[55] Grupta GP, Massague J. Cancer metastasis: Building a framework. *Cell* 2006;127:679-695.

[56] Saito Y, et al. Specific activation of microRNA-127 with down-regulation of the protooncogene BCL6 by chromatin-modifying drugs in human cancer cells. *Cancer Cell* 2006;9:435-443.

[57] Lehmann U, et al. Epigenetic inactivation of microRNA gene has-mir-9-1 in human breast cancer. *J. Pathol.* 2008;214:17-24.

[58] Garofalo M, Di Leva G, Romano G, Nuovo G, et al. miR-221 and 222 regulate TRAIL resistance and enhance tumorigenicity through PTEN and TIMP3 down-regulation. *Cancer Cell* 2009;16:498-509.

[59] Liu X, Sempere LF, Ouyang H, Memoli VA, Andrew AS, et al. MicroRNA-31 functions as an oncogenic microRNA in mouse and human lung cancer cells by repressing specific tumor suppressors. *J. Clin. Invest.* 2010;120:1298-1309.

[60] Schaefer U, Boloshanenko O, Willen D, Walczak H. TRAIL: a multifunctional cytokine. *Front Biosc*i. 2007;12:3813-3824.

[61] Weiss GJ, Bemis LT, Nakajima E, Sugita M, et al. EGFR regulation by microRNA in lung cancer: correlation with clinical response and survival to gefitinib and EGFR expression in cell lines. *Ann. Oncol.* 2008;19:1053-1059.

[62] Chan JA, Krichevsky AM, Kosik KS. MicroRNA-21 is an anti-apoptotic factor in human glioblastoma cells. *Cancer Res.* 2005;65:6029-33.

[63] Kota J, Chivukula RR, O;onnell KA, et al. therapeutic microRNA delivery suppresses tumorigenesis in a murine liver cancer model. *Cell* 2009;137:1005-17.

[64] Weiler J, Hunziker J, Hall J. Anti-miRNA oligonucleotides (AMOs): ammunition to target miRNAs implicated in human disease? *Gene. Ther.* 2006;13:496-502.

[65] Ebert MS, Neilson JR, Sharp PA. MicroRNA sponges: competitive inhibitors of small RNAs in mammalian cells. *Nat. Methods* 2007;4:721-6.

[66] Jeong SH, Wu HPark WY. LIN28B confers radio-resistance through the posttranscriptional control of KRAS. *Exp. Mol. Med.* 2009;41:912-8.

[67] Kawamata T, Tomari Y. Making RISC. *Trends Biochem. Sci.* 2010;35:368-76.

[68] Iki T, Yoshikawa M, Nshikiori M et al. In vitro assembly of plant RNA-induced silencing complexes facilitated by molecular chaperone HSP90. *Mol. Cell* 2010;39:282-291.

[69] Krek A, Grun D, Poy MN et al. Combinatorial microRNA target predictions. *Nat. Genet.* 2005;37:495-500.

[70] Pii RS, Artus CG, Filipowicz W. Tethering of human Ago proteins to mRNA mimics the miRNA mediated repression of protein synthesis. *RNA* 2004;10:1518-25.

[71] Humphreys T, Westman BJ, Martin DI, Preiss T. MicroRNAs control transition initation by inhibiting eukaryotic initiation factor 4E/cap and poly(A) tail function. *Proc. Natl. Acad. Sci. USA* 2005;102: 16961-6.

[72] Pillai RS, Bhattacharyya SN, Artus CG et al. Inhibition of translational initiation by let-7. MicroRNA in human cells. *Science* 2005;447:875-8.

[73] Bates GJ, Nicol SM, Wilson BJ et al. The DEAD box protein p68: a novel transcriptional coactivator of the tumor suppressor. *EMBO J.* 2005;24:543-53.

[74] Liu J, Valenca-Sanchez MA, Hannon GJ, Parker R. MicroRNA-dependent localization of targeted mRNAs to mammalian P-bodies. *Nat. Cell Biol.* 2005;7:719-23.

[75] Paroo Z, Ye X, Chen S, Liu Q. Phosphorylation of the human microRNA-generating complex mediates MAPK/Erk signaling. *Cell* 2009;139:112-22.

In: Signaling, Gene Regulation and Cancer
Editors: Shree Ram Singh and Manoj K. Mishra
ISBN: 978-1-61942-088-5

Chapter 18

Histone Posttranslational Modifications and Cancer Epigenomics

Chhavi Chauhan*
Mouse Cancer Genetics Program, National Cancer Institute (NCI)
National Institutes of Health, Frederick, MD, US

Abstract

The genetic material in each cell is stored as the deoxy ribonucleic acid (DNA), which is further bound by histone proteins and compacted into a structure called the chromatin. These histone protein tails can undergo covalent modifications either strengthening or weakening the DNA histone contacts, usually leading to gene repression and active transcription, respectively. These modifications are referred to as epigenetic modifications, the changes that do not affect DNA sequence but cause stable heritable phenotypes by chromatin modification. These epigenetic changes depend on both the internal and the external cellular milieu and are required for a cell to maintain its identity. In this chapter, I have attempted to discuss histone posttranslational modifications (PTM) and the factors that might affect/ promote these modifications. There have been many reports in the recent past implicating epigenetic mis-regulation of key genes, resulting in cancer. The later half of the chapter, therefore, focuses on the relevance of these PTMs in context of a complex human disease like cancer.

Abbreviations

ADP	Adenosine diphosphate
APC	Adenomatosis polyposis coli
ChIP	Chromatin Immunoprecipitation
DNA	Deoxy ribonucleic acid
H	Histone
HATs	Histone acetyltransferases

* E-mail: chhavi.chauhan@nih.gov.

HDACs	Histone deacetylases
HP1	Heterochromatin Protein 1
H3K4me3	Histone3 lysine 4 trimethyl mark (and so on)
MLL	Mixed lineage leukemia
NCI	National Cancer Institute
NETs	Neuroendocrine tumors
NIH	National Institutes of Health
PTM	Posttranslational modification
Rb	Retinoblastoma
RNA	Ribonucleic acid
VHL	Von Hippel-Lindau

1. Introduction

Despite the fact that the deoxy ribonucleic acid (DNA) in every single cell in our body is the same, each cell performs unique functions. This ability of the cells to perform diverse yet exclusive functions is possible due to the epigenetic information stored in each cell.

Epigenetic modifications refer to stable heritable phenotypes that result from changes in the chromosome without alterations in the DNA sequence. Either this epigenetic modification can affect the DNA or the proteins that are directly associated with the DNA, like the histone proteins. Two most commonly studied epigenetic modifications include the 1) Methylation of DNA and 2) Covalent modification of the histones that are associated with the DNA. In this chapter, I will discuss the covalent modifications of histones as well as how these modifications epigenetically control and regulate gene expression.

1.1. Chromatin, Nucleosomes and Histones

In most eukaryotic cells, in order to fit the DNA into the nucleus, it is compacted with histone proteins to give rise to a structure called the chromatin. There are two categories of histone proteins: the highly conserved core histone proteins that include Histone 2-4 (H2-4) and the less conserved linker histone proteins that include H1 and H5. The dimers of core histone proteins H2A, H2B, H3 and H4 assemble into an octamer with approximately 147 base pairs of DNA wound around this octamer to form a structure called the nucleosome. The DNA that connects two nucleosomes is called the linker DNA. The histone protein associated with this linker DNA is called the linker histone, H1/5. The linker histone proteins are not only less conserved evolutionarily but also often have more than one form within a species.

The DNA-histone interactions make the chromatin highly compact and stable, limiting the accessibility of DNA to various other DNA binding proteins. In order to allow transcription to happen, the DNA-histone interactions must be modified. Therefore, cells regulate the expression of specific genes in a tissue specific manner by controlling the DNA-histone interaction, which in turn dictates DNA-DNA binding protein interactions. The DNA-histone interactions are governed by the covalent modifications of the various lysine residues in the histone tails as discussed in section 1.2.

1.2. Histones Regulate DNA Transcription by Undergoing Posttranslational Modifications (PTMs)

The four core histones: H2A, H2B, H3 and H4 can undergo a range of posttranslational modifications (PTMs). These PTMs include acetylation, methylation, citrullination, phosphorylation, SUMOylation, O-GlcNac modification, ubiquitination and adenosine diphosphate (ADP) ribosylation [1-3]. Although, a lot is known about transcription-induced changes to core histones, not much is known about how linker histones are affected. In fact, these histone modifications have led to the postulation of the "Histone Code Hypothesis" which states "PTMs function as a signal platform to recruit effecter modules to local chromatin, and its effectors/ readers that ultimately determine the functional outcome of certain PTMs" [1, 4].

These PTMs directly influence the chromatin structure by altering the DNA-histone association to allow for the recruitment of other DNA binding proteins that regulate gene transcription [5]. For example, the acetylation of the lysine residues can reduce the positive charge of the histones. Since the DNA is negatively charged, the acetylation of lysine residues can weaken interaction between the DNA and the histones, making the nucleosome more mobile to allow access to other DNA binding proteins [6, 7]. In fact, it has been demonstrated that the acetylation of even a single residue (H4K16) can alter the compaction level of the nucleosome arrangement [8].

1.3. Effectors/Readers Dictate PTM Read-outs

In the past several years, it has been demonstrated that PTM of histones plays a key role in the regulation of chromatin structure and function. This further affects important cellular processes ranging from mitosis, replication, and transcription, to DNA repair and apoptosis [9].

The functional outcome of a specific PTM in the given chromatin context is determined by effectors and readers. The proteins that recognize histone modification site specifically and thus "affect" gene transcription process are known as the "effectors". These PTMs seem to indirectly influence the outcome from a distance by recruiting effectors to distinct genomic regions [1, 4, 10]. Heterochromatin protein 1 (HP1) is one such effector protein [11, 12].The domains that can "read" a specific PTM are referred to as the "readers". Some examples of readers include the bromo and the chromo domains [13, 14].

1.4. Functional Readouts of PTM

Yun et al (came up with a classification of various readers that can dictate the functional readouts of particular PTMs [15]. The readers are divided into the following categories:

1. Chromatin architectural proteins
2. Chromatin remodelers
3. Chromatin modifiers
4. Adaptors

1.4.1. Chromatin Architectural Proteins

Chromatin architectural proteins include the protein complexes that simultaneously bind to multiple nucleosomes and can affect chromatin structure by either inducing chromatin compaction or by acting as physical shields to block access to underlying DNA. These architectural proteins can often spread across a large region through self-propagation and oligomerization [16]. One such example of the chromatin architectural protein is the SIR complex that targets hypoacetylated and H3K79-un-methylated regions [17].

1.4.2. Chromatin Remodelers

As the name suggests, chromatin remodelers include chromatin remodeling complexes that can either move nucleosomes to different positions or increase the accessibility of the nucleosomal DNA [6]. This category of readers do not stay bound to the site of action and are often not enriched at the target site, for example, the yeast Isw2 complex [18].

1.4.3. Chromatin Modifiers

Chromatin modifiers refer to the group of proteins that can modify the chromatin by recruiting secondary modifiers, without directly affecting the chromatin structure. For example, Rpd3S is recruited by K36me3 mark and RpdS3 in turn deacetylates transcribed chromatin [19].

1.4.4. Adaptors

In addition to the above mentioned categories of proteins, PTM readers can function as adaptors to recruit other factors that can affect DNA activities such as DNA transcription, recombination, DNA damage repair, replication and RNA processing [15].

1.5. PTMs Are Identified Using Chromatin Immunoprecipitation (ChIP)

Histone modifications are commonly detected by using a technique called Chromatin Immunoprecipitation (ChIP). ChIP relies on the principle of cross-linking DNA with associated proteins such as histones followed by the detection of the associated proteins using antibodies specific for various histone and histone PTMs [2, 20]. Of the various PTMs, the most studied modifications include histone acetylation and methylation, which are discussed in more details in section 1.6.

1.6. Histone Acetylation and Methylation Are the Most Widely Studied Histone PTMs

Histone acetylation and methylation are the two most widely studied histone PTMs. The key enzymes that affect histone acetylation status include histone acetyltransferases (HATs), that acetylate lysine residues and histone deacetylases (HDACs) that remove these acetyl marks [21]. On the other hand, histone methyltransferases carry out the methylation of histone tails [21]. Similar to the role of histone deacetylases, histone demethylases remove the methyl marks put by the methyltransferases [21].

Histone lysine residues can be mono-, di- or tri-methylated and each modification has a unique effect, depending on both the residue modified and the amount of methylation. For instance, H3K4 and H3K36 methylation is usually associated with euchromatin (open chromatin) structure and hence transcriptional activation. On the other hand, H3K9 and H3K27 methylation is associated with heterochromatin (closed chromatin) structure and hence gene silencing. However, recently there have been reports of bivalent domains that make the situation more complex [22, 23]. For instance, chromatin regions marked by both H3K4 trimethylation (active transcription mark) and H3K27 trimethylation (gene repression mark). It is possible that genes that are silent yet poised for transcription carry these bivalent marks and can be identified on the basis of the presence of these bivalent marks [20, 23]. There appears to be a functional link between PTM of histones and DNA methylation. For example, methyl-CpG binding proteins recruit histone deacetylases to methylated DNA. It has been proposed that DNA methylation, and histone PTMs are specifically recognized by nucleosome remodeling complexes. For example, nucleosome remodeling complexes recognize trimethylated H3K4 and methyl CpG binding proteins and move nucleosomes in an adenosine triphosphate (ATP)- dependent manner [2]. In addition, recent studies have suggested an imperative interplay between various epigenetic marks such as histone modifications, DNA methylation along with microRNAs [3].

Since the histone modifications are epigenetic modifications that can be inherited by daughter cells to maintain lineage specific transcription profiles [7]; understanding the histone modifications and how they affect DNA properties has become a central focus of the chromatin field [24].

2. Cancer Epigenomics – Epigenetic Modifications Are Associated with Cancer

There have been extensive studies that have identified various epigenetic modifications associated with cancer. In fact, the very first epigenetic change that was described in cancer cells was global hypomethylation. However, recent studies have identified that both global as well as gene-specific hypomethylation and hypermethylation are involved in cancer. In addition, other widespread chromatin modifications are also associated with cancer [25, 26]. Studies have shown that in tumors, chromosome instability results from global hypomethylation; and further results in transcriptional activation of proto-oncogenes and oncogenes (e.g. *ras*). All these chromosomal changes result in the onset of cancer or oncogenesis [27]. In addition, it has been demonstrated that there is an activation of growth promoting genes through hypomethylation resulting in oncogenesis. Evidence suggests that anhydrase IX activation and S100 calcium-binding protein A4 upon hypomethylation in tumors results in renal cell cancer and colon cancer, respectively [28, 29]. Alternately, promoter specific hypermethylation of CpG islands of tumor suppressors results in gene silencing, as in the cases of *Retinoblastoma (Rb), p16, von Hippel-Lindau (VHL), MLH1, adenomatosis polyposis coli (APC),* and *E-cadherin* [25, 30, 31].

Several studies have shown that alterations in the chromatin and histones also play an important role in the development of cancer. For e.g., histone changes resulting in gene silencing have been reported in various cancers such as prostate, colorectal and lymphomas

[32, 33]. These studies reemphasize the fact that epigenetic modifications play a causative role not only in regulating gene expression but also in the read-out of these regulations in terms of the development of the phenotype.

It is extremely important to study the various epigenetic modifications of chromatin. In the recent past, over twenty epigenetic marks have been reported. However, further research needs to be done to understand the functional consequences of these epigenetic marks. It is possible that some of these epigenetic marks are more relevant compared to others. However, there is not much evidence to gauge the importance of various epigenetic marks. In addition, it is highly likely that we have not yet identified all the different epigenetic marks. Once new epigenetic marks are identified, we need to determine their functions as well. The epigenetic marks are fundamentally different in various tissues. Therefore, in order to study a particular tissue, we need to define the various epigenetic marks in that tissue and analyze the functions of these marks.

Recently, there has been exponential growth in the number of assays and methods to describe epigenetic marks. However, more research needs to be done to compare the various assays and improve our understanding and usage of these epigenomic assays. In addition, researchers are developing ways and techniques to understand higher order chromatin organization and the functions associated with these modifications.

As mentioned earlier, all somatic cells that arise from a common progenitor contain almost identical genotype. However, during the course of development, they differentiate to perform diverse biological functions because of differential gene regulation (gene activation or repression) that is dependent on the cellular milieu. During development, these epigenetic marks are maintained to impart identity to a specific cell type.

The epigenetic changes in a specific cell type / tissue are heritable, are likely to cause gene expression changes, and are linked with disease development. These marks help maintain cellular identity through the organization and maintenance of specific chromatin marks (signature). In addition, the nature of chromatin (euchromatin vs. heterochromatin) and its interaction with chromatin remodeling factors can play crucial roles in preventing or promoting both genomic instability and tumorigenesis. Further, we need to establish powerful tools using our knowledge of genetics, systems biology, epigenomics and genomics to understand the progression of complex human diseases such as cancer.

2.1. Genes and Epigenetic Marks Associated with Cancer

There are varieties of genes that are linked with cancer due to the mis-regulation of their epigenetic marks. *Men1* gene that encodes for a transcription factor Menin, is associated with familial endocrine tumors and is the most common genetic lesion in neuroendocrine tumors (NETs) (both heritable and sporadic), including 43% of the PanNETs reported in Jiao and co-workers' study [34]. Menin recruits Mixed Lineage Leukemia (MLL) complex with H3K4me3 histone methyltransferases, MLL2 and MLL3. *MLL2* and *MLL3* are often found mutated in childhood medulloblastomas [35].

There have been a lot of studies in the recent past that have identified mutations associated with pathways that are involved in either the methylation or demethylation of marks like H3K27 (gene silencing mark) [36-38].

Identification and functional characterization of the various histone PTMs using genome-wide analysis techniques will further advance diagnostics and treatment of a range of human cancers. In fact, studies are ongoing to understand the role of epigenetic changes in human cancers. Research is underway at the National Institutes of Health (NIH) and the National Cancer Institute (NCI) to develop novel epigenomics assays, make reference epigenomic maps, and illustrate disease-specific epigenomic changes (epigenomics@nih.gov). It is a challenging process since unlike the DNA, the epigenome is influenced by different environmental factors and can potentially mediate the interaction between the genotype, internal and external environmental factors [39, 40].

Conclusion

Each cell draws its identity not just from the heritable genetic material in the nucleus, but also from the epigenetic changes in this heritable material that are determined and influenced by both internal and external environmental factors. Complex human diseases such as cancer have found to be associated with the misregulation of various epigenetic modifications of certain key genes. It is therefore important to elucidate the various epigenetic marks and to characterize them functionally to fully understand their role in oncogenesis.

References

[1] Jenuwein, T. and C.D. Allis, Translating the histone code. *Science,* 2001. 293(5532): p. 1074-80.

[2] Berger, S.L., The complex language of chromatin regulation during transcription. *Nature,* 2007. 447(7143): p. 407-12.

[3] Guil, S. and M. Esteller, DNA methylomes, histone codes and miRNAs: tying it all together. *Int. J. Biochem. Cell Biol.*, 2009. 41(1): p. 87-95.

[4] Strahl, B.D. and C.D. Allis, The language of covalent histone modifications. *Nature,* 2000. 403(6765): p. 41-5.

[5] Li, B., M. Carey, and J.L. Workman, The role of chromatin during transcription. *Cell,* 2007. 128(4): p. 707-19.

[6] Workman, J.L. and R.E. Kingston, Alteration of nucleosome structure as a mechanism of transcriptional regulation. *Annu. Rev. Biochem.*, 1998. 67: p. 545-79.

[7] Ahmad, K. and S. Henikoff, Epigenetic consequences of nucleosome dynamics. *Cell,* 2002. 111(3): p. 281-4.

[8] Shogren-Knaak, M., et al., Histone H4-K16 acetylation controls chromatin structure and protein interactions. *Science*, 2006. 311(5762): p. 844-7.

[9] Iizuka, M. and M.M. Smith, Functional consequences of histone modifications. *Curr. Opin. Genet. Dev.*, 2003. 13(2): p. 154-60.

[10] Daniel, J.A., M.G. Pray-Grant, and P.A. Grant, Effector proteins for methylated histones: an expanding family. *Cell Cycle*, 2005. 4(7): p. 919-26.

[11] James, T.C., et al., Distribution patterns of HP1, a heterochromatin-associated nonhistone chromosomal protein of Drosophila. *Eur. J. Cell Biol.*, 1989. 50(1): p. 170-80.

[12] Eissenberg, J.C., et al., Mutation in a heterochromatin-specific chromosomal protein is associated with suppression of position-effect variegation in Drosophila melanogaster. *Proc. Natl. Acad. Sci. U S A,* 1990. 87(24): p. 9923-7.

[13] Dhalluin, C., et al., Structure and ligand of a histone acetyltransferase bromodomain. *Nature,* 1999. 399(6735): p. 491-6.

[14] Jacobson, R.H., et al., Structure and function of a human TAFII250 double bromodomain module. *Science*, 2000. 288(5470): p. 1422-5.

[15] Yun, M., et al., Readers of histone modifications. *Cell Res.* 21(4): p. 564-78.

[16] Buhler, M. and S.M. Gasser, Silent chromatin at the middle and ends: lessons from yeasts. *Embo J.*, 2009. 28(15): p. 2149-61.

[17] Martino, F., et al., Reconstitution of yeast silent chromatin: multiple contact sites and O-AADPR binding load SIR complexes onto nucleosomes in vitro. *Mol. Cell,* 2009. 33(3): p. 323-34.

[18] Gelbart, M.E., et al., Genome-wide identification of Isw2 chromatin-remodeling targets by localization of a catalytically inactive mutant. *Genes Dev.*, 2005. 19(8): p. 942-54.

[19] Carrozza, M.J., et al., Histone H3 methylation by Set2 directs deacetylation of coding regions by Rpd3S to suppress spurious intragenic transcription. *Cell*, 2005. 123(4): p. 581-92.

[20] Downs, J.A., M.C. Nussenzweig, and A. Nussenzweig, Chromatin dynamics and the preservation of genetic information. *Nature*, 2007. 447(7147): p. 951-8.

[21] Razin, A., CpG methylation, chromatin structure and gene silencing-a three-way connection. *Embo J.*, 1998. 17(17): p. 4905-8.

[22] Cloos, P.A., et al., Erasing the methyl mark: histone demethylases at the center of cellular differentiation and disease. *Genes Dev.*, 2008. 22(9): p. 1115-40.

[23] Mikkelsen, T.S., et al., Genome-wide maps of chromatin state in pluripotent and lineage-committed cells. *Nature,* 2007. 448(7153): p. 553-60.

[24] Kouzarides, T., Chromatin modifications and their function. *Cell,* 2007. 128(4): p. 693-705.

[25] Feinberg, A.P., Phenotypic plasticity and the epigenetics of human disease. *Nature*, 2007. 447(7143): p. 433-40.

[26] Feinberg, A.P. and B. Vogelstein, Hypomethylation distinguishes genes of some human cancers from their normal counterparts. *Nature,* 1983. 301(5895): p. 89-92.

[27] Ehrlich, M., DNA methylation in cancer: too much, but also too little. *Oncogene*, 2002. 21(35): p. 5400-13.

[28] Danesh, F.R. and Y.S. Kanwar, Modulatory effects of HMG-CoA reductase inhibitors in diabetic microangiopathy. *Faseb J.,* 2004. 18(7): p. 805-15.

[29] Wilson, A.S., B.E. Power, and P.L. Molloy, DNA hypomethylation and human diseases. *Biochim. Biophys. Acta*, 2007. 1775(1): p. 138-62.

[30] Greger, V., et al., Epigenetic changes may contribute to the formation and spontaneous regression of retinoblastoma. *Hum. Genet.*, 1989. 83(2): p. 155-8.

[31] Jones, P.A. and S.B. Baylin, The fundamental role of epigenetic events in cancer. *Nat. Rev. Genet.*, 2002. 3(6): p. 415-28.

[32] Varambally, S., et al., The polycomb group protein EZH2 is involved in progression of prostate cancer. *Nature,* 2002. 419(6907): p. 624-9.

[33] Fraga, M.F., et al., Loss of acetylation at Lys16 and trimethylation at Lys20 of histone H4 is a common hallmark of human cancer. *Nat. Genet.*, 2005. 37(4): p. 391-400.

[34] Jiao, Y., et al., DAXX/ATRX, MEN1, and mTOR pathway genes are frequently altered in pancreatic neuroendocrine tumors. *Science*. 331(6021): p. 1199-203.

[35] Parsons, D.W., et al., The genetic landscape of the childhood cancer medulloblastoma. *Science*. 331(6016): p. 435-9.

[36] Dalgliesh, G.L., et al., Systematic sequencing of renal carcinoma reveals inactivation of histone modifying genes. *Nature*. 463(7279): p. 360-3.

[37] van Haaften, G., et al., Somatic mutations of the histone H3K27 demethylase gene UTX in human cancer. *Nat. Genet.,* 2009. 41(5): p. 521-3.

[38] Morin, R.D., et al., Somatic mutations altering EZH2 (Tyr641) in follicular and diffuse large B-cell lymphomas of germinal-center origin. *Nat. Genet*. 42(2): p. 181-5.

[39] Bjornsson, H.T., M.D. Fallin, and A.P. Feinberg, An integrated epigenetic and genetic approach to common human disease. *Trends Genet.*, 2004. 20(8): p. 350-8.

[40] Wilson, A.G., Epigenetic regulation of gene expression in the inflammatory response and relevance to common diseases. *J. Periodontol.,* 2008. 79(8 Suppl): p. 1514-9.

In: Signaling, Gene Regulation and Cancer
Editors: Shree Ram Singh and Manoj K. Mishra

ISBN: 978-1-61942-088-5

Chapter 19

EPIGENETIC REGULATION OF CANCER

Rakesh Kumar Singh**, Rohit Seth and Akash Gunjan

Department of Biomedical Sciences, College of Medicine,
Florida State University, Tallahassee, FL, US

ABSTRACT

Cancer is a class of diseases characterized by uncontrolled cell division. Under normal conditions cell division is very tightly controlled and several layers of inbuilt safety mechanisms have evolved to ensure that cell division takes place only when and where it is required. However, under certain conditions a particular cell may acquire a set of mutations that may initiate carcinogenesis by breaking free of the normal regulatory mechanisms. There are multiple layers of barriers that must be breached before a cell can successfully establish itself as a cancer cell. Typically, the loss of tumor suppressor genes and/or activation of oncogenes could lead to uncontrolled cell division. In last few decades epigenetic regulation that is DNA sequence independent has emerged as a major player in carcinogenesis. In eukaryotes, the genomic DNA is organized into chromatin with the help of histone proteins that also regulate access to the genetic information contained within DNA. Histone post-translational modifications such as methylation, phosphorylation, acetylation etc., combined with DNA methylation play very important roles in deciding whether a set of genes will be expressed or silenced. These epigenetic modifications have been found to be altered in various kinds of cancers. Unlike the genetic mutations, most of these epigenetic modifications are reversible and therefore provide excellent targets for therapeutic intervention. Since alterations in some of the epigenetic modifications appear very early in carcinogenesis, they may also prove valuable for early diagnosis of cancer which is very critical for successful therapeutic intervention. In this chapter, we will discuss these epigenetic regulatory mechanisms in cancer and recent developments in the area of epigenetic therapy of cancer.

* E-mail:rakesh.singh@med.fsu.edu.

1. INTRODUCTION

According to the World Health Organization, cancer is a leading cause of death worldwide. The disease accounted for 7.9 million deaths (or around 13% of all deaths worldwide) in 2007. Deaths from cancer worldwide are projected to continue rising, with an estimated 12 million deaths in 2030. Contrary to public perception, cancer (malignant neoplasia) is not a single disease. It is a group of hundreds of diseases that are unified by the overarching property that they all result from uncontrolled cell division. Under normal conditions, multicellular organisms have many layers of regulation to prevent such uncontrolled cell division. Hence, in order for cells to become cancerous, they need to overcome several barriers. In a landmark review in the journal Cell, Hanahan and Weinberg (2000) enumerated at least six essential alterations in cell physiology for the development of cancer. These included self-sufficiency in growth signals, insensitivity to growth-inhibitory signals, evasion of programmed cell death, limitless replicative potential, sustained angiogenesis, tissue invasion and metastasis (Hanahan and Weinberg, 2000). These authors have now proposed revised hallmarks (Hanahan and Weinberg, 2011) after taking into account the spectacular progress made in the first decade of the twenty first century. They have added two new emerging hallmarks to the list; reprogramming of energy metabolism and evading immune destruction. Each of these alterations result in the breach of specific cellular defense mechanisms. Since a cell must breach several defensive barriers before it can successfully establish itself as cancer, carcinogenesis has been defined as a multistep process. This in part explains why cancer is an aging related disorder and the frequency of cancer occurrence increases with the age. Tumors also exhibit another layer of complexity by recruiting 'normal' cells that help in the cancer development by providing specific 'tumor microenvironment'.

Two classes of genes that have been extensively studied in the carcinogenesis field are the oncogenes (genes that promote carcinogenesis) and the tumor suppressor genes (genes that prevent carcinogenesis). Genetic studies have played a pivotal role in deciphering the molecular mechanisms of carcinogenesis and have been the prime focus of study to discover preventive and therapeutic options against cancer for most of the twentieth century. The completion of Human Genome Project (HGP) in the beginning of the 21st century has opened up numerous opportunities for the researchers to perform genome wide studies to find differences between cancerous and normal cells. The past decade has also seen a surge of interest in the scientific community to discover additional layers of regulation, such as epigenetic mechanisms, that may be involved in regulating carcinogenesis. Historically, epigenetics was invoked to explain phenotypic variations that could not be explained by genetics. The father of epigenetics, Conrad Waddington (Waddington, 1942), defined epigenetics as "the branch of biology which studies the causal interactions between genes and their products, which bring the phenotype into being". Epigenetics may be considered as a bridge between genotype and phenotype; a phenomenon that changes the final outcome of a locus or chromosome without changing the underlying DNA sequence. More specifically, Goldberg et al. (2007) recently defined epigenetics as the study of any potentially stable and, ideally, heritable change in gene expression or cellular phenotype that occurs without changes in Watson-Crick base-pairing of DNA. Simply put, all multicellular organisms start their life as a single cell (zygote) and hence all the different cells in an organism share the same genome but have very different physiological roles to play. These differences between the hundreds of different kinds of cells and tissue types in an individual are the result of epigenetic regulation. This epigenetic regulation is largely derived from two sources – a variety of histone modifications and DNA methylation.

All eukaryotic cells have their DNA wrapped around highly basic proteins called histones. Two molecules of each of the core histones (H2A, H2B, H3 and H4) interact to form an octameric protein core around which 147 bp of DNA is wrapped in about 1.8 turns to make a nucleosome that forms the monomeric subunit of the nucleoprotein polymer called chromatin (van Holde, 1988). The piece of DNA between the two nucleosomes is called as linker DNA that makes numerous contacts with the histone H1 (or linker histone) in higher eukaryotes. Initially, histone proteins were simply considered to be the packaging material for DNA. However, this view has completely changed in past three decades as it has been found that histones not only package the DNA, but also regulate access to the DNA. Hence, all the biochemical processes that require access to DNA (DNA replication, transcription, recombination and repair), must find a way to overcome the barrier created by the histones. These core histone molecules can be extensively modified by the occurrence of posttranslational modifications such as phosphorylation, methylation, acetylation, ADP ribosylation, sumoylation and ubiquitylation (Fischle et al., 2003). Different histone modifications could signal several different downstream events including switching off and on of key genes. Apart from histone modifications, the DNA itself can be methylated at CpG sequences. Cytosine is methylated to 5-methyl cytosine which has emerged as a key regulatory modification in DNA (Newell-Price et al., 2000). Thus, a combination of histone modifications and DNA methylation provides an additional layer of regulation for gene expression. Many of these modifications can be passed on to the daughter cells and are therefore hereditary. However, the major difference between these modifications and DNA sequence alterations is that the former are reversible. Since these modifications are reversible, they provide a powerful avenue for therapeutic intervention against disease conditions in which they may be playing a significant role. In recent years, it has become increasingly clear that misregulation of either or both the main components of epigenetic inheritance namely DNA methylation and the histone modifications by the deregulation of factors that mediate 'writing', 'erasing' and 'reading' of these marks actively contribute in human cancers. In this chapter we will present an overview of these modifications and the promises and possibilities of epigenetic therapy of cancer.

2. DNA Methylation and Cancer

DNA methylation is the process of enzymatic addition of a methyl group on to the DNA, primarily on the cytosine pyrimidine ring at the number 5 carbon. The most frequent sequence to be methylated is the dinucleotide sequence CpG. This is achieved by a group of enzymes known as DNA methyltransferases (DNMTs) capable of transferring the methyl group from S-adenosylmethionine (SAM) to cytosine. There are two kinds of methyltransferases, namely *de novo* methyltransferases and maintenance methyltransferases. DNMT3a and DNMT3b are considered to be *de novo* methyl transferases (Lyko et al., 1999) as they can catalyze methylation reaction at previously unmethylated sequences, whereas

DNMT1 is considered to be the maintenance methyltransferase (Lei et al., 1996) as this enzyme can catalyze methylation reaction immediately following DNA replication using the methylated parental strand as the template. DNA methylation is generally considered a signal for long term gene silencing. DNA methylation patterns established during the early embryonic development seem to persist for a very long time. Although the different types of cells and tissues originate from the same zygotic cell and have the same genetic material, they differ dramatically in their gene expression profiles. DNA methylation is one of the primary mechanisms by which cells generate differential temporal, spatial, tissue specific and developmental stage specific gene expression profiles. It is clear from the above description that DNA methylation is an essential process and it has been linked to several key processes such as genetic imprinting, dosage compensation and aging related disorders including carcinogenesis. Early estimates suggested that approximately 70-80% of CpG dinucleotides are methylated in mammalian cells (Bird and Taggart, 1980) and majority of the unmethylated CpG dinucleotides are grouped in clusters known as CpG islands. These CpG islands are primarily located in the promoter regions of many genes that are crucial for their proper expression. In many kinds of cancers, these CpG islands have been found to be abnormally methylated thereby silencing key regulatory tumor suppressor genes. Since DNA methylation can provide a long term signal, this might result in heritable transcriptional silencing. The methylation of CpG islands can lead to transcriptional silencing either directly by preventing the binding of different transcription factors or it may induce formation of heterochromatin (Fujita et al., 2003) via binding of methyl-CpG-binding domain proteins (MBDs). MBDs help heterochromatinization by recruiting histone deacetylase and other chromatin remodeling enzymes.

When compared to a normal genome, the DNA methylation pattern of cancer cells has been found to be altered. This disturbance in normal methylation patterns could be either by hypermethylation of unmethylated sequences or conversely by hypomethylation of normally methylated sequences. These disturbances have been found to be associated with early stages of carcinogenesis and might even initiate it (Gaudet et al., 2003). These two generally opposing events (hypomethylation and hypermethylation) are known to coexist in cancer cells. Whereas global hypomethylation has been found to be associated with several types of cancers, hypermethylation of discrete areas is known to coexist with the above mentioned global hypomethylation (Riggs and Jones, 1983; Feinberg and Vogelstein, 1983). Overall, hypomethylation takes place predominantly in repetitive DNA and has been linked to the generation of chromosomal instability, oncogene activation, transposon activation, loss of imprinting and nuclear disorganization (Ehrlich, 2002). DNA hypomethylation can be the direct result of reduced activity of DNMTs, or indirectly due to DNA repair where methylated sequences are replaced by unmethylated DNA. There is evidence to suggest that DNA repair mechanisms can replace methylated cytosine and incorporate unmethylated cytosine, thereby resulting in hypomethylation (Barreto *et al.*, 2007). On the other hand, hypermethylation occurs in the CpG islands located in the promoters of certain tumor suppressor genes such as p16 ^{INK4a}, BRCA1, or MLH1 (Baylin and Herman, 2006). These hypermethylated sites can be targeted by DNA-demethylating agents, thereby opening up new avenues for cancer treatment. The methylation-associated silencing of these genes affects several cellular pathways, and often has a tumor-type-specific profile and could therefore serve as a specific tumor marker. Environmental influences such as availability of SAM that is synthesized via the folate cycle can greatly affect methylation levels. It has been demonstrated in the Apc^{Min}

mouse model for colon cancer that the knockout of DNMT1, in combination with a low-folate diet, results in hypomethylation and reduced tumorigenesis (Trasler et al., 2003). However, mice carrying a hypomorphic DNMT1 allele, which reduces DNMT1 expression levels to 10% of wild-type levels, leads to the development of aggressive T-cell lymphomas (Gaudet et al., 2003). These findings indicate that a stable DNA-methylation pattern is crucial for cancer prevention and that the contribution of DNA-methylation changes to tumorigenesis depends on both cell type and environmental influences.

2.1. DNA Methylation and Cancer Therapy

Most of the currently known demethylating agents act by inhibiting DNMT activity. The rationale for the use of these agents was provided by the discovery that genetic inactivation of DNMT1 and DNMT3b led to demethylation of all known hypermethylated tumor suppressor genes and slower tumor growth (Rhee et al., 2002; Paz et al., 2003). DNMTs have binding sites for both the acceptor cytosine residue and the methyl donor SAM. Any molecule that can potentially block either one of the two sites can reduce methylation by acting as a competitive inhibitor.

The first analog tested to determine whether it can act as an inhibitor of DNA methylation was 5-azacytidine (Jones and Taylor, 1980) which is incorporated into the DNA during replication and RNA during transcription. It subsequently attaches covalently to DNMTs, thereby inactivating them and resulting in cytotoxicity. As such, tumors with increased levels of these enzymes are expected to be more sensitive to this drug. Treatment with equimolar amounts of both cytidine and 5-azacytidine inhibits the incorporation of the latter into nucleic acids, resulting in no alteration of the cell cycle either *in vivo* or *ex vivo*. The cytidine and 2'-deoxycytidine analogs are the most extensively studied members of this class. All these compounds only inhibit DNMTs when incorporated into double-stranded DNA, but the exact mechanism differs slightly depending on the presence of a hydrogen atom at the 5' position of the pyrimidine ring. In analogs that lack this atom, the intermediate of the methylation reaction cannot undergo elimination to release the enzyme, and so the latter becomes covalently trapped. In other cases, for instance 5, 6-dihydro-5-aza-cytidine, the enzyme forms a very stable complex with the analog, reducing its turnover. In all cases, nucleoside or deoxynucleoside analogs are transformed into the corresponding deoxynucleotide triphosphate inside the cell. Nucleosides are first converted into diphosphate nucleotides, which are then reduced and, after phosphorylation, are finally incorporated into DNA during the S phase. Cells lacking any of the enzymes involved in the synthesis of triphosphate deoxynucleotide derivatives are resistant to these compounds, and analogs that do not bind any of these enzymes cannot be incorporated into DNA, because they are inactive.

The analog 5-aza-2'-deoxycytidine (Decitabine) is one of the most commonly used demethylating drugs in assays with cultured cells (Daskalakis et al., 2007). It overcomes the problem of incorporation of 5-azacytidine into RNA and reduces its side effects. Indeed, 5-aza-2'-deoxycytidine is only incorporated into DNA. Decitabine was shown to have significant anti chronic myelogenous leukemia (CML) activity (Kantarjian et al., 2003) when given to CML patients.

Zebularine [1-(beta-D-ribofuranosyl)-1,2-dihydropyrimidin- 2-one] is another cytidine analog that has recently been developed (Zhou et al., 2002; Billam et al., 2010). Similar to 5-azacytidine, it was first tested as an anticancer agent. Initially designed as a cytidine deaminase inhibitor, it has been shown to form a covalent complex with DNA methyltransferases, to deplete human DNMTs. Most importantly, it leads to actual DNA demethylation and induces the reactivation of hypermethylated genes in yeast models as well as $p16^{INK4a}$ gene in bladder cancer cells (Cheng et al., 2003). Zebularine was also found to be a promising antitumor agent in mouse xenograft models (Cheng et al., 2003). The novel feature of this DNA demethylating agent, compared with 5-azacytidine and 5-aza-2'-deoxycytidine, is that it is chemically stable with very low toxicity, and hence it is the first drug in its class that can be given orally. However in mouse models, it required high doses, up to 1g/kg body weight, making its use in human clinical trials very difficult. To overcome this problem Zebularine may be used as a combination drug with other DNA demethylating drugs, because it has been shown that sequential treatment with 5-aza-2'-deoxycytidine followed by Zebularine hindered the remethylation of tumor-suppressor genes (Cheng et al., 2004).

The use of the nucleoside analogs in clinical trials has been limited by their side effects that are probably caused by the cytotoxic effects associated with the incorporation of the drugs into DNA independently of their DNA-methylation status. This mechanistic finding has led to the search for inhibitors of DNA methylation that cannot be incorporated into DNA. The FDA approved drug Procainamide used for the treatment of cardiac arrhythmias has been proposed as a nonnucleoside inhibitor of DNA methylation (Cornacchia et al., 1988). Procainamide restores expression of the detoxifier gene GSTP1 in prostate cancer cells silenced by hypermethylation by causing global DNA hypomethylation. This action is thought to be mediated by binding of procainamide to GC-rich DNA sequences. Another FDA approved drug, Procaine, has also been found to have DNA hypomethylation and growth-inhibitory activity. Procaine acts as an inhibitor of DNA methylation in breast cancer cells, causing global genomic DNA hypomethylation and demethylation and reactivation of tumor- suppressor genes with hypermethylated CpG islands. This effect is associated with, and possibly mediated by, strong Procaine binding to CpG-rich DNA. Procaine also suppresses growth in these breast cancer cells simultaneously with the occurrence of DNA demethylation (Villar-Garea et al., 2003).

3. Histone Modifications and Cancer

The core histone proteins are subjected to a plethora of post-translational modifications. All core histone proteins have a globular histone fold domain and an unstructured N-terminal tail and both of these, particularly the latter, are heavily modified under different conditions. Lysine, arginine, serine, threonine and tyrosine residues are frequently modified. Lysine and arginine residues can be methylated whereas lysines can also be acetylated, ubiquitylated and SUMOylated. Serine, threonine and tyrosine residues are frequently phosphorylated. Many of these modifications are known to exhibit extensive cross talk and hence the dynamic equilibrium of various modifications have been proposed to generate a 'histone code' that can potentially determine whether a particular locus will be transcriptionally active or silent

(Jenuwein and Allis, 2000). Histone acetylation and methylation reactions are catalyzed by the histone acetyltransferases (HATs) and Histone methyltransferases (HMTs) respectively. These enzymes are also referred to as code 'writers'. Histone deacetylases (HDACs) and histone demethylases (HDMs) remove acetyl and mcthyl groups respectively and they are sometimes referred to as code 'erasers'. A third group of proteins that recognize these histone modification marks and affect the gene expression patterns are known as code 'readers'. The biological outcome of histone modifications may be either mediated by an altered chromatin structure or by direct signaling pathways mediated by proteins that recognize histone modification patterns (Strahl and Allis, 2000). We will focus mainly on the role of histone phosphorylation, ubiquitylation, methylation and demethylation in cancer in this chapter. A detailed discussion on the role of histone acetylation and deacetylation in cancer is available in another chapter.

3.1. Histone Phosphorylation and Cancer

All four core histones as well as linker histone H1 have been shown to be phosphorylated at serine and threonine residues, while histones H3 and H2A.X have also been shown to be phosphorylated at tyrosine residues. Histone phosphorylation leads to a reduction of total positive charge on histone and thereby leads to reduced affinity for DNA. Histone serine and threonine phosphorylation have been associated with mitosis (H3S10, H3S28), DNA damage response (H4 S1 and H2A.X S139), control of gene expression (H3S10 and H3T11), apoptosis (H2B S10) and meiosis (Cheung et al., 2003; Krishnamoorthy et al., 2006; Metzger et al., 2008 and Shimada et al., 2008). Histone tyrosine phosphorylation has been reported only very recently (Singh and Gunjan, 2011) and it has been shown to play a role in cellular response to DNA damage (H2A.X Y142) (Xiao et al., 2009; Cook et al., 2009 and Krishnan et al., 2009) and chromatin architecture/oncogenesis (H3 Y41) in mammalian cells (Dawson et al., 2009), as well as histone turnover in yeast (H3 Y99) (Singh et al., 2009). Since processes like mitosis, apoptosis and DNA damage response are very critical in genome maintenance, any deregulation of kinases and or phosphatases regulating these histone phosphorylation events may contribute to carcinogenesis.

Phosphorylation of the linker histone H1 has been shown to be involved in cell cycle progression and it has been considered a marker for mitotic cells (Boulikas, 1995). CDK2 is the major kinase that is known to phosphorylate H1 *in vivo* during mitosis peaking at metaphase, when chromosomes are maximally condensed. Elevated H1 phosphorylation has been observed in cancer cells and tumors (Deshpande et al., 2005). In a recent study Zheng et al. (2010) implicated H1 phosphorylation in transcriptional regulation by RNA polymerases I and II in the interphase nucleus. They also found that this phosphorylation plays an important role in the control of ribosome biogenesis and cell growth.

Among the known histone phosphorylations, H2A(X) phosphorylation is probably the most studied and recognized in the context of DNA damage. In mammals, this phosphorylation occurs at serine 139 of histone H2A variant H2AX in the C-terminal SQ motif (Rogakou et al., 1998). In the budding yeast, where there are no H2A variants, phosphorylation occurs at serine 129 (Downs et al., 2000). This unique modification is catalyzed by phosphoinositide-3 kinases (PI3K) such as Ataxia Telangiectasia-Mutated (ATM), (Ataxia telangiectasia and Rad3 related) (ATR) and DNA-dependent protein kinase

(DNA-PK) in mammals and Mec1/Tel1 in budding yeast and is often referred as gamma H2A or γH2A(X). In yeast, H2A phosphorylation can be detected as far as 50Kb on either side of a DNA double strand break (DSB) (Shroff et al., 2004), whereas it spreads over 2 Mbps surrounding a DNA break in mammalian cells (Rogakou et al., 1998). γH2A(X) is believed to be a very early marker of the DNA damage and contributes to the initiation of the DNA damage response signaling cascade. The phosphate on serine 139 is recognized by BRCA1 C-Terminus (BRCT) domain of the adaptor protein, mediator of DNA damage checkpoint protein 1 (MDC1), which interacts with ATM as well (Stucki et al., 2005). γH2A(X) and MDC1 recruit Mre11, Rad50 and Nbs1 (MRN) complex which in turn helps recruit more mediators and subsequent repair effectors (Chapman and Jackson, 2008). Mutations in many of these enzymes are related to human disease, including Ataxia telangiectasia (AT), Blooms syndrome (BS), and severe combined immunodeficiency (SCID). These patients exhibit high levels of chromosome rearrangements and are susceptible to cancer. Double mutant mice defective in both H2A.X and tumor suppressor p53 were found to develop lymphomas associated with inappropriate recombination in B cells (Bassing and Alt, 2004). Recently, the Allis group discovered that Y142 is constitutively phosphorylated under normal physiological conditions and becomes dephosphorylated upon DNA damage (Xiao et al., 2009). This phosphorylation was carried out by WSTF (Williams-Beuren Syndrome transcription factor), which is a component of WICH complex (WSTF-ISWI ATP-dependent chromatin-remodeling complex) and has intrinsic tyrosine kinase activity by virtue of its N-terminal domain. Following DNA damage, WSTF was found to dissociate from chromatin resulting in a decrease in Y142 phosphorylation, which in turn promoted the maintenance of phosphorylation at S139 as well as the γH2A.X dependent recruitment of MDC1 and other repair complexes. Two additional groups independently described the phosphorylation of H2A.X Y142. Cook et al., (2009) found that during embryonic development of mouse kidney, H2A.X Y142 phosphorylation was reversible. They identified EYA (named after the Drosophila mutant "eyes absent") as the main tyrosine phosphatase responsible for this and found an EYA dependent reduction of Y142 phosphorylation following DNA damage. Similar findings were published by Krishnan et al. (2009) implicating EYA as the tyrosine phosphatase responsible for dephosphorylation at Y142. Together these studies suggest that the balance between the opposing activities of WSTF and EYA will be the crucial determinant of the outcome of DNA damage response, since H2A.X S139 phosphorylation seems to be dependent on the phosphorylation status of Y142. These studies indicated that phosphorylation status of Y142 may serve as an important determinant of cell fate after DNA damage. Quite possibly, when low amounts of damage occur and repair is feasible, phospho-H2A.X Y142 is dephosphorylated, while S139 phosphorylation accumulates and leads to the formation of repair foci by recruiting repair factors such as MDC1. Under extensive damage conditions, phospho-H2A.X Y142 persists, while phospho-S139 is attenuated and this leads to the recruitment of c-Jun N-terminal protein kinase-1 (JNK1) leading to apoptosis. The reciprocal regulation of DNA repair and apoptosis by the reversible phosphorylation of H2A.X Y142 is a very attractive hypothesis to explain how the cell selects the most appropriate response following DNA damage. Since H2AX Y142 phosphorylation seems to be very important in determining cell fate following DNA damage, it will be interesting to study if cancer cells have subverted this mechanism in favor of their survival. If future studies do indeed find evidence for the subversion of this pathway in cancer cells, these regulatory kinases and or phosphatases may become very important therapeutic targets.

Histone H2B has been shown to be phosphorylated by the Ste20 kinase in yeast and its homolog Mst1 in humans (Ahn et al., 2005). Mst1 kinase has been shown to phosphorylate H2B in vertebrates from frogs to mammals at serine 14 and is believed to play a role in chromatin compaction during apoptosis (Cheung et al., 2003). This apoptotic pathway was shown to exist even in unicellular budding yeast where Ste20, the yeast homolog of Mst1, was found to phosphorylate H2B at serine10 in a hydrogen peroxide induced cell death pathway (Ahn et al., 2005). Drosophila Mst1 phosphorylates Drosophila Inhibitor of Apoptosis 1 (DIAPI), and regulates the expression of both DIAPI and cyclin E (Jia et al., 2003). Since histone H2B phosphorylation is highly conserved from budding yeast to humans and is known to play a critical role in cell proliferation and cell death pathways, its role in carcinogenesis can hardly be overemphasized. Moreover, in a recent study, it was found that adenosine monophosphate–activated protein kinase (AMPK) directly associated with chromatin and phosphorylated H2B at serine 36 (Bungard et al., 2010). Since AMPK is a central regulator of cellular energy homeostasis, the S36 phosphorylation of H2B appears to play a direct regulatory role in the transcriptional control of AMPK dependent genes. Cancer cells are usually under a fair amount of stress and generally subvert energy metabolism and therefore this study directly brings H2B phosphorylation at the forefront of epigenetic signaling in cancer.

In mammalian cells, phosphorylation of H3 at S10 and S28 has been associated with chromatin condensation during mitosis and transcription (De Souza et al., 2000; Hsu et al., 2000). These phosphorylations are mediated by Aurora B kinase and mitogen- and stress-activated protein kinase 1 (MSK1) respectively(Peterson and Laniel, 2004). Phosphorylation of histone H3 at Ser10 (H3S10p) is critical during cellular transformation (Kim et al., 2008; Choi et al., 2005), and H3S10 phosphorylation levels are elevated in oncogene-transformed cells as well as human tumor cell lines (Chadee et al., 1999). Furthermore, increased H3S10 phosphorylation levels due to MSK1 and Aurora B kinases lead to chromosomal instability, which is a hallmark of cancer (Espino et al., 2009). Hence, it has been suggested that regulation of histone H3 phosphorylation levels may be a possible target for cancer treatment.

Phosphorylation of H3 at T3 is mediated by the Haspin kinase (Dai et al., 2005). This modification has been shown to be directly recognized by the Baculovirus IAP Repeat (BIR) domain of Survivin, a member of the chromosomal passenger complex (CPC). This leads to the recruitment of the CPC to chromosomes and activation of its kinase subunit Aurora B (Kelly et al., 2010). Consistently, modulation of the kinase activity of Haspin, which phosphorylates H3T3, leads to defects in the Aurora B–dependent process of spindle assembly and inhibition of nuclear reformation. These findings establish a direct cellular role for mitotic histone H3T3 phosphorylation, which is read and translated by the CPC to ensure accurate cell division.

H3Y41 was recently shown to be directly phosphorylated by the protein tyrosine kinase JAK2 (Janus Kinase 2) in human hematopoietic cell lines (Dawson et al., 2009). The phosphorylation status of H3Y41 was shown to modulate the binding of HP1α (heterochromatin protein 1 alpha) to histone H3. The binding of HP1α was destabilized by H3Y41 phosphorylation and inhibition of JAK2 led to reduced expression of leukemia oncogene *LMO2*, probably due to the increased binding of HP1α as a result of reduced H3Y41 phosphorylation. Thus, H3Y41 phosphorylation may regulate chromatin architecture around certain gene promoters. Aberrant JAK2 activity may result in high levels of H3Y41 phosphorylation leading to the loss of HP1α binding and an upregulation of target genes such as *LMO2*, which is likely to promote oncogenesis of hematopoietic cells. Interestingly, overexpression of HP1 reversed the leukemic phenotype resulting from excessive JAK signaling in fruit flies (Shi et al., 2006). These findings make JAK2 an attractive therapeutic target for certain human hematopoietic malignancies.

Using a combination of biochemistry, yeast genetics and site-directed mutagenesis, Singh et al., (2009) showed that excess histone H3 is phosphorylated at tyrosine 99 (H3Y99) and this phosphorylation of H3 presumably by the DNA damage checkpoint kinase Rad53 was crucial for the ubiquitylation and degradation of this histone. Due to steric limitations, H3Y99 is likely to be accessible for phosphorylation only when it is not in association with histone H4 (Singh et al., 2009). Since H3 is tightly associated with H4 in the nucleosome, this may be an important safety mechanism against inadvertent degradation of nucleosomal histones. Due to the cytotoxicity of excess histones (Singh et al., 2010), cells have to rapidly eliminate any histones that have not been incorporated in the chromatin. Whereas budding yeast has only two copies of each core histone gene, human cells have about two dozen copies of each core histone gene and are even more likely to face the adverse effects of excess histone accumulation. The tumor suppressor CHEK2 (checkpoint kinase 2) is the human counterpart of the Rad53 kinase, while CHEK1 is an essential kinase closely related to CHEK2 (Bartek and Lukas, 2003). Not surprisingly, both CHEK1 and CHEK2 often exhibit overlapping substrate specificities and are known to play an important role in the maintenance of genome stability. It remains to be determined if these proteins are also involved in the regulation of histone metabolism in human cells and if so, this function of these proteins may be contributing to genome maintenance and tumor suppression independent of their roles in the DNA damage response.

3.2. Histone Methylation and Cancer

Histone methylation is chemical modification of histone that involves addition of one, two or three methyl groups on the lysine residues or up to two symmetric or asymmetric methyl marks on arginine residues. Histone methyl transferases (HMTs) use SAM as a cosubstrate for the transfer of methyl group. Methylation influences the basicity as well as hydrophobicity of histones and their affinity towards certain proteins, such as transcription factors. Human genome is believed to encode over 70 methyltransferases, many of which have been implicated in cancer. HMTs are characterized by a conserved SET (su(var)3-9, enhancer-of-zeste, trithorax) domain. Depending on the homology within the SET domain, HMTs have been divided into five groups (Keppler and Archer, 2008) which are SUV39, SET1, SET2, RIZ1 (Retinoblastoma protein-interacting zinc-finger) and SMYD3 (SET- and MYND-domain containing protein-3). Disruptor of telomeric silencing-1 (DOT1) is a non-SET domain containing histone lysine methyltransferase. Each subgroup contains several HMTs with the capacity to methylate histones on lysine residues. Lysine methylation can lead to either gene activation or repression depending on the residue and degree of methylation. Generally H3K9 or H3K27 or H4 K20 methylation is associated with gene silencing whereas H3K4, H3K36 and H3K79 methylation correlates with transcriptional activation (**Volkel and Angrand, 2007**). Since histone methylation can either activate or repress genes depending on the degree of methylation and the residue involved, it can influence carcinogenesis by either activating oncogenes or repressing tumor suppressor genes. For example, misregulation of

H3K4 methylation has been reported in human lymphoid and myloid leukemia. MLL gene rearrangement is the most common form of chromosomal anomaly in human leukemia accounting for over 80% infant leukemia (Hess 2004). Chromosomal translocation of MLL1, the H3K4 HMT, leads to upregulation of several Hox genes and plays key role in leukemic progression (Argiropoulos and Humphries, 2007; Milne et al., 2002; Nakamura et al., 2002). In yet another example, enhancer of zeste 2 (EZH2) (a subunit of polycomb repressor complex 2 (PRC2)) catalyzes histone H3 methylation on gene promoters thus repressing genes that induce stem cell differentiation to maintain an embryonic stem cell signature. EZH2 has been found to be significantly upregulated in metastatic forms of prostate cancer and its downregulation leads to reduced growth of prostate cancer cells (Varambally et al., 2002). Metastatic breast cancer has also been shown to have significantly higher levels of EZH2 compared to normal breast tissue (Kleer et al., 2003).

G9a is the predominant histone H3K9 methyltransferase in euchromatin, required for overall mono- and dimethylation of H3K9 (Tachibana et al., 2002). G9a has been implicated in the silencing of tumor suppressor genes as well as in the targeting of *de novo* DNA methylation in embryonic stem cells (Tachibana et al., 2008). Knockout mouse embryos die around day 10 and show severe growth retardation, thus establishing an essential function of G9a in normal cells. Overexpression of G9a has been observed in several cancer cell lines such as prostate, breast, lung, and colon cancer cell lines (Kondo et al., 2007).

Histone methylation is a reversible posttranslational mark on histones. The methyl group can be removed by two families of histone demethylases namely, lysine-specific demethylase-1 (LSD1/KDM1A) and the Jumanji domain containing enzymes. LSD1 was discovered by Shi et al., (2004) and catalyzes lysine demethylation in a flavin adenine dinucleotide (FAD)-dependent manner. It can demethylate mono- and di-methyl (but not tri-methyl) marks on histone H3 at K4 and K9. LSD1 requires protonated nitrogen as a hydrogen donor. Since the nitrogen of a tri-methylated lysine is not protonated it cannot be demethylated by LSD1. LSD1 is essential during normal developmental processes and targeted deletion in mice results in embryonic lethality (Wang et al., 2009a). LSD1 is overexpressed in estrogen receptor (ER) negative breast cancers (Lim et al., 2010). Further, it has also been shown to be a part of NuRD complex that may serve to inhibit metastasis in breast cancer (Wang et al., 2009b) and as such H3K4 methylation inhibitions may have therapeutic potential in this disease. LSD1 was also found to be playing an important role in regulating the pluripotency of stem cells by regulating H3K4 demethylation at bivalent chromatin domains that carry both H3K4 di/trimethylation and H3K27 trimethylation marks and are found at the regulatory regions of several genes involved in development (Adamo et al., 2011). This study may pave the way for therapeutic intervention at the level of cancer stem cells in the future.

The demethylation reaction carried out by Jumanji C (JmjC) domain proteins requires the presence of Fe(II) and α-ketoglutarate to generate formaldehyde and succinate and allows removal of methyl marks from mono-, di- and tri-methylated lysines. The first JmjC domain-containing histone demethylase to be reported was KDM2A/JHDM1A (JmjC domain-containing histone demethylation protein 1A)/FBXL11 (F-box and leucine-rich repeat protein (Tsukada et al., 2006). Mammalian KDM2A and KDM2B/JHDM1B/FBXL10 have been demonstrated to catalyze H3K36 demethylation. Cell culture studies suggest that KDM2 proteins might have both pro- and anti-oncogenic functions in different contexts. KDM2A can suppress the growth of colon cancer cells (Lu et al., 2010), while KDM2B can inhibit proliferation of tumor cells by repressing rRNA genes (Frescas et al., 2007). KDM2B has been shown to prevent premature senescence in MEFs (mouse embryonic fibroblasts) most probably by repressing the Ink4b gene (He et al., 2008; Pfau et al., 2008). In addition to their functions in proliferation, KDM2 proteins also play diverse roles in the maintenance of genomic stability. Several other KDM families (KDM3-KDM6) have been reported and many more will hopefully be reported in the near future. Since the discovery of the first histone demethylase in 2004 (Shi et al., 2004), tremendous progress has been made in this field. However, further studies are required to clearly define the roles of the multiple demethylases before we can fully exploit the therapeutic potential of histone demethylases. At the very least, the mechanism of recruitment of these demethylases to their targets and the importance of their demethylase activities in regulating various signaling pathways need to be addressed before we can begin to determine which of these histone demethylases may be suitable targets for therapeutic drugs.

3.2.1. Modulation of Histone Methylation/Demethylation in Cancer Therapy

Since histone methylation/demethylation are enzymatic processes, they offer opportunities for therapeutic interventions. Several enzymes involved in modulating histone methylation are ideal therapeutic targets as their enzymatic activity can be potentially modified using small molecule inhibitors. At present EZH2 seems to be the best-characterized histone methyltransferase target for tumor therapy. But several important questions with respect to tumor specificity of EZH2 expression and function still need to be answered. The first attempt at the pharmacological inhibition of EZH2 by 3-deazaneoplanocin A (DZNep) has been reported (Tan et al., 2007). The major concern is that it does not specifically inhibit EZH2, but rather induces the degradation of EZH2 and associated proteins (Tan et al., 2007). A high-throughput screen for G9a inhibitors led to identification of BIX-01294, which caused a moderate reduction in histone H3 lysine 9 dimethylation in bulk histone preparations. Treatment of different cell lines with twice the IC_{50} concentrations of BIX-01294 did not lead to detectable morphological changes, and global histone methylation changes appeared small (Kubicek et al., 2007). Hence, additional work will be required to establish G9a inhibitors as drug candidates for further development.

Inhibition of LSD1 by polyamine analogs has been found to reverse aberrant gene silencing (Huang et al., 2007). Another class of polyamine analogues, the oligoamines, were found to be competitive inhibitors of recombinant LSD1, suggesting that the oligoamines may compete directly with the substrate at the active site *in vivo* (Huang et al., 2009). This inhibitor of histone demethylases was found to block colon tumor-cell growth coincident with histone methylation and re-expression of silenced genes. Interestingly, it also markedly potentiated the activity of DNA hypomethylating agents *in vitro* and *in vivo*, suggesting that histone demethylase inhibitors may synergize with DNA methylation inhibitors, and thus represent a class of potential epigenetic therapeutic agents that can be used in combination therapy to treat human disease.

CONCLUSION

Epigenetic alterations in cancer cells have been widely reported. Unlike genetic mutations, epigenetic alterations such as DNA methylation and/or histone modifications are reversible. Therein lies the great potential for epigenetic drugs based therapeutic intervention in various types of cancer. Tumor suppressor genes that have been rendered useless via epigenetic silencing during carcinogenesis can be brought back to life by the use of 'epidrugs'. Upon reactivation, these genes could potentially halt tumor development simply by reinforcing the cellular control on the division of tumor cells. Alternatively, these rapidly dividing tumor cells could be pushed towards differentiation, which might result in substantial therapeutic gains. 'Epidrugs' can also be potentially used to silence the oncogenes leading to clinical gain by retarding the growth of tumors. The latest technological developments in the field of proteomics, genomics and epigenomics are set to accelerate the development of novel epigenetic therapeutics. Systematic mapping of histone modification marks in cancer cells compared to normal cells are expected to shed more light on the epigenetic mechanisms involved in carcinogenesis. The first comprehensive genome-scale histone methylation mapping has already suggested novel functions of known histone methylation marks in T-cells (Barski et al., 2007). Another study mapped genome-wide distribution of histone trimethylation marks in mouse embryonic stem cells and lineage-committed cells (Mikkelsen et al., 2007). Advancement in genome-sequencing technologies and epigenome-mapping projects are likely to revolutionize the field of epigenomics and 'epidrug' development. Epigenetic modifications are known to crosstalk with each other extensively and a detailed understanding of these combinatorial interactions will greatly aid in the discovery of novel targets for therapeutic intervention. For example, DNMTs and HDACs interact through methyl-CpG binding proteins and not surprisingly, DNA demethylation combined with inhibition of histone deacetylation led to re-expression of silenced genes in cancer (Cameron et al., 1999). These early experimental studies certainly suggest that crosstalk between epigenetic silencing systems is worth investigating. More recently, another example of crosstalk between a DNMT and a histone-modifying enzyme has been reported where it was shown that the targeted deletion of the LSD1 caused progressive loss of DNA methylation (Wang et al, 2009b). Further, since histone- and DNA-modifying enzymes exhibit moderate to high substrate specificity and appear to be targeted to specific loci in the genome, it is possible that combinations of different small-molecule modifiers of these enzymes would permit a powerful and flexible method to combat diverse human disorders in the future, based on the same basic premise of modulating epigenetics. Thus, it can be safely argued that even though epigenetic therapy of cancer is still in its infancy, it holds great promise for our fight against cancer in the future. With high throughput studies of epigenetic modifications becoming reality, epigenetic signatures of various kinds of cancers can be defined in the next few years. Specific therapeutic strategies based on the epigenetic signature of individual cancers could be developed on a case by case basis. This is expected to usher in the era of 'personalized' medicine. Identification and targeting tumor specific interactions between epigenetic pathways will also become a reality in the future. However, there is a lot of work that remains to be done before we can start reaping all the potential benefits of epigenetic therapies in cancer treatment.

References

[1] Adamo A, Sesé B, Boue S, Castaño J, Paramonov I, Barrero MJ, Izpisua Belmonte JC. (2011) LSD1 regulates the balance between self-renewal and differentiation in human embryonic stem cells. *Nat. Cell Biol.* 13:652-660.

[2] Ahn SH, Cheung WL, Hsu JY, Diaz RL, Smith MM, Allis CD. (2005) Sterile 20 kinase phosphorylates histone H2B at serine 10 during hydrogen peroxide-induced apoptosis in S. cerevisiae. *Cell.* 120:25-36.

[3] Argiropoulos B, Humphries RK. (2007) Hox genes in hematopoiesis and leukemogenesis. *Oncogene* 26, 6766–6776.

[4] Barreto G, Schäfer A, Marhold J, Stach D, Swaminathan SK, Handa V, Döderlein G, Maltry N, Wu W, Lyko F, Niehrs C. (2007) Gadd45a promotes epigenetic gene activation by repair-mediated DNA demethylation, *Nature* 445:671–675.

[5] Barski A, Cuddapah S, Cui K, Roh TY, Schones DE, Wang Z, Wei G, Chepelev I, Zhao K. (2007) High-resolution profiling of histone methylations in the human genome. *Cell.* 129:823-837.

[6] Bartek J, Lukas J. (2003) Chk1 and Chk2 kinases in check¬point control and cancer. *Cancer Cell.* 3:421-429.

[7] Bassing CH, Alt FW. (2004) The cellular response to general and programmed DNA double strand breaks. *DNA Repair* (Amst). 3:781-796.

[8] Baylin SB, Herman JG. (2000) DNA hypermethylation in tumorigenesis: epigenetics joins genetics. *Trends Genet.* 16:168-174.

[9] Billam M, Sobolewski MD, Davidson NE. (2010) Effects of a novel DNA methyltransferase inhibitor zebularine on human breast cancer cells. *Breast Cancer Res. Treat.*120:581-592.

[10] Bird AP, Taggart MH. (1980) Variable patterns of total DNA and rDNA methylation in animals. *Nucleic Acid Res.* 8:1485-1497.

[11] Boulikas T. (1995) The phosphorylation connection to cancer. *Int. J. Oncol.* 6:271-278.

[12] Bungard D, Fuerth BJ, Zeng PY, Faubert B, Maas NL, Viollet B, Carling D, Thompson CB, Jones RG, Berger SL. (2010) Signaling kinase AMPK activates stress-promoted transcription via histone H2B phosphorylation. *Science*. 329:1201-1205.

[13] Cameron EE, Bachman KE, Myöhänen S, Herman JG, Baylin SB. (1999) Synergy of demethylation and histone deacetylase inhibition in the re-expression of genes silenced in cancer. *Nat. Genet.* 21:103-107.

[14] Chadee DN, Hendzel MJ, Tylipski CP, Allis CD, Bazett-Jones DP, Wright JA, Davie JR. (1999) Increased Ser-10 phosphorylation of histone H3 in mitogen-stimulated and oncogene-transformed mouse fibroblasts. *J. Biol. Chem.* 274:24914-24920.

[15] Cheng JC, Matsen CB, Gonzales FA, Ye W, Greer S, Marquez VE, Jones PA, Selker EU. (2003) Inhibition of DNA methylation and reactivation of silenced genes by zebularine. *J. Natl. Cancer Inst.* 95:399-409.

[16] Cheng JC, Weisenberger DJ, Gonzales FA, Liang G, Xu GL, Hu YG, Marquez VE, Jones PA. (2004) Continuous zebularine treatment effectively sustains demethylation in human bladder cancer cells. *Mol. Cell Biol.* 24:1270-1278.

[17] Cheung WL, Ajiro K, Samejima K, Kloc M, Cheung P, Mizzen CA, Beeser A, Etkin LD, Chernoff J, Earnshaw WC, Allis CD. (2003) Apoptotic phosphorylation of histone H2B is mediated by mammalian sterile twenty kinase. *Cell.* 113:507-517.

[18] Choi HS, Choi BY, Cho YY, Mizuno H, Kang BS, Bode AM, Dong Z. (2005) Phosphorylation of histone H3 at serine 10 is indispensable for neoplastic cell transformation. *Cancer Res.* 65:5818-5827.

[19] Cook PJ, Ju BG, Telese F, Wang X, Glass CK, Rosenfeld MG. (2009) Tyrosine dephosphorylation of H2A.X modulates apoptosis and survival decisions. *Nature.* 458:591-596.

[20] Cornacchia E, Golbus J, Maybaum J, Strahler J, Hanash S, Richardson B. (1988) Hydralazine and procainamide inhibit T cell DNA methylation and induce autoreactivity. *J. Immunol.* 140:2197-2200.

[21] Dai J, Sultan S, Taylor SS, Higgins JM. (2005) The kinase haspin is required for mitotic histone H3 Thr 3 phosphorylation and normal metaphase chromosome alignment. *Genes Dev.* 19:472-488.

[22] Daskalakis M, Blagitko-Dorfs N, Hackanson B. (2010) Decitabine. *Recent Results Cancer Res.* 184:131-157.

[23] Dawson MA, Bannister AJ, Göttgens B, Foster SD, Bartke T, Green AR, Kouzarides T. (2009) JAK2 phosphorylates histone H3Y41 and excludes HP1alpha from chromatin. *Nature.* 461:819-822.

[24] De Souza CP, Osmani AH, Wu LP, Spotts JL, Osmani SA. (2000) Mitotic histone H3 phosphorylation by the NIMA kinase in Aspergillus nidulans. *Cell.* 102:293-302.

[25] Deshpande A, Sicinski P, Hinds PW. (2005) Cyclins and cdks in development and cancer: A perspective. *Oncogene* 24:2909–2915.

[26] Downs JA, Lowndes NF, Jackson SP. (2000) A role for Saccharomyces cerevisiae histone H2A in DNA repair. *Nature.* 408:1001-1004.

[27] Ehrlich M. (2003) Expression of various genes is controlled by DNA methylation during mammalian development. *J. Cell Biochem.* 88:899-910.

[28] Ehrlich M. (2002) DNA methylation in cancer: too much, but also too little. *Oncogene* 21:5400–5413.

[29] Espino PS, Pritchard S, Heng HH, Davie JR. (2009) Genomic instability and histone H3 phosphorylation induction by the Ras-mitogen activated protein kinase pathway in pancreatic cancer cells. *Int. J. Cancer.* 124:562-567.

[30] Feinberg AP, Vogelstein B. (1983) A technique for radiolabeling DNA restriction endonuclease fragments to high specific activity. *Anal. Biochem.* 132:6-13.

[31] Fischle W, Wang Y, Allis CD. (2003) Histone and chromatin cross-talk. *Curr. Opin. Cell Biol.* 15:172-183.

[32] Frescas D, Guardavaccaro D, Bassermann F, Koyama-Nasu R, Pagano M. (2007) JHDM1B/FBXL10 is a nucleolar protein that represses transcription of ribosomal RNA genes. *Nature.* 450:309-313.

[33] Fujita N, Watanabe S, Ichimura T, Tsuruzoe S, Shinkai Y, Tachibana M, Chiba T, Nakao M. (2003) Methyl-CpG binding domain 1 (MBD1) interacts with the Suv39h1-HP1heterochromatic complex for DNA methylation-based transcriptional repression. *J. Biol. Chem.* 278:24132-24138.

[34] Gaudet F, Hodgson JG, Eden A, Jackson-Grusby L, Dausman J, Gray JW, Leonhardt H, Jaenisch R. (2003) Induction of tumors in mice by genomic hypomethylation. *Science.* 300:489-492.

[35] Goldberg AD, Allis CD, Bernstein E. (2007) Epigenetics: a landscape takes shape. *Cell.* 128:635-638.

[36] Hanahan D, Weinberg RA. (2000) The hallmarks of cancer. *Cell* 100:57-70.

[37] Hanahan D, Weinberg RA. (2011) Hallmarks of cancer: the next generation. *Cell* 144:646-674.

[38] He J, Kallin EM, Tsukada Y, Zhang Y. The H3K36 demethylase Jhdm1b/Kdm2b regulates cell proliferation and senescence through p15(Ink4b). (2010) *Nat. Struct. Mol. Biol.* 15:1169-1175.

[39] Hess JL. (2004) MLL: a histone methyltransferase disrupted in leukemia. *Trends Mol. Med.* 10:500–507.

[40] Hsu JY, Sun ZW, Li X, Reuben M, Tatchell K, Bishop DK, Grushcow JM, Brame CJ, Caldwell JA, Hunt DF, Lin R, Smith MM, Allis CD. (2000) Mitotic phosphorylation of histone H3 is governed by Ipl1/aurora kinase and Glc7/PP1 phosphatase in budding yeast and nematodes. *Cell.* 102:279-291.

[41] Huang Y, Greene E, Murray Stewart T, Goodwin AC, Baylin SB, Woster PM, Casero RA Jr. (2007) Inhibition of lysine-specific demethylase 1 by polyamine analogues results in reexpression of aberrantly silenced genes. *Proc. Natl. Acad. Sci. USA* 104:8023–8028.

[42] Huang Y, Stewart TM, Wu Y, Baylin SB, Marton LJ, Perkins B, Jones RJ, Woster PM, Casero RA Jr. (2009) Novel oligoamine analogues inhibit lysine-specific demethylase 1 and induce reexpression of epigenetically silenced genes. *Clin. Cancer Res.* 15:7217–7228.

[43] Jenuwein T, Allis CD. (2001) Translating the histone code. *Science.* 293:1074-1080.

[44] Jia J, Zhang W, Wang B, Trinko R, Jiang J. (2003) The Drosophila Ste20 family kinase dMST functions as a tumor suppressor by restricting cell proliferation and promoting apoptosis. *Genes Dev.* 17:2514-2519.

[45] Jones PA, Taylor SM. (1980) Cellular differentiation, cytidine analogs and DNA methylation. *Cell.* 20:85-93.

[46] Kantarjian, H. M., O'Brien, S., Cortes, J., Giles, F. J., Faderl, S., Issa, J.-P., Garcia-Manero, G., Rios, M. B., Shan, J., Andreeff, M., Keating, M. and Talpaz, M. (2003), Results of decitabine (5-aza-2'deoxycytidine) therapy in 130 patients with chronic myelogenous leukemia. *Cancer* 98: 522–528.

[47] Kelly AE, Ghenoiu C, Xue JZ, Zierhut C, Kimura H, Funabiki H. (2010) Survivin reads phosphorylated histone H3 threonine 3 to activate the mitotic kinase Aurora B. *Science.* 330:235-239.

[48] Keppler BR, Archer TK. (2008) Chromatin-modifying enzymes as therapeutic targets--Part 1. *Expert. Opin. Ther Targets.* 12(10):1301-1312.

[49] Kim HG, Lee KW, Cho YY, Kang NJ, Oh SM, Bode AM, Dong Z. (2008) Mitogen- and stress-activated kinase 1-mediated histone H3 phosphorylation is crucial for cell transformation. *Cancer Res.* 68:2538-2547.

[50] Kleer CG, Cao Q, Varambally S, Shen R, Ota I, Tomlins SA, Ghosh D, Sewalt RG, Otte AP, Hayes DF, Sabel MS, Livant D, Weiss SJ, Rubin MA, Chinnaiyan AM. (2003) EZH2 is a marker of aggressive breast cancer and promotes neoplastic transformation of breast epithelial cells. *Proc. Natl. Acad. Sci. U S A.* 100:11606-11611.

[51] Kondo Y, Shen L, Suzuki S, Kurokawa T, Masuko K, Tanaka Y, Kato H, Mizuno Y, Yokoe M, Sugauchi F, Hirashima N, Orito E, Osada H, Ueda R, Guo Y, Chen X, Issa JP, Sekido Y. (2007) Alterations of DNA methylation and histone modifications contribute to gene silencing in hepatocellular carcinomas. *Hepatol. Res.* 37: 974–983.

[52] Krishnan N, Jeong DG, Jung SK, Ryu SE, Xiao A, Allis CD, Kim SJ, Tonks NK. (2009) Dephosphorylation of the C-terminal tyrosyl residue of the DNA damage-related histone H2A.X is mediated by the protein phosphatase eyes absent. *J. Biol. Chem.* 284:16066-16070.

[53] Kubicek S, O'Sullivan RJ, August EM, Hickey ER, Zhang Q, Teodoro ML, Rea S, Mechtler K, Kowalski JA, Homon CA, Kelly TA, Jenuwein T. (2007) Reversal of H3K9me2 by a small-molecule inhibitor for the G9a histone methyltransferase. *Mol. Cell* 25: 473–481.

[54] Lei H, Oh SP, Okano M, Jüttermann R, Goss KA, Jaenisch R, Li E. (1996) De novo DNA cytosine methyltransferase activities in mouse embryonic stem cells. *Development.* 122:3195-3205.

[55] Lim S, Janzer A, Becker A, Zimmer A, Schüle R, Buettner R, Kirfel J. (2010) Lysine-specific demethylase 1 (LSD1) is highly expressed in ER-negative breast cancers and a biomarker predicting aggressive biology. *Carcinogenesis.* 31:512-520.

[56] Lu T, Jackson MW, Wang B, Yang M, Chance MR, Miyagi M, Gudkov AV, Stark GR. (2010) Regulation of NF-kappaB by NSD1/FBXL11-dependent reversible lysine methylation of p65. *Proc. Natl. Acad. Sci. U S A.* 107:46-51.

[57] Lyko F, Ramsahoye BH, Kashevsky H, Tudor M, Mastrangelo MA, Orr-Weaver TL and Jaenisch R. (1999) Mammalian (cytosine-5) methyltransferases cause genomic DNA methylation and lethality in Drosophila. *Nat. Genet.* 23:363-366.

[58] Mikkelsen TS, Ku M, Jaffe DB, Issac B, Lieberman E, Giannoukos G, Alvarez P, Brockman W, Kim TK, Koche RP, Lee W, Mendenhall E, O'Donovan A, Presser A, Russ C, Xie X, Meissner A, Wernig M, Jaenisch R, Nusbaum C, Lander ES, Bernstein BE. (2007) Genome-wide maps of chromatin state in pluripotent and lineage-committed cells. *Nature.* 448:553-560.

[59] Milne TA, Briggs SD, Brock HW, Martin ME, Gibbs D, Allis CD, Hess JL. (2002) MLL targets SET domain methyltransferase activity to Hox gene promoters. *Mol. Cell.* 10:1107–1117.

[60] Nakamura T, Mori T, Tada S, Krajewski W, Rozovskaia T, Wassell R, Dubois G, Mazo A, Croce CM, Canaani E. (2002) ALL-1 is a histone methyltransferase that assembles a supercomplex of proteins involved in transcriptional regulation. *Mol. Cell.* 10:1119–1128.

[61] Newell-Price J, Clark AJ, King P. (2000) DNA methylation and silencing of gene expression. *Trends Endocrinol. Metab.* 11:142-148.

[62] Paz MF, Wei S, Cigudosa JC, Rodriguez-Perales S, Peinado MA, Huang TH, Esteller M. (2003) Genetic unmasking of epigenetically silenced tumor suppressor genes in colon cancer cells deficient in DNA methyltransferases. *Hum. Mol. Genet.* 12:2209-2219.

[63] Peterson CL, Laniel MA. (2004) Histones and histone modifications. *Curr. Biol.* 14:R546-551.

[64] Pfau R, Tzatsos A, Kampranis SC, Serebrennikova OB, Bear SE, Tsichlis PN. (2008) Members of a family of JmjC domain-containing oncoproteins immortalize embryonic fibroblasts via a JmjC domain-dependent process. *Proc. Natl. Acad. Sci. U S A.* 105:1907-1912.

[65] Rhee I, Bachman KE, Park BH, Jair KW, Yen RW, Schuebel KE, Cui H, Feinberg AP, Lengauer C, Kinzler KW, Baylin SB, Vogelstein B. (2002) DNMT1 and DNMT3b cooperate to silence genes in human cancer cells. *Nature*. 416:552-556.

[66] Riggs AD, Jones PA. (1983) 5-methylcytosine, gene regulation, and cancer. *Adv. Cancer Res.* 40:1-30.

[67] Rogakou EP, Pilch DR, Orr AH, Ivanova VS, Bonner WM. (1998) DNA double-stranded breaks induce histone H2AX phosphorylation on serine 139. *J. Biol. Chem.* 273:5858-5868.

[68] Shi S, Calhoun HC, Xia F, Li J, Le L, Li WX. (2006) JAK signaling globally counteracts heterochromatic gene silencing. *Nat. Genet.* 38:1071-1076.

[69] Shi Y, Lan F, Matson C, Mulligan P, Whetstine JR, Cole PA, Casero RA, Shi Y. (2004) Histone demethylation mediated by the nuclear amine oxidase homolog LSD1. *Cell.* 119:941-953.

[70] Shroff R, Arbel-Eden A, Pilch D, Ira G, Bonner WM, Petrini JH, Haber JE, Lichten M. (2004) Distribution and dynamics of chromatin modification induced by a defined DNA double-strand break. *Curr. Biol.* 14:1703-1711.

[71] Singh RK, Gunjan A. (2011) Histone tyrosine phosphorylation comes of age. *Epigenetics*. 6:153-160.

[72] Singh RK, Kabbaj MH, Paik J, Gunjan A. (2009) Histone levels are regulated by phosphorylation and ubiquitylation-dependent proteolysis. *Nat. Cell Biol.* 11:925-933.

[73] Singh RK, Liang D, Gajjalaiahvari UR, Kabbaj MH, Paik J, Gunjan A (2010). Excess histone levels mediate cytotoxicity via multiple mechanisms. *Cell Cycle*. 9:4236-4244.

[74] Strahl BD, Allis CD. (2000) The language of covalent histone modifications. *Nature* 403:41–45.

[75] Stucki M, Clapperton JA, Mohammad D, Yaffe MB, Smerdon SJ, Jackson SP. (2005) MDC1directly binds phosphorylated histone H2AX to regulate cellular responses to DNA double-strand breaks. *Cell.* 123:1213-26.

[76] Tachibana M, Matsumura Y, Fukuda M, Kimura H, Shinkai Y. (2008) G9a/GLP complexes independently mediate H3K9 and DNA methylation to silence transcription. *EMBO J.* 27: 2681–2690.

[77] Tachibana M, Sugimoto K, Nozaki M, Ueda J, Ohta T, Ohki M, Fukuda M, Takeda N, Niida H, Kato H, Shinkai Y. (2002) G9a histone methyltransferase plays a dominant role in euchromatic histone H3 lysine 9 methylation and is essential for early embryogenesis. *Genes Dev*. 16:1779–1791.

[78] Tan J, Yang X, Zhuang L, Jiang X, Chen W, Lee PL, Karuturi RK, Tan PB, Liu ET, Yu Q. (2007) Pharmacologic disruption of Polycomb-repressive complex 2-mediated gene repression selectively induces apoptosis in cancer cells. *Genes Dev*. 21:1050-1063.

[79] Trasler J, Deng L, Melnyk S, Pogribny I, Hiou-Tim F, Sibani S, Oakes C, Li E, James SJ, Rozen R. (2003) Impact of Dnmt1 deficiency, with and without low folate diets, on tumor numbers and DNA methylation in Min mice. *Carcinogenesis.* 24:39-45.

[80] Tsukada Y, Fang J, Erdjument-Bromage H, Warren ME, Borchers CH, Tempst P, Zhang Y. (2006) Histone demethylation by a family of JmjC domain-containing proteins. *Nature*. 439:811-816.

[81] van Holde, KE (1988) *Chromatin*. Springer-Verlag, New York.

[82] Villar-Garea A, Fraga MF, Espada J, Esteller M. (2003) Procaine is a DNA-demethylating agent with growth-inhibitory effects in human cancer cells. *Cancer Res*. 63:4984-4989.

[83] Völkel P, Angrand PO. (2007) The control of histone lysine methylation in epigenetic regulation. *Biochimie*. 89:1-20

[84] Waddington CH. (1942) *The epigenotype. Endeavour*, 1, 18-20.

[85] Wang J, Hevi S, Kurash JK, Lei H, Gay F, Bajko J, Su H, Sun W, Chang H, Xu G, Gaudet F, Li E, Chen T. (2009a) The lysine demethylase LSD1 (KDM1) is required for maintenance of global DNA methylation. *Nat. Genet*. 41:125-129.

[86] Wang Y, Zhang H, Chen Y, Sun Y, Yang F, Yu W, Liang J, Sun L, Yang X, Shi L, Li R, Li Y, Zhang Y, Li Q, Yi X, Shang Y. (2009b) LSD1 is a subunit of the NuRD complex and targets the metastasis programs in breast cancer. *Cell*. 138:660-672.

[87] Xiao A, Li H, Shechter D, Ahn SH, Fabrizio LA, Erdjument-Bromage H, Ishibe-Murakami S, Wang B, Tempst P, Hofmann K, Patel DJ, Elledge SJ, Allis CD. (2009) WSTF regulates the H2A.X DNA damage response via a novel tyrosine kinase activity. *Nature*. 457:57-62.

[88] Zheng Y, John S, Pesavento JJ, Schultz-Norton JR, Schiltz RL, Baek S, Nardulli AM, Hager GL, Kelleher NL, Mizzen CA. (2010) Histone H1 phosphorylation is associated with transcription by RNA polymerases I and II. *J. Cell Biol*. 189:407-415.

[89] Zhou L, Cheng X, Connolly BA, Dickman MJ, Hurd PJ, Hornby DP. (2002) Zebularine: a novel DNA methylation inhibitor that forms a covalent complex with DNA methyltransferases. *J. Mol. Biol*. 321:591-599.

[90] Krishnamoorthy T, Chen X, Govin J, Cheung WL, Dorsey J, Schindler K, Winter E, Allis CD, Guacci V, Khochbin S, Fuller MT, Berger SL. (2006) Phosphorylation of histone H4 Ser1 regulates sporulation in yeast and is conserved in fly and mouse spermatogenesis. *Genes Dev*. 20:2580-2592.

[91] Metzger E, Yin N, Wissmann M, Kunowska N, Fischer K, Friedrichs N, Patnaik D, Higgins JM, Potier N, Scheidtmann KH, Buettner R, Schüle R. (2008) Phosphorylation of histone H3 at threonine 11 establishes a novel chromatin mark for transcriptional regulation. Nat Cell Biol. 2008 Jan;10(1):53-60. Epub 2007 Dec 9. Erratum in: *Nat. Cell Biol*. 10:53-60.

[92] Shimada M, Niida H, Zineldeen DH, Tagami H, Tanaka M, Saito H, Nakanishi M. (2008) Chk1 is a histone H3 threonine 11 kinase that regulates DNA damage-induced transcriptional repression. *Cell*. 132:221-232.

[93] Chapman JR, Jackson SP. (2008) Phospho-dependent interactions between NBS1 and MDC1 mediate chromatin retention of the MRN complex at sites of DNA damage. *EMBO Rep*. 9:795-801.

[94] Varambally S, Dhanasekaran SM, Zhou M, Barrette TR, Kumar-Sinha C, Sanda MG, Ghosh D, Pienta KJ, Sewalt RG, Otte AP, Rubin MA, Chinnaiyan AM. (2002) The polycomb group protein EZH2 is involved in progression of prostate cancer. *Nature*. 419:624-629.

In: Signaling, Gene Regulation and Cancer
Editors: Shree Ram Singh and Manoj K. Mishra
ISBN: 978-1-61942-088-5

Chapter 20

An Overview of Cell Cycle Regulation and Tumor Viruses

Shuntele N. Burns
Department of Mathematics and Science
Alabama State University, Montgomery, AL, US

Abstract

Cell division is vital for normal growth and development. The highly coordinated sequence of events that ultimately leads to cell replication in eukaryotes is called the cell cycle. Both external and internal signals influence the cycle. Several checkpoints exist within the cycle to ensure that cellular processes have been properly completed before the cell advances to the next step. The activities of cyclin-dependent kinases (Cdk's) are central to cell cycle regulation, and fluctuations in Cdk activity are largely due to changes in the concentration of cyclins. Different kinds of cyclins are manufactured during specific periods of the cycle, resulting in the construction of special cyclin-Cdk complexes that stimulate particular cellular events. For example, progression beyond G_1 of the cycle normally depends on the inactivation of the retinoblastoma tumor suppressor gene product by cyclin-Cdk complexes.

Disruptions in the cell cycle control system can lead to the development of cancer. Approximately 20% of human cancers are induced by viruses, in that viral infections are responsible for obstructing normal cell cycle regulation. RNA viruses primarily induce cancer by creating oncogenes, while DNA viruses promote excessive cell division by encoding for proteins that inhibit tumor suppressor genes. Research related to tumor viruses has provided valuable information concerning the basic principles of cell cycle control and the roles of major genes in regulation.

Introduction

The cell cycle involves a complex series of events that includes interphase, mitosis, and cytokinesis. During the cycle, a cell increases in mass, replicates its DNA, and eventually separates into two cells. Cell division is vital to the formation of tissues and organs and the replacement of worn and damaged cells. Proto-oncogenes (genes that promote cell division) and tumor suppressor genes (genes that suppress cell division) are critical in regulating the number of cells produced. Progression of the cell through the cell cycle is influenced by an array of extracellular and intracellular chemical signals and is monitored at certain checkpoints. Protein complexes composed of cyclins and kinases are critical in detecting whether specific cellular processes have occurred at the appropriate time and whether or not the cell cycle should advance beyond certain points.

Cancer occurs when mechanisms that normally regulate cell division malfunction. A healthy cell may undergo a series of changes that releases it from the normal control mechanisms responsible for limiting the number of cells produced. These mechanisms can become faulty when cellular genes incur mutations or when proteins manufactured during a viral infection alter control system function. This chapter will analyze the principle components of the cell cycle, including mechanisms that normally regulate cell proliferation, and will examine various ways in which DNA and RNA tumor viruses usurp the cell cycle control system to transmute healthy cells to cancer cells.

THE CELL CYCLE

Each individual originates from a one-celled fertilized egg, which, through cell division, becomes trillions. Cell division is essential for development. In adults, the role of cell division moves from growth and specialization to primarily replacement of cells that have died naturally or were lost to injury (Craig et al., 2010). There are relatively few types of cells that continually divide once a person reaches maturity, and the production of these cells is guided by the need for regular replacement (Hanson, 2005). For example, the body is constantly generating thousands of new skin cells and cells lining the digestive and respiratory tracts (Mader et al., 2011). It is important that cell division is strictly regulated to ensure the proper make up and function of tissues and organs. The loss of control can lead to an unrestricted production of cells, a feature exhibited by cancer cells (Craig et al., 2010).

Cell division is a component of the cell cycle, which is an ordered set of events encompassing cell growth, genetic duplication, and the eventual separation of the cell into two daughter cells. When repeated over and over, the cycle is a constant source for the manufacturing of cells. External signals, like growth factors, normally provide the appropriate stimuli to induce a human cell to enter the cell cycle (Mader et al., 2011). The two main phases of the cycle are interphase and mitotic phase. Though the duration of the cycle varies significantly from one cell type to another, a typical animal cell spends approximately 24 hours completing the cycle, with about 90% of the time engaged in activities associated with interphase—growth and the duplication of chromosomes—and the relative brief mitotic phase devoted to the separation of duplicated chromosomes and the ultimate division of the cell into two (Krogh, 2002).

Interphase dominates the cell cycle through a series of events that involves the cell increasing in size, the cytoplasmic components doubling in number, and the DNA replicating. Interphase proceeds in the following order: G_1 ("gap-one"), S, and G_2 ("gap-two"). During G_1, the cell grows as certain organelles, such as ribosomes and mitochondria, multiply, and chemicals necessary for DNA replication are produced. At this point, each chromosome within the cell consists of a single chromatid (Mader et al., 2011). The period of time in which the cell remains in G_1 can vary considerably with different cell types. Some cells, like neurons of the brain, are arrested at G_1 and will not typically complete the cell cycle; these cells are metabolically active but have entered the quiescent G_0 state (Sadava et al., 2008). Liver and kidney cells are also normally detained in G_0; however, appropriate external cues can initiate cell proliferation to replace those that have been lost as a result of injury (Cooper, 2000).

The synthesis of DNA takes place during the S subphase, in that each chromosome becomes duplicated, i.e., at the conclusion of this subphase, the chromosome possesses two identical sister chromatids that are joined together. During G_2, the cell produces proteins that will be required for cell division; proteins associated with the spindle apparatus, for instance, will be needed to move chromatids to opposite poles of the dividing cell (Mader et al., 2011). The mitotic phase, the second of the two main phases of the cell cycle, involves the division of nuclear and cytoplasmic material and the ensuing separation of the cell into two cells. The mitotic phase has two primary components: mitosis and cytokinesis. Mitosis entails a sequence of stages (prophase, metaphase, anaphase, and telophase) that leads to the separation of sister chromatids of each duplicated chromosome. During this process, the sister chromatids become daughter chromosomes that are allocated to two daughter nuclei. Finally, the cell enters cytokinesis, which includes the division of cytoplasm and the subsequent splitting of the cell, resulting in two identical daughter cells. The cycle repeats with the two daughter cells shifting back into interphase (Starr and Taggart, 2004).

CONTROLLING THE CELL CYCLE

Extracellular and intracellular signals are integral in monitoring and directing the different processes that occur during the cell cycle. External cues called growth factors are proteins that can initiate or suppress cell division (Mader et al., 2011). A stimulatory growth factor binds in a lock-and-key fashion to a specific receptor on the plasma membrane of a targeted cell. This binding prompts the production of proteins that participate in different reactions involved in triggering cell division (Raven et al., 2009). In mammals, growth factors are critical in triggering early cell cycle reactions (Mader et al., 2011). Various types of growth factors exist, and cell proliferation is dependent upon the appropriate growth factor required for the particular cell type. For example, when an injury results in the loss of blood, platelets not only initiate blood clotting at the wound, but they also produce and release a chemical called platelet-derived growth factor that binds to adjacent skin cells and induces them to divide and heal the lesion. Interleukins are other growth factors, produced by white blood cells, that induce cell division in other cells vital to the immune system (Sadava et al., 2008). As noted earlier in this chapter, some cells typically remain in the quiescent G_0 subphase, unless stimulated to reenter the cell cycle by external factors (Cooper, 2000). Liver cells, for instance, can be driven to enter G_1 with several key growth factors—hepatocyte growth factor, epidermal growth factor, and transforming growth factor-α—in order to restore damaged or lost cells (Fausto, 2000).

The membrane-bound protein identified as Ras has an intimate connection with growth factors and is the product of a proto-oncogene, which promotes cell division and inhibits apoptosis (programmed cell death). Once a growth factor binds to a cell membrane receptor, the Ras protein becomes activated. Next, a series of molecules are successively activated, eventually sending a chemical signal to the nucleus, which then promotes gene expression for cell division (Mader et al., 2011). The Ras protein is involved in the production of cyclin D, a protein that is critical in cell cycle progression (Weinberg, 2007b) and whose specific role will be discussed later in the chapter. Proto-oncogenes, like the *RAS* gene, have the potential to cause cancer; mutated proto-oncogenes that have become cancer-inducing are called oncogenes. The *RAS* oncogene produces a protein that is permanently activated; therefore, no growth factor is required, and the cell is stimulated to divide uncontrollably (Medema and Bos, 1993). A mutated Ras protein has been detected in about 25% of all human tumors (Mader et al., 2011).

A group of proteins called cyclin-dependent kinases (Cdk's) are important components of the internal control of the cell as it progresses through the cell cycle; these proteins monitor the progress of the cell through the cycle and prompt the cell to enter subsequent stages when conditions are favorable. A kinase is an enzyme that transfers a phosphate group to a molecule, thereby altering its shape and function. The Cdk's are important in initiating steps in the cell cycle by catalyzing the phosphorylation of particular proteins. The kinase does not work alone but functions in concert with a protein called cyclin, which activates the kinase by changing its shape and displaying its active site, the region where the substrate binds (Sadava et al., 2008). In the late 1980s, Hartwell and Weinert demonstrated that a Cdk is required in order for yeast cells to proceed from G_1 to S. Further investigation has revealed that several cyclin-Cdk complexes exist at checkpoints throughout the human cell cycle (Sadava et al., 2008).

The cell cycle is temporarily halted at specific times, checkpoints, to detect whether conditions favor cell division. Checkpoints ensure that the cell advances through the cycle in the correct order and that each step is properly concluded (Starr, 2005). The cyclin-Cdk complexes, along with a host of other molecules, serve as the basis of processes that advance, postpone, or prevent the movement of the cell within the cycle. For example, the cyclin A-Cdk2 complex initiates DNA duplication during the S subphase, and cyclin B-Cdk1 prompts the cell to enter mitosis (Sadava et al., 2008). A major role of checkpoints is to gauge DNA damage (Cagle, 2008). The following proteins play major roles in detecting genetic damage in humans: ATM, ATR, Rad17, Rad9, Rad1, Hus1. Once damage is identified, a signal is transmitted to target molecules associated with the cell cycle that are responsible for halting cell cycle progression. Proteins BRCA1, CLASPIN, Chk1, and Chk2 serve as mediators, and several participate in DNA repair. For example, Chk1 and Chk2 phosphorylate the target protein Cdc25, which then inactivates cyclin E-Cdk2, a cyclin-Cdk complex that is critical in moving the cell from G_1 to S subphase (Weinberg, 2007b). Research has uncovered various pathways that are responsible for correcting DNA mutations. Most of these procedures involve the removal of defective DNA and the subsequent insertion of nucleotides, using the unharmed complementary DNA strand as a template. Following DNA repair, chemical cues stimulate the cell to continue through the cycle. However, if DNA damage is irreparable, apoptosis may ensue. Checkpoints arrest the cycle for DNA repair or set the path for cell death in order to eliminate the distribution of damaged DNA to daughter cells. Hence, abnormalities within checkpoints contribute to the onset of cancer—the uncontrolled propagation of mutated cells (Cagle, 2008).

The restriction point is a checkpoint within G_1 that determines whether the cell will make the transition to the S subphase. The restriction point involves several cyclin-Cdk complexes and tumor suppressor genes, which encode proteins necessary for inhibiting the proliferation of cells. When activated, the retinoblastoma protein (RB), a tumor suppressor, blocks the production of molecules that are critical for numerous biological processes, including DNA replication and mitosis (Ishida et al., 2001). However, when RB is inactivated by the phosphorylation of cyclin-Cdk complexes, the cell moves beyond the restriction point and is committed to enter the S subphase (Sadava et al., 2008).

The events leading to the successful passage of the cell past the restriction point illustrate a sophisticated cascade of biochemical reactions. In its active state, RB binds to the transcription factor E2F, a protein that controls the expression of several genes necessary for the transition of the cell into the S subphase. DNA synthesis cannot take place as long as RB and E2F are bound. However, when RB is phosphorylated by cyclin D-Cdk4/Cdk6 complexes, the RB protein changes shape and then releases E2F. Once freed from RB, E2F triggers its own gene expression, as well as that of other genes related to the cycle. These genes include cyclin E, proteins vital to the initiation of DNA replication, and molecules needed later in the cycle. The binding of cyclin E to Cdk2 forms a complex that also phosphorylates RB, thus inactivating it and allowing the cell to advance past the restriction point (Ravin and Johnson, 2001). The Cdk's become inactive when cyclin breaks down (Weinberg, 2007b).

The RB protein can also operate with other tumor suppressor proteins within G_1 to prevent cells with abnormal DNA from progressing past the restriction point. For example, proteins p53 and p21 can work together to prevent cell division (Koshland, 1993). If damaged DNA is detected in G_1, there is an increased production of the p53 protein. These proteins serve as transcription factors that activate *p21* gene expression. The p21 protein subsequently inhibits the cyclin-Cdk complexes—cyclin D-Cdk4/Cdk6 and cyclin E-Cdk2—by binding to them; therefore, the cycle is halted to allow for DNA repair. Once DNA has been repaired, the p21 protein breaks down, which then releases cyclin-Cdk complex inhibition, thereby allowing the cell to continue its progress through the cycle (He et al., 2005). However, if DNA injury is too severe for repair, p53 induces apoptosis with enzymes called caspases (Harris and Levine, 2005). About half of all human cancers have a defective *p53* gene, leading to ineffective cell cycle controls (Mader et al., 2011). Cancer arises from inappropriate cell replication, and the interruption in normal cyclin-Cdk activities can be found in these abnormal cells. For instance, abnormalities in the *p53* gene of some breast cancer cells allow for elevated levels of cyclin D, which overstimulate Cdk's and subsequently lead to excessive cell division (Sadava et al., 2008).

Protein p16 is another key molecule that can suppress cell cycle progression into the S subphase. Genetic abnormalities can stimulate p16 to bind to Cdk4 and Cdk6, thereby inhibiting cyclin D activity. As a result, the cyclin D-Cdk4/Cdk6 complexes are not available to phosphorylate RB (Meyerson and Harlow, 1994).

THE BASICS OF CANCER AND THE DISRUPTION IN CELL CYCLE CONTROLS

The examples provided above illustrate some of the regulator proteins involved in keeping cell division under control. However, when genes that code for these proteins become mutated, the proteins may not function properly, and continuous cell division can occur, thereby transferring defective genes to daughter cells. Tissue resulting from unregulated cell growth is usually non-functional, does not exhibit the proper form, and cannot interact properly with different cell types (Raven et al., 2009).

Neoplasms are abnormal masses of cells that no longer have control over cell division. Neoplasms can be benign (noncancerous) or malignant (cancerous). Benign neoplasms, like ordinary skin moles, grow very slowly and usually do not migrate. However, malignant growths can disturb neighboring tissues, both physically and metabolically (Starr, 2005).

There are over 100 different types of cancers, which generally comprise three main groups: carcinomas, sarcomas, and leukemias/lymphomas (Cooper, 2000). About 85% of cancers are carcinomas and involve malignances of epithelial cells (Pecorino, 2008). Sarcomas are tumors of connective tissues, like muscle and bone. Leukemia is a malignancy of the blood cells in the bone marrow, while lymphoma is a cancer of the lymphocytes, typically in the lymph nodes. The tissue of origin (e.g., lung carcinoma) and the cell type (e.g., erythroid leukemia) can be used to further classify malignant growths (Cooper, 2000).

Environment and heredity are the primary factors contributing to the onset of cancer, with environmental influences accounting for the vast majority of cancers (Weinberg, 2007a). Environmental factors include diet, tobacco, radiation, and infectious organisms (Anand et al., 2008).

Hanahan and Weinberg (2000) have identified six general characteristics of cancer cells. First, normal growth factor signaling is not required to initiate cell division. For example, many cancer cells have the ability to produce specialized growth factors that promote their own proliferation, leading to continuous autostimulation of cell division; the cancer cells are less dependent on growth factors synthesized by normal cells (Fedi et al., 1997). As noted earlier in this chapter, when the *RAS* proto-oncogene becomes mutated, its protein product may be constantly activated without stimulation from growth factors (Medema and Bos, 1993).

Second, growth inhibitory signals that normally suppress cell proliferation are not recognized; therefore, cells grow unchecked. Alterations in tumor suppressor genes, such as *RB*, can encode proteins that no longer inhibit cell grow. The RB protein may not normally bind to E2F, and gene expression for proteins necessary for the cell to advance through the G_1 subphase takes place (Raven et al., 2009).

Third, cancer cells are insensitive to apoptotic signals. A component of the apoptotic signaling pathway is the p53 tumor suppressor protein. When the protein is inactivated, the cell may proceed through the cell cycle regardless of damaged DNA. Programmed cell death is essential in preventing abnormal cells from growing relentlessly (Symonds et al., 1994).

Fourth, there is unlimited replicative potential in cancer cells. In a non-cancerous cell, the chromosomal ends, telomeres, shorten each time the cell divides. These DNA "caps" prevent the deterioration of genes at the ends of the chromosomes; however, over time, the telomeres become short enough to prevent cell division, and the cell eventually dies. In a cancerous cell, however, the enzyme telomerase maintains telomeric DNA sequences, allowing the cell to divide without restraint (Sadava et al., 2008).

Fifth, cancer cells encourage angiogenesis, the formation of new blood vessels. Chemical cues induce the growth of blood vessels which provide the tumor cells with oxygen and nutrients for growth. Furthermore, these vessels deprive normal cells of essential molecules (Mader et al., 2011).

Sixth, tissue invasion and metastasis are characteristics of malignant tumors. Cancer cells generally lose adhesive properties that allow the cells to have close interaction. The reduction in molecules that permit adhesion contributes to the invasion and metastasis of cancer cells. In addition, reduced adhesiveness causes morphological and cytoskeletal changes. For example, many cancer cells are rounder than normal cells, primarily because malignant cells are not securely attached to neighboring cells (Cooper, 2000).

Cancer arises from the accumulation of multiple mutations (Loeb and Loeb, 2000), and abnormalities in proto-oncogenes and tumor suppressor genes have been identified in many malignant cells. Over 100 proto-oncogenes have been identified, and at least 15 tumor suppressor genes have been discovered (Pecorino, 2008). Proto-oncogenes normally promote cell division; however, once they mutate to become cancer-causing genes, oncogenes, excessive cell proliferation can occur in a number of ways: (1) Mutations lead to the over-manufacturing of growth factors. (2) A defective cell membrane receptor does not require the binding of a growth factor to initiate cell division. (3) Mutant transcription factors inappropriately induce cell division by becoming permanently bound to the promoter of a target gene. (4) Proteins that participate in signal transduction pathways are always activated when mutated; therefore, an external stimulus is not needed to induce cell division (Sadava et al., 2008).

Tumor suppressor genes produce proteins that normally inhibit cell division; however, when mutations inactivate these proteins, cell division is uncontrollable: (1) Mutant adhesion/recognition proteins prevent neighboring cells from interacting; therefore, cells spread to different areas of the body (Hirohashi and Kanai, 2003). (2) Proteins that are normally involved in repairing DNA do not make corrections when they are defective, resulting in the accumulation of DNA mutations. These alterations, in turn, can impair other tumor suppressor genes (Markowitz, 2000). (3) Tumor suppressor proteins involved in halting the cell cycle in G_1 become damaged to the point that they are inactive, resulting in the cell incorrectly advancing past the restriction point and entering subphase S (Sadava et al., 2008). (4) Defective proteins no longer induce apoptosis in irreparably damaged cells (Weinberg, 2007b).

Characteristics of Viruses and Historical Perspective of Research Involving Tumor Viruses

Viruses are not composed of cells; therefore, they cannot be grouped under either of the three domains of the living world—Bacteria, Archaea, and Eukarya. In fact, most viruses are significantly smaller than bacteria. In addition, viruses do not perform some of the basic functions of living things, such as executing metabolic activities—taking in nutrients and expelling waste products. However, viruses have the capability of reproducing within host cells. Viruses that infect animals can be very diverse. Some consist of nucleic acid enclosed by particles of proteins (capsid), and others possess a membrane originating from the plasma membrane of a host cell. Animal viruses have either DNA or RNA as genetic material, and the viral genome primarily codes for just a few proteins (Sadava et al., 2008).

There are five principal steps during which a typical animal virus multiples within the host cell: attachment, penetration, replication and protein synthesis, assembly, and release. First, the virus binds to a plasma membrane receptor of the host cell, and either the intact virus or just its nucleic acid—DNA or RNA—enters the cell. Next, viral genetic material instructs the host cell to manufacture many copies of the virus's genome and proteins. The manufactured viral components are then put together to produce new clones of the virus, which can eventually leave the host cell to infect other cells (Starr, 2005).

Hundreds, even thousands of new viruses can be released from a host cell. Depending on its life cycle, a virus can be virulent or temperate. A virulent virus ultimately leads to the destruction of the host cell, while a temperate virus allows the cell to live for an extended period with viral genome integrated into the host cell's DNA (Weinberg, 2007a).

Viruses are among the many factors that have been found to increase the risk of developing certain types of cancer. The relationship between viruses and cancer was a major discovery in cancer investigation. These tumor viruses make genetic alterations in host cells that cause them to become cancerous. Approximately 20% of all human cancers are induced by viruses (Crawford, 2005).

Toward the end the nineteenth century, Louis Pasteur and Robert Koch discovered the pathogenic agents that were responsible for diseases such as cholera, rabies, and dysentery. These disease-causing agents were placed in two major categories, bacteria and viruses. Infectious agents that were large enough to be collected in the pores of filters were described as bacteria, while those that were not retained in filters were considered viruses. Therefore, diseases like smallpox and rabies were identified as viruses (Weinberg, 2007a).

Recognition of the association between viruses and cancer developed over many years with a number of critical experiments. In 1876 a Russian veterinarian reported the first tumor to be transplanted experimentally; tumor tissue from a dog was transferred to a second dog, and a tumor later developed. These findings demonstrated to many researchers that cancer could be transmissible (Das and Das, 2000). In 1908 Wilhelm Ellermann and Oluf Bang demonstrated that a filterable agent (virus) from leukemia cells of chickens could induce the disease in other birds upon transmission. However, at that time, leukemia was not considered a form of cancer, and, unfortunately, this finding was primarily overlooked (Shors, 2009). Peyton Rous continued with this line of study, and in 1911 he reported that a virus obtained from a sarcoma in the breast muscle of hens initiated tumor growths in healthy chickens following injections of the virus; furthermore, an elevated level of viruses was later recovered from injected chickens, far more than was originally introduced. Subsequent work by Rous uncovered two other viruses associated with chickens. Nonetheless, the importance of Rous's discovery that cancer could be induced by a virus was not fully appreciated for decades, after which other novel experiments revealed the existence of additional tumor viruses. Rous eventually received the Nobel Prize in Medicine and Physiology in 1966, more than 50 years after his original discovery (Weinberg, 2007a).

Rous's studies of tumor viruses, along with those of several other key scientists, laid the foundation for the discovery of oncogenes. The Rous Sarcoma Virus (RSV) encodes for three genes (*env*, *pol*, and *gag*) that are required for viral proliferation and production of virus particles. In the 1970s a fourth gene, *src*, was found to induce healthy cells to become cancerous—oncogenic transformation. Viral genes responsible for causing cancer in normal cells were described as viral tumor-causing genes or oncogenes (Shors, 2009). In 1976 J. Michael Bishop and Harold Varmus reported that a normal gene with a homologous sequence to *src* was found in healthy chickens. Further inquiry revealed that this gene can be found in humans and many other animals (Pecorino, 2008), and in 1989 Bishop and Varmus each received a Nobel Prize for their groundbreaking investigation (Shors, 2009). These and other findings established that oncogenes are mutated forms of normal genes, proto-oncogenes (Pecorino, 2008).

RNA Tumor Viruses

The actions of many RNA and DNA viruses have been shown to promote cancer development in different groups of animals, from amphibians to primates, and certain human cancers are virally induced (Pandey and Madhuri, 2010). Studies of retroviruses, which consist of an RNA genome and include the well-known Rous Sarcoma Virus, have provided valuable insights into cancer biology, in particular our comprehension of oncogenes (Macdonald et al., 2004). RNA tumor viruses rarely infect humans but are usually associated with chickens, cats, and mice. The human T-cell leukemia viruses (HTLV-1 and HTLV-2) and the human immunodeficiency virus (HIV) are retroviruses known to induce cancers in people (Raju and Madala, 2005).

The retrovirus has a simple genome, typically consisting of three regions—*gag*, *pol*, and *env*—that encode for the core protein, the enzyme reverse transcriptase, and the envelope protein, respectively. The two ends of the viral genome consist of promoter and enhancer sequences, identified as long terminal repeats (LTRs), which control the transcription of genes (Welch and Chrysogelos, 2002).

The reproductive cycle of the retrovirus is quite unique. Once the virus enters the host, a single-stranded viral RNA is converted to a double-stranded DNA copy. This procedure is directed by the viral enzyme reverse transcriptase. The DNA copy is eventually integrated into the host chromosome; the introduced DNA is called a provirus. This latent strategy helps the provirus elude recognition from the host's immune system. New viruses are not manufactured during latency; however, certain external factors, such as ultraviolet radiation, can trigger the production of viruses (Mader et al., 2011). During this process, viral DNA is transcribed to mRNA by the host cell's RNA polymerase. The mRNA molecule is then translated by cellular ribosomes. The resulting proteins are assembled with the viral mRNA to produce progeny viruses that will eventually leave the host cell to infect new cells (Shors, 2009).

The introduction of a retrovirus into a host chromosome may cause a gene in the host to come under the control of novel regulatory pathways, thus setting the stage for unregulated cell proliferation. The transformation of proto-oncogenes to oncogenes by retroviruses can lead to abnormalities in growth factors, growth factor receptors, intracellular signal transducers, and nuclear transcription factors, along with other proteins involved in cell division (Pecorino, 2008).

Transduction and insertional mutagenesis are two major mechanisms used by retroviruses to transform a proto-oncogene to an oncogene in non-humans (Welch and Chrysogelos, 2002). In transduction, the viral genome is integrated within the host's DNA, and recombination takes place between sequences, resulting in the incorporation of a segment of the host cell's DNA into the viral genome (Raju and Madala, 2005). When the cell divides, the resulting daughter cells will each receive a copy of the normal genome of the cell, along with a copy of the introduced provirus (Sadava et al., 2008). Occasionally transduction will result in the provirus acquiring a portion or all of a host cell's proto-oncogene. The transduced gene sequence is now under the control of the active viral promoter and has become oncogenic, leading to unrestrained cell division. Oncogenic activity can also take place if the acquired gene incurs structural alterations during transduction and generates abnormal protein that contributes to cell proliferation (Welch and Chrysogelos, 2002).

Insertional mutagenesis is another major mechanism by which retroviruses disrupt normal gene expression in non-humans (Pecorino, 2008). Insertional mutagenesis entails a more direct method of altering host genes that normally promote cell division. During viral genome integration with the host, the retrovirus inserts its DNA close to a proto-oncogene of the host, resulting in continuous gene expression driven by the local viral promoter of the LTRs (Macdonald et al., 2004).

Retroviruses that infect humans can induce oncogenic transformation of host cells by trans-activation. This mechanism is displayed in human T-cell leukemia virus type 1 (HTLV-1), which was the first discovered human retrovirus. Adult T-cell leukemia/lymphoma and HTLV-associated myclopathy (neurological disorder) are caused by HTLV-1. In addition to the structural and enzymatic genes *gag*, *pol*, and *env*, HTLV-1 possesses two partially overlapping regulatory genes, *tax* and *rex*. The encoded protein from the *tax* gene fosters transcription of proviral DNA, as well as initiates the expression of various proto-oncogenes. The protein product of the *rex* gene serves as a post-transcriptional activator that escalates the movement of viral mRNA molecules from the host's nucleus to the cytoplasm (Ryan et al., 2010). The retrovirus HTLV-1 does not have a transduced oncogene in its genetic material, nor does it continuously activate a local proto-oncogene (Coffin et al., 1997). Instead, with trans-activation, a protein expressed from a viral gene, such as the *tax* gene, initiates the production of cellular proteins essential for cell division (Ryan et al., 2010).

DNA Tumor Viruses

DNA tumor viruses are very diverse, comprising different structures, genetic organization, and replication strategies (Shors, 2009). Notable DNA viruses that are carcinogenic to humans include papovaviruses, adenoviruses, and hepatitis B (Pecorino, 2008). Studies are incomplete regarding the manner in which mammalian DNA viruses assemble their genetic material into the viral capsid. While some viruses produce their own proteins, others take advantage of the host cell's mechanism. For instance, papovaviruses employ certain histones from the host to assemble their genetic material (Allison, 2007).

Contingent on the type of host cell, there are two modes of replication that take place in DNA tumor viruses. In permissive host cells, the entire viral genome is expressed, viral particles are assembled to form viral progeny which are released from the cell, and the host cell eventually dies (Matigan et al., 2000). In nonpermissive cells, the viral DNA is usually integrated into the host cell's genome at random locations. The viral DNA is replicated with the host's DNA; however, a portion of the genetic material of the virus is expressed. Since structural proteins associated with the virus are not produced, no viral progeny are released. At some point, the host cell becomes altered to where unregulated cell division takes place, eventually leading to a tumor (Allison, 2007).

Protein-protein interaction is the primary mechanism by which DNA tumor viruses convert normal cells into cancerous cells. Viral proteins called tumor antigens and cellular proteins interact in a manner such that cellular proteins are removed from their normal functional positions, which then sets the path for abnormal cell proliferation (Raju and Madala, 2005). The DNA tumor virus generally inhibits normal activity of protein products of host tumor suppressor genes, such as the *p53* and *RB* genes, thereby altering normal cell cycle progression. The inactivation of tumor suppressors is essential for productive viral replication (Shors, 2009).

Almost all cervical cancers are associated with the human papillomavirus (HPV), a member of the viral family papova. Of the 130 different types of HPV, only a small number are considered high risk for cervical cancer. After infection occurs, the cancer may take at least ten years to develop. Genital HPV is one of the most common sexually acquired viral infections. The HPV proteins E6 and E7 are critical in promoting cancer by interacting with RB, which normally prevents the host cell from advancing to subphase S of the cell cycle (Pecorino, 2008). Protein E7 binds to RB, causing it to release E2F, and the cell progresses forward in the cycle. The E7 protein of high-risk cancer HPV's binds to RB with greater affinity than the protein of low-risk HPV's, facilitating an increase in cell transformation (Jenkins and Hoffman, 2001). Protein E6 binds to ubiquitin ligase, which then complexes with p53. Once ubiquitinated, p53 undergoes degradation, in that the half-life of p53 is decreased from 3 hours to 20 minutes with a corresponding reduction in its biological activity. The loss of p53 by E6 blocks the host cell from repairing its DNA or undergoing apoptosis. At this point, the cell is unable to prevent further viral injury and is very vulnerable to additional genetic abnormalities (Stubenrauch and Iftner, 2006). As with E7, the E6 of high-risk cancer HPV's binds more effectively to p53 than the E6 of low-risk viruses (Jenkins and Hoffman, 2001). Furthermore, E6 has been found to promote telomerase activity, which functions in replenishing the ends of chromosomes, thereby extending the number of times a cell will divide (Klingelhutz et al., 1996).

Though the HPV genome is integrated randomly in cellular DNA, the *E1* and *E2* genes are consistently disrupted. Proteins encoded by these genes suppress the transcription of *E6* and *E7* genes. Therefore, disruption of *E1* and *E2* leads to the unregulated production of E6 and E7 proteins, increasing the risk of developing cancer (Jenkins and Hoffman, 2001).

CONCLUSION

In multicellular organisms, cell division is essential for growth and repair of tissues. Cell division is a principle component of the cell cycle, which is an ordered progression of events including cell growth, genetic replication, and, finally, division into two daughter cells. Eukaryotic cells have an elaborate mechanism for initiating and coordinating different processes in the cell cycle. External and internal factors are important in controlling the cycle. Cyclin-Cdk complexes are key to regulating the progression of the cell through the phases of the cycle. For example, genetic damage will activate a series of steps involving the combination of cyclin-Cdk complexes and tumor suppressor proteins so that the cycle is arrested while repairs are made. Once DNA repair is complete, protein complexes disassociate, and the cell advances through the cycle. However, when proper regulatory controls malfunction, the cell may divide uncontrollably to form cancer cells.

Approximately 20% of human cancers are attributed to viruses. Researchers have identified various ways in which RNA and DNA tumor viruses interfere with the normal control mechanisms of the cell cycle. RNA viruses mainly promote the development of cancer by creating oncogenes, and DNA viruses induce elevated cell production by altering tumor suppressor genes. Studies conducted over the years to ascertain the methods by which tumor viruses alter cell division have been invaluable in providing insights into the mysteries surrounding the genetic basis of human cancers, including those that have no relationship to viral infections.

REFERENCES

Allison, L. A. (2007). *Fundamental Molecular Biology*. Malden, MA: Blackwell Publishing.

Anand, P., Kunnumakara, A. B., Sundaram, C., Harikumar, K. B., Tharakan, S. T., Lai, S. C., Sung, B., and Aggarwa, B. B. (2008). Cancer is a preventable disease that requires major lifestyle changes. *Pharmaceutical Research*, 25, 2097-2116.

Cagle, P. T., Jagirdar, J., and Popper, H. H. (2008). Molecular genetics of lung and pleural neoplasms. In T. F. Tomashefski, P. T. Cagle, C. F. Farver, and A. E. Fraire (Eds.), *Dail and Hammar's Pulmonary Pathology: Neuroplastic Lung Disease* (3rd edition, pp. 47-157). New York, NY: Springer Science and Business Media.

Coffin, J. M., Hughes, S. H., and Varmus, H. E. (1997). *Retroviruses*. Woodbury, NY: Cold Spring Harbor Laboratory Press.

Cooper, G. M. (2000). *The Cell: A Molecular Approach* (2nd edition). Sunderland, MA: Sinauer Associates.

Craig, A. A., and Barry, S. A. (2010). *Cancer: Basic Science and Clinical Aspects*. Hoboken, NY: John Wiley and Sons.

Crawford, D. H. An introduction to viruses and cancer [online]. 2005 [2011 April 11]. Available from: www.sgm.ac.uk/pubs/micro_today/pdf/ 080501.pdf.

Das, U., and Das, A. K. (2000). Review of canine transmissible venereal sarcoma. *Veterinary Research Communications*, 24, 545-556.

Fausto, N. (2000). Liver regeneration. *Journal of Hepatology*, 32, 19-31.

Fedi, P., Tronick, S. R., and Aaronson, S. A. (1997). Growth factors. In J. F. Holland, R. C. Bast, D. L. Morton, E. Frel, D. W. Kufe, and R. R. Weichselbaum (Eds.), *Cancer Medicine* (pp. 41-64). Baltimore, MD: Williams and Wilkins.

Hanahan, D., and Weinberg, R. A. (2000). The hallmarks of cancer. *Cell,* 100, 57-70.

Hanson, B. A. (2005). *Understanding Medicinal Plants: Their Chemistry and Therapeutic Action*. Binghamton, NY: Haworth Herbal Press.

Harris, S. L., and Levine, A. J. (2005). The p53 pathway: positive and negative feedback loops. *Oncogene*, 24: 2899-2908.

He, G., Siddik, Z. H., Huang, Z., Wang, R., Koomen, J., Kobayashi, R., Khokhar A. R., and Kuang, J. (2005). Induction of p21 by p53 following DNA damage inhibits both cdk4 and cdk2 activities. *Oncogene*, 24: 2929-2943.

Hirohashi, S., and Kanai, Y. (2003). Cell adhesion system and human cancer morphogenesis. *Cancer Science*, 94: 575-581.

Ishida, S., Huang, E., Zuzan, H., Spang, R., Leone, G., West, M., and Nevins, J. R. (2001). Role for E2F in control of both DNA replication and mitotic functions as revealed from DNA microarray analysis. *Journal of Molecular Cell Biology*, 21: 4684-4699.

Jenkins, F. J., and Hoffman, L. J. (2001). Viral carcinogenesis. In K. Bland, J. Daly, and C. Karakousis (Eds.), *Surgical Oncology: Contemporary Principles and Practice* (pp. 92-100). New York, NY: The McGraw-Hill Companies, Inc.

Klingelhutz, A. J., Foster, S. A., and McDougall, J. K. (1996). Telomerase activation by the E6 gene product of human papillomavirus type 16. *Nature*, 380: 79-82.

Koshland, D. E. (1993). Molecule of the year. *Science*, 262: 1953.

Krogh, D. (2002). *Biology: A Guide to the Natural World* (2nd edition). Upper Saddle River, NJ: Prentice-Hall, Inc.

Loeb, K. R., and Loeb, L. A. (2000). Significance of multiple mutations in cancer. *Carcinogenesis*, 21: 379-385.

Macdonald, F., Ford, C. H. J., and Casson, A. G. (2004). *Molecular Biology of Cancer* (2nd edition). New York, NY: Garland Science/BIOS Scientific Publishers.

Mader, S. S., Isaacson, J. A., Lyle-Ippolito, K. G., and Storfer, A. T. (2011). *Inquiry into Life* (13th edition). New York, NY: The McGraw-Hill Companies, Inc.

Markowitz, S. (2000). DNA repair defects inactivate tumor suppressor genes and induce hereditary and sporadic colon cancers. *Journal of Clinical Oncology*, 18: 75-80.

Matigan, M. T., Martinko, J. M., and Parker, J. (2000). *Brock Biology of Microorganisms.* New York, NY: Garland Science/BIOS Scientific Publishers.

Medema, R. H., and Bos, J. L. (1993). The role of p21 ras in receptor tyrosine kinase signaling. *Critical Reviews in Oncogenesis*, 4, 615-661.

Meyerson, M., and Harlow, E. (1994). Identification of G1 kinase activity for cdk6, a novel cyclin D partner. *Molecular Cell Biology*, 14, 2077-2086.

Pandey, G., and Madhuri, S. Oncogenic DNA and RNA viruses causing the cancer pathogenesis [online]. 2010 [2011 April 22]. Available from: globalresearchonline.net/journalcontents/ volume5issue3/Article-021.pdf.

Pecorino, L. (2008). *Molecular Biology of Cancer: Mechanisms, Targets, and Therapeutics* (2nd edition). New York, NY: Oxford University Press Inc.

Raju, S. M., and Madala, B. (2005). *Illustrated Medical Biochemistry*. New Delhi, India: Jaypee Brothers Medical Publishers.

Raven, P. H., and Johnson, G. B. (2001). *Biology* (4th edition). New York, NY: The McGraw-Hill Companies, Inc.

Raven, P. H., Johnson, G. B., Mason, K. A., Losos, J., and Singer, S. (2009). *Biology* (8th edition). New York, NY: The McGraw-Hill Companies, Inc.

Ryan, K. J., Ray, C. G., Ahnad, N., Drew, W. L., and Plorde, J. J. (2010). *Sherris Medical Microbiology* (5th edition). New York, NY: The McGraw-Hill Companies, Inc.

Sadava, D., Heller, C., Orlans, G., Purves, B., and Hills, D. (2008). *Life: The Science of Biology* (8th edition). Sunderland, MA: Sinauer Associates, Inc.

Shors, T. (2009). *Understanding Viruses.* Sudbury, MA: Jones and Bartlett Publishers.

Starr, C. (2005). *Biology: Today and Tomorrow.* Belmont, CA: Thomas Learning, Inc. Starr, C., and Taggart, R. (2004). *Biology: The Unity and Diversity of Life* (10th edition). Belmont, CA: Thomas Learning, Inc.

Stubenrauch, F., and Iftner, T. (2006). Molecular basis of cervical neoplasm. In Jordan, J. A., Singer, A., Jones, H. W., Shafi, M. I. (Eds.), *The Cervix* (2nd edition, pp. 277-288). Malden, MA: Blackwell Publishing, Inc.

Symonds, H., Krall, L., Remington, L., Saenz-Robies, M., Lowe, G., Jacks, T., and Van Dyke, T. (1994). p53-dependent apoptosis suppresses tumor growth and progression in vivo. *Cell*, 78: 703-711.

Weinberg, R. A. (2007a). *The Biology of Cancer.* New York, NY: Garland Science, Taylor and Francis Group.

Weinberg, R. A. (2007b). Cancer—principles and overview. In B. Lewin, L. Cassimeris, V. R. Lingappa, and G. Plopper. (Eds.), *Cells* (pp. 561-586). Sudbury, MA: Jones and Bartlett Publishers.

Welch, J. N., and Chrysogelos, S. A. (2002). Positive mediators of cell proliferation. In W. B. Coleman and G. T. Tsongalis (Eds.), *The Molecular Basis of Human Cancer* (pp. 65-79). Totown, NJ: Humana Press Inc.

In: Signaling, Gene Regulation and Cancer
Editors: Shree Ram Singh and Manoj K. Mishra
ISBN: 978-1-61942-088-5

Chapter 21

BIODEGRADABLE POLYMERIC NANO-BIOMATERIALS FOR CANCER THERAPEUTIC APPLICATIONS

Vinoy Thomas[1], Bini Mathew[3], Derrick R. Dean[2] and Elijah Nyairo[4]
[1]Center for Nanoscale Materials and Biointegration (CNMB), Department of Physics and
[2]Department of Material Science and Engineering University of Alabama at Birmingham (UAB), Birmingham, AL, US
[3]Department of Organic Chemistry, Drug Discovery Division, Southern Research Institute (SRI), Birmingham, Alabama, US
[4]Center for Nanobiotechnology Research (CNBR), Alabama State University, AL, US

ABSTRACT

Nano-biotechnology has the potential to impact the detection, treatment and monitoring of cancer significantly and to revolutionize cancer therapies. Nanoparticles are particles with dimensions in the 1-1000 nm range, the scale at which materials have unique physico-chemical and structural properties including ultra small size, large surface to mass ratio, high reactivity, and unique interactions with biological systems. This chapter aims to overview briefly nano-particulate and nanofibrous therapeutic delivery systems based on biocompatible, biodegradable polymeric materials. These multi-functionalized nanocarriers designated as 'magic bullets' as postulated by Paul Ehrlich, the Founder of chemotherapy are presently being developed with the capability of delivering the cancer drugs of interest and bio-imaging for the detection of cancer.

1. Introduction

Cancer is a major human health problem worldwide. It is the second leading cause of human mortality, and accounts 7 million deaths per year (12.5% of deaths worldwide) [1-3]. In the U.S.A, it is estimated that one out of four deaths is cancer-related. Since the early 1970's, the survival rate has improved only by ~10% as more advanced methods and imaging tools have been introduced for early cancer detection [3]. Although the recent reports from US national cancer institute (NCI) indicate a decline in the cancer incidence rates as well as in the death rates related to cancer, the absolute number of individuals diagnosed with cancer will continue because of aging population, leading to increased demands for cancer-related medical services through the spectrum of diagnosis, active treatment, and post treatment management [4]. Nanomedicine in cancer therapy is an emerging field that deals with the effective therapeutic delivery and diagnosis utilizing functional structures on the nanometer scale (≤100 nm). At the nanoscale, the properties of materials often differ drastically from those of the corresponding macro-sized bulk materials (eg. nanoparticles, nanofibers, nanotubes, nanorods, quantum dots). There are numerous applications for a combination of nanotechnology, biotechnology, and biomedical technology in cancer therapies (chemotherapy, photodynamic therapy [5], hyperthermia [6] and electrochemotherapy [7]) and other medical disciplines. Among them, applications involved in the controlled and targeted delivery of therapeutics for treatment and diagnosis or monitoring biological systems together has recently been referred by the term 'theranostics'. The nanomaterials used as carriers/probes in these applications can be broadly grouped into inorganic nanoparticles and polymeric nanoparticles, although, all of them are not in particulate form though.

Inorganic nanoparticles explored for diagnostic-imaging to date include, iron oxide nanocrystals, gold nanoshells, quantum dots, nanodiamonds and nanosilica. Iron oxide nanocrystals [8-10] with super paramagnetic properties (5–100 nm in diameter) are used as contrast agents in magnetic resonance imaging (MRI) of cancer tissues. Gold nanoshells (10–300 nm in diameter) which are unique nanoparticles consisting of a silica-core surrounded by a thin gold-shell have been approved by the FDA for hyperthermia treatment for many cancers [11-13]. Quantum dots (QDs) are attractive as fluorescent imaging agents compared to conventional fluorophores (organic dyes) due to their strong resistance to photo-bleaching [14-19]. They are fine semiconductor nanocrystals (3-10 nm) composed of an elemental core (e.g. cadmium, zinc) surrounded by a metal shell. Nanodiamonds [20, 21] are carbon nanoparticles (2-8 nm) produced by explosion synthesis (detonation). They are commercially available with a large number of organic functional groups (for drug conjugation), with nitrogen vacancy (NV) centers for florescence emission, and they could be an alternative to quantum dots in terms of non-toxicity and photo-blinking. Organically modified nanosilica particles (ORMOSIL) are another class of inorganic nanoparticles having potential in cancer therapeutic applications [3]. These non-biodegradable, inorganic nanoparticles can be synthesized in the nano size range (5-30 nm) suitable for unimpeded circulation in the bloodstream and extravasation into tumor tissue [3].

By definition, organic nanoparticles or polymeric nanoparticles are colloidal systems of submicron size (10-1000 nm) [22, 23] that can be prepared from a large variety of biodegradable and biocompatible polymeric materials in a large variety of combinations. Polymeric nanoparticles find applications in cancer therapy mainly for the encapsulation and delivery of therapeutic drugs rather than cancer diagnostics [22-27]. This chapter is arranged to review briefly on the cancer-therapuetic applications of major biodegradable polymeric biomaterials, approved by the FDA for various medical applications. Attempts to use electrospun biomaterials based on nanofibers of biodegradable polymers in cancer is discussed towards the end of the chapter. The advantages of using polymeric nanoparticles as carriers in drug delivery systems are many-fold. An 'ideal nanomedicine-based therapeutic device' should hold the anti-cancer drug in the blood for a long time ('long circulating nanoparticles') and allow a controlled drug release at the cancer tissue only, without toxic side-effects on other normal functional tissues. It is well-known that a majority of anti-cancer drugs are highly toxic to normal functional cells. Conventional treatment modalities in chemotherapy bear high risk due to the severe side effects resulting from the non-specificity of the drugs to cancer tissues, poor bio-availability and bio-distribution of drugs, and eventual multi-drug resistance (MDR) [28, 29] by the cancer cells. MDR, one of the most challenging threats to survival and a common cause to tumor persistence despite the invasive chemotherapy, refers to a cross-resistance to structurally and functionally unrelated drugs, thereby rendering the tumor unresponsive to most chemotherapeutics. The strategic use of multifunctional polymeric nanoparticles as targeting drug delivery vehicles for anticancer therapeutics has great potential to modulate drug resistance [30]. Commonly used polymer based organic nanoparticles are in the form of a single polymeric particle, or core-shell particle, polymeric micelles, dendrimers, and liposomes [1, 24, 31, 32]. Although micelles, liposomes and dendrimers are classical examples for polymer based colloidal nanostructures, this review is focused mainly on the polymeric nanoparticles (nanospheres and nanocapsules) only.

2. Biodegradable Polymeric Biomaterials for Nanoparticle Synthesis

The first issue with regard to tissue engineering and drug delivery is the choice of the suitable scaffolding/carrier materials. The drug-carrier materials require desirable characteristics such as biodegradability (degradable into nontoxic products), biocompatibility (elicit no unwanted immune reactions), and the right surface chemistry to prevent agglomerations and to resist cell-adhesive protein adsorption. Both natural and synthetic materials have been considered for making promising polymeric carriers in nanomedicne formulation [24, 33] and scaffolds in regenerative medicine [34]. The major attractions of synthetic polymers over natural biopolymers are their availability with a wide range of physico-chemical and mechanical properties. The properties can be tailored for specific applications by optimizing composition and structure-property relations. However, mild toxicity of the intermediate acidic degradation products is a concern for their use in tissue engineering (TE) [35]. Biopolymers have advantages in terms of good biocompatibility, but lack mechanical and dimensional integrity in physiological media. Moreover, because of the native-origin, they suffer batch-to-batch variation in properties, molecular weight, purity etc. and are easily degradable without much control on the rate of degradation.

Polymers used in cancer therapies are classified as either biodegradable or non-biodegradable polymers. While biodegradable polymers find use in therapeutic-delivery of drugs, non-biodegradable polymers have found to use in functionalizing aforementioned inorganic diagnostic agents. Most of the synthetic biodegradable polymers involved in drug delivery systems (DDS) are polyesters (polymer with large number of hydrolytically labile ester groups formed by the condensation polymerization of corresponding di-acids and diols) such as polyglycolide (PGA), polylactide (PLA), poly-(lactide- co-glycolide) (PLGA), and poly(ε-caprolactone) (PCL) or polyanhydrides (formed by the condensation of acid monomers) [33-36]. These polymers have sterilizability, favorable biocompatibility, controlled biodegradability and the degradation by-products can be metabolized and removed from the body through normal metabolic pathways. Most widely used native biopolymers in nanoparticle preparation include gelatin, chitosan, alginate etc.

Synthetic polymers have the advantage of a sustained release of the therapeutic agent over a period of days to several weeks as compared with natural polymers with a relatively shorter duration of drug release. Biodegradation behavior of these polymers can be altered by changing the block copolymer composition and molecular weight [36]. Hence, the release of an encapsulated therapeutic agent can be adjusted to last from days to months. Alternatively, several crosslinking methods can be adapted to improve the biostability/degradability characteristics of native polymers [37, 38]. The polymers selected must meet several requirements such as drug compatibility, suitable degradation kinetics as well as biocompatibility and ease of processing [33-39]. Table 1 gives the physico-chemical and degradation characteristics of the most commonly used biodegradable and biocompatible polymers in cancer-drug/nanoparticle formulations.

Table 1. Biodegradable polymers studied extensively in cancer-nanotechnology

Polymer	Characteristics	Biodegradation
PGA	Highly crystalline hydrophilic (T_g=35-40 °C, T_m=225-230 C)	2-3 months
PLA	Amorphous hydrophobic (T_g =55-60 °C)	12-16 months
PLGA	Amorphous hydrophobic (T_g =45-50 °C)	6-12 months
PCL	Semi-crystalline hydrophobic (T_g =-55 °C)	16-24 months
Gelatin	Protein biopolymer, denatured collagen, hydrophilic	Fast degradation
Chitosan	Carbohydrate biopolymer, hydrophilic	Fast degradation

PLA and its copolymer (PLGA) are the most investigated polymeric nanoparticles for drug delivery [40, 41]. They are synthesized by ring-opening polymerization of cyclic monomers. PLA is extensively used for the encapsulation of many therapeutic agents due to its high hydrophobicity, biodegradability, biocompatibility, low toxicity, strong mechanical strength and slow drug release [41]. PLGA, the most successfully used synthetic biodegradable polymer for the development of nanomedicines, undergoes hydrolysis in the body and produces the biodegradable metabolite monomers, lactic acid and glycolic acid

(Figure 1). PLGA is generally preferred over PLA and PGA (polyglycolide) as it can be dissolved by a range of solvents including tetrahydrofuran (THF), acetone, dichloromethane (DCM) etc.

The degradation profile of this polymer can be easily controlled by selecting the optimum ratio of monomers (lactic acid and glycolic acid) in a polymerization reaction to get the optimum molecular weight, hydrophilicity etc. As a result, PLGA (85:15) showed slower drug-release as compared to the other varieties of PLGA such as PLGA (75:25), PLGA (50:50). Increased entrapment efficiency and drug-loading was observed for particles of PLGA (85:15) than the PLGA with lower L-lactide moiety [42]. Further, PLGA (85:15) produced nanoparticles of lower polydispersity index (PDI) [42] and relatively smaller size compared to other varieties of PLGA, in a given preparation technique [43].

PLA and PLGA nanoparticles have been mostly prepared by emulsification–diffusion [44], emulsion–solvent evaporation [45], interfacial deposition [46] and nano-precipitation methods [47]. Excellent reviews on the preparation of polymeric nanoparticles and size control parameters [48], surface properties [49] and efficient drug incorporation methods [46, 49] have been reported elsewhere.

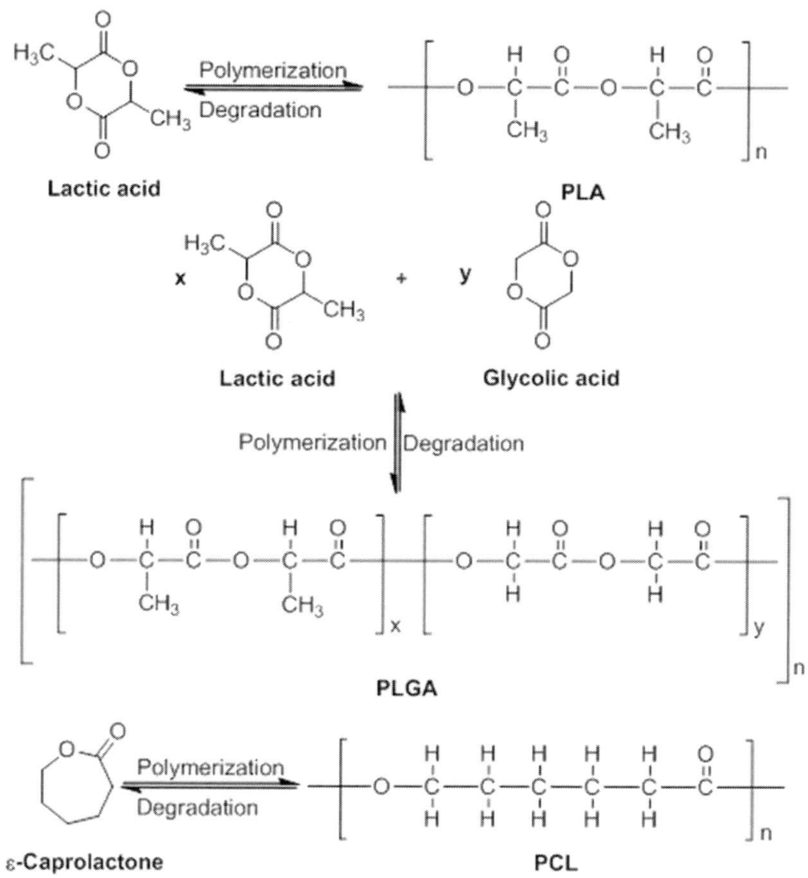

Figure 1. Schematic representation of chemical structures of PLA, PLGA, and PCL their monomers.

Double emulsion- solvent evaporation, a very widely used method for the preparation of PLGA nanoparticles, is illustrated as an example in polymeric nanoparticle preparation section. Depending on the polymer used as carrier, the drug loading and encapsulation may vary. For example, Paclitaxel (anticancer drug) nanomedicine has 100% encapsulation efficiency on PLGA as reported by Mu et al. [50], while having only 20% on PCL nanoparticles as reported by Kim et al. [51]. Several other anticancer drugs including doxorubicin, 5-fluorouracil and dexamethasone have been successfully formulated using PLGA.

Polycaprolactone (PCL) is semi-crystalline biodegradable polyester, especially interesting for the preparation of degradable implants and sutures due to its slow degradation behavior compared to PLGA. PCL is degraded by hydrolysis of its ester linkages in physiological conditions and has also received attention in drug delivery. PLCL copolymers synthesized from ε-caprolactone and DL-lactide yield materials with a more rapid degradation rate than PCL. Nanoparticles of PCL are prepared mostly by nano-precipitation, solvent displacement and solvent evaporation. For example, Devalapally et al. [52] have developed biodegradable PEG–PCL nanoparticles for the delivery of Paclitaxel and ceramide (CER) to overcome drug resistance in ovarian cancer. PEG–PCL amphiphilic polymeric nanospheres containing Taxol reported by Kim et al. [51] showed promising anticancer activity.

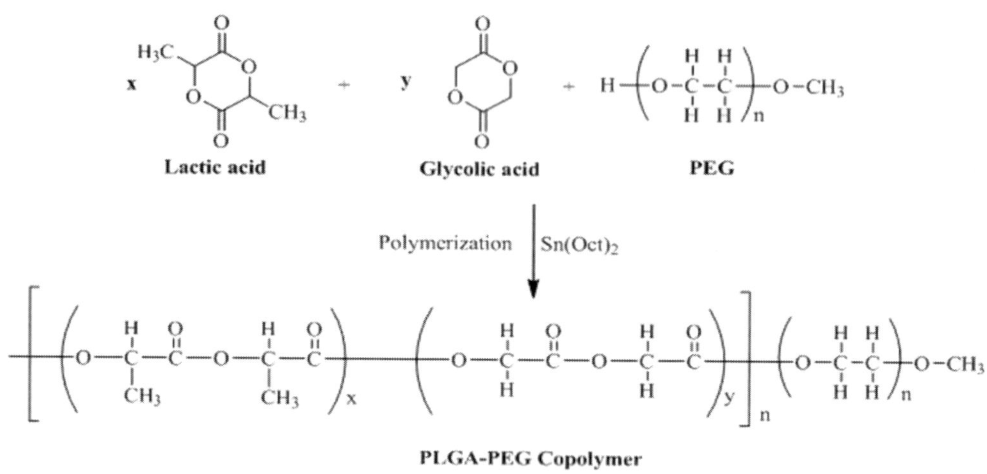

Figure 2. Schematic representation of PLGA-PEG formation from monomers.

Polyethylene glycol (PEG) is a non-toxic, water-soluble polymer with proven biocompatibility. PEGlylation (covalent grafting or adsorption of PEG chains onto nanoparticles) gives a hydrophilic surface to the synthetic hydrophobic nanoparticles, reduces macrophage recognition and complement activation, thereby prolonging their circulation-time in blood. Moreover, PEGlylated nanoparticles with very low interfacial energy in an aqueous environment prevent their unwanted aggregation due to secondary interactions between nanoparticles. A majority of the biodegradable homo polyesters (PLA, PCL, PLGA, PLCL) are strongly hydrophobic and limit their practical drug-formulations. To add hydrophilicity and long-term circulating ability, PEG has been co-polymerized with these polyesters. For example, block copolymers consisting of a hydrophobic PLGA segment and a hydrophilic

PEG segment have attracted significant attention due to their biodegradability, biocompatibility, and tailor-made properties.

A wide variety of drug formulations, such as micro/nano-particles micelles, hydrogels, and injectable drug delivery systems have been developed using PLGA-PEG block copolymers for pharmaceutical applications [53-57]. PLGA-PEG block copolymers with unique properties, such as micro-phase separation, crystallinity, water-solubility, and biodegradability can be classified according to their block structure as AB diblock [58] ABA or BAB [59] or triblock, multi-block [60] branched block [61] star-shaped block [62] and graft block [63] copolymers (where A is a polyester block and B is a PEG block). The synthetic route of a PLGA-PEG copolymer using stannous octoate as catalyst is given in Figure 2.

3. Biodegradable Biopolymers for Nanoparticle Synthesis

Besides the commonly used synthetic polymers, naturally occurring hydrophilic polymers are also used in the preparation of nanoparticles for drug delivery. Most commonly used biopolymers in pharmaceutical applications are polysaccharides such as chitosan, alginate, etc and proteins such as gelatin, albumin etc. Nanoparticles prepared from these biopolymers have hydrogel characteristics (hydrophilicity and extremely high water content) and are commonly known as 'nanogels' [64].

Chitosan (α(1-4)-2-amino-2-deoxy β-D-glucan), is an abundant polysaccharide prepared by deacetylation of chitin from crustacean shells (Figure 3). Chitosan has received attention as a biomaterial for biomedical as well as therapeutic applications due to its well-documented biocompatibility [65-67]. Being a positively charged polyelectrolyte at physiological conditions, chitosan interacts easily with negatively charged molecules and other polyanions. These interactive forces and the resulting sol–gel transition stages have been exploited for nano-encapsulation purposes [68-71]. Chitosan and thiolated chitosan have documented ability of adhering to the mucosal surfaces, a property essential for mucosal drug delivery and mucosal vaccination [72, 73].

Different methods involving emulsion, gelation, and precipitation can be adopted to prepare chitosan nanoparticles. Chitosan nanospheres immobilized with 5-fluorouracil (5-FU), an anticancer drug by a water-in-oil (w/o) emulsion method followed by glutaraldehyde (Glu) crosslinking of the amino groups in chitosan was reported by Ohya et al. [64].

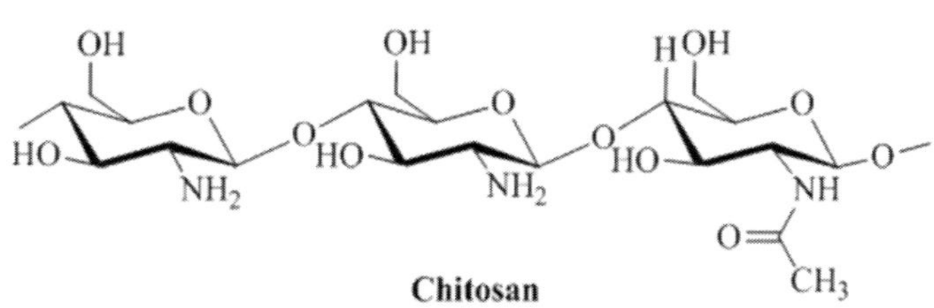

Figure 3. Schematic representation of chemical structure of chitosan.

Since chitosan has the ability to gel upon contact with polyanions (ionotropic gelation) as previously mentioned, nanoparticles of chitosan have been developed based on ionotropic gelation of chitosan with tripolyphosphate (TPP) [74–76] by the addition of an alkaline phase (pH = 7–9) containing TPP into an acidic phase (pH = 4–6) containing chitosan. Nanoparticles are formed immediately upon mixing of the two phases through inter and intra molecular linkages created between TPP phosphates and chitosan amino groups. Chitosan nanoparticles prepared by the ionic gelation of TPP indicated particle-size ranging between 20 and 200 nm with spherical shapes [77].

Another possibility for the synthesis of chitosan nanoparticles is the use of de-solvating or precipitating agents. For example, Berthold et al. [78] have used sodium sulfate as a precipitating agent to form chitosan particles by a drop-wise addition of sodium sulfate into a solution of chitosan and a stabilizer (polysorbate 80 used as stabilizer for the suspension) under both stirring and ultrasonication to desolvate chitosan in a particulate form. Additionally, chitosan-DNA particles prepared using the complex coacervation technique [79 80] exhibited the particle size at 100–250 nm range with a narrow distribution. The chitosan-DNA particles could partially protect the encapsulated plasmid DNA from nuclease degradation. Drug-chitosan nanoparticles have shown potent cytotoxic effects on various tumor cell lines both in vitro and in vivo [81-83].

Gelatin, denatured collagen, is another extensively studied protein-biomaterial in drug delivery applications due to its nontoxic, biodegradable, and bio-absorbable properties [84, 85]. Gelatin is chemically similar to collagen and it is widely used as food-additive and in other medical products. Gelatin nanoparticles can be prepared by the desolvation/coacervation method [84]. In the desolvation/coacervation process, a homogeneous solution of charged macromolecules undergoes liquid–liquid phase separation, giving rise to a polymer-rich dense phase as the bottom-layer and a transparent solution as the top-layer. The addition of natural salt or alcohol normally promotes coacervation and the control of turbidity/crosslinking that results in nanoparticle formation [84]. Many drugs have been successfully encapsulated into gelatin nanoparticles. For example, Paclitaxel-loaded gelatin nanoparticles have been reported by the desolvation method [85]. A rapid drug release from gelatin nanoparticles was observed in PBS with ~90% released at 37 °C after 2 h. The paclitaxel-loaded gelatin nanoparticles were found effective against human bladder transitional cancer cells as reported by Lu et al. [85].

Carbohydrate polymers such as alginate and cyclodextrins (CDs) are also used in drug delivery applications [87-89]. Alginate is an anionic biopolymer consisting of linear chains of α-L-glucuronic acid and β-D-mannuronic acid with properties such as a high degree of aqueous solubility, biocompatibility, and non-toxicity [86]. Crosslinking and formation of polymeric gel-networks result in particle formation upon the addition of counter-ions (e.g. calcium ions) to alginate solution. Cyclodextrins are cyclic oligomers having a unique amphiphilic structure consisting of a hydrophilic cup-shape exterior and a hydrophobic interior. CDs are water-soluble and form inclusion complexes with hydrophobic guest (e.g. adamantine). Several CD-based gene carriers have shown stable polyplexing with DNA with high transfection efficiency and low toxicity [87-89].

4. Preparation of Polymeric Nanoparticles and Drug Release

As mentioned earlier, polymeric nanoparticles can be nanospheres or nanocapsules, depending on the method of preparation. The nanocapsule has a core, shell and surface. The core determines the particle size and generally consists of the inactive matrix, the active agent itself or an active agent-containing matrix. The acts as a base to which surface molecules, that sometimes include active agents, are anchored or bound with or without intermediate spacers and in some cases, it acts as an active agent-dispersed matrix. The surface molecules strongly affect the bio-distribution of the nanoparticles that usually consist of site-, tissue-, cell- and/or receptor-specific molecules (targeting agents). PEG molecules attached to the shell or surface molecules by physical adsorption or covalent bonding with or without intermediate spacers, possibly, induce a bound water layer on the particle surface. The drug can be dispersed homogenously in the polymer matrix, encapsulated/entrapped in core-shell fashion, or covalently conjugated/attached with the polymer molecules. Diffusion, polymer matrix swelling, polymer erosion and biodegradation mechanisms cause the release of the drugs formulated in polymeric devices

Polymeric nanoparticles have been synthesized into nanospheres or nanocapsules using various methods [46-48], depending on applications and type of drugs to be encapsulated. There are many preparation methods such as solvent evaporation, solvent displacement, salting out, and solvent diffusion for formulating drug-loaded nanoaprticles of PLA, PLGA, and PCL. In a typical solvent evaporation method, the polymer is dissolved in an organic solvent (DCM, chloroform, or ethyl acetate), followed by dissolution/dispersion of the drug of interest. This is then added to an aqueous phase containing an emulsifying agent (PVA, gelatin, poloxamer-188, polysorbate-80, etc) by continuous stirring or sonication PVA solution to form an oil-in-water emulsion (O/W). The water-oil-water (W/O/W) double emulsion technique has also been widely used to prepared drug-loaded nanocapsules of polyesters [90-91]. For example, development of biodegradable polymer based tamoxifen citrate loaded nanoparticles was reported for breast cancer therapy by double-emulsion and solvent evaporation techniques [90]. In a typical experiment, the specified amount of the polymer PLGA (250 mg) was dissolved in dichloromethane and tamoxifen citrate was solubilized in methanol. The drug (5-10 mg) and polymer solutions were mixed in a 10 mL beaker and then the aqueous phase containing polyvinyl alcohol (PVA) with specific concentration was added drop-wise into the oil phase and homogenized in a high speed homogenizer at a specific optimized speed (16000 rpm) for 4 minutes. The primary emulsion (W/O) was formed. The primary emulsion was then added drop-wise into the aqueous phase containing PVA with specific concentrations and homogenized in the same speed for 6 minutes and a double emulsion (W/O/W) was formed. The prepared emulsion was placed on a magnetic stirrer and stirred for 12 hours at room temperature for the evaporation of organic solvent such as dichloromethane and methanol. Nanoparticles (200 nm) obtained were filtered, washed in a cooling centrifuge at 4°C and kept in an ultra-low freezer at −80°C and dried in a freeze drier at −42°C for 8 hours. The salting out procedure is based on the separation of a water miscible solvent from aqueous solution by adding a salting out agent like magnesium chloride, calcium chloride, etc. The main advantage of the salting out procedure is that it minimizes stress to protein encapsulants. Detailed procedures of various methods used for preparing nanoparticles (emulsification–diffusion [44], emulsion–solvent evaporation [45], interfacial deposition [46] and nano-precipitation methods [47]) from synthetic polymers and biopolymers have been reported by many researchers [26, 46, 48, 49, 77, 78, 84, 85].

Nanoparticle analyzers based on light scattering methods and microscopic techniques such as scanning electron microscopy (SEM), transmission electron microscopy (TEM) and atomic force microscopy (AFM) are the commonly used for the characterization of size-distribution, zeta potential and surface morphology of polymeric nanoparticles. An SEM image of PLGA nanoparticles prepared by double emulsion is given in Figure 4.

A highly efficient polymeric nanoparticle system must have high loading capacity to reduce the quantity of the polymer required for drug-administration. Drug loading into nanoparticles is achieved either by incorporating the drug at the time of nanoparticle preparation or by adsorbing the drug after the formation of nanoparticles by incubating them in the drug solution [49]. Drugs can be also loaded onto nanoparticles by adding them to the reaction mixture during the polymerization process [49]. These methods provide solid solution for the drug encapsulation in the polymer, dispersion of the drug in the polymer, adsorption of the drug onto the surface of the nanoparticles and chemical binding of the drug to the polymer. The amount of bound-drug and the type of interaction of drug and nanoparticles depend on the chemical nature and structure of the drug as well as the polymer and the conditions of drug loading [92]. The determination of the adsorption isotherm is one possible way to detect the type of binding and the binding rate (mg drug/mg nanoparticle). So far two dozen plus nanotechnology-based therapeutics are approved by US-FDA for clinical use or clinical trial [30]. Two well-known examples of nanomedicine formulations approved by the FDA for cancer therapy are albumin bound Paclitaxel (Abraxane®) and liposomal-doxorubicin (Doxil®).

Paclitaxel Taxol® (Figure 5a), a diterpenoid drug first identified by the US National Cancer Institute in 1967 was extracted from the bark of Pacific yew trees. It has been found to be a potent anticancer drug against a wide range of cancers, including head and neck, ovarian, lung, breast, and colon cancers. Paclitaxel interferes with the normal function of microtubule breakdown by binding with a subunit of tubulin. This drug promotes the polymerization of tubulin causing cell death by disrupting the dynamics necessary for cell division.

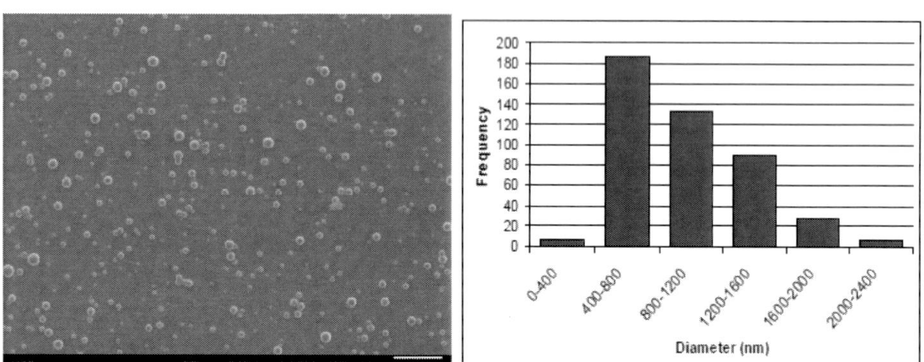

Figure 4. Nano/micro particle of PLGA prepared by a double-emulsion method and the corresponding particle size distribution.

The currently available commercial formulation, Taxol® is formulated with Cremophor EL (polyethoxylated caster oil) and ethanol in a 1:1 ratio. Cremophor EL can induce severe side effects such as hyper sensitivity reactions and is nephrotoxic and neurotoxic. The practical chemotherapeutic application of paclitaxel has been limited also by its high hydrophobicity and poor solubility in water (less than 0.5 mg/L). A nanoparticle formulation of paclitaxel, in which serum albumin is included as a carrier [nanometer-sized (130 nm) albumin bound paclitaxel approved by FDA in 2005) has been applied in the clinic for the treatment of metastatic breast cancer [93].

Doxorubicin (DOX) (Figure 5b) is one of the most widely used anticancer-drug, but its clinical application is still limited by its severe toxic side effects. PEGylated liposomal DOX formulation (Doxil ®, approved by FDA FDA in 1995) Doxil is used for treating advanced ovarian cancer. It is also used with another medicine to treat a type of bone marrow cancer (multiplemyeloma) in certain patients. Doxil is an antineoplastic antibiotic. It works by killing cancer cells. Liposomal DOX coated with PEG exhibits prolonged systemic circulation, prevents opsonization and avoids monocyte phagocytic system (MPS) clearance, and reduces cardiotoxicity and gastrointestinal toxicity [94].

Cisplatin, the *cis*-isomer of square planar Pt(II) complex (*cis*-$PtCl_2(NH_3)_2$), is another very potent drug used in cancer medicine, but the full therapeutic exploitation of cisplatin is limited due to its toxicity [95]. The selective delivery (targeting) of cisplatin to tumor cells would significantly reduce drug toxicity, and improve its therapeutic index. Cisplatin has been encapsulated on PLGA–mPEG nanoparticles prepared by double emulsion methods. PLGA methoxy(polyethylene glycol) (mPEG) nanaoparticles revealed prolonged drug residence in blood upon intravenous administration [96]. However, at the targeted site PLGA–mPEG nanoparticles encapsulated with cisplatin exhibited rapid degradation and sustained release. Very recently, a study on the targeted delivery of a cisplatin prodrug using PLGA-PEG nanoparticles (150±15 nm encapsulating ~5% wt/wt Pt(IV) prodrug) for safer and more effective prostate cancer therapy in vivo was documented by Dhar et al. [97].

The mechanism of drug release from nanoparticles varies with the strategy of their biodistribution and delivery. Typically, the active agents are released from nanoparticles by a desorption/diffusion or by an erosion/biodegradation mechanism in to the blood, the interstitial fluid, in tumor tissue, the tumor cell surface, or in to the intracellular space.

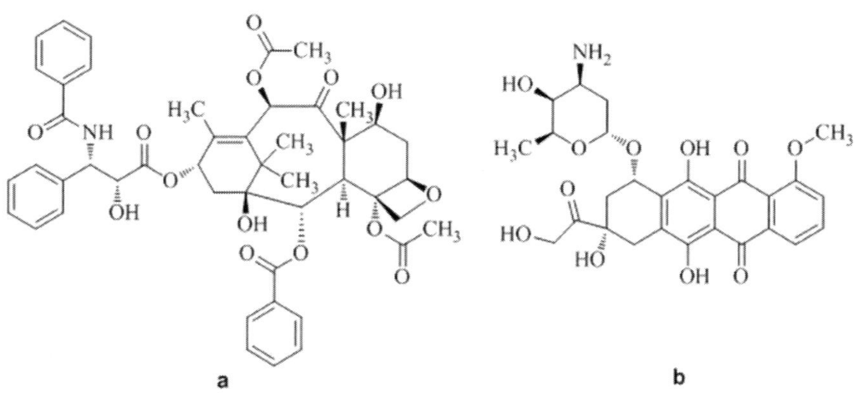

Figure 5. Chemical structure of Paclitaxel (a) and Doxorubicin (b).

The endocytosed particles are transported to the lysosomes and subsequently degraded. Kumari et al. [26] have pointed out five possible methods for drug release: (a) desorption of drug bound to the surface, (b) diffusion through the nanoparticle matrix, (c) diffusion through the polymer wall of nanocapsules, (d) nanoparticle matrix erosion, or (e) a combined erosion–diffusion process [98]. The kinetic analysis of drug release from nanoparticles can be expressed by a mathematical function [26], $C = Ae^{-\alpha t} + B^{-\beta t}$, where C is the concentration of drug remaining in the nanoparticles at time t, A and B are system characteristic constants (A is used for diffusion control system and B for erosion control system), and α and β are rate constants that can be obtained from semi-logarithmic plots [98]. In general, the drug release rate depends upon solubility, diffusion and biodegradation of the drug/matrix materials. The release-mechanism can be tuned by the choice of polymer matrices. Drug release also depends upon the loading efficiency of drug as well as the size of nanoparticles. Larger particles have a smaller initial burst-release than smaller particles [26].

5. Passive Delivery of Drug by Polymeric Nanoparticles

The efficiency of any cancer-therapeitc is directly related to its ability to locate selectively the cancer cells. Majority of anticancer drugs used in conventional chemotherapy have no tumor selectivity and are randomly distributed in the body, resulting in a relatively low therapeutic index. The passive targeting phenomenon, first identified in 1986 by Matsumura et al. [99], has been called the ‘enhanced permeation and retention (EPR) effect.’ Fast-growing cancer cells demand the recruitment of new vessels (neovascularization) or rerouting of existing vessels near the tumor mass to supply them with oxygen and nutrients [100]. The imbalance of angiogenic regulators such as growth factors and matrix metalloproteinases makes tumor vessels highly disorganized or leaky with numerous pores or enlarged gap-junctions between endothelial cells and poor lymphatic drainage [100]. These unique anatomical and patho-physiological abnormality features “the EPR effect”, which controls the passive delivery of drugs by nanoparticles. Numerous studies have shown that the EPR effect results in passive accumulation of macromolecules and nanosized particulates (e.g. polymer conjugates, polymeric micelles, dendrimers, and liposomes) in solid tumor tissues, increasing the therapeutic index while decreasing side effects.

Passive targeting by ‘EPR effect’ takes the advantage of the inherent size and surface properties of nanoparticles and also the tumor environment. Normal healthy tissues have tight, continuous vasculature-walls that allow small molecules to penetrate easily in contrast to large molecules. The size of the pores ranges from approximately 9 nm to 50 nm in diameter; however the larger pores are 100-1000 times fewer in number. Tumor tissues have discontinuous leaky walls with a large number of pores that are approximately 100 nm in diameter due to the absence of basal lamina. This leads to enhanced penetration of 50-100 nm particles in tissues and organs that have discontinuous capillary walls [9]. Since the kidneys are capable of filtering particles of size $\leq$ 10 nm and the liver can capture particles of size $\geq$ 100 nm, the ideal nanocarrier size is somewhere between 10 and 100 nm.

In addition to their size, the surface characteristics such as surface charge and surface hydrophilicity of nanoparticles are also an important factor determining their life span and fate during circulation. Clearance organs (such as the kidneys) can filter nanocarriers not only based on their size, but also by their surface charge. Since positively charged particles can bind easily cells in the body, including endothelial cells, before reaching the target tumor, therefore, it is important to maintain the nanocarriers as either neutral or anionic for successful evasion of renal elimination. Nanoparticles should ideally have a hydrophilic surface to escape from macrophage and monocyte phagocytic systems (MPS) capture [38]. Particles of size below 200 nm with highly hydrophilic surfaces also tend to exhibit an improved EPR effect, which has been attributed to the increased residence time of the carrier in blood. This hydrophilicity can be achieved in two ways, (1) coating the surface of nanoparticles with a hydrophilic polymer (e.g. PEG) and (2) preparing nanoparticles from block copolymers having hydrophilic PEG. The second method of covalent coupling of amphiphilic copolymers (PLA, PCL, PLGA chemically bonded to PEG) is generally preferred, as it avoids aggregation and ligand desorption effectively when in contact with blood components. Other polymers such as poloxamines, poloxamers, polysaccharides (chitosan, alginate, hyaluronic acid, pullulan etc.) can also be coated onto nanoparticles to provide the necessary hydrophilicity, and to circumvent their macrophage clearance. Nanoparticles that satisfy the size and surface characteristics requirements described above for escaping reticuloendothelial system (RES) capture have the ability to circulate for longer times in the bloodstream and a greater chance of reaching the targeted tumor tissues. By tuning the size and surface characteristics of such nanocarriers, an improved biodistribution and activity, as well as a prolonged in vivo half-life, can be reached.

Another contributor to passive targeting is the unique micro-environment surrounding the tumor cells, which is different from that of normal cells. Hyper-proliferative cancer cells show a high metabolic rate (the supply of oxygen and nutrients is usually not sufficient for them to maintain) and therefore, tumor cells use glycolysis to obtain extra energy, resulting in an acidic environment [104]. Additionally, cancer cells over-express and release unique enzymes called matrix metalloproteinases (MMP) [105]. The strategy is to utilize this unique tumor environment in a scheme is called 'tumor-activated prodrug therapy'. In this case, the drug conjugated to a tumor-specific molecule remains inactive until it reaches the target ('pH triggered' delivery). For example, the pH-sensitive liposomes are designed to be stable at a physiologic pH of 7.4, but release active drug in target tissues in which the pH is less than physiologic values (such as in the acidic environment of tumor cells) [106]. An example of albumin-bound form of doxorubicin incorporating a matrix metalloproteinase- 2–specific octapeptide sequence (Gly-Pro-Leu-Gly-Ile-Ala-Gly-Gln) between the drug and the carrier was observed to be efficiently and specifically cleaved by matrix metalloproteinase-2 (MMP-2) in an in vitro study [107]. Localized delivery approach is another route in passive delivery to deliver therapeutic agent directly to the target site and to minimize systemic toxicity and increase the drug-concentration at the site of action [40].

6. Polymeric Nanoparticles in Active Targeted Delivery

Liposomes, micelles, and other polymeric nanoparticles based nanomedicine formulations which depend on the passive delivery of drugs are the first generation nanotherapeutics. Although the first generation nanomedicines provide many advantages such as prolonging drug half-life, improving solubility of hydrophobic drugs, reducing potential immunogenicity, other side effects, and/or releasing drugs in a sustained or 'pH triggered' fashion compared to conventional therapies, they inevitably face intrinsic limitations to its specificity. It is widely believed that active targeting, through the modification of nanoparticles with ligands has the potential to enhance the therapeutic efficacy and reduce the side effects related to conventional therapeutics [108]. The ability to actively target specific cells rather than tissues also allows ligand-conjugated nanocarriers to outperform first generation, non-targeted nanocarriers. The presence of targeting ligands can greatly enhance the retention and cellular uptake of nanoparticles via receptor-mediated endocytosis, even though tumor accumulation is largely determined by the physicochemical properties of nanoparticles [108, 109]. This can then lead to higher intracellular drug concentration and increase in therapeutic activity.

Active targeting to the tumor can be achieved by specific recognition of cancer cells via interactions such as lectin-carbohydrate, ligand–receptor, antibody–antigen couples, or by targeting through aptamers [110]. Lectin-carbohydrate is one of the old examples of targeted drug delivery [111]. Lectins are proteins of non-immunological origin, capable of recognizing and binding to glycoproteins expressed on cell surfaces. However, drug delivery systems (DDS) based on lectin-carbohydrate have mainly been developed to target whole organs [112], which can pose harm to normal cells. Cancer cells often display exclusive proteins on cell surface (tumor-specific antigens) or increased expression of certain proteins (tumor-associated antigens) compared to normal healthy cells. Active drug targeting can be achieved by conjugating a nanoparticle to a targeting component that strongly interacts with these specific antigens or over-expressed receptors by the target tissue, leading to preferential accumulation of the drug in the targeted organ, tissue, or cells. The use of a targeting moiety not only decreases adverse side effects by allowing the drug to be delivered to the specific site of action, but also facilitates cellular uptake of the drug by receptor-mediated endocytosis. Briefly, the specificity and characteristics of targeting moiety are important for circulation time, cellular uptake, affinity and extravasation.

Antibody-mediated targeting is regarded as a promising strategy to enhance the specific delivery of nanoparticles and imaging of cancer due to the high specificity of antigen-antibody reaction. For example, Solbrig et al. [113] have optimized a polymer preparation method to deliver both defined tumor-associated proteins and the complex mixtures of tumor-antigens present in tumors using PLGA(50/50) copolymer. Immunogenicity of antibodies is a serious problem in the application for therapeutic purposes. Antibodies modifier with PEG has shown both reduced antigenicity or immunogenicity and increased targeting efficiency. But, both PEG density and chain length [114, 115] may influence the activity of PEG grafted antibody. For example, a reduction in antibody activity has been observed when multiple PEG chains are conjugated to a single antibody [116].

Transferrin (TfR) is a β1-glycoprotein and facilitates the transport of ferric ions through transferrin receptors on the plasma membrane. The transferrin expression on tumors is above 10 times higher in comparison to non-tumor cells [117] and, therefore, transferrin has been used as a ligand for site-specific delivery to tumors. Kobayashi et al. illustrated TfR-targeted liposomes to deliver doxorubicin and showed that a significant delivery (3.5 fold higher) of encapsulated DOX in both SBC-3 and SBC-3/ADM cells compared to free doxorubicin [118].

The folate receptor is one of the most highly studied targets for cancer therapeutics. Folic acid (FA) is highly over-expressed in a number of human tumors including ovarian, lung, brain, renal cell, and breast cancers, whereas in normal tissue, its expression is significantly lower or absent. Folate mediated targeting has many advantages over other ligands. It is non-immunogenic, easy to conjugate to nanoparticles, has high binding affinity, has high specificity to tumors and is stable in storage as well as in blood circulation [119, 120]. For example, Zhang et al. [121] have synthesized doxorubicin loaded, folate-decorated PLGA-vitamin E nanoparticles for targeted chemotherapy. In another example by Yoo and Park [122], polymeric micelles from PLGA-PEG were prepared to achieve folate-targeted delivery of doxorubicin. Recently, a new folate receptor–targeted nanoparticle formulation of Paclitaxel using heparin as a carrier [heparin-folate-Taxol, HFT] was formulated and tested using nude mouse animal models [123].

Aptamers are oligonucleotides (DNA or RNA) that bear unique 3d-conformations capable of binding to target antigens with high affinity and specificity [124]. They have been applied in DDS as a ligand to enhance selectivity [125]. Aptamers offer advantages over antibodies as they can be engineered to obtain the desirable storage properties, they are easy to characterize and modify, elicit little or no immunogenicity and exhibit high specificity and affinity for their target antigens. The combination of targeted delivery of nanoparticles and EPR effect, effective and controlled release of drug at the site of cancer will result in next generation "smart nano therapeutics" that are safer than what is available today.

7. Potential Use of Electrospun Polymeric Nanofibers in Cancer Therapy

Electrospinning is a simple and fascinating nanotechnology-enabled fiber spinning technique used to produce polymeric fibers in nanoscale dimensions for tissue engineering and drug delivery applications [34, 126-132]. In this process, a polymer jet is initiated when the electrostatic charge overcomes the surface tension of the pendent droplet from a capillary nozzle. As the electrospinning jet moves away from the nozzle, it is stretched by whipping and bending instabilities [133]. Typically, the bending instability causes the length of a jet to elongate by more than 10,000 times in a very short time period (50 ms or less). The elongation rate during the bending instability is extremely high (up to $1x10^6$ s^{-1}) [133, 134]. This huge elongation rate can effectively stretch the macromolecular chains and closely align them along the nanofiber axis. In addition, different combinations of polymer blends, composite fibers with ceramic nanoparticles or carbon nanotubes can be readily incorporated into fibers with alignment along the fiber axis [34, 37, 135, 136]. Because of the nature of electrospinning process, when nanofibers are produced in an electrostatic field they are deposited randomly, layer by layer/over and over with the formation of variously sized pores (ranging from 100 nm to 25 μm) [137-139]. Various parameters such as concentration (viscosity) of the polymer solution, solution feeding rate, conductivity of the solution, electric potential, nozzle diameter, etc. affect the formation as well as the diameter of the nanofibers. Anisotropic structural organization including preferred fiber alignment can be obtained by using a rotating target or one with a specific geometry [140]. The fiber-diameter is reduced further due to the additional stretching exerted from the rotational alignment [141, 142].

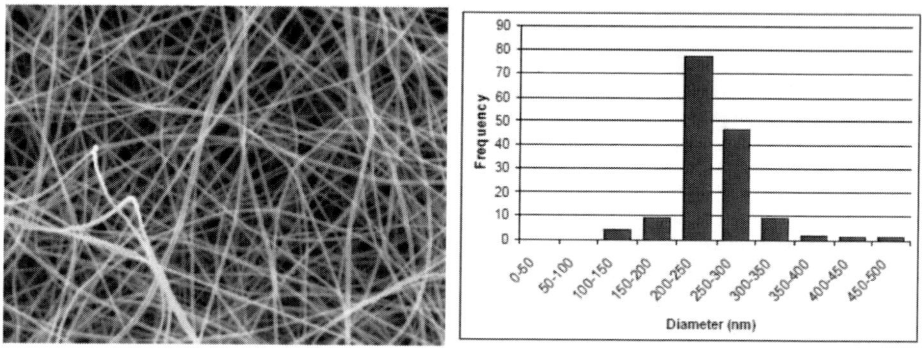

Figure 6. SEM image of nanofibrous structure of electrospun PLGA (left) and diameter distribution (%) (right).

Electrospun polymeric nanofibers are highly attractive and promising as tissue engineering scaffolds due to the fact that the morphological nano-feature (fiber-diameter) and chemical composition of electrospun fibers mimic closely those of extracellular matrix (ECM) proteins [128-140]. The nanofibrous structure has a high surface area/volume ratio, providing a favorable substrate for cell attachment (and therefore, a higher cell density per unit of space) compared to conventional scaffold structures (Figure 6). The high surface area/volume ratio of electrospun fibers allows the efficient localized delivery of a loaded drug. Various drugs, proteins and other growth factors have been successfully introduced (i.e., maintain their structure and bioactivity) into nanofiber scaffolds, which can potentially be used as drug delivery vehicles [126, 127]. A number of different drug loading methods such as coating, embedding, and encapsulation of the drug (coaxial and emulsion electrospinning) as well as covalent-linking can be used. The first method to incorporate therapeutic drugs into nanofibers involves blending/dispensing the drug into the polymer solution to be spun and both drugs and polymers are electrospun into drug-loaded ultrafine fibers [143]. In this method, drugs could minimally load inside the fiber but, easily deposit on the fiber-surface because of the rapid evaporation of the solvent and high ionic strength in solution. Therefore, fibrous mats prepared by blending method will most probably result in a burst-effect in the initial stage of drug release [144-146]. An initial burst-release is necessary to eliminate the inflammatory bacteria before they begin to proliferate in the case of delivery of antibiotics. For example, Biteral leaded PCL exhibited a burst release of 80% after 3 h under in vitro conditions [145]. Using this method, a loading efficiency of 90% into PDLA nanofibers was achieved for the antibiotic drug Mefoxin by blending and co-spinning [144]. Covalent conjugation of the drug to polymers before or after spinning into fibers represents another method to immobilize the drug and to modulate its release [147].

Modified spinning techniques such as co-axial electrospinning or emulsion electrospinning have also been employed to control and tune the release-kinetics of drugs. Using a coaxial dual-capillary spinneret, one component of the fiber can easily be encapsulated inside another component leading to what is known as a core-shell structure. This structure could be of high interest in the sustained-delivery of drugs and growth factors. The shell could not only shield and conserve the activity of drug but also control its release-kinetics. The core-shell electrospun fibers were fabricated via coaxial electrospinning from both water (core) and organic solvent (shell) soluble polymers. Jiang et al. [148] disclosed first the spinning of core-shell nanofibers containing therapeutic agents such as drugs, polypeptides, growth factors and DNAs for delivery applications. These core-shell fiber structures are advantageous in protecting the bioactive agents from denaturation or deactivation and in controlling the release rate from the nanofibers, and thus avoiding the burst release phenomenon For example, Huang et al. used co-axial electrospinning to prepare core-shell nanofibers for controlled release of multi drugs. Polycaprolactone was used as the shell and two medically pure drugs, Resveratrol and Gentamycin Sulfate, were used as the cores. The drugs were released in a controlled way without any initial burst effect [149].

Emulsion electrospinning involves preparation of core-sheath fibers from a stabile polymer-emulsion. Zhang et al. investigated the feasibility of encapsulating a non-electrospinnable protein along with a water-soluble polymer, PEG, into biodegradable PCL nanofibers via the coaxial electrospinning technique [50]. Xu *et al.* have successfully encapsulated poly(ethylene oxide) (PEO) and water-soluble doxorubicin hydrochloride in the shell material of PEG-PLLA [151, 152]. Incorporation of the anticancer drug BCNU into electrospun PEG-PLLA mats and subsequent cell responses was reported by the same group previously [153]. Both the polymer and the BCNU were dissolved in chloroform and electrospun, giving BCNU embedded in the PEG-PLLA fibers. The average fiber diameters depended on drug loading with values ranging from 690 to 1350 nm for drug loadings of 5 and 30 wt%, respectively. Subsequent studies on the effect of BCNU release from electrospun PEG-PLLA mats on the growth of rat Glioma C6 cells showed that BCNU loaded PEG-PLLA mats exhibited anticancer activity over a period of 72 h. Xie and Wang have used electrospun PLGA mat to encapsulate Paclitaxel, a well-known mitotic inhibitor and radiosensitizing anti-cancer drug [154] and determined that encapsulation efficiency of the Paclitaxel in electrospun PLGA mats as more than 90%. An in vitro drug release study demonstrated controlled release from both PLGA nano- and microfibers over a period of 61 days. Nanofibers had a faster release rate, releasing approximately 80% of the drug over the 61-day release period as compared to approximately 60% for the microfibers. An in vitro cytotoxicity study demonstrated that depending on the drug loading, up to 70% of C6 Glioma cells could be killed after 72 h. Later, Ranganath and Wang [155] conducted both in vitro and in vivo studies of electrospun PLGA (85/15) and PLGA (50/50) fibrous scaffolds loaded with Paclitaxel. Under in vitro conditions, the micro and nanofiberous scaffolds revealed a sustained release of 26% and 46% of the drug over 80 days period. The apoptosis studies via caspase-3 activity in vitro using C6 Glioma cells and in vivo subcutaneous implantation in nude mice (C6 Glioma tumor) revealed signs of higher activity compared to controls.

Electrospun mats have the ability to overcome drug loading limitations seen with other drug delivery devices, such as micelles and liposomes, currently used for targeting tumors. The drug release through different pathways such as diffusion, erosion, and biodegradation of fibers depends on physico-chemical (amorphous or crystalline, hydrophilic or hydrophobic), compositional and morphological characteristics (blends/composites porous/nonporous, and diameter) of polymer fibers. Cui et al. [156] investigated the effect of fiber diameter and drug loading on the release behavior in an electrospun polymer drug delivery system prepared using blending method, and they found that electrospun fibers with larger diameter exhibited longer period of nearly zero order release, and higher drug encapsulation resulted in a more significant burst release. As a modification to the co-spinning of drug blended composite fibers or co-axial spinning of core-shell fibers, we and others are focused on a combined synergistic approach of drug encapsulated/conjugated nanoparticles (polymeric and inorganic particles and nanodiamond) incorporated into the electrospun fibers for enhanced and effective delivery of drugs. Figure 7 is an example of Cisplatin loaded PLGA nanoparticle co-electrospun PVA nanofibers. At present, both degradable and non-degradable polymers are under investigation to be developed as drug carriers for local delivery of antibiotics and anticancer drugs.

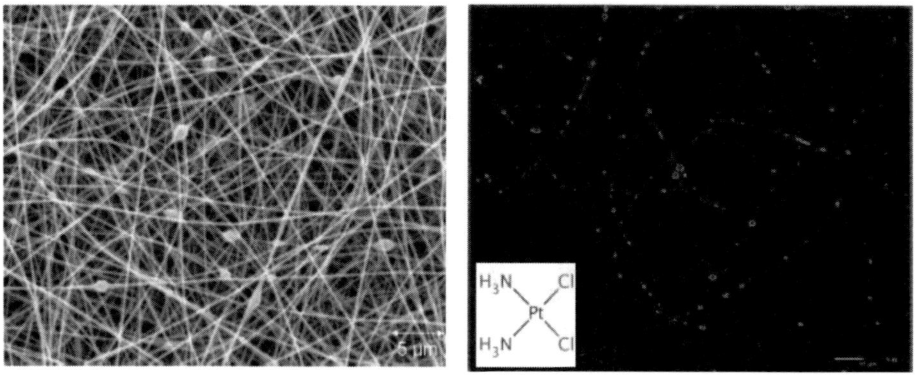

Figure 7. SEM (left) and CLSM (right) images of nanofibrous structure of electrospun PVA fibers incorporated PLGA nanoparticles loaded with cisplatin. Inset is chemical structure of cisplatin.

Conclusion

In summary, cancer nano-biotechnology is emerging as a new field of interdisciplinary research, cutting across the disciplines of medicine, biology, chemistry, materials science and engineering and is expected to lead to major advances in cancer detection, diagnosis, and treatment [157, 158]. Cancer nanomedicines have the potential to provide break-through and cutting-edge solutions for current limitations of cancer therapies. In contrast to conventional chemotherapy, nanotechnology-based drugs can be of several different drugs (combo-drugs) and simultaneously support complex treatments through localized release of multiple drugs. It has been proven that targeted nanoparticles can recognize and attach to cancer cells expressing specific receptors and thus increase the specificity and therapeutic efficiency. The use of polymeric nanoparticles in biological application has been rapidly advancing toward practical applications in human cancer diagnosis and therapy.

Polymeric nanoparticles are biodegradable and flexible in surface functionalities for drug-conjugation, therefore, can be used for controlled release of encapsulated or conjugated chemotherapeutics. But, these nanoparticles are of relatively large size (200–500 nm) and thus, the long circulating stability of these particles on systemic administration (as opposed to topical delivery) may be compromised.Ultra-fine inorganic nanoparticles (5-30 nm) act as image contrast agents (gold nanoshells, quantum dots (QDs, and iron-oxide nanoparticles) [159] and they are suitable for unimpeded circulation in the bloodstream and extravasation into tumor tissues. Depending on their size and composition, QDs have unique properties (exhibit broad excitation spectra, narrow emission spectra, a longer fluorescence life time, and a strong resistance to photo-bleaching) that make them attractive as fluorescent imaging agents against conventional fluorescent dyes. However, their practical application has been limited by problems with toxicity (heavy metal-toxicity), instability, and lack of drug encapsulation efficiency and selectivity for the disease site.

In order to reduce the toxic effects and to enhance the drug-conjugation or entrapmentvarious attempts have been made by physically or chemically anchoring biocompatible polymers on the surfaces of inorganic diagnostic nanoparticles. Hydrophilic polymers such as PEG or its copolymers and other biopolymers have been widely used for surface modification of nanoparticles such as iron oxide or QD to overcome their limitations [160]. In addition, surface coating or covalent surface modification of nanoparticles with hydrophilic polymers may minimize recognition by proteins and cells in the body, allowing them to circulate in the blood for a longer period of time and thereby increasing the possibility that the drug will reach the target site. Moreover, conjugation of PEGylated nanoparticles with cationic bovine serum albumin (CBSA) has been developed to allow crossing of the intact blood-brain-barrier (BBB) and reach therapeutic levels in the brain. [161]. Fluorescent, super-paramagnetic nanospheres for drug storage, targeting, and imaging using multifunctional nanocarrier system for cancer diagnosis and treatment has been reported recently [162]. For example, quantum dots with emissions in the near-IR range (~ 800 nm) were conjugated onto the surface of a spherical polystyrene matrix (~ 150 nm) having internally embedded high fraction of super-paramagnetic Fe_3O_4 nanoparticles (~ 10 nm). Thereafter, a chemotherapeutic agent (paclitaxel) was loaded onto the surfaces of these composite multifunctional nanocarriers by using a layer of PLGA. Hence, the particle targeting efficiency may be enhanced even further by conjugating the appropriate ligands or targeting moieties to chemically reactive functional groups on the polymer-coated nanoparticles, thus improving the utility of therapeutic and diagnostic nanomedicines ("theranostics"). Recent advances such as 'click chemistry' [163-165] and development of 'fluorescent nanodiamonds' [166, 167] opens new opportunities for the development of multifunctional nanoparticles or 'magic bullets' capable of effective targeting, delivery, and imaging in cancer-therapies without any side effect.

Acknowledgments

We acknowledge support from the National Science Foundation Grant Number NSF-HRD-0734232. We thank Moncy Jose and Alisha Peterson for the help received during the nano-biomaterials preparation and characterizations. We sincerely apologize to those authors whose works may have been relevant to this review but was not mentioned or cited due to space limitation.

References

[1] G. Orive, R. M. Harnandez, A. R. Gascon, and J. L. Pedraz. Micro and nano drug delivery systems in cancer therapy. *Cancer Therapy*, 3:131–138, 2005.

[2] E. M. Bilensoy, I. Vural, A. Bochot, et al. Tamoxifen citrate loaded amphiphilic ß-cyclodextrin nanoparticles; in-vitro characterization and cytotoxicity. *J. Control Release.*, 104:489–496, 2005

[3] K. T. Yong, I. Roy, M. T. Swihart, and P. N. Prasad. Multifunctional nanoparticles as biocompatible targeted probes for human cancer diagnosis and therapy. *J. Mater Chem.*, 19: 4655–4672, 2009.

[4] B. A. Kohler, E. Ward, B. J. McCarthy, M. J. Schymura, L. A. G. Ries, C. Eheman, A. Jemal, R. N. Anderson, U. A. Ajani, and B. K. Edwards. Annual report to the nation on the status of cancer, 1975–2007, featuring tumors of the brain and other nervous system. *J. Natl. Cancer Inst.*, 103:1–23, 2011.

[5] Y. E. K. Lee, and R. Kopelman. Polymeric nanoparticles for photodynamic therapy. In S. J. Hurst, editor, *Biomedical Nanotechnology: Methods and Protocols, Methods in Molecular Biology*, vol 726, Springer Science Business Media, 2011.

[6] P. Wust, B. Hildebrandt, G. Sreenivasa, B. Rau, J. Gellermann, H. Riess, R. Felix and P. M. Schlag. *The Lancet Oncology*, 3(8): 487-497, 2002.

[7] T. Jarm, M. Cemazar, D. Miklavcic, and G. Sersa. Antivascular effects of electrochemotherapy: implications in treatment of bleeding metastases. *Expert Rev. Anticancer Ther.*, 10:729–746, 2010.

[8] Y. K. Kim, H. S. Kwak, C. S. Kim, G. H. Chung, Y. M. Han, J. M. Lee. Hepatocellular carcinoma in patients with chronic liver disease: comparison of SPIO-enhanced MR imaging and 16-detector row CT. *Radiology*, 238:531–41, 2006.

[9] M. G. Harisinghani, J. Barentsz, P. F. Hahn, W. M. Deserno, S. Tabatabaei, C. H. van de Kaa CH, et al. Noninvasive detection of clinically occult lymph-node metastases in prostate cancer. *N. Engl. J. Med.*, 348:2491–9, 2003.

[10] A. Ito, H. Hond,a and T. Kobayashi, Cancer diagnosis and therapy using magnetite nanoparticles. *Chem. Eng. Japan,* 67: 692–695, 2003.

[11] C. Loo, A. Lowery, N. Halas, J. West, and R. Drezek. Immunotargeted nanoshells for integrated cancer imaging and therapy. *Nano Lett.*, 5:709–11, 2005.

[12] C. L. Nehl, N. K. Grady, G. F. Goodrich, F. Tam, N. J. Halas, J. H. Hafner. Scattering spectra of single gold nanoshells. *Nano Lett.,* 4:2355–2359, 2004.

[13] L. R. Hirsch, R. J. Stafford, J. A. Bankson, S.R. Sershen, B. Rivera, R. F. Price, J. D. Hazle, N. J. Halas, and J.L. West. Nanoshell-mediated near-infrared thermal therapy of tumors under magnetic resonance guidance. *Proc. Natl. Acad. Sci. USA,* 100:13549-13554, 2003.

[14] X. Gao, L. Yang, J. A. Petros, F. F. Marshall, J. W. Simons, S. Nie, In vivo molecular and cellular imaging with quantum dots. *Curr. Opin. Biotechnol.*, 16:63-72, 2005.

[15] M. E. Akerman,W. C. Chan, P. Laakkonen, S. N. Bhatia, and E. Ruoslahti, Nanocrystal targeting in vivo, *Proc. Natl. Acad. Sci. USA,* 99:12617–12621, 2002.

[16] P. Alivisatos. The use nanocrystals in biological detection. *Nature Biotechnol.*, 22(1):47-52, 2004.

[17] W. Cai, D. W. Shin, K. Chen, O. Gheysens, Q. Cao, S. X. Wang, S. S. Gambhir, X. Chen. Peptide labeled near infrared quantum dots for imaging tumour vasculature in living subjects. *Nano Lett.*, 6:669–676, 2006.

[18] D. R. Larson, W. R. Zipfel, R. M. Williams, S. W. Clark, M. P. Bruchez, F. W. Wise, W. W.Webb.Water-soluble quantum dots for multiphoton fluorescence imaging in vivo *Science*, 300:1434–1436, 2003.

[19] F Erogbogbo, KT Yong, I Roy, GX Xu. P. N. Prasad, and M. T. Swihart. Biocompatible luminescent silicon quantum dots for imaging of cancer cells. *ACS Nano*, *2*(5):873–878, 2008.

[20] M.F. Weng, S.Y. Chiang, N.S. Wang, H. Niu. Fluorescent nanodiamonds for specifically targeted bioimaging: Application to the interaction of transferrin with transferrin receptor. *Diamond and Related Materials*, 18:587–591, 2009.

[21] C. Fu, H.Y. Lee, K. Chen, T.S. Lim, H.Y.Wu, P.K. Lin, P.K.Wei, P.H. Tsao, H.C. Chang,W. Fann. Characterization and application of single fluorescent nanodiamonds as cellular biomarkers. *Proc. Natl. Acad. Sci. USA* 104: 727-732, 2007.

[22] J. Kreuter. Nanoparticles. In: J. Kreuter, editor, *Colloidal drug delivery systems*, vol. 66.. p. 219–342, New York: Marcel Dekker; 1994.

[23] P. Couvreur. Polyalkylcyanoacrylates as colloidal drug carriers. *Crit. Rev. Ther. Drug Carr. Syst.*,5:1–20, 1988.

[24] J. H Park, S. Lee, J. H. Kim, K. Park, K. Kim, and I. C. Kwon. Polymeric nanomedicine for cancer therapy. *Prog. Polym. Sci.*, 33:113–137, 2008.

[25] K. Cho, X. Wang, S. Nie, Z. Chen, and D.M. Shin Therapeutic nanoparticles for drug delivery in cancer. *Clin. Cancer Res.,* 14:1310-1316, 2008.

[26] A. Kumari, S. K. Yadav, S. C. Yadav. Biodegradable polymeric nanoparticles based drug delivery systems. *Colloids and Surfaces B: Biointerfaces* 75: 1–18, 2010.

[27] Y. Fukumorei, andH. Ichikawa. Nanoparticles for cancer therapy and diagnosis. *Advanced Powder Technol.,*17(1):1–28, 2006.

[28] R. Krishna, L. D. Mayer. Multidrug resistance (MDR) in cancer—mechanisms, reversal using modulators of MDR and the role of MDR modulators in influencing the pharmacokinetics of anticancer drugs. *Eur. J. Pharm. Sci.* 11:265–83, 2000.

[29] M. Links, R. Brown. Clinical relevance of the molecular mechanisms of resistance to anticancer drugs. *Expert Rev. Mol. Med.* 1999:1–21, 1999.

[30] J. Shi, A.R. Votruba, O. C. Farokhzad, and R. Langer. Nanotechnology in drug delivery and tissue engineering: From discovery to applications. *Nano Lett.,* 10: 3223–3230, 2010.

[31] D. Peer, J. M. Karp, S. Hong, O. C. Farokhzad, R. Margalit, and R. Langer Nanocarriers as an emerging platform for cancer therapy. *Nature Nanotech.* 2: 751–760, 2007.

[32] C. Vauthier, and K. Bouchemal. Methods for the preparation and manufacture of polymeric nanoparticles, *Pharmaceutical Research*, 26(5): 1025-1058, 2009.

[33] J. C. Middleton, A. J. Tipton Synthetic biodegradable polymers as orthopedic devices, *Biomaterials* 21: 2335-2346, 2000.

[34] V. Thomas, R. R. Dean, Y. K. Vohra. Nanostructured biomaterials in regenerative medicine. *Curr. Nanosci.* 2: 155-, 2006.

[35] P. A. Gunatillake, and R. Adhikari. Biodegradable synthetic polymers for tissue engineering *European cells and Materials*, 5: 1-16, 2003.

[36] L.S. Nair and C.T. Laurencin. Polymers as biomaterials for tissue engineering and controlled drug delivery, *Adv. Biochem. Engin./Biotechnol.*, 102: 47–90, 2006.

[37] M. V. Jose, V. Thomas, D. R. Dean, and E. Nyairo, Fabrication and characterization of aligned nanofibrous PLGA/Collagen blends as bone tissue scaffolds. *Polymer,* 50:3778–3785, 2009.

[38] M. C. Bottino, V. Thomas, M. V. Jose, D. R. Dean, G. M. Janowski, Acellular dermal matrix graft: Synergistic effect of rehydration and natural crosslinking on mechanical properties. *J. Biomed. Mater Res. Part B: Appl. Biomater.* 95B: 276–282, 2010.

[39] R. A. Jain.The manufacturing techniques of various drug loaded biodegradable poly(lactide-co-glycolide) (PLGA) devices. *Biomaterials,* 21:2475–2490, 2000.

[40] S. Parveen, S.K. Sahoo. Polymeric nanoparticles for cancer therapy *Journal of Drug Targeting, 16(2):* 108–123, 2008.

[41] J. M. Anderson, and M. S. Shive, Biodegradation and biocompatibility of PLA and PLGA microspheres, *Advanced Drug Delivery Reviews,* 28: 5-24, 1997.

[42] G. Mittal., D. K. Sahana, V. Bhardwaj, et al. Estradiol loaded PLGA nanoparticles for oral administration: Effect of polymer molecular weight and copolymer composition on release behavior in vitro and in vivo. *J. Control. Release,*119(1):77–85, 2007.

[43] V. R. Sinha, and A. Trehan. Formulation, characterization, and evaluation of ketorolac tromethamine-loaded biodegradable microspheres. *Drug Delivery*, 12:133–139, 2005.

[44] D.K. Sahana, G. Mittal, V. Bhardwaj, and M.N.V. Ravikumar. PLGA nanoparticles for oral delivery of hydrophobic drugs: influence of organic solvent on nanoparticle formation and release behavior in vitro and in vivo using estradiol as a model drug, *J. Pharm. Sci.* 97 (4):1530–1542, 2008.

[45] M.F. Zambaux, F. Bonneaux, R. Gref, E. Dellacherie, and C. Vigneron. Preparation and characterization of protein C-loaded PLA nanoparticles, *J. Control. Release* 60:179–188, 1999.

[46] C. P. Reis, R. J. Neufed, A. J. Ribeiro, and F. V. Veiga. Nanoencapsulation I. Methods for preparation of drug loaded polymeric nanoparticles, *Nanomedicine*, 2 (1): 8–21, 2006.

[47] J.M. Barichello, M. Morishita, K. Takayama, and T. Nagai. Encapsulation of hydrophilic and lipophilic drugs in PLGA nanoparticles by the nanoprecipitation method, *Drug Dev. Ind. Pharm.* 25 (4):471–476, 1999.

[48] J. P. Raoa, and K. E. Geckeler Polymer nanoparticles: Preparation techniques and size-control parameters. *Progress in Polymer Science* 36: 887–913, 2011

[49] K.S. Soppimatha, T.M. Aminabhavia, A. R. Kulkarnia, and W.E. Rudzinski Biodegradable polymeric nanoparticles as drug delivery devices. *Journal of Controlled Release* 70: 1–20, 2001.

[50] L. Mu, and S.S. Feng, A novel controlled release formulation for the anticancer drug paclitaxel (Taxol): PLGA nanoparticles containing vitamin E TPGS. *J. Control. Release*, 86 (1): 33–48, 2003.

[51] S.Y. Kim, and Y.M. Lee, Taxol-loaded block copolymer nanospheres composed of methoxy poly(ethylene glycol) and poly(epsilon-caprolactone) as novel anticancer drug carriers. *Biomaterials.* 22 (13): 1697–1704, 2001.

[52] Devalapally H, Duan Z, Seiden MV, and Amiji MM. Paclitaxel and ceramide co administration in biodegradable polymeric nanoparticulate delivery system to overcome drug resistance in ovarian cancer. *Int. J. Cancer*, 121: 1830 –1880, 2007.

[53] M. Morlock, T. Kissel, Y. X. Li, H. Koll, and G. Winter. Erythropoietin loaded microspheres prepared from biodegradable LPLG-PEO-LPLG triblock copolymers: protein stabilization and in-vitro release properties. *J. Control. Rel.*, 56:105-115, 1998.

[54] R. T. Liggins, and H. M. Burt. Polyether-polyester diblock copolymers for the preparation of paclitaxel loaded polymeric miecelle formations. *Adv. Drug Delivery Rev.*, 54:191-20, 2002.

[55] B. Jeong, Y. H. Bae, D. S. Lee, and S.W. Kim. Biodegradable block copolymers as injectable drug- delivery systems. *Nature*, 388:860-862, 1997.

[56] Y. Ikada, and H. Tsuji. Biodegradable polyesters for medical and ecological applications. *Macromol. Rapid Commun.*, 21:117-132, 2000.

[57] M. Chaubal. Polylactides/glycolides-excipients for injectable drug delivery and beyond. *Drug Delivery Technology*, 2:34-36, 2002.

[58] T. Kissel, Y. Li, and F. Unger. ABA-triblock copolymers from biodegradable polyester A-blocks and hydrophilic poly(ethylene oxide) B-blocks as a candidate for in situ forming hydrogel delivery systems for proteins. *Adv. Drug Delivery Rev.*, 54:99-134, 2002.

[59] K. M. Huh,and Y. H. Bae. Synthesis and characterization of poly(ethylene glycol)/poly(-lactic acid) alternating multiblock copolymers. *Polymer,* 40:6147-6155, 1999.

[60] Y. H. Bae, K. M. Huh, Y. Kim, and K. H.Park. Biodegradable amphiphilic multiblock copolymers and their implications for biomedical applications. *J. Control. Rel.,* 64:3-13, 2000.

[61] J. S. Hrkach, M. T. Peracchia, A. Domb, N. Lotan, and R.Langer. Nanotechnology for biomaterials engineering: structural characterization of amphiphilic polymeric nanoparticles by 1H NMR spectroscopy. *Biomaterials*, 18:27-30, 1997.

[62] Y. Li, and T. Kissel. Synthesis, characteristics and in vitro degradation of star-block copolymers consisting of L-lactide, glycolide, and branched multi-arm poly(ethylene oxide). *Polymer*, 39:4421-4427, 1998.

[63] B. Jeong, M. R. Kibbey, J. C. Birnbaum, Y. Y. Won, and A. Gutowska. Thermogelling biodegradable polymers with hydrophilic backbones: PEG-g-PLGA. *Macromolecules*, 33:8317-8322, 2000.

[64] M. Hamidi, A. Azadi, and P. Rafiei. Hydrogel nanoparticles in drug delivery. *Advanced Drug Delivery Reviews,* 60:1638–1649, 2008.

[65] J. Knapczyk, L. Krowczynski, J. Krzck, M. Brzeski, E. Nirnberg, D. Schenk, and H. Struszcyk. Requirements of chitosan for pharmaceutical and biomedical applications, In: G. Skak-Braek, T. Anthonsen, and P. Sandford, Editors, Chitin and Chitosan: *Sources, Chemistry, Biochemistry, Physical Properties and Applications*, pp. 657–663, Elsevier, London, 1989,

[66] S. Hirano, H. Seino, I. Akiyama, and I. Nonaka. Chitosan: a biocompatible material for oral and intravenous administration, In: C.G. Gebelein, and R.L. Dunn, Editors, *Progress in Biomedical Polymers*, pp. 283–289, Plenum Press, New York, 1990.

[67] SA Agnihotri, NN Mallikarjuna, and T. M. Aminabhavi. Recent advances on chitosan based micro- and nanoparticles in drug delivery, *Journal of Controlled Release*, 100: 5-28, 2004.

[68] T.G. Shutava, and Y.M. Lvov. Nano-engineered microcapsules of tannic acid and chitosan for protein encapsulation, *J. Nanosci. Nanotechnol.*, 6:1655–1661, 2006.

[69] Y.F. Fan, Y.N.Wang, Y.G. Fan, and J.B. Ma. Preparation of insulin nanoparticles and their encapsulation with biodegradable polyelectrolytes via the layer-by-layer adsorption, *Int. J. Pharm.,* 324: 158–167, 2006.

[70] R. Pandey, Z. Ahmad, S. Sharma, and G.K. Khuller. Nano-encapsulation of azole antifungals: potential applications to improve oral drug delivery, *Int. J. Pharm.*, 301: 268–276, 2005.

[71] C.M. Lehr, J.A. Bowstra, E.H. Schacht, and H.E. Junginger. In vitro evaluation of mucoadhesive properties of chitosan and some other natural polymers, *Int. J. Pharm.* 78: 43–48, 1992.

[72] K.A. Janes, P. Calvo, and M.J. Alonso. Polysaccharide colloidal particles as delivery systems for macromolecules, *Adv. Drug Del. Rev.*, 47: 83–97, 2001.

[73] P. Artursson, T. Lindmark, S.S. Davis, and L. Illum. Effect of chitosan on the permeability of monolayers of intestinal epithelial cells (Caco-2), *Pharm. Res.* 11 :1358–1361, 1994.

[74] Y. Kawashima, T. Handa, H. Takenaka, S.Y. Lin, and Y. Ando. Novel method for the preparation of controlled-release theophylline granules coatedwith a polyelectrolyte complex of sodium polyphosphate-chitosan, *J. Pharm. Sci.,* 74 : 264–268, 1985.

[75] T.H. Dung, S.R. Lee, S.D. Han, S.J. Kim, Y.M. Ju, M.S. Kim, and H. Yoo. Chitosan-TPP nanoparticle as a release system of antisense oligonucleotide in the oral environment, *J. Nanosci. Nanotechnol.,* 7: 3695–3699, 2007.

[76] X.Z. Shu, and K.J. Zhu. A novel approach to prepare tripolyphosphate/chitosan complex beads for controlled release drug delivery, *Int. J. Pharm.,* 201: 51–58, 2000.

[77] Y. Xu, and Y. Du. Effect of molecular structure of chitosan on protein delivery properties of chitosan nanoparticles, *Int. J. Pharm.*, 250: 215–226, 2003.

[78] A. Berthold, K. Cremer, and J. Kreuter. Preparation and characterization of chitosan microspheres as drug carrier for prednisolone sodium phosphate as model for anti-inflammatory drugs, *J. Control. Release*, 39: 17–25, 1996.

[79] V. Dodane, and D. Vilivalam. Pharmaceutical applications of chitosan, *Pharm. Sci. Technol. Today*, 16 :246–253, 1998.

[80] K. Ray, H.Q. Mao, K.Y. Lin, S.K. Huang, and K.W. Leong. Oral immunization with DNAchitosan nanoparticles, *Proc. Int. Symp. Control. Release Mater.,* 26: 348–349, 1999.

[81] V. Dodane, and V. D. Vivivalam. Pharmaceutical applications of chitosan. *Pharm. Sci. Technol. Today*, 1:246–253, 1998.

[82] Illum L. Chitosan and its use as a pharmaceutical excipient. *Pharm. Res.,* 15:1326–1331,1998.

[83] L. Qi, Z. Xu, and M. Chen. In vitro and in vivo suppression of hepatocellular carcinoma growth by chitosan nanoparticles. *Eur. J. Cancer,* 43:184–193, 2007.

[84] R.C. Oppenhiem. Paclitaxel loaded gelatin nanoparticles for intravesical bladder cancer therapy, *Int. J. Pharm.*, 8 :217-234, 1981.

[85] Z. Lu, T.K. Yeh, M. Tsai, J. L. S. Au, and M. G Wientjes. Paclitaxel-loaded gelatin nanoparticles for intravesical bladder cancer therapy, *Clin. Cancer Res.* 10:7677-7684, 2004.

[86] H.H. Tonnesen, and J. Karlsen, Alginate in drug delivery systems, *Drug Dev. Ind. Pharm.* 28 : 621–630, 2002.

[87] H. Gonzalez, S. J. Hwang, and M. E. Davis. New class of polymers for the delivery of macromolecular therapeutics. *Bioconjug. Chem.*, 10:1068–74, 1999.

[88] S. H. Pun, N. C. Bellocq, A. Liu, G. Jensen, T. Machemer, E. Quijano. Cyclodextrin-modified polyethylenimine polymers for gene delivery. *Bioconjug. Chem.*, 15: 831–40, 2004.

[89] D. W. Pack, A. S. Hoffman, S. Pun S, and P. S. Stayton. Design and development of polymers for gene delivery. *Natl. Rev. Drug Discov.*, 4:581–93, 2005.

[90] B. Sahana, K. Santra, S. Basu, and B. Mukherjee. Development of biodegradable polymer based tamoxifen citrate loaded nanoparticles and effect of some manufacturing process parameters on them: a physicochemical and in-vitro evaluation. *Int. J. Nanomedicine*, 5: 621–630, 2010.

[91] M.F. Zambaux, F. Bonneaux, R. Gref, P. Maincent, E. Dellacherie, M.J. Alonso, P. Labrude, and C. Vigneron. Influence of experimental parameters on the characteristics of poly(lactic acid) nanoparticles prepared by double emulsion method. *J. Control. Rel.* 50: 31–40, 1998.

[92] D.B. Shenoy, and M.M. Amiji, Poly(ethylene oxide)-modified poly (epsiloncaprolactone) nanoparticles for targeted delivery of tamoxifen in breast cancer. *Int. J. Pharm.*, 293 (1–2): 261–270, 2005.

[93] W. J. Gradishar,S. Tjulandin, N. Davidson, H. Shaw, N. Desai, P. Bhar, M. Hawkins, and J. O'Shaughnessy. Phase III trial of nanoparticle albumin-bound paclitaxel compared with polyethylated castor oil-based paclitaxel in women with breast cancer. *J. Clin Oncol.*, 23:7794-803, 2005.

[94] M. B. Bally, R. Nayar, D. Masin, M. J. Hope, P. R Cullis, and L. D. Mayer. Liposomes with entrapped doxorubicin exhibit extended blood residence time *Biochim. Biophys. Acta* 1023:133-139, 1990.

[95] B. Rosenberg, Fundamental studies with cisplatin, *Cancer,* 55 (10): 2303–2306, 1985.

[96] K. Avgoustakis, A. Beletsi, Z. Panagi, P. Klepetsanis, A.G. Karydas, and D.S. Ithakissios. PLGA–mPEGnanoparticles of cisplatin: in vitro nanoparticledegradation, in vitro drug release and in vivo drug residence in blood properties, *J. Control. Release* 79 :123–135, 2002.

[97] S. Dhar, N. Kolishetti, S. J. Lippard, and O. C. Farokhzad. Targeted delivery of a cisplatin prodrug for safer and more effective prostate cancer therapy in vivo. *Proc. Natl. Acad. Sci. U S A.* 108(5): 1850–1855, 2011.

[98] K. Ringe, C. Walz, B. Sabel, Nanoparticle drug delivery to the brain, in: H.S. Nalwa, Editor, *Encyclopedia of Nanoscience and Nanotechnology*, vol. 7, American Scientific Publishers, New York, 2004.

[99] Y. Matsumura, and H. Maeda. A new concept for macromolecular therapeutics in cancer chemotherapy: mechanism of tumoritropic accumulation of proteins and the antitumor agent smancs. *Cancer Res.* 46: 6387–92, 1986.

[100] P. Carmeliet, and R. K. Jain. Angiogenesis in cancer and other diseases. *Nature*, 407:249-57, 2000.

[101] H. Ringsdorf. Structure and properties of pharmacologically active polymers. *J. Polym. Sci. Polym. Symp.* 51:135–53, 1975.

[102] F. F. Davis. The origin of pegnology. *Adv. Drug Deliv. Rev.* 54:457–58, 2002.

[103] S. M. Moghimi, and A. C. Hunter. Poloxamers and poloxamines in nanoparticle engineering and experimental medicine. *Trends Biotechnol.*, 18:412–20, 2000.

[104] H. Pelicano, D. S. Martin, R. H. Xu, and Huang P. Glycolysis inhibition for anticancer treatment. *Oncogene*, 25:4633-46, 2006.

[105] E. I. Deryugina, and J. P. Quigley. Matrix metalloproteinases and tumor metastasis. *Cancer Metastasis Rev.*, *25*:9-34, 2006.

[106] M. B. Yatvin, W. Kreutz, B. A. Horwitz, and M. Shinitzky. pH-sensitive liposomes: possible clinical implications. *Science,* 210:1253-5, 1980.

[107] A. M. Mansour, J. Drevs, N. Esser, F. M. Hamada, O. A. Badary, C. Unger, I. Fichtner, and F. A. Kratz. A new approach for the treatment of malignant melanoma: enhanced antitumor efficacy of an albumin-binding doxorubicin prodrug that is cleaved by matrix metalloproteinase 2. *Cancer Res.*, 63:4062-6, 2003.

[108] D. W. Bartlett, H. Su, I. J. Hildebrandt, W. A. Weber, and M. E. Davis. Impact of tumor-specific targeting on the biodistribution and efficacy of siRNA nanoparticles measured by multimodality *in vivo* imaging. *Proc. Natl. Acad. Sci. U.S.A.* 104: 15549–15554, 2007.

[109] K. F. Pirollo, and E. H. Chang. Does a targeting ligand influence nanoparticle tumor localization or uptake? *Trends Biotechnol.*, 26, 552–558, 2008.

[110] Y. Liu, J. Sun, J. Han, and Z. He. Long circulating targeted nanoparticles for cancer therapy, *Current Nanosci.,* 6:347-354, 2010.

[111] R. Kannagi, M. Izawa, T. Koike, K. Miyazaki, and N. Kimura N. Carbohydrate mediated cell adhesion in cancer metastasis and angiogenesis. *Cancer Sci.,* 95:377–84, 2004.

[112] N. Yamazaki, S. Kojima, N. V. Bovin, S. Andre, S. Gabius, and H. J. Gabius. Endogenous lectins as targets for drug delivery. *Adv. Drug Deliv. Rev.,* 43:225–44, 2000.

[113] C. M. Solbrig, J. K. Saucier-Sawyer, V. Cody, W. M. Saltzman, and D. J. Hanlon. Polymer nanoparticles for immunotherapy from encapsulated tumor-associated antigens and whole tumor cells. *Mol. Pharm.,* 4:47–57, 2007.

[114] A.P Chapman. PEGylated antibodies and antibody fragments for improved therapy. *Adv. Drug Deliv Rev.*, 54:531-545, 2002.

[115] H. Hatakeyama, H. Akita, E. Ishida, K. Hashimoto, H. Kobayashi, T. Aoki, J. Yasuda, K. Obata, H. Kikuchi, T. Ishida, H. Kiwada, and H. Harashima. Tumor targeting of doxorubicin by anti-MTI-MMP antibody modified PEG liposomes. *Int. J. Pharm.*, 342: 194-200, 2007.

[116] K. Kitamura, T. Takahashi, T. Yamaguchi, A. Noguchi, K. Takashima, H. Tsurumi, M. Inagake, T. Toyokuni, and S. hakomori. Chemical engineering of the monoclonal antibody A7 by polyethylene glycol for targeting cancer chemotherapy, *Cancer Res.*, 51:4310-4315, 1991.

[117] A. Agarwal, S. Saraf, A Asthana, U. Gupta, V Gajbhiye, and N K. Jain Ligand based dendritic systems for tumor targeting. *Int. J. Pharm.,* 350: 3-13, 2008.

[118] T. Kobayashi, T. Ishida, Y. Okada, S. Ise, H. Harashima, and H. Kiwada. Effect of transferrin receptor- targeted liposomal doxorubicin in P-glycoprotein-mediated drug resistant tumor cells. *Int. J. Pharm.,* 329:94-102, 2007.

[119] J. Sudimack, and R. J. Lee. Targeted drug delivery via the folate receptor. *Adv. Drug Deilv. Rev.*, 41: 147-162, 2000.

[120] P. S. Low, and A. C. Antony. Folate receptor-targeted drugs for cancer and inflammatory diseases. *Adv. Drug Deilv. Rev.*, 56:1055-1058, 2004.

[121] Z. Zhang, L. S. Huey, and S. S. Feng. Folate-decorated poly(lactideco- glycolide)-vitamin E TPGS NPs for targeted drug delivery. *Biomaterials*, 28:1889–1899, 2007.

[122] H. S. Yoo, and T. G. Park. Folate receptor targeted biodegradable polymeric doxorubicin micelles. *J. Control. Release*, 96:273–283, 2004.

[123] K. Cho, X. Wang, S. Nie, Z. Chen, and D. M. Shin. Therapeutic Nanoparticles for Drug Delivery in Cancer. *Clin. Cancer Res.*, 14:1310-1316, 2008.

[124] B. J. Hicke, A. W. Stephens, T. Gould, Y. F. Chang, C. K. Lynott, J. Heil et al. Tumor targeting by an aptamer. *J. Nucl. Med.*, 47: 668-78, 2006.

[125] O. C. Farokhzad, S. Jon, A. Khademhosseini, T. N. Tran, D. A. Lavan, and R. Langer. Nanoparticle-aptamer bioconjugates: a new approach for targeting prostate cancer cells. *Cancer Res.*, 64:7668-72, 2004.

[126] M. Hadjiargyrou, and J. B. Chiu. Enhanced composite electrospun nanofiber scaffolds for use in drug delivery. *Expert Opin. Drug Deliv.*, 5(10):1093-106, 2008.

[127] D. G. Yu, L.M. Zhu, K.White, and C. Branford-White. Electrospun nanofiber-based drug delivery systems. *Health*, 1 (2): 67-75, 2009.

[128] Q. P. Pham, U. Sharma and A. G. Mikos, Electrospinning of Polymeric Nanofibers for Tissue Engineering Applications: A Review. *Tissue Eng.,* 12:1197-1211, 2006.

[129] C.P. Barnes, S.A. Sell, E.D. Boland, D.G. Simpson, and G.L. Bowlin, Nanofiber technology: designing the next generation of tissue engineering scaffolds. *Adv Drug Del Rev,* 59:1413-1433, 2007.

[130] T.J. Sill and H.A. von Recum. Electrospinning: Applications in drug delivery and tissue engineering. *Biomaterials*, 29: 1989-2006, 2008.

[131] V. Thomas, T. Donahoe, E. Nyairo, D. R. Dean, and Y.K. Vohra, Electrospinning of Biosyn based tubular conduits: Stuctural, morphological and mechanical characterizations. *Acta Biomaterialia*, 7: 2070-2079, 2011.

[132] V. Thomas, D. R. Dean, M. V. Jose, B. Mathew, S. Chowdhury, and Y. K. Vohra. Nanostructured biocomposite scaffolds based on collagen co-electrospun with nanohydroxyapatite. *Biomacromolecules*, 8: 631-637, 2007.

[133] D.H. Reneker, and A. L. Yarin, Electrospinning jets and polymer nanofibers. *Polymer*, 49: 2387-2425, 2008.

[134] S. Agarwal, A. Greiner, and J.H. Wendorff. Electrospinning of manmade and biopolymer nanofibers Progress in techniques, materials, and applications. *Adv. Funct. Mater*, 19: 2863-2879, 2009.

[135] H. Deshpande, C. Schindler, D. Dean, W.Clem, S. Bellis, E. Nyairo, M. Mishra, and V. Thomas. Nanocomposite scaffolds based on electrospun polycaprolactone/modified carbon nanofiber/ nanohydroxyapatite by electrosphoretic deposition. *J. Biomaterials and Tissue Engineering*. 2011, (In Press).

[136] V. Thomas, S. Jagani, K. Johnson, M.V. Jose, D.R. Dean, Y.K. Vohra, and E. Nyairo, Electrospun bioactive nanocomposite scaffolds based on polycaprolactone and nanohydroxyapatite for bone tissue engineering. *J. Nanosci. Nanotechnol.* 6:487-493, 2006.

[137] W.J. Li, C.T. Laurencin, E.J. Caterson, R.S. Tuan, and F.K. Ko, Electrospun nanofibrous structure: a novel scaffold for tissue engineering. *J. Biomed. Mater Res.*, 60: 613-621, 2002.

[138] M. C. Bottino, V. Thomas, and G. M. Janowski. A novel spatially designed and functionally graded electrospun membrane for periodontal regeneration. *Acta Biomaterialia* 7:216–224, 2011.

[139] V. Thomas. X. Zhang, S. A. Catledge, and Y. K. Vohra. Functionally -graded electrospun tubular scaffolds with tunable mechanical properties for vascular tissue regeneration. *Biomedical Materials*, 2: 224-232.

[140] R. Murugan, and S. Ramakrishna. Design strategies of tissue engineering scaffolds with controlled fiber orientation. *Tissue Eng.*, *13* (8): 1845-1886, 2007

[141] V. Thomas, M. V. Jose, S. Chowdhury, J. F. Sullivan, D. R. Dean, and Y. K. Vohra, Aligned nanofibrous scaffolds of polycaprolactone fabricated by electrospinning. *J.* Biomater. Sci. Polymer *Edn.* 17: 969-984 (2006).

[142] A. Baji, Y.W. Mai, S.C. Wong, M. Abtahi, and P. Chen, Electrospinning of polymer nanofibers: Effects on oriented morphology, structures and tensile properties. *Composites Science and Technology* 70: 703–718, 2010.

[143] E. R. Kenawy, G. L. Bowlin, K. Mansfield, J. Layman, D. G. Simpson, E. H. Sanders, and G. E. Wnek. Release of tetracycline hydrochloride from electrospun poly(ethylene-co-vinylacetate), poly(lactic acid), and a blend. *J. Contr. Release*, 81: 57-64, 2002.

[144] X. H. Zong, K. Kim, D. F. Fang, S.F. Ran, B.S. Hsiao, and B. Chu. Structure and process relationship of electrospun bioabsorbable nanofiber membranes. *Polymer*, 43:4403-4412, 2002.

[145] N. Bolgen, I. Vargel, P. Korkusuz, Y. Z. Menceloglu and E. Piskin. In vivo performance of antibiotic embedded electrospun PCL membranes for prevention of abdominal adhesions. *J. Biomed. Mater Res. B Appl. Biomater.,* 81(2):530-43, 2007.

[146] K. Kim, Y. K. Luu, C. Chang, D. Fang, B. S. Hsiao, B. Chu B. Incorporation and controlled release of a hydrophilic antibiotic using poly(lactide-coglycolide)- based electrospun nanofibrous scaffolds. *J. Control. Release*, 98(1):47-56, 2004.

[147] M. Thanou and R. Duncan. Polymer-protein and polymer- drug conjugates in cancer therapy. *Curr. Opin. Invest. Drugs,* 4: 701-709, 2003.

[148] H. Jiang, Y. Hu, P. Zhao, Y. Li, and K. Zhu K. Modulation of protein release from biodegradable core-shell structured fibers prepared by coaxial electrospinning. *J. Biomed. Mater Res. B Appl. Biomater.*, 79(1):50-7, 2006.

[149] Z. M. Huang, C. L. He, A. Yang, Y. Zhang, X. J. Han, J. Yin, and Q. Wu. Encapsulating drugs in biode-gradable ultrafine fibers through co-axial electrospinning. *J. Biomed. Mater. Res. Part A,* 77: 169-179, 2006.

[150] Y. Zhang, X. Wang, Y. Feng, J. Li, C. Lim, and S. Ramakrishna. Coaxial electrospinning of (fluorescein isothiocyanate-conjugated bovine serum albumin)-encapsulated poly(ε-caprolactone) nanofibers for sustained release. *Biomacromolecules*, 7: 1049-1057, 2006.

[151] X. Xu, X. Zhuang, X. Chen, X. Wang, L. Yang, and X. Jing X Preparation of Core-Sheath Composite Nanofibers by Emulsion Electrospinning. Macromol. Rapid Commun., 27:1637–1642, 2006.

[152] X. L. Xu, L.X. Yang, X.Y. Xu, X. Wang, X.S. Chen, Q.Z. Liang, J. Zeng, J. and X. B. Jing. Ultrafine medicated fibers electrospun from W/O emulsions, *J. Control. Release*, 108:33-42, 2005.

[153] X. Xu, X. Chen, X. Xu, T. Lu, X. Wang, L. Yang, et al. BCNU-loaded PEGPLLA ultrafine fibers and their in vitro antitumor activity against Glioma C6 cells. *J. Control. Release* 114:307-16, 2006.

[154] J. Xie, and C. H. Wang. Electrospun micro- and nanofibers for sustained delivery of paclitaxel to treat C6 glioma in vitro. *Pharm. Res.*, 23:1817-26, 2006.

[155] S.H. Ranganath, C. H. Wang, Biodegradable microfiber implant delivering paclitaxel for post surfigical chemotherapy against malignant glioma. *Biomaterials* 29: 2996-3003, 2008.

[156] W. Cui, X. Li, X. Zhu, G. Yu, S. Zhou, and J. Weng. Investigation of drug release and matrix degradation of electrospun poly(DLlactide) fibers with paracetamol inoculation *Biomacromolecules* 7: 1623-1629, 2006.

[157] M. Ferrari. Cancer nanotechnology: opportunities and challenges. *Nat. Rev. Cancer* 5:161–71, 2005.

[158] P.R. Srinivas, P. Barker, and S. Srivastava Nanotechnology in early detection of cancer. *Lab. Invest.* 82:657–62, 2002.

[159] K. T. Yong, I. Roy, M.T. Swihart, and P.N. Prasad. Multifunctional nanoparticles as biocompatible targeted probes for human cancer diagnosis and therapy. *J. Mater Chem.* 1; 19: 4655–4672, 2009.

[160] S.A. Meenach, J. Zach Hilt, and K. W. Anderson. Poly(ethylene glycol)-based magnetic hydrogel nanocomposites for hyperthermia cancer therapy, *Acta Biomaterialia,* 6(3):1039-1046, 2010.

[161] A. K. Kumagai, J. B. Eisenberg, and W.M. Pardridge. Absorptive-mediated endocytosis of cationized albumin and a beta-endorphin-cationized albumin chimeric peptide by isolated brain capillaries. Model system of blood-brain barrier transport. *J. Biol. Chem. 262:15214-9, 1987.*

[162] H. S. Cho, Z. Dong, G.M. Pauletti, J. Zhang, H. Xu,H. Gu, L. Wang, R. C. Ewing, C. Huth, F. Wang, and D. Shi. Fluorescent, Superparamagnetic nanospheres for drug storage, targeting, and imaging: A multifunctional nanocarrier system for cancer diagnosis and treatment *ACS Nano, 4* (9): 5398–5404, 2010.

[163] G. Maltzahn, Y. Ren, J. H. Park, D.H. Min, V. R. Kotamraju, J. Jayakumar, V. Fogal, M. J. Sailor, E. Ruoslahti, and S. N. Bhatia. In vivo tumor cell targeting with "click" nanoparticles, *Bioconjugate Chem.* 19: 1570–1578, 2008.

[164] J. Lu, M. Shi, and M. S. Shoichet. Click chemistry functionalized polymeric nanoparticles target corneal epithelial cells through RGD-cell surface receptors. *Bioconjugate Chem.,* 20:87–94, 2009.

[165] M. Mueller, A. Schallon, A. Walther, R. Freitag, and A.H. E. Mueller. Clickable, biocompatible, and fluorescent hybrid nanoparticles for intracellular delivery and optical imaging. *Biomacromolecules,* 11:390–396, 2010.

[166] R. Lam, M. Chen, E. Pierstorff, H. Huang, E. Osawa, and Dean Ho Nanodiamond-embedded microfilm devices for localized chemotherapeutic elution. *ACS Nano,* 2(10):2095–2102, 2008.

[167] H. Huang, E. Pierstorff, E. Osawa, and D. Ho. Active nanodiamond hydrogels for chemotherapeutic delivery. *Nano Letters,* 7:3305–3314, 2007.

In: Signaling, Gene Regulation and Cancer
Editors: Shree Ram Singh and Manoj K. Mishra
ISBN: 978-1-61942-088-5

Chapter 22

CISPLATIN IN CANCER MANAGEMENT: NEW APPROACHES FOR BYPASSING NEPHROTOXICITY

Rohit Seth**, Rakesh Kumar Singh and Alka Seth
College of Medicine, Florida State University, Tallahassee, FL, US

ABSTRACT

Cisplatin (CIS) is an anticancer agent widely used in clinical practice for treating variety of malignant tumors. The antineoplastic activity of this compound occurs from DNA cross-links and adducts, in addition to the generation of superoxide radicals. CIS's irreversible protein binding in plasma presents serious dose-limiting side effects like nephrotoxicity, ototoxicity and neurotoxicity. Recently, lot of research focus is centralized around role of CIS on the signaling pathways leading to tubular cell death and inflammation. Various renal protective approaches are being discovered, however, it is unclear if these approaches would limit the anticancer effects of CIS in tumors. Various combinations like biochemical, herbal and nano particle tagged CIS are proposed as an alternative to direct application of CIS. New formulations like, liposomal CIS, lipoplatin, have been successfully tested thoroughly in preclinical (*in vitro*) and phase I, II and III trials. In this chapter we will focus our discussion on different strategies adopted for limiting CIS induced nephrotoxicity. It is our hope that newer combinational therapies with CIS will improve overall survival in patients with malignant tumors.

INTRODUCTION

First report of the cytotoxic properties of certain platinum coordination complexes by Rosenberg and colleagues (Rosenberg B 1969) caused widespread interest in the evaluation of these compounds for cancer chemotherapy. Cisplatin (CIS) is an inorganic platinum agent

* Corresponding author: Rohit Seth, Ph.D. - College of Medicine, Florida State University. Tallahassee, Fl- 32306-4300. Phone: 850-644-8025. Fax: 850-644-5781. E-mail: rohitseth123@gmail.com.

(cis-diamminedichloroplatinum) with antineoplastic activity. CIS forms highly reactive, charged, platinum complexes which bind to nucleophilic groups such as GC-rich sites in DNA, inducing intrastrand and interstrand DNA cross-links, as well as DNA-protein cross-links. These cross-links result in apoptosis and cell growth inhibition thereby preventing the growth of cancer cells and cause their death. For quite long CIS has been used for therapeutic management of several tumors, such as those of the ovary, testes, and the head and neck (Prestayko AW 1979).

Even though the repertoire of knowledge has increased about CIS's mechanism of action, a lot still remains to be explored to define its mode of action in various combinations and in different experimental settings. This understanding is also critical for elucidating mechanisms underlying the drug-resistant phenotype, which radically limits the clinical utility of CIS (Siddik 2003). In this chapter we are highlighting mechanism of action of CIS and the role and importance of CIS as a therapeutic agent for treatment of various cancers as well as combinational approaches that are used for bypassing nephrotoxic effects of CIS.

Role of CIS in Cancer Treatment

CIS is approved to be used together with other drugs to treat advanced ovarian cancer, metastatic ovarian cancer (Polyzos A 2009; Echarri Gonzalez MJ 2011) or testicular cancer (Chung P 2011) in patients who have already had surgery or radiotherapy. CIS is found useful in treatment of locally advanced squamous cell carcinoma of head and neck (SCCHN) (Ghadjar, Simcock et al. 2011), late stage cervical cancer (Punushapai U 2010), malignant mesothelioma (Jakobsen and Sørensen 2011) and locally advanced, advanced or metastatic non-small cell lung cancer (NSCLC) that cannot be treated with surgery (Chang, Ahn et al. 2011; Sen, Saglam et al. 2011). CIS-based chemotherapy during pelvic radiation therapy improves long-term progression-free and overall survival among locally advanced cervical cancer patients collectively and for stage IIB and III disease, individually and there was no observed increase in late toxicity with CIS-based chemoradiotherapy (Rose, Ali et al. 2007). CIS plus gemcitabine improves survival of patients with advanced biliary tract cancer (Valle, Wasan et al. 2010). S-1, a fourth-generation oral fluoropyrimidine plus CIS holds promise of becoming a standard first-line treatment for patients with advanced gastric cancer (Koizumi, Narahara et al. 2008). In nasopharyngeal carcinoma, chemotherapy using CIS led to a small, but significant, benefit for overall survival and event-free survival. This benefit was essentially observed when chemotherapy was administered concomitantly with radiation therapy (Baujat, Audry et al. 2006). In a clinical trial whole-abdominal irradiation and doxorubicin-CIS chemotherapy were compared in women with stage III or IV endometrial carcinoma. It is noted that chemotherapy with doxorubicin-CIS significantly improved progression-free and overall survival compared with whole-abdominal irradiation (Randall, Filiaci et al. 2006). CIS combination chemotherapy seems to be the cornerstone for treatment of many cancers. Initial platinum responsiveness is high but the majority of cancer patients will eventually relapse with CIS-resistant disease. It is also proposed that CIS-induced apoptosis lead to overexpression of Bcl-2, anti-apoptotic factor which rendered the cells partially resistant to CIS (Takeda, Kobayashi et al. 1997; Zhan Y 1999). Some common mechanisms for CIS resistance that are now proposed include changes in cellular uptake and efflux of the drug and increased detoxification of the drug. Other drug resistance pathways include deactivation of platinum drugs by thiols, increased nucleotide excision repair of Pt-DNA adducts, decreased mismatch repair, and defective apoptosis result in resistance to platinum therapy (Ahmad 2010).

MOLECULAR MECHANISMS OF CIS ACTION

CIS is transported into cells by the copper transporter Ctr1 (Ishida, Lee et al. 2002). Once inside the cell the chloride ions dissociate from the platinum due to the low intracellular chloride concentrations. The positively charged platinum ion binds cellular nucleophiles in DNA, RNA and proteins (Cohen SM 2001). Several cellular pathways get activated in response to this interaction, which include recognition by high-mobility group (HMG) and repair proteins, translesion synthesis by polymerases, and induction of apoptosis. The apoptotic process is regulated by activation of caspases, p53 gene, and several proapoptotic and antiapoptotic proteins. HMG1 proteins binds specifically to DNA containing CIS d(GpG) or d(ApG) intrastrand cross-links, which unwind and bend DNA in a specific manner (Pil and Lippard 1992). Detailed amino-terminal sequence analysis identified 26.5- and 28-kDa proteins that were identical to HMG proteins HMG-2 and HMG-1, respectively (Hughes, Engelsberg et al. 1992). Recently, the role of Lys-2 and 81 as target sites for acetylation in full-length HMG1 and truncated tail-less protein, respectively, has been studied by mutation analysis for the abilities of these proteins to bind and bend DNA. It appears that Lys-81 is critical for the DNA bending ability of truncated HMG1 and Lys-2 and Lys-81 are both necessary for the protein's binding to CIS-modified DNA (Elenkov I 2011). Another report suggests importance of two cysteine residues in HMG1 domain A which form a reversible disulfide bond under mildly oxidizing conditions. The reduced domain A protein binds to a 25-bp DNA probe containing a central 1,2-d(GpG) intrastrand cross-link, the major platinum-DNA adduct, with a 10-fold greater binding affinity than the oxidized domain A, this study reveal that the cellular redox environment can influence the interaction of HMG1 with the platinated DNA (Park and Lippard 2011). Thiols such as the sulfur of glutathione (GSH) can bind to the platinum molecule of CIS, replacing one of the chloride ions and preventing binding to other cellular nucleophiles (Berners-Price and Kuchel 1990).

Nephrotoxicity with Cisplatin

Nephrotoxicity is a major concern in patients using CIS for treating any kind of cancer. Tubular necrosis is derived from epithelial cell death by apoptosis and necrosis in the proximal and distal renal tubules (Sancho-Martínez, Piedrafita et al. 2011). In mice CIS injections, resulted in renal functional and structural abnormalities consistent with nephrotoxicity, which was accompanied by elevated circulating and renal levels of TNF-α and IL-1β, subunits of NADPH oxidase, thiobarbituric acid-reactive substances, and PGE(2) (Jia, Wang et al. 2011). CIS treatment also caused an increase in renal IL-10R1 expression and STAT3 phosphorylation (Tadagavadi and Reeves 2010). In rats CIS injection resulted in an increase in malondialdehyde, 8-isoprostane, and tumor necrosis factor-α level of kidney and significantly deranged renal function (Tuzcu, Sahin et al. 2010). There is a growing body of evidence that activation of p53 and oxidative stress plays a critical role in CIS-induced renal apoptotic injury. We are going to touch on each of these mechanisms in little detail in the following section.

Apoptotic Mechanisms

CIS is known to induce apoptosis in renal tubular epithelial cells (RTEC) and is emerging as a final common pathway in response to a variety of cellular stresses applied at intensity below the threshold for necrosis (Ueda, Kaushal et al. 2000; Levine 2001). This observation especially holds true for CIS nephrotoxicity, in which necrotic cell death is encountered with higher doses whereas lower concentrations induce apoptosis (Lau 1999). Inhibition of apoptosis may represent a novel and powerful therapeutic strategy for the prevention and treatment of CIS nephrotoxicity. Various cell culture models have been used to validate this idea, CIS was first shown to cause apoptosis in cultured mouse proximal tubular cells (Lieberthal, Triaca et al. 1996; Fukuoka, Takeda et al. 1998) and RTEC (Kruidering M 1998; Kaushal, Kaushal et al. 2001; Park, De Leon et al. 2002), human proximal tubular cells (Razzaque, Koji et al. 1999; Cummings and Schnellmann 2002; Nowak 2002), and collecting duct cells (Hwa Lee, Mi Song et al. 2001). Caspase inhibitors such as ZVAD.fmk and DEVD.CHO, are effective in protecting against CIS-induced DNA damage and cell death (Zhan Y 1999; Kaushal, Kaushal et al. 2001). CIS nephrotoxicity also upregulates caspase 9 and to a lesser extent caspase 8 (Park, De Leon et al. 2002). It was noticed that CIS induce apoptosis in cultured RTEC and inhibitor of executioner caspase3 blocks only about 50% of apoptosis (Cummings and Schnellmann 2002; Nowak 2002) suggesting remaining 50% cell death occurs in a caspase independent manner.

Activation of Death Receptor Pathways

CIS induced apoptosis in tumor cells leads to activation of the Fas pathway (Friesen C 1999). In response to CIS cultured human proximal tubular epithelial cells underwent apoptosis which was temporally correlated with an increased expression of Fas (Razzaque, Koji et al. 1999). In primary cultures of proximal tubular cells from wild-type mice, CIS upregulated expression of Fas mRNA and protein, increased caspase 8 activity, and apoptosis, all these effects were blunted in primary cultures of proximal tubular cells from a Fas mutant mice (Tsuruya, Ninomiya et al. 2003). Similarly, RTEC cultures from wild-type mice exhibited increased TNF-α secretion following CIS exposure, and RTEC from TNFR1-deficient mice displayed an attenuated functional and apoptotic response to CIS (Ramesh and Reeves 2002).

Activation of Mitochondrial Pathways

Various mitochondrial factors that participate in either apoptosis or free radical generation get activated in renal cells in response to CIS. CIS nephrotoxicity causes and increased renal expression of Bax *in vivo* (Huang, Dunn et al. 2001), and translocation of Bax from the cytosol to a membrane fraction in cultured cells (Hwa Lee, Mi Song et al. 2001). CIS-induced oxidative stress in renal mitochondria is also known to cause the release of cytochrome c into the cytosol and activation of caspase 9 (Nishikawa, Nagatomi et al. 2001). Apoptosis in RTEC by CIS is also regulated by p53-mediated caspase-2 activation and translocation of apoptosis inducing factor (AIF) from mitochondria to cytosol and nucleus, causing DNA fragmentation and cell death (Seth, Yang et al. 2005). Gene silencing of AIF with small interfering RNA (siRNA) suppressed CIS-induced AIF expression and provided a marked protection against cell death (Seth, Yang et al. 2005).

CIS-Induced Activation of p53

CIS can induce a rapid activation and nuclear translocation of p53 in kidneys (Miyaji T 2001) and cultured RTEC (Cummings and Schnellmann 2002). Inhibition of p53 prior to CIS exposure blunted the apoptotic response by 50%. CIS-induced DNA damage as well as CIS-induced oxidative stress can activate p53 (Müller, Wilder et al. 1998; Chandel NS 2000), which in turn can activate both Bax as well as the Fas-FADD axis (Burns and El-Deiry 1999). In part p53 can also regulate expression of both p21 and 14-3-3σ (Hermeking, Lengauer et al. 1997) which are over-expressed in terminally differentiating epithelia. Both are required for proper coordination of the cell cycle, and the absence of either of these factors can accelerate apoptosis (Megyesi, Andrade et al. 2002). Recently, p53 was shown to regulate transcriptional control of the caspase-6 and -7 genes and their functional implications in CIS nephrotoxicity both *in vitro* and *in vivo* (Yang, Kaushal et al. 2007). In this study the executioner caspase-6 and -7 but not caspase-3 were identified as transcriptional targets of p53. Primary cultures of p53 (-/-) RTEC treated with CIS did not induce caspase-6 and -7 expression and subsequent activation; however, caspase-3 (-/-) RTEC, inhibition of caspase-6 and -7 activations markedly prevented cisplatin-induced cell death. Silencing of p53 by siRNA completely abrogate hypersensitivity not only to CIS but also to genotoxic inducers of p53 such as the Mdm2 antagonist Nutlin-3 and the proteasome inhibitor Bortezomib (Gutekunst, Oren et al. 2011).

CIS Induced Oxidative Stress

Cytotoxic action of CIS on renal proximal tubular cells is mediated by reactive oxygen species (ROS) formation, decline of mitochondrial membrane potential and lysosomal membrane leakiness before cell lysis ensued. Studies have suggested that ROS, mainly generated in mitochondria, play a central role in CIS-induced renal damage (Santos, Bezerra et al. 2008). CIS is known to induce ROS in RTEC primarily by decreasing the activity of antioxidant enzymes and by depleting intracellular concentrations of GSH (Sadzuka Y 1994; Huang, Dunn et al. 2001). In rats, administration of orgotein, a pharmaceutical preparation of Zn-Cu superoxide dismutase (SOD), lowered the nephrotoxicity induced by CIS injections (McGinness JE 1978). Apart from SOD, dimethylthiourea, and GSH have been shown to reduce the degree of renal failure and tubular cell damage when administered simultaneously with CIS in rats (McGinness JE 1978; Matsushima, Yonemura et al. 1998). It is shown that

CYP2E1 involved in CIS oxidative stress cytotoxicity mechanism and intracellular nitric oxide enhancement protects the RTEC against the CIS-induced cytotoxicity. It appears that CIS nephrotoxicity is associated with mutual mitochondrial/lysosomal potentiation (cross-talk) of oxidative stress in RPT cells. This cross-talk finally results in release of lysosomal digestive proteases and phospholipases and mitochondrial MPT pore opening leading to cytochrome c release and activation of caspase cascade which signal apoptosis (Pourahmad, Hosseini et al. 2010).

Role of Meprins in CIS Nephrotoxicity

In a recent study a critical role of zinc-dependent metalloproteinase meprins is reported in CIS induced injury to RTEC. Meprins are highly expressed in the brush-border membranes of both the kidney and the intestines and are capable of proteolytically degrading extracellular matrix proteins, proteolytically processing bioactive proteins, and play a role in inflammatory processes (Herzog, Seth et al. 2007). Meprin, composed of α and/or β subunits, consist of membrane-bound and secreted forms that are abundantly expressed in proximal tubules of the kidney. Meprin β is expressed in rat glomerular epithelial cells and suggest a potential involvement of meprin β in glomerulonephritis (Oneda, Lods et al. 2008). Role of meprins is studied in endotoxemic acute renal failure using meprin α knockout (αKO), meprin β knockout (βKO), and wild-type (WT) mice. Meprin αKO mice displayed lower blood urea nitrogen levels and serum cytokine profiles compared with WT and meprin βKO mice. This work suggest that meprin A contributes to the renal and urogenital pathogenesis of endotoxicity (Yura, Bradley et al. 2009). The excretion of meprin β fragment is seen to rise even before the rise in serum creatinine and blood urea nitrogen (BUN) suggesting its usefulness as a biomarker for AKI (Herzog, Seth et al. 2007).

Combinatorial Regimen for Effective Treatment

Combination therapy shows some promising results for using CIS, bypassing nephrotoxicity without any compromise with its anti-neoplastic activity. Several combinations have been proposed in this regards, for convenience we have categorized them in four major groups namely, biochemical combinations, herbal combinations, liposomal CIS and the most recent one using nanoparticles in combination with CIS.

Biochemical Combinations

Several biochemical combinations are used along with CIS that can minimize nephrotoxic effects of CIS. In a recent work, losartan, an angiotensin receptor blocker, was tested against oxidative stress induced by CIS. Losartan has protective effects against CIS-induced nephrotoxicity as evidenced by restoration of normal serum levels of creatinine and BUN, and LDH leakage. It significantly counteracted CIS-induced lipid peroxidation and GSH depletion (Saleh S 2009). With idea of testing antioxidative, antiapoptotic and anti-inflammatory properties of acetyl L-carnitine (ALCAR), this drug was used in combination with CIS. ALCAR administration resulted in an improvement in kidney function tests along with decreased apoptotic cell death and caspase-3, 8 and 9 expression (Tufekci O 2009). Carvedilol, an antihypertensive with strong antioxidant properties, is suggested as a potential drug for the adjuvant nephroprotective therapy along with CIS chemotherapy (Carvalho Rodrigues, Rodrigues et al. 2010), it is shown to prevent the mitochondrial oxidative stress and apoptosis in kidney of rats treated with CIS (Rodrigues, Rodrigues et al. 2011). Selective agonists of cannabinoid-2 (CB(2)) receptors also exert anti-inflammatory and tissue-protective effects in various disease models. Effect of CIS was tested using CB(2) receptor agonist HU-308 and CB(2) knockout mice, while CB(2) agonist attenuated the CIS-induced inflammatory response, oxidative/nitrosative stress, and cell death in the kidney and improved renal function, CB(2) knockouts developed enhanced inflammation and tissue injury (Mukhopadhyay, Rajesh et al. 2010).

Protective effects of selenium (Se) on hematological and oxidative stress parameters were tested in rats. Se and CIS co-treatment stimulated erythropoiesis and significant recovery of reduced GSH levels (Marković, Djačić et al. 2010). Use of synthetic organoselenium compound diphenylmethyl selenocyanate is shown to prevent against the nephrotoxic damage induced by CIS in Swiss albino mice (Chakraborty, Roy et al. 2011). Another compound zinc picolinate is shown to be protective in CIS induced nephrotoxicity. A single dose of CIS resulted in an increase in malondialdehyde, 8-isoprostane, and tumor necrosis factor-α (TNF-α) levels of kidney and significantly deranged renal function. Zinc picolinate treatment significantly reduced BUN, creatinine, malondialdehyde, 8-isoprostane, and TNF-α levels suggesting that zinc picolinate may be a potential preventive agent in CIS-induced renal injury through decreasing oxidative stress and inflammation (Tuzcu, Sahin et al. 2010).

Some enzymes and inhibitors of biochemical pathways are successfully tested as possible combinations with CIS, for example protein tyrosine kinase inhibitors tyrphostins was tested in the prophylaxis and treatment of acute kidney injury (AKI), it appears to be nephroprotective in animal models of AKI (Gonçalves S 2010). Another multi-targeted receptor tyrosine kinase inhibitor sunitinib is reported to significantly decrease CIS-induced changes in serum creatinine, BUN, creatinine clearance and regulate micro total protein in urine, renal malondialdehyde levels and reduced GSH contents. These findings suggest that sunitinib could ameliorate the CIS-evoked toxicity as well as to improve the chemotherapeutic effect (Suddek 2011). Role of proteasome inhibitors, PS-341 and MG-132 has been tested in preventing mitochondrial release of AIF in CIS-exposed RTEC. PS-341 and MG-132 blocked CIS-induced Mcl-1 depletion and markedly prevented mitochondrial release of AIF (Liu, Yang et al. 2010). One recent study documents the protective effect of cilastatin, a specific inhibitor of renal dehydrodipeptidase1, cilastatin prevents hydrolysis of imipenem and its accumulation in the proximal tubule. While CIS increased cell death, apoptotic-like morphology, caspase activation, and mitochondrial injury in proximal tubular cells in a dose- and time-dependent way. Concomitant treatment with cilastatin reduced all CIS-induced changes in renal tubular function (Camano, Lazaro et al. 2010). Fructose-1,6-bisphosphate is also reported as renal protective when given along with CIS in rats. The drugs when administered intraperitoneally as a single dose presented a significantly lower level of creatinine and BUN compared to the CIS group. The anti-inflammatory effect of fructose-1,6-bisphosphate confirms its protective effect in cases of cellular injury (Azambuja, Lunardelli et al. 2011). A glutathione disulfide mimetic with chemoprotective activity, NOV-002 combined with CIS was tested in clinical trials and shown to significantly improve survival and decreased tumor progression rates in non-small cell lung (NSCLC) and ovarian cancer patients. NOV-002-treated mice had significantly lower levels of plasma creatinine compared to mice treated with CIS alone. NOV-002 protected the kidneys from CIS mediated proximal tubule damage. Since CIS-induced nephrotoxicity can be mediated by a glutathione-platinum conjugate catalyzed by gamma-glutamyl-transpeptidase and GSH is an endogenous substrate of GGT, the protective effect of NOV-002 in the kidney may be attributed to its ability to act as a competitive substrate for the enzyme (Jenderny, Lin et al. 2010). Recently cimetidine, an H_2 receptor antagonist is shown to inhibit nephrotoxicity induced by CIS on osteosarcoma both, *in vitro* and *in vivo* when the serum concentration of cimetidine was maintained over 20 μg/ml for 4 h by bolus and continuous intravenous infusion in Wistar rats (Katsuda, Yamashita et al. 2010). A recent study used pravastatin, a HMG-CoA reductase inhibitor along with CIS. Findings of this study suggest that pravastatin has a partial protective effect against CIS nephrotoxicity via antioxidant activity as well as attenuation of the p53 response, and lipid-lowering effects (Fujieda, Morita et al. 2011). Pravastatin showed significant protection as evidenced by the decrease of elevated serum creatinine and BUN, and improvement of histopathological injury induced by CIS. The formation of kidney malondialdehyde with a concomitant reduction of reduced GSH was inhibited by pravastatin, while the activities of kidney SOD, catalase (CAT) and glutathione peroxidase (GPx) were increased. The over expressions of kidney induced nitric oxide synthase (iNOS) and nitrotyrosine (3-NT) were suppressed by pravastatin. Pravastatin also suppressed CIS-induced p38 mitogen-activated protein kinase (MAPK) activation in the kidney of mice (An, Xin et al. 2011).

Another study suggests a dose related effect of sildenafil, an inhibitor of cGMP-specific phosphodiesterase type 5, in combating renal toxicity by CIS. Sildenafil given at a dose of 0.4 mg/ kg/day, i.p. in rats for 5 days with CIS, showed a dramatic improvement in renal histopathology, reduction in N-acetyl-β-D-glucosaminidase and increase in renal blood flow (Ali, Abdelrahman et al. 2011). Inhibitors of histone deacetylases, including suberoylanilide hydroxamic acid (SAHA) and trichostatin A (TSA), were tested for cytoprotective effects. SAHA is reported to ameliorate Bax translocation and cytochrome c release. Mechanistically, SAHA inhibited and TSA delayed p53 phosphorylation, acetylation, and activation during CIS incubation suggesting anti-apoptotic role for these inhibitors. In HCT116 colon cancer cells, SAHA suppressed CIS-induced p53 activation, but enhanced apoptosis (Dong, Luo et al. 2010).

Prostaglandin E1 analogue misoprostol is reported to reduce CIS nephrotoxicity on account of its ROS scavenging activity. Misoprostol treatment after CIS application protected the renal tissue (Ozer, Asci et al. 2011). CIS nephrotoxicity can be ameliorated by genetic or pharmacologic blockade of prostaglandin synthesis (Jia, Wang et al. 2011). Inflammation, oxidative stress, and arachidonate metabolism associated with renal failure were all diminished in microsomal prostaglandin E synthase-1 (mPGES-1) null mice (Jia, Wang et al. 2011). Pituitary adenylate cyclase-activating polypeptide (PACAP) has been shown to protect kidneys from ischemic injury and light chain-induced damage by modulating inflammation. HK-2 human renal cells were treated with CIS in the presence or absence of either PACAP38 or p53 siRNA. PACAP markedly inhibited CIS-induced apoptosis of renal tubules. While exposure to CIS significantly suppressed the expression of fibronectin and collagens I and IV, and altered the integrin repertoire of human renal tubule cells, PACAP partially reversed these reductions. Further, use of PACAP receptor antagonists and siRNA silencing of p53 conferred renal protection which was mediated by the PAC(1) receptor and through both p53-dependent and -independent suppression of apoptosis. PACAP also prevented the rise in BUN and creatinine in mice treated with CIS (Li, Balamuthusamy et al. 2010).

Herbal Combinations

Recently lot of focus is on using herbal combinations along with CIS for reducing nephrotoxic effects of CIS. Although in most cases mechanism of action of these herbal combinations is not known, still it hold promise for future investigations. Some herbal combinations that are successfully used are mentioned here.

A recent report suggests use of *Astragalus* injection (AI) with CIS for nephroprotection. CIS treatment significantly increased the number of TUNEL-positive cells and the Bax/Bcl-2 ratio in the kidneys of CIS treated mice. High-dose AI pretreatment significantly reduced the elevated number of TUNEL-positive cells and the Bax/Bcl-2 ratio. It is reported that CIS plus AI led to enhanced tumor growth inhibition as compared to CIS alone suggesting that AI attenuates CIS-induced nephrotoxicity without compromising the anti-tumor efficacy of CIS (Liu L 2010). The rhizome of turmeric, *Curcuma longa* (CL), is another herbal medicine which is tested for genome wide mRNA expression profiles of HEK 293 -cells together with CIS. The results of microarray analysis and real-time PCR revealed that NFκB pathway-related genes and apoptosis-related genes were down-regulated in CL-treated HEK 293 cells. This study suggests therapeutic role for CL that can be used for the prevention or treatment of CIS-induced renal disorders. (Sohn, Ko et al. 2010).

Quercetin, a plant-derived flavonoid found in fruits, vegetables, leaves and grains partially prevented nephrotoxic effects of CIS, while not affecting its anti-tumour activity (Sanchez-Gonzalez PD 2011). Another flavone found in medicinal herbs and plants, Luteolin, has been reported to exhibit anti-inflammatory, antioxidant and anticarcinogenic activities. Treatment of mice with CIS resulted in renal damage, showing an increase in BUN and creatinine levels, tubular damage, oxidative stress and apoptosis. Treatment of CIS-treated mice with luteolin significantly improved renal dysfunction, reducing tubular cell damage, oxidative stress and apoptosis. This study suggests that luteolin ameliorates the CIS-mediated nephrotoxicity through down-regulation of p53-dependent apoptotic pathway in the kidney (Kang, Park et al. 2011).

Resveratrol, (3,5,4'-trihydroxy-trans-stilbene) a stilbenoid, and a phytoalexin produced naturally by several plants when under attack by pathogens was tested for ameliorating CIS-induced kidney damage. It was reported that SIRT1 activation by resveratrol decreased CIS-induced apoptosis while improving the glomerular filtration rate. SIRT1 activation causes deacetylation of p53 which in turn plays an important role to attenuate CIS-induced kidney injury (Kim DH 2011). In another study epigallocatechin- gallate and resveratrol were tested together for their capacity to ameliorate CIS -induced disruption of renal glomerular filtration rate in rats. Contrary to previous study this study showed that EGCG but not resveratrol can protect against CIS-induced nephrotoxicity. This study further confirms that CIS triggers a high level of oxidative stress and systemic inflammation, events that were all abrogated with EGCG; better than resveratrol or quercetin. (El-Mowafy AM 2011)

Phenolic extract of soybean (PESB) was tested with CIS in rats. Rats pretreated with PESB showed a decreased BUN, serum creatinine and urinary N-acetyl-beta-D-glucosaminidase. The extract also significantly increased renal myeloperoxidase activity. PESB also decreased renal xanthine oxidase activity and serum nitrate/nitrite in the CIS-treated rats. In addition, PESB significantly improved liver histology and increased the activities of the antioxidant enzymes measured [SOD, CAT, glutathione-S-transferase], prevented GSH depletion and decreased malondialdehyde level following CIS treatment (Ekor, Emerole et al. 2010). The extracts of medicinal plants *Hemidesmus indicus L.* (Apocynaceae) and *Acorus calamus L.* (Araceae) protected the renal tissue effectively from CIS-induced toxicity. Treatment of CIS-administered animals with the plant extracts of *H. indicus* and *A. calamus* significantly lowered BUN and serum creatinine levels and restored renal pathology (Sandeep D 2010). Nephroprotective role of PTY, prepared from methanolic extract of tubers of *Pueraria tuberosa* D.C., was studied in rats. PTY prevented the rise in serum creatinine, BUN and decline in GSH content as well as activities of SOD and CAT. PTY treatment also prevented DNA damage, tubular swelling, cellular necrosis and protein cast deposition as compared to CIS treated rat kidneys (Nagwani and Tripathi 2010). Similarly, ethanol extract of *Momordica dioica* fruit extract was studied for its nephroprotective and curative activities in rats. While CIS alone could induce nephrotoxicity as indicated by rise in BUN and serum creatinine, reduced GSH and increase in malondialdehyde production. *M. dioica* fruit extract administration showed significant recovery in the levels of these biochemicals, suggesting a nephroprotective and curative effect of *M. dioica* fruit extract (Jain A 2010). Ethanolic extract of *Bauhinia variegata* (Linn.) exhibited significant and comparable nephroprotective potential to that of the standard polyherbal drug cystone (Panda, Sahoo et al. 2011).

Sulforaphane (SFN), a isothiocyanate produced by cruciferous vegetables, is known to induce cytoprotective enzymes through the transcription factor Nrf2. Preincubation of pig RTEC with SFN for 24 h was able to prevent CIS-induced cell death in a dose dependent manner. SFN also attenuated CIS-induced renal dysfunction, structural damage, oxidative/nitrosative stress, GSH depletion, enhanced urinary hydrogen peroxide excretion and the decrease in antioxidant enzymes (CAT, GPx and glutathione-S-transferase) in rats (Guerrero-Beltrán, Calderón-Oliver et al. 2010). Protective effect of pomegranate seed extract (PSE) was tested against oxidative stress caused by CIS induced damage to the kidneys and liver in rats. PSE administration for 15 days by gavage along with a single injection of CIS showed significant protection against CIS injury. PSE significantly attenuated nephrotoxicity and hepatotoxicity by the way of its antioxidant, radical-scavenging, and anti-apoptotic effects, suggesting the use of PSE as a dietary supplement in patients receiving chemotherapy (Çayır, Karadeniz et al. 2011). Similarly *Glycine max* and *Chrysanthemum indicum* extracts exhibited a promising protective effect against CIS-induced nephrotoxicity (Pongjit, Ninsontia et al. 2011). Lycopene, which is a carotenoid found in tomatoes is a potent antioxidant it was tested in combination with CIS. Rats treated with tomato lycopene complex + CIS showed lower serum creatinine and BUN levels compared with rats treated with CIS alone, suggesting a protective effects against CIS-induced nephrotoxicity in rats (Dogukan A 2011). Royal jelly (RJ) and green tea (GT) extracts are also tested along with CIS in adult albino mice. The mice treated with only CIS exhibited an increase in serum BUN and creatinine levels and oxidative damage resulting in decrease in GSH levels and a significant increase in MDA levels in kidneys. RJ and GT supplementation attenuated elevation in serum creatinine and BUN levels and restored GSH content and MDA production levels in the kidney tissue following CIS treatment (Yapar, Çavuşoğlu et al. 2009). Another study reported similar observation in rats. In this study adult male Wistar albino rats receiving RJ in association with CIS, showed improvement in some oxidative stress parameters and certain other biochemical parameters, pre-treatment with RJ is suggested to be more effective in coping with oxidative stress resulting from CIS treatment (Silici, Ekmekcioglu et al. 2011).

The root of *Angelica gigas Nakai* contains two major coumarins, which have been previously identified as decursin and decursinol. Decursin has been demonstrated to exhibit potent anti-cancer activity both *in vitro and in vivo*. Decursin showed protective effects of CIS-induced damage in normal human primary renal epithelial cells (HRCs). Decursin significantly blocked CIS-induced cytotoxicity in HRCs. Decursin also inhibited sub-G1 and cell death by suppression of cleavage of caspase-3, -9 and poly(ADP-ribose) polymerase (PARP) induced by CIS treatment in HRCs. Decursin effectively restored the activities of Cu/Zn SOD, CAT and GPx in CIS-treated HRCs, suggesting this combination suitable to be used with other anti-cancer drugs (Kim, Jeong et al. 2010).

Recently in southern India a terpene which is isolated from fruiting bodies of *Ganoderma lucidum* was tested along with CIS in mice. While IP administration of CIS resulted in significant nephrotoxicity in mice and resulted in increased BUN, creatinine and alkaline phosphtase (ALP) levels. The renal antioxidant defense system such as SOD, CAT, GPx and concentration of reduced GSH were affected by CIS treatment. The oral administration of terpenes at a dose of 100mg/kg body weight prevented increase in BUN, creatinine levels and ALP activity and also maintained the renal antioxidant defense. The *Ganoderma* terpenes also imparted protection against CIS induced renal tissue lipid peroxidation (Pillai, John et al. 2011). Silymarin from seeds of milk thistle, *Silybum marianum* has been shown to protect against CIS-induced cell death in a dose-dependent manner in renal cells and could be beneficial for the development of this considerably safe compound as a renoprotective agent (Ninsontia C 2011).

Use of Liposomal CIS

Recently, a new formulation for CIS, liposomal CIS has been introduced and tested thoroughly in preclinical (*in vitro*) and phase I, II and III trials. Experiments in animals showed that lipoplatin is less toxic than CIS and that it produces tumor reduction. No nephrotoxicity was observed in a phase I-II study using a combination of lipoplatin and gemcitabine in pretreated patients with advanced pancreatic cancer. Up to now, the data on lipoplatin treatment in malignant tumors are quite impressive, because of the negligible toxicity and because it is equal if not superior to CIS with regard to response rate (Stathopoulos 2010). Combination of 5-fluorouracil/lecovorin with the administration of lipoplatin shows minimal nephrotoxicity and neutropenia. A net improvement in the performance status was recorded 2 months after the end of therapy suggesting its suitability for the treatment of gastric cancer (Koukourakis, Giatromanolaki et al. 2010). In another trial study patients with inoperable non-small-cell lung cancer, liposomal CIS in combination with paclitaxel has been shown to be much less toxic than the original CIS combined with paclitaxel. Nephrotoxicity in particular was negligible after liposomal CIS administration. TTP and survival were similar in both treatment arms. (Stathopoulos, Antoniou et al. 2010).

CIS in Combination with Nanoparticles

Recently lot of focus is on using CIS in combination with nanoparticles, the hypothesis is that rapid cytoplasmic release profile from nanoparticles would potentiate the anticancer activity of CIS. Some early reports suggest the entrapment of CIS into biodegradable colloidal systems (polyalkylcyanoacrylates) (Egea MA 1994). CIS-loaded nanoparticles with pH-responsive poly[2-(N,N-diethylamino)ethyl methacrylate] (PDEA) cores was more effective in reducing tumor burdens in mice with some incidences of cellular pyknosis (Xu, Van Kirk et al. 2006). Similarly CIS adsorbed magnetic nanoparticles (cis-MNPs) used for minimal invasive cancer treatments exhibit advantageous features for a facilitated desorption of CIS in biological media and the heating potential is adequate for hyperthermic treatments (Kettering and et al. 2009). Different doses of blank and CIS-loaded PLGA-mPEG nanoparticles were tested on SCID mice xenografted with colorectal adenocarcinoma HT 29 cells. This study suggests that mice treated with CIS-loaded nanoparticles exhibited higher survival rate compared to the free CIS group (Mattheolabakis, Taoufik et al. 2009). Rational engineering of a novel polymer, glucosamine-functionalized polyisobutylene-maleic acid, in unique platinum to polymer ratio, self-assembles into a nanoparticle complex that releases CIS in a pH-dependent manner. The nanoparticles exhibited significantly improved antitumor efficacy in terms of tumor growth delay in breast and lung cancers and tumor regression in a K-ras(LSL/+)/Pten(fl/fl) ovarian cancer model (Paraskar, Soni et al. 2010). Same group reports use of a novel polyethylene glycol-functionalized poly-isobutylene-maleic acid copolymer complexed with CIS. This complex exhibits significantly improved antitumor efficacy in a 4T1 breast cancer model *in vivo*, with limited nephrotoxicity (Paraskar and et al. 2011). Another nanocombination uses hyperbranched polyglycerols (HPG's) with hydrophobic cores and derivatized with methoxy poly (ethylene glycol) and functionalized with carboxylate groups to bind and deliver CIS. Using this combination, release of CIS into artificial urine was considerably faster than into buffer. Carboxylated HPGs demonstrated good biocompatibility, and drug-loaded HPGs effectively inhibited proliferation of KU-7-luc bladder cancer cells (Ye, Letchford et al. 2010). *In vitro* cell viability study, using four human tumor cell lines proved that the core-shell type star polymer bearing carboxylate functions used as a nanocarrier of CIS conjugates, exhibited lower cytotoxicity compared to the free agent (Kowalczuk, Stoyanova et al. 2011). Antivascular tumor therapeutic using the F3 peptide to target CIS-loaded nanoparticles shows high binding specificity to both human ovarian tumor cells and tumor endothelial cells *in vitro*, however, it only show cytotoxic activity against the tumor endothelial cells (Winer, Wang et al. 2010). Similarly, poly(ethylene glycol)-b-poly(acrylic acid) (PEG-b-PAA) block copolymer nanoparticles containing CIS were released by erosion of the hydrogels were shown to inhibit the growth of human bladder carcinoma EJ cells without any cytotoxicity (Zhu, Li et al. 2010).

Drug delivery is very important consideration with regards to CIS administration. Use of nanoparticles appears to significantly improve drug delivery. In a recent study dextran-magnetite conjugated CIS (DM-Cis) complex was used for thermal ablation and delivery of chemotherapeutic drug (Sonoda A 2010). Similarly, acid-responsive therapeutic nanoparticles (NPs) consisting of polymer-CIS conjugates are shown to readily precipitate to form sub-100 nm NPs. This approach holds the promise to suppress cancer cell chemoresistance by rapidly releasing a high dose of chemotherapy drugs inside the tumor cells, thereby improving the therapeutic efficacy of the drug payload. (Aryal, Hu et al. 2009). In more recent works codelivered CIS and docetaxel (Dtxl) to prostate cancer cells was tested. This work reveals the potential of a single, programmable nanoparticle to blend and deliver a combination of drugs for cancer treatment (Kolishetti, Dhar et al. 2010). Another work suggests using targeted, multifunctional siRNA dclivery system for cancer therapy. For this superparamagnetic iron oxide nanoparticles and poly(propyleneimine) generation 5 dendrimers were used to cooperatively provoke siRNA. The developed siRNA delivery system is reported to be capable of sufficiently enhancing *in vivo* antitumor activity of CIS (Taratula O 2011). Other technologies as liposomes, dendrimers, polymers and nanotubes, are all showing promise for the delivery of platinum compounds CIS and related carboplatin and oxaliplatin. The main aim is improving drug targeting to reduce side-effects.(Harper, Krause-Heuer et al. 2010)

Route of Drug Administration

Route of drug administration with regards to CIS administration is a very important factor in determining the outcome of treatment. In a comprehensive study that evaluates the patient-reported outcomes associated with IP versus IV therapy, it appears that IP arm experienced more health-related quality-of-life disruption, abdominal discomfort and neurotoxicity compared with patients receiving conventional IV therapy (Wenzel LB 2007). In another study using cervical cancer patient's CIS chemotherapy led to a small, but significant, benefit for overall survival and event-free survival. This benefit was essentially observed when chemotherapy was administered concomitantly with RT (Long, Bundy et al. 2005). So depending on site of cancer route of drug administration needs to be adjusted accordingly.

Conclusion

The high incidence of nephrotoxicity caused by CIS stresses the importance of undertaking research to identify platinum complexes that would feature antitumor properties with less nephrotoxicity. Research during the last few years has gained significant insights into the signaling pathways that are responsible for CIS nephrotoxicity. However, how the various pathways are integrated to induce a remarkable renal pathology remains largely unknown. Along with the mechanistic studies, numerous renal protective approaches have been documented; nevertheless, the demonstrated protection is mostly partial and combinatorial strategies may offer better protective effects. As we move forward, it is our dream to also focus on validating combination therapies, putative molecular markers of prognosis, drug sensitivity, and drug resistance. New promising technologies will be very helpful to guide patient selection and treatment decisions in the future and will help in refining the therapeutic approaches.

REFERENCES

Ahmad, S. (2010). "Platinum–DNA Interactions and Subsequent Cellular Processes Controlling Sensitivity to Anticancer Platinum Complexes." *Chemistry and Biodiversity* 7(3): 543-566.

Ali, B. H., A. M. Abdelrahman, et al. (2011). "The Effect of Sildenafil on Cisplatin Nephrotoxicity in Rats." *Basic and Clinical Pharmacology and Toxicology:* no-no.

An, Y., H. Xin, et al. (2011). "Amelioration of cisplatin-induced nephrotoxicity by pravastatin in mice." *Experimental and Toxicologic Pathology* 63(3): 215-219.

Aryal, S., C.-M. J. Hu, et al. (2009). "Polymer−Cisplatin Conjugate Nanoparticles for Acid-Responsive Drug Delivery." *ACS Nano* 4(1): 251-258.

Azambuja, A., A. Lunardelli, et al. (2011). "Effect of Fructose-1,6-bisphosphate on the Nephrotoxicity Induced by Cisplatin in Rats." *Inflammation* 34(1): 67-71.

Baujat, B., H. Audry, et al. (2006). "Chemotherapy in locally advanced nasopharyngeal carcinoma: An individual patient data meta-analysis of eight randomized trials and 1753 patients." *International Journal of Radiation Oncology, Biology, Physics* 64(1): 47-56.

Berners-Price, S. J. and P. W. Kuchel (1990). "Reaction of cis- and trans-[PtCl2(NH3)2] with reduced glutathione inside human red blood cells, studied by 1H and 15N-{1H} DEPT NMR." *Journal of Inorganic Biochemistry* 38(4): 327-345.

Burns, T. F. and W. S. El-Deiry (1999). "The p53 pathway and apoptosis." Journal of Cellular Physiology 181(2): 231-239.

Camano, S., A. Lazaro, et al. (2010). "Cilastatin Attenuates Cisplatin-Induced Proximal Tubular Cell Damage." *Journal of Pharmacology and Experimental Therapeutics* 334(2): 419-429.

Carvalho Rodrigues, M. A., J. L. Rodrigues, et al. (2010). "Carvedilol protects against the renal mitochondrial toxicity induced by cisplatin in rats." *Mitochondrion* 10(1): 46-53.

Çayır, K., A. Karadeniz, et al. (2011). "Pomegranate Seed Extract Attenuates Chemotherapy-Induced Acute Nephrotoxicity and Hepatotoxicity in Rats." *Journal of Medicinal Food* 0(May 6. [Epub ahead of print]): null.

Chakraborty, P., S. S. Roy, et al. (2011). "Amelioration of cisplatin-induced nephrotoxicity in mice by oral administration of diphenylmethyl selenocyanate." *Free Radical Research* 45(2): 177-187.

Chandel NS, V. H. M., Thompson CB, Schumacker PT. (2000). "Redox regulation of p53 during hypoxia." *Oncogene* 19(34): 3840-3848.

Chang, G.-C., M.-J. Ahn, et al. (2011). "Comparative effectiveness of bevacizumab plus cisplatin-based chemotherapy versus pemetrexed plus cisplatin treatment in East Asian non-squamous non-small cell lung cancer patients applying real-life outcomes." *Asia-Pacific Journal of Clinical Oncology* 7: 34-40.

Chung P, W. P. (2011). "Testicular cancer: seminoma." *Clin. Evid.* (Online). (Jan 25, 2011): pii: 1807.

Cohen SM, L. S. (2001). "Cisplatin: from DNA damage to cancer chemotherapy." *Prog. Nucleic Acid Res. Mol. Biol.*(67): 93-130.

Cummings, B. S. and R. G. Schnellmann (2002). "Cisplatin-Induced Renal Cell Apoptosis: Caspase 3-Dependent and-Independent Pathways." *Journal of Pharmacology and Experimental Therapeutics* 302(1): 8-17.

Dogukan A, T. M., Agca CA, Gencoglu H, Sahin N, Onderci M, Ozercan IH, Ilhan N, Kucuk O, Sahin K. (2011). "A tomato lycopene complex protects the kidney from cisplatin-induced injury via affecting oxidative stress as well as Bax, Bcl-2, and HSPs expression." *Nutr. Cancer* 63((3)): 427-434.

Dong, G., J. Luo, et al. (2010). "Inhibitors of histone deacetylases suppress cisplatin-induced p53 activation and apoptosis in renal tubular cells." *American Journal of Physiology - Renal Physiology* 298(2): F293-F300.

Echarri Gonzalez MJ, G. R., Muggia FM (2011). "Intraperitoneal drug delivery for ovarian cancer: why, how, who, what, and when?" *Oncology* (Williston Park) 25(2): 156-165.

Egea MA, G. F., Valero J, Garcia ME, Garcia ML. (1994). "Entrapment of cisplatin into biodegradable polyalkylcyanoacrylate nanoparticles." *Farmaco.* 49(3): 211-217.

Ekor, M., G. O. Emerole, et al. (2010). "Phenolic extract of soybean (Glycine max) attenuates cisplatin-induced nephrotoxicity in rats." *Food and Chemical Toxicology* 48(4): 1005-1012.

El-Mowafy AM, S. H., Al-Gayyar MM, El-Mesery ME, El-Azab MF. (2011). "Evaluation of renal protective effects of the green-tea (EGCG) and red grape resveratrol: role of oxidative stress and inflammatory cytokines." *Nat. Prod. Res.* 25((8)): 850-856.

Elenkov I, P. P., Ugrinova I, Takahashi M, Pasheva E (2011). "The DNA Binding and Bending Activities of Truncated Tail-less HMGB1 protein are Differentially Affected by Lys-2 and Lys-81 Residues and Their Acetylation." *Int. J. Biol. Sci.* 7((6)): 691-699.

Friesen C, F. S., Debatin KM (1999). "Cytotoxic drugs and the CD95 pathway." *Leukemia* 13(11): 1854-1858.

Fujieda, M., T. Morita, et al. (2011). "Effect of pravastatin on cisplatin-induced nephrotoxicity in rats." *Human and Experimental Toxicology* 30(7): 603-615.

Fukuoka, K., M. Takeda, et al. (1998). "Distinct interleukin-1[beta]-converting enzyme family proteases mediate cisplatin- and staurosporine-induced apoptosis of mouse proximal tubule cells." *Life Sciences* 62(12): 1125-1138.

Ghadjar, P., M. Simcock, et al. (2011). "Concomitant Cisplatin and Hyperfractionated Radiotherapy in Locally Advanced Head and Neck Cancer: 10-Year Follow-up of a Randomized Phase III Trial (SAKK 10/94)." *International Journal of Radiation Oncology, Biology, Physics* In Press, Corrected Proof.

Gonçalves S, F.-S. R., Sanchez-Niño MD, Tejedor A, Neria F, Egido J, Ruiz-Ortega M, Ortiz A. (2010). "Tyrphostins as potential therapeutic agents for acute kidney injury." *Curr. Med. Chem.* 17((10)): 974-986.

Guerrero-Beltrán, C. E., M. Calderón-Oliver, et al. (2010). "Sulforaphane protects against cisplatin-induced nephrotoxicity." *Toxicology Letters* 192(3): 278-285.

Gutekunst, M., M. Oren, et al. (2011). "p53 Hypersensitivity Is the Predominant Mechanism of the Unique Responsiveness of Testicular Germ Cell Tumor (TGCT) Cells to Cisplatin." *PLoS ONE* 6(4): e19198.

Harper, B. W., A. M. Krause-Heuer, et al. (2010). "Advances in Platinum Chemotherapeutics." *Chemistry – A European Journal* 16(24): 7064-7077.

Hermeking, H., C. Lengauer, et al. (1997). "14-3-3Ã Is a p53-Regulated Inhibitor of G2/M Progression." *Molecular cell* 1(1): 3-11.

Herzog, C., R. Seth, et al. (2007). "Role of meprin A in renal tubular epithelial cell injury." *Kidney Int.* 71(10): 1009-1018.

Huang, Q., R. T. Dunn, et al. (2001). "Assessment of Cisplatin-Induced Nephrotoxicity by Microarray Technology." *Toxicological Sciences* 63(2): 196-207.

Hughes, E. N., B. N. Engelsberg, et al. (1992). "Purification of nuclear proteins that bind to cisplatin-damaged DNA. Identity with high mobility group proteins 1 and 2." *Journal of Biological Chemistry* 267(19): 13520-13527.

Hwa Lee, R., J. Mi Song, et al. (2001). "Cisplatin-induced apoptosis by translocation of endogenous Bax in mouse collecting duct cells." *Biochemical Pharmacology* 62(8): 1013-1023.

Ishida, S., J. Lee, et al. (2002). "Uptake of the anticancer drug cisplatin mediated by the copper transporter Ctr1 in yeast and mammals." *Proceedings of the National Academy of Sciences* 99(22): 14298-14302.

Jain A, S. A. (2010). "Nephroprotective activity of Momordica dioica Roxb. in cisplatin-induced nephrotoxicity." *Nat. Prod. Res.* 24((9)): 846-854.

Jakobsen, J. and J. Sørensen (2011). "Review on clinical trials of targeted treatments in malignant mesothelioma." *Cancer Chemotherapy and Pharmacology*: 1-15.

Jenderny, S., H. Lin, et al. (2010). "Protective effects of a glutathione disulfide mimetic (NOV-002) against cisplatin induced kidney toxicity." *Biomedicine and Pharmacotherapy* 64(1): 73-76.

Jia, Z., N. Wang, et al. (2011). "Amelioration of cisplatin nephrotoxicity by genetic or pharmacologic blockade of prostaglandin synthesis." *Kidney Int.* 79(1): 77-88.

Kang, K. P., S. K. Park, et al. (2011). "Luteolin ameliorates cisplatin-induced acute kidney injury in mice by regulation of p53-dependent renal tubular apoptosis." *Nephrology Dialysis Transplantation* 26(3): 814-822.

Katsuda, H., M. Yamashita, et al. (2010). "Protecting Cisplatin-Induced Nephrotoxicity with Cimetidine Does Not Affect Antitumor Activity." *Biological and Pharmaceutical Bulletin* 33(11): 1867-1871.

Kaushal, G. P., V. Kaushal, et al. (2001). "Role and regulation of activation of caspases in cisplatin-induced injury to renal tubular epithelial cells." *Kidney Int.* 60(5): 1726-1736.

Kettering, M. and et al. (2009). "Characterization of iron oxide nanoparticles adsorbed with cisplatin for biomedical applications." *Physics in Medicine and Biology* 54(17): 5109.

Kim DH, J. Y., Lee JE, Lee AS, Kang KP, Lee S, Park SK, Han MK, Lee SY, Ramkumar KM, Sung MJ, Kim W. (2011). "SIRT1 activation by resveratrol ameliorates cisplatin-induced renal injury through deacetylation of p53." *Am. J. Physiol. Renal Physiol.* 2011 May 18. [Epub ahead of print].

Kim, J. H., S.-J. Jeong, et al. (2010). "Decursin Prevents Cisplatin-Induced Apoptosis via the Enhancement of Antioxidant Enzymes in Human Renal Epithelial Cells." *Biological and Pharmaceutical Bulletin* 33(8): 1279-1284.

Koizumi, W., H. Narahara, et al. (2008). "S-1 plus cisplatin versus S-1 alone for first-line treatment of advanced gastric cancer (SPIRITS trial): a phase III trial." *The Lancet Oncology* 9(3): 215-221.

Kolishetti, N., S. Dhar, et al. (2010). "Engineering of self-assembled nanoparticle platform for precisely controlled combination drug therapy." *Proceedings of the National Academy of Sciences* 107(42): 17939-17944.

Koukourakis, M. I., A. Giatromanolaki, et al. (2010). "Concurrent Liposomal Cisplatin (Lipoplatin), 5-Fluorouracil and Radiotherapy for the Treatment of Locally Advanced Gastric Cancer: A Phase I/II Study." *International Journal of radiation oncology, biology, physics* 78(1): 150-155.

Kowalczuk, A., E. Stoyanova, et al. (2011). "Star-shaped nano-conjugates of cisplatin with high drug payload." *International Journal of Pharmaceutics* 404(1-2): 220-230.

Kruidering M, v. d. W. B., Zhan Y, Baelde JJ, Heer E, Mulder GJ, Stevens JL, Nagelkerke JF (1998). "Cisplatin effects on F-actin and matrix proteins precede renal tubular cell detachment and apoptosis in vitro." *Cell Death Differ.* Jul;5((7)): 601-614.

Lau, A. H. (1999). "Apoptosis induced by cisplatin nephrotoxic injury." *Kidney Int.* 56(4): 1295-1298.

Levine, J., Lieberthal, W (2001). "Terminal pathways to cell death." In: Molitoris, BA.; Finn, WF., editors. *AcuteRenal Failure*. W.B. Saunders; Philadelphia: 30-59.

Li, M., S. Balamuthusamy, et al. (2010). "Pituitary adenylate cyclase-activating polypeptide ameliorates cisplatin-induced acute kidney injury." *Peptides* 31(4): 592-602.

Lieberthal, W., V. Triaca, et al. (1996). "Mechanisms of death induced by cisplatin in proximal tubular epithelial cells: apoptosis vs. necrosis." *American Journal of Physiology -Renal Physiology* 270(4): F700-F708.

Liu, L., C. Yang, et al. (2010). "Proteasome inhibitors prevent cisplatin-induced mitochondrial release of apoptosis-inducing factor and markedly ameliorate cisplatin nephrotoxicity." *Biochemical Pharmacology* 79(2): 137-146.

Liu L, Z. J., He R, Zhou L, Zhang J. (2010). "[Astragalus injection ameliorates cisplatin-induced nephrotoxicity in mice]." *Zhongguo Zhong Yao Za Zhi* 35((20)): 2736-2740.

Long, H. J., B. N. Bundy, et al. (2005). "Randomized Phase III Trial of Cisplatin With or Without Topotecan in Carcinoma of the Uterine Cervix: A Gynecologic Oncology Group Study." *Journal of Clinical Oncology* 23(21): 4626-4633.

Marković, S., D. Djačić, et al. (2010). "Effects of Acute In vivo Cisplatin and Selenium Treatment on Hematological and Oxidative Stress Parameters in Red Blood Cells of Rats." *Biological Trace Element Research*: 1-11.

Matsushima, H., K. Yonemura, et al. (1998). "The role of oxygen free radicals in cisplatin-induced acute renal failure in rats." *The Journal of laboratory and clinical medicine* 131(6): 518-526.

Mattheolabakis, G., E. Taoufik, et al. (2009). "In vivo investigation of tolerance and antitumor activity of cisplatin-loaded PLGA-mPEG nanoparticles." *European Journal of Pharmaceutics and Biopharmaceutics* 71(2): 190-195.

McGinness JE, P. P., Demopoulos HB, Hokanson JA, Kirkpatrick DS. (1978). "Amelioration of cis-platinum nephrotoxicity by orgotein (superoxide dismutase)." *Physiol. Chem. Phys.* 10(3): 267-277.

Megyesi, J., L. Andrade, et al. (2002). "Coordination of the cell cycle is an important determinant of the syndrome of acute renal failure." *American Journal of Physiology - Renal Physiology* 283(4): F810-F816.

Miyaji T, K. A., Yasuda H, Fujigaki Y, Hishida A. (2001). "Role of the Increase in p21 in Cisplatin-Induced Acute Renal Failure in Rats." *Journal of the American Society of Nephrology* 12(5): 900-908.

Müller, M., S. Wilder, et al. (1998). "p53 Activates the CD95 (APO-1/Fas) Gene in Response to DNA Damage by Anticancer Drugs." *The Journal of Experimental Medicine* 188(11): 2033-2045.

Mukhopadhyay, P., M. Rajesh, et al. (2010). "Cannabinoid-2 receptor limits inflammation, oxidative/nitrosative stress, and cell death in nephropathy." *Free Radical Biology and Medicine* 48(3): 457-467.

Nagwani, S. and Y. B. Tripathi (2010). "Amelioration of cisplatin induced nephrotoxicity by PTY: A herbal preparation." *Food and Chemical Toxicology* 48(8-9): 2253-2258.

Ninsontia C, P. K., Chaotham C, Chanvorachote P. (2011). "Silymarin selectively protects human renal cells from cisplatin-induced cell death." *Pharm. Biol.* 2011 May 18. [Epub ahead of print].

Nishikawa, M., H. Nagatomi, et al. (2001). "Targeting Superoxide Dismutase to Renal Proximal Tubule Cells Inhibits Mitochondrial Injury and Renal Dysfunction Induced by Cisplatin." *Archives of Biochemistry and Biophysics* 387(1): 78-84.

Nowak, G. y. (2002). "Protein Kinase C-α and ERK1/2 Mediate Mitochondrial Dysfunction, Decreases in Active Na+ Transport, and Cisplatin-induced Apoptosis in Renal Cells." *Journal of Biological Chemistry* 277(45): 43377-43388.

Oneda, B., N. Lods, et al. (2008). "Metalloprotease Meprinβ in Rat Kidney: Glomerular Localization and Differential Expression in Glomerulonephritis." *PLoS ONE* 3(5): e2278.

Ozer, M. K., H. Asci, et al. (2011). "Effects of misoprostol on cisplatin-induced renal damage in rats." *Food and Chemical Toxicology* 49(7): 1556-1559.

Panda, P., S. Sahoo, et al. (2011). *Nephroprotective effect of <i>Bauhinia variegata</i> (linn.) whole stem extract against cisplatin-induced nephropathy in rats.*

Paraskar, A. and et al. (2011). "Rationally engineered polymeric cisplatin nanoparticles for improved antitumor efficacy." *Nanotechnology* 22(26): 265101.

Paraskar, A. S., S. Soni, et al. (2010). "Harnessing structure-activity relationship to engineer a cisplatin nanoparticle for enhanced antitumor efficacy." *Proceedings of the National Academy of Sciences* 107(28): 12435-12440.

Park, M. S., M. De Leon, et al. (2002). "Cisplatin Induces Apoptosis in LLC-PK1 Cells via Activation of Mitochondrial Pathways." *Journal of the American Society of Nephrology* 13(4): 858-865.

Park, S. and S. J. Lippard (2011). "Redox State-Dependent Interaction of HMGB1 and Cisplatin-Modified DNA." *Biochemistry* 50(13): 2567-2574.

Pil, P. and S. Lippard (1992). "Specific binding of chromosomal protein HMG1 to DNA damaged by the anticancer drug cisplatin." *Science* 256(5054): 234-237.

Pillai, T. G., M. John, et al. (2011). "Prevention of cisplatin induced nephrotoxicity by terpenes isolated from Ganoderma lucidum occurring in Southern Parts of India." *Experimental and Toxicologic Pathology* 63(1-2): 157-160.

Polyzos A, T. N., Gogas H, Lagadas A, Polyzos K, Giannakopoulos K, Felekouras E, Tsigris C, Karatzas T, Papadopoulos O, Giannopoulos A. (2009). "Cisplatin-Ifosfamide-gemcitabine as salvage chemotherapy in ovarian cancer patients pretreated with platinum compounds and Paclitaxel." *Anticancer Res.* 29(7): 2681-2686.

Pongjit, K., C. Ninsontia, et al. (2011). "Protective effect of glycine max and chrysanthemum indicum extracts against cisplatin-induced renal epithelial cell death." *Human and Experimental Toxicology*.

Pourahmad, J., M.-J. Hosseini, et al. (2010). "Mitochondrial/lysosomal toxic cross-talk plays a key role in cisplatin nephrotoxicity." *Xenobiotica* 40(11): 763-771.

Prestayko AW, D. A. J., Issell BF and Crooke ST (1979). "Cisplatin (cis-diamminedichloroplatinum II)." *Cancer Treat. Rev.* 6(1): 17–39.

Punushapai U, Y. P., Chumworathayi B, Luanratanakorn S, Udomthavornsuk B. (2010). "Weekly cisplatin 20 mg/m2 in patients with carcinoma of cervix receiving pelvic radiotherapy at Srinagarind Hospital: a randomized controlled trial." *Asian Pac. J. Cancer Prev.* 11(1): 201-207.

Ramesh, G. and W. B. Reeves (2002). "TNF-α mediates chemokine and cytokine expression and renal injury in cisplatin nephrotoxicity." *The Journal of Clinical Investigation* 110(6): 835-842.

Randall, M. E., V. L. Filiaci, et al. (2006). "Randomized Phase III Trial of Whole-Abdominal Irradiation Versus Doxorubicin and Cisplatin Chemotherapy in Advanced Endometrial Carcinoma: A Gynecologic Oncology Group Study." *Journal of Clinical Oncology* 24(1): 36-44.

Razzaque, M. S., T. Koji, et al. (1999). "Cisplatin-induced apoptosis in human proximal tubular epithelial cells is associated with the activation of the Fas/Fas ligand system." *Histochemistry and Cell Biology* 111(5): 359-365.

Rodrigues, M. A. C., J. L. Rodrigues, et al. (2011). "Carvedilol protects against cisplatin-induced oxidative stress, redox state unbalance and apoptosis in rat kidney mitochondria." *Chemico-Biological Interactions* 189(1-2): 45-51.

Rose, P. G., S. Ali, et al. (2007). "Long-Term Follow-Up of a Randomized Trial Comparing Concurrent Single Agent Cisplatin, Cisplatin-Based Combination Chemotherapy, or Hydroxyurea During Pelvic Irradiation for Locally Advanced Cervical Cancer: A Gynecologic Oncology Group Study." *Journal of Clinical Oncology* 25(19): 2804-2810.

Rosenberg B, V. L., Trosko JE and Mansour VH (1969). "Platinum compounds: a new class of potent antitumour agents." *Nature* (Lond) (222): 385-386.

Sadzuka Y, S. Y., Takino Y, Hirota S. (1994). "Protection against cisplatin-induced nephrotoxicity in the rat by inducers and an inhibitor of glutathione S-transferase." *Biochem. Pharmacol.* 48((3)): 453-459.

Saleh S, A.-S. A., El-Demerdash E, Khalef MM. (2009). "Protective effects of the angiotensin II receptor blocker losartan on cisplatin-induced kidney injury." *Chemotherapy* 55((6)): 399-406.

Sanchez-Gonzalez PD, L.-H. F., Perez-Barriocanal F, Lopez-Novoa JM. (2011). "Quercetin reduces cisplatin nephrotoxicity in rats without compromising its anti-tumour activity." *Nephrol. Dial Transplant.* 2011 May 19. [Epub ahead of print].

Sancho-Martínez, S. M., F. J. Piedrafita, et al. (2011). "Necrotic concentrations of cisplatin activate the apoptotic machinery but inhibit effector caspases and interfere with the execution of apoptosis." *Toxicological Sciences.*

Sandeep D, K. N. C. (2010). "Amelioration of cisplatin-induced nephrotoxicity by extracts of Hemidesmus indicus and Acorus calamus." *Pharm. Biol.* 48((3)): 290-295.

Santos, N., C. Bezerra, et al. (2008). "Hydroxyl radical scavenger ameliorates cisplatin-induced nephrotoxicity by preventing oxidative stress, redox state unbalance, impairment of energetic metabolism and apoptosis in rat kidney mitochondria." *Cancer Chemotherapy and Pharmacology* 61(1): 145-155.

Sen, F., E. Saglam, et al. (2011). "Weekly docetaxel and cisplatin with concomitant radiotherapy in addition to surgery and/or consolidation chemotherapy in stage III non-small cell lung cancer." *Cancer Chemotherapy and Pharmacology*: 1-9.

Seth, R., C. Yang, et al. (2005). "p53-dependent Caspase-2 Activation in Mitochondrial Release of Apoptosis-inducing Factor and Its Role in Renal Tubular Epithelial Cell Injury." *Journal of Biological Chemistry* 280(35): 31230-31239.

Siddik, Z. H. (2003). "Cisplatin: mode of cytotoxic action and molecular basis of resistance." *Oncogene* 22(47): 7265-7279.

Silici, S., O. Ekmekcioglu, et al. (2011). "The protective effect of royal jelly against cisplatin-induced renal oxidative stress in rats." *World Journal of Urology* 29(1): 127-132.

Sohn, S.-H., E. Ko, et al. (2010). "The genome-wide expression profile of Curcuma longa-treated cisplatin-stimulated HEK293 cells." *British Journal of Clinical Pharmacology* 70(4): 547-556.

Sonoda A, N. N., Nitta-Seko A, Ohta S, Takamatsu S, Ikehata Y, Nagano I, Jo J, Tabata Y, Takahashi M, Matsui O, Murata K. (2010). "Complex comprised of dextran magnetite and conjugated cisplatin exhibiting selective hyperthermic and controlled-release potential." *Int. J. Nanomedicine.* 5: 499-504.

Stathopoulos, G. P. (2010). "Liposomal cisplatin: a new cisplatin formulation." *Anti-Cancer Drugs* 21(8): 732-736 710.1097/CAD.1090b1013e32833d 32839adf.

Stathopoulos, G. P., D. Antoniou, et al. (2010). "Liposomal cisplatin combined with paclitaxel versus cisplatin and paclitaxel in non-small-cell lung cancer: a randomized phase III multicenter trial." *Annals of Oncology* 21(11): 2227-2232.

Suddek, G. (2011). "Sunitinib improves chemotherapeutic efficacy and ameliorates cisplatin-induced nephrotoxicity in experimental animals." *Cancer Chemotherapy and Pharmacology* 67(5): 1035-1044.

Tadagavadi, R. K. and W. B. Reeves (2010). "Endogenous IL-10 Attenuates Cisplatin Nephrotoxicity: Role of Dendritic Cells." *The Journal of Immunology* 185(8): 4904-4911.

Takeda, M., M. Kobayashi, et al. (1997). "Cisplatin-induced apoptosis of immortalized mouse proximal tubule cells is mediated by interleukin-1β converting enzyme (ICE) family of proteases but inhibited by overexpression of Bcl-2." *Archives of Toxicology* 71(10): 612-621.

Taratula O, G. O., Savla R, Wang YA, He H, Minko T. (2011). "Multifunctional nanomedicine platform for cancer specific delivery of siRNA by superparamagnetic iron oxide nanoparticles-dendrimer complexes." *Curr. Drug Deliv.* 8((1)): 59-69.

Tsuruya, K., T. Ninomiya, et al. (2003). "Direct involvement of the receptor-mediated apoptotic pathways in cisplatin-induced renal tubular cell death." *Kidney Int.* 63(1): 72-82.

Tufekci O, G. D., Ozoğul C, Kolatan E, Altun Z, Yilmaz O, Aktaş S, Erbayraktar Z, Kirkim G, Mutafoğlu K, Soylu A, Serbetçioğlu B, Güneri EA, Olgun N. (2009). "Evaluation of the effect of acetyl L-carnitine on experimental cisplatin nephrotoxicity." *Chemotherapy* 55((6)): 451-459.

Tuzcu, M., N. Sahin, et al. (2010). "Protective Role of Zinc Picolinate on Cisplatin-Induced Nephrotoxicity in Rats." Journal of renal nutrition : *the official Journal of the Council on Renal Nutrition of the National Kidney Foundation* 20(6): 398-407.

Ueda, N., G. P. Kaushal, et al. (2000). "Apoptotic mechanisms in acute renal failure." *The American Journal of medicine* 108(5): 403-415.

Valle, J., H. Wasan, et al. (2010). "Cisplatin plus Gemcitabine versus Gemcitabine for Biliary Tract Cancer." *New England Journal of Medicine* 362(14): 1273-1281.

Wenzel LB, H. H., Armstrong DK, Walker JL, Cella D (2007). "Health-related quality of life during and after intraperitoneal versus intravenous chemotherapy for optimally debulked ovarian cancer: a Gynecologic Oncology Group Study." *Journal of Clinical Oncology* 25(4): 437-443.

Winer, I., S. Wang, et al. (2010). "F3-Targeted Cisplatin-Hydrogel Nanoparticles as an Effective Therapeutic That Targets Both Murine and Human Ovarian Tumor Endothelial Cells In vivo." *Cancer Research* 70(21): 8674-8683.

Xu, P., E. A. Van Kirk, et al. (2006). "Anticancer Efficacies of Cisplatin-Releasing pH-Responsive Nanoparticles." *Biomacromolecules* 7(3): 829-835.

Yang, C., V. Kaushal, et al. (2007). "Transcriptional activation of caspase-6 and -7 genes by cisplatin-induced p53 and its functional significance in cisplatin nephrotoxicity." *Cell Death Differ.* 15(3): 530-544.

Yapar, K., K. Çavuşoğlu, et al. (2009). "Protective Effect of Royal Jelly and Green Tea Extracts Effect Against Cisplatin-Induced Nephrotoxicity in Mice: A Comparative Study." *Journal of Medicinal Food* 12(5): 1136-1142.

Ye, L., K. Letchford, et al. (2010). "Synthesis and Characterization of Carboxylic Acid Conjugated, Hydrophobically Derivatized, Hyperbranched Polyglycerols as Nanoparticulate Drug Carriers for Cisplatin." *Biomacromolecules* 12(1): 145-155.

Yura, R. E., S. G. Bradley, et al. (2009). "Meprin A metalloproteases enhance renal damage and bladder inflammation after LPS challenge." *American Journal of Physiology - Renal Physiology* 296(1): F135-F144.

Zhan Y, v. d. W. B., Wang Y, Stevens JL (1999). "The roles of caspase-3 and bcl-2 in chemically-induced apoptosis but not necrosis of renal epithelial cells." *Oncogene* 18.((47)): 6505-6512.

Zhu, W., Y. Li, et al. (2010). "Supramolecular Hydrogels from Cisplatin-Loaded Block Copolymer Nanoparticles and α-Cyclodextrins with a Stepwise Delivery Property." *Biomacromolecules* 11(11): 3086-3092.

INDEX

#

A

B

C

D

E

F

G

I

J

K

L

M

N

O

P

Q

R

S

T

U

V

W

X

Y

Z